AF575312

Dynamics of Atmospheric Re-Entry

Dynamics of Atmospheric Re-Entry

Frank J. Regan
Satya M. Anandakrishnan
Naval Surface Warfare Center
Dahlgren Division, White Oak Detachment
Silver Spring, Maryland

EDUCATION SERIES
J. S. Przemieniecki
Series Editor-in-Chief
Air Force Institute of Technology
Wright-Patterson Air Force Base, Ohio

Published by
American Institute of Aeronautics and Astronautics, Inc.,
370 L'Enfant Promenade, SW, Washington, DC 20024-2518

American Institute of Aeronautics and Astronautics, Inc., Washington, DC

Library of Congress Cataloging-in-Publication Data

Regan, Frank J.
Dynamics of atmospheric re-entry / Frank J. Regan, Satya M. Anandakrishnan.
p. cm.—(AIAA education series)
Includes index.
1. Space vehicles—Atmospheric re-entry. 2. Space vehicles—Aerodynamics. I. Anandakrishnan, Satya M. II. Title. III. Series.
TL1060.R42 1993 629.4'15-dc20 92-33727
ISBN 1-56347-048-9

Texts Published in the AIAA Education Series

Re-Entry Vehicle Dynamics
Frank J. Regan, 1984
Aerothermodynamics of Gas Turbine and Rocket Propulsion
Gordon C. Oates, 1984
Aerothermodynamics of Aircraft Engine Components
Gordon C. Oates, Editor, 1985
Fundamentals of Aircraft Combat Survivability Analysis and Design
Robert E. Ball, 1985
Intake Aerodynamics
J. Seddon and E. L. Goldsmith, 1985
Composite Materials for Aircraft Structures
Brian C. Hoskins and Alan A. Baker, Editors, 1986
Gasdynamics: Theory and Applications
George Emanuel, 1986
Aircraft Engine Design
Jack D. Mattingly, William Heiser, and Daniel H. Daley, 1987
An Introduction to the Mathematics and Methods of Astrodynamics
Richard H. Battin, 1987
Radar Electronic Warfare
August Golden Jr., 1988
Advanced Classical Thermodynamics
George Emanuel, 1988
Aerothermodynamics of Gas Turbine and Rocket Propulsion, Revised and Enlarged
Gordon C. Oates, 1988
Re-Entry Aerodynamics
Wilbur L. Hankey, 1988
Mechanical Reliability: Theory, Models and Applications
B. S. Dhillon, 1988
Aircraft Landing Gear Design: Principles and Practices
Norman S. Currey, 1988
Gust Loads on Aircraft: Concepts and Applications
Frederic M. Hoblit, 1988
Aircraft Design: A Conceptual Approach
Daniel P. Raymer, 1989
Boundary Layers
A. D. Young, 1989
Aircraft Propulsion Systems Technology and Design
Gordon C. Oates, Editor, 1989
Basic Helicopter Aerodynamics
J. Seddon, 1990
Introduction to Mathematical Methods in Defense Analyses
J. S. Przemieniecki, 1990

Space Vehicle Design
Michael D. Griffin and James R. French, 1991
Inlets for Supersonic Missiles
John J. Mahoney, 1991
Defense Analyses Software
J. S. Przemieniecki, 1991
Critical Technologies for National Defense
Air Force Institute of Technology, 1991
Orbital Mechanics
Vladimir A. Chobotov, 1991
Nonlinear Analysis of Shell Structures
Anthony N. Palazotto and Scott T. Dennis, 1992
Optimization of Observation and Control Processes
Veniamin V. Malyshev, Mihkail N. Krasilshikov, and Valeri I. Karlov, 1992
Aircraft Design: A Conceptual Approach
Second Edition
Daniel P. Raymer, 1992
Rotary Wing Structural Dynamics and Aeroelasticity
Richard L. Bielawa, 1992
Spacecraft Mission Design
Charles D. Brown, 1992
Introduction to Dynamics and Control of Flexible Structures
John L. Junkins and Youdan Kim, 1993
Dynamics of Atmospheric Re-Entry
Frank J. Regan and Satya M. Anandakrishnan, 1993

Published by
American Institute of Aeronautics and Astronautics, Inc., Washington, DC

Foreword

In 1984, the publication of *Re-Entry Vehicle Dynamics* by Frank J. Regan inaugurated the AIAA Education Series of textbooks and monographs in aeronautics and astronautics. The series now contains over 30 titles and serves well its original goal to provide textbooks and supplemental materials for professional work. This latest addition to the Series, *Dynamics of Atmospheric Re-Entry* by Frank J. Regan and Satya M. Anandakrishnan, is an extension of the 1984 text, and it includes an expanded coverage of the subject.

The new text presents a comprehensive treatise on the dynamics of atmospheric entry. All mathematical concepts are fully explained in the text so that there is no need for any additional reference materials. The first half of the text deals with fundamental concepts and practical applications of the atmospheric model, Earth's gravitational field and form, axis transformations, force and moment equations, Keplerian motion, and re-entry mechanics. The second half includes special topics such as re-entry decoys, maneuvering re-entry vehicles, angular motion, flowfields around re-entering bodies, error analysis, and inertial guidance.

The AIAA Education Series embraces a broad spectrum of theory and application of different disciplines in aerospace, including aerospace design practice, and more recently the Series has been expanded to include defense science, engineering, and technology. The basic philosophy for the Series is to develop both teaching texts for students and reference materials for practicing engineers and scientists.

J. S. Przemieniecki
Editor-in-Chief
AIAA Education Series

Preface

This is the second book that I have been invited to prepare for the AIAA Education Series. The first book, *Re-Entry Vehicle Dynamics,* was prepared from the notes that accompanied a course I gave first at the Naval Surface Weapons Center (now Naval Surface Warfare Center) and later at other institutions. The present volume, *Dynamics of Atmospheric Re-Entry,* is an extension of the original book. The "I" now becomes "we," as the present volume has two authors.

A strong effort has been made to present a broad coverage of topics. The only justification for any apparent imbalance in the selections is that we have found the included material to have great utility. Our work has been largely with modeling the dynamics of re-entry vehicles (including at times the presence of an active defense in the vicinity of the impact area). We have included chapters on atmospheric and gravity modeling (Chapters 2 and 3), because such models are fundamental to any study of re-entry dynamics.

Chapters 4 and 5 contain discussions of force and moment representations and general mathematical methods for resolving these vectors into reference frames. In Chapter 6 we derive and present closed-form expressions for exoatmospheric or Keplerian motion; we then use the derived expressions to evaluate the effect of initial conditions on position and velocity errors particularly at impact. Chapter 7 provides some closed-form expressions that are useful in evaluating the performance of a broad class of re-entry vehicles, including those capable of maneuvering. Chapter 8 discusses modeling of decoys or, in a general sense, determining the likelihood that the measurables of an observed object fall within some category definition or preset bounds.

In Chapter 9 we have extended the ideas of Chapter 7 on maneuvering re-entry vehicles, including some simple guidance and targeting algorithms. Since a maneuvering re-entry vehicle may encounter an active defense, we also include some interceptor dynamics and guidance laws.

In Chapter 10 we return to the Keplerian environment of Chapter 6 and extend the work from particle mechanics to rigid bodies and hence angular motion as well. In *Re-Entry Vehicle Dynamics,* aerodynamics loads were limited to Newtonian expressions; in Chapter 11 we have examined the aerodynamic flowfield in greater detail from the free molecular flow regime to the continuum regime. Having highlighted some of the special features of hypersonic flowfields, we present some simple computational methods for calculating aerodynamic loads that we have found useful in preliminary design.

In Chapter 12 we extend the angular motion discussed in Chapter 10 to the atmospheric segment of the trajectory. In Chapter 13 we return to error analysis and present some ideas for modeling trajectories in the presence of modeling errors. Chapter 14 might seem a bit out of place in a discussion of re-entry vehicles, but in our modeling work the inertial navigator has been a principal part of the effort, particularly strapdown implementations. For example, the angular velocities could be taken from the moment equations, possibly altered with an error model, and then used in an updating algorithm. Since the strapdown system is computationally intensive we have included some of the updating algorithms that we have found useful.

We wish to acknowledge the particular help that has been given in bringing this work to published form. Our thanks go to John Vamos, Branch Head, Reentry Systems Branch, and C. A. Fisher, Head of Weapons Dynamics Division, both of Naval Surface Warfare Center (NSWC), White Oak Laboratory, who have provided encouragement for this effort. Also special thanks go to W. C. Lyons for his invaluable suggestions and for proofreading some of the chapters.

No originality is claimed for any of the included material. Every effort has been made to cite original sources in the preparation of the chapters. Nevertheless, some compilations are invaluable. For example, in Chapter 7 the extensive work of J. J. Martin will be recognized in the organization of this chapter. The extensive work of John Allwine of NSWC is evident throughout Chapter 9, and many of the positive results come from his imaginative application of computer modeling.

Chapter 12 makes use of the extensive work of N. S. Vinh (University of Michigan) as well as the wind tunnel studies of D. C. Freeman of NASA. The error analyses of Chapter 13 were the result of many discussions with and impromptu tutorials from A. Gorechlad of NSWC. Chapter 14 drew on help that extends over 15 years from G. Schmidt of C. Stark Draper Laboratory (CSDL) in both the software and hardware of the inertial navigator, particularly in strapdown. The strapdown algorithms come largely from the extensive work of W. VanderVelde of MIT, former teacher of the first author. Thanks also go to P. G. Savage of Strapdown Associates, Inc., who has provided extensive information on instruments and has permitted our use of his graphic presentations.

We would both like to acknowledge the extensive help of Jeanne Godette of AIAA in bringing this work to completion as well as the understanding and patience of our wives, Kay Regan and May Maniam, enduring what must have seemed like an interminable effort.

Frank J. Regan
Satya M. Anandakrishnan
October 1992

Table of Contents

1
Introduction

"I could more easily believe two Yankee Professors would lie than that stones would fall from heaven."

—Thomas Jefferson (President of the American Philosophical Society) commenting upon a theory of two New England astronomers that a meteorite found in Weston, Connecticut, was of extraterrestrial origin (1807)

1.0 Background

The dynamics of re-entry bodies may be regarded as an application of rational mechanics, the oldest branch of physics, to the high-speed motion of an object in a planetary atmosphere. The foundation of classical, or nonrelativistic, rigid-body mechanics was essentially completed by the close of the nineteenth century, although other areas of mechanics such as the dynamics of fluids and elastic bodies are still under active development. The continual evolution of hypersonic and hypervelocity gasdynamics has been motivated largely by the design of manned and unmanned vehicles which enter the atmosphere at velocities which are a significant fraction of the speeds associated with orbital motion.

At the lower speeds of atmospheric flight vehicles, the flowfield primarily imparts a distributed, shear, and normal pressure load to the vehicle. Control is then effected by making local alterations to this pressure field, often through configurational changes such as the deflection of a flap. At the much higher speeds of atmospheric entry, the support and control of these distributed loads is a great engineering problem. In addition, the flowfield surrounding the entry object adds a distributed thermal load to the pressure load. Coping with this thermal load presents an even more formidable engineering challenge than that encountered in sustaining the pressure loads.

Most of the emphasis in this book is on the dynamics associated with the atmospheric re-entry of aerospace vehicles from near-orbital speeds. Nevertheless, it might be of some interest to consider the magnitude of the problem presented by the heat load. For example, a satellite in a circular orbit at 320 km above the Earth's surface has a specific kinetic energy of 3.11×10^7 J/kg (1.3162×10^4 Btu/lb). An interplanetary vehicle (or natural space debris) on a parabolic orbit would have twice as high a specific kinetic energy. Although the re-entry vehicles considered in this book are for the most part on suborbital trajectories, the specific energy of such vehicles can approach that of a circular orbit. The significance of a specific energy of 3.11×10^7 J/kg becomes clear when we consider the specific energies of vaporization of water and carbon. Water will vaporize at a specific energy of 2.32×10^6 J/kg (1000 Btu/lb), and carbon, with one of the highest specific heats of vaporization, will vaporize at a specific energy of 6.03×10^7 J/kg (2.6×10^4 Btu/lb).

The problem of dissipating this energy or converting it into a form that does not prove catastrophic to the vehicle is a central engineering concern. Clearly, if all of the body's kinetic energy were to be converted into thermal energy, the object would vaporize unless made almost entirely of carbon. The survival of natural atmospheric entry bodies such as meteorites indicates that a significant fraction of the mechanical energy associated with entry must be dissipated by phenomena other than thermal absorption. The energy fraction absorbed by an artifact, or re-entry body, as thermal energy depends upon both the design (body shape) and operation (trajectory). For a natural entry body (meteor or meteorite) the entry speeds are usually higher than those encountered by re-entry bodies; meteorite materials vary widely from silicon to nickel-iron.

1.1 Meteorites—Nature's (Re-)Entry Bodies

We are concerned, in this book, primarily with the dynamics associated with the passage of an object (usually an artifact) from exoatmospheric conditions through the atmosphere to ground impact. Let us then first consider natural objects which experience atmospheric entry. Meteorites have been called by scientists the "poor man's space probe"; the re-entry engineer might well call meteorites the "poor man's re-entry body."

The first atmospheric entry bodies (if not re-entry bodies) were meteorites, objects of extraterrestrial origin which somehow penetrated the atmosphere and, depending upon their size, came to rest more or less intact on the Earth's surface. During the eighteenth-century Age of Enlightenment, the intellectual establishment began to respond to reports of rocks from the sky. Jefferson, the most prominent American adherent of European enlightenment, provided the response given above to the meteorite hypothesis. Nearly all of his European intellectual counterparts shared his conviction.

Before examining meteors further, let's first establish some definitions. A *meteoroid* is any natural object which might enter the atmosphere, a *meteor* is observed in transit through the atmosphere, and a *meteorite* strikes the Earth (or, more precisely, it is recovered and identified). A meteorite that is recovered after an observed impact is called a *fall*, whereas a meteorite which cannot be associated with any observed trajectory is called a *find*.

In spite of Jefferson's considered opinion, the first recorded fall occurred on November 16, 1492, coincidentally only a month after Columbus discovered America and 250 years before Jefferson was born. At the time, the name given to this natural atmospheric entry body was *donnerstein*, or "thunderstone." One of the recovered meteorites is still preserved in the city hall of Ensisheim in the Alsace region of present-day France. The fall preserved in Ensisheim is a Chondrite (stone or silicon-rich) meteor.

Another possible fall is recorded in the biblical Acts of the Apostles where the Earth-goddess Diana is identified with a metallic meteorite recovered at Ephesus.

> Is there anybody alive who does not know that the city of the Ephesians is the guardian of the temple of great Diana and of her statue that fell from heaven? [Acts 19:36]

The association with a fertility goddess is probably due to the benippled appearance characteristic of an iron-nickel or simply *iron* meteorite, particularly one which maintains a fixed orientation during entry. The windward side of such a meteor is often heavily dimpled due to surface ablation. The ablative pattern remains fixed in the surface after the meteor cools during the terminal phase of the trajectory. The hills and valleys of this wavy pattern, characteristic of iron meteorites, are called *regmaglypts*. Not surprisingly, a similar gouged pattern has been observed on the nose tips of manmade re-entry bodies. Of course, *iron finds* supplied the primitive metallurgist with this important metal, although its origin was seldom recognized.

The scientific community began to accept the extraterrestrial nature of meteorites in 1794. A book published by E. F. F. Chladni contained reasoned arguments that observed fireballs often drop meteorites and that certain large masses of iron were prehistoric meteorites. Four years earlier, in 1790, a scientist P. Bertholon commented regarding the 1790 fall of a meteorite, "... how sad it is that the entire municipality enters folk tales upon the official record."[1] Also in 1790, A. Stutz, in discussing the fall of a meteorite near Hraschina, Croatia, remarked, "... the straightforward manner with which everything is accounted, the agreement among witnesses who had no grounds to agree so completely regarding a lie.... [Nevertheless] in our times it would be unfortunate to hold such tales to be probable."[1]

It should be mentioned that the quotation attributed to Jefferson at the beginning of this chapter is contested by some, and in truth the source has not been confirmed directly. In fairness to Jefferson, his letter to Daniel Solomon on February 15, 1808 provides a much more balanced assessment: "... its [the recovered meteorite in Connecticut] descent from the atmosphere presents so much difficulty as to require careful examination.... We are certainly not to deny whatever we cannot account for.... The actual fact... is the thing to be established."[1]

The turning point in Europe occurred about five years earlier in 1803, when J. B. Biot, a member of the French Academy, accepted reports of a meteoritic stone that had fallen from the sky at L'Aigle, France.

In a sense Jefferson's letter to Solomon might be taken as no less than the beginning of the Space Age, for at this point America's foremost intellectual and natural philosopher began to seriously consider that there was some contact between the celestial world of the astronomer and the terrestrial world of humans. (A European might insist that Chladni's book ushered in the scientific aspect of the Space Age; to be sure, Chladni's statements were more positive than

Jefferson's, and they predate Jefferson's thoughts by nearly two decades.) Of course, impacted meteorites had been recovered and their origin identified much earlier, but always by nonscientists. The scientific establishment much before 1808 could not accept the extraterrestrial origin of the meteorites. Thirty years later, however, few scientists would deny that meteorites were indeed "stones from heaven." A detailed tracing of the evolution of the hypothesis of the extraterrestrial origin and atmospheric entry of meteors is contained in Wasson's excellent book.[1] Within a decade or two of Jefferson's letter, and certainly by the end of the first third of the nineteenth century, the extraterrestial origin of the meteorite was considered an established scientific fact.

The occasional atmospheric entry of meteors has little obvious impact on everyday conditions on the Earth. However, there have been spectacular examples. In the plateau country of northern Arizona (about 25 km west of Winslow) is the Meteor Crater (sometimes associated with Barringer, its initial investigator). This crater is about 1200 m across and about 200 m deep. Somewhere between 25,000 to 50,000 years ago, a nickel-iron meteor of mass 4×10^9 kg and radius 50 m impacted the Coconino Sandstone of the Diablo Canyon.[1] The impact velocity, clearly in the hypervelocity range, has been estimated at about 20 km/s. A projectile of this size would scarcely be affected by the atmosphere. The impact was doubtlessly explosive, as only a few tenths of metric tons (megagrams) of debris have ever been recovered, in spite of extensive drilling in both the crater floor and in the south rim. (From the shape of the crater, the meteor is assumed to have impacted from the north.)

On June 30, 1908, at about 07:17, local time, a meteor or comet maybe 30 to 90 m in diameter on a grazing orbit exploded over a remote area of northern Siberia near the Tunguska River. Seismographs at several locations recorded the air pressure wave. The explosion has been estimated to contain energy as great as a 100-megaton fusion bomb. The burned vegetation indicates a surface temperature of something like 75°C. Regardless of the energy content, the impact produced no crater, and no meteorite fragments have ever been found. Clearly, the object exploded or fragmented in the atmosphere; the entry velocity has been widely estimated to lie within the range of 28 to 47 km/s.[1]

Several other meteors have been observed and the craters from hundreds of others which impacted in prehistoric times have been identified. Space limitations do not permit a discussion of these impacts here. However, literature which discusses historic and prehistoric meteors, their dates, crater sizes and locations, masses and velocities, and effects on the biosphere is readily available. Among such sources are Dodd,[2] Heide,[3] Kriznov,[4] and Erickson.[5] Kriznov's book, originally published in Russia, discusses the Tunguska meteor in great detail.

Consider a case for which detailed measurements of a meteor were made. Most of the information we have concerning meteorites is based on the impact crater and the recovered meteor or fragments thereof. According to Dodd[2] there are now 1000 asteroids with a diameter of at least 1 km that cross the Earth's orbit. The crossing frequency is lower for greater sizes, but some calculations show that an asteroid with a diameter of 10 km should impact the Earth every

40 million years. At this point we might want to consider some kind of relationship between the frequency of atmospheric entry and the size of a meteoroid. Figure 1.1 indicates that there is an inverse relationship between the frequency of encounter and the mass of the meteoroid. Dodd's conclusions are also supported by the earlier work of Grieve.[7] An increase in size by a factor of 10 corresponds to a decrease in frequency by a factor of 10. It would seem that a 1.1×10^9 metric ton iron or stone-iron meteorite should strike the Earth once every 5.0×10^7 years. Such an object would be about 300 m in diameter.

The smallest meteors that can be recovered have a mass of about 1 g. About 1.6×10^{12} objects of that size or larger strike the Earth each year (about 15 per km^2).[2] Crater-forming meteors are about 325 metric tons. Such meteorites are not affected much by the atmosphere and therefore impact the Earth with a kinetic energy equal to an equivalent mass of TNT. According to Fig. 1.1, on the average, five such meteors strike the Earth's atmosphere each year; a meteor of the size that caused the Arizona meteor crater strikes the Earth's atmosphere every 60 years or so. Fortunately, most strike the oceans and most of the remainder are stone meteors that break up in the atmosphere. A nickel-iron meteor similar to the one that left its mark at Diablo Canyon should visit the Earth once every 5000 years or so. One of the largest craters on record is the

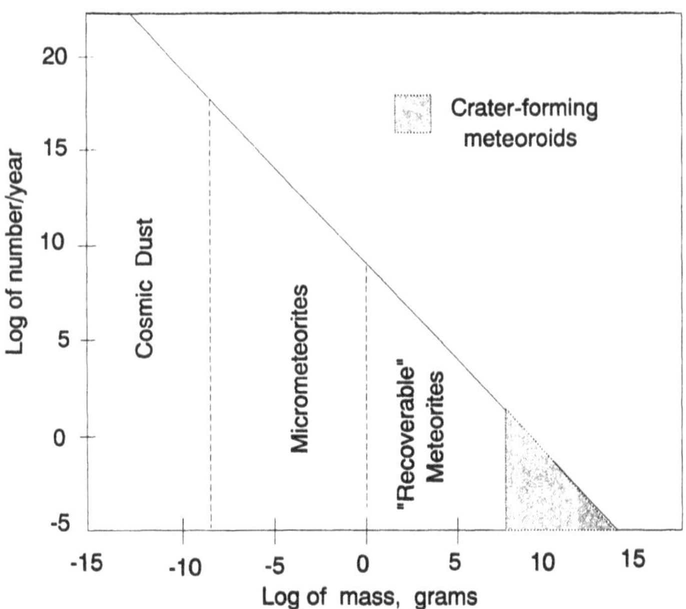

Fig. 1.1 Frequency of meteoroid occurrence vs meteoroid size. (Reprinted by permission of the publishers of *Thunderstones and Shooting Stars* by R. T. Dobb, Harvard University Press, Cambridge, MA, copyright 1986 by R. T. Dobb.)

circular crater at the Manicouagan River in Quebec, Canada. The crater is 60 km in diameter with an estimated age of more than 200 million years. Many of the oldest craters were recognized only after satellite observations of the Earth became routine. Erosion and vegetation growth makes identification of craters and estimation of their sizes difficult.

Figure 1.2 gives a relationship between age and crater diameter. Note that the largest craters are two orders of magnitude greater than the Barringer meteor crater in Arizona. The craters at Vredefort, South Africa (1.97×10^9 years old), and Sudbury, Ontario (1.89×10^9 years old), are about 40 km wide (although estimates in some sources are as large as three times this figure).

Nobel laureate Alvarez hypothesizes that mass extinctions could have been caused by the impact of a large object with the Earth.[6] (See also Erickson.[5]) About 15 or so mass extinctions have been identified by paleontologists as having occurred during the past 600 million years.[6,7] One of the most noted of these extinctions occurred about 100 million years ago at the end of the Cretaceous period. Alvarez postulates that an extraterrestrial body about 10 km in diameter with a specific gravity of about 3 struck the Earth at a speed of 20 km/s, releasing an energy equivalent to 10^8 megatons (100 million megatons) of TNT, orders of magnitude greater than the combined energy of all nuclear arsenals. One kiloton of TNT equals about 4.0×10^{12} Joules and is equivalent to a very small nuclear weapon. If this object struck land it would have caused a crater 100 km wide. Speculations as to the consequences of this event are contained in Refs. 2, 5, and 6–9.

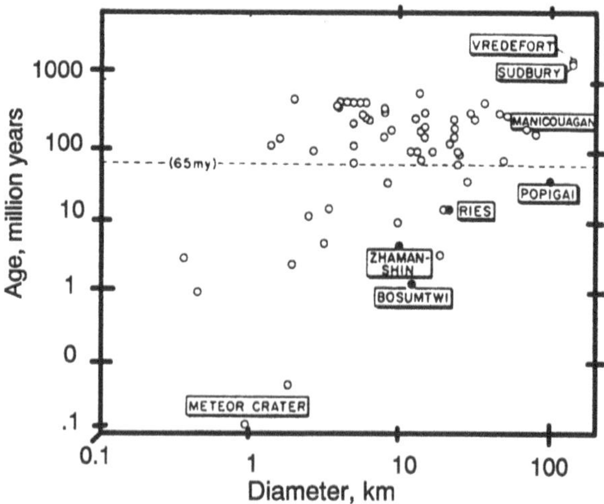

Fig. 1.2 Crater age vs crater diameter. (Reprinted by permission of the publishers of *Thunderstones and Shooting Stars* by R. T. Dobb, Harvard University Press, Cambridge, MA, copyright 1986 by R. T. Dobb.)

Meteors have played a negligible role in the physical composition of the Earth.[10,11] However, they may have had a profound effect on the evolution of biological forms. Arguably, the acceptance of the extraterrestrial origin of the meteor about 175 years ago by the scientific community established the mindset that could conceive and then develop a meteor of human origin—the re-entry body.

Let's look now at a portion of a very limited body of data concerning meteors whose trajectories have been measured—measurements which are similar to those that might be made for a re-entry body. An orthographic network is in place which can photographically record fireballs from widely separated sites. Each station possesses an array of cameras with rotating shutters that are triggered by the illumination of a fireball. To date, three meteor trajectories have been recorded. Consider how the velocity magnitude obtained from a set of measurements compares to theoretical predictions for the Přibram H5, a stone or Chrondite meteor. The theoretical predictions come from Chapter 7 of this book or from Buchwald.[13]

In Eq. (7.32) of this book, the magnitude of the deceleration, a, of a re-entry body is given as

$$a = \frac{\mathrm{d}V}{\mathrm{d}t} = -\left(\frac{\rho_0 S C_D}{2m}\right) V^2 e^{-h/H} + g \sin\gamma \tag{1.1}$$

where C_D is the drag coefficient, S the reference area, γ the flight path angle, and m the mass. The terms ρ_0 and H are parameters in an exponential atmosphere model [Eq. (2.32)]. Equation (1.1) cannot be integrated in closed form, but it would seem to indicate that under some circumstances there would be a terminal velocity associated with a meteor [drag balance's gravity for $(\mathrm{d}V/\mathrm{d}t) = 0$]. In other words, for a meteor below a certain size limit, the impact speed at the Earth's surface would not be influenced by the speed with which the meteor entered the atmosphere. Before examining this issue further, we might consider how the velocity changes with altitude when the acceleration is significantly greater than one g.

If we change the variable of integration from time t to altitude h and regard the gravitational acceleration as negligible, we obtain the following expression for the velocity magnitude:

$$V = V_E \exp\left[-\frac{\rho_0 H e^{-h/H}}{2\beta \sin(\gamma_E)}\right] \tag{1.2}$$

where V_E is the velocity at entry, γ_E is the flight path angle (the angle the velocity vector makes with the horizontal) at entry, and β is defined and then evaluated for a spherical-shaped meteor as follows:

$$\beta = \frac{m}{C_D S} \approx \frac{\rho_V \frac{4}{3}\pi R_b^3}{C_D \pi R_b^2} = \left(\frac{\frac{3}{4} C_D}{\rho_V R_b}\right)^{-1} = (K \sin\gamma_E)^{-1}$$

where ρ_V is the average material density, C_D is the drag coefficient, and K is a parameter Buchwald[13] uses to characterize the meteor mass and aerodynamic drag. Clearly,

$$K = 1/\beta \sin(\gamma_E) \tag{1.3}$$

contains both a shape/composition term (β) and an operational term (γ_E).

Whereas β has units of kg/m^2, K has units of m^2/kg, or in Buchwald's usage, cm^2/g. The Přibram H5 Chondrite meteor has a value of K equal to 8.25×10^{-3} cm^2/g, or 8.25×10^{-4} m^2/kg. For a vertical trajectory the Přibram meteor has a ballistic coefficient β of 1212.1 kg/m^2, or 248.24 lb/ft^2. Assuming a drag coefficient of 0.5 and a density of 3.0 g/cm^3, the radius of the body would be about 15.0 cm.

Note from Fig. 1.3 that this meteor loses virtually all of its entry velocity before impact. For a much larger meteorite, most of the entry velocity magnitude is carried to impact. For example, consider the meteor which created the Arizona Meteor Crater. It had a mass of 4×10^9 kg, a radius of 50 m, an estimated drag coefficient C_D of 1.0, and a density of 7.0 g/cm^3 (estimated from recovered fragments) corresponding to a Buchwald constant K of 2.2×10^{-6} assuming vertical re-entry. From Fig. 1.3, such an object will carry to impact nearly all of its entry kinetic energy.

1.2 Artifacts—Manmade Re-Entry Bodies

In the previous section we argued that the Space Age began when the scientific establishment began to accept the extraterrestrial origin of meteorites. However, it took another 150 years before technology could support the engi-

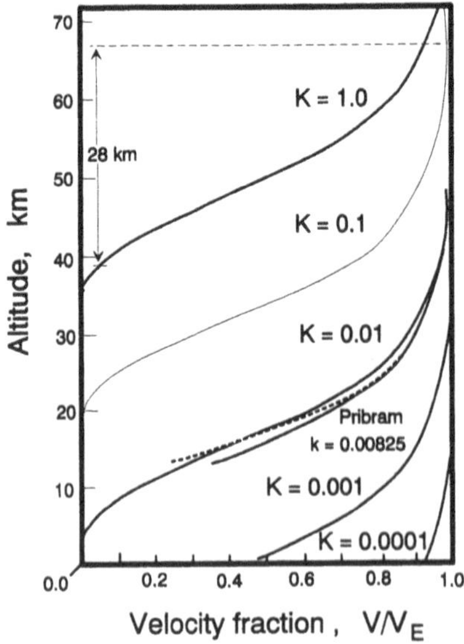

Fig. 1.3 Altitude vs velocity fraction.

neering part of the Space Age: the design, construction, launch, and recovery of an artifact from orbit.

On August 10, 1960 the *Discovery 13* satellite completed 17 orbits and re-entered the Earth's atmosphere. During its descent Hawaiian radar tracked it to an oceanic impact. The recovery of the remnants of this satellite by U.S. Navy frogmen gave to it a unique distinction: *Discovery 13* was the first manmade object to be recovered from space.[11]

Discovery 13 was not designed to survive an atmospheric entry well enough to maintain integrity to impact. However, an appropriately designed artifact can maintain functional integrity to a particular altitude or to impact. Certain re-entry bodies might have additional capabilities, such as the ability to develop lifting or maneuvering loads. A re-entry vehicle might use lift to increase accuracy, to evade an active defense, or to soft land at a selected site. Other re-entry vehicles might have to be designed to survive not only atmospheric re-entry but also penetration of the Earth to a certain depth.

Most of the emphasis in this book will be on the dynamics associated with atmospheric re-entry. Nevertheless, it might be of some interest to consider the magnitude of the problem presented by the heat load. Selecting an appropriate external shape is a major design response to controlling the heat flux into a vehicle. Vehicle shape also has a major impact on the stability and controllability of a vehicle. First, let's consider the magnitude of the thermal problem facing a re-entering vehicle as well as the evolution of re-entry vehicle morphology.

As pointed out earlier, a satellite in a circular orbit 320 km above the Earth's surface has a specific kinetic energy of 3.11×10^7 J/kg (1.316×10^9 Btu/lb). Clearly, if all of the body's kinetic energy were converted into thermal energy and this energy were entirely absorbed by the body, there would be enough heat to vaporize most materials. (The specific heat of vaporization of carbon is 6.03×10^7 J/kg.) The survival of meteorites indicates that even for an object of random "design," not all of the kinetic energy is converted into thermal energy and certainly not all of the thermal energy is absorbed by the vehicle. It becomes a major engineering problem to keep to a minimum the absorption of thermal energy by the body. The fraction of the energy absorbed by the body depends upon materials, particularly in the nosetip region, as well as the body shape and the trajectory traversed during re-entry.

Of course, the roots of re-entry body dynamics and aerodynamics are at the high-performance end of atmospheric vehicle engineering. Since drag minimization is a primary goal in the design of such vehicles, it might seem obvious to design a re-entry body to meet some kind of minimum drag condition. Not surprisingly, therefore, many of the earlier shapes proposed for re-entry bodies had fairly pointed vertices. All of this was to change. The change from the pointed, streamlined shape to the blunt shape is a result of the aerodynamic research at NACA's Ames Laboratory, particularly the work of H. J. Allen and coworker A. J. Eggers.

In 1952 Allen of NACA Ames proposed that re-entry from high-speed exoatmospheric conditions required a blunt high-drag body.[15] Allen's conjecture, followed later by an exact analytic effort coauthored with Eggers,[16] is considered by many to be the breakthrough paper in the engineering of re-entry vehicles.

It is not too much of an exaggeration to identify this paper of Allen and Eggers as complementary to the letter of Jefferson to Solomon, partially quoted earlier. Jefferson's letter set the stage for the scientific investigation of atmospheric entry; Allen and Eggers initiated the engineering aspects of re-entry. More than 35 years have passed since their seminal paper was published, yet it remains a classic of engineering literature: it is readable and a model of clarity, the analysis is to the point, and the conclusions are stated unequivocally. The following paragraphs discuss Allen and Egger's findings.

Supersonic speeds are characterized by the formation of a shock wave at the vertex of the body. If the body has a sharp vertex, then the shock wave is attached; if the body is blunt, the shock wave is detached ahead of the body. Figure 1.4 shows a blunt body as well as some of the salient flow characteristics. As Allen states, "The bow shock is normal to the stagnation streamline and converts the supersonic flow ahead of the shock to a low subsonic speed at high static temperature downstream of the shock."[16] Allen and Eggers then treat the nose, or vertex, section as if it were a segment of a sphere of radius σ in a subsonic flowfield. They show that the maximum heating rate at the stagnation point, $\mathrm{d}H_s/\mathrm{d}t$, may be expressed as

$$\frac{\mathrm{d}H_s}{\mathrm{d}t} = K\left(\frac{\rho}{\sigma}\right)^{1/2} V^3 \tag{1.4}$$

where K is a constant, ρ is the atmospheric density, σ is the nose radius, and V is the air speed. Obviously, the heating rate is inversely proportional to the square root of the nose radius. The conclusion is that sharp-nosed re-entry vehicles might have low drag, but they will also experience severe heating problems. If we replace the airspeed V by the expression given in Eq. (1.2), the heating

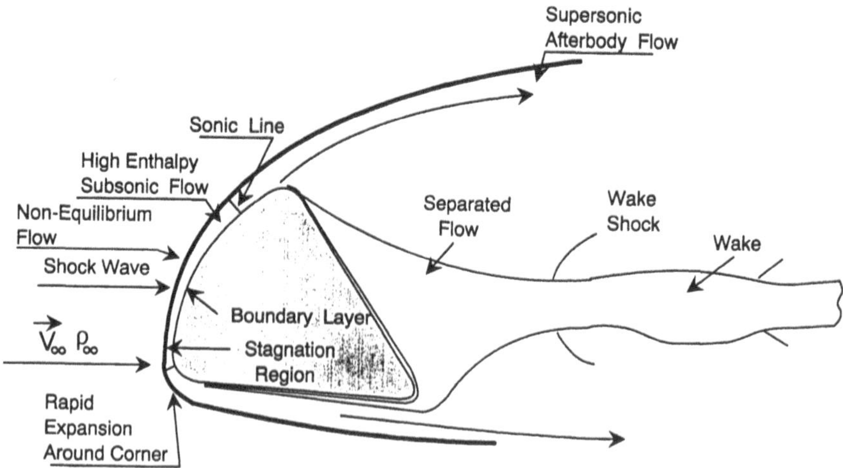

Fig. 1.4 Flow characteristics around a blunt body.

rate is given by the following function of entry conditions V_E and γ_E, ballistic coefficient β, and altitude h:

$$\frac{dH_s}{dt} = K\left(\frac{\rho_0}{\sigma}\right)^{1/2} V_E^3 \exp\left(-\frac{h}{2H}\right)\exp\left[\frac{-3\rho_0 H}{2\beta\sin(\gamma_E)}e^{-h/H}\right] \tag{1.5}$$

The maximum heat transfer rate occurs at an altitude of

$$h_{\max} = H\,\ell n\left[\frac{3\rho_0 H}{2\beta\sin(\gamma_E)}\right] \tag{1.6}$$

with a corresponding velocity magnitude $V_{\max}$ of

$$V_{\max} = V_E \exp\left(-\tfrac{1}{6}\right) \approx 0.85\, V_E$$

As an aside, Eq. (1.6) shows that for micrometeorites (less than a few grams), the altitude for the maximum heat flux is very high, but since the density is very low, the actual heat flux is very low; consequently, a micrometeorite usually survives to ground impact.

Of more immediate interest, Allen and Eggers show that Q, the total heat absorbed by the entering object, is given as:

$$Q = \frac{1}{4}\left(\frac{C_f S_w}{C_D A}\right) m V_E^2 \left\{1 - \exp\left[\frac{-\rho_0 H}{\beta\sin(\gamma_E)}\right]\right\} \tag{1.7}$$

where C_f is the equivalent skin-friction coefficient, S_w the wetted area, A the reference area, and m the mass. For most re-entry vehicles of interest the term in the braces is nearly equal to unity, so the following approximation is useful:

$$Q \approx \frac{1}{4} m V_E^2 \left(\frac{C_f S_w}{C_D A}\right) \tag{1.8}$$

Equation (1.8) represents a pivotal concept in the engineering analysis of the re-entry phenomenon and leads to a rather surprising nonintuitive result: total heating is reduced by increasing the total drag coefficient C_D (assuming that an increase in C_D is not accompanied by an increase in the skin-friction coefficient C_f).

We can therefore conclude that the total heat load, Q, is reduced by selecting a blunt shape. If $(V/V_E)^2$ is the fraction of entry kinetic energy at any altitude, Eq. (1.2) indicates that a smaller β corresponds to a lesser fraction of remaining kinetic energy. The "lost" kinetic energy must then have been converted into heat. However, the amount of this heat that is absorbed by the object will be less for the blunter shape.

It should be emphasized that Eq. (1.8) contradicts what might at first appear to be a valid intuitive argument: a lower-drag vehicle converts less kinetic energy into heat than a blunt, high-drag vehicle and should therefore absorb less heat. Nevertheless, Eq. (1.8) indicates that a blunt vehicle will absorb less heat than a low-drag vehicle. (The heat not absorbed by the vehicle is of course absorbed by the atmosphere.) Note that the drag increase must not be attained by significantly increasing the effective skin-friction coefficient. The ballistic coefficient, β, does appear in Eq. (1.7), but for vehicles of engineering interest (β around 7500 kg/m^2 or less) the magnitude of β is of secondary importance in determining the heat load. If Eq. (1.8) is an acceptable approximation to Eq. (1.7) for determining the total heat load, then the ballistic coefficient β does not have a major influence on heat absorption. Before leaving Allen and Eggers, we will include a statement from their classic paper:

> It seems unlikely that a pointed nose will be of practical interest for high-speed missiles since not only is the local heat-transfer rate exceedingly large in this case, but the capacity for heat retention is small. . . . Body shapes . . . would more probably, then, be those with nose shapes having nearly hemispherical tips.

The most fundamental, or minimum, re-entry body is a BRB—ballistic re-entry body. The only intended loads acting on this object are gravity and aerodynamic, and the aerodynamic loads should be entirely along the negative of the velocity vector. Asymmetric ablation might lead to trim angles of attack, resulting in transverse or lifting forces. Of course, for a BRB, such loads would be regarded as spurious and unintended. Typically, the maximum stagnation pressures encountered by an unmanned BRB are about 150 atm. The maximum heat flux in the sonic region behind the shock wave would be about 3.0×10^8 J/(m^2-s) (3.5×10^4 Btu/ft^2-s); the total heat load in the sonic flow region would typically be 2.5×10^9 J/m^2 (3.0×10^5 Btu/ft^2).

A typical manned ballistic re-entry body is shown in Fig. 1.5. This vehicle is the *Mercury* capsule, which had to meet such requirements as minimum heat transfer to the vehicle, low maximum axial acceleration, and low terminal velocity. For a *Mercury* capsule the peak heating loads were much less than the typical unmanned BRB [a peak flux of about 8.0×10^5 J/(m^2-s) and a total heat load (heat pulse) of about 8.0×10^7 J/kg]. The *Mercury* capsule is a museum piece; it is virtually impossible that a single-seat mission will ever be undertaken again. It is interesting to note, though, that this vehicle made use of all three of the basic thermal protection systems: reradiation, heat sink, and ablation.

As the name implies, the reradiation system provides a thermodynamic balance between the absorbed and radiated heat energy. The *Mercury* capsule used shingles of René 41, a nickel-coated alloy, on the cone. The recovery canister, which experienced higher heating rates, used a heat sink of beryllium. A heat sink, as the name implies, absorbs more heat than can be reradiated to the atmosphere. The maximum value of specific heat that can be absorbed by a practical material is on the order of 2.5×10^6 J/kg (1000 Btu/lb). Copper and beryllium are the only materials which have found much application as heat sinks. Most of the heat load was taken by the glass-reinforced charring ablator. The word *ablation* originally had a surgical meaning, as in cutting or removing; later the term was used by atmospheric physicists to describe the mass loss of meteors entering the atmosphere. Unlike a heat sink, an ablator has a low thermal conductivity. Organic plastics are the only ablators which have found

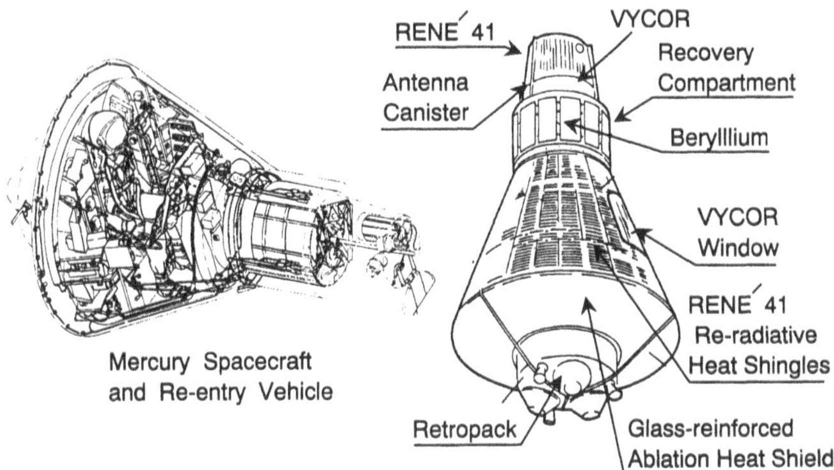

Fig. 1.5 Mercury re-entry vehicle.

much engineering application; the heat load liquifies or vaporizes some of the material of the ablation shield to minimize heat soak into the vehicle.

The capability of a re-entry body may be extended considerably by providing some method for effecting maneuvering. A maneuvering re-entry vehicle (MaRV) can use maneuvering to offset the effects of positional/velocity errors acquired at boost termination. For example, we will show in Chapter 6 that the sensitivity of range error to velocity error might be expressed as

$$\frac{\partial R}{\partial V} = \frac{2R_E}{V_0}\left\{\sin(\theta_i) + \cos(\gamma_0)[1 - \cos(\theta_i)]\right\} \tag{1.9}$$

where R_E is the radius of the Earth, V_0 and γ_0 are the velocity and flight path angle at boost termination, and θ_i is central Earth angle subtended by the trajectory segment from burnout to impact. As discussed in Chapter 6, a typical value of $\partial R/\partial V$ is about 4.5 km/(m/s). This means that a 1-m/s velocity error can result in a range error at impact of 4.5 km. A maneuvering vehicle, which has a means of detecting position (say, at atmospheric entry) would be able to reduce this ballistic error to an acceptably small value. Maneuvering can also be used for other purposes, such as avoiding an active defense (see Chapter 9).

Most of the methods proposed for providing maneuvering loads are based upon creating an asymmetry in the near flowfield. Obvious methods include movable flaps which can provide one, two, or three degrees of freedom (pitch, yaw, and roll). Another aerodynamic approach to developing maneuvering loads is the injection of material into the boundary layer or simply jet interaction. Jet interaction seems to be suited only to steering out navigational errors, however, not for defensive maneuvering.

Control can also be effected by moving a mass laterally in the vehicle to offset the center of gravity. The resulting mass asymmetry is equivalent to an aerody-

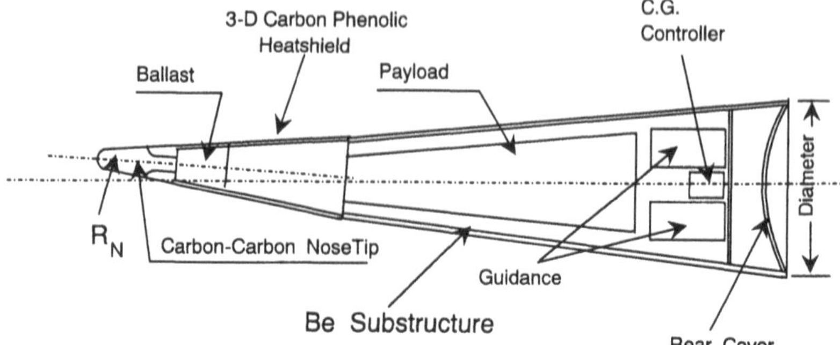

Fig. 1.6 Typical bent body MaRV configuration.

namic asymmetry. Usually the capacity to maneuver is enhanced by "building in" an aerodynamic asymmetry; the moving mass is then positioned to control the direction of the lift force that is generated by the configurational asymmetry. The *Gemini* capsule, for example, had a more elaborate control system than did *Mercury*; *Gemini* also had a slight mass offset to provide a trim angle of attack which produced a lift whose direction and magnitude could be controlled by the astronaut.

An alternative to the moving-mass controller is the pure aerodynamic control. Figure 1.6 shows a typical bent-body moving-mass control system, and Fig. 1.7 shows the aerodynamic split-windward flap approach. A more exotic

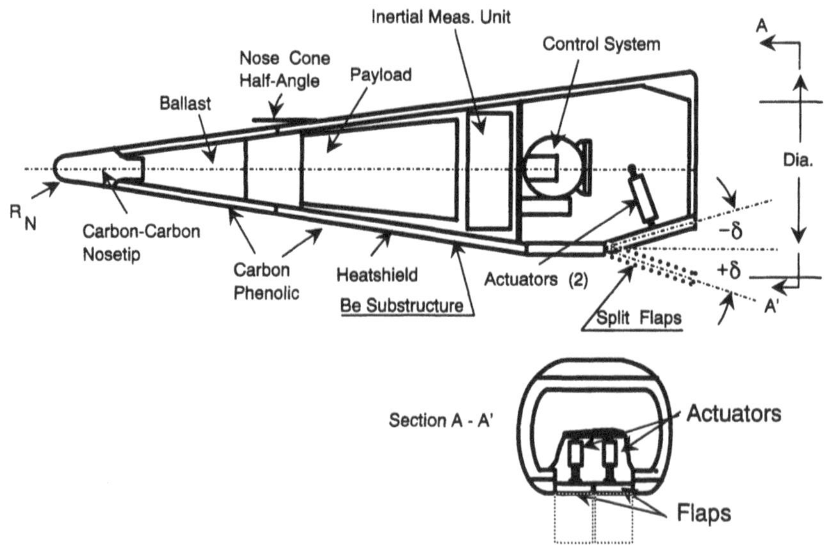

Fig. 1.7 Re-entry body with split-windward flap.

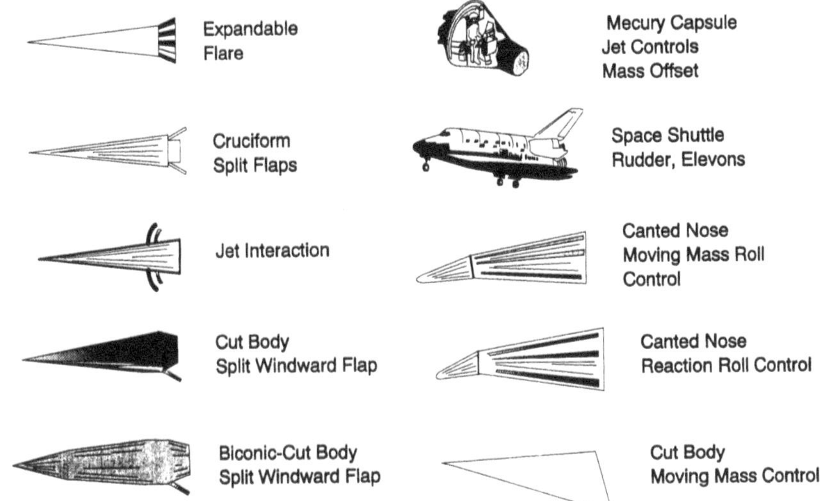

Fig. 1.8 Maneuvering re-entry vehicle control concepts.

method is to alter the pressure field of the ionized gas flow in the vicinity of the vehicle by the onboard generation of an electromagnetic field. A survey of several types of re-entry vehicles is presented in Fig. 1.8.

In addition to the thermal loads generated by atmospheric entry, there are very important dynamic effects. For a manned re-entry vehicle the trim angle of attack must be strictly controlled: a vehicle with a lift-to-drag ratio, C_L/C_D, that is modest by the standards of atmospheric vehicles can result in excessive side loads at re-entry if the trim angle of attack is not severely restricted.

Beyond even load considerations, at least for hyperbolic and near parabolic orbits, is the controlling of load to cause capture of a re-entry body by a planet. To ensure capture of an RB, the flight path angle at entry must be very tightly controlled to effect capture and to limit aerodynamic loads.

The seminal paper in atmospheric entry analysis was contributed by Chapman.[17] Chapman defined the perigee parameter F_p as

$$F_p = \frac{\rho_p}{2\beta}(R_p H)^{1/2} \tag{1.10}$$

where ρ_p is the atmospheric density at perigee and R_p is the distance from the planet center to the perigee (the point of closest approach). Figure 1.9 shows a plot of the maximum acceleration (in g) versus the perigee parameter for several values of the entry velocity ratio

$$\lambda_E = \frac{V_E}{\sqrt{gR_0}}$$

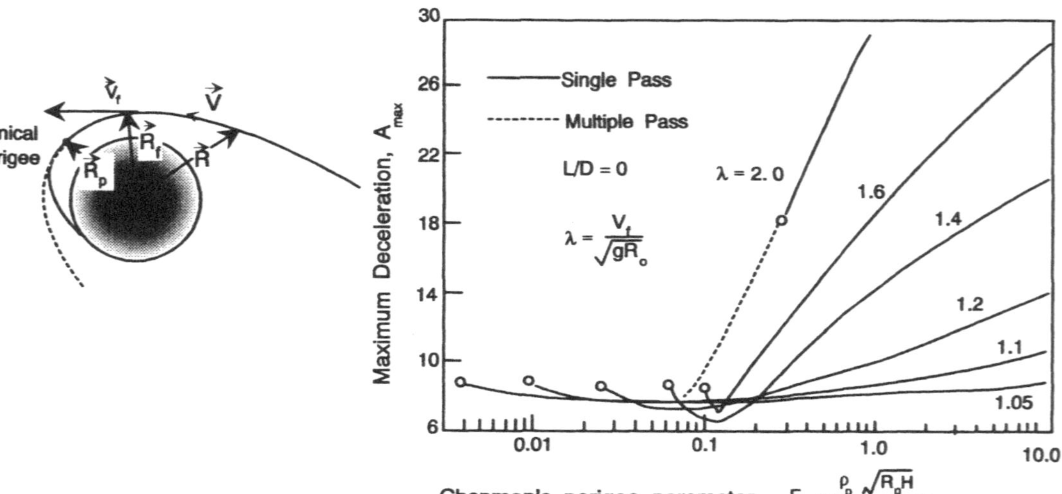

Fig. 1.9 Maximum deceleration of nonlifting vehicles vs Chapman's perigee parameter.

where R_0 is the radius of the planet. Note the rise in maximum acceleration with increasing values of entry velocity.

1.3 Overview

We shall not be concerned in this book with the engineering design of either vehicle structures to survive inertial and pressure loads or heat shields to survive thermal loads. It might be sufficient to recognize that the details of heat load distribution and intensity vary with vehicle configuration and re-entry trajectory. Presumably both the vehicle shape (design) and trajectory (operation) are under the control of the mission planner.

A direct cause-and-effect relationship between vehicle configuration and trajectory (the cause) and thermal and pressure distributions (the effect) is a bit of an oversimplification. It is equally true that the existing pressure/thermal distributions have an influence on subsequent configurations and therefore on pressure loads and ultimately the trajectory. For example, the thermal characteristics of the near flowfield will determine the gas constituents and properties in that field, thereby affecting the moment and drag loads on the body. Obviously, heat-shield shape changes resulting from asymmetric ablation can "freeze in" an asymmetry, bringing about a nonzero trim angle, which in turn will generate an aerodynamic load normal to the trajectory tangent.

In spite of the complex interactions among the thermal and pressure distributions, the configuration, and the trajectory, we shall place most of the emphasis in this book on vehicle dynamics. The distributed pressure field will be represented by various "lumped" parameters known as aerodynamic coefficients and stability derivatives. These parameters will vary significantly over the entire trajectory. However, such variations are less important when compared to changes in atmospheric density, which will vary by several orders of magnitude from entry to impact.

If the re-entry body can be regarded as a rigid body immersed in and moving through a gas and subjected to a gravitational field, a vectorial representation of the loads might take the following form:

$$m\boldsymbol{a}^b = \boldsymbol{F}_a^b + \boldsymbol{F}_g^b$$

where m is the body mass, $\boldsymbol{a}^b$ is the acceleration of the body, $\boldsymbol{F}_a^b$ is the aerodynamic load, and $\boldsymbol{F}_g^b$ is the gravitational load. The superscript b indicates that all loads are resolved in some body-fixed frame. The above equation could be regarded as a generalized force equation where m could also be the inertia tensor, $\boldsymbol{a}$ the angular acceleration, and $\boldsymbol{F}_a$ the generalized moments. An important part of this book discusses the forms that this force equation takes in various coordinate frames. For example, the vector $\boldsymbol{F}_g^b$ represents the resolution of the gravitational vector in a body frame; however, gravity is defined more fundamentally in a local Earth-frame. Thus, considerable attention is devoted to the transformation of vector and tensor entities (and their derivatives) from one reference frame to another.

At this point it might be of interest to summarize the remaining chapters of the text. Since the most interesting and critical engineering problems associated with re-entry bodies occur within the atmosphere, we consider various atmospheric models in Chapter 2. Throughout its trajectory, both exo- and endoatmospheric, the re-entry body is subjected to the Earth's gravitational field. Therefore in Chapter 3 we consider both the Earth's shape and gravitational field, with emphasis on deviations from the simple inverse-square law. Because of the fundamental importance of transferring vector quantities among various frames, we will examine various ways of dealing with axis transformation geometry in Chapter 4. The fundamental physics supporting the force and moment equations is given in Chapter 5. Keplerian, or exoatmospheric, motion is covered in Chapter 6 for particles and in Chapter 10 for rigid bodies. For endoatmospheric motion, particle motion is covered in Chapter 7, and maneuvering body motion is covered in Chapter 9. Some consideration is given to decoy masking of re-entry bodies in Chapter 8. An adequate treatment of the gasdynamics associated with re-entry would require several volumes dedicated to this subject alone. In Chapter 11 we consider the lumped parameter representation of aerodynamic loads and provide a few computer programs for calculating loads on certain restricted shapes. In Chapter 12 we again return to the re-entry analysis of Chapter 7 and extend the treatment to angular motion. In Chapter 13, we identify various error sources and consider in great detail various ways of quantitatively assessing the effects of such errors. Finally, in Chapter 14 we discuss the analytical background surrounding inertial navigators.

References

[1]Wasson, J. T., *Meteorites—Their Record of the Early Solar System History*, W. H. Freeman, Salt Lake City, UT, 1985.

[2]Dodd, R. T., *Thunderstones and Shooting Stars—The Meaning of Meteorites*, Harvard University Press, Cambridge, MA, 1986.

[3]Heide, F., *Meteorites*, Chicago University Press, Chicago, IL, 1962.

[4]Kriznov, E. L., *Giant Meteorites* (translated by J. S. Romankiewica), Pergammon Press, Elmsford, NY, 1966.

[5]Erickson, J., *Target Earth*, TAB Books, Summit, PA, 1991.

[6]Alvarez, L. W., et al., "Impact Theory of Mass Extinctions and Invertebrate Fossil Record," *Science*, Vol. 223, 1984, pp. 1135–1140.

[7]Grieve, R. A. F., "The Record of Impact on Earth—Implications for a Major Cretaceous/Tertiary Impact Event," Special Paper 190, *Geological Implications of Impacts of Large Asteroids and Comets on the Earth*, edited by L. T. Silver and P. H. Schultz, Geological Society of America, Boulder, CO, 1982, pp. 25–38.

[8]Walback, W. S., Lewis, R. S., and Anders, E., "Cretaceous Extinctions: Evidence of Wildfires and the Search for Meteoric Material," *Science*, Vol. 230, 1985, pp. 167–170.

[9]Tappan, H., "Extinction or Survival: Selectivity and Causes of Phanerozoic Crises," Paper 190, *Geological Implications of Impacts of Large Asteroids and Comets on the Earth*, edited by L. T. Silver and P. H. Schultz, Geological Society of America, Boulder, CO, 1982, pp. 265–276.

[10]Nininger, H. H., *Out of the Sky*, Dover Publications, Mineola, NY, 1952.

[11]Wasson, J. T., *Meteorites: Classification and Properties*, Springer Publishing, New York, NY, 1974.

[12]Opik, E. J., *Physics of Meteor Flight in the Atmosphere*, Interscience Tracts on Physics and Astronomy, No. 6, Interscience Publishers, Wiley, Somerset, NJ, 1958.

[13]Buchwald, V. F., *Iron Meteorites*, University of California Press, Berkeley, CA, 1975.

[14]McDougall, W. P., *. . . The Heavens and the Earth—a Political History of the Space Age*, BASIC Books, Harper-Row, Scranton, PA, 1985.

[15]Baker, D., *The History of Manned Space Flight*, published in the United States by Crown Publishers, New York, NY, 1981.

[16]Allen, H. J., and Eggers, A. J., "A Study of Motion and Aerodynamic Heating of Ballistic Missiles Entering the Earth's Atmosphere at High Supersonic Speeds," NACA Rept. 1381, 1958.

[17]Chapman, D. R., "An Approximate Analytic Method for Studying Entry into Planetary Atmospheres," NACA-TN-4246, May 1958.

2
Atmospheric Model

2.1 Introduction

The loads imparted to a re-entry vehicle (RV) during its trajectory may be conveniently divided into body forces (gravitational) and contact or surface forces (aerodynamic). While gravitational forces are present throughout the trajectory, the aerodynamic forces dominate at altitudes below 50 km. It will be shown subsequently that peak aerodynamic decelerations, or the axial component of the specific aerodynamic forces, can be as high as 100 *g*; certainly, axial loads in excess of 20 *g* can be present over most of the trajectory segment below 30 km. Furthermore, even an axially symmetric maneuvering re-entry vehicle (MaRV) can generate transverse loads of at least 10 *g*. Uncertainties in the aerodynamic loads can cause significant errors at impact. Thus, it is important to be able to represent the atmosphere adequately for computer or analytical modeling of RV trajectories.

An atmospheric model must be able to represent the vertical run of pressure, density, and temperature. (From pressure, density, and temperature there are many additional derived quantities such as dynamic viscosity, molecular mean free path length, and molecular collision frequency.) At first glance the atmosphere might be described as a thermodynamic medium bounded by a gravitational field and in hydrostatic equilibrium set by solar radiation. Since solar radiation and atmospheric reradiation must vary diurnally and annually, any single (i.e., standard) atmospheric model cannot be more than an approximation. We also expect atmospheric properties to vary not only with time, but also with location; i.e., we would expect such properties to vary with longitude and latitude. Furthermore variations in solar radiation threaten the premise of hydrostatic equilibrium by creating atmospheric circulations. Nevertheless, we will limit our ambitions here to representing the atmosphere by a single model with sufficient adjustment points to match such variations that might be appropriate to a particular time or location. An underlying premise of this chapter is that such a standard atmospheric model can have utility in the evaluation of the performance of a re-entry vehicle or components of such a vehicle.

2.2 Standard Atmospheres

The purpose of a Standard Atmosphere has been defined by the World Meteorological Organization (WMO)[1]:

A hypothetical vertical distribution of atmosphere temperature, pressure, and density which by international agreement for historical reasons is roughly representative of year-round mid-latitude conditions. Typical usages are as a basis for pressure altimeter calibrations, aircraft performance calculations, aircraft and rocket design, ballistic tables.... The air is assumed to obey the perfect gas law and the hydrostatic equation, which taken together relate temperature, pressure, and density with geopotential [altitude]. Only one standard atmosphere should be specified at a given time.... The atmosphere shall also be considered to rotate with the Earth and be an average over a diurnal cycle, semiannual variations, and the range of conditions from active to quiet geomagnetic and active to quiet sunspot conditions.

In essence, Standard Atmosphere is a single mapping of primary properties (pressure and temperature) and derived properties (density, viscosity, speed of sound propagation, mean free path length, etc.) with altitude. Consider an accepted standard: the COESA (Committee on Extension to the Standard Atmosphere) *U.S. Standard Atmosphere 1976.*[2] In discussing this model we draw a distinction between kinetic and molecular temperatures (T and T_M, respectively) and geometric and geopotential altitudes (Z and h, respectively). Kinetic temperature is the measure of molecular kinetic energy and for all practical purposes is identical to thermometer temperature at low altitudes; molecular temperature accepts the fiction that the molecular weight of the air at any altitude, M, remains constant at its sea-level value, M_0. The two temperatures are related as follows:

$$T_M = \left[\frac{M_0}{M}\right] T \tag{2.1}$$

It will be seen that maintaining molecular weight invariant with altitude leads to attractive mathematical simplifications; the penalty, of course, is the introduction of the artificial molecular temperature T_M. The geopotential altitude h is based upon the assumption that the gravitational acceleration is constant with altitude, rather than obeying the familiar inverse-square law [see Eq. (3.71a)]. The geometric, or measurable, altitude includes the familiar variation of gravitational acceleration with the square of the reciprocal of the geocentric distance. Thus, h and Z are related as follows:

$$h = \left[\frac{R_E}{(R_E + Z)}\right] Z \tag{2.2a}$$

or, in differential form,

$$\begin{aligned} \mathrm{d}h &= \left[\frac{R_E}{(R_E + Z)}\right]^2 \mathrm{d}Z \\ &= \left(\frac{g}{g_0}\right) \mathrm{d}Z \end{aligned} \tag{2.2b}$$

The *U.S. Standard Atmosphere 1976*[2] is an atmospheric model that meets the conditions set forth by WMO. The model covers the atmosphere from the Earth's surface to an altitude of 1000 km at a latitude of 45°N. The defining atmospheric elements are sea-level values of temperature and pressure and segmented layers over which the molecular temperature has a defined altitude profile. The atmospheric region from sea level to 86 km is divided into seven strata. Within each strata, the molecular temperature is represented as a linear function of geopotential altitude as

$$T_M = T_{M_i} + L_{h_i}(h - h_i) \tag{2.3a}$$

where the subscript i identifies the layer ($0 \leq i \leq 7$) and

$$L_{h_i} = \left.\frac{\mathrm{d}T_M}{\mathrm{d}h}\right|_i \tag{2.3b}$$

where L_{h_i} is known here as the *thermal lapse rate*. Linear variation of T_M with geometric altitude Z uses the following related definition:

$$T_M = T_{M_i} + L_{Z_i}(Z - Z_i) \tag{2.4a}$$

where

$$L_{Z_i} = \left.\frac{\mathrm{d}T_M}{\mathrm{d}Z}\right|_i \tag{2.4b}$$

It will be shown subsequently how the thermal lapse rate, L_Z or L_h, enters into the computer model of the atmosphere. At this time, however, we must accept (at least tentatively) that it is appropriate to divide the atmosphere into distinct strata; within each strata the temperature varies uniquely with altitude. In most cases this variation is linear; linear variation of temperature with altitude means that any given layer can be characterized by its lapse rate and the value of the temperature at the lowest altitude of that layer. The lapse rate can be positive, negative, or zero. Layers having a zero lapse rate are identified as being *isothermal*.

In our computer modeling efforts we use a sort of hybrid atmosphere, the 1976 Standard Atmosphere, from sea level to a geometric altitude of 86 km, and the 1962 Standard Atmosphere above 86 km. The seven-layer 1976 atmosphere[2] below 86 km is given in Table 2.1. In Table 2.1, note that there are two strata (1–2 and 4–5) over which the temperature remains constant, i.e., $L_Z = L_h = 0$, and two layers (2–3 and 3–4) over which the temperature increases.

The 1976 model's layers above an altitude of 86 km are given in Table 2.2. Unfortunately, the 1976 model has two layers (8–9 and 10–11) over which the temperature variation is nonlinear. For layer 8–9 the temperature varies elliptically with altitude, whereas for layer 10–11 the variation is exponential. The appropriate derivatives (i.e., the thermal lapse rates L_Z or L_h) can easily be calculated, but such rates are not constant over the layer.

Table 2.1 Seven-layer atmosphere

Layer index	Geopotential altitude h, km	Geometric altitude Z, km	Molecular temperature T_M, K	Lapse rate L_h, K/km
0	0.0	0.0	288.150	−6.5
1	11.0	11.0102	216.650	0.0
2	20.0	20.0631	216.650	+1.0
3	32.0	32.1619	228.650	+2.8
4	47.0	47.3501	270.650	0.0
5	51.0	51.4125	270.650	−2.8
6	71.0	71.8020	214.650	−2.0
7	84.8520	86.0	186.946	

For our simple model we have eschewed the complexities of the 1976 atmosphere at the higher altitudes (above an 86-km geometric altitude), replacing it with an earlier 1962 atmosphere.[2] The thermal lapse rates are given in Table 2.3 from 86 km up to an altitude of 700 km. While the atmosphere has been modeled for altitudes above 86 km, two conditions must be held before the analyst.

First, the properties of the atmosphere over the altitude range of 86 to 700 km will vary widely because at such altitudes the atmosphere is greatly influenced by solar radiation. Above 86 km or so, the assumption of hydrostatic equilibrium is no longer valid. Because of diffusion and vertical motion of the various constituent gas species, the atmosphere above 86 km is highly dynamic. As suggested by Jursa[1] it is necessary to include diffusive separation of the gas constituents (see Chapter 14). Such elaborate modeling, although essential for many geophysical studies, is tangential to our main interests; the reader may pursue such considerations in Refs. 1, 4, and 5.

Second, the utility of an atmospheric model in re-entry vehicle studies lies in the computation of loads on a re-entry vehicle. Aerodynamic loads are nearly always negligible fractions of vehicle weight at altitudes above 80 km. Thus, using an approximate but computationally tractable model is entirely appropriate for the intended use of the model. Temperature variations with altitude for the 1962 and 1976 Standard Atmospheres models are shown in Fig. 2.1.

2.3 Atmospheric Description

A literal description of the atmospheric strata is not as finely divided as the numerical description which is based upon changes in the thermal lapse rate. For the literal description, a change of strata is indicated by an altitude at which the atmosphere becomes isothermal, i.e., the thermal lapse rate becomes zero. For instance, the lowest atmospheric layer ends at an altitude where the

Table 2.2 1976 standard atmosphere strata from 86 km to 1000 km

Geometric altitude range: 86.0 km to 91.0 km (Index 7–8)

$$\frac{dT}{dZ} = 0.0 \text{ K/km}$$

$$T_8 = T_7$$

Geometric altitude range: 91.0 km to 110.0 km (Index 8–9)

$$T = T_C + A\left[1 - \left(\frac{Z - Z_8}{a}\right)^2\right]^{1/2}$$

$$\frac{dT}{dZ} = -\frac{A}{a}\left(\frac{Z - Z_8}{a}\right)^2\left[1 - \left(\frac{Z - Z_8}{a}\right)^2\right]^{-1/2}$$

$$T_c = 263.1902 \text{ K}$$

$$A = -76.3232 \text{ K}$$

$$a = -19.9429 \text{ km}$$

Geometric altitude range: 110.0 km to 120.0 km (Index 9–10)

$$T = T_9 + L_Z(Z - Z_9)$$

$$\frac{dT}{dZ} = +12.0 \text{ K/km}$$

Geometric altitude range: 120.0 km to 1000.0 km (Index 10–11)

$$T = T_\infty - (T_\infty - T_{10})\exp(-\lambda\xi)$$

$$\frac{dT}{dZ} = \lambda(T_\infty - T_{10})\left[\frac{R_E + Z_{10}}{R_E + Z}\right]^2 \exp(-\lambda\xi)$$

$$\xi = (Z - Z_{10})\left[\frac{R_E + Z_{10}}{R_E + Z}\right]$$

$$\lambda = 0.01875/\text{km}$$

$$R_E = 6.356766 \times 10^3 \text{ km}$$

$$T_\infty = 1000 \text{ K}$$

Table 2.3 1962 standard atmosphere from 86 km to 700 km

Layer index	Geometric altitude, km	Molecular temperature, K	Kinetic temperature, K	Molecular weight	Lapse rate, K/km
7	86.0	186.946	186.946	28.9644	+ 1.6481
8	100.0	210.65	210.02	28.88	+ 5.0
9	110.0	260.65	257.00	28.56	+10.0
10	120.0	360.65	349.49	28.08	+20.0
11	150.0	960.65	892.79	26.92	+15.0
12	160.0	1110.65	1022.2	26.66	+10.0
13	170.0	1210.65	1103.4	26.49	+ 7.0
14	190.0	1350.65	1205.4	25.85	+ 5.0
15	230.0	1550.65	1322.3	24.70	+ 4.0
16	300.0	1830.65	1432.1	22.65	+ 3.3
17	400.0	2160.65	1487.4	19.94	+ 2.6
18	500.0	2420.65	1506.1	16.84	+ 1.7
19	600.0	2590.65	1506.1	16.84	+ 1.1
20	700.0	2700.65	1507.6	16.17	

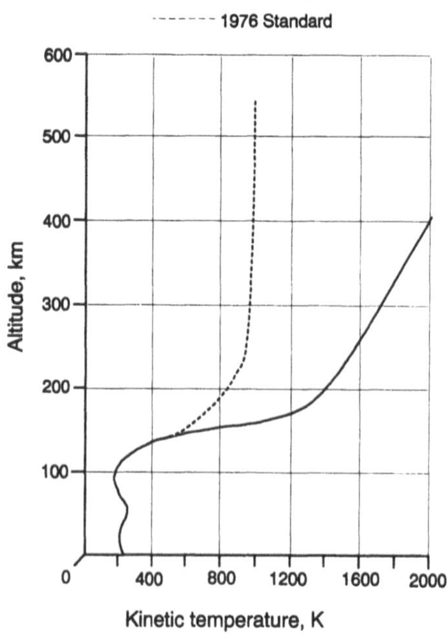

Fig. 2.1 Temperature variations for 1962 and 1976 standard atmospheres.

atmosphere becomes isothermal. According to Table 2.1, this shift occurs at an altitude of 11 km. Refer to Fig. 2.2.

The region of the atmosphere between sea level and 11 km is known as the *troposphere*. It will be shown that the *first e-folding* of the atmosphere occurs at about 6.7 km. This means that the density ratio (density at altitude of 6.7 km to density at sea level) is $e^{-1} = 1/e = 0.3679$. Thus 63% of the atmosphere lies below an altitude of 6.7 km. The layer above the troposphere is called the *stratosphere;* the dividing line is called the *tropopause*. The stratosphere extends to the next isothermal layer, which starts at an altitude of 47 km. The region above the stratosphere is called the *mesosphere,* with the dividing line called the *stratopause*. The mesosphere extends from roughly 47 km to 86 km or so.

Above an altitude of 86 km, hydrodynamic equilibrium is no longer a valid assumption. At an altitude of 500 km, the temperature is set at the so-called exospheric temperature. The region between 86 km and 500 km is called the *thermosphere;* the region above the thermosphere (i.e., above 500 km) is called the *exosphere*. The exospheric temperature is constant from 500 km throughout the remaining atmosphere.

There is some ambiguity as to the value of the exospheric temperature. Reference 1 gives the average exospheric temperature used by the various models. These are summarized in Table 2.4. Clearly there is some variation in the value of the exospheric temperature. For Table 2.3 we selected a value of 1500 K in accordance with the 1962 Standard Atmosphere model. The kinetic temperature variations with altitude for the various models has been shown in Fig. 2.1. Some of the questions concerning the setting of the exospheric temperature are discussed at length in Ref. 1. Since the atmosphere above 100 km is of minor importance in the dynamics of re-entry vehicles, no further discussion will be

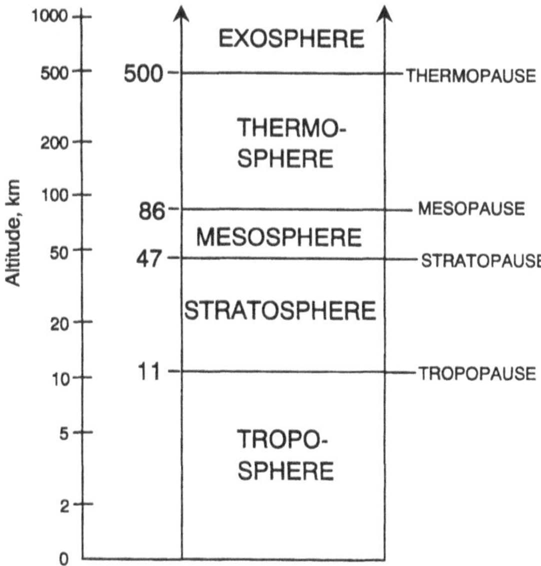

Fig. 2.2 Atmosphere strata definitions.

Table 2.4 Average exospheric temperature models

Model	Average exospheric temperature
U.S. Standard Atmosphere 1962	1500 K
Mean circa 1965	1200 K
Mean circa 1972	1000 K
U.S. Standard Atmosphere 1976 (1.5 solar cycles)	1000 K

given here. Obviously an analyst who takes exception to setting the exospheric temperature at 1500 K may make an adjustment in accordance with evolving upper atmospheric physical models or solar activity at the time of re-entry. However, setting the exospheric temperature at 1000 K will not bring the model into strict compliance with the 1976 standard model because the 1976 model uses nonlinear thermal distributions in two strata.

2.4 Physical Foundations of an Atmospheric Model

Before considering the physical foundations of an atmospheric model, the first concern is the intended use of such a model. For our purposes the primary concern is to have available the pressure, temperature, and derived density as a function of altitude. We wish to have a computer program for which altitude is an input and pressure, temperature, density, and various quantities that are derived from these three are output. Certainly, other atmospheric models may be developed in which other quantities may be considered primary (such as wind magnitude and direction, electron density, gas constituents, etc.).

The first fundamental relationship is the atmospheric equilibrium equation, which relates pressure P and density ρ. Figure 2.3 depicts an atmospheric element in equilibrium under pressure and gravitational forces. Summing both pressure and gravitational forces gives

$$-\rho g A\,\mathrm{d}Z + [P - (P + \mathrm{d}P)]A = 0$$

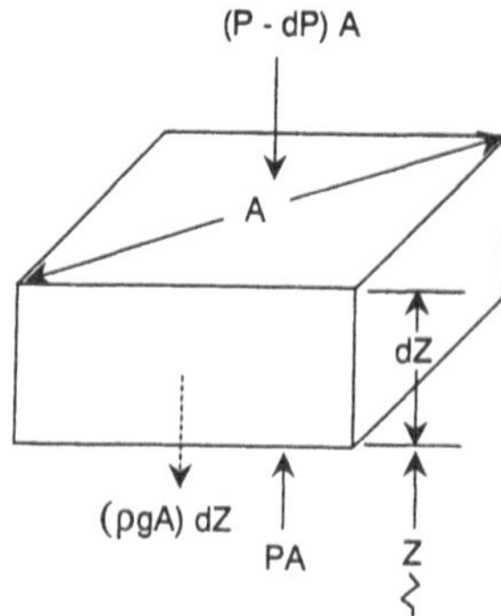

Fig. 2.3 Atmospheric element under equilibrium conditions.

or

$$\frac{dP}{dZ} = -\rho g \tag{2.5}$$

In addition to assuming force equilibrium, we assume the atmosphere to be a thermodynamic fluid. A satisfactory relationship among pressure, density, and temperature takes the form of the thermal equation of state of an ideal gas, i.e.,

$$PV = NR^{*}T \tag{2.6}$$

where the intensive properties, pressure P and temperature T, are related to the volume V, the number of moles present N, and the universal gas constant R^{*}.

In this section numerical constants are presented sparingly; the reader should refer to Table 2.5 for the needed numerical values. Equation (2.6) may be written in terms of intensive properties if we restrict our attention to a particular gas or, more to the point, a gas whose constituents are fixed. (At higher altitudes—say, above 100 km—we will have to modify this assumption of fixed gas constituency.) First, we must divide Eq. (2.6) by the volume V and recognize that

$$N = m/M \qquad \text{and} \qquad \rho = m/V \tag{2.7}$$

Table 2.5 Atmospheric constants

Definition	Symbol	Value	Units
Sea-level pressure	P_0	1.013250×10^5	N/m^2
Sea-level temperature	T_0	288.15	K
Sea-level density	ρ_0	1.225	kg/m^3
Avagadro's number	Na	6.0220978×10^{23}	/kg-mole
Universal gas constant	R^*	8.31432×10^3	J/kg-mole-K
Gas constant (air)	R	287.0	J/kg-K
Adiabatic polytropic constant	γ	1.405	
Sea-level molecular weight	M_0	28.96643	
Sea-level gravity acceleration	g_0	9.806	m/s^2
Radius of Earth (equator)	R_e	6.3781×10^6	m
Thermal constant	β	1.458×10^{-6}	kg/(m-s-K$^{1/2}$)
Sutherland's constant	S	110.4	K
Collision diameter	σ	3.65×10^{-10}	m

Note: K: degrees Kelvin; s: seconds; kg: kilograms; m: meters; J: joules.

where M is the molecular weight (of the gas under consideration), m is the mass of the gas, and ρ is the mass per unit volume (or density) of the gas. Equation (2.6) then becomes

$$P = \rho R^* T/M \tag{2.8}$$

We can express pressure and density in a form known as the *polytropic equation of state* as follows:

$$P/\rho^n = \text{constant} = P_0/\rho_0^n \tag{2.9}$$

where the subscript 0 refers to a reference condition. Equation (2.9) is useful in revealing some of the physics underlying the pressure-density relationship with altitude; however, if we use Eq. (2.9) [rather than Eq. (2.8)] we must come to terms with the exponent n and its variation with altitude.

Equations (2.8) and (2.9) provide two equations for three unknowns, namely pressure, temperature, and density. In our simple atmosphere model we resort to what is essentially an empiricism; namely, we impose a temperature-altitude relationship. In Tables 2.1 and 2.3 the assumption is made that temperature is a linear function of altitude. Table 2.2 indicates that at the higher altitudes (above 86 km) a more accurate temperature-altitude relationship may require a non-linear relationship. If we impose at most a linear variation of temperature with altitude, then the altitudes at which the thermal gradients change can be designated as *break points*. The altitude break points for the thermal atmosphere are given in Table 2.1 for altitudes up to 86 km and in Table 2.3 for altitudes above 86 km. The linear temperature profiles have already been given in Eqs. (2.3) and (2.4); these relationships are repeated below:

$$T_M = T_{M_i} + L_{Z_i}(Z - Z_i) \tag{2.10a}$$

$$T_M = T_{M_i} + L_{h_i}(h - h_i) \tag{2.10b}$$

where the subscript i refers to the layer over which the relationship is appropriate. The subscript M refers to the molecular temperature, which is defined in Eq. (2.1) and is an artificial temperature based upon the assumption that the molecular weight of air remains constant at its sea-level value throughout the whole atmosphere. Unfortunately, this assumption is not tenable for altitudes above 100 km, where the gas constituents change markedly from the sea-level mixture. Thus we will be required to include in the atmospheric model variations of the molecular weight M with altitude.

Figure 2.4 depicts the variations of both the temperature (molecular) and molecular weight with geometric altitude. We now consider how *geopotential* and *geometric* altitudes differ. To do this we must first recognize the relationship between the altitude and the gravitational force. An adequate relationship for our purposes is given in Eqs. (2.2).

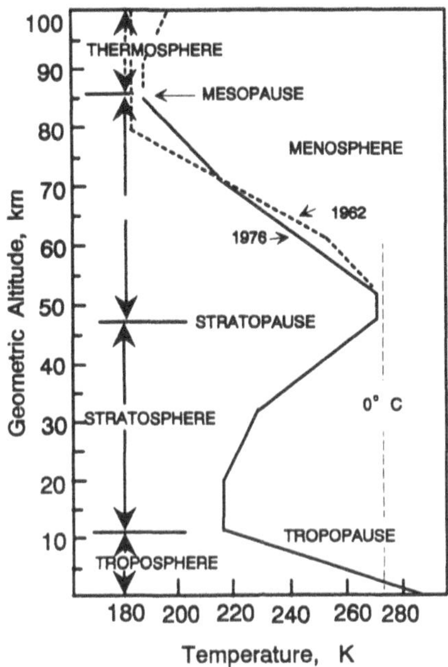

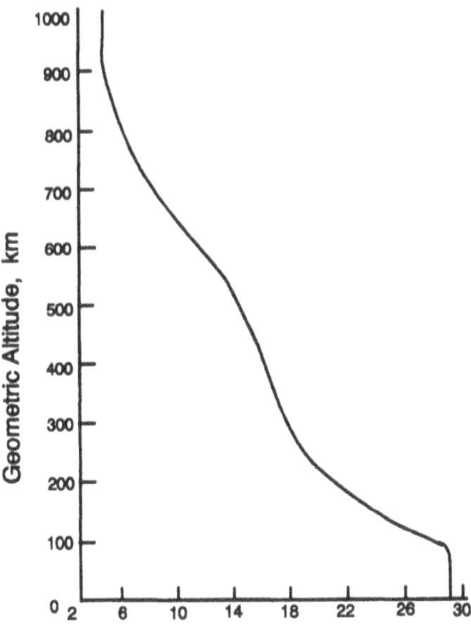

Fig. 2.4 Temperature and molecular weight vs altitude.

The geopotential altitude h is found by assuming a constant sea-level value for the gravitational acceleration throughout the entire atmosphere. It is related to the more physically realistic geometric altitude Z by the following equation:

$$\mathrm{d}h = (g/g_0)\,\mathrm{d}Z \tag{2.11}$$

The polytropic density-pressure relationship is an interesting way to "explain" the thermal gradient. First, rewrite both Eqs. (2.8) and (2.9) in terms of logarithmic differentials as

$$\frac{\mathrm{d}P}{P} = \frac{\mathrm{d}\rho}{\rho} + \frac{\mathrm{d}T}{T} \tag{2.12}$$

and

$$\frac{1}{n}\frac{\mathrm{d}P}{P} = \frac{\mathrm{d}\rho}{\rho} \tag{2.13}$$

respectively. By eliminating $\mathrm{d}\rho/\rho$ between Eqs. (2.12) and (2.13) we get

$$\frac{\mathrm{d}P}{P} = \left[\frac{n}{(n-1)}\right]\frac{\mathrm{d}T}{T}$$

or

$$\frac{\mathrm{d}P}{\mathrm{d}Z} = \left[\frac{n}{(n-1)}\right]\frac{P}{T}\frac{\mathrm{d}T}{\mathrm{d}Z} \tag{2.14}$$

From Eqs. (2.5) and (2.8) we get

$$\rho = -\frac{1}{g}\frac{\mathrm{d}P}{\mathrm{d}Z} = \frac{PM}{TR^*} \tag{2.15}$$

Insertion of Eq. (2.15) into Eq. (2.14) gives

$$\frac{\mathrm{d}T_M}{\mathrm{d}Z} = -\frac{(n-1)}{n}\left(\frac{gM_0}{R^*}\right) \tag{2.16}$$

where the kinetic temperature has been replaced by the molecular temperature according to Eq. (2.1). Integration of Eq. (2.16) is complicated by the altitude variation of g. As pointed out previously, use of the geopotential altitude avoids this problem by assuming a constant sea-level value for g.

The main interest in Eq. (2.16) is that it points out a simple relationship between the temperature gradient and the polytropic exponent n. An interesting comment has been made by H. Schlichting[6] concerning the exponent, n:

> The temperature gradient dT/dZ determines the stability of the stratification in the stationary atmosphere. The stratification is more stable when the temperature decrease with increasing height becomes smaller. For $dT/dZ = 0$ when $n = 1$, the atmosphere is isothermal and has a very stable stratification. For $n = \gamma = 1.405$, the stratification is adiabatic (isentropic) with $dT/dZ = -0.98$ K per 100 m [-9.8 K per km]. This stratification is indifferent because an air volume moving upward for a certain distance cools off through expansion at just the same rate as the temperature drops with height. The air volume maintains the temperature of the ambient air and is, therefore, in an indifferent equilibrium at every altitude. Negative temperature gradients of a larger magnitude than 0.98 K per 100 m result in unstable stratification.

It is of interest to see how the polytropic exponent varies with altitude and use it as a measure of the stability of each atmospheric layer. Table 2.6 provides a comparison of the thermal lapse rate L_h and the polytropic exponent n for the first 86 km of geometric altitude.

According to Schlichting[6] all layers should be stable when the polytropic exponent is between isothermal ($n = 1.0$) and adiabatic ($n = 1.405$). Of course, layers having a polytropic coefficient less than 1.0 are layers of thermal inversion in that the temperature increases with altitude. Such layers are also stable in the sense that an air mass will not tend to rise in such a thermal gradient. Note that there are no layers which pe̊rmit an adiabatic expansion of an air mass.

In developing an analytical model of the atmosphere, we eliminate the density in the equilibrium equation [Eq. (2.5)] by using the ideal gas equation [Eq. (2.8)] to give the following relationship between the proportional pressure change and the geometric altitude increment dZ:

$$\frac{dP}{P} = -\frac{M_0 g\, dZ}{R^* T_M(Z)} = -\frac{g\, dZ}{R T_M(Z)} \tag{2.17a}$$

Table 2.6 Variation of the polytropic exponent with altitude

Geopotential altitude h, km	Thermal lapse rate L_h, K/km	Polytropic exponent n
0.0		
	−6.5	1.2350
11.0		
	0.0	1.0000
20.0		
	+1.0	0.9716
32.0		
	+2.8	0.9242
47.0		
	0.0	1.0000
51.0		
	−2.8	1.0893
71.0		
	−2.0	1.0622
84.8520		

In terms of the geopotential altitude h, the relationship is as follows:

$$\frac{dP}{P} = -\frac{M_0 g\,dh}{R^* T_M(h)} = -\frac{g_0 dh}{RT_M(h)} \tag{2.17b}$$

where the gas constant for air has replaced the universal gas constant ($R = R^*/M_0$). We then insert the appropriate linear relationship for temperature [Eqs. (2.3) or (2.4)]. The next step is to integrate the above equation from the beginning of the ith layer to some point within the layer but below the $(i+1)$th layer.

We shall carry out the integration for two cases: 1) the isothermal atmosphere where L_Z is zero throughout the layer and 2) the nonisothermal layer where L_Z is nonzero. In both cases we consider the geometric altitude as the running, or independent, variable. There seems little point in using the geopotential altitude as the dependent variable when microcomputers are readily available to perform the calculations.

First we must express the gravitational acceleration in a form convenient for integration. We can represent the spherical Earth gravity field as

$$g = g_0\left[\frac{R_E^2}{(R_E + Z)^2}\right] \tag{2.18}$$

The above expression may be expanded to give the following satisfactory approximation:

$$g \approx g_0\left[1 - \left(\frac{2}{R_E}\right)Z\right] = g_0[1 - bZ] \tag{2.19}$$

where $b = 3.139 \times 10^{-7}/\text{m}$. Inserting Eq. (2.19) into Eq. (2.17a) we get

$$\int_{P_i}^{P} \frac{dP}{P} = -\frac{g_0}{R}\int_{Z_i}^{Z} \frac{(1 - bZ)dZ}{T_{M_i} + L_Z(Z - Z_i)} \tag{2.20}$$

For $L_Z = 0$ (the isothermal case) the above equation becomes

$$P = P_i \exp\left\{-\left[\frac{g_0(Z - Z_i)}{RT_{M_i}}\right]\left[1 - \frac{b}{2}(Z - Z_i)\right]\right\} \tag{2.21a}$$

$$T_M = T_{M_i} \tag{2.21b}$$

$$\rho = \rho_i \exp\left\{-\left[\frac{g_0(Z - Z_i)}{RT_{M_i}}\right]\left[1 - \frac{b}{2}(Z - Z_i)\right]\right\} \tag{2.21c}$$

For the nonisothermal layers we have, from Eq. (2.20),

$$\int_{P_i}^{P} \frac{\mathrm{d}P}{P} = -\frac{g_0}{RL_{Z_i}} \int_{Z_i}^{Z} \frac{(1 - bZ)\mathrm{d}Z}{[T_{M_i}/L_{Z_i} + (Z - Z_i)]} \tag{2.22}$$

The above equation integrates to

$$P = P_i \left[\left(\frac{L_{Z_i}}{RT_{M_i}} \right)(Z - Z_i) + 1 \right]^{-\left\{ \left(\frac{g_0}{RL_{Z_i}} \right) \left[1 + b\left(\frac{T_{M_i}}{L_{Z_i}} - Z_i \right) \right] \right\}} \exp\left\{ \left(\frac{g_0 b}{RL_{Z_i}} \right)(Z - Z_i) \right\} \tag{2.23a}$$

and the temperature is expressed linearly with geometric altitude as

$$T = T_{M_i} + L_{Z_i}(Z - Z_i) \tag{2.23b}$$

Density follows from the temperature and pressure by means of the ideal gas equation [Eq. (2.8)] as

$$\rho = \rho_i \left[\left(\frac{L_{Z_i}}{RT_{M_i}} \right)(Z - Z_i) + 1 \right]^{-\left\{ \left(\frac{g_0}{RL_{Z_i}} \right) \left[\frac{RL_{Z_i}}{g_0} + 1 + b\left(\frac{T_{M_i}}{L_{Z_i}} - Z_i \right) \right] \right\}} \exp\left\{ \left(\frac{g_0 b}{RL_{Z_i}} \right)(Z - Z_i) \right\} \tag{2.23c}$$

Equations (2.21) are used for zero lapse rate (i.e., for isothermal) layer calculations and Eqs. (2.23) are used for the nonzero lapse rate layers. These equations are coded in Appendix A in TRUEBASIC.

This section has presented some of the supporting physics for a so-called Standard Atmosphere model. No attempt will be made to account for daily or annual variations in the defining parameters or to make adjustments for variations with latitude. Such questions are discussed further in the volume edited by Jursa.[1]

It appears that there is no satisfactory analytical representation of atmospheric winds that is consistent with the idea of a standard atmosphere. It seems impossible to simply make some ground-based wind measurements and possibly a few additional measurements at altitude to provide an accurate wind-vector interpolator. However, there are some useful quantities which can be derived from a knowledge of pressure, temperature, and density. These topics will be discussed briefly in the next section of this chapter.

2.5 Derived Atmospheric Quantities

It was shown in the previous section that given pressure, temperature, and molecular weight at sea level and molecular temperature at set altitude intervals, it is possible to define an atmospheric model that is closely representative of an international standard. In Appendix A a computer program is listed which makes the model available for re-entry vehicle trajectory analysis. This standard model provides not only the primary atmospheric quantities such as pressure, temperature, and density, but also other atmospheric quantities which are readily derived from these primary quantities. Interesting summaries of many of these secondary atmospheric properties are available.[1,7] In the remainder of this section some of the derived quantities are discussed briefly.

Air Particle Speed

The mean air particle speed, which we designate as V, is the average of the distribution of speeds of all air particles within a sample coupon of air. Of course, V is a statistical variable and hence has meaning only if a sufficient number of particles are available in the coupon. It is assumed that variations in both pressure and density are negligible within the volume. From the kinetic theory of gases it can be shown that

$$V = \left(\frac{8R^*}{\pi M}\right)^{1/2} T^{1/2} = \left[\left(\frac{8R^*}{\pi M_0}\right) T_M\right]^{1/2} \tag{2.24}$$

Collision Frequency

The kinetic theory of gases defines the collision frequency F as the frequency with which a gas molecule collides with other molecules. All colliding particles are characterized by an effective collision diameter, designated as σ. The numerical value assigned to σ is considered small compared to the mean free path length λ. The collision frequency F may be written as

$$F = \sqrt{2}\pi\sigma^2 V n \tag{2.25}$$

where n is the number of molecules in a coupon of air.

Mean Free Path Length

The mean free path length λ is another statistical quantity. The mean free path length is the average distance traveled by a gas particle before collision with another particle. Since the reciprocal of F is the average time between collisions,

$$\lambda = V/F \tag{2.26}$$

Viscosity

The viscosity is an essential quantity for calculating the viscosity similarity parameter known as the Reynolds number. Further discussion of this quantity will be deferred until aerodynamic loads are discussed in a later chapter. the dynamic viscosity μ is calculated from the Sutherland equation as

$$\mu = \beta T^{3/2}/(T + S) \tag{2.27}$$

where the thermal constant β and the Sutherland constant S are parameters adjusted for a particular gas. The appropriate numerical values for air are given in given in Table 2.5. The kinematic viscosity ν is simply the ratio of the dynamic viscosity to the density.

$$\nu = \mu/\rho \tag{2.28}$$

Equations (2.27) and (2.28) are without value above about 80 km and are questionable above an altitude of 40 km because of the low value of density at these altitudes.

Thermal Conductivity

The thermal conductivity parameter k, a function of the temperature, quantifies the thermal conduction properties of a material. The parameter k quantifies the ability of a material slab to conduct heat; it has units of J/s-m-K. The thermal conductivity for air is calculated from a relationship similar to Sutherland's law [Eq. (2.27)] for viscosity.

$$k = \frac{B_1 T^{3/2}}{T + S_1\left(10^{-12/T}\right)} \tag{2.29a}$$

where $B_1 = 2.64638 \times 10^{-3}$ J/s-m-K$^{1/2}$ and $S_1 = 245.4$ K.

The thermal conductivity is related to the viscosity as follows:

$$k = \mu c_p/\mathrm{Pr} \tag{2.29b}$$

where c_p is the specific heat at constant pressure, Pr is the Prandtl number (for most gases Pr is independent of temperature and pressure), and μ may be determined from Eq. (2.27).

Speed of Sonic Wave Propagation

The speed of sound, or the speed of sonic wave propagation, is solely a function of temperature T; for a perfect gas the speed of sound is given by

$$C = (\gamma R T)^{1/2} \tag{2.30}$$

The speed of sound appears in aerodynamic load calculations where it is divided into the velocity of the RV relative to the atmosphere to obtain the Mach number M.

2.6 Exponential Atmosphere

In the preceding discussion of the Standard Atmosphere model, we divided the atmosphere into a number of layers and labeled these layers according to whether the temperature varied linearly or remained constant over the layer. As simple as such a model may appear, it still requires the use of a computer. For closed-form solutions a simpler representation of the atmosphere is needed.

The starting point for developing a simple analytical model is Eq. (2.21c), which gives a relationship between the density ρ and geometric altitude Z with the assumption that the layer is isothermal. We then further simplify Eq. (2.21c) by replacing the geometric altitude Z with the geopotential altitude h. By using h as the altitude variable we neglect gravity variations with altitude; consequently, the parameter which accounts for such variations, namely $b = 2/R_E$, is set to zero. Finally we assume that the atmosphere is entirely isothermal; thus, we can set $T_{M_i} = T_0$, the sea-level value of the atmospheric temperature. We may therefore rewrite Eq. (2.21c) as

$$\rho = \rho_0 \exp\left\{ -\frac{h g_0}{(R T_{M_0})} \right\} \tag{2.31}$$

According to Table 2.5, the parameter ρ_0, which we can identify as the first atmospheric parameter, equals 1.225 kg/m^3. The lumped term RT_M/g_0, which is known as the atmosphere scale height H is 8.434×10^3 m. Clearly, the atmosphere is not isothermal through most of its altitude range; however, we can adjust the two parameters to give a better fit over the altitude range of interest, say, from 5 to 40 km. An acceptable two-parameter atmosphere model might be written as

$$\rho = \rho_0 e^{-h/H} \tag{2.32}$$

where

$$\rho_0 = 1.752 \text{ kg/m}^3$$

and

$$H = 6.7 \times 10^3 \text{ m}$$

Equation (2.32) is very useful in obtaining some closed-form solutions of various special re-entry vehicle trajectories. A graphical comparison of Eq. (2.32) with the *U.S. Standard Atmosphere 1976* is given in Fig. 2.5.

2.7 Planetary Atmospheres

The study of the physics of planetary atmospheres is in a fairly preliminary stage. In spite of limited data from direct measurements some important work has been done. For example, a preliminary study of the Jovian atmosphere has been done using data from *Pioneer 10* radio occultation.[4] The thermal profile of the Venusian atmosphere has also followed from radio occultation measurements.[4] In the case of Venus, the interplanetary vehicle was *Mariner 5*, with in situ measurements from *Venera 8* probes during and after atmospheric descent. The

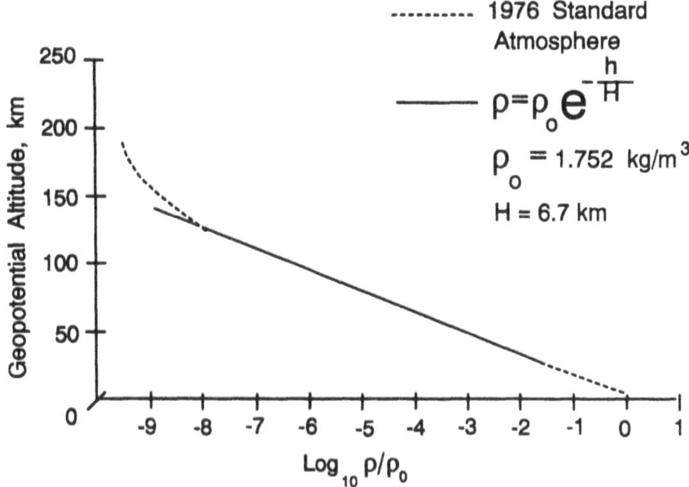

Fig. 2.5 Geopotential altitude vs density ratio ρ/ρ_0 for standard and exponential altitudes.

Martian surface measurements were carried out by the Viking landers and with optical and occultation measurements by the *Mariner 6, 7,* and *9* vehicles.[4,8,9]

Before the two *Viking* soft landings on Mars in 1976 and the four *Pioneer* probes to Venus in 1978, information about the atmospheres of both planets was greatly in error. According to Seiff,[10] in 1960 it was believed that the surface pressure on Mars was about 80 mb (mille-bars) and on Venus something like 4 bar (4 Earth atmospheres). Subsequent measurements have shown that the Martian estimate was 10 to 12 times too high and the Venusian estimate 25 times too low. The surface pressure on Mars is about 8 mb (less than 4 mb in the region of the poles during winter) and on Venus about 100 bar. Clearly, it is difficult to design re-entry bodies, especially those intended for soft landings, without fairly accurate knowledge of the density variation with altitude.

In 1963 the decision was made to send probes into the atmospheres of both Mars and Venus. As a consequence the Planetary Atmospheric Experiments Test (PAET) probe was developed. Tests were carried out in the Earth's atmosphere in 1971. In 1976 the *Viking* lander carried out the first test of the atmospheric profile of another planet—Mars. In 1978 the U.S. *Pioneer* Venus orbiter contained four dedicated PAETs for entry into the atmosphere at four widely separated points in the hemisphere facing the Earth.

Of all of the planets, Mars alone seems suited for surface exploration, at least during the next century. Nevertheless, the atmospheres of Venus and Jupiter are of great interest. In Fig. 2.6 we may compare the thermal profiles of the atmospheres of Jupiter, Mars, Venus, and Earth.

Chamberlin[4] points out that much of the data on the Jovian atmosphere is still conjectural. The adiabatic lapse rate in the troposphere is uncertain, since the mix of H_2 and He is uncertain. With a ratio of specific heat γ estimated to be approximately 1.6, the lapse rate is about 2.9 K/km. The tropopause minimum is about 100 K or maybe 120 K. A stratosphere (region of constant temperature)

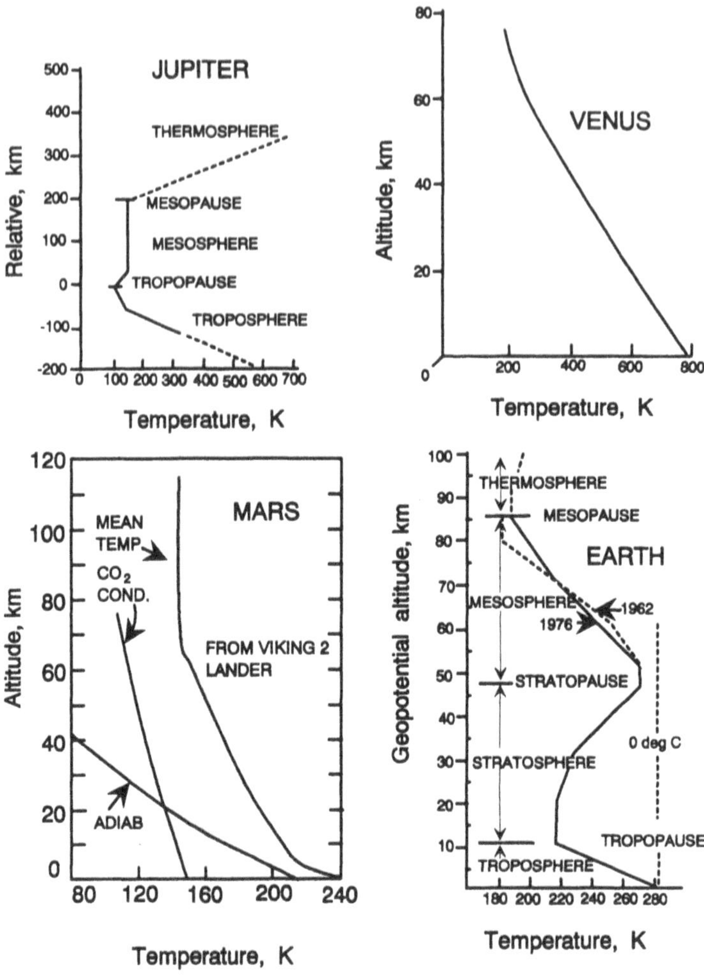

Fig. 2.6 Comparison of planetary thermal profiles.

is not indicated since the constancy of temperature above the tropopause is not established. Originally it was thought that in the thermosphere the temperature increase was only about 15 K. However, radio occultation experiments on *Pioneers 10* and *11* showed that in the thermosphere the temperature rises to about 750 K.

Atmospheric measurements of Venus were first obtained from radio occultation experiments aboard *Mariner 5* and in situ sensing on the *Venera 8* probes. The temperature profile is essentially adiabatic up to about 50 km. Horizontal temperature gradients are quite low, indicating strong atmospheric dynamics. Polarization studies show that there is a layer of clouds composed of sulfuric acid. Above the clouds no stratosphere has been discovered. Throughout the thermosphere (above 70 to 80 km) the temperature increases again to about 300 K.

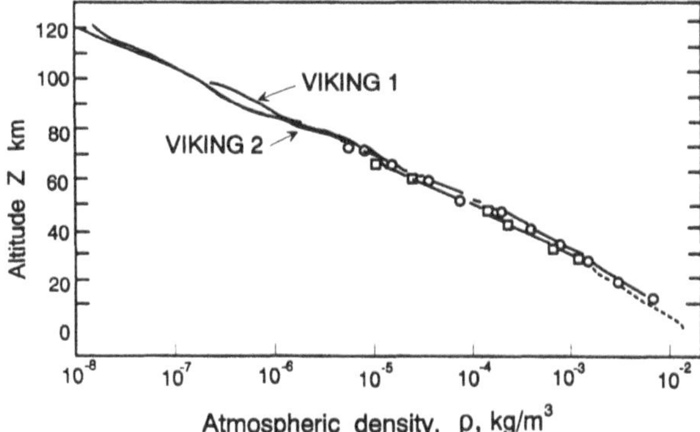

Fig. 2.7a Martian atmospheric density structure from Viking lander measurements.

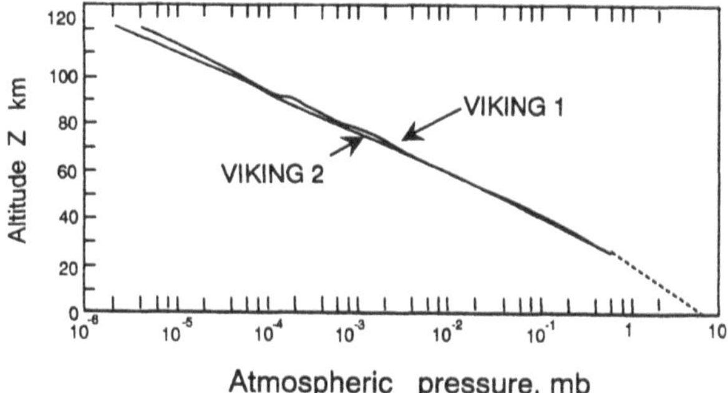

Fig. 2.7b Martian atmospheric pressure structure from Viking lander measurements.

The temperature profiles of the Martian atmosphere vary with surface temperature (which is something like 245 K). There is also a significant variation of temperature profiles with season (and dust loading), especially at extreme latitudes. The temperature lapse rates dT/dZ are smaller than the adiabatic lapse rate, resulting in a stable atmosphere. Except for a surface boundary layer, little or no thermal convection should take place. As indicated in Fig. 2.6, temperatures in the atmosphere are above the CO_2 condensation boundary. However, condensation does occur overnight in the winter at high latitudes.

The density profile of the Martian atmosphere is given in Fig. 2.7a. These data were taken from the *Viking 1* and *2* landers. The separate measurements agree quite well. Both *Viking* landers provided density measurements over a range from the surface to an altitude of 120 km, with an accuracy that would have been impossible to accomplish from Earth. Pressures derived from the temperature and density profiles are given in Fig. 2.7b. The pressure at the surface is about 7.5 mb, although at the poles in the winter the pressure might be half this value. At an altitude of 120 km, the pressure is about 10^{-6} mb.

An atmospheric probe measures acceleration (actually deceleration) during entry; from these measurements the density is obtained from the following relationship:

$$\rho = -\left(\frac{2m}{C_D A}\right)\left(\frac{a}{V^2}\right) \tag{2.33}$$

where m is the mass of the probe, C_D is the drag coefficient, A is the reference area, V is the velocity magnitude, and a is the acceleration (which in this case is negative because deceleration is being measured). A derivation of this relationship is deferred until Chapter 7. Obviously, a ground-based calibration of the probe is a necessary part of the process. Integration of the vertical component of the velocity vector provides altitude. Then, using the hydrostatic equation [Eq. (2.5)], we may obtain the pressure as a function of altitude. The equation of state of an ideal gas [Eq. (2.6)] then provides temperature. In order to use the state equation we need to know the molecular weight of the constituent atmospheric gases. For Mars the molecular weight was provided by the Upper Atmosphere Mass Spectrometer; for Venus, it was provided by the Orbiter Mass Spectrometer.

Unfortunately, the central goal of the *Viking* lander was to conduct an ill-considered (and unsuccessful) search for life; planetary studies would have been much more appropriate.[11] Nevertheless, some atmospheric experiments were included: a mass spectrometer to determine composition of the upper atmosphere and a retarding potential analyzer for measurements of the ionosphere. The atmospheric pressure uncertainty was between 4 and 10 mb during the design of the soft lander. An uncertainty of this magnitude is significant because the Martian pressure, as we have seen, is around 7 or 8 mb. Nevertheless, the lander did soft land without damage. Further details of the Viking lander may be found in Seiff's paper.[10]

The *Pioneer* Venus mission sent four probes into the atmosphere over widely separated points of the Earth-side hemisphere. Figure 2.8 presents the thermal

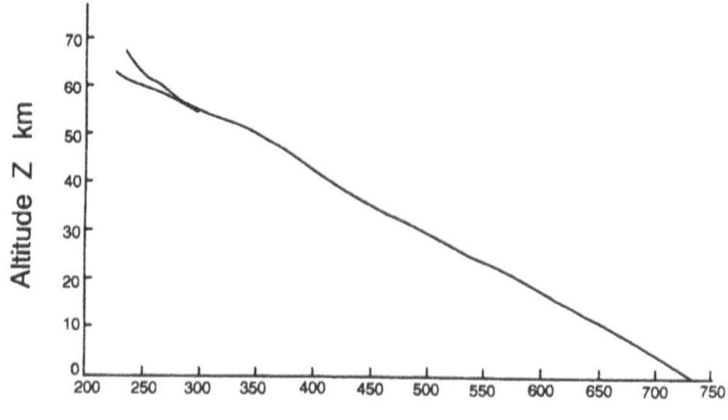

Fig. 2.8 Average Venusian atmospheric temperature profile from four widely separated probes.

profile of Venus, and the pressure profile is given in Fig. 2.9. The curves represent averages of the data from all four probes. Further discussion of the atmospheres of Venus and Mars is available (see Hunten[12]).

The composition of the Venusian atmosphere is 96.5% CO_2 and 3.5% N_2, with trace amounts of SO_2, Ar, CO, and O_2. Interestingly, the bulk of the atmospheres of both Venus and Mars are similar. Water is in far less abundance than on Earth. Winds are from east to west, decreasing from a peak magnitude of 110 m/s at the cloud tops to about 1 m/s at the surface.

For Eq. (2.31) we identified the atmosphere scale height as

$$H = \frac{RT_M}{g_0} = \frac{R^* T_M}{M_0 g_0} \tag{2.34}$$

While it is true in the case of the Earth that we "adjusted" the above constant to increase the utility of the two-parameter atmosphere model, we will retain the computed terrestrial value. Using numbers from Table 2.5 for Earth, we get

$$H = (287.0)(288.15)/(9.806) = 8.434 \text{ km}$$

Table 2.7 permits comparisons to be made among Earth, Mars, Venus, and Jupiter.

It is interesting to note that the scale heights of the four planets are fairly close to one another. In comparing Mars and Earth we see that the high molecular

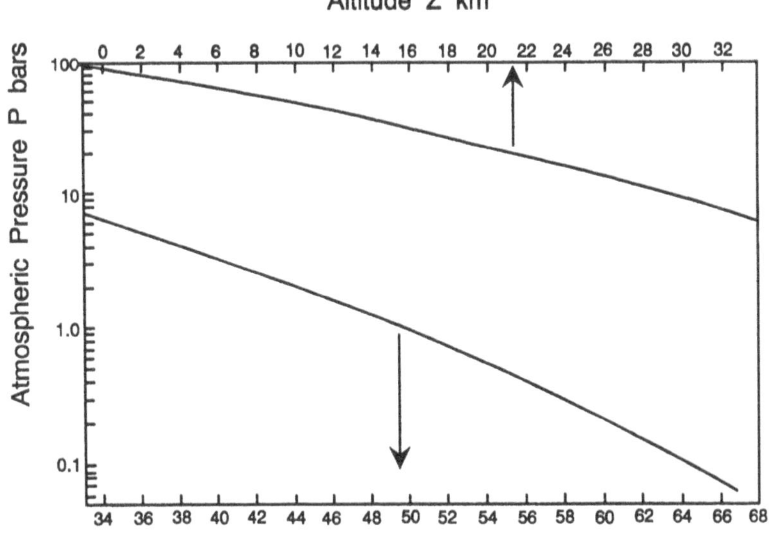

Fig. 2.9 Average Venusian atmospheric pressure profile from four widely separated probes.

Table 2.7 Scale heights of planetary atmospheres

Planet	Atmosphere composition	Molecular weight M_0	Gravity acceleration g_e, m/s^2	Average surface temperature T_{M_0}, K	Scale height H, km	P_0 Earth ATM	L_Z, K/km
Earth	N_2, O_2	28.9	9.81	288	8.5	1.0	− 9.8
Venus	CO_2	44.0	8.88	700	14.9	100	−10.7
Mars	CO_2	44.0	3.73	210	10.6	6×10^{-3}	− 4.5
Jupiter	H_2, He	2.0	26.20	160	25.3	(?)	−20.5

weight of the CO_2 on Mars is counteracted by the low gravitational acceleration with respect to Earth. On Venus the high molecular weight of CO_2 is offset by the high surface temperature. Jupiter has both a high surface temperature and a high gravitational acceleration (about three times that of the Earth), but these values are offset by the low molecular weight of H_2.

The information contained in Table 2.7 can be used to construct a two-parameter atmospheric model for Venus and Mars. The value of the atmospheric density at the surface may be calculated from the ideal gas equation as

$$\rho = MP_0/R^*T_0 \tag{2.35}$$

Calculations of aerodynamic re-entry loads into a Martian or Venusian atmosphere using such a two-parameter model may be adequate for rough estimations of aerodynamic drag loads; space does not permit the preparation of planetary atmospheric models comparable in precision to that given in Appendix A. However, a computer model like that given in Appendix A could be prepared for Venus or Mars if the thermal strata could be identified with appropriate numerical values for the lapse rates.

References

[1]Jursa, A. S. (ed.), *Handbook of Geophysics and the Space Environment,* Air Force Geophysics Laboratory, Air Force Systems Command, Cambridge, MA, 1985.

[2]COESA (Committee on Extension to the Standard Atmosphere), *U.S. Standard Atmosphere 1962,* U.S. Government Printing Office, Washington, DC, 1962.

[3]COESA, *U.S. Standard Atmosphere 1976,* U.S. Government Printing Office, Washington, DC, 1976.

[4]Chamberlin, J. W., *The Theory of Planetary Atmospheres—An Introduction to Their Physics and Chemistry,* Academic Press, New York, 1978.

[5]Craig, R. A., *The Upper Atmosphere—Meteorology and Physics*, International Geophysics Series, Academic Press, Niles, IL, 1965.

[6]Schlichting, H., *Aerodynamics of the Airplane,* McGraw-Hill, Princeton, NJ, 1979.

[7]*Ballistic Missile Series,* Headquarters, U.S. Army Material Command, Rept. CP-706-283, Washington, DC, 1965.

[8]Gierason, P. J., and Goody, R. M., "The Effect of Dust on the Temperature of the Martian Atmosphere," *Journal of Atmospheric Sciences,* Vol. 29, March 1972, pp. 400–402.

[9]Kliore, A., "Occultation Experiment: Results of the First Direct Measurement of Mars' Atmosphere and Ionosphere," *Science,* Vol. 149, No. 3684, 1965, pp. 1243–1248.

[10]Seiff, A., "Atmospheres of Earth, Mars and Venus as Defined by Entry Probe Experiments," *Journal of Spacecraft and Rockets,* AIAA, Vol. 28, No. 3, 1991, pp. 265–275.

[11]Cooper, H. S. F., *The Search for Life on Mars: Evolution of an Idea,* H. Holt Rinehart & Winston, Orlando, FL, 1984.

[12]Hunten, D. M., Colin, L., Donahue, T. M., and Moroz, V. I. (eds.), *Venus,* University of Arizona Press, Tucson, AZ, 1983.

Bibliography

Avduevsky, V. D., "A Tentative Model of the Venus Atmosphere Based on the Measurements of Veneras 5 and 6," *Journal of Atmospheric Sciences,* July 1970, pp. 561–568.

Goody, R. M., and Walker, J. C. G., *Atmospheres*, Prentice-Hall, Old Tappan, NJ, 1972.

Houghton, J. T., *The Physics of Atmospheres*, Cambridge University Press, Port Chester, NY, 1977.

Wallace, J. M., and Hobbs, P. V., *Atmosphere Science—An Introductory Survey*, Academic Press, Niles, IL, 1977.

3
Earth's Form and Gravitational Field

"It cannot be seen, cannot be felt,
Cannot be heard, cannot be smelt,
It lies behind stars and under hills,
And empty holes it fills."

—J. R. R. Tolkien, *The Hobbit*

3.1 Introduction

One of the fundamental principles of rational mechanics—indeed, of physics itself—is Newton's Law of Gravitation. This gravitational law does not "explain" gravity in terms of more fundamental speculations, but rather relates this pervasive force to certain metrical quantities such as the mass of the objects producing the field and the distance of their separation. Newton recognized the fundamental scope and limitations of his description of gravity when he wrote, "Gravity must be caused by an agent acting constantly according to certain laws; but whether this agent be material or immaterial, I have left to the consideration of my readers." Nevertheless, Newton indicated that he thought the agent was material when, in a 1692 letter to Richard Bentley, he wrote

> [T]hat one body may act on another through a vacuum, without the mediation of anything else, by and through which their action and force may be conveyed from one to another, is to be so great an absurdity, that I believe no man who has . . . a competent faculty of thinking can ever fall into it.

Space does not permit an extended discussion of the implications of this most fundamental of physical concepts. An interesting and extensive discussion of the gravitational law applied to a Sun/Earth/Moon system is given by Eisberg and Lerner.[1]

For our purposes we may represent the Law of Gravitation as an interaction between a re-entry vehicle of mass m and Earth of mass M_E, both treated as mass particles in the following expression:

$$F_G = \frac{GM_E m}{R^2} = \frac{GM_E m}{(R_E + h)^2} \tag{3.1}$$

where G is the universal gravitational constant and R is the distance separating the "particles" Earth (E) and re-entry vehicle (RV). In the second expression we have replaced R with the sum of the radius of the Earth, R_E, and the altitude h of the RV above the surface of the Earth. It would seem that we are violating our assumption that the Earth may be regarded as a particle when we introduce a finite Earth radius R_E. It can be shown that if the Earth were a sphere of uniform density, then the external gravitational field of the Earth would be indistinguishable from that of a (dimensionless) particle whose mass is equal to that of the Earth. However, the Earth's form deviates from that of a sphere, and its density is certainly not constant.

The purpose of this chapter is to consider two distinct but related questions: 1) What is the shape of the Earth? and 2) What is an adequate analytical representation of the Earth's external gravitational field? In the next section we combine these two questions by determining the shape of an equipotential surface that may be used as an approximation of the Earth's surface.

3.2 Geoid and Reference Ellipsoid

Two reference surfaces are used in geometric geodesy: the *geoid* and the *reference ellipsoid*. The geoid is an equipotential surface of the Earth's gravitational field that coincides with the undisturbed mean sea level. This surface is continued under the continents. The direction of the local gravity field is everywhere along the normal to the geoid. The size of the geoid is set by mean sea level; the shape is set by the requirement that the local gravitational acceleration or gravitational force be normal to the geoid at all points.

The reference ellipsoid is a simple mathematical figure that closely matches the geoid. The reference ellipsoid is a surface of revolution described by the rotation of an ellipse about its minor axis. This minor axis is coincident with the Earth's rotational axis. The result is an oblate spheroid, or an ellipsoid of revolution. Points on the ellipsoid are given by the geodetic latitude L, the angle which the normal to the ellipsoid makes with respect to the equatorial plane and the geodetic longitude l, which is positive eastward from the Greenwich meridian. These two angles L and l determine the direction of the local normal and hence provide a good (usually acceptable) approximation of the direction of the gravitational force. The reference ellipsoid is discussed in this section and the following.

Because the reference ellipsoid is an approximation, precise work may require some definition of the deflection of the gravitational field within and out of the meridian. Such precision is not needed for most re-entry body studies. However, we present some of the definitions of the deflection of the vertical in Section 3.8.

Newton's gravitational law describes the gravitational attraction between two point masses. If we are willing to accept the approximation that both the Earth and RV may be treated as mass particles, then Eq. (3.1) provides a simple relationship between RV altitude and the gravitational force. For some simple analyses such a relationship is entirely adequate and in keeping with even more approximate representations of atmospheric properties. However, if our interest

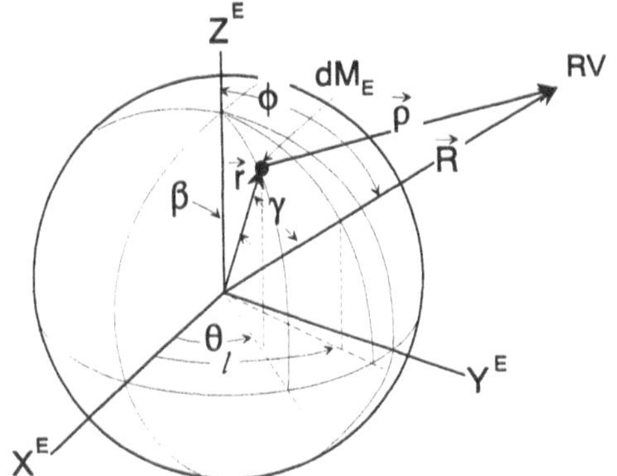

Fig. 3.1 Variables associated with the Earth's gravity potential.

extends to relatively low altitude (i.e., $h \ll R_E$) and highly accurate trajectories, the Earth can then no longer be regarded as a particle. Figure 3.1 shows an RV acted upon by the gravitational field of a mass element $\mathrm{d}M_E$, where $\mathrm{d}M_E$ and the RV are separated by a distance $\boldsymbol{\rho}$; the RV is at a distance $\boldsymbol{R}$ from the center of mass of the Earth.

The gravitational potential at a point R (relative to the Earth's center) due to a mass element $\mathrm{d}M_E$ at point r is given by

$$\mathrm{d}U_g = \frac{G\mathrm{d}M_E}{\rho} \tag{3.2}$$

where

$$\boldsymbol{\rho} = \boldsymbol{R} - \boldsymbol{r} \tag{3.3}$$

This relationship permits the calculation of the differential force $\mathrm{d}\boldsymbol{F}_g$ due to the differential mass element $\mathrm{d}M_E$. The potential due to the entire Earth is given by the integral of the above differential potential over the entire Earth. The potential due to the entire Earth may be expressed as

$$U_g = G\int \frac{\mathrm{d}M_E}{\rho} = G\int \frac{\mathrm{d}M_E}{|\boldsymbol{R} - \boldsymbol{r}|} \tag{3.4}$$

In this section we confine our attention to a point on the Earth's surface because we are seeking the analytical form of an equipotential surface that may be used as a close approximation of the Earth's surface. A point on the Earth's surface (or a point which remains stationary above a point fixed on the Earth's surface) is subjected to both centrifugal and gravitational forces. The so-called centrifugal force is more properly identified as a force of constraint required to maintain an object at a fixed position within a rotating or non-Newtonian reference frame. The force (per unit mass) potential due to centrifugal force is

$$U_{\mathrm{rot}} = \tfrac{1}{2}R^2\omega_{e/I}^2 = \tfrac{1}{2}\left(X^2 + Y^2\right)\omega_{e/I}^2 \tag{3.5}$$

where $\boldsymbol{R} = X\boldsymbol{i} + Y\boldsymbol{j}$ and $\omega_{e/I}$ is the rotation rate of the Earth. Clearly, the derivative of U_{rot} with respect to radial distance R gives the radial force due to the rotation rate $\omega_{e/I}$. Combining both potentials U_g and U_{rot} gives the total potential

$$\begin{aligned} U &= U_g + U_{\mathrm{rot}} \\ &= G\int \frac{\mathrm{d}M_E}{|\boldsymbol{R}-\boldsymbol{r}|} + \frac{1}{2}(X^2+Y^2)\omega_{e/I}^2 \end{aligned} \tag{3.6}$$

It should be emphasized here that our goal is to obtain an expression for the equipotential surface; thus, a test point is first located on the surface of the Earth. Later we will consider the potential at a point removed from the Earth, e.g., a point which locates an RV. In this latter case, we will omit the centrifugal contribution since an axially symmetric Earth would develop the same gravitational field whether rotating or not. (Actually, there is a very weak dependence of the Earth's field on longitude, which will be discussed briefly at the end of this chapter.)

Returning to Eq. (3.6), we may write

$$\frac{1}{|\boldsymbol{R}-\boldsymbol{r}|} = \frac{1}{[R^2+r^2-2Rr\cos(\gamma)]^{1/2}} \tag{3.7}$$

where γ is the angle between the vector $\boldsymbol{r}$ which locates the mass element $\mathrm{d}M_E$ and the vector $\boldsymbol{R}$ which locates the field point. Expanding the denominator in the above equation (subject to the condition that $R > r$) gives

$$\begin{aligned} \frac{1}{|\boldsymbol{R}-\boldsymbol{r}|} = \frac{1}{R}\bigg\{ & 1 + \frac{r}{R}\cos(\gamma) + \frac{1}{2}\left(\frac{r}{R}\right)^2[3\cos^2(\gamma)-1] \\ & + \frac{1}{2}\left(\frac{r}{R}\right)^3\cos(\gamma)[5\cos^2(\gamma)-3] + \cdots \bigg\} \end{aligned} \tag{3.8}$$

or, in terms of Legendre polynomials (LP),

$$\begin{aligned} \frac{1}{|\boldsymbol{R}-\boldsymbol{r}|} = \frac{1}{R}\bigg\{ & P_0[\cos(\gamma)] + \left(\frac{r}{R}\right)P_1[\cos(\gamma)] + \left(\frac{r}{R}\right)^2 P_2[\cos(\gamma)] \\ & + \left(\frac{r}{R}\right)^3 P_3[\cos(\gamma)] + \cdots \left(\frac{r}{R}\right)^n P_n[\cos(\gamma)] \bigg\} \end{aligned} \tag{3.9}$$

where

$$\begin{aligned} P_0[\cos(\gamma)] &= 1 \\ P_1[\cos(\gamma)] &= \cos(\gamma) \\ P_2[\cos(\gamma)] &= \tfrac{1}{2}[3\cos^2(\gamma)-1] = \tfrac{1}{4}[3\cos(2\gamma)+1] \\ P_3[\cos(\gamma)] &= \tfrac{1}{2}[5\cos^3(\gamma)-3\cos(\gamma)] = \tfrac{1}{8}[5\cos(3\gamma)+3\cos(\gamma)] \end{aligned} \tag{3.10}$$

In general, for an LP of order n, we may identify the following generating function according to Rodrigues[2]:

$$P_n(v) = \frac{1}{2^n n!}\frac{\mathrm{d}^n}{\mathrm{d}v^n}(v^2 - 1) \tag{3.11a}$$

where $v = \cos(\gamma)$. Another generating function carries out the differentiation to give[3]

$$P(v) = \sum_{k=0}^{n/2} \frac{1 \cdot 3 \cdot 5 \cdots (2n - 2k - 1)(-1)^k v^{n-2k}}{2^k k!(n - 2k)!} \tag{3.11b}$$

The following expression gives a recursive relationship of more utility in computations[4]:

$$P_{n+1}(v) = 2vP_n(v) - P_{n-1}(v) - \left[\frac{vP_n(v) - P_{n-1}(v)}{(n + 1)}\right] \tag{3.11c}$$

Referring to Fig. 3.1 we can write the following expressions for the unit vectors along $\boldsymbol{r}$ and $\boldsymbol{R}$, i.e., $\boldsymbol{u}_r$ and $\boldsymbol{u}_R$:

$$\begin{aligned} \boldsymbol{u}_r &= [\sin(\beta)\cos(\theta), \sin(\beta)\sin(\theta), \cos(\beta)]^{\mathrm{T}} \\ \boldsymbol{u}_R &= [\cos(L_c)\cos(l), \cos(L_c)\sin(l), \sin(L_c)]^{\mathrm{T}} \end{aligned} \tag{3.12}$$

where the latitude L_c is related to ϕ the colatitude by $\phi = \pi/2 - L_c$. It follows that

$$\begin{aligned} \cos(\gamma) &= \boldsymbol{u}_r \cdot \boldsymbol{u}_R \\ &= \sin(\beta)\cos(\theta)\cos(L_c)\cos(l) \\ &\quad + \sin(\beta)\sin(\theta)\cos(L_c)\sin(l) + \cos(\beta)\sin(L_c) \end{aligned} \tag{3.13a}$$

or, alternatively,

$$\cos(\gamma) = \sin(\beta)\cos(L_c)\cos(\theta - l) + \cos(\beta)\sin(L_c) \tag{3.13b}$$

Next we insert Eq. (3.9) into Eq. (3.6) to get

$$\begin{aligned} U = \frac{G}{R}\Bigg\{&\int_{M_E} \mathrm{d}M_E + \frac{1}{R}\int_{M_E} P_1[\cos(\gamma)]r\,\mathrm{d}M_E \\ &+ \frac{1}{R^2}\int_{M_E} P_2[\cos(\gamma)]r^2\mathrm{d}M_E + \cdots\Bigg\} + \frac{1}{2}\left(X^2 + Y^2\right)\omega_{e/I}^2 \end{aligned} \tag{3.14}$$

The first integral gives M_E, the mass of the Earth. The second integral reduces to a series of first moments about the axes through O, the origin of the Earth-centered system. Since we are assuming that the origin O is at the center of mass of the Earth, this second integral vanishes. The third integral may be written as

$$\int_{M_E} P_2[\cos(\gamma)]r^2 \mathrm{d}M_E = \int_{M_E} \left[\frac{3}{2}\cos^2(\gamma) - \frac{1}{2}\right] r^2 \mathrm{d}M_E$$

By using Eq. (3.13a) to replace $\cos(\gamma)$, we obtain, after some manipulation,

$$\begin{aligned}\int_{M_E} [\cos(\gamma)]r^2 \mathrm{d}M_E &= \frac{3}{2}\left[\frac{I^e_{xx} + I^e_{yy} - 2I^e_{zz}}{2}\right]\left[\sin^2(L_c) - \frac{1}{3}\right] \\ &\quad + \frac{3}{4}\left(I^e_{yy} - I^e_{xx}\right)\cos^2(L_c)\cos(2l) \end{aligned} \tag{3.15}$$

where I^e_{xx}, I^e_{yy}, I^e_{zz} are the moments of inertia of the Earth about the X^e, Y^e, and Z^e Earth-fixed axes. Numerical evaluation of these moments of inertia is not important here. However, we do note that the third integral of Eq. (3.14) does include both latitude L_c and longitude l. We further note (as expected) that if the Earth were a homogeneous body whose axis of symmetry is coincident with the axis of rotation (the Z^e-axis), the longitude l would not appear in the gravitational potential U, because I^e_{xx} would equal I^e_{yy}. Finally, we truncate the series representing the potential expansion after the third term. The potential U given in Eq. (3.14) may now be written as

$$\begin{aligned} U = \frac{M_E G}{R}\Bigg\{ & 1 + \frac{K}{2R^2}[1 - 3\sin^2(L_c)] + \frac{3(I^e_{yy} - I^e_{xx})}{4M_E R^2}\cos^2(L_c)\cos(2l) \\ & + \frac{\omega^2_{e/I} R^3}{2M_E G}\cos^2(L_c)\Bigg\} \end{aligned} \tag{3.16}$$

where

$$K = \frac{2I^e_{zz} - (I^e_{xx} + I^e_{yy})}{2M_E} \approx \frac{(I^e_{zz} - I^e_{xx})}{M_E} \tag{3.17}$$

Note that we have assumed that $I^e_{xx} = I^e_{yy}$, which is consistent with ignoring longitudinal variations.

Our goal now is to look at the first-order correction to the potential U to account for nonsphericity in the gravitational field. This can be done by making the above assumption, namely that $I^e_{xx} = I^e_{yy}$; the consequence is the elimination of the third term in Eq. (3.16) to give

$$U = \frac{M_E G}{R}\left\{1 + \frac{K}{2R^2}[1 - 3\sin^2(L_c)] + \frac{\omega^2_{e/I} R^3}{2M_E G}\cos^2(L_c)\right\} \tag{3.18}$$

Now we can introduce the requirement that the Earth's surface is an equipotential surface. This may be done by letting $U = U_0$ and solving for $R = R_E$. A glance at Eq. (3.18) shows that solving for R (in terms of U_0) would be a tedious task and certainly difficult to express in a relatively simple closed form. However, we also know that the second and third terms are small in comparison to unity (the first term) if for no other reason than the Earth being "nearly" spherical in shape and mass distribution. Thus, recognizing that Eq. (3.18) is an approximation anyway, we may set R equal to the Earth's equatorial radius R_e. When this is done, the solution for R_E, the distance from the Earth's mass center to a point on its equipotential surface, is

$$R_E = \frac{M_E G}{U_0}\left\{1 + \frac{K}{2R_e^2}\left[1 - 3\sin^2(L_c)\right] + \frac{\omega_{e/I}^2 R_e^3}{2M_E G}\cos^2(L_c)\right\} \tag{3.19}$$

To put the rotational term (the third term) in some perspective we set d as the ratio of the centripetal acceleration at the equator to the gravitational attraction at the equator, that is,

$$d = \frac{R_e\omega_{e/I}^2}{(M_E G/R_e^2)} = \frac{R_e^3\omega_{e/I}^2}{M_E G} \tag{3.20}$$

Equation (3.19) then becomes

$$\begin{aligned} R_E &= \frac{M_E G}{U_0}\left\{1 + \frac{K}{2R_e^2}\left[1 - 3\sin^2(L_c)\right] + \frac{d}{2}\cos^2(L_c)\right\} \\ &\approx \frac{M_E G}{U_0}\left\{\left(1 + \frac{K}{2R_e^2} + \frac{d}{2}\right)\left[1 - \left(\frac{3K}{2R_e^2} + \frac{d}{2}\right)\sin^2(L_c)\right]\right\} \end{aligned} \tag{3.21}$$

where we have assumed that $[1 + (K/2R_e^2) + (d/2)]$ is nearly unity. We can numerically evaluate d as

$$d \approx \frac{R_e\omega_{e/I}^2}{M_E G/R_e^2} \approx \frac{(6.378 \times 10^6)(7.272 \times 10^{-5})^2}{9.80} \approx 3.44 \times 10^{-3} \tag{3.22}$$

where $\omega_{e/I}$ is the Earth's rotational rate.

With the X and Y moments of inertia equal, we can rewrite Eq. (3.17) as

$$K \approx \frac{I_{zz}^e - I_{xx}^e}{M_E} \qquad \text{or} \qquad \frac{K}{R_e^2} \approx \frac{I_{zz}^e - I_{xx}^e}{R_e^2 M_E} \tag{3.23}$$

Inserting some numerical values we have

$$\frac{K}{R_e^2} \approx \frac{I_{zz}^e - I_{xx}^e}{I_{zz}^e/0.3309} \approx (0.3309)(3.2729 \times 10^{-3}) = 1.08 \times 10^{-3}$$

where the inertia ratio identified by Garland[3] as the *mechanical ellipticity* is given by

$$\frac{I_{zz}^e - I_{xx}^e}{I_{zz}^e} = 3.2729 \times 10^{-3}$$

and where the polar moment of inertia I_{zz}^e is approximated as

$$I_{zz}^e \approx 0.3309 M_E R_e^2$$

These numerical values for K/R_e^2 and d may be used in a simplified and rearranged version of Eq. (3.21) as follows:

$$R = R_E = \frac{M_E G}{U_0}\left(1 + \frac{K}{2R_e^2} + \frac{d}{2}\right)\left[1 - \left(\frac{3K}{2R_e^2} + \frac{d}{2}\right)\sin^2(L_c)\right] \tag{3.24}$$

which is of the form

$$R = R_E = R_e\left[1 - f\sin^2(L_c)\right] \tag{3.25}$$

where f is the "flattening" of the sphere. Equations (3.24) or (3.25) provide us with an equation representing a surface which closely approximates the geoid; this surface will be shown in Section 3.3 to be nearly spheroidal in shape. (A spheroid is defined as an ellipsoid of revolution.) The constant f may be written as

$$f = \frac{3K}{2R_e^2} + \frac{d}{2} \tag{3.26}$$

The geoid is the actual geopotential surface at mean sea level; however, it is sometimes identified as the approximate analytically described surface. The potential U evaluated at the equator, $U = U_e = U_0$, is the constant potential used in the definition of this approximate geopotential surface; that is,

$$U_0 = U_e = GM_E/R_e \tag{3.27}$$

We can now return to Eq. (3.18) and find g_E, the variation of the gravitational acceleration with latitude, as follows:

$$g_E = -\frac{\partial U}{\partial R} = \frac{M_E G}{R^2}\left\{1 + \frac{3K}{2R^2}\left[1 - 3\sin^2(L_c)\right] + d\cos^2(L_c)\right\} \tag{3.28}$$

Since the gravitational acceleration is to be determined on the surface of the Earth, we must evaluate the derivative in the above expression at $R = R_E$. It should be emphasized that R_E is the distance from the origin (mass center of

Earth) to a point on the surface of the Earth. Equation (3.24) may be inserted into Eq. (3.28) to give

$$g_E \approx \frac{U_0^2}{M_E G}\left(1 + \frac{K}{2R_e^2} - 2d\right)\left[1 + \left(2d - \frac{3K}{2R_e^2}\right)\sin^2(L_c)\right] \tag{3.29}$$

Equation (3.29) is of the form

$$g_E = g_e[1 + B\sin^2(L_c)] \tag{3.30}$$

where

$$B = 2d - 3K/2R_e^2 \tag{3.31}$$

and g_e is the gravitational acceleration at the equator. Or, using Eq. (3.26), we can rewrite Eq. (3.31) to give

$$B = \tfrac{5}{2}d - f \tag{3.32}$$

where B is the proportional difference between the equatorial and polar gravitational accelerations. By using previously given values for d and K/R_e^2 we obtain the following value for B:

$$\begin{aligned} B &= 2(d) - 1.5K/R_e^2 \\ &= 2\left(3.4 \times 10^{-3}\right) - 1.5\left(1.08 \times 10^{-3}\right) = 5.18 \times 10^{-3} \end{aligned}$$

Equation (3.30) should not be used for precise calculations. For numerical work an improvement to Eq. (3.30) is

$$g_E = g_e\left[1 + F_2\sin^2(L_c) + F_4\sin^4(L_c) + F_6\sin^6(L_c)\right] \tag{3.33}$$

where the F-values are given in Table 3.1.

3.3 Geocentric Position Vector

In the preceding section we provided some physical justification for at least the form of the Earth's surface. In this section we assume that the Earth's form is that of a spheroid, and we arrive at a representation that is nearly identical to that given in Eq. (3.25).

First we must recall that the geoid defines an equipotential surface which is coincident with mean sea level. If the Earth were a fluid body but still retained its mass distribution and rotation, the resulting figure would define the geoid. We assume that such a surface may be represented adequately by a spheroid, as shown in Fig. 3.2. The polar and equatorial radii, R_p and R_e, respectively, are given in Table 3.1. There are two ways of describing the deviation of an ellipse from a circle: the ellipticity e, defined as

$$e = (R_e - R_p)/R_e = 1 - (R_p/R_e) \tag{3.34a}$$

Table 3.1 Earth constants

Symbol	Definition	Value	Uncertainty	Units
GM_E	Gravitational constant	3.986005×10^{14}	--	m^3/s^2
$\omega_{e/I}$	Earth rotation rate	$7.292115147 \times 10^{-5}$	--	rad/s
R_e	Equatorial radius	6.378135×10^6	± 5.0	m
R_p	Polar radius	6.356750×10^6	± 5.0	m
e	Ellipticity	$1/298.257$	$\pm 6.0 \times 10^{-8}$	--
K	Eccentricity	8.18192×10^{-2}	--	--
g_e	Gravitational acceleration at equator	9.780326771	$\pm 1.8 \times 10^{-9}$	m/s^2
F_2	Grav-latitude coefficient	$5.27904138 \times 10^{-3}$	--	--
F_4	Grav-latitude coefficient	$3.27179493 \times 10^{-4}$	--	--
F_6	Grav-latitude coefficient	1.2621789×10^{-6}	--	--
J_2	2nd-order Jeffery constant	1.08263×10^{-3}	$\pm 2.0 \times 10^{-7}$	--
J_3	3rd-order Jeffery constant	2.532153×10^{-7}	$\pm 1.0 \times 10^{-7}$	--
J_4	4th-order Jeffery constant	1.6109876×10^{-7}	$\pm 1.0 \times 10^{-7}$	--

or the eccentricity K, defined as

$$K = \left[1 - \left(\frac{R_p}{R_e}\right)^2\right]^{1/2} = \left[2e\left(1 - \frac{e}{2}\right)\right]^{1/2} \tag{3.34b}$$

The eccentricity K should not be confused with the inertia to mass ratio of Eq. (3.17).

In replacing a spherical Earth with a spheroidal Earth, we introduce a complication: the normal to the Earth's surface no longer passes through the center of

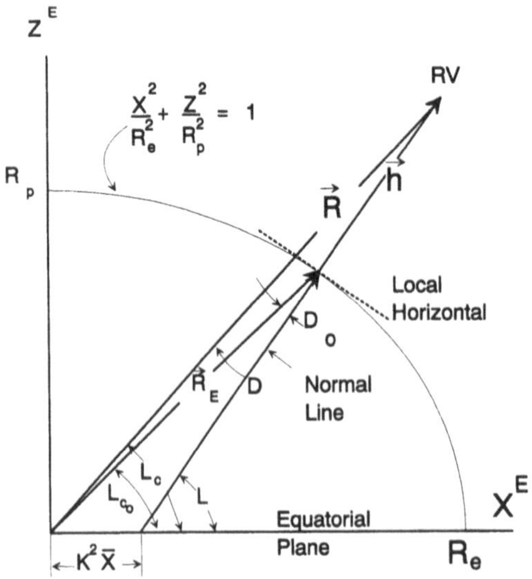

Fig. 3.2 Earth meridian plane.

the Earth. Since the geoid is an equipotential surface, the gravitational specific force must be directed along the normal to the geoid. As a consequence, "up" and "down" are no longer geocentric. A planar section through the spheroid that contains the axis of symmetry defines the elliptical Earth contour. This sectional view also displays, in edge view, the Earth's equatorial plane. The equation of an ellipse may be written as

$$X^2/R_e^2 + Z^2/R_p^2 = 1 \tag{3.35a}$$

where R_e and R_p are the equatorial and polar radii, respectively. The slope of the ellipse at the generic point $(\bar{X}, \bar{Z})$ may be written as

$$\left.\frac{\mathrm{d}Z}{\mathrm{d}X}\right|_{\bar{X},\bar{Z}} = -\left[\frac{\bar{X}}{\bar{Z}}\right]\left(\frac{R_p}{R_e}\right)^2 \tag{3.35b}$$

The slope of the normal is the negative reciprocal of $\mathrm{d}Z/\mathrm{d}X$. Let (X_n, Z_n) be any point along this normal. Using the point-slope representation for a straight line we may write for this normal

$$\frac{Z - Z_n}{X - X_n} = \left(\frac{\bar{Z}}{\bar{X}}\right)\left(\frac{R_e}{R_p}\right)^2 \tag{3.36}$$

The intersection of the normal with the X-axis is found by setting Z_n in the above equation to zero and solving for X_n as

$$X_n = \bar{X}\left[1 - \left(\frac{R_p}{R_e}\right)^2\right] = K^2\bar{X} \tag{3.37}$$

where K as used in this equation should not be confused with the K used in Eq. (3.17).

We can introduce a geographic reference frame (sometimes known as a *geodetic frame* or, in some applications, a *navigation frame*) such that the Z-axis is down along the local normal, the X-axis is north, and the Y-axis is east. Note that Fig. 3.2 depicts three vectors: $\boldsymbol{R_E}$, which locates the generic point on the Earth; $\boldsymbol{h}$, which locates the re-entry vehicle above the Earth's surface and along the local normal; and $\boldsymbol{R}$, which locates the RV from the center of the Earth. Also note that

$$\boldsymbol{R} = \boldsymbol{R_E} + \boldsymbol{h} \tag{3.38}$$

where

$$\begin{aligned}
\boldsymbol{R}_E^g &= [-R_E\sin(D_0), 0, -R_E\cos(D_0)]^{\mathrm{T}} \\
\boldsymbol{h}^g &= [0, 0, -h]^{\mathrm{T}} \\
\boldsymbol{R}^g &= [-R\sin(D), 0, -R\cos(D)]^{\mathrm{T}} \\
&= [-R_E\sin(D_0), 0, -R_E\cos(D_0) - h]^{\mathrm{T}}
\end{aligned}$$

and where the angle between the normal and the geocentric direction is given by D_0 or D (depending on whether the point is on or above the Earth). We must now relate the magnitudes of $\boldsymbol{R}$, $\boldsymbol{R}_E$, and $\boldsymbol{h}$. It is easy to show that

$$R^2 = R_E^2 + h^2 + 2R_E h \cos(D_0) \tag{3.39}$$

Upon completing the square we get

$$R = \left\{(R_E + h)^2 - 2R_E h[1 - \cos(D_0)]\right\}^{1/2} \tag{3.40a}$$

or

$$R = (R_E + h)\left\{1 - \frac{2hR_E[1 - \cos(D_0)]}{(R_E + h)^2}\right\}^{1/2} \tag{3.40b}$$

D_0 is a small angle in the sense that

$$1 - \cos(D_0) \approx D_0^2/2$$

It will be shown later that at a latitude of 45°, D_0 equals the ellipticity, which is about 1/297 rad (about 0.192 deg). Consequently, we may write for R the following expression:

$$R = (R_E + h) - \frac{R_E h D_0^2}{2(R_E + h)} - \frac{R_E^2 h^2 D_0^4}{8(R_E + h)^2} \tag{3.41}$$

By using the numerical quantities in Table 3.1, it is easy to show that

$$R \approx (R_E + h) - h\left(5.62 \times 10^{-6}\right) \tag{3.42}$$

So, with h equal to 3.0×10^5 m, we get about 1.68 m for the magnitude of the third term in Eq. (3.41). Therefore, we seem justified in ignoring the deviation of the normal in calculating geocentric distances.

3.4 Deflection of the Vertical

The succeeding development follows closely the work of Britting.[5] As pointed out earlier, modeling the Earth as a spheroid introduces an ambiguity in the definition of the vertical. From a point on the Earth (or above the Earth), the spheroidal normal is not geocentric. The angle between the normal and the geocentric direction, designated by D_0 or D (depending upon whether the generic point is on or above the Earth), is given by

$$D_0 = L - L_{c_0}, \qquad D = L - L_c \tag{3.43}$$

where L and L_c are the geographic (or geodetic) and geocentric latitudes of the generic point, respectively.

From the law of Sines we may write, after referring to Fig. 3.2 and Eq. (3.37),

$$\sin(D)/K^2\bar{X} = \sin(\pi - L)/R \tag{3.44}$$

where $\bar{X}$ continues to be the X-coordinate of a generic point on the Earth's surface (i.e., the equatorial projection of the Earth radius vector $\boldsymbol{R}_E$) and may be written as

$$\bar{X} = R_E \cos(L_{c_0}) \tag{3.45}$$

By using Eq. (3.43) we may rewrite the above equation as

$$\bar{X} = R_E[\cos(L)\cos(D_0) + \sin(L)\sin(D_0)] \tag{3.46}$$

Upon substituting Eqs. (3.34b), (3.45), and (3.46) into Eq. (3.44) and making use of the approximation given by Eq. (3.42) we get

$$\sin(D) = \left(\frac{eR_E}{R_E + h}\right)\left(1 + \frac{e}{2}\right)\sin(2L)\cos(D_0) + 2e\left(1 - \frac{e}{2}\right)\left(\frac{R_E}{R_E + h}\right)\sin^2(L)\sin(D_0) \tag{3.47}$$

The above equation is identical to that given by Britting.[5] If Eq. (3.47) is evaluated at the surface of the Earth, i.e., at $h = 0$ and $D = D_0$, we get

$$\sin(D_0) = e\left(1 - \frac{e}{2}\right)\sin(2L) + 2e\left(1 - \frac{e}{2}\right)\sin^2(L)\sin(D_0)$$

or

$$\sin(D_0)\left[1 - 2e\left(1 - \frac{e}{2}\right)\sin^2(L)\right] = e\left(1 - \frac{e}{2}\right)\sin(2L)$$

Accepting the small angle approximation for D_0 (i.e., $D_0 \leq e$), we get as a valid approximation the following expression:

$$D_0 = e\sin(2L) + \Delta_0 \tag{3.48a}$$

where

$$\Delta_0 = -\left(\frac{e}{2}\right)^2 \sin(2L) + 2e^2\sin(2L)\sin^2(L) + \cdots \tag{3.48b}$$

which has a value no larger than 5.6 μrad (1.16 arc-s).[5] Again following the work of Britting,[5] we get

$$D = e \sin(2L) + \Delta \tag{3.49a}$$

where

$$\Delta = -e \sin(2L)\left(\frac{e}{2} + \frac{h}{R_E}\right) + \cdots \tag{3.49b}$$

If re-entry begins at 3×10^5 m, then we can show that Δ is about 163 μrad, or 34 arc-s. The first term in either Eq. (3.48a) or Eq. (3.49a) is about 700 arc-s, so there is some justification for ignoring Δ and Δ_0.

3.5 Earth's Radius

In Section 3.2 we assumed that the Earth's shape could be approximated by an equipotential surface, an approximation to the geoid. An acceptable approximation for the Earth's radius R_E as a function of geocentric latitude was given in Eq. (3.25). In this section the Earth's form is approximated by a spheroid [see Eq. (3.35a)]. We show here that the geoid is nearly identical to a spheroid or, equivalently, that a section through the geoid that views the equatorial plane edge view has the form of an ellipse. If altitude must be included in the approximation (say, for the purpose of obtaining the altitude rate $\mathrm{d}h/\mathrm{d}t$), then Britting[5] suggests the following expression for the deviation of the normal, D, to include altitude, h:

$$D = e\left(1 - \frac{h}{R_E}\right)\sin(2L) \tag{3.49c}$$

Again taking $(\bar{X}, \bar{Z})$ to be a generic point on the Earth's surface, we may write

$$\bar{X} = R_E \cos(L_{c_0}), \qquad \bar{Z} = R_E \sin(L_{c_0})$$

Inserting the above relationships into Eq. (3.35a) we get

$$R_E = \frac{R_p}{\left\{1 - \left[1 - (R_p/R_e)^2\right]\cos^2(L_{c_0})\right\}^{1/2}} \tag{3.50}$$

The quantity in the inner brackets is recognized as the square of the eccentricity of the ellipse. Hence, we may expand the denominator as

$$R_E = R_p\left[1 + \frac{K^2}{2}\cos^2(L_{c_0}) + \frac{3K^4}{8}\cos^4(L_{c_0}) + \frac{5K^6}{16}\cos^6(L_{c_0}) + \cdots\right] \tag{3.51}$$

Taking

$$\frac{K^2}{2} = e\left(1 - \frac{e}{2}\right) \approx e$$

and using numerical values for e, we can show that a satisfactory approximation to Eq. (3.51) is formed by ignoring the third and higher terms to give

$$\begin{aligned} R_E &\approx R_p\left[1 + (K^2/2)\cos^2(L_c)\right] \\ &\approx R_p\left[1 + e\cos^2(L_{c_0})\right] \end{aligned} \tag{3.52a}$$

Britting[5] also shows that the expression

$$L_{c_0} = L - D_0$$

along with Eq. (3.48a) for D_0 yield an alternate form of Eq. (3.52a) as follows:

$$\begin{aligned} R_E &= R_e\left\{1 - \frac{e}{2}[1 - \cos(2L)] + \frac{5}{16}e^2[1 - \cos(4L)] - \cdots\right\} \\ R_E &\approx R_e[1 - e\sin^2(L)] \end{aligned} \tag{3.52b}$$

The above expression is accurate to within about 50 m. So, unless the uncertainty in the equatorial radius R_e is less than 1 m, the above expression should be adequate.

Equation (3.52b) is identical to Eq. (3.25) if f is taken to be equal to e. This fact seems to indicate that the Earth's shape is nearly that of a rotating molten sphere. Note in Eq. (3.26) that the Earth's rotation appears directly in the constant d as well as indirectly in the constant K/R_e^2, since the rotation causes the moments of inertia I_{zz}^e and I_{xx}^e to be unequal.

3.6 Earth's Gravitational Potential

In the preceding sections we directed our attention toward justifying an elliptical, or nearly elliptical, shape of the Earth's meridian. In this section we derive an expression for the potential of the Earth's external gravitational field. In doing so, we include effects due to a nonspherical mass distribution, although we consider the pole axis to be the axis of symmetry. We ignore the Earth's rotation because such rotation has no gravitational effect on a body removed from the surface. Rotation, as we have seen, contributes to the total body force experienced by an object on the Earth's surface. We can thus write Eq. (3.14) in a more compact form:

$$U = \frac{G}{R}\int_V \sum_{k=0}^{\infty} P_k[\cos(\gamma)]\left(\frac{r}{R}\right)^k \mathrm{d}M_E \tag{3.53}$$

where $P_k[\cos(\gamma)]$ is the Legendre polynomial defined in Eq. (3.11a) or alternately by using Rodrigues' generating function with $v = \cos(\gamma)$. That is,

$$P_k(v) = \frac{1}{2^k k!}\frac{\mathrm{d}^k(v^2 - 1)}{\mathrm{d}v^k} \tag{3.54a}$$

For completeness we will include the associated Legendre functions P_k^j also using Rodriques' generating function.

$$P_k^j = (\nu^2 - 1)^{j/2} \frac{\mathrm{d}^j P_k(\nu)}{\mathrm{d}\nu^j} \tag{3.54b}$$

The mass element $\mathrm{d}M_E$ must also be replaced by the appropriate spherical variables as follows:

$$\mathrm{d}M_E = D(r, \beta, \theta) r^2 \sin(\beta) \mathrm{d}r \mathrm{d}\beta \mathrm{d}\theta \tag{3.55}$$

where $D(r, \beta, \theta)$ is the density function.

We can now rewrite Eq. (3.13b) in the following form:

$$\cos(\gamma) = \sin(\beta)\sin(\phi)\cos(\theta - l) + \cos(\beta)\cos(\phi) \tag{3.56}$$

where we have replaced L_{c_0}, the geocentric latitude, by its complement, ϕ.

The procedure then is to replace $\cos(\gamma)$ using the above relationship along with a knowledge of the density function $D(r, \beta, \theta)$ of Eq. (3.55). The potential U, according to Eq. (3.53), becomes a summation of a series of integrals. The analytical manipulations would obviously be formidable even if the density function were a constant. However, with an imperfect knowledge of the density function, there would be limited utility to a closed-form or even a numerical solution. Therefore, the procedure that we follow is to use the form of Eq. (3.53) for the potential, but to assign to the constants numerical values based upon measurements rather than attempt some kind of solution. We will now outline how this might be done.

We may use the addition formula of spherical harmonics (associated Legendre polynomials) to write the following replacement for $P_k[\cos(\gamma)]$[6]:

$$\begin{aligned} P_k[\cos(\gamma)] &= P_k[\cos(\phi)] P_k[\cos(\beta)] \\ &\quad + 2\sum_{j=1}^{k} \left\{ \frac{(k-1)!}{(k+1)!} \cos[j(l-\theta)] P_k^j[\cos(\phi)] P_k^j[\cos(\beta)] \right\} \end{aligned} \tag{3.57}$$

where P_k^j are the associated Legendre polynomials (more appropriately called "functions") of the first kind of degree k and order j; the relationship between the two types of Legendre polynomials is given by the following:

$$\begin{aligned} P_k^j(\nu) &= \left(1 - \nu^2\right)^{j/2} \frac{\mathrm{d}^j}{\mathrm{d}\nu^j} [P_k(\nu)] \\ &= \frac{\left(1 - \nu^2\right)^{j/2}}{2^k k!} \frac{\mathrm{d}^{j+k}}{\mathrm{d}\nu^{j+k}} \left(\nu^2 - 1\right)^k \end{aligned} \tag{3.58}$$

A brief but very readable development of associated Legendre functions is given in Appendix G of Ref. 4.

It is appropriate at this point to note that since we are assuming a spheroidal shape (i.e., a shape which possesses rotational symmetry about the polar axis), we may write

$$D(r, \beta, \theta) = D(r, \beta) \tag{3.59}$$

and the summation terms in Eq. (3.57) may be ignored. In more complete representations useful in some Earth science studies, the θ-dependent terms produce periodic functions of longitude l in the potential U. For re-entry studies such precision is not needed. As a consequence the potential U may be written as

$$U = \frac{1}{R}\sum_{k=0}^{\infty}\frac{A_k}{R^k}P_k[\cos(\phi)] \tag{3.60}$$

where

$$A_k = G\int_V \left[r^k P_k \cos(\beta)\right] D(r, \beta) r^2 \sin(\beta) \mathrm{d}r\, \mathrm{d}\beta\, \mathrm{d}\theta \tag{3.61}$$

Equation (3.60) may be rewritten in terms of k for $k = 0$, $k = 1$, and $k \geq 2$ as follows:

$$U(R, \phi) = \frac{G}{R}\int_{M_E} \mathrm{d}M_E + \frac{G}{R^2}\cos(\phi)\int_{M_E} r\cos(\beta)\mathrm{d}M_E + \sum_{k=2}^{\infty}\frac{A_k}{R^{k+1}}P_k[\cos(\phi)] \tag{3.62}$$

The first term provides the mass of the Earth. The second term is identically zero because $r\cos(\phi)$ is the distance in the equatorial plane from the origin of the geocentric axis system to the elementary mass $\mathrm{d}M_E$; because the axis system is geocentric, this first moment of the mass must vanish. Finally, we have

$$U(R, \phi) = \frac{GM_E}{R} + \sum_{k=2}^{\infty}\frac{A_k}{R^{k+1}}P_k[\cos(\phi)] \tag{3.63}$$

If we let

$$\frac{R_e^k}{R_e^k R^{k+1}} = \frac{1}{RR_e^k}\left(\frac{R_e}{R}\right)^k \tag{3.64}$$

we may rewrite Eq. (3.63) as

$$U(R, \phi) = \frac{GM_E}{R}\left\{1 - \sum_{k=2}^{\infty}\left(\frac{R_e}{R}\right)^k J_k P_k[\cos(\phi)]\right\} \tag{3.65}$$

where

$$J_k = -\frac{A_k}{R_e^k}\left(\frac{1}{GM_E}\right) = \frac{A_k}{\mu R_e^k} \tag{3.66}$$

where the J-terms are sometimes known as the *Jeffery constants*. Numerical (or, in principle, analytical) integrations would in the past have produced the A_k terms given in Eq. (3.61). The issue is avoided and the J-constants are evaluated today using satellite observations.

How these constants are determined is tangential to our efforts. However, a very readable presentation of such measurements is given by Garland.[3] Useful values of some of the J-constants are given in Table 3.1.

It is possible to give some physical meaning to the J_2 term by setting Eq. (3.65) equal to Eq. (3.18) to get

$$\frac{K}{2R^2} = \left(\frac{R_e}{R}\right)^2 J_2$$

When combined with Eq. (3.17), the above expression leads to

$$J_2 = \frac{I_{zz}^e - I_{xx}^e}{M_E R_e^2} = \frac{K}{2R_e^2} \tag{3.67}$$

Thus, J_2 represents the difference between the polar and equatorial moments of inertia. By using Eq. (3.26) with $\omega_{e/I} = 0 = d$, we can show that

$$J_2 \approx \tfrac{1}{3}f \approx \tfrac{1}{3}e \tag{3.68}$$

which indicates that J_2 is closely related to the form factor f (or, through Eq. (3.52b), the ellipticity e). A quick calculation of $e/3$ gives 1.176×10^{-3}, close to the value of J_2. Because of this relationship between J_2 and the ellipticity, J_2 is often identified as the *oblateness* term; that is, J_2 quantifies Earth oblateness. Another designation is the *dynamical form factor.*

Details regarding the measurement of J_2 from a satellite in Earth orbit are given in Chapter 11 of Ref. 3 and in various places in Refs. 7 and 8. The effect of the flattening of the Earth from a sphere to a spheroid is to cause the orbit of a low-altitude Earth satellite to precess at about 8 deg per day for a near-equatorial orbit to about 4 deg per day for an orbit inclined 60 deg to the equator. The term *precession* here means that there is a westward retreat of the argument of the ascending node of the orbit. The precessional rate $\mathrm{d}\Omega/\mathrm{d}t$ is given as follows:

$$\frac{\mathrm{d}\Omega}{\mathrm{d}t} = \left(\frac{3}{2}\right) J_2 \cos(i) \left(\frac{R_e}{R}\right)^2 (g_e)^{1/2} \left(\frac{R_e}{R^{3/2}}\right) \tag{3.69}$$

where i is the inclination of the orbit from the Earth's equatorial plane.

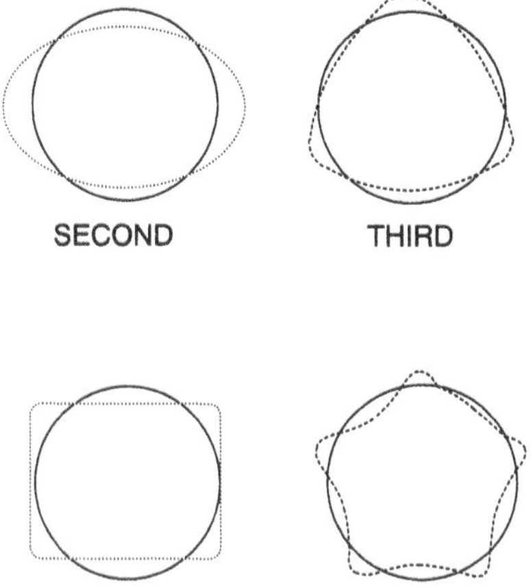

Fig. 3.3 Form of Earth due to the second through fifth harmonics.

For an orbit with an altitude of 8.0×10^5 m, the above equation indicates that the orbit will precess westward at a rate of about 6.2 deg per day, a quantity which is easily measurable to better than 0.1%. Insertion of the measured value of precessional rate, along with orbital information such as altitude and inclination, permits a straightforward calculation for J_2. For highly eccentric orbits (where there is significant variation in altitude), J_2 may be measured by tracking the perigee, because J_2 causes the perigee to shift several degrees per day. Orbital precession causes the perigee to shift, say, from north to south (and of course back again). The differences in these north-south perigee altitudes (about 10 km) are used to calculate J_3, often called the pear-shaped term. A sketch of the various harmonics is shown in Fig. 3.3. These sketches are obtained by using the program given in Appendix B. It should be pointed out that J_2 is one of the most accurately measured of all of the Earth's geophysical constants.

For interest, the gravitational potential may be expanded to the fourth J-term as follows:

$$U = \frac{GM_E}{R}\left\{1 - \frac{J_2}{2}\left(\frac{R_e}{R}\right)^2 [3\cos^2(\phi) - 1] - \frac{J_3}{2}\left(\frac{R_e}{R}\right)^3 (5\cos^3(\phi) - 3\cos(\phi))\right.$$
$$\left. - \frac{J_4}{8}\left(\frac{R_e}{R}\right)^4 [35\cos^4(\phi) - 30(\cos^2(\phi) + 3] - \cdots\right\} \tag{3.70}$$

The gravitational field in spherical coordinates may be found by inserting U from the above expression into the gradient operator as follows:

$$\boldsymbol{g} = g_R \boldsymbol{i}_R + g_\phi \boldsymbol{i}_\phi = -\nabla U = -\left[\frac{\partial}{\partial R}(U)\boldsymbol{i}_R + \frac{\partial}{R\partial\phi}(U)\boldsymbol{i}_\phi\right]$$

to give

$$g_R = -\frac{GM_E}{R^2}\left\{1 - \frac{3}{2}J_2\left(\frac{R_e}{R}\right)^2[3\cos^2(\phi) - 1]\right.$$
$$- 2J_3\left(\frac{R_e}{R}\right)^3\cos(\phi)[5\cos^2(\phi) - 3]$$
$$\left. - \frac{5}{8}J_4\left(\frac{R_e}{R}\right)^4[35\cos^4(\phi) - 30\cos^2(\phi) + 3]\right\} \tag{3.71a}$$

and

$$g_\phi = \frac{3GM_E}{R^2}\left(\frac{R_e}{R}\right)^2\sin(\phi)\cos(\phi)\left\{J_2 + \frac{1}{2}J_3\left(\frac{R_e}{R}\right)\sec(\phi)[5\cos^2(\phi) - 1]\right.$$
$$\left. + \frac{5}{6}J_4\left(\frac{R_e}{R}\right)^2[7\cos^2(\phi) - 1]\right\} \tag{3.71b}$$

3.7 Gravitational Field in an Inertial Frame

Equations (3.71) provide the specific gravitational force vector in the geocentric direction and along the normal to the geocentric direction within the meridian plane. The geocentric angle ϕ is the complement of the geocentric latitude L_c. Thus we may write the following expression for the components of the gravitational acceleration in the geocentric frame:

$$\boldsymbol{g}^c = [-g_\phi, 0, -g_R]^{\mathrm{T}} \tag{3.72}$$

For a definition of the various coordinate frames, refer to Appendix F.

The vector $\boldsymbol{g}^c$ may now be evaluated in an inertial frame as

$$\boldsymbol{g}^I = C_c^I \boldsymbol{g}^c$$

which yields

$$\boldsymbol{g}^I = \begin{bmatrix} g_\phi \sin(L_c)\cos(\lambda) + g_R\cos(L_c)\cos(\lambda) \\ g_\phi \sin(L_c)\sin(\lambda) + g_R\cos(L_c)\sin(\lambda) \\ -g_\phi\cos(L_c) + g_R\sin(L_c) \end{bmatrix} = \begin{bmatrix} g_x^I \\ g_y^I \\ g_z^I \end{bmatrix} \tag{3.73}$$

It is shown in Appendix C how the latitude L_c and the longitude may be calculated from a knowledge of the inertial coordinates of the re-entry vehicle.

The vector $\boldsymbol{g}^I$ may be rewritten slightly. First, we note that λ is the celestial longitude. As an aside, we note that the terrestrial longitude l and the celestial longitude λ are related as follows:

$$l = l_0 + \lambda - \omega_{e/I}t \tag{3.74}$$

where l_0 is the terrestrial longitude at time $t = 0$. Initialization of the celestial longitude λ is arbitrary. For re-entry body trajectory work we might assume that λ is zero at burnout (trajectory initialization) so that l_0 is the terrestrial longitude at burnout.

The radius vector in an inertial frame (Z-axis coincident with the Earth's polar axis) is given by

$$X = R\cos(L_c)\cos(\lambda) \tag{3.75a}$$

$$Y = R\cos(L_c)\sin(\lambda) \tag{3.75b}$$

$$Z = R\sin(L_c) \tag{3.75c}$$

By using Eq. (3.73) we can write, according to Britting,[5]

$$g_x^I = [g_R + g_\phi \tan(L_c)]X/R \tag{3.76a}$$

$$g_y^I = [g_R + g_\phi \tan(L_c)]Y/R \tag{3.76b}$$

$$g_z^I = [g_R - g_\phi \cot(L_c)]Z/R \tag{3.76c}$$

Now we may insert the expressions for g_R and g_ϕ from Eqs. (3.71) into Eqs. (3.76) to give

$$g_x^I = -\frac{\mu}{R^2}\left\{1 + \frac{3}{2}J_2\left(\frac{R_E}{R}\right)^2\left[1 - 5\left(\frac{Z}{R}\right)^2\right]\right\}\frac{X}{R} \tag{3.77a}$$

$$g_y^I = -\frac{\mu}{R^2}\left\{1 + \frac{3}{2}J_2\left(\frac{R_e}{R}\right)^2\left[1 - 5\left(\frac{Z}{R}\right)^2\right]\right\}\frac{Y}{R} \tag{3.77b}$$

$$g_z^I = -\frac{\mu}{R^2}\left\{1 + \frac{3}{2}J_2\left(\frac{R_e}{R}\right)^2\left[3 - 5\left(\frac{Z}{R}\right)^2\right]\right\}\frac{Z}{R} \tag{3.77c}$$

In Eqs. (3.77) we have of course omitted the J_3 and J_4 terms which are present in Eqs. (3.71). For most re-entry trajectory modeling, retention of the J_2 term should be sufficient. The maximum error incurred by omitting J_3 and J_4 is about 1.2×10^{-5} g for the X/R and Y/R terms. The error may be as high as 2.0×10^{-5} g for the Z/R term. The criterion which must be applied here is the accuracy of the accelerometer. If the accelerometer is capable of measurements in the 10-μg range, then omission of the J_2 and higher Jeffery constants seems to be acceptable.

Equations (3.77) provide the resolution of the $\boldsymbol{g}$ vector in a nonrotating inertial frame; in a geocentric Earth-fixed frame, we must "correct" Eqs. (3.77) for centrifugal acceleration. Thus we have in the Earth-fixed frame

$$\boldsymbol{g}^e = \boldsymbol{g}^I - \Omega_{e/I}^e \Omega_{e/I}^e \boldsymbol{R}^e \tag{3.78}$$

where $\boldsymbol{g}^e$ is the gravity vector in the Earth-fixed frame, $\Omega_{e/I}$ is the rotation rate of the Earth relative to inertial space, and $\boldsymbol{R}$ is the geocentric position vector.

In the next section we will briefly consider variations of the gravitational field from that associated with the geoid, i.e., the field given by Eqs. (3.71). From Appendix G we define the geocentric frame (or c-frame as having the Z^c-axis) opposite the geocentric position vector $\boldsymbol{R}$, the X^c-axis in the local meridian, and the Y^c-axis in the direction of east. Thus, we may write the following expression for the gravity vector in the geocentric frame:

$$\boldsymbol{g}^c = [-g_\phi, 0, -g_R]^{\mathrm{T}} \tag{3.79}$$

Of course, as Eq. (3.78) shows, a measurement of the gravitational acceleration at the surface of the Earth cannot separate the centrifugal contribution. Einstein's equivalency principle means that the centrifugal and gravitational accelerations cannot be separated through measurement; therefore, the centrifugal contribution must be removed computationally.

Of some interest are the components of the gravitational attraction in the navigation frame. In the navigation frame, the Z^n-axis is normal to the reference ellipsoid and deviates from the geocentric position vector by the angle D discussed earlier. The X^n-axis is in the meridian and directed north; the Y^n-axis is directed east. From Appendix G it is easily shown that

$$g_N = -g_\phi \cos(D) - g_R \sin(D) \tag{3.80a}$$

$$g_E = 0 \tag{3.80b}$$

$$g_D = g_\phi \sin(D) - g_R \cos(D) \tag{3.80c}$$

The deviation of the normal angle D is given in Eqs. (3.43). It is easy to show that the components of the centrifugal acceleration in an Earth frame are given by

$$\boldsymbol{\Omega}^e_{e/I}\boldsymbol{\Omega}^e_{e/I}\boldsymbol{R}^e = \begin{bmatrix} R\omega^2_{e/I}\cos(L_c)\cos(l) \\ R\omega^2_{e/I}\cos(L_c)\sin(l) \\ 0 \end{bmatrix} \tag{3.81a}$$

By using the DCM (Directional Cosine Matrix) C^n_e to transfer to the navigation frame, we obtain the component of the centrifugal acceleration in a navigation frame as

$$\begin{bmatrix} R\omega^2_{e/I}\sin(L)\cos(L_c) \\ 0 \\ R\omega^2_{e/I}\cos(L)\cos(L_c) \end{bmatrix} \tag{3.81b}$$

Next we must express the gravitational vector in the navigation frame. First, the transformation between the geocentric and the navigation frame is as follows:

$$g^n = C_c^n g^c$$

Thus, the gravitational vector becomes

$$g^n = \begin{bmatrix} g_N \\ g_E \\ g_D \end{bmatrix} = \begin{bmatrix} \cos(D) & 0 & \sin(D) \\ 0 & 1 & 0 \\ -\sin(D) & 0 & \cos(D) \end{bmatrix} \begin{bmatrix} -g_\phi \\ 0 \\ -g_R \end{bmatrix} \tag{3.82a}$$

giving

$$\begin{aligned} g_N &= -g_\phi \cos(D) - g_R \sin(D) \\ g_E &= 0 \\ g_D &= g_\phi \sin(D) - g_R \cos(D) \end{aligned} \tag{3.82b}$$

Now the gravitational acceleration in the navigation frame becomes

$$g^n = \begin{bmatrix} g_N - R\omega_{e/I}^2 \sin(L)\cos(L_c) \\ g_E \\ g_D - R\omega_{e/I}^2 \cos(L)\cos(L_c) \end{bmatrix} \tag{3.83}$$

3.8 Gravitational Anomalies and Deflection of the Vertical

Up to this point we have modeled the Earth as a body of revolution. Consequently, the gravitational vector has remained within the meridian plane, although the vector was not (usually) geocentric. As might be expected, there will be some discrepancies in the magnitude and direction of the gravitational vector based upon the spheroidal model and the actual value. A discrepancy between a local measurement of the gravity vector and that predicted by the model is identified as a *gravitational anomaly* when referring to the magnitude and as a *deflection of the vertical* when referring to the direction. Figure 3.4 provides an illustration of the deflection of the vertical; the angle ξ measures the deflection in the meridian plane, and η measures the deflection in a plane normal to the meridian. Common values of ξ are about -10 to $+4$ arc-s whereas η might vary between -14 and $+8$ arc-secs. However, such values are included here only to give some idea of the size of this effect; the actual values can vary significantly over a distance as short as 50 km.

Let's designate the gravitational acceleration associated with the reference ellipsoid by g_e, the gravity anomaly by Δg, and the effect due to deflection of

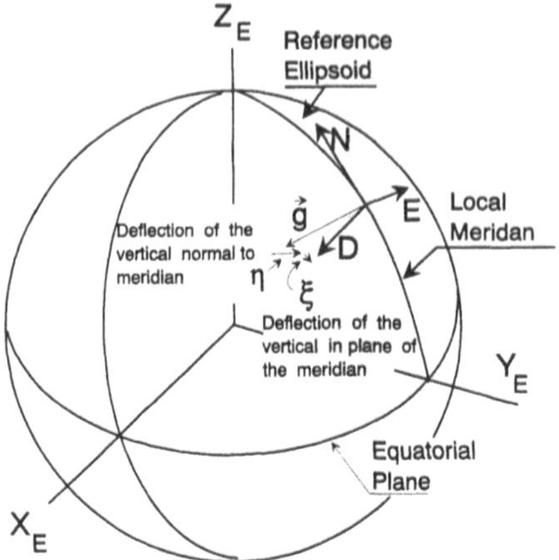

Fig. 3.4 Deflection of the vertical.

the vertical by ξg_e and ηg_e. Thus we may write the following for the gravity in the navigation frame:

$$\boldsymbol{g}^n = \begin{bmatrix} \xi g_e \\ -\eta g_e \\ g_e + \Delta g \end{bmatrix} \tag{3.84}$$

Once again we note that in the geographic (navigation) frame, the X-axis points north and is in the meridian, the Y-axis is directed east (a positive η rotates the gravity vector west), and the Z-axis is down normal to the geoid (and not usually geocentric).

We might separate the anomalies into a Δ-vector as

$$\boldsymbol{g}^n = \begin{bmatrix} 0 \\ 0 \\ g_e \end{bmatrix} + \begin{bmatrix} \xi g_e \\ -\eta g_e \\ \Delta g \end{bmatrix} = \begin{bmatrix} 0 \\ 0 \\ g_e \end{bmatrix} + \boldsymbol{\Delta g}^n \tag{3.85}$$

Since g_e is the gravitational acceleration for the reference ellipsoid, we have, from Eq. (3.83),

$$g_e = g_D - R\omega_{e/I}^2 \cos(L)\cos(L_c) \tag{3.86a}$$

and, from Eq. (3.80c),

$$g_D = -g_R \cos(D) + g_\phi \sin(D) \tag{3.86b}$$

Since D is small, we have, from Eq. (3.49a),

$$g_D \approx -g_R + g_\phi e \sin(2L) \tag{3.86c}$$

where we have replaced $\sin D$ by D.

Next, we replace g_R from Eq. (3.71) to get:

$$g_D = \frac{\mu}{4^2}\left\{1 - \frac{3}{4}J_2[1 - 3\cos(2L)]\right\} + \epsilon \tag{3.87}$$

where ϵ contains higher-order terms. Britting estimates the error term ϵ to be on the order of 20 μg.[5] Of course, additional Jeffery constants might be included, but if so the approximation leading to Eqs. (3.86) might not be justified.

Finally, the expression for the gravitational acceleration associated with the reference geoid may be written as

$$g_e \approx \frac{\mu}{R^2}\left\{1 - \frac{3}{4}J_2[1 - 3\cos(2L)]\right\} - R\omega_{e/I}^2 \cos(L)\cos(L - D) \tag{3.88}$$

Note that the radius R in the above expression varies according to the Earth's shape. Of course, if we are interested in the gravitational acceleration at the surface of the Earth, then we should replace R with R_E. If Eq. (3.88) is to be used to compute the gravitational acceleration at the Earth's surface, we must replace R as follows:

$$R = R_E = R_e[1 - e\sin^2(L)] \tag{3.89}$$

or by using some other approximation [e.g., Eqs. (3.50),(3.51), or (3.52)].

3.9 Longitudinal Dependencies

Up to this point in the development of the gravitational potential U, we have consistently ignored variations with longitude. Longitude terms were presented in Eq. (3.53) but were subsequently ignored when we assumed the Earth shape to be spheroidal. In this section we briefly consider the modeling of longitudinal dependencies. A model of the Earth's potential which includes longitudinal variations is possibly overly elaborate for re-entry vehicle studies. However, it is included here for completeness.

The gravitational potential U, which includes longitudinal terms, may be written as follows[4]:

$$U(R, l, \phi) = \frac{GM_E}{R}\left\{1 - \sum_{k=2}^{\infty}\left(\frac{R_e}{R}\right)^k J_k P_k[\cos(\phi)] + \sum_{k=2}^{\infty}\sum_{j=1}^{k}\left(\frac{R_e}{R}\right)^k [C_{kj}\cos(kl) + S_{kj}\sin(Kl)]\, P_k^j \cos(\phi)\right\} \tag{3.90}$$

where R_e is, of course, the radius of the Earth at the equator (given in Table 3.1). The terms P_k^j are the associated Legendre functions of degree k and order j, first encountered in this book in Eq. (3.54b). Note that the Legendre function P_k^o is equal to the Legendre polynomial P_k. The coefficients C_{kj} and S_{kj}, apparently the coefficients of a Fourier series in longitude l, are called the *Tesseral harmonic coefficients* when j is not equal to k and the *Sectoral harmonic coefficients* when j is equal to k. Appendix G of Ref. 4 provides a pictorial representation of the various harmonics. Although we have identified the J_k-terms as the Jeffery constants, an alternate nomenclature identifies them as the *Zonal harmonic coefficients*. Thus, for example, J_2 might be designated as the *Second-order zonal harmonic*. The C_{kj} and S_{kj} coefficients are numerically determined from satellite observations; the representation of Eq. (3.90) is used mostly to help provide the ephemeri of new satellites.

Equation (3.90) is useful because the gravitational acceleration may be calculated as a function of altitude, colatitude, and longitude as

$$\boldsymbol{g}^c = -\nabla U = -\left[\frac{\partial U}{\partial R}\boldsymbol{i}_R + \frac{1}{R}\frac{\partial U}{\partial \phi}\boldsymbol{i}_\phi + \frac{1}{R\sin(\phi)}\frac{\partial U}{\partial l}\boldsymbol{i}_l\right] \tag{3.91}$$

The above equation is a more elaborate version of Eqs. (3.71); the result is a longitudinally dependent version of Eq. (3.72). This line of development will not be pursued further here, however, because the potential given by Eq. (3.90) is usually not needed for most re-entry studies and simulations.

3.10 Gravity Gradient

For some types of re-entry problems the gradient of the gravitational force over the vehicle is important (see Appendix F). Also, in the error analysis of an inertial system, an error in position means an error in the gravitational term which must be added to the specific force to obtain the kinematic acceleration. This gravitational error is expressed directly in terms of the gravity gradient.

The gravity gradient is given as the divergence of the vector $\boldsymbol{g}$ with respect to the position vector $\boldsymbol{R}$ as follows:

$$\boldsymbol{G} = \frac{\mathrm{d}\boldsymbol{g}}{\mathrm{d}\boldsymbol{R}} \tag{3.92}$$

where $\boldsymbol{G}$ is the gradient matrix. For a spherical gravity field, $\boldsymbol{g}$ is defined as

$$\boldsymbol{g} = -\mu \boldsymbol{R}/R^3 \tag{3.93}$$

The vector derivative used in Eq. (3.92) is defined as

$$\frac{\mathrm{d}(\,)}{\mathrm{d}\boldsymbol{R}} \doteq \boldsymbol{i}_1\frac{\mathrm{d}(\,)}{\mathrm{d}X_1} + \boldsymbol{i}_2\frac{\mathrm{d}(\,)}{\mathrm{d}X_2} + \boldsymbol{i}_3\frac{\mathrm{d}(\,)}{\mathrm{d}X_3} \tag{3.94}$$

Inserting Eq. (3.93) into Eq. (3.92) gives

$$\boldsymbol{G} = -\mu\left\{(\boldsymbol{R}^{\mathrm{T}}\boldsymbol{R})^{-3/2}\frac{\mathrm{d}\boldsymbol{R}}{\mathrm{d}\boldsymbol{R}} + \frac{\mathrm{d}}{\mathrm{d}\boldsymbol{R}}\left[\left(\boldsymbol{R}^{\mathrm{T}}\boldsymbol{R}\right)^{-3/2}\right]\boldsymbol{R}\right\} \tag{3.95}$$

First, we note that

$$\frac{\mathrm{d}\boldsymbol{R}}{\mathrm{d}\boldsymbol{R}} = \boldsymbol{I} \tag{3.96a}$$

and

$$\frac{\mathrm{d}}{\mathrm{d}\boldsymbol{R}}\left[\left(\boldsymbol{R}^{\mathrm{T}}\boldsymbol{R}\right)^{-3/2}\right]\boldsymbol{R} = -\frac{3}{2}\left(\boldsymbol{R}^{\mathrm{T}}\boldsymbol{R}\right)^{-5/2}\frac{\mathrm{d}}{\mathrm{d}\boldsymbol{R}}\left[\left(\boldsymbol{R}^{\mathrm{T}}\boldsymbol{R}\right)\right]\boldsymbol{R} \tag{3.96b}$$

but

$$\frac{\mathrm{d}}{\mathrm{d}\boldsymbol{R}}\left(\boldsymbol{R}^{\mathrm{T}}\boldsymbol{R}\right) = 2\boldsymbol{R}^{\mathrm{T}} \tag{3.96c}$$

The result given in Eq. (3.96c) is easily demonstrated. First, note that

$$\boldsymbol{R}^{\mathrm{T}}\boldsymbol{R} = X_1^2 + X_2^2 + X_3^2$$

Next, by applying Eq. (3.94) we get

$$\begin{aligned}\frac{\mathrm{d}}{\mathrm{d}\boldsymbol{R}}\left(\boldsymbol{R}^{\mathrm{T}}\boldsymbol{R}\right) &= \frac{\mathrm{d}}{\mathrm{d}\boldsymbol{R}}\left(X_1^2 + X_2^2 + X_3^2\right) = [2X_1, 2X_2, 2X_3]\\ &= 2\boldsymbol{R}^{\mathrm{T}}\end{aligned}$$

With Eq. (3.96c) substituted into Eq. (3.96b), Eq. (3.95) can be rewritten as

$$\boldsymbol{G} = -\mu\left[\frac{1}{R^3}\boldsymbol{I} - \frac{3}{R^5}\boldsymbol{R}\boldsymbol{R}^{\mathrm{T}}\right] \tag{3.97a}$$

An alternate form may be obtained by normalizing by $\boldsymbol{R}$ to give

$$\boldsymbol{G} = -\frac{\mu}{R^3}\left[\boldsymbol{I} - 3\boldsymbol{u}_R\boldsymbol{u}_R^{\mathrm{T}}\right] \tag{3.97b}$$

where $\boldsymbol{u}_R$ is the unit vector in the R-direction.

A simplified application of Eq. (3.97) might be to take the $\boldsymbol{r}$ vector along the 3-axis. If Δr is also along the 3-axis (along the gravity vector), then

$$\boldsymbol{r}_r = \boldsymbol{i}_3$$

$$\Delta \boldsymbol{r} = \Delta r \boldsymbol{u}_r = \Delta r \boldsymbol{i}_3$$

or

$$\boldsymbol{u}_r \boldsymbol{u}_r^{\mathrm{T}} = \begin{bmatrix} 0 & 0 & 0 \\ 0 & 0 & 0 \\ 0 & 0 & 1 \end{bmatrix} \tag{3.98}$$

Since $\boldsymbol{g}$ is taken as negative, Eq. (3.97b) indicates that while $\boldsymbol{g}$ remains negative as the radius vector increases by an amount $\Delta \boldsymbol{r}$, the new gravitational acceleration is diminished in magnitude by the following amount:

$$\frac{2\mu}{r^3}\Delta r$$

We may also obtain the gravity gradient for the nonspherical gravity field where we include the oblateness coefficient J_2. First, we return to Eqs. (3.77), where we have expressed the components of the gravitational acceleration in an inertial frame. As we pointed out, these equations are valid without change in an Earth-fixed geocentric frame except that the centrifugal acceleration must be added.

The gravity gradient for the spherical field is given in Eqs. (3.97). Thus, we need calculate only an additional gradient contribution due to oblateness and add this term to that of the spherical gravity field. We designate the contribution of the oblateness to the gravitational acceleration and gradient, respectively, by Δg_0 and ΔG_0.

From Eqs. (3.77) we may write in vector form the following:

$$\Delta \boldsymbol{g}_0 = -\frac{\mu}{R^2}\frac{3}{2}J_2\left(\frac{R_e}{R}\right)^2\frac{\boldsymbol{R}}{R} \tag{3.99}$$

Following the steps that led to Eqs. (3.97), we have

$$\Delta \boldsymbol{G}_0 = -\frac{3}{2}\mu J_2 R_e^2\left[\boldsymbol{I}\left(\frac{1}{R^5}\right) - \frac{5}{R^6}\boldsymbol{R}\frac{\boldsymbol{R}^{\mathrm{T}}}{R}\right] \tag{3.100a}$$

or in terms of the unit vector $\boldsymbol{u_R}$

$$\Delta \boldsymbol{G}_0 = -\frac{3}{2}\mu J_2\left(\frac{R_e^2}{R^5}\right)[\boldsymbol{I} - 5\boldsymbol{u}_R\boldsymbol{u}_R^{\mathrm{T}}] \tag{3.100b}$$

References

[1]Eisberg, R. M., and Lerner, L. S., *Physics—Foundations and Applications*, McGraw-Hill, Hightstown, NJ, 1981.

[2]Sneddon, I. N., *Special Functions of Mathematical Physics and Chemistry*, Oliver and Boyd, distributors in U.S. Interscience Publishers, New York, 1976.

[3]Garland, G. D., *Introduction to Geophysics—Core and Crust*, Sanders, Philadelphia, PA, 1979.

[4]Wertz, J. R. (ed.), *Spacecraft Attitude Determination and Control*, Vol. 73, Astrophysics and Space Science Library, D. Reidel, distributed by Kluwer Boston, Hingham, MA, 1986.

[5]Britting, K. R., *Inertial Navigation Systems Analysis*, Wiley Interscience, Wiley, Somersest, NJ, 1971.

[6]Korn, G. A., and Korn, T. M., *Mathematical Handbook for Scientists and Engineers*, McGraw-Hill, Hightstown, NJ, 1968.

[7]Heiskanen, W. A., *The Earth and Its Gravity Field*, McGraw-Hill, Hightstown, NJ, 1958.

[8]King-Hele, D., "The Shape of the Earth," *Science Magazine*, Vol. 192, No. 4246, June 25, 1976.

4
Axis Transformations

4.1 Background

> "*Quaternion* was defined by an American school girl to be an 'ancient religious ceremony.' The ancients . . . knew not and did not worship quaternions. . . . A quaternion is neither a scalar nor a vector, but a sort of combination of both. It has no physical representatives, but is a highly abstract mathematical."
>
> — Oliver Heaviside[1]

While the reader may ultimately concur with Heaviside's opinion regarding quaternions, we should first emphasize the main topic of this chapter. This topic can be put in the form of a question: if we should have a vector with components (numerical or symbolic) available in one coordinate system, how might we represent the components of the same vector in a second system? Three mutually orthogonal unit vectors with a common origin will be taken here to define a coordinate system. In the nontrivial formulation of this problem, the frames associated with each of the systems are misaligned with respect to each other; in other words the (1, 2, 3) or (X, Y, Z) axes of one system are not collinear with the corresponding axes of the second system. Our concern in this chapter is how to manipulate the components of a vector $\boldsymbol{X}$ available in an a-frame, i.e., $\boldsymbol{X}^a$, to produce the same vector with components in the b-frame, i.e., $\boldsymbol{X}^b$. As will be seen, there are many ways of manipulating $\boldsymbol{X}^a$ to obtain $\boldsymbol{X}^b$, although there must be some analytical equivalency among such methods. Probably the most fundamental approach to this problem is to multiply the vector $\boldsymbol{X}^a$ by a matrix C to obtain $\boldsymbol{X}^b$, i.e.,

$$\boldsymbol{X}^b = C\boldsymbol{X}^a \tag{4.1a}$$

Since the matrix C transforms (or carries) vector components *from* the a-frame *to* the b-frame, we should attach symbols to the matrix C to indicate this "from-to" process. Unfortunately there is no universally accepted standard. Kane,[2] Hughes,[3] Wertz,[4] and Britting[5] differ in symbolic representations. The symbolism used by Britting[5] is preferred here. Using Britting's symbolism, the transformation *from* the a-frame *to* the b-frame will be represented with subscripts and superscripts as follows:

$$X^b = C_a^b X^a \tag{4.1b}$$

The a-frame–to–b-frame misalignment is contained in the matrix C_a^b.

As suggested by the quotation at the beginning of this chapter, some methods for vector transformation are not satisfying because of limited physical or geometric appeal. A reader studying the topic of frame transformations must not only deal with rather geometrically abstract concepts but also be prepared to carry out extensive algebraic manipulations. In addition, to be able to read some of the literature, the student must become conversant in a number of symbolic dialects.

Several methods for reference-frame manipulation have been developed over the past 200 years. The controversy surrounding the various approaches to the problem are discussed at length by Crowe.[1] Some methods which were intensely developed during the mid-nineteenth century languished during the early twentieth century only to be "rediscovered" during the last third of the twentieth century. The reason for such renewed interest in methods which appeared to be obsolete was that computational algorithms using such methods had clear advantages over algorithms based upon the more conventional methods. These computational algorithms were used in a technological setting that could not have been imagined during the gestation period of any of the methods of frame transformation.

4.2 Directional Cosine Matrix

Probably the most fundamental of all frame transformation methods is the directional cosine matrix, or DCM. Consider a vector $\boldsymbol{X}$ with components (X_1^a, X_2^a, X_3^a) and (X_1^b, X_2^b, X_3^b) in the a- and b-frames, respectively. Both vector and matrix representations are given here, although only the vector form will be used initially. The vector form is

$$\begin{aligned} \boldsymbol{X} &= X_1^a \boldsymbol{i}_1^a + X_2^a \boldsymbol{i}_2^a + X_3^a \boldsymbol{i}_3^a = \sum_{j=1}^{3} X_j^a \boldsymbol{i}_j^a \\ \boldsymbol{X} &= X_1^b \boldsymbol{i}_1^b + X_2^b \boldsymbol{i}_2 + X_3^b \boldsymbol{i}_3^b = \sum_{i=1}^{3} X_i^b \boldsymbol{i}_i^b \end{aligned} \tag{4.2a}$$

The matrix form is

$$X^a = \left[X_1^a, X_2^a, X_3^a\right]^{\mathrm{T}}, \qquad X^b = \left[X_1^b, X_2^b, X_3^b\right]^{\mathrm{T}} \tag{4.2b}$$

Note that the superscript designating the frame (a or b) is used only for the matrix form; it is not needed for the vector form.

The vector components are shown in Fig. 4.1. The first step in relating the vector components in the a- and b-frames makes use of the unit triads. Finding the components of the b-triad expressed in terms of the a-triad (see Fig. 4.2)

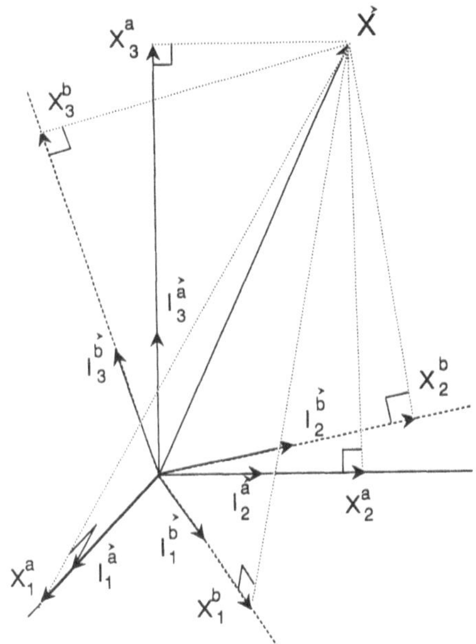

Fig. 4.1 Components of a vector in two Cartesian frames.

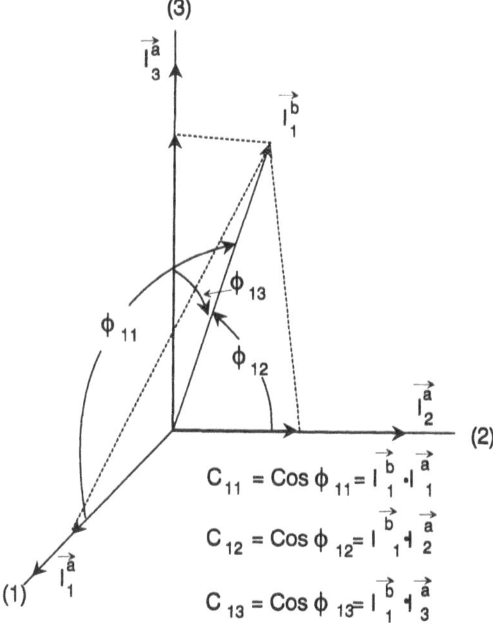

Fig. 4.2 Resolution of the I_1^b-vector in the a-frame.

gives the following expression for the unit vector along the 1-direction in the b-frame:

$$\boldsymbol{i}_1^b = \left(\boldsymbol{i}_1^b \cdot \boldsymbol{i}_1^a\right)\boldsymbol{i}_1^a + \left(\boldsymbol{i}_1^b \cdot \boldsymbol{i}_2^a\right)\boldsymbol{i}_2^a + \left(\boldsymbol{i}_1^b \cdot \boldsymbol{i}_3^a\right)\boldsymbol{i}_3^a$$

or, in general,

$$\boldsymbol{i}_i^b = \sum_{j=1}^{3} \left(\boldsymbol{i}_i^b \cdot \boldsymbol{i}_j^a\right)\boldsymbol{i}_j^a \tag{4.3}$$

Note here that a superscript is used for the unit vector. This exception to the above rule of vector formalism is necessary because the symbol i is used for unit vector designation regardless of the frame. The frame unit triads (e.g., $\boldsymbol{i}_i^b$) are embedded in the frame; hence, the frame designation is necessary.

Clearly, the inner product of two unit vectors is the cosine of the angle between the vectors, producing

$$\boldsymbol{i}_i^b = \sum_{j=1}^{3} \left(\boldsymbol{i}_i^b \cdot \boldsymbol{i}_j^a\right)\boldsymbol{i}_j^a = \sum_{j=1}^{3} \cos(\phi_{ij})\boldsymbol{i}_j^a = \sum_{j=1}^{3} C_{ij}\boldsymbol{i}_j^a \tag{4.4a}$$

where the nine cosines are given by

$$C_{ij} = \cos(\phi_{ij}) = \boldsymbol{i}_i^b \cdot \boldsymbol{i}_j^a \qquad \text{for} \qquad i = 1, 2, 3; \qquad j = 1, 2, 3 \tag{4.4b}$$

The inner product of the $\boldsymbol{i}_1^b$ vector in the a-frame is shown in Fig. 4.2.

The coordinates of the vector $\boldsymbol{X}$ may be written in terms of both vector triads as

$$X_1^b = \boldsymbol{X} \cdot \boldsymbol{i}_1^b, \qquad X_1^a = \boldsymbol{X} \cdot \boldsymbol{i}_1^a$$

or, in general,

$$X_i^b = \boldsymbol{X} \cdot \boldsymbol{i}_i^b, \qquad X_j^a = \boldsymbol{X} \cdot \boldsymbol{i}_j^a$$

We may now replace the b-frame triad with the a-frame triad by using Eqs. (4.4):

$$X_1^b = \boldsymbol{X} \cdot \boldsymbol{i}_1^b = \boldsymbol{X} \cdot \sum_{j=1}^{3} C_{1j}\boldsymbol{i}_j^a = \sum_{j=1}^{3} C_{1j}\boldsymbol{X} \cdot \boldsymbol{i}_j^a$$

$$X_1^b = \sum_{j=1}^{3} C_{1j}X_j^a = C_{11}X_1^a + C_{12}X_2^a + C_{13}X_3^a$$

In general, then,

$$X_i^b = \sum_{j=1}^{3} C_{ij} X_j^a \tag{4.5}$$

The above equation may now be written in matrix form as

$$X^b = \begin{bmatrix} X_1^b \\ X_2^b \\ X_3^b \end{bmatrix} = \begin{bmatrix} C_{11} & C_{12} & C_{13} \\ C_{21} & C_{22} & C_{23} \\ C_{31} & C_{32} & C_{33} \end{bmatrix} \begin{bmatrix} X_1^a \\ X_2^a \\ X_3^a \end{bmatrix} \tag{4.6}$$

The square matrix C transforms the set of vector components in an a-frame to components in a b-frame. To emphasize the purpose of C we use the notation of Eq. (4.1b) to give

$$X^b = C_a^b X^a \tag{4.7}$$

Note that X^b is represented here as a column matrix. The superscript indicates the appropriate frame. In the above usage C_a^b is a direction cosine matrix which transforms vector components in an a-frame (given in the subscript) to vector components in a b-frame (given in the superscript).

Since a and b are general designations for nonaligned frames, we may write

$$X^a = C_b^a X^b \tag{4.8}$$

which upon substitution of Eq. (4.7) gives

$$X^a = C_b^a C_a^b X^a = I X^a \tag{4.9}$$

If we repeat the steps leading to Eq. (4.5) but instead start with the a-frame, we can show that

$$X_j^a = \sum_{i=1}^{3} C_{ji} X_i^b$$

which would lead to the equivalent of Eq. (4.7) [i.e., Eq. (4.8)]. The conclusion is that

$$C_b^a = \text{Transpose}\left(C_a^b\right) \doteq \left(C_a^b\right)^{\mathrm{T}} \tag{4.10}$$

and from Eq. (4.9) we would get

$$\left(C_a^b\right)^{\mathrm{T}} C_a^b = I = C_b^a \left(C_b^a\right)^{\mathrm{T}} \tag{4.11a}$$

Thus,

$$C_a^b = (C_b^a)^{-1} \tag{4.11b}$$

Equations (4.11) are the result of the following mathematical property: the inner product of a row with itself is unity, and the inner product of one row of the DCM matrix with either of the other rows must be zero. Similarly, the inner product of any column with itself is unity, while the inner product of any column with either of the other two columns is zero. These results are given in Eqs. (4.14) below. However, note first that the determinant of C_a^b (or C_b^a) must always be unity.

$$\det\left[C_b^a C_a^b\right] = \det[I] = 1 = \det\left[C_b^a\right]\det\left[C_a^b\right] \tag{4.12}$$

which leads to

$$\det\left[C_a^b\right] = 1 = \det\left[C_b^a\right] \tag{4.13}$$

From Eq. (4.11) we get

$$\begin{bmatrix} C_{11} & C_{21} & C_{31} \\ C_{12} & C_{22} & C_{32} \\ C_{13} & C_{23} & C_{33} \end{bmatrix} \begin{bmatrix} C_{11} & C_{12} & C_{13} \\ C_{21} & C_{22} & C_{23} \\ C_{31} & C_{32} & C_{33} \end{bmatrix} = \begin{bmatrix} 1 & 0 & 0 \\ 0 & 1 & 0 \\ 0 & 0 & 1 \end{bmatrix}$$

which leads to the following relationships among the coefficients (sometimes called the *redundancy relationships*):

$$\begin{aligned} C_{11}^2 + C_{12}^2 + C_{13}^2 &= 1 = C_{11}^2 + C_{21}^2 + C_{31}^2 \\ C_{21}^2 + C_{22}^2 + C_{23}^2 &= 1 = C_{12}^2 + C_{22}^2 + C_{32}^2 \\ C_{31}^2 + C_{32}^2 + C_{33}^2 &= 1 = C_{13}^2 + C_{23}^2 + C_{33}^2 \end{aligned} \tag{4.14a}$$

$$\begin{aligned} C_{11}C_{12} + C_{21}C_{22} + C_{31}C_{32} &= 0 = C_{11}C_{21} + C_{12}C_{22} + C_{13}C_{23} \\ C_{21}C_{31} + C_{22}C_{32} + C_{23}C_{33} &= 0 = C_{12}C_{13} + C_{22}C_{23} + C_{32}C_{33} \\ C_{11}C_{31} + C_{12}C_{32} + C_{13}C_{33} &= 0 = C_{11}C_{13} + C_{21}C_{23} + C_{31}C_{33} \end{aligned} \tag{4.14b}$$

All of the above relationships follow from the orthonormal property of the DCM (i.e., the product of the DCM with its transpose is the identity matrix).

The redundancy relationships [Eqs. (4.14)] have some utility as a check on the orthogonality of the DCM and might be used as part of an algorithm to bring the DCM back into an orthonormal condition. As we will see, the change in the misalignment between frames means that the DCM representing that misalignment must have time-dependent elements; this is equivalent to saying that elements of the DCM must be solutions of differential equations. While some of the elements of the DCM might be updated from solving the differential equations, others may be found from the algebraic relationships given in Eqs. (4.14). For example, the element C_{11} may be obtained from the first or third relationship in Eqs. (4.14b) as follows:

$$C_{11} = \frac{-(C_{21}C_{22} + C_{31}C_{32})}{C_{12}} = \frac{-(C_{12}C_{22} + C_{13}C_{23})}{C_{21}} \tag{4.15}$$

If the element in the denominator (say, C_{12}) is zero, then C_{11} could not be obtained from that relationship. Remember that C_{12} is the cosine of the angle between the unit vectors in the 1-direction in the b-frame and the 2-direction in the a-frame. Thus, if C_{12} is zero, then these two unit vectors are orthogonal. From a computational point of view, when C_{12} is at the limit of precision, then the misalignment of the two frames becomes imprecise. A final point to be emphasized regarding Eqs. (4.14) is that Eqs. (4.14a) have limited utility due to the sign uncertainty resulting from the root operation.

A more useful set of redundancy relationships follows again from the orthogonality of the unit triads. Because both the a- and b-frames are orthogonal, we can write

$$\boldsymbol{i}_1^b = \boldsymbol{i}_2^b \times \boldsymbol{i}_3^b, \qquad \boldsymbol{i}_2^b = \boldsymbol{i}_3^b \times \boldsymbol{i}_1^b, \qquad \boldsymbol{i}_3^b = \boldsymbol{i}_1^b \times \boldsymbol{i}_2^b \tag{4.16}$$

Along with Eq. (4.4a), the first of the above relationships yields

$$\begin{aligned} \boldsymbol{i}_1^b &= C_{11}\boldsymbol{i}_1^a + C_{12}\boldsymbol{i}_2^a + C_{13}\boldsymbol{i}_3^a \\ &= (C_{21}\boldsymbol{i}_1^a + C_{22}\boldsymbol{i}_2^a + C_{23}\boldsymbol{i}_3^a) \times (C_{31}\boldsymbol{i}_1^a + C_{32}\boldsymbol{i}_2^a + C_{33}\boldsymbol{i}_3^a) \end{aligned} \tag{4.17}$$

By extending the second and third terms of Eqs. (4.16) as we did for the first term in Eq. (4.17), we get some useful redundancy relationships:

$$\begin{aligned} C_{11} &= C_{22}C_{33} - C_{32}C_{23} \\ C_{12} &= C_{23}C_{31} - C_{21}C_{33} \\ C_{13} &= C_{21}C_{32} - C_{31}C_{22} \\ C_{21} &= C_{32}C_{13} - C_{12}C_{33} \\ C_{22} &= C_{11}C_{33} - C_{31}C_{13} \\ C_{23} &= C_{31}C_{12} - C_{11}C_{32} \\ C_{31} &= C_{12}C_{23} - C_{22}C_{13} \\ C_{32} &= C_{21}C_{13} - C_{11}C_{23} \\ C_{33} &= C_{11}C_{22} - C_{21}C_{12} \end{aligned} \tag{4.18}$$

Remember that the adjoint of a matrix is the transpose of the matrix formed by replacing each element in the original matrix by its signed minor (cofactor). From Eqs. (4.18) it is obvious that

$$\left(C_a^b\right)^{\mathrm{T}} = \operatorname{adj}\left(C_a^b\right) \tag{4.19}$$

Note also that

$$\begin{aligned}\left(C_a^b\right)^{-1} &= \frac{\operatorname{adj}\left(C_a^b\right)}{\det\left(C_a^b\right)} \\ &= \operatorname{adj}\left(C_a^b\right) = \left(C_a^b\right)^{\mathrm{T}}\end{aligned} \tag{4.20}$$

Thus, Eqs. (4.18) are a consequence of the inverse of an orthonormal DCM being equal to its transpose. We can bypass the manipulations of Eq. (4.17) and write Eqs. (4.18) by inspection. Simply stated, each element in any orthonormal coordinate transformation matrix is equal to its own cofactor.

4.3 Updating the DCM

Since the DCM represents the angular misalignment between two frames, we should expect that the elements of a DCM would be time-dependent if the misalignment is time-dependent. Let's assume that one frame does not rotate and designate this frame as the I-frame, or inertial frame. A second frame, the b-frame, is fixed to a re-entry vehicle. To an observer fixed in the I-frame, the re-entry vehicle appears to be rotating. A vector measured in the b-frame can be transferred to the I-frame by the DCM C_b^I. Because of the frame rotation, the elements of the DCM must be time-dependent. We would expect to obtain the elements of C_b^I as follows:

$$C_b^I(t) = \int^t \frac{\mathrm{d}}{\mathrm{d}t}\left[C_b^I\right] \mathrm{d}\tau \tag{4.21}$$

Our goal in this chapter is to obtain the derivative shown in Eq. (4.21). From Eq. (4.4a) we may write

$$\boldsymbol{i}_k^I = \sum_{j=1}^{3} C_{kj} \boldsymbol{i}_j^b \tag{4.22}$$

where k and j are independent axis labels (1, 2, 3).

We now must look at the derivatives of the I- and b-frame triads with respect to the I-frame.

$$\left.\frac{\mathrm{d}\boldsymbol{i}_k^I}{\mathrm{d}t}\right|_I = 0, \qquad \left.\frac{\mathrm{d}\boldsymbol{i}_j^b}{\mathrm{d}t}\right|_I = \boldsymbol{\omega}_{b/I} \times \boldsymbol{i}_j^b \tag{4.23}$$

The first equation states that the unit vectors of the I-frame do not change direction with respect to the I-frame. The second equation states that the unit vectors of the b-frame change direction at right angles to both the vector and the angular velocity vector. The symbolic protocol used here follows that developed by Britting[5]; $\boldsymbol{\omega}_{b/I}$ is the angular velocity vector representing the rotation of the b-frame relative to the I-frame.

Thus, upon differentiation, Eq. (4.23) becomes

$$\left.\frac{\mathrm{d}\boldsymbol{i}_k^I}{\mathrm{d}t}\right|_I = \frac{\mathrm{d}}{\mathrm{d}t}\left(\sum_{j=1}^{3} C_{kj}\boldsymbol{i}_j^b\right) = \sum_{j=1}^{3}\left(\frac{\mathrm{d}C_{kj}}{\mathrm{d}t}\right)\boldsymbol{i}_j^b + \sum_{j=1}^{3} C_{kj}\frac{\mathrm{d}\boldsymbol{i}_j^b}{\mathrm{d}t}$$

$$0 = \sum_{j=1}^{3}\left(\frac{\mathrm{d}C_{kj}}{\mathrm{d}t}\right)\boldsymbol{i}_j^b + \sum_{j=1}^{3} C_{kj}\left(\boldsymbol{\omega}_{b/I} \times \boldsymbol{i}_j^b\right) \tag{4.24}$$

$$0 = \sum_{j=1}^{3}\left(\frac{\mathrm{d}C_{kj}}{\mathrm{d}t}\right)\boldsymbol{i}_j^b + \boldsymbol{\omega}_{b/I} \times \sum_{j=1}^{3} C_{kj}\boldsymbol{i}_j^b$$

If we set $k = 1$ we get

$$\frac{\mathrm{d}\boldsymbol{i}_1^I}{\mathrm{d}t} = 0 = \frac{\mathrm{d}C_{11}}{\mathrm{d}t}\boldsymbol{i}_1^b + \frac{\mathrm{d}C_{12}}{\mathrm{d}t}\boldsymbol{i}_2^b + \frac{\mathrm{d}C_{13}}{\mathrm{d}t}\boldsymbol{i}_3^b + \det\begin{bmatrix} \boldsymbol{i}_1^b & \boldsymbol{i}_2^b & \boldsymbol{i}_3^b \\ \omega_1 & \omega_2 & \omega_3 \\ C_{11} & C_{12} & C_{13} \end{bmatrix} \tag{4.25}$$

Note that we have taken

$$\boldsymbol{\omega}_{b/I} = \omega_1\boldsymbol{i}_1^b + \omega_2\boldsymbol{i}_2^b + \omega_3\boldsymbol{i}_3^b \tag{4.26}$$

Expansion of the cross-product determinant in Eq. (4.25) followed by term collection gives three equations relating C_{11}, C_{12}, C_{13}, their derivatives, and the components of the angular velocity vector. If we then repeat the procedure for $k = 2$ and $k = 3$ we will obtain six more first-order differential equations. The differential equations for all nine elements of the DCM are as follows:

$$\begin{aligned}
\dot{C}_{11} &= C_{12}\omega_3 - C_{13}\omega_2, & \dot{C}_{12} &= C_{13}\omega_1 - C_{11}\omega_3, & \dot{C}_{13} &= C_{11}\omega_2 - C_{12}\omega_1 \\
\dot{C}_{21} &= C_{22}\omega_3 - C_{23}\omega_2, & \dot{C}_{22} &= C_{23}\omega_1 - C_{21}\omega_3, & \dot{C}_{23} &= C_{21}\omega_2 - C_{22}\omega_1 \\
\dot{C}_{31} &= C_{32}\omega_3 - C_{33}\omega_2, & \dot{C}_{32} &= C_{33}\omega_1 - C_{31}\omega_3, & \dot{C}_{33} &= C_{31}\omega_2 - C_{32}\omega_1
\end{aligned} \tag{4.27}$$

Alternately, the above equations may be written more compactly in matrix notation as

$$\frac{\mathrm{d}}{\mathrm{d}t}\left[C_b^I\right] \doteq \dot{C}_b^I = \begin{bmatrix} \dot{C}_{11} & \dot{C}_{12} & \dot{C}_{13} \\ \dot{C}_{21} & \dot{C}_{22} & \dot{C}_{23} \\ \dot{C}_{31} & \dot{C}_{32} & \dot{C}_{33} \end{bmatrix}$$

$$= \begin{bmatrix} C_{11} & C_{12} & C_{13} \\ C_{21} & C_{22} & C_{23} \\ C_{31} & C_{32} & C_{33} \end{bmatrix} \begin{bmatrix} 0 & -\omega_3 & \omega_2 \\ \omega_3 & 0 & -\omega_1 \\ -\omega_2 & \omega_1 & 0 \end{bmatrix} \tag{4.28}$$

The 3×3 angular velocity matrix in the above expression is called the *skew-symmetric* form of the angular velocity vector. It will be shown that this form is useful in providing the matrix equivalent of a vector cross product. Thus, the angular velocity may be written in either a vector (column) form or a matrix (skew-symmetric) form as follows:

$$\boldsymbol{\omega}_{b/I}^b = \begin{bmatrix} \omega_1 \\ \omega_2 \\ \omega_3 \end{bmatrix}, \qquad \Omega_{b/I}^b = \begin{bmatrix} 0 & -\omega_3 & \omega_2 \\ \omega_3 & 0 & -\omega_1 \\ -\omega_2 & \omega_1 & 0 \end{bmatrix} \tag{4.29}$$

It will be left as an exercise for the reader to show that

$$\left(\Omega_{b/I}^b\right)^{\mathrm{T}} = -\Omega_{b/I}^b, \qquad \Omega_{b/I}^b \boldsymbol{\omega}_{b/I}^b = \begin{bmatrix} 0 \\ 0 \\ 0 \end{bmatrix} = \left[\left(\boldsymbol{\omega}_{b/I}^b\right)^{\mathrm{T}} \Omega_{b/I}^b\right]^{\mathrm{T}}$$

Equation (4.28) may be written in the following compact form:

$$\frac{\mathrm{d}}{\mathrm{d}t}\left(C_b^I\right) = C_b^I \Omega_{b/I}^b \tag{4.30}$$

The above expression is fundamental. In a simulation, the angular velocity matrix is assigned numerical values from the integration of the moment equations; in a strap-down navigator, the angular velocity comes from a gyro triad.

It is important to emphasize the symbolic usage here. The C_b^I symbol indicates a matrix which will transform a vector *from* the b-frame *to* the I-frame (body-to-inertial). The subscript b/I in the angular velocity matrix indicates a rotation of the b-frame relative to the I-frame. The superscript b indicates that the elements of the angular velocity (skew-symmetric) matrix are components of the angular velocity in the body frame.

Although the expression given in Eq. (4.30) is fundamental in flight dynamics, the equation may be generalized to any a- and b-frames as follows:

$$\frac{\mathrm{d}}{\mathrm{d}t}\left(C_a^b\right) = C_a^b \Omega_{a/b}^a \tag{4.31}$$

By taking the transpose of both sides we have

$$\dot{C}_b^a = \left(\dot{C}_a^b\right)^{\mathrm{T}} = \left(\Omega_{a/b}^a\right)^{\mathrm{T}}\left(C_a^b\right) = -\Omega_{a/b}^a C_b^a = \Omega_{b/a}^a C_b^a$$

or

$$\dot{C}_I^b = \Omega_{I/b}^b C_I^b = -\Omega_{b/I}^b C_I^b \tag{4.32}$$

Since a and b are entirely arbitrary, we may rewrite the above equation to get

$$\frac{\mathrm{d}}{\mathrm{d}t}\left(C_a^b\right) = \Omega_{a/b}^b C_a^b \qquad \text{or} \qquad \frac{\mathrm{d}}{\mathrm{d}t}\left(C_b^I\right) = \Omega_{b/I}^I C_b^I \tag{4.33}$$

If we compare Eqs. (4.30) and (4.33) we see that by premultiplying rather than postmultiplying by the skew-symmetric matrix, the frame of angular velocity resolution has changed from the b-frame to the I-frame.

Clearly, we have

$$C_a^b C_b^a = I$$

or

$$\dot{C}_a^b C_b^a + C_a^b \dot{C}_b^a = 0 = \Omega_{a/b}^b C_a^b C_b^a + C_a^b \Omega_{b/a}^a C_b^a$$

which gives

$$\Omega_{a/b}^b = C_a^b \Omega_{a/b}^a C_b^a, \qquad \Omega_{b/I}^I = C_b^I \Omega_{b/I}^b C_I^b \tag{4.34}$$

It is important to recognize that $\Omega_{b/I}^b$ and $\Omega_{b/I}^I$ are skew-symmetric forms of the same vector, namely the rotation rate of the b-frame relative to the I-frame; the superscript indicates the frame of vector resolution. Equation (4.34) indicates how the angular velocity in skew-symmetric form transforms from the b-frame to the I-frame (or the a-frame to the b-frame); the equivalent vector form of the transformation is of course

$$\boldsymbol{\omega}_{a/b}^b = C_a^b \boldsymbol{\omega}_{a/b}^a, \qquad \boldsymbol{\omega}_{b/I}^I = C_b^I \boldsymbol{\omega}_{b/I}^b \tag{4.35}$$

4.4 Euler Angles

One of the drawbacks of the DCM method is that we must deal with a total of nine elements, each with its own differential equation. Since a rigid body has only three rotational degrees of freedom, we recognize that DCMs are highly redundant. Although the Euler angle method is entirely compatible and equivalent to the DCM method, the Euler angle method requires only three parameters.

Figure 4.3 illustrates rotations about each of the orthogonal coordinate axes. The prime-superscript is used to denote a frame after its initial rotation. The

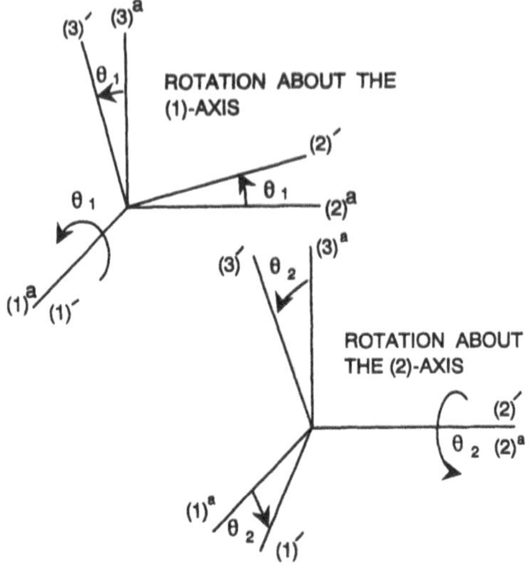

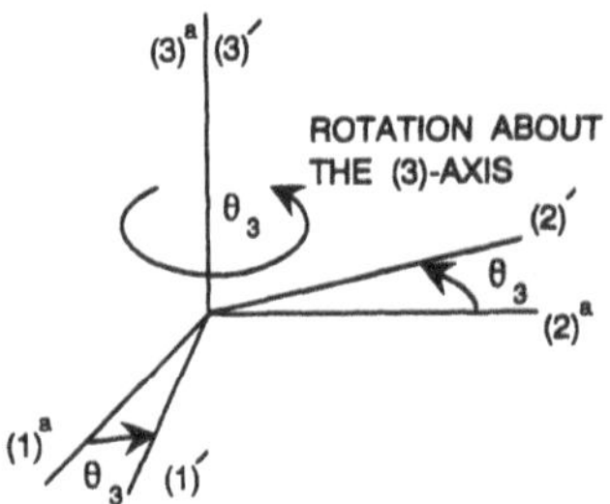

Fig. 4.3 Principal axis rotations.

corresponding matrices which relate measurements in the initial frames to the prime frames are given below, with the subscripts indicating the axis about which rotation takes place.

$$C_1(\theta_1) = \begin{bmatrix} 1 & 0 & 0 \\ 0 & \cos(\theta_1) & \sin(\theta_1) \\ 0 & -\sin(\theta_1) & \cos(\theta_1) \end{bmatrix} \tag{4.36a}$$

$$C_2(\theta_2) = \begin{bmatrix} \cos(\theta_2) & 0 & -\sin(\theta_2) \\ 0 & 1 & 0 \\ \sin(\theta_2) & 0 & \cos(\theta_2) \end{bmatrix} \tag{4.36b}$$

$$C_3(\theta_3) = \begin{bmatrix} \cos(\theta_3) & \sin(\theta_3) & 0 \\ -\sin(\theta_3) & \cos(\theta_3) & 0 \\ 0 & 0 & 1 \end{bmatrix} \tag{4.36c}$$

The above rotations are designated as the *principal rotations*.

We now consider a sequence of axis rotations. Since in general we need three parameters, we must expect three rotations; succeeding rotations must be about different axes. Rotation from the a-frame to the b-frame requires rotation through two intermediate frames which we designate as the prime and double-prime frames, using $'$ and $''$ symbols, respectively. We can tolerate this symbolic awkwardness because the prime frames are only a sort of "scaffolding" which will be removed. The rotation angle θ has a numerical subscript to indicate the order of the sequence. To keep the symbols in agreement with those used by Hughes,[3] we let the first rotation angle be θ_3, the second θ_2, and the final rotation θ_1. The rotation from the a-frame to the prime frame is

$$X' = C_i(\theta_3)X^a \tag{4.37a}$$

where i may be equal to 1, 2, or 3 to indicate the axis of rotation. The double-prime frame is related to the a-frame as follows:

$$X'' = C_j(\theta_2)X' = C_j(\theta_2)C_i(\theta_3)X^a \tag{4.37b}$$

where j may be equal to 1, 2, or 3 but may not equal whatever axis number was selected for i. (If $i = 3$, then j could only equal 1 or 2.) Finally, the vector X in the b-frame is related to the vector X in the a-frame through double-prime and prime rotations as follows:

$$X^b = C_k(\theta_1)C_j(\theta_2)C_i(\theta_3)X^a = C_a^bX^a \tag{4.37c}$$

where k cannot equal j but can, of course, equal i.

Again, the rotation sequence is as follows: a rotation about the i-axis through the angle θ_3 to form the prime frame; the frame undergoes a rotation about the j-axis in the prime frame through the angle θ_2 to form the double-prime frame; finally, the frame undergoes a rotation through the angle θ_1 about the k-axis in the double-prime frame. Obviously, there are 12 possible sequences that can be followed. Sometimes each particular sequence is designated as an *Euler angle scheme*. All combinations have been carried out and are given in Appendix C.

The question arises as to which sequence is the "best." The answer is problem-dependent, as might be the selection of the type of coordinate system (e.g., cylindrical or Cartesian). Suppose we choose the sequence $C_1C_2C_1$ (rotation about the 1-axis in the a-frame, followed by a rotation about the 2-axis in the prime frame, followed by a rotation about the 1-axis in the double-prime frame). Before considering the details of this sequence, let's consider what happens when one of the angles is zero. If, for example, θ_2 is effectively zero, we would get the following for the sequence indicated by Eq. (4.37c):

$$C_1C_2C_1 = \begin{bmatrix} 1 & 0 & 0 \\ 0 & \cos(\theta_1 + \theta_3) & \sin(\theta_1 + \theta_3) \\ 0 & -\sin(\theta_1 + \theta_3) & \cos(\theta_1 + \theta_3) \end{bmatrix} \tag{4.38}$$

Clearly, since θ_1 and θ_3 appear only as part of a sum, we can replace this sum by a single angle, indicating that when θ_2 is zero, the above sequence appears to be a single rotation about the 1-axis. It will be shown later that under these circumstances there is some difficulty in relating Euler angle rates to components of angular velocity available in the b-frame.

An Euler sequence convenient for relating a re-entry vehicle frame, i.e., a b-frame, to a wind or velocity frame, i.e., a ν-frame, is the sequence $C_1C_2C_1$, which we just used above. The rotational sequences are given in Fig. 4.4. The 1^ν-axis is along the velocity vector, which is not necessarily horizontal; the

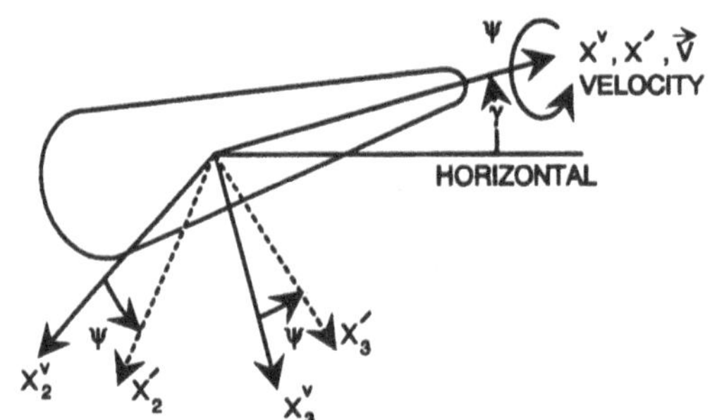

FIRST SEQUENCE - ORIENT ANGLE OF ATTACK PLANE

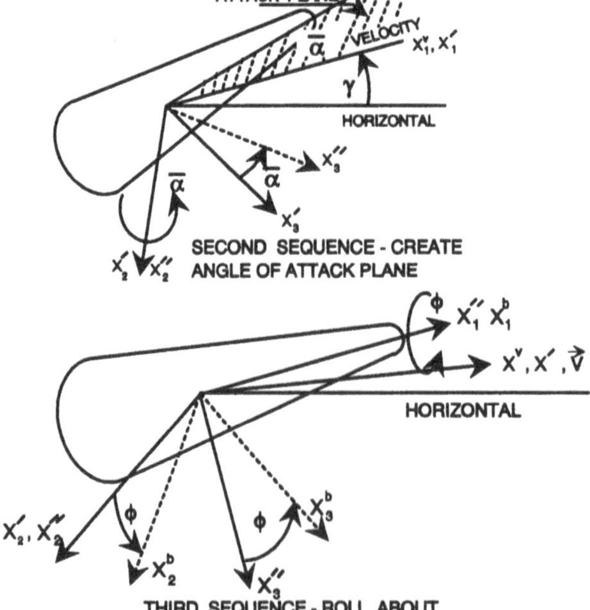

THIRD SEQUENCE - ROLL ABOUT
ANGLE OF ATTACK PLANE

Fig. 4.4 Angular sequences for re-entry bodies.

2^v-axis is horizontal and, of course, normal to the 1^v-axis (and the velocity vector); the 3^v-axis completes the right-handed triad.

We assume that the RV is nearly a body of rotational symmetry, but we allow for the possibility that there is some contour irregularity which can cause some aerodynamic roll sensitivity. The body frame, or b-frame, is assumed to be aligned initially with the velocity frame, i.e., $1^b, 2^b$, and 3^b are assumed to be aligned with $1^v, 2^v$, and 3^v. The body is first rotated about the 1^v-axis through the angle $\theta_3 = \psi$; next, the body is rotated about the new $2'$-axis through the angle $\theta_2 = \bar{\alpha}$. Thus, the new $1''$-axis along the RV's axis of symmetry makes an angle α with the 1^v-axis (which, of course, is in the direction of the positive velocity vector). Since the angle $\bar{\alpha}$ measures the angular separation between the RV's axis of symmetry (zero lift line) and the velocity vector, it is called the *total angle of attack*. The angle ψ is a measure of the angle between the vertical and the angle-of-attack plane. If ψ is zero, then the angle-of-attack plane is vertical. (We would not expect the aerodynamic loads under any circumstance to be a function of the angle ψ.) The final rotation is about the $1''$-axis (which is also the 1^b-axis) through the roll angle $\theta_1 = \phi$. This angle is a measure of the rotation of the RV relative to the angle-of-attack plane. If the RV is a body of revolution (i.e., if the 1^b-axis is an axis of symmetry), the aerodynamic loads will not depend upon ϕ.

The rotations described above are given below.

$$C_v^b = C_1(\theta_1)C_2(\theta_2)C_1(\theta_3) \tag{4.39a}$$

$$C_v^b = \begin{bmatrix} C\bar{\alpha} & S\bar{\alpha}S\psi & -S\bar{\alpha}C\psi \\ S\phi s\bar{\alpha} & C\phi C\psi - S\phi C\bar{\alpha}S\psi & C\phi S\psi + S\phi C\bar{\alpha}C\psi \\ C\phi S\bar{\alpha} & -S\phi C\psi - C\phi C\bar{\alpha}S\psi & -S\phi S\psi + C\phi C\bar{\alpha}C\psi \end{bmatrix} \tag{4.39b}$$

where C and S are shorthand for cosine and sine, respectively, when they appear in elements of matrices. The above sequence of operations is taken from one of 12 such sequences in Appendix C.

4.5 Updating Euler Angles

The DCM, whose elements are expressed in terms of Euler angles, is a function of time when the Euler angles are functions of time. Thus, we would expect the Euler angles to be expressible as ordinary differential equations. Numerical solutions of such equations will then permit an updating of elements of the DCM.

Let's assume that we are interested in the relative alignment between a body- or b-frame, and an inertial- or I-frame. Equation (4.30) gave the following expression:

$$\frac{\mathrm{d}}{\mathrm{d}t}\left(C_b^I\right) = C_b^I \Omega_{b/I}^b \tag{4.30}$$

which may be rewritten as

$$\left(\dot{C}_b^I\right)^{\mathrm{T}} = \left(C_b^I \Omega_{b/I}^b\right)^{\mathrm{T}} = \left(\Omega_{b/I}^b\right)^{\mathrm{T}} \left(C_b^I\right)^{\mathrm{T}}$$

$$\dot{C}_I^b = -\Omega_{b/I}^b C_I^b$$

Solving for the angular velocity in a body frame (skew-symmetric form) gives

$$\Omega_{b/I}^b = -\dot{C}_I^b \left(C_I^b\right)^{-1} = -\dot{C}_I^b \left(C_I^b\right)^{\mathrm{T}} \tag{4.40}$$

As we indicated in the previous section, the matrix C_I^b is defined through a three-step rotational sequence as follows:

$$C_I^b = C_{II}^b C_I^{II} C_I^I \doteq C_i(\theta_1) C_j(\theta_2) C_k(\theta_3) \tag{4.41}$$

Upon inserting Eq. (4.41) into Eq. (4.40) we get

$$\Omega_{b/I}^b = -\left(\frac{\mathrm{d}}{\mathrm{d}t} C_i C_j C_k\right) [C_i C_j C_k]^{\mathrm{T}}$$

which gives, upon expansion,

$$\Omega_{b/I}^b = -\left(\dot{C}_i C_j C_k + C_i \dot{C}_j C_k + C_i C_j \dot{C}_k\right)\left(C_k^{\mathrm{T}} C_j^{\mathrm{T}} C_i^{\mathrm{T}}\right)$$

Carrying out the indicated multiplications gives

$$\Omega_{b/I}^b = -\dot{C}_i C_i^{\mathrm{T}} - C_i \dot{C}_j C_j^{\mathrm{T}} C_i^{\mathrm{T}} - C_i C_j \dot{C}_k C_k^{\mathrm{T}} C_j^{\mathrm{T}} C_i^{\mathrm{T}} \tag{4.42}$$

The principal rotations for any sequence set (i, j, k) have been given in Eqs. (4.36). Once a sequence is selected, the appropriate matrices can be differentiated, transposed, and multiplied to give the angular velocity in skew-symmetric form. As one might expect, the amount of symbolic manipulation is not pleasant to contemplate, much less to carry out for all 12 varieties of Euler angle schemes. We may assume correctly that such manipulations have already been carried out; except for practice or some enlightenment, such steps need not be repeated. With the angular velocity available (in vector form) we can relate this vector to the derivatives of the Euler angles as

$$\begin{bmatrix} \omega_1 \\ \omega_2 \\ \omega_3 \end{bmatrix} = E \begin{bmatrix} \dot{\theta}_1 \\ \dot{\theta}_2 \\ \dot{\theta}_3 \end{bmatrix} \tag{4.43a}$$

where E is a matrix transforming the Euler angle rates into the angular velocity components [see Eq. (4.52)].

Now the three Euler angles are taken about three axes which are not orthogonal; consequently, the matrix E is not orthogonal and its inverse is therefore not equal to its transpose. Nevertheless, complexity notwithstanding, the inverse of E is important, since we usually have $(\omega_1, \omega_2, \omega_3)$ available from the moment equations when carrying out a simulation or from a gyro triad when performing inertial navigation. More often than not, we have available the components of the angular velocity but wish to obtain the derivatives of the Euler angles.

$$\begin{bmatrix} \dot{\theta}_1 \\ \dot{\theta}_2 \\ \dot{\theta}_3 \end{bmatrix} = E^{-1} \begin{bmatrix} \omega_1 \\ \omega_2 \\ \omega_3 \end{bmatrix} \tag{4.43b}$$

As an example let's look at the rotation given in Eq. (4.36a):

$$C_1 = \begin{bmatrix} 1 & 0 & 0 \\ 0 & \cos(\theta_1) & \sin(\theta_1) \\ 0 & -\sin(\theta_1) & \cos(\theta_1) \end{bmatrix}$$

If we form the product shown on the left below we can obtain

$$-\dot{C}_1 C_1^{\mathrm{T}} = \begin{bmatrix} 0 & 0 & 0 \\ 0 & 0 & -\dot{\theta}_1 \\ 0 & \dot{\theta}_1 & 0 \end{bmatrix} \tag{4.44}$$

which is the skew-symmetric form of the rotation vector $\mathrm{d}\theta_1/\mathrm{d}t$, i.e.,

$$\frac{\mathrm{d}}{\mathrm{d}t}\boldsymbol{\theta}_1 = \begin{bmatrix} \dot{\theta}_1 \\ 0 \\ 0 \end{bmatrix} = \dot{\theta}_1 \boldsymbol{u}_1 \tag{4.45}$$

where $\boldsymbol{u}_1$ is a unit vector along the 1-axis. If we want to write the vector $(\mathrm{d}\theta_1/\mathrm{d}t)\boldsymbol{u}_1$ in skew-symmetric form, we can use the star $(*)$ superscript as follows:

$$-\dot{C}_1 C_1^{\mathrm{T}} = \left(\dot{\theta}_1 \boldsymbol{u}_1\right)^* \tag{4.46}$$

(We use the star superscript to indicate the applicable vector in skew-symmetric form; however, where appropriate, we will continue to use the upper case to indicate the skew-symmetric matrix and the lower case to indicate column or vector form.)

We may now rewrite Eq. (4.42) as follows:

$$\Omega^b_{b/i} = \left(\dot{\theta}_1 \boldsymbol{u}_i\right)^* + C_i \left(\dot{\theta}_2 \boldsymbol{u}_j\right)^* C_i^{\mathrm{T}} + C_i C_j \left(\dot{\theta}_3 \boldsymbol{u}_k\right)^* \left(C_i C_j\right)^{\mathrm{T}} \tag{4.47}$$

The above relationship may be written as a sum of skew-symmetric matrices by using the relationship given in Eq. (4.34), rewritten below as

$$\Omega^b_{b/I} = C^b_I \Omega^I_{b/I} \left(C^b_I\right)^{\mathrm{T}} = \left(C^b_I \omega^i_{b/I}\right)^*$$

or, in general,

$$\Omega^b_{b/a} = C^b_a \Omega^a_{b/a} \left(C^b_a\right)^{\mathrm{T}} = C^b_a \left(\omega^a_{b/a}\right)^* \left(C^b_a\right)^{\mathrm{T}} = \left(C^b_a \omega^a_{b/a}\right)^* \tag{4.48}$$

Equation (4.47) now becomes the sum of skew-symmetric matrices:

$$\Omega^b_{b/I} = \left(\boldsymbol{u}_i \dot{\theta}_1\right)^* + \left(C_i \boldsymbol{u}_j \dot{\theta}_2\right)^* + \left(C_i C_j \boldsymbol{u}_k \dot{\theta}_3\right)^* \tag{4.49}$$

By way of review, the matrix C_k transforms the original i-frame to the prime (or intermediate) frame; the matrix C_j then transforms the prime frame to the double-prime frame, and C_i transforms the double-prime frame to the final (in this case) b-frame. Equation (4.49) can be reformatted in a vector form as

$$\boldsymbol{\omega}^b_{b/I} = \boldsymbol{u}_i \dot{\theta}_1 + C_i \boldsymbol{u}_j \dot{\theta}_2 + C_i C_j \boldsymbol{u}_k \dot{\theta}_3 \tag{4.50}$$

Suppose we apply the preceding expression to the sequence often used by aircraft or re-entry vehicles having planar symmetry (rather than the more usual case of rotational symmetry). First, the object is rotated around the 3-axis through the angle $\theta_3 = \psi$, then about the (new) 2-axis through $\theta_2 = \theta$, and finally about the 1-axis through the angle $\theta_1 = \phi$. Again, the first angular rotation is always θ_3, followed by θ_2 and finally by θ_1. Equation (4.50) becomes

$$\boldsymbol{\omega}^b_{b/I} = \boldsymbol{u}_1 \dot{\phi} + C_1 \boldsymbol{u}_2 \dot{\theta} + C_1 C_2 \boldsymbol{u}_3 \dot{\psi} \tag{4.51}$$

In the preceding expression we should note that the first rate, $\mathrm{d}\theta_3/\mathrm{d}t = \mathrm{d}\psi/\mathrm{d}t$, is about the 3-axis and is therefore unaffected by the C_3 rotation matrix; however, this angular-rate vector must be transformed by the C_2 and C_1 matrices. The second rate, $\mathrm{d}\theta_2/\mathrm{d}t = \mathrm{d}\theta/\mathrm{d}t$, is transformed only by the C_1 matrix because this second rotation does not affect the direction of the second rate. The final rotation rate, $\mathrm{d}\theta_1/\mathrm{d}t = \mathrm{d}\phi/\mathrm{d}t$, is also along the final axis (the 1-axis in this case), so this rate needs no further transformation. If the transformation were from an inertial I-frame to a body b-frame, $\boldsymbol{u}_3$ would be the unit vector $\boldsymbol{K}^I$ and $\boldsymbol{u}_1$ the unit vector $\boldsymbol{I}^b$. The C_1 and C_2 matrices would follow from Eqs. (4.36b) and (4.36c). The ψ, θ, and ϕ angles of Eq. (4.51) are the Euler angles appropriate to the sequence [first about the 3-axis (inertial), then about the 2-axis (prime), and then about the 1-axis (body)]. The result is a matrix between the angular rate with components in the b-frame and the derivatives of the Euler angles:

$$\boldsymbol{\omega}_{b/i}^{b} = E[\dot{\phi}, \dot{\theta}, \dot{\psi}]^{\mathrm{T}} = E[\dot{\theta}_1, \dot{\theta}_2, \dot{\theta}_3]^{\mathrm{T}}$$

where

$$E = \begin{bmatrix} 1 & 0 & -\sin(\theta) \\ 0 & \cos(\phi) & \sin(\phi)\cos(\theta) \\ 0 & -\sin(\phi) & \cos(\phi)\cos(\theta) \end{bmatrix} = \begin{bmatrix} 1 & 0 & -\sin(\theta_2) \\ 0 & \cos(\theta_1) & \sin(\theta_1)\cos(\theta_2) \\ 0 & -\sin(\theta_1) & \cos(\theta_1)\cos(\theta_2) \end{bmatrix} \tag{4.52}$$

where the matrix E was first introduced in Eq. (4.43a).

In many, if not most, simulation problems of interest, the components of the angular rate are expressed in the body frame because the aerodynamic moments are available in this frame. Consequently, we are usually required to find the Euler angle rates from known values of angular velocity rates in the body frame. With these Euler angle rates available and integrated, we can then recompute (or update) the directional cosine matrix. Thus, the inverse of the E-matrix given in Eq. (4.52) is needed. This inverse is given in Table C.2 of Appendix C for all possible sequences of Euler angles.

A glance at Table C.2 of Appendix C indicates that any scheme of Euler angles will have a singularity somewhere. For the 1 2 1 sequence of interest to the rotationally symmetric RV, we have a singularity at

$$\sin(\theta_2) = \sin(\bar{\alpha}) = 0$$

$$\bar{\alpha} = 0$$

In Eq. (4.38) we showed that when $\bar{\alpha} = 0$ (i.e., when $\theta_2 = 0$), ψ and $\theta(\theta_3$ and $\theta_1)$ add directly, and the sequence $C_1 \quad C_2 \quad C_3$ becomes a single rotation matrix. Thus, if

$$\boldsymbol{\omega}_{b/I}^{b} = \begin{bmatrix} \omega_1 \\ \omega_2 \\ \omega_3 \end{bmatrix} = E^{-1} \begin{bmatrix} \dfrac{\mathrm{d}\phi}{\mathrm{d}t} \\ \dfrac{\mathrm{d}\bar{\alpha}}{\mathrm{d}t} \\ \dfrac{\mathrm{d}\psi}{\mathrm{d}t} \end{bmatrix} \tag{4.53}$$

we can show that, at $\bar{\alpha} = 0$,

$$\omega_2 = \frac{\mathrm{d}\bar{\alpha}}{\mathrm{d}t}\cos(\phi) + \frac{\mathrm{d}\psi}{\mathrm{d}t}\sin(\phi)$$

and that ω_1 and ω_3 are indeterminate.

4.6 Axis/Angle Parameters

An interesting way of describing the relative angular displacement of two frames is to make use of a theorem credited to Euler, which states that if one frame remains fixed while a second frame, initially coincident with the first frame, rotates in some arbitrary fashion, there is always a fixed axis through the point of common origin about which the rotation can be assumed to have taken place. Regardless of what sequence of rotations actually took place, the final condition of misalignment between the two frames is always equivalent to a rotation through a finite angle about some axis. Describing frame misalignment in such a fashion requires four parameters: the angle of rotation and the three directional cosines associated with the axis of rotation. If we identify the axis of rotation by the unit vector $\boldsymbol{a}$, then the directional cosines associated with this axis of rotation are the components of vector $\boldsymbol{a}$ in both the rotating and nonrotating frames. These four parameters will be shown to be derived from the elements of the DCM. Given the four axis/angle parameters, we will be able to obtain the elements of the DCM describing the frame misalignment.

The moving frame is designated by m and the nonmoving, or fixed, frame by f. Since the axis vector $\boldsymbol{a}$ is transformed through the DCM describing the misalignment, we have what should be a relationship between the coordinates of $\boldsymbol{a}$ in both the m- and f-frames. That is,

$$\boldsymbol{a}^m = C_f^m \boldsymbol{a}^f \tag{4.54}$$

However, because the $\boldsymbol{a}$ vector is along the axis of frame rotation (by definition), $\boldsymbol{a}$ must have the same direction cosines in both the m- and f-frames. Thus,

$$\boldsymbol{a}^m = (\lambda I)\boldsymbol{a}^f \tag{4.55}$$

where λ is a constant accounting for the fact that the magnitude of $\boldsymbol{a}$, although taken as unity, is arbitrary, because any vector of arbitrary length along $\boldsymbol{a}$ could serve as the axis. Furthermore, as it turns out, $\boldsymbol{a}$ is the only real eigenvector of the DCM; the other two eigenvectors must have complex or imaginary components, so it is likely that the value of λ corresponding to these vectors is also complex.

Combining Eqs. (4.54) and (4.55) gives

$$\left(C_f^m - \lambda I\right)\boldsymbol{a}^f = 0$$

Since $\boldsymbol{a}^f$ is not zero, the inverse of the bracketed term above must be undefined. Consequently, the determinant of this term must be zero, i.e.,

$$\det\left(C_f^m - \lambda I\right) = 0$$

The three values of λ must be either all real or one real and two complex conjugates such that

$$\lambda_1\lambda_2\lambda_3 = 1$$

The most general form for the λ must be

$$\lambda_1 = 1, \qquad \lambda_2 = e^{j\phi}, \qquad \lambda_3 = e^{-j\phi} \tag{4.56}$$

The λ are the eigenvalues of the matrix C_m^f. The trace of C_m^f must equal the sum of the eigenvalues, i.e.,

$$\sigma = \text{tr}\left[C_f^m\right] = C_{11} + C_{22} + C_{33} = \lambda_1 + \lambda_2 + \lambda_3 = 1 + 2\cos(\phi) \tag{4.57}$$

Solving for ϕ we get

$$\cos(\phi) = \tfrac{1}{2}(\sigma - 1) \tag{4.58}$$

where ϕ is the angle through which the m-frame rotates about the axis whose direction is described by the unit vector $\boldsymbol{a}$. Thus, given the three diagonal elements of C_f^m, we can calculate the rotation angle.

Initially the f- and m-frames are aligned; rotation of the m-frame about the axis vector $\boldsymbol{a}$ through the angle ϕ results in the present misalignment between the two frames. This misalignment is described by elements of a DCM, by Euler angles, or now by the direction cosines of the axis vector $\boldsymbol{a}$ and the angle through which the rotation takes place. Refer to Fig. 4.5. It is assumed that the generic vector $\boldsymbol{v}$ remains fixed in the f-frame. To an observer in the m-frame, the vector $\boldsymbol{v}$ appears to move in a direction opposite to that of the frame rotation. Thus, the vector $\boldsymbol{v}$ is designated here by the appropriate superscript to indicate the frame in which its components are evaluated (i.e., assigned literal or numerical values).

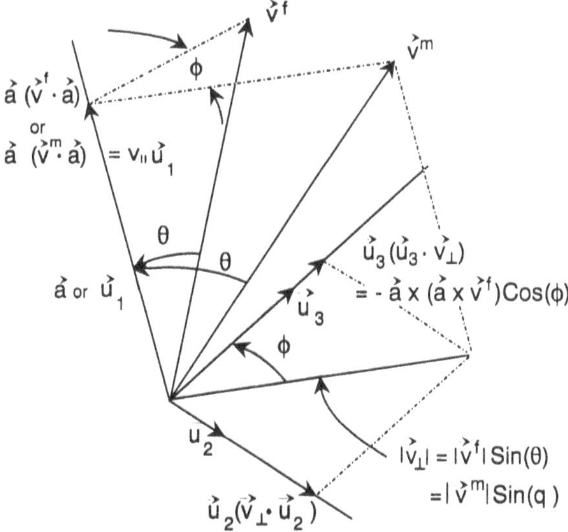

Fig. 4.5 Geometric interpretation of the axis/angle method.

The next step is to construct a triad for resolving the vector $\boldsymbol{v}$ into components. The first axis of the triad is along the axis vector $\boldsymbol{a}$, given by the unit vector $\boldsymbol{u}_1$; the component of $\boldsymbol{v}^f$ along the axis vector $\boldsymbol{a}$ (or $\boldsymbol{u}_1$) is given as $\boldsymbol{v}_\parallel$:

$$\boldsymbol{v}_\parallel = \left(\boldsymbol{v}^f \cdot \boldsymbol{a}\right)\boldsymbol{a} \tag{4.59a}$$

The second and third axes, designated by unit vectors $\boldsymbol{u}_2$ and $\boldsymbol{u}_3$, are

$$\boldsymbol{u}_2 = \langle \boldsymbol{v}^f \times \boldsymbol{a} \rangle = \frac{\boldsymbol{v}^f \times \boldsymbol{a}}{|\boldsymbol{v}^f|\sin(\theta)} \tag{4.59b}$$

$$\boldsymbol{u}_3 = \langle \boldsymbol{a} \times \left(\boldsymbol{v}^f \times \boldsymbol{a}\right) \rangle = \frac{\boldsymbol{a} \times \left(\boldsymbol{v}^f \times \boldsymbol{a}\right)}{|\boldsymbol{v}^f|\sin(\theta)} \tag{4.59c}$$

The vector $\boldsymbol{v}^m$ can now be projected into the $(\boldsymbol{u}_2, \boldsymbol{u}_3)$ plane to get $\boldsymbol{v}_\perp$,

$$\boldsymbol{v}_\perp = \boldsymbol{v}^m - \boldsymbol{a}(\boldsymbol{v}^m \cdot \boldsymbol{a}) \tag{4.60a}$$

The magnitude of $\boldsymbol{v}_\perp$ is

$$|\boldsymbol{v}_\perp| = |\boldsymbol{v}^f|\sin(\theta) = |\boldsymbol{v}^m|\sin(\theta) \tag{4.60b}$$

The vector $\boldsymbol{v}$ as well as the unit vector triad $(\boldsymbol{u}_1, \boldsymbol{u}_2, \boldsymbol{u}_3)$ are illustrated in Fig. 4.5. The components of $\boldsymbol{v}$ along the $\boldsymbol{u}_2$ and $\boldsymbol{u}_3$ directions are the components of $\boldsymbol{v}_\perp$ along both of these axes. That is,

$$\begin{aligned}
\boldsymbol{u}_2(\boldsymbol{u}_2 \cdot \boldsymbol{v}_\perp) &= \boldsymbol{u}_2|\boldsymbol{v}^f|\sin(\theta)\sin(\phi) \\
&= |\boldsymbol{v}^f|\sin(\theta)\frac{\boldsymbol{v}^f \times \boldsymbol{a}\sin(\phi)}{|\boldsymbol{v}^f|\sin(\theta)} \\
&= -(\boldsymbol{a} \times \boldsymbol{v}^f)\sin(\phi)
\end{aligned} \tag{4.61a}$$

$$\begin{aligned}
\boldsymbol{u}_3(\boldsymbol{u}_3 \cdot \boldsymbol{v}_\perp) &= \boldsymbol{u}_3|\boldsymbol{v}^f|\sin(\theta)\cos(\phi) \\
&= \frac{\boldsymbol{a} \times \left(\boldsymbol{v}^f \times \boldsymbol{a}\right)}{|\boldsymbol{v}^f|\sin(\theta)}|\boldsymbol{v}^f|\sin(\theta)\cos(\phi) \\
&= -\boldsymbol{a} \times \left(\boldsymbol{a} \times \boldsymbol{v}^f\right)\cos(\phi)
\end{aligned} \tag{4.61b}$$

We can now add all three components of the vector $\boldsymbol{v}^m$ to provide the components of the vector $\boldsymbol{v}$ as seen in the m-frame.

$$\boldsymbol{v}^m = \left(\boldsymbol{v}^f \cdot \boldsymbol{a}\right)\boldsymbol{a} - \left(\boldsymbol{a} \times \boldsymbol{v}^f\right)\sin(\phi) - \boldsymbol{a} \times \left(\boldsymbol{a} \times \boldsymbol{v}^f\right)\cos(\phi) \tag{4.62a}$$

or, by using the triple cross-product identity, we get

$$\boldsymbol{v}^m = \boldsymbol{a}\left(\boldsymbol{a} \cdot \boldsymbol{v}^f\right) - \sin(\phi)\left(\boldsymbol{a} \times \boldsymbol{v}^f\right) - \cos(\phi)\left[\boldsymbol{a}\left(\boldsymbol{a} \cdot \boldsymbol{v}^f\right) - (\boldsymbol{a} \cdot \boldsymbol{a})\,\boldsymbol{v}^f\right] \tag{4.62b}$$

The above relationship may be written in matrix form as

$$v^m = \left\{ aa^{\mathrm{T}} - \sin(\phi)a^* - \cos(\phi)aa^{\mathrm{T}} + [\cos(\phi)]\, I \right\} v^f \tag{4.63a}$$

where $a \cdot a$ or $a^{\mathrm{T}}a = 1$ because a is a unit vector. The star superscript indicates that the vector is written in skew-symmetric form, providing the matrix equivalent of the vector cross-product operation. The term contained within the braces in Eq. (4.63a) acts like a directional cosine matrix in that it transforms the components of vector v from the f- to the m-frame. Useful rearrangements of the above expression are

$$v^m = \left\{ aa^{\mathrm{T}} + \left[\left(I - aa^{\mathrm{T}}\right)\cos(\phi)\right] - \left[a^* \sin(\phi)\right]\right\} v^f \tag{4.63b}$$

$$v^m = \left\{ [\cos(\theta)]\, I + [1 - \cos(\phi)]\, aa^{\mathrm{T}} - [\sin(\phi)]\, a^* \right\} v^f \tag{4.63c}$$

Either of the above forms also transforms a vector v^f (vector v with components in the f-frame) into v^m (vector v with components in the m-frame). Thus the term in the braces in Eqs. (4.63) is equivalent to a DCM; that is,

$$C_f^m = [\cos(\phi)]\, I + [1 - \cos(\phi)]\, aa^{\mathrm{T}} - [\sin(\phi)]\, a^* \tag{4.64}$$

In the preceding analysis we assumed that the vector v was fixed in the f-frame and that the observation was made from the m-frame. By having v available in the f-frame and knowing the axis/angle parameters $(\phi, a) = (\phi, a_1, a_2, a_3)$, we can obtain the components of the same vector in the m-frame. If we assume that v is fixed in the m-frame (and shares that frame's rotation), we could obtain the vector v^f as follows:

$$\begin{aligned} v^f &= [C_f^m]^{-1} v^m = C_m^f v^m \\ &= \left\{ [\cos(\phi)]\, I + [1 - \cos(\phi)]\, aa^{\mathrm{T}} + \left[\sin(\phi)a^*\right]\right\} v^m \end{aligned} \tag{4.65}$$

The inverse is formed by replacing the rotation angle ϕ by its negative.

Note that in the above expression we have depicted v^f and v^m as matrices instead of vectors. When using matrix notation, a vector is indicated by a superscript which indicates the frame of resolution.

If we carry out the manipulations of Eq. (4.64), we can relate the elements of the DCM to the four axis/angle parameters. The results are given in Table 4.1.

Subtracting the matrix C_f^m [Eq. (4.64)] from its inverse [or its transpose, Eq. (4.65)] gives

$$a^* = \left[\left(C_m^f\right)^{\mathrm{T}} - C_m^f \right] / 2\sin(\phi)$$

$$a = \begin{bmatrix} a_1 \\ a_2 \\ a_3 \end{bmatrix} = \frac{1}{2\sin\phi} \begin{bmatrix} C_{23} - C_{32} \\ C_{31} - C_{13} \\ C_{12} - C_{21} \end{bmatrix} \tag{4.66}$$

The a-parameters are given in Table 4.2 along with singularity conditions.

Table 4.1 DCM elements in terms of axis/angle parameters

$$
\begin{aligned}
C_{11} &= (1 - \cos\phi)a_1^2 + \cos\phi \\
C_{12} &= (1 - \cos\phi)a_1 a_2 + a_3 \sin\phi \\
C_{13} &= (1 - \cos\phi)a_1 a_3 - a_2 \sin\phi \\
C_{21} &= (1 - \cos\phi)a_2 a_1 - a_3 \sin\phi \\
C_{22} &= (1 - \cos\phi)a_2^2 + \cos\phi \\
C_{23} &= (1 - \cos\phi)a_2 a_3 + a_1 \sin\phi \\
C_{31} &= (1 - \cos\phi)a_3 a_1 + a_2 \sin\phi \\
C_{32} &= (1 - \cos\phi)a_3 a_2 - a_1 \sin\phi \\
C_{33} &= (1 - \cos\phi)a_3^2 + \cos\phi
\end{aligned}
$$

Upon examining Eq. (4.66) there is an obvious singularity in the determination of the components of $\boldsymbol{a}$ when $\phi = \pm n\pi$, where n is an integer. When $\phi = 2n\pi$, the final orientation of the m-frame is indistinguishable from the condition of no rotation. With no rotation the vector $\boldsymbol{a}$ must remain undefined. Under such circumstances the trace of the DCM must be equal to 3 [see Eq. (4.57)]. When $\phi = \pm(2n+1)\pi$, the rotation is equivalent to a rotation angle of π (or $-\pi$) regardless of the value of n. For this case the trace of the DCM is -1. When $\sigma = -1$, the root leaves an ambiguity of sign, which may be resolved by computing the products

$$
a_1 a_2 = c_{12}/2, \qquad a_2 a_3 = c_{23}/2, \qquad a_1 a_3 = c_{13}/2
$$

Table 4.2 Axis directional numbers

	$\sigma \neq -1,\ 3$	$\sigma = -1$
a_1	$\dfrac{C_{23} - C_{32}}{2\sin\phi}$	$\left[\dfrac{(1 + C_{11})}{2}\right]^{1/2}$
a_2	$\dfrac{C_{31} - C_{13}}{2\sin\phi}$	$\left[\dfrac{(1 + C_{22})}{2}\right]^{1/2}$
a_3	$\dfrac{C_{12} - C_{21}}{2\sin\phi}$	$\left[\dfrac{(1 + C_{33})}{2}\right]^{1/2}$

The angle ϕ has already been evaluated in terms of the trace, σ, in Eq. (4.58). It is given by

$$\begin{aligned}\cos\phi &= \tfrac{1}{2}(\sigma - 1)\\ &= \tfrac{1}{2}(C_{11} + C_{22} + C_{33} - 1)\end{aligned} \tag{4.67}$$

Thus, from a knowledge of the elements of a DCM, the angle of rotation ϕ and the components of the vector $\boldsymbol{a}$ may be found from Eqs. (4.66) and (4.67). Conversely, if the axis/angle parameters are known, then the elements of the DCM may be calculated from Table 4.1.

The preceding discussion of the axis/angle parameters is an expansion of the development given by Hughes.[3] An alternate analytical description is given by Kane,[2] and a more geometric description is given by Gelman.[6–8]

4.7 Updating the Axis/Angle Parameters

Since the misalignment between the m- and f-frames will not remain constant, we expect that the elements of the C_f^m DCM will be time-dependent. Equivalently, we expect the four axis/angle parameters to be time-dependent. An excellent treatment of time-varying axis/angle parameters is given by Setterlund.[9]

Equation (4.67) may be rewritten as

$$\cos\phi = \left[\mathrm{tr}\left(C_f^m\right) - 1\right]/2 \tag{4.68}$$

Differentiating the previous expression gives

$$\begin{aligned}\frac{\mathrm{d}\phi}{\mathrm{d}t} &= -\frac{1}{2\sin\phi}\left\langle \frac{\mathrm{d}}{\mathrm{d}t}\left[\mathrm{tr}\left(C_f^m\right)\right]\right\rangle\\ &= -\frac{1}{2\sin\phi}\left\langle \mathrm{tr}\left[\frac{\mathrm{d}}{\mathrm{d}t}\left(C_f^m\right)\right]\right\rangle\end{aligned} \tag{4.69}$$

Since the trace and differentiation operations are linear, we may exchange their order. From Eq. (4.30), we may write, after letting $b = m$ and $I = f$,

$$\frac{\mathrm{d}C_m^f}{\mathrm{d}t} = C_m^f \Omega_{m/f}^m$$

After transposing both sides and recognizing the properties of the skew-symmetric matrix, we get

$$\frac{\mathrm{d}C_f^m}{\mathrm{d}t} = -\Omega_{m/f}^m C_f^m \tag{4.70}$$

Equation (4.69) becomes

$$\frac{\mathrm{d}\phi}{\mathrm{d}t} = -\frac{1}{2\sin\phi}\left[\mathrm{tr}\left(\Omega^m_{m/f} C^m_f\right)\right] = -\frac{1}{2\sin\phi}\left\{\mathrm{tr}\left[\Omega^m_{m/f}\left(C^f_m\right)^{\mathrm{T}}\right]\right\}$$

$$= -\frac{1}{2\sin\phi}\mathrm{tr}\left\{\begin{bmatrix} 0 & -\omega_3 & \omega_2 \\ \omega_3 & 0 & -\omega_1 \\ -\omega_2 & \omega_1 & 0 \end{bmatrix}\begin{bmatrix} C_{11} & C_{12} & C_{13} \\ C_{21} & C_{22} & C_{23} \\ C_{31} & C_{32} & C_{33} \end{bmatrix}^{\mathrm{T}}\right\}$$

$$= \frac{1}{2\sin\phi}\left[\omega_1(C_{23} - C_{32}) + \omega_2(C_{31} - C_{13}) + \omega_3(C_{12} - C_{21})\right]$$

The above expression can be rewritten using Eq. (4.66) as

$$\frac{\mathrm{d}\phi}{\mathrm{d}t} = \frac{1}{2\sin\phi}\left(\boldsymbol{\omega}^m_{m/f}\right)^{\mathrm{T}}\begin{bmatrix} C_{23} - C_{32} \\ C_{31} - C_{13} \\ C_{12} - C_{21} \end{bmatrix} = \left(\boldsymbol{\omega}^m_{m/f}\right)^{\mathrm{T}}\boldsymbol{a} \tag{4.71}$$

The above result is fairly intuitive since it says that the component of the angular velocity along the axis vector is equal to the rotation rate about the axis.

Next we must find an expression for the derivative of the unit axis vector $\boldsymbol{a}$. It was pointed out earlier that the axis vector $\boldsymbol{a}$ has the same directional numbers in both the m- and f-frames (moving and fixed). We can restate Eq. (4.54) as

$$C^m_f \boldsymbol{a}^f = \boldsymbol{a}^m = \boldsymbol{a}^f$$

Differentiating the above gives

$$\frac{\mathrm{d}C^m_f}{\mathrm{d}t}\boldsymbol{a}^f + C^m_f\frac{\mathrm{d}\boldsymbol{a}^f}{\mathrm{d}t} = \frac{\mathrm{d}\boldsymbol{a}^f}{\mathrm{d}t}$$

$$\frac{\mathrm{d}C^m_f}{\mathrm{d}t}\boldsymbol{a}^f = \left(I - C^m_f\right)\frac{\mathrm{d}\boldsymbol{a}^f}{\mathrm{d}t}$$

From Eq. (4.32), with $m = b$ and $f = I$, the above relationships become

$$-\Omega^m_{m/f} C^m_f \boldsymbol{a}^f = \left(I - C^m_f\right)\frac{\mathrm{d}\boldsymbol{a}^f}{\mathrm{d}t}$$

but from Eq. (4.54)

$$-\Omega^m_{m/f}\boldsymbol{a}^f = \left(I - C^m_f\right)\frac{\mathrm{d}\boldsymbol{a}^f}{\mathrm{d}t}$$

After inserting Eq. (4.64) for C_f^m we get

$$-\Omega_{m/f}^{m}\boldsymbol{a} = \left[I - \langle \boldsymbol{a}\boldsymbol{a}^{\mathrm{T}} + \cos\phi(I - \boldsymbol{a}\boldsymbol{a}^{\mathrm{T}}) - a^{*}\sin\phi\rangle\right]\frac{\mathrm{d}\boldsymbol{a}}{\mathrm{d}t} \tag{4.72}$$

Note that we have dropped the frame superscript on the axis vector since this vector has the same directional cosines in both the m- and f-frames. Further simplification is possible since,

$$\frac{\mathrm{d}\boldsymbol{a}}{\mathrm{d}t}\cdot\boldsymbol{a} = \frac{\mathrm{d}\boldsymbol{a}^{\mathrm{T}}}{\mathrm{d}t}\boldsymbol{a} = \boldsymbol{a}^{\mathrm{T}}\dot{\boldsymbol{a}} = 0$$

$$-\Omega_{m/f}^{m}\boldsymbol{a} = \left[(1-\cos\phi)I + a^{*}\sin\phi\right]\frac{\mathrm{d}\boldsymbol{a}}{\mathrm{d}t} \tag{4.73}$$

or

$$\frac{\mathrm{d}\boldsymbol{a}}{\mathrm{d}t} = \left[(1-\cos\phi)I + a^{*}\sin\phi\right]^{-1}\left(-\Omega_{m/f}^{m}\boldsymbol{a}\right)$$

$$\frac{\mathrm{d}\boldsymbol{a}}{\mathrm{d}t} = \left[(1-\cos\phi)I + a^{*}\sin\phi\right]^{-1}a^{*}\boldsymbol{\omega}_{m/f}^{m} \tag{4.74}$$

where we made use of the identity

$$-\Omega_{m/f}^{m}\boldsymbol{a} = a^{*}\boldsymbol{\omega}_{m/f}^{m}$$

The inverse indicated in Eq. (4.74) may be written as

$$[(1-\cos\phi)I + \sin\phi a^{*}]^{-1} = \frac{1}{1-\cos\phi}\left[I - \frac{1}{2}\sin\phi a^{*} + \frac{1}{2}(1-\cos\phi)(\boldsymbol{a}\boldsymbol{a}^{\mathrm{T}} - 1)\right] \tag{4.75}$$

and checked by multiplying this inverse by the original matrix. Equation (4.74) now becomes

$$\frac{\mathrm{d}\boldsymbol{a}}{\mathrm{d}t} = \frac{1}{1-\cos\phi}\left[a^{*} - \frac{1}{2}\sin\phi(a^{*})^{2} - \frac{1}{2}(1-\cos\phi)a^{*}\right]\boldsymbol{\omega}_{m/f}^{m} \tag{4.76}$$

Note that

$$\boldsymbol{a}\boldsymbol{a}^{\mathrm{T}}a^{*} = \boldsymbol{a}(\boldsymbol{a}^{\mathrm{T}}a^{*}) = 0$$

$$(\boldsymbol{a}\boldsymbol{a}^{\mathrm{T}} - I) = a^{*}a^{*} \doteq (a^{*})^{2}$$

Thus,

$$\frac{\mathrm{d}\boldsymbol{a}}{\mathrm{d}t} = \left[\frac{1}{2}a^* - \frac{1}{2}\cos\left(\frac{\phi}{2}\right)(\boldsymbol{a}\boldsymbol{a}^{\mathrm{T}} - I)\right]\boldsymbol{\omega}^m_{m/f} \tag{4.77}$$

where we made use of the identity

$$\cos(\phi/2) = \sin\phi/(1 - \cos\phi)$$

We now have differential equations for ϕ [Eq. (4.71)] and $\boldsymbol{a}$ [Eq. (4.77)]. As Setterlund[9] points out, these equations are difficult to solve. One reason is that the vector $\boldsymbol{a}$ is defined only after a rotation has taken place; consequently, there is some problem in setting an initial value for $\boldsymbol{a}$.

4.8 Euler Four-Parameter Method (Quaternions)

The axis/angle method discussed in the last section is a four-parameter system involving ϕ, a_1, a_2, and a_3. A closely related four-parameter system has alternate designations as *quaternions* or the *Euler four parameters*. Quaternion algebra was a precursor to modern vector analysis and the Euler Four Parameters could be manipulated using this algebra.[1] With the rapid development of strapdown inertial measurement units (IMU), interest in quaternions has been renewed because this four-parameter method has certain computational advantages compared to other analytically equivalent methods.

First, we can give the quaternion Q geometric significance (of sorts) by relating the four elements of Q to the four axis/angle parameters as follows:

$$Q = \begin{bmatrix} q_1 \\ q_2 \\ q_3 \\ --- \\ q_4 \end{bmatrix} = \begin{bmatrix} a_1\sin(\phi/2) \\ a_2\sin(\phi/2) \\ a_3\sin(\phi/2) \\ ------ \\ \cos(\phi/2) \end{bmatrix} = \begin{bmatrix} \boldsymbol{q} \\ --- \\ q_4 \end{bmatrix} = \begin{bmatrix} \boldsymbol{a}\sin(\phi/2) \\ ------ \\ \cos(\phi/2) \end{bmatrix} \tag{4.78a}$$

It would appear that the first three quaternions are vector components and are related to the unit axis vector $\boldsymbol{a}$ as follows:

$$\boldsymbol{q} = \boldsymbol{a}\sin(\phi/2)$$

$$q_1 = a_1\sin(\phi/2), \qquad q_2 = a_2\sin(\phi/2), \qquad q_3 = a_3\sin(\phi/2) \tag{4.78b}$$

$$q_4 = \cos(\phi/2)$$

from which follows the useful normalizing relationship

$$q_1^2 + q_2^2 + q_3^2 + q_4^2 = 1 = \boldsymbol{q}^{\mathrm{T}}\boldsymbol{q} + q_4^2$$

Since $\boldsymbol{a}$ has identical components in either the m- or f-frames, so does $\boldsymbol{q}$, the vector part of the quaternion.

First, we endeavor to express the DCM C_f^m, which transforms vector components in the fixed frame to components in the moving frame. As an example we can take the element C_{11} from Table 4.1.

$$C_{11} = \cos\phi + a_1^2(1 - \cos\phi)$$

but

$$\cos\phi = 2\cos^2(\phi/2) - 1 = 1 - 2\sin^2(\phi/2)$$

By inserting the above identities into the expression for C_{11}, we get

$$C_{11} = 2\cos^2(\phi/2) - 1 + 2a_1^2\sin^2(\phi/2)$$

By using the definition of quaternion Q given in Eq. (4.78a), we can easily show that

$$C_{11} = 2q_4^2 - 1 + 2q_1^2$$

or, by using the normalization property of the quaternion, we can obtain the following alternative to the above expression:

$$C_{11} = q_1^2 - q_2^2 - q_3^2 + q_4^2 = 1 - 2\left(q_2^2 + q_3^2\right)$$

Table 4.3 gives expressions for all of the elements of the DCM in terms of quaternions.

Table 4.3 DCM elements in terms of quaternions

$$C_{11} = 1 - 2(q_2^2 + q_3^2)$$
$$C_{12} = 2(q_1q_2 + q_3q_4)$$
$$C_{13} = 2(q_1q_3 - q_2q_4)$$
$$C_{21} = 2(q_2q_1 - q_3q_4)$$
$$C_{22} = 1 - 2(q_1^2 + q_3^2)$$
$$C_{23} = 2(q_2q_3 + q_1q_4)$$
$$C_{31} = 2(q_1q_3 + q_2q_4)$$
$$C_{32} = 2(q_2q_3 - q_1q_4)$$
$$C_{33} = 1 - 2(q_1^2 + q_2^2)$$

We must now take steps analogous to those taken earlier in the development of the DCM, Euler angle, and axis/angle update schemes [given by Eqs. (4.30), (4.53), (4.71), and (4.77)]. Again, this development closely follows the approach used by Setterlund.[9] First, we differentiate the quaternion vector $\boldsymbol{q}$ given by Eq. (4.78b) as follows:

$$\frac{\mathrm{d}\boldsymbol{q}}{\mathrm{d}t} = \frac{1}{2}\cos\left(\frac{\phi}{2}\right)\boldsymbol{a}\frac{\mathrm{d}\phi}{\mathrm{d}t} + \sin\left(\frac{\phi}{2}\right)\frac{\mathrm{d}\boldsymbol{a}}{\mathrm{d}t} \tag{4.79}$$

We can replace $\mathrm{d}\phi/\mathrm{d}t$ and $\mathrm{d}\boldsymbol{a}/\mathrm{d}t$ by using Eqs. (4.71) and (4.77) to get

$$\frac{\mathrm{d}\boldsymbol{q}}{\mathrm{d}t} = \left[\left(\boldsymbol{\omega}^m_{m/f}\right)^{\mathrm{T}}\boldsymbol{a}\right]\frac{1}{2}\cos\left(\frac{\phi}{2}\right)\boldsymbol{a} + \sin\left(\frac{\phi}{2}\right)\left[\frac{1}{2}a^* - \frac{1}{2}\cot\left(\frac{\phi}{2}\right)(\boldsymbol{a}\boldsymbol{a}^{\mathrm{T}} - I)\right]\boldsymbol{\omega}^m_{m/f}$$

This easily reduces to

$$\frac{\mathrm{d}\boldsymbol{q}}{\mathrm{d}t} = \frac{1}{2}\sin\left(\frac{\phi}{2}\right)a^*\boldsymbol{\omega}^m_{m/f} + \frac{1}{2}\cos\left(\frac{\phi}{2}\right)\boldsymbol{\omega}^m_{m/f}$$

or, using the definition of the quaternion vector $\boldsymbol{q}$ and scalar q_4, we get from Eqs. (4.78) the following:

$$\frac{\mathrm{d}\boldsymbol{q}}{\mathrm{d}t} = -\frac{1}{2}\Omega^m_{m/f}\boldsymbol{q} + \frac{1}{2}q_4\boldsymbol{\omega}^m_{m/f} \tag{4.80}$$

Differentiation of the quaternion scalar q_4 gives

$$\frac{\mathrm{d}q_4}{\mathrm{d}t} = -\frac{1}{2}\sin\left(\frac{\phi}{2}\right)\frac{\mathrm{d}\phi}{\mathrm{d}t}$$

or, using Eqs. (4.71) and (4.78b),

$$\begin{aligned}\frac{\mathrm{d}q_4}{\mathrm{d}t} &= -\frac{1}{2}\sin\left(\frac{\phi}{2}\right)\left(\boldsymbol{\omega}^m_{m/f}\right)^{\mathrm{T}}\boldsymbol{a}\\ &= -\frac{1}{2}\sin\left(\frac{\phi}{2}\right)\boldsymbol{a}^{\mathrm{T}}\boldsymbol{\omega}^m_{m/f} = -\frac{1}{2}\boldsymbol{q}^{\mathrm{T}}\boldsymbol{\omega}^m_{m/f}\end{aligned} \tag{4.81}$$

Equations (4.80) and (4.81) may be combined into a matrix format as follows:

$$\frac{\mathrm{d}Q}{\mathrm{d}t} = \begin{bmatrix}\dfrac{\mathrm{d}\boldsymbol{q}}{\mathrm{d}t}\\ \hline \dfrac{\mathrm{d}q_4}{\mathrm{d}t}\end{bmatrix} = \begin{bmatrix}-\frac{1}{2}\Omega^m_{m/f} & \frac{1}{2}\boldsymbol{\omega}^m_{m/f}\\ -\frac{1}{2}(\boldsymbol{\omega}^m_{m/f})^{\mathrm{T}} & 0\end{bmatrix}\begin{bmatrix}\boldsymbol{q}\\ \hline q_4\end{bmatrix} \tag{4.82a}$$

or

$$\frac{\mathrm{d}Q}{\mathrm{d}t} = \frac{\mathrm{d}}{\mathrm{d}t}\begin{bmatrix} q_1 \\ q_2 \\ q_3 \\ q_4 \end{bmatrix} \doteq \begin{bmatrix} \dot{q}_1 \\ \dot{q}_2 \\ \dot{q}_3 \\ \dot{q}_4 \end{bmatrix} = \frac{1}{2}\begin{bmatrix} 0 & \omega_3 & -\omega_2 & \omega_1 \\ -\omega_3 & 0 & \omega_1 & \omega_2 \\ \omega_2 & -\omega_1 & 0 & \omega_3 \\ -\omega_1 & -\omega_2 & -\omega_3 & 0 \end{bmatrix}\begin{bmatrix} q_1 \\ q_2 \\ q_3 \\ q_4 \end{bmatrix} \tag{4.82b}$$

or, finally, in a more compact form,

$$\frac{\mathrm{d}Q}{\mathrm{d}t} = \frac{1}{2}AQ \tag{4.82c}$$

The elements of matrix A are obvious from Eq. (4.82b). A slight rearrangement of Eq. (4.82b) gives

$$\begin{bmatrix} \frac{\mathrm{d}q_1}{\mathrm{d}t} \\ \frac{\mathrm{d}q_2}{\mathrm{d}t} \\ \frac{\mathrm{d}q_3}{\mathrm{d}t} \\ \frac{\mathrm{d}q_4}{\mathrm{d}t} \end{bmatrix} = \frac{\mathrm{d}Q}{\mathrm{d}t} = \frac{1}{2}\begin{bmatrix} q_4 & -q_3 & q_2 \\ q_3 & q_4 & -q_1 \\ -q_2 & q_1 & q_4 \\ -q_1 & -q_2 & -q_3 \end{bmatrix}\begin{bmatrix} \omega_1 \\ \omega_2 \\ \omega_3 \end{bmatrix} \tag{4.82d}$$

or, in a compact form similar to Eq. (4.82c), we have

$$\frac{\mathrm{d}Q}{\mathrm{d}t} = \frac{1}{2}B\boldsymbol{\omega}_{m/f}^{m} \tag{4.82e}$$

There are other forms for the quaternion time derivative which frequently appear. For example, we may express the time derivatives of the vector and scalar parts of the quaternion separately as

$$\frac{\mathrm{d}\boldsymbol{q}}{\mathrm{d}t} = \frac{1}{2}\left[q^* + q_4 I\right]\boldsymbol{\omega}_{m/f}^{m} \tag{4.82f}$$

and

$$\dot{q}_4 = -\tfrac{1}{2}\boldsymbol{q}^{\mathrm{T}}\boldsymbol{\omega}_{m/f}^{m} \tag{4.82g}$$

Table 4.4a Parameter options

Name	Number of parameters	Relevant parameters	Relevant form of the DCM
Direction cosines	9	C_{ij}	$C_a^b = [C_{ij}]$
Euler angles	3	$\theta_1, \theta_2, \theta_3$	$C_i(\theta_1) \cdot C_j(\theta_2) \cdot C_k(\theta_3)$
Axis/angle	4	$\boldsymbol{a}, \phi$	$I \cos\phi + (1 - \cos\phi)\boldsymbol{a}\boldsymbol{a}^{\mathrm{T}} - \sin\phi$
Quaternions	4	q_1, q_2, q_3, q_4	$(q_4^2 - \boldsymbol{q}^{\mathrm{T}}\boldsymbol{q})i + 2\boldsymbol{q}\boldsymbol{q}^{\mathrm{T}} - 2q_4\boldsymbol{q}^*$

Table 4.4b Parameter options

Name	Angular velocity	Time derivatives of the parameters
Direction cosines	$\Omega_{a/b}^b = \dot{C}_a^b (C_a^b)^{\mathrm{T}}$	$\dot{C}_a^b = \Omega_{a/b}^b C_a^b$
	$\Omega_{a/b}^a = (C_a^b)^{\mathrm{T}} \dot{C}_a^b$	$\dot{C}_a^b = C_a^b \Omega_{a/b}^a$
Euler angles	$\boldsymbol{\omega} = [1, C_i, C_i C_j] \dfrac{\mathrm{d}\boldsymbol{\theta}}{\mathrm{d}t}$	$\dfrac{\mathrm{d}\boldsymbol{\theta}}{\mathrm{d}t} = E^{-1}[\theta_1, \theta_2]\boldsymbol{\omega}$ (see Appendix C)
Axis/angle	$\boldsymbol{\omega} = \phi\boldsymbol{a} - (1 - \cos\phi)a^*\boldsymbol{a} + \boldsymbol{a}\sin\phi$	$\dfrac{\mathrm{d}\boldsymbol{a}}{\mathrm{d}t} = \dfrac{1}{2}\left[a^* - \cos\left(\dfrac{\phi}{2}\right)a^*a^*\right]\boldsymbol{\omega}$; $\dfrac{\mathrm{d}\phi}{\mathrm{d}t} = \boldsymbol{a}^{\mathrm{T}}\boldsymbol{\omega}$
Quaternions	$\boldsymbol{\omega} = 2\left(q_4 \dfrac{\mathrm{d}\boldsymbol{q}}{\mathrm{d}t} - \dfrac{\mathrm{d}q_4}{\mathrm{d}t}\boldsymbol{q}\right) - 2\boldsymbol{q}^* \dfrac{\mathrm{d}\boldsymbol{q}}{\mathrm{d}t}$	$\dfrac{\mathrm{d}q_4}{\mathrm{d}t} = -\dfrac{1}{2}\boldsymbol{q}^{\mathrm{T}}\boldsymbol{\omega}$; $\dfrac{\mathrm{D}Q}{\mathrm{D}t} = \dfrac{1}{2}(q^* + q_4 I)\boldsymbol{\omega}$; $\dfrac{\mathrm{D}Q}{\mathrm{D}t} = \dfrac{1}{2}AQ$; $\dfrac{\mathrm{D}Q}{\mathrm{D}t} = \dfrac{1}{2}B\boldsymbol{\omega}$

For what value it may have, we state without further development the inverse of Eq. (4.82b), from which the angular velocity components may be obtained if the quaternion elements are given:

$$\boldsymbol{\omega}^{m}_{m/f} = 2\left[q_4\frac{\mathrm{d}\boldsymbol{q}}{\mathrm{d}t} - \frac{\mathrm{d}q_4}{\mathrm{d}t}\boldsymbol{q}\right] - 2q^{*}\frac{\mathrm{d}\boldsymbol{q}}{\mathrm{d}t} \tag{4.83}$$

The usual procedure is to obtain the angular velocity vector $\boldsymbol{\omega} = [\omega_1, \omega_2, \omega_3]^{\mathrm{T}}$ from the moment equations (simulation) or from a gyro cluster (navigation/flight test) and update the quaternions by using Eq. (4.82d). With the new (or updated) quaternions, the DCM can be updated using the entries of Table 4.3.

4.9 Summary

This chapter has considered four systems for describing frame orientation. The presentation has by no means been exhaustive. Other systems such as Cayley-Klein and Euler-Rodrigues parameters could be discussed; however, further cataloging of methods would be tangential to the primary goals of this book. In Tables 4.4a and 4.4b, we summarize some of the results from this chapter.

References

[1]Crowe, M. J., *A History of Vector Analysis—The Evolution of the Idea of a Vectoral System,* Dover, Moneola, NY, 1985.

[2]Kane, T. R., Litkins, P. W., and Levinson, D. A., *Space Dynamics,* McGraw-Hill, Hightstown, NJ, 1983.

[3]Hughes, P. C., *Spacecraft Attitude Dynamics,* Wiley, Somerset, NJ, 1986.

[4]Wertz, J. R. (ed.), *Spacecraft Attitude Determination and Control,* distributed by Kluwer Boston, Hingham, MA, 1978 (later reprint 1985).

[5]Britting, K. R., *Inertial Navigation Systems Analysis,* Wiley-Interscience, Wiley, Somerset, NJ, 1971.

[6]Gelman, H., "The Second Orthogonality Condition in the Theory of Proper and Improper Rotations: I—Derivation of the Conditions and of Their Main Consequences," *Journal of Research,* National Bureau of Standards, Bulletin Mathematical Sciences, No. 3, 1968, pp. 229–237.

[7]Gelman, H., "The Second Orthogonality Condition in the Theory of Proper and Improper Rotations: II—The Intrinsic Vector," *Journal of Research,* National Bureau of Standards, Bulletin Mathematical Sciences, No. 2, 1969, pp. 125–138.

[8]Gelman, H., "A Note on the Time Dependence of the Effective Axis and Angle of Rotation," *Journal of Research,* National Bureau of Standards, Bulletin Mathematical Sciences, Nos. 3 and 4, 1971.

[9]Setterlund, R. H., "Novel Method for In-Flight Alignment of Master-Slave Inertial Platforms," M.Sc. Thesis, Dept. of Aeronautics and Astronautics, Massachusetts Inst. of Technology, Cambridge, MA, 1974.

5
Force and Moment Equations

5.1 Newton's Second Law of Motion

"The change of motion is proportional to the motive force impressed; and is made in the direction of the right line in which that force is impressed."

—Isaac Newton, *Mathematical Principles of Natural Philosophy*

In Chapter 3 we discussed Newton's Law of Gravitation and the extensions of that law in order to describe the nonspherical gravitational field of the Earth. In addition to his speculations concerning gravity, Newton also laid the foundations of rational mechanics in what has become known as *Newton's Three Laws of Motion*. In this chapter we are concerned with the second law, which relates *force*, the agent that causes motion, to the kinematic entity called *acceleration*, essentially what Newton identifies as "change of motion." The constant of proportionality between applied force and the resulting acceleration is identified as the mass. An interesting speculation on the axiomatic nature of Newton's Second Law is given by Lindsay.[1] Lindsay shows that while acceleration can be given a kinematic definition, force and mass are fundamentally related only through Newton's Second Law. Thus either force or mass must be axiomatic with the other defined through the Second Law.

A natural or artificial atmospheric entry vehicle is subjected to a variety of forces, some of which may include aerodynamic, magnetic, terrestrial and nonterrestrial gravitational fields, and perhaps other effects. Over certain segments of the trajectory some forces may be negligibly small, only to dominate over a later segment. For example, the motion of a re-entry body moving along an exoatmospheric trajectory is determined almost entirely by the terrestrial gravitational field; even the nonspherical terms discussed in Chapter 3 are more influential than nonterrestrial fields, solar winds, or atmospheric gas residuals. However, during the later endoatmospheric trajectory, the aerodynamic pressure field may cause drag loads on the order of 100 g. Under such circumstances, we may completely ignore the gravitational field and still retain the essential features of the trajectory. Obviously, there must be a boundary of sorts between the exotrajectory (gravity-dominated) and the endotrajectory (aerodynamics-dominated). In Chapter 6 we will examine exoatmospheric (or Keplerian) motion, and in Chapter 7 we will consider some elementary trajectory representations for the much more complex endoatmospheric trajectory.

Before trying to represent the applied forces in any detail at all, we must first learn to apply Newton's Second Law to situations where data are provided by an observer who may be rotating and accelerating.

5.2 Vector Differentiation

From a historical point of view, we might identify mechanics as the most fundamental of all divisions of classical physics. Certainly one of the major concerns of physicists even up to the end of the nineteenth century was providing a quantitative description of particle and rigid-body motion—i.e., dynamics. Newton's Second Law, stated at the beginning of this chapter, can be expressed in a more analytical form.

Assume that a vector $\boldsymbol{R}$ represents a directed distance from an observer at O to a re-entry body. We assume that the observer is capable of making measurements of both the magnitude and direction of $\boldsymbol{R}$ at regular and arbitrarily small time intervals Δt. Newton's Second Law states that

$$m \left.\frac{\mathrm{d}^2\boldsymbol{R}}{\mathrm{d}t^2}\right|_I = \boldsymbol{F} \tag{5.1}$$

where m is the mass of the re-entry body and $\boldsymbol{F}$ is the vector sum of the forces acting on the re-entry body. The acceleration of the re-entry body, represented by the second derivative of $\boldsymbol{R}$, must be a second derivative of the directed distance $\boldsymbol{R}$ observed in a frame that neither rotates nor accelerates due to nonfield forces. Unfortunately, most re-entry body observations discussed in this book are made from an Earth-fixed station; thus, Eq. (5.1) is not directly applicable. The subscript I designates this nonrotating, nonaccelerating observational frame. (To avoid confusion from overuse of the symbol I, the inertial frame is henceforth referred to as an f-frame, where f stands for "fixed.")

Figure 5.1 depicts directed distance measurements of a re-entry vehicle at times t, $t + \Delta t$, and $t + 2\Delta t$, i.e., it depicts $\boldsymbol{R}_t$, $\boldsymbol{R}_{t+\Delta t}$, and $\boldsymbol{R}_{t+2\Delta t}$. We can first define velocity vectors $\boldsymbol{V}$ as

$$\boldsymbol{V}_t = \lim_{\Delta t \to 0} \frac{(\boldsymbol{R}_{t+\Delta t} - \boldsymbol{R}_t)}{\Delta t} = \lim_{\Delta t \to 0} \frac{\Delta \boldsymbol{R}_t}{\Delta t} \doteq \left.\frac{\mathrm{d}\boldsymbol{R}}{\mathrm{d}t}\right|_t \tag{5.2a}$$

$$\boldsymbol{V}_{t+\Delta t} = \lim_{\Delta t \to 0} \frac{(\boldsymbol{R}_{t+2\Delta t} - \boldsymbol{R}_{t+\Delta t})}{\Delta t} = \lim_{\Delta t \to 0} \frac{\Delta \boldsymbol{R}_{t+\Delta t}}{\Delta t} = \left.\frac{\mathrm{d}\boldsymbol{R}}{\mathrm{d}t}\right|_{t+\Delta t} \tag{5.2b}$$

and an acceleration as

$$\frac{\mathrm{d}^2\boldsymbol{R}}{\mathrm{d}t^2} \doteq \lim_{\Delta t \to 0} \left[\frac{(\boldsymbol{V}_{t+\Delta t} - \boldsymbol{V}_t)}{\Delta t}\right] \tag{5.2c}$$

If the observation frame [i.e., the frame in which the vectors $\boldsymbol{R}$ and $\boldsymbol{V}$ are quantized, or (as Britting[2] says) "componentized"] is an inertial frame, then

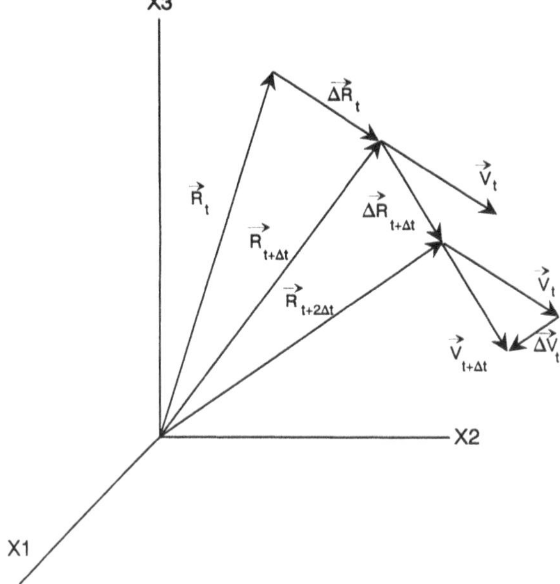

Fig. 5.1 Vector differentiation.

we can multiply the acceleration given in Eq. (5.2c) by the mass of the RV to determine the magnitude and direction of the force which caused the observed motion.

The observed motion is, of course, given by the measured components of the vector $\boldsymbol{R}$ at time intervals Δt. The analyst must then become an applied geophysicist or aerodynamicist to "explain" the origin of the observed force. The force is not "observed" as such but is deduced from positional measurements (or their equivalent) via Newton's Second Law. (The word *observed* should not be taken literally even when applied to positional measurement; the vectorial change in the directed distance from an origin to a point interior to a RV has meaning, although observation of such a point may be entirely impractical.)

The use of measurements from a noninertial frame in Eq. (5.1) will lead to apparent "forces" which cannot be justified physically. For example, applying Eq. (5.1) directly to Earth-fixed observations of an RV in exoatmospheric motion will yield forces which cannot be justified by the gravitational field model given in Chapter 3.

Figure 5.2 provides an illustration of both an inertial observation frame at O′ and a noninertial frame at O. Positions of the RV relative to each frame are given by the directed distances $\boldsymbol{R}$ and $\boldsymbol{r}$ from the inertial and moving (noninertial) frames, respectively. From Fig. 5.2 it is clear that the following vector additive rule applies:

$$\boldsymbol{R} = \boldsymbol{R}_0 + \boldsymbol{r} \tag{5.3}$$

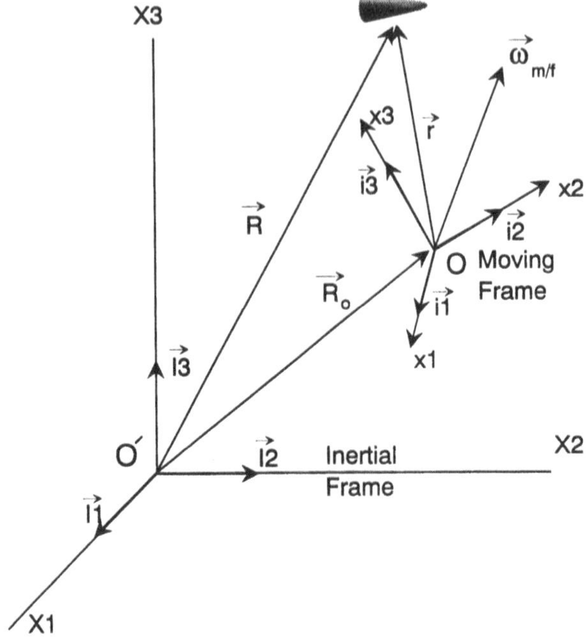

Fig. 5.2 Inertial (fixed) and moving observation frames.

Since changes in $\boldsymbol{R}$ and $\boldsymbol{R}_0$ can be "observed" in the inertial frame at O′, the inertial change in $\boldsymbol{r}$ has meaning as the vector difference of changes in $\boldsymbol{R}$ and $\boldsymbol{R}_0$. However, $\boldsymbol{r}$ is also the directed distance of the RV as observed from the moving frame at O. The vectorial changes in $\boldsymbol{r}$ as observed from the moving frame at O will not be equal to the vectorial changes in $\boldsymbol{r} = \boldsymbol{R} - \boldsymbol{R}_0$ as observed from the inertial frame at O′. The position, velocity, and acceleration of $\boldsymbol{r}$ as observed from the moving frame at O may be expressed as

$$\boldsymbol{r} = x_1\boldsymbol{i}_1 + x_2\boldsymbol{i}_2 + x_3\boldsymbol{i}_3 \tag{5.4a}$$

$$\left.\frac{\mathrm{d}\boldsymbol{r}}{\mathrm{d}t}\right|_m = \dot{x}_1\boldsymbol{i}_1 + \dot{x}_2\boldsymbol{i}_2 + \dot{x}_3\boldsymbol{i}_3 \tag{5.4b}$$

$$\left.\frac{\mathrm{d}^2\boldsymbol{r}}{\mathrm{d}t^2}\right|_m = \ddot{x}_1\boldsymbol{i}_1 + \ddot{x}_2\boldsymbol{i}_2 + \ddot{x}_3\boldsymbol{i}_3 \tag{5.4c}$$

The vector $\boldsymbol{R}$, which locates the RV from the inertial frame, and the vector $\boldsymbol{R}_0$, which locates the origin of the moving frame (or m-frame) relative to the inertial frame, may be expressed in component form in the inertial frame as follows:

$$\boldsymbol{R} = \sum_{k=1}^{3} X_k\boldsymbol{I}_k, \qquad \boldsymbol{R}_0 = \sum_{k=1}^{3} X_{0_k}\boldsymbol{I}_k \tag{5.5}$$

The vector $\boldsymbol{r}$, the directed distance from the origin of the moving frame to the re-entry body, is most conveniently expressed in terms of the m-frame coordinates as

$$\boldsymbol{r} = \sum_{k=1}^{3} x_k \boldsymbol{i}_k \tag{5.6a}$$

but could be expressed in terms of the inertial frame (or f-frame) as

$$\boldsymbol{r} = \sum_{k=1}^{3} (X_k - X_{0_k}) \boldsymbol{I}_k \tag{5.6b}$$

In usual formulations of this problem we would have available the coordinates (x_1, x_2, x_3) of the re-entry body with respect to the m-frame, and $(\omega_1, \omega_2, \omega_3)$, the angular velocity of the m-frame relative to the f-frame.

From Eqs. (5.3) and (5.5) we can write the vector $\boldsymbol{R}$ as

$$\boldsymbol{R} = \sum_{k=1}^{3} X_k \boldsymbol{I}_k = \sum_{k=1}^{3} X_{0_k} \boldsymbol{I}_k + \sum_{k=1}^{3} x_k \boldsymbol{i}_k \tag{5.7}$$

Differentiation of $\boldsymbol{R}$ with respect to inertial space gives

$$\left.\frac{\mathrm{d}\boldsymbol{R}}{\mathrm{d}t}\right|_f = \sum_{k=1}^{3} \left.\frac{\mathrm{d}X_{0_k}}{\mathrm{d}t}\right|_f \boldsymbol{I}_k + \sum_{k=1}^{3} \left.\frac{\mathrm{d}x_k}{\mathrm{d}t}\right|_m \boldsymbol{i}_k + \sum_{k=1}^{3} x_k \left.\frac{\mathrm{d}\boldsymbol{i}_k}{\mathrm{d}t}\right|_f \tag{5.8}$$

Because the m-frame rotates relative to the f-frame, the unit vectors that span the m-frame are time-varying when observed in the f-frame. Note again that we are designating the inertial frame as the f-frame. From Eq. (4.23) we have

$$\left.\frac{\mathrm{d}\boldsymbol{i}_k}{\mathrm{d}t}\right|_f = \boldsymbol{\omega}_{m/f} \times \boldsymbol{i}_k \tag{5.9}$$

Equation (5.8) then becomes

$$\left.\frac{\mathrm{d}\boldsymbol{R}}{\mathrm{d}t}\right|_f = \left.\frac{\mathrm{d}\boldsymbol{R}_0}{\mathrm{d}t}\right|_f + \sum_{k=1}^{3} \left.\frac{\mathrm{d}x_k}{\mathrm{d}t}\right|_m \boldsymbol{i}_k + \boldsymbol{\omega}_{m/f} \times \sum_{k=1}^{3} x_k \boldsymbol{i}_k$$

Collecting terms then gives

$$\left.\frac{\mathrm{d}\boldsymbol{R}}{\mathrm{d}t}\right|_f = \left.\frac{\mathrm{d}\boldsymbol{R}_0}{\mathrm{d}t}\right|_f + \left.\frac{\mathrm{d}\boldsymbol{r}}{\mathrm{d}t}\right|_m + \boldsymbol{\omega}_{m/f} \times \boldsymbol{r} \tag{5.10}$$

Note that the differential chain rule provides the second and third terms in Eq. (5.8). The second term is the velocity of the RV as measured by an observer fixed in the m-frame, and the third term accounts for frame rotation. The subscript m on the second term in Eq. (5.10) indicates the rate of change of $\boldsymbol{r}$ as observed in the m-frame, and the third term accounts for the velocity of $\boldsymbol{r}$ (relative to inertial space) due to frame rotation.

The second derivative of $\boldsymbol{R}$ with respect to the f-frame follows from Eq. (5.9):

$$\begin{aligned}
\left.\frac{\mathrm{d}^2\boldsymbol{R}}{\mathrm{d}t^2}\right|_f &= \left.\frac{\mathrm{d}^2\boldsymbol{R}_0}{\mathrm{d}t^2}\right|_f + \left.\frac{\mathrm{d}}{\mathrm{d}t}\right|_f \sum_{k=1}^{3}\left.\frac{\mathrm{d}x_k}{\mathrm{d}t}\right|_m \boldsymbol{i}_k + \left.\frac{\mathrm{d}}{\mathrm{d}t}\right|_f \left(\boldsymbol{\omega}_{m/f} \times \sum_{k=1}^{3} x_k \boldsymbol{i}_k\right) \\
&= \left.\frac{\mathrm{d}^2\boldsymbol{R}_0}{\mathrm{d}t^2}\right|_f + \sum_{k=1}^{3}\left.\frac{\mathrm{d}^2x_k}{\mathrm{d}t^2}\right|_m \boldsymbol{i}_k + \boldsymbol{\omega}_{m/f} \times \sum_{k=1}^{3}\left.\frac{\mathrm{d}x_k}{\mathrm{d}t}\right|_m \boldsymbol{i}_k \\
&\quad + \frac{\mathrm{d}\boldsymbol{\omega}_{m/f}}{\mathrm{d}t} \times \sum_{k=1}^{3} x_k \boldsymbol{i}_k + \boldsymbol{\omega}_{m/f} \times \sum_{k=1}^{3}\left.\frac{\mathrm{d}x_k}{\mathrm{d}t}\right|_m \boldsymbol{i}_k \\
&\quad + \boldsymbol{\omega}_{m/f} \times \left(\boldsymbol{\omega}_{m/f} \times \sum_{k=1}^{3} x_k \boldsymbol{i}_k\right)
\end{aligned}$$

Collecting terms gives

$$\left.\frac{\mathrm{d}^2\boldsymbol{R}}{\mathrm{d}t^2}\right|_f = \left.\frac{\mathrm{d}^2\boldsymbol{R}_0}{\mathrm{d}t^2}\right|_f + \left.\frac{\mathrm{d}^2\boldsymbol{r}}{\mathrm{d}t^2}\right|_m + \frac{\mathrm{d}\boldsymbol{\omega}_{m/f}}{\mathrm{d}t} \times \boldsymbol{r} + 2\boldsymbol{\omega}_{m/f} \times \left.\frac{\mathrm{d}\boldsymbol{r}}{\mathrm{d}t}\right|_m + \boldsymbol{\omega}_{m/f} \times (\boldsymbol{\omega}_{m/f} \times \boldsymbol{r}) \tag{5.11}$$

The first and second vector derivatives are given in Eqs. (5.10) and (5.11). We may also rewrite these equations in matrix form. The matrix form has utility in that it can easily be converted into computer code. In the vector formulation no specific reference frames are specified beyond noting relative frame rotation and linear acceleration. In our presentation so far we have specified one frame as inertial (nonrotating and nonaccelerating) and a second frame rotating and accelerating relative to the inertial frame. If we were not interested in applying Newton's Second Law, then we might consider only relative frame rotation and might have no requirement that one frame be inertial.

In order to rewrite Eqs. (5.10) and (5.11) in matrix form we assume that at any instant of time there is a directional cosine matrix (DCM) between the f- and m-frames. By following the notation of Eq. (4.30) we may write

$$\boldsymbol{R}^f = \boldsymbol{R}_0^f + \boldsymbol{r}^f = \boldsymbol{R}_0^f + C_m^f \boldsymbol{r}^m \tag{5.12}$$

Differentiation of Eq. (5.12) gives the following matrix equivalent of Eq. (5.10):

$$\frac{\mathrm{d}\boldsymbol{R}^f}{\mathrm{d}t} = \frac{\mathrm{d}\boldsymbol{R}_0^f}{\mathrm{d}t} + C_m^f\left(\frac{\mathrm{d}\boldsymbol{r}^m}{\mathrm{d}t} + \Omega_{m/f}^m \boldsymbol{r}^m\right) \tag{5.13}$$

A second differentiation yields the following matrix equivalent of Eq. (5.11):

$$\frac{d^2\boldsymbol{R}^f}{dt^2} = \frac{d^2\boldsymbol{R}_0^f}{dt^2} + C_m^f\left(\frac{d^2\boldsymbol{r}^m}{dt^2} + \frac{\Omega_{m/f}^m}{dt}\boldsymbol{r}^m + 2\Omega_{m/f}^m\frac{d\boldsymbol{r}^m}{dt} + \Omega_{m/f}^m\Omega_{m/f}^m\boldsymbol{r}^m\right) \tag{5.14}$$

Up to now the first and second derivatives of the position vector $\boldsymbol{R}$ have been expressed in terms of the position vectors $\boldsymbol{R}_0$ and $\boldsymbol{r}$ and their derivatives. In Eq. (5.10) we may regard $d\boldsymbol{r}/dt|_m$ as the change in the directed distance of the re-entry body as observed in the moving m-frame. The term $\boldsymbol{\omega}_{m/f} \times \boldsymbol{r}$ must be regarded as a correction term to account for rotation of the m-frame. Thus, the symbol $\boldsymbol{v}$ will be used to represent the velocity of the RB relative to a frame whose origin is at O but which does not share the m-frame's rotation. We have

$$\boldsymbol{v} \doteq \left.\frac{d\boldsymbol{r}}{dt}\right|_f = \left.\frac{d\boldsymbol{r}}{dt}\right|_m + \boldsymbol{\omega}_{m/f} \times \boldsymbol{r} \tag{5.15a}$$

and

$$\boldsymbol{V}_0 \doteq \left.\frac{d\boldsymbol{R}_0}{dt}\right|_f \tag{5.15b}$$

where $\boldsymbol{V}_0$ is the velocity of the origin of the m-frame. Consequently, Eqs. (5.10) and (5.11) become

$$\left.\frac{d\boldsymbol{R}}{dt}\right|_f \doteq \boldsymbol{V} = \boldsymbol{V}_0 + \boldsymbol{v} \tag{5.16a}$$

$$\begin{aligned}\frac{d^2\boldsymbol{R}}{dt^2} \doteq \left.\frac{d\boldsymbol{V}}{dt}\right|_f &= \left.\frac{d\boldsymbol{V}_0}{dt}\right|_f + \left.\frac{d\boldsymbol{v}}{dt}\right|_f \\ &= \left.\frac{d\boldsymbol{V}_0}{dt}\right|_f + \left.\frac{d\boldsymbol{v}}{dt}\right|_m + \boldsymbol{\omega}_{m/f} \times \boldsymbol{v}\end{aligned} \tag{5.16b}$$

We may carry over the above ideas into the matrix formulation by writing

$$\boldsymbol{V}^f = \boldsymbol{V}_0^f + C_m^f\boldsymbol{v}^m \tag{5.17a}$$

and

$$\boldsymbol{A}^f \doteq \frac{d\boldsymbol{V}^f}{dt} = \frac{d\boldsymbol{V}_0^f}{dt} + C_m^f\left(\frac{d\boldsymbol{v}^m}{dt} + \Omega_{m/f}^m\boldsymbol{v}^m\right) \tag{5.17b}$$

Since the velocity $\boldsymbol{V}_0$ might be expressed in the m-frame, an alternate form of Eqs. (5.17a) and (5.17b) might be

$$V^f = C^f_m [V^m_0 + v^m] \tag{5.17c}$$

$$\frac{dV^f}{dt} = C^f_m \left[\frac{dV^m_0}{dt} + \Omega^m_{m/f} V^m_0\right] + C^f_m \left[\frac{dv^m}{dt} + \Omega^m_{m/f} v^m\right] \tag{5.17d}$$

$$A^f \doteq \frac{dV^f}{dt} = C^f_m A^m$$

Table 5.1 summarizes the velocity and acceleration expressions using both vector and matrix formulations.

Table 5.1 Symbol definitions

Vector	Matrix	Description
$\boldsymbol{r}$	$\boldsymbol{r}^m$	Vector location of RV coordinates from m-frame origin
$\boldsymbol{R}_0$	$\boldsymbol{R}^f_0$	Vector location of m-frame origin in f-frame coordinates
$\boldsymbol{R}$	$\boldsymbol{R}^f$	Vector location of RV in f-frame coordinates
$\left.\frac{d\boldsymbol{r}}{dt}\right\rvert_m$	$\dot{\boldsymbol{r}}^m$	Positional time derivative, or velocity, as observed in m-frames (in m-frame coordinates)
$\boldsymbol{\omega}_{m/f}$	$\boldsymbol{\omega}^f_{m/f}$	Angular velocity of m-frame relative to f-frame (in f-frame coordinates)
$\boldsymbol{\omega}_{m/f} \times \boldsymbol{r}$	$\Omega^m_{m/f}\boldsymbol{r}^m$	Linear velocity due to rotation of the m-frame relative to the f-frame (in m-frame coordinates)
$\boldsymbol{V}_0, \left.\frac{d\boldsymbol{R}_0}{dt}\right\rvert_f$	$\boldsymbol{V}^f_0, \dot{\boldsymbol{R}}^f_0$	Velocity of the m-frame origin relative to the f-frame origin (in f-frame coordinates)
$\left.\frac{d^2\boldsymbol{r}}{dt^2}\right\rvert_m$	$\ddot{\boldsymbol{r}}^m$	Second positional time derivative, or acceleration, as observed in m-frame (in m-frame coordinates)
$\frac{d\boldsymbol{\omega}_{m/f}}{dt} \times \boldsymbol{r}$	$\dot{\Omega}^m_{m/f}\boldsymbol{r}^m$	Linear acceleration of observations in m-frame due to angular acceleration (in m-frame coordinates)
$2\boldsymbol{\omega}_{m/f} \times \left.\frac{d\boldsymbol{r}}{dt}\right\rvert_m$	$2\Omega^m_{m/f}\dot{\boldsymbol{r}}^m$	Linear acceleration of observations in m-frame due to Coriolis effect (in m-frame coordinates)
$\boldsymbol{\omega}_{m/f} \times (\boldsymbol{\omega}_{m/f} \times \boldsymbol{r})$	$\Omega^m_{m/f}\Omega^m_{m/f}\boldsymbol{r}^m$	Linear acceleration of observations in m-frame due to centrifugal acceleration (in m-frame coordinates)

5.3 Force Equations

Figure 5.3 shows a generic mass element m_i embedded in a re-entry vehicle. This element is being "observed" from a fixed frame (with origin at O′) and from a moving frame (with an origin at O). The moving frame is embedded in the RV, sharing all of the RV's angular and linear motions; any mass element fixed in the RV is stationary when observed from the m-frame. The forces acting on the generic mass element m_i are separated into an external force $\boldsymbol{F}_i$ and an internal force $\boldsymbol{F}_{ij}$, the force exerted on the element m_i by the element m_j. If a force is internal, according to Newton's Third Law, the element m_i must exert an equal and opposite force on m_j; furthermore, the line of action of these internal forces must be along the line $\boldsymbol{r}_{ij}$ connecting the mass elements. The total forces acting on m_i must equal the mass m_i times the acceleration of the element relative to inertial space. We write the acceleration in vector form but give the final result in both vector and matrix forms. From Newton's Second Law we have

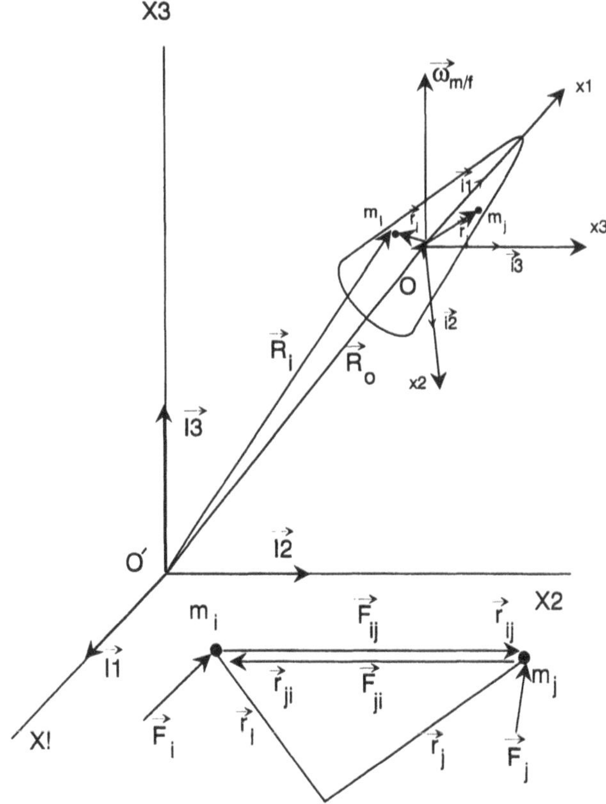

Fig. 5.3 Mass element shown in both inertial (fixed) and body (moving) frames.

$$
\begin{aligned}
\boldsymbol{F}_i + \sum_j \boldsymbol{F}_{ij} \frac{\boldsymbol{r}_{ij}}{|\boldsymbol{r}_{ij}|} = m_i \Bigg(& \left.\frac{\mathrm{d}\boldsymbol{V}_0}{\mathrm{d}t}\right|_m + \boldsymbol{\omega}_{m/f} \times \boldsymbol{V}_0 + \left.\frac{\mathrm{d}^2\boldsymbol{r}_i}{\mathrm{d}t^2}\right|_m + \frac{\mathrm{d}\boldsymbol{\omega}_{m/f}}{\mathrm{d}t} \times \boldsymbol{r}_i \\
& + 2\boldsymbol{\omega}_{m/f} \times \left.\frac{\mathrm{d}\boldsymbol{r}_i}{\mathrm{d}t}\right|_m + \boldsymbol{\omega}_{m/f} \times (\boldsymbol{\omega}_{m/f} \times \boldsymbol{r}_i) \Bigg)
\end{aligned} \tag{5.18}
$$

If we take the sum of the above equations over all generic mass elements, we must first acknowledge the disappearance of all internal forces; the internal forces $\boldsymbol{F}_{ij}$ and $\boldsymbol{F}_{ji}$ are of all equal magnitude and, by Newton's Third Law, of opposing direction, i.e.,

$$
\boldsymbol{F}_{ij} \frac{\boldsymbol{r}_{ij}}{|\boldsymbol{r}_{ij}|} = \boldsymbol{F}_{ji} \frac{-\boldsymbol{r}_{ji}}{|\boldsymbol{r}_{ji}|}
$$

Therefore

$$
\begin{aligned}
\boldsymbol{F} = \sum_i \boldsymbol{F}_i = & \left.\frac{\mathrm{d}\boldsymbol{V}_0}{\mathrm{d}t}\right|_m \sum_i m_i + \boldsymbol{\omega}_{m/f} \times \boldsymbol{V}_0 \sum_i m_i + \left.\frac{\mathrm{d}^2}{\mathrm{d}t^2}\right|_m \sum_i (\boldsymbol{r}_i m_i) \\
& + \frac{\mathrm{d}\boldsymbol{\omega}_{m/f}}{\mathrm{d}t} \times \sum_i m_i \boldsymbol{r}_i + 2\boldsymbol{\omega}_{m/f} \times \left.\frac{\mathrm{d}}{\mathrm{d}t}\right|_m \sum_i (\boldsymbol{r}_i m_i) \\
& + \boldsymbol{\omega}_{m/f} \times \left[\boldsymbol{\omega}_{m/f} \times \sum_i (m_i \boldsymbol{r}_i) \right]
\end{aligned} \tag{5.19}
$$

Let

$$
\sum_i m_i = m \qquad \boldsymbol{r}_c = \sum_i m_i \boldsymbol{r}_i / m \tag{5.20}
$$

where m is the total mass and $\boldsymbol{r}_c$ is a vector which locates the center of mass from the origin of the m-frame. We use the symbol m to indicate total mass because M will be used for the pitching moment (or, with a superscript, for the moment vector). Thus, the total external force acting on the RV may be written in vector form as

$$
\begin{aligned}
\boldsymbol{F} = m \Bigg[& \left.\frac{\mathrm{d}\boldsymbol{V}_0}{\mathrm{d}t}\right|_m + \boldsymbol{\omega}_{m/f} \times \boldsymbol{V}_0 + \left.\frac{\mathrm{d}^2\boldsymbol{r}_c}{\mathrm{d}t^2}\right|_m + \frac{\mathrm{d}\boldsymbol{\omega}_{m/f}}{\mathrm{d}t} \times \boldsymbol{r}_c \\
& + 2\boldsymbol{\omega}_{m/f} \times \left.\frac{\mathrm{d}\boldsymbol{r}_c}{\mathrm{d}t}\right|_m + \boldsymbol{\omega}_{m/f} \times (\boldsymbol{\omega}_{m/f} \times \boldsymbol{r}_c) \Bigg]
\end{aligned} \tag{5.21}
$$

and in matrix form as

$$
\begin{aligned}
\boldsymbol{F}^m = m \Bigg[& \frac{\mathrm{d}\boldsymbol{V}_0^m}{\mathrm{d}t} + \Omega_{m/f}^m \boldsymbol{V}_0^m + \frac{\mathrm{d}^2\boldsymbol{r}_c^m}{\mathrm{d}t^2} + \frac{\mathrm{d}\Omega_{m/f}^m}{\mathrm{d}t} \boldsymbol{r}_c^m \\
& + 2\Omega_{m/f}^m \frac{\mathrm{d}\boldsymbol{r}_c^m}{\mathrm{d}t} + \Omega_{m/f}^m \Omega_{m/f}^m \boldsymbol{r}_c^m \Bigg]
\end{aligned} \tag{5.22}
$$

Of course, with the origin of the m-frame fixed at the center of gravity (i.e., $\boldsymbol{r}_c = 0$, $\mathrm{d}\boldsymbol{r}_c/\mathrm{d}t = 0$ and $\mathrm{d}^2\boldsymbol{r}_c/\mathrm{d}t = 0$), Eqs. (5.21) and (5.22) become greatly simplified.

$$\boldsymbol{F} = m\left(\left.\frac{\mathrm{d}\boldsymbol{V}_0}{\mathrm{d}t}\right|_m + \boldsymbol{\omega}_{m/f} \times \boldsymbol{V}_0\right) = m\left.\frac{\mathrm{d}\boldsymbol{V}_0}{\mathrm{d}t}\right|_f \tag{5.23a}$$

$$\boldsymbol{F}^m = m\left(\frac{\mathrm{d}\boldsymbol{V}_0^m}{\mathrm{d}t} + \Omega^m_{m/f}\boldsymbol{V}_0^m\right) = m\boldsymbol{A}^m \tag{5.23b}$$

Equations (5.23) will be extensively applied in Chapter 7.

Application of the Force Equations—Example 1

Figure 5.4 illustrates an RV with a moving mass controller. A detailed discussion of the moving mass controller is given later in this chapter. For now we wish only to estimate the forces acting on this mass, which is confined in a tube normal to the axis of symmetry of the RV. The forces considered here are both those causing motion of the mass and those constraining the mass to remain within the tube. Since the tube is fixed within the RV, the tube must share all angular and linear motion of the RV.

For this example let's use the matrix formulation. We assume that the re-entry body's angular velocity vector has a constant magnitude of 12 rad/s along the X_1-axis (axis of symmetry) of the re-entry vehicle. The linear velocity is also

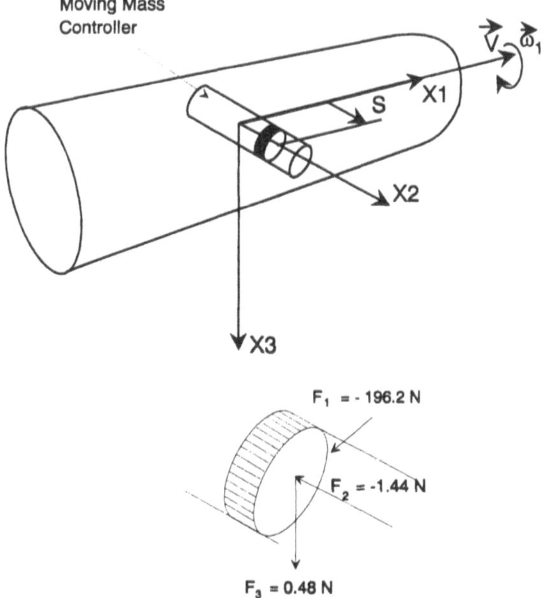

Fig. 5.4 Moving mass controller.

along the X_1-axis and has a magnitude of 4000 m/s; the linear acceleration is along the negative X_1-axis and of magnitude 20 g. The moving mass is confined to motion along a tube aligned with the X_2-axis. Consider the case where the moving mass has a displacement of $S = +0.1$ m from the X_1-axis and is moving relative to the RV at a constant velocity of 0.2 m/s in the positive X_2-direction. We want to find the forces acting on the moving mass if its mass is 0.1 kg.

From Eq. (5.22) we have the following:

$$\boldsymbol{V}^m = [4000, 0, 0]^{\mathrm{T}} \qquad \frac{\mathrm{d}\boldsymbol{V}_0^m}{\mathrm{d}t} = [-196.2, 0, 0]^{\mathrm{T}} \tag{5.24a}$$

$$\Omega^m_{m/f} = \begin{bmatrix} 0 & 0 & 0 \\ 0 & 0 & -12 \\ 0 & 12 & 0 \end{bmatrix} \qquad \frac{\mathrm{d}\Omega^m_{m/f}}{\mathrm{d}t} = \begin{bmatrix} 0 & 0 & 0 \\ 0 & 0 & 0 \\ 0 & 0 & 0 \end{bmatrix} \tag{5.24b}$$

$$\boldsymbol{r}^m = [0, 0.1, 0]^{\mathrm{T}} \qquad \frac{\mathrm{d}\boldsymbol{r}^m}{\mathrm{d}t} = [0, 0.2, 0]^{\mathrm{T}} \qquad \frac{\mathrm{d}^2\boldsymbol{r}^m}{\mathrm{d}t^2} = [0, 0, 0]^{\mathrm{T}} \tag{5.24c}$$

Equation (5.22) may now be rewritten as

$$\boldsymbol{F}^m = \begin{bmatrix} F_1 \\ F_2 \\ F_3 \end{bmatrix} = (0.1)\left(\begin{bmatrix} -196.2 \\ 0 \\ 0 \end{bmatrix} + \begin{bmatrix} 0 & 0 & 0 \\ 0 & 0 & -12 \\ 0 & -12 & 0 \end{bmatrix} \begin{bmatrix} 4000 \\ 0 \\ 0 \end{bmatrix} \right.$$
$$\left. + 2\begin{bmatrix} 0 & 0 & 0 \\ 0 & 0 & -12 \\ 0 & 12 & 0 \end{bmatrix} \begin{bmatrix} 0 \\ 0.2 \\ 0 \end{bmatrix} + \begin{bmatrix} 0 & 0 & 0 \\ 0 & 0 & -12 \\ 0 & 12 & 0 \end{bmatrix} \begin{bmatrix} 0 & 0 & 0 \\ 0 & 0 & -12 \\ 0 & -12 & 0 \end{bmatrix} \begin{bmatrix} 0 \\ 0.1 \\ 0 \end{bmatrix} \right) \tag{5.25}$$

Carrying out the indicated operations gives

$$F_1 = -196.2 \text{ N}$$

$$F_2 = -1.44 \text{ N}$$

$$F_3 = 0.48 \text{ N}$$

The above results indicate that if the mass is at a position of 0.1 m from the axis of symmetry and is moving at a constant speed (relative to the RV) of 0.2 m/s, then a force of 1.44 N must be applied in a direction opposite to the direction of motion in order to maintain this constant speed. Forces F_1 and F_3 are forces of constraint. Force F_1 acts on the mass in the negative X_1-direction; the mass, of course, exerts a force of the same magnitude in the positive X_1-direction. A little reflection will show that all forces F_1, F_2, and F_3 are internal forces.

Application of the Force Equations—Example 2

At a point P on the surface of the Earth we have erected a coordinate system such that the X_1-axis is positive to the east, the X_2-axis is positive to the north, and the X_3-axis positive up; a second coordinate frame is geocentric and inertial. Assume that we are launching a small rocket at an elevation angle E and an azimuth angle A. We assume that the performance of the rocket is limited, so that the maximum achievable altitude is a negligible fraction of the Earth's radius. Thus, without further justification, we accept a "flat Earth" model in the vicinity of the point P. The moving and inertial frames are shown in Fig. 5.5.

For this problem we will use Eq. (5.11) as a starting point. First, let's rewrite the first term on the right side as follows:

$$\left.\frac{\mathrm{d}^2\boldsymbol{R}_0}{\mathrm{d}t^2}\right|_f = \left.\frac{\mathrm{d}^2\boldsymbol{R}_0}{\mathrm{d}t^2}\right|_m + \frac{\mathrm{d}\boldsymbol{\omega}_{m/f}}{\mathrm{d}t} \times \boldsymbol{R}_0 + 2\boldsymbol{\omega}_{m/f} \times \left.\frac{\mathrm{d}\boldsymbol{R}_0}{\mathrm{d}t}\right|_m + \boldsymbol{\omega}_{m/f} \times (\boldsymbol{\omega}_{m/f} \times \boldsymbol{R}_0) \tag{5.26}$$

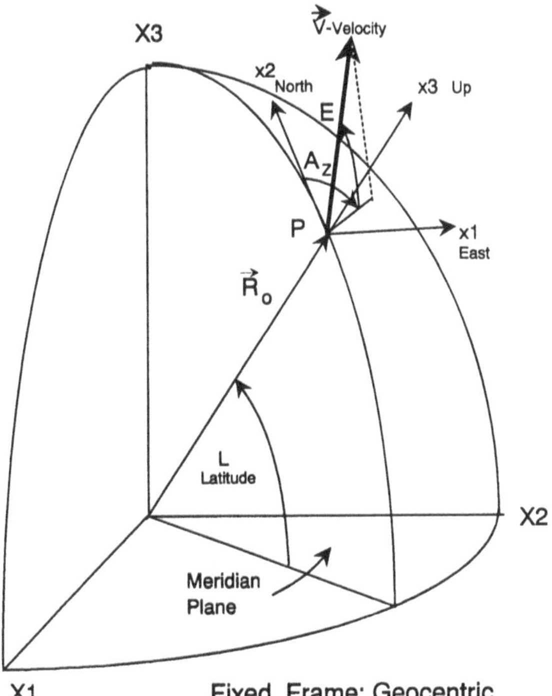

Fig. 5.5 Inertial and moving frames for rocket launch example.

The rotational rate of the moving frame at P relative to inertial space is the sidereal rotation rate of the Earth, designated as Ω_E. This angular velocity may be written as components in the m-frame as follows (see Fig. 5.6):

$$\boldsymbol{\omega}_{m/f} = [\Omega_E \cos(L)]\, \boldsymbol{i}_2 + [\Omega_E \sin(L)]\, \boldsymbol{i}_3 \tag{5.27a}$$

Of course, this rate is constant, i.e.,

$$\frac{\mathrm{d}\boldsymbol{\omega}_{m/f}}{\mathrm{d}t} = 0 \tag{5.27b}$$

Assuming a spherical Earth we have

$$\boldsymbol{R}_0 = R_E \boldsymbol{i}_3 \tag{5.28a}$$

[We could have assumed an elliptical Earth meridian through P and used Eq. (3.25) to express R_E as a function of latitude L.] Because we are assuming a spherical Earth, we can write

$$\left.\frac{\mathrm{d}\boldsymbol{R}_0}{\mathrm{d}t}\right|_m = \frac{\mathrm{d}}{\mathrm{d}t}(R_E)\boldsymbol{i}_3 = 0 \tag{5.28b}$$

Consequently, Eq. (5.26) becomes

$$\begin{aligned} \left.\frac{\mathrm{d}^2\boldsymbol{R}_0}{\mathrm{d}t^2}\right|_f &= \boldsymbol{\omega}_{m/f} \times (\boldsymbol{\omega}_{m/f} \times \boldsymbol{R}_0) = \boldsymbol{\omega}_{m/f}(\boldsymbol{\omega}_{m/f} \cdot \boldsymbol{R}_0) - \boldsymbol{R}_0 \omega^2_{m/f} \\ &= \boldsymbol{i}_2 \left[\frac{1}{2}\Omega_E^2 R_E \sin(2L)\right] - \boldsymbol{i}_3 \left[\Omega_E^2 R_E \cos(L)\right] \end{aligned} \tag{5.29}$$

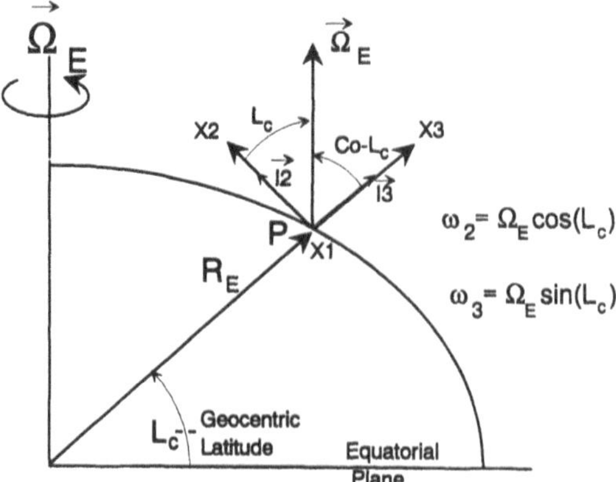

Fig. 5.6 Resolution of Earth's angular velocity in a moving frame.

Consider Eq. (5.11), and let

$$
\begin{aligned}
\boldsymbol{r} &= x_1\boldsymbol{i}_1 + x_2\boldsymbol{i}_2 + x_3\boldsymbol{i}_3 \\
\left.\frac{\mathrm{d}\boldsymbol{r}}{\mathrm{d}t}\right|_m &\doteq \boldsymbol{v} = \dot{x}_1\boldsymbol{i}_1 + \dot{x}_2\boldsymbol{i}_2 + \dot{x}_3\boldsymbol{i}_3 \\
\left.\frac{\mathrm{d}^2\boldsymbol{r}}{\mathrm{d}t^2}\right|_m &\doteq \boldsymbol{a} = \ddot{x}_1\boldsymbol{i}_1 + \ddot{x}_2\boldsymbol{i}_2 + \ddot{x}_3\boldsymbol{i}_3
\end{aligned}
\tag{5.30}
$$

where $\boldsymbol{v}$ and $\boldsymbol{a}$ are the velocity and acceleration, respectively, as observed from the moving frame. Of course, the product of the mass of the rocket and $\boldsymbol{a}$ may not be equated to the acting forces since $\boldsymbol{a}$ is the acceleration of the RV observed in a noninertial frame.

We recognize that the left side of Eq. (5.11) is the acceleration of the rocket relative to inertial space and, hence, by Newton's Second Law, is the force per unit mass associated with the rocket; i.e.,

$$
\left.\frac{\mathrm{d}^2\boldsymbol{R}}{\mathrm{d}t^2}\right|_f = -G\boldsymbol{i}_3 + \frac{(T-D)}{m}\left(\frac{\boldsymbol{v}}{|\boldsymbol{v}|}\right) \tag{5.31}
$$

where G is the gravitational acceleration (which we will assume to be a constant), T is the thrust, and D is the drag. Note that we have assumed that both T and D act in the same direction as the vector $\boldsymbol{v}$; dividing $\boldsymbol{v}$ by its magnitude provides a unit velocity vector. Equation (5.11) may then be written as

$$
\begin{aligned}
-G\boldsymbol{i}_3 + \frac{(T-D)}{m}\left(\frac{\boldsymbol{v}}{|\boldsymbol{v}|}\right) &= \boldsymbol{\omega}_{m/f} \times (\boldsymbol{\omega}_{m/f} \times \boldsymbol{R}_0) + \boldsymbol{a} \\
&\quad + 2\boldsymbol{\omega}_{m/f} \times \boldsymbol{v} + \boldsymbol{\omega}_{m/f} \times (\boldsymbol{\omega}_{m/f} \times \boldsymbol{r})
\end{aligned}
\tag{5.32}
$$

The second term on the right has been expanded in Eq. (5.29). Note that we can ignore the fourth term in comparison to the first term because we have assumed that $\boldsymbol{r} \ll \boldsymbol{R}_0$.

Upon examining Eq. (5.30) we can see that it would be relatively straightforward to replace $\boldsymbol{a}$ and $\boldsymbol{v}$ with derivatives of positional variables in the m-frame. The result would be three second-order differential equations in terms of x_1, x_2, and x_3. These three equations might be solved numerically, providing great practice in computer coding. However, an approximate solution may be obtained by assuming the elementary "nonrotating Earth" value for the velocity. Such an approximation has some validity for a no-thrust, no-drag trajectory. Thus, by assuming $T = 0$, $D = 0$, and $\boldsymbol{\omega}_{m/f} = 0$ we get

$$
\ddot{x}_1 = 0, \qquad \ddot{x}_2 = 0, \qquad \ddot{x}_3 = -G \tag{5.33a}
$$

which integrate to yield the following velocity magnitude components:

$$\dot{x}_1 = V\cos(E)\sin(A) \qquad \dot{x}_2 = V\cos(E)\cos(A) \qquad \dot{x}_3 = V\sin(E) - Gt \tag{5.33b}$$

Remember that V, A, and E are the initial values of velocity magnitude, azimuth, and elevation (see Fig. 5.7). The x_3 term integrates to

$$x_3 = [V\sin(E)]\,t - \tfrac{1}{2}Gt^2 \tag{5.33c}$$

which may be solved for the time-of-flight, $t = t_f$, which is given by

$$t_f = 2V\sin(E)/G \tag{5.33d}$$

Equations (5.33b) may be inserted into the third term on the right of Eq. (5.32) to give the following approximation:

$$\begin{aligned}2\boldsymbol{\omega}_{m/f} \times \boldsymbol{v} \approx{}& 2\{[\Omega_E\cos(L)]\boldsymbol{i}_2 + [\Omega_E\sin(L)]\boldsymbol{i}_3\} \times \{[V\cos(E)\sin(A)]\boldsymbol{i}_1 \\ &+ [V\cos(E)\cos(A)]\boldsymbol{i}_2 + [V\sin(E) - Gt]\boldsymbol{i}_3\} \\ ={}& 2\boldsymbol{i}_1\{(V\Omega)_E)[\cos(L)\sin(E) - \cos(E)\cos(A)\sin(L)] \\ &- [G\Omega_E\cos(L)]t\} \\ &+ 2\boldsymbol{i}_2\{(V\Omega_E)[\sin(L)\cos(E)\sin(A)]\} \\ &+ 2\boldsymbol{i}_3\{(V\Omega_E)[-\cos(E)\sin(A)\cos(L)]\}\end{aligned} \tag{5.34}$$

Suppose we have available the initial values of V (velocity magnitude), L (latitude), A (azimuth), and E (elevation). We are interested only in determining the impact point of the rocket. More precisely, we are interested in the difference

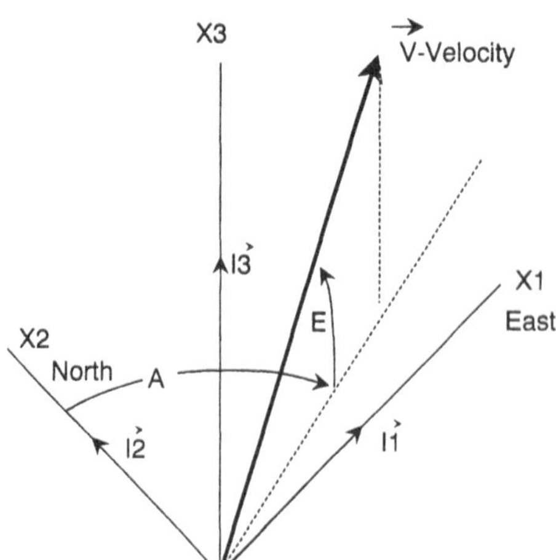

Fig. 5.7 Resolution of velocity in moving frame.

between two impact points; one impact point is calculated by taking into account the Earth's rotation rate, and the other is calculated by ignoring the rotation. It is of course this Earth rotation which prevents the frame at P from being an inertial frame. Inserting Eqs. (5.29), (5.30), and (5.34) into Eq. (5.32) gives

$$\ddot{x}_1 + 2V\Omega_E[\cos(L)\sin(E) - \sin(L)\cos(E)\cos(A)] - [2G\Omega_E\cos(L)]t = 0$$

$$\ddot{x}_2 + \tfrac{1}{2}\Omega_E^2 R_E \sin(2L) + 2V\Omega_E \sin(L)\cos(E)\sin(A) = 0 \qquad (5.35)$$

We can integrate Eqs. (5.35), ignoring the initial conditions given in Eqs. (5.33b). The resulting coordinates x_1 and x_2 will give the location of the rotation impact point relative to the no-rotation impact point. If Eqs. (5.35) are integrated and evaluated at $t = t_f$, we get the following relative displacement of the rotation impact point:

$$x_1 = -\Omega_E\left\{V[\cos(L)\sin(E) - \sin(L)\cos(E)\cos(A)]\,t_f^2 - G\cos(L)\frac{t_f^3}{3}\right\}$$

$$x_2 = -\tfrac{1}{4}[\Omega_E^2 R_E \sin(2L)]t_f^2 - [V\Omega_E \sin(L)\cos(E)\sin(A)]t_f^2 \qquad (5.36)$$

Suppose an object were launched toward the east under the following conditions:

$$A = 90°, \qquad G = 9.8\ \text{m/s}^2, \qquad R_E = 6.378 \times 10^6\ \text{m}$$

$$E = 45°, \qquad V = 300\ \text{m/s}, \qquad L = 45°$$

$$\Omega_E = 7.2722 \times 10^{-5}\ \text{rad/s}$$

From Eq. (5.33d) the time of flight t_f is

$$t_f = 2V\sin(E)/G = (2)(300)\sin(45)/9.81 = 43.3\ \text{s}$$

Equations (5.36) then give

$$x_1 = -6.8\ \text{m}, \qquad x_2 = -36.2\ \text{m}$$

Thus, if the impact point is calculated under the assumption that Earth rotation may be ignored, the actual impact point will lie about 36 m to the south and 7 m to the west (see Fig. 5.8). Of course, including the presence of atmospheric drag will reduce this difference, principally by reducing the time of flight. However, this exercise does indicate that improper assumptions concerning the existence of an inertial frame can result in fairly significant errors.

5.4 Moment Equations

In this section we derive the equations for the angular momentum of a rigid body as shown in Fig. 5.3. We have an m-frame (fixed to the RV) and an inertial

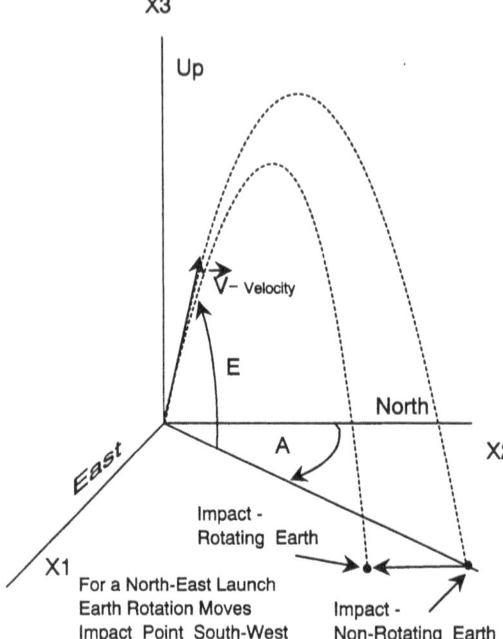

Fig. 5.8 Impact points for nonrotating and rotating Earth.

frame for describing motion. The vector $\boldsymbol{R}_0$ is the directed distance to the origin of the body frame; $\boldsymbol{R}_i$ locates a generic mass element m_i in the inertial frame, and $\boldsymbol{r}_i$ locates the same element in the moving frame. The relationship among these three vectors is

$$\boldsymbol{R}_i = \boldsymbol{R}_0 + \boldsymbol{r}_i \tag{5.37}$$

The angular momentum of the element m_i is the moment of linear momentum, which is given by

$$\boldsymbol{H}_i = \boldsymbol{R}_i \times m_i \left.\frac{\mathrm{d}\boldsymbol{R}_i}{\mathrm{d}t}\right|_f \tag{5.38}$$

Inserting Eq. (5.37) into Eq. (5.38) and applying Eq. (5.10) gives

$$\begin{aligned} \boldsymbol{H}_i = {} & (\boldsymbol{R}_0 \times \boldsymbol{V}_0)m_i + \boldsymbol{R}_0 \times m_i \left.\frac{\mathrm{d}\boldsymbol{r}_i}{\mathrm{d}t}\right|_m + \boldsymbol{R}_0 \times (\boldsymbol{\omega}_{m/f} \times m_i \boldsymbol{r}_i) \\ & + m_i \boldsymbol{r}_i \times \boldsymbol{V}_0 + \boldsymbol{r}_i \times m_i \left.\frac{\mathrm{d}\boldsymbol{r}_i}{\mathrm{d}t}\right|_m + \boldsymbol{r}_i \times (\boldsymbol{\omega}_{m/f} \times m_i \boldsymbol{r}_i) \end{aligned} \tag{5.39}$$

In Chapter 6 we will regard the RV to be a point mass; under such circumstances, $\boldsymbol{R}_0$ is the distance from the Earth's center to the RV, and all terms but the first vanish from Eq. (5.39). In this chapter, however, the vector $\boldsymbol{R}_0$ is arbitrary; hence, all terms containing $\boldsymbol{R}_0$ are arbitrary. Thus, we can set $\boldsymbol{R}_0$ to zero; consequently, the first three terms of Eq. (5.39) vanish. The angular momentum $\boldsymbol{H}$ is the summation over all the mass elements of the last three terms of Eq. (5.39), i.e.,

$$\boldsymbol{H} = \sum_i \boldsymbol{H}_i = \sum_i [(m_i \boldsymbol{r}_i) \times \boldsymbol{V}_0] + \sum_i \left[m_i \boldsymbol{r}_i \times \left.\frac{\mathrm{d}\boldsymbol{r}_i}{\mathrm{d}t}\right|_m \right] + \sum_i [\boldsymbol{r}_i \times (\boldsymbol{\omega}_{m/f} \times m_i \boldsymbol{r}_i)] \tag{5.40}$$

Note that if the RV is rigid then the second term vanishes; if the origin of the axis system is at the center of mass, then the first term vanishes as well.

Before applying Eq. (5.40) we must show that the moment is equal to the time derivative of the angular momentum relative to inertial space. If $\boldsymbol{M}$ is the moment, we prove our point through the following sequence:

$$\begin{aligned} \boldsymbol{M} &= \sum_i \boldsymbol{R}_i \times \boldsymbol{F}_i = \sum_i \boldsymbol{R} \times m_i \left.\frac{\mathrm{d}^2\boldsymbol{R}_i}{\mathrm{d}t^2}\right|_f = \sum_i \left(\boldsymbol{R}_i \times m_i \left.\frac{\mathrm{d}^2\boldsymbol{R}_i}{\mathrm{d}t^2}\right|_f \right. \\ &\quad \left. + m_i \left.\frac{\mathrm{d}\boldsymbol{R}_i}{\mathrm{d}t}\right|_f \times \left.\frac{\mathrm{d}\boldsymbol{R}_i}{\mathrm{d}t}\right|_f \right) = \frac{\mathrm{d}}{\mathrm{d}t}\left(\boldsymbol{R}_i \times m_i \left.\frac{\mathrm{d}\boldsymbol{R}_i}{\mathrm{d}t}\right|_f \right) \\ &= \left.\frac{\mathrm{d}}{\mathrm{d}t}\right|_f \sum \boldsymbol{H}_i = \left.\frac{\mathrm{d}\boldsymbol{H}}{\mathrm{d}t}\right|_f = \left.\frac{\mathrm{d}\boldsymbol{H}}{\mathrm{d}t}\right|_m + \boldsymbol{\omega}_{m/f} \times \boldsymbol{H} \end{aligned} \tag{5.41}$$

Obviously, applying Eq. (5.41) with all three terms in Eq. (5.40) would result in a lengthy expression. Rather than indulging in such an exercise, lets look at some cases of major interest.

Application of the Moment Equations—Example 1

Assume that an RV is controlled by means of a moving mass. When the control mass is moved a distance S along the X_2-axis, the center of mass of the RV moves to a position e units normal to the plane of configurational symmetry. To simplify calculations we assume that at null conditions the moving mass is located at the vehicle's center of mass. Let's also assume that the RV has no angular velocity. This situation is depicted in Fig. 5.9 and may be expressed as follows:

$$\sum_i m_i \boldsymbol{r}_i = m\boldsymbol{r}_c = me\boldsymbol{i}_2$$

$$\boldsymbol{\omega}_{m/f} = 0$$

$$\left.\frac{\mathrm{d}\boldsymbol{r}_i}{\mathrm{d}t}\right|_m = 0$$

where m is the total vehicle mass.

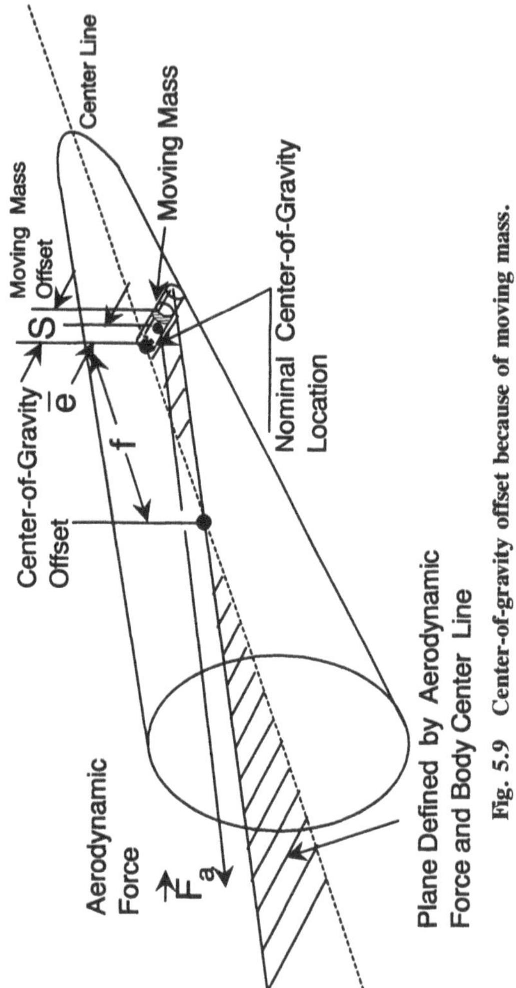

Fig. 5.9 Center-of-gravity offset because of moving mass.

Equation (5.40) becomes

$$\boldsymbol{H} = me\boldsymbol{i}_2 \times \boldsymbol{V}_0$$

Taking the derivative of the angular momentum according to Eq. (5.41) gives the moment acting at the origin of the moving (body-fixed) frame. The origin of the moving frame is at the center of mass, with the control mass in a null position (on the RV's axis of symmetry), so

$$\boldsymbol{M} = me\boldsymbol{i}_2 \times \boldsymbol{a}_0 \tag{5.42}$$

where

$$\boldsymbol{a}_0 \doteq \left.\frac{d\boldsymbol{V}_0}{dt}\right|_f = \left.\frac{d\boldsymbol{V}_0}{dt}\right|_m + \boldsymbol{\omega}_{m/f} \times \boldsymbol{V}_0 = \left.\frac{d\boldsymbol{V}_0}{dt}\right|_m$$

because $\boldsymbol{\omega}_{m/f} = 0$. We first recognize that movement of the control mass changes the mass distribution of the RV but does not affect the shape (and aerodynamic pressure distribution). Therefore, the aerodynamic moment may be represented as a force acting at a center of pressure. The center of pressure remains on the axis of symmetry or within the plane of symmetry of the re-entry vehicle. Thus,

$$\boldsymbol{M} = (-f\boldsymbol{i}_1) \times \boldsymbol{F}_a \tag{5.43}$$

where f locates the center of pressure from the null center of gravity.

Let the applied force and acceleration be resolved in the body frame as follows:

$$\boldsymbol{a}_0 = a_1\boldsymbol{i}_1 + a_2\boldsymbol{i}_2 + a_3\boldsymbol{i}_3 \tag{5.44a}$$

$$\boldsymbol{F}_a = F_1\boldsymbol{i}_1 + F_2\boldsymbol{i}_2 + F_3\boldsymbol{i}_3 \tag{5.44b}$$

Inserting Eqs. (5.44) into Eqs. (5.42) and (5.43) and then equating the result gives

$$me\boldsymbol{i}_2 \times (a_1\boldsymbol{i}_1 + a_2\boldsymbol{i}_2 + a_3\boldsymbol{i}_3) = (-f\boldsymbol{i}_1) \times (F_1\boldsymbol{i}_1 + F_2\boldsymbol{i}_2 + F_3\boldsymbol{i}_3)$$

The result is

$$\boldsymbol{i}_1(mea_3) = 0, \qquad \boldsymbol{i}_2(fF_3) = 0, \qquad \boldsymbol{i}_3(-mea_1 + fF_2) = 0$$

We conclude that

$$a_3 = 0, \qquad F_3 = 0, \qquad e/f = F_2/ma_1 = F_2/F_1 \tag{5.45}$$

Under conditions of trim (i.e., zero net moment acting on the vehicle) the aerodynamic force must pass through both the center of pressure and the center of gravity. Since the RV's center of mass is moved a distance e units along the positive i_2-axis, the aerodynamic force no longer is along the axis of symmetry (or in the plane of symmetry, as the case may be). It might be of interest to calculate the lateral force (expressed in terms of a coefficient) that can be achieved by a given lateral center-of-mass offset in an RV which retains configurational symmetry. For an angle of attack, α, the yawing moment coefficient may be written as

$$C_m = C_{m_0} + \left[\frac{dC_m}{d\alpha}\bigg|_{\alpha=0} \alpha \right] = 0$$

where we have set $C_m = 0$ because the RV is assumed to be trimmed. Strictly speaking, we should use β for the aerodynamic yaw angle and C_n for the yawing moment, but for a body of revolution, pitching and yawing are equivalent. Solving for the trim angle of attack α_T gives

$$\alpha_T = -\left[C_{m_0} \bigg/ \left(\frac{dC_m}{d\alpha} \right) \right] \tag{5.46}$$

C_{m_0} is the moment coefficient that exists when the angle of attack is zero. This moment (taken about the center of gravity) is

$$C_{m_0} = C_D(e/L) \tag{5.47}$$

where C_D is the drag coefficient and where L is the length of the RV and also the reference length in the moment coefficient definition. C_m, C_D, and C_n may be expressed as follows:

$$C_m = \frac{M}{\frac{1}{2}\rho V^2 SL}, \qquad C_D = \frac{D}{\frac{1}{2}\rho V^2 S}, \qquad C_n = \frac{N}{\frac{1}{2}\rho V^2 SL} \tag{5.48}$$

where D and N are the drag force and yawing moment, respectively, where ρ is the atmospheric density, V is the velocity magnitude, and S is the reference area. It will be shown in a later chapter that

$$\frac{dC_m}{d\alpha} = \frac{dC_N}{d\alpha}\left(\frac{f}{L}\right) \tag{5.49}$$

where C_N is the normal force coefficient. Thus, inserting Eqs. (5.47) and (5.49) into Eq. (5.46) gives

$$\alpha_T = \left[C_D \bigg/ \left(\frac{dC_N}{d\alpha} \right) \right]\left(\frac{e}{f}\right) \tag{5.50a}$$

Typical values of these parameters are as follows:

$$e = 0.003 \text{ m} \qquad f = 0.015 \text{ m}$$

$$\frac{\mathrm{d}C_N}{\mathrm{d}\alpha} = 1.5/\text{rad} \qquad C_D = 0.10$$

The trim angle of attack for these values would be

$$\alpha_T = \left(\frac{0.003}{0.015}\right)\left(\frac{0.1}{1.5}\right) = 0.0133 \text{ rad} = 0.764 \text{ deg}$$

The normal force coefficient C_N follows from Eq. (5.50a) as

$$C_N = C_D(e/f) \tag{5.50b}$$

The normal force that is available for lateral acceleration is proportional to the drag coefficient and the ratio of the center-of-mass offset to the center of pressure, e/f. Equation (5.50b) illustrates the classical compromise that usually exists between maneuverability and stability. A large value of f indicates large directional stability. However, to achieve a given normal force coefficient, a large value of f requires a large value of e. Increasing the value of e requires either increasing the control mass or the distance through which the mass must be moved. In either case, increasing e means increasing the control actuation time.

To complete the moment equations we must return to Eq. (5.40). At this point we make the assumption that the re-entry vehicle is rigid and that the origin of the body-fixed axis system is at the center of mass. Hence, the first two terms of Eq. (5.40) vanish, leaving

$$\begin{aligned} \boldsymbol{H} &= \sum_i \boldsymbol{H}_i = \sum_i \boldsymbol{r}_i \times (\boldsymbol{\omega}_{m/f} \times m_i \boldsymbol{r}_i) \\ &= \sum_i \left[\boldsymbol{\omega}_{m/f}(r_i^2) - \boldsymbol{r}_i(\boldsymbol{r}_i \cdot \boldsymbol{\omega}_{m/f})\right] m_i \end{aligned} \tag{5.51a}$$

or in matrix form

$$\boldsymbol{H}^m = \left[\sum m_i \left(\boldsymbol{r}_i^{\mathrm{T}} \boldsymbol{r}_i \boldsymbol{\nu} - \boldsymbol{r}_i \boldsymbol{r}_i^{\mathrm{T}}\right)\right] \boldsymbol{\omega}_{m/f}^m \tag{5.51b}$$

where $\boldsymbol{\nu}$ is the identity matrix. The bracketed term will be shown to be the inertia tensor. Now the location of a generic mass particle m_i is given as

$$\boldsymbol{r}_i = x_{1_i} \boldsymbol{i}_1 + x_{2_i} \boldsymbol{i}_2 + x_{3_i} \boldsymbol{i}_3 \tag{5.52a}$$

and the angular velocity may be represented in vector form as

$$\boldsymbol{\omega}_{m/f} = \omega_1 \boldsymbol{i}_1 + \omega_2 \boldsymbol{i}_2 + \omega_3 \boldsymbol{i}_3 \tag{5.52b}$$

Inserting Eqs. (5.52) into Eq. (5.51a) gives

$$\begin{aligned}\boldsymbol{H} = &\left[\omega_1 \sum_i \left(x_{2_i}^2 + x_{3_i}^2\right) m_i - \omega_2 \sum_i x_{1_i} x_{2_i} m_i - \omega_3 \sum_i x_{1_i} x_{3_i} m_i\right] \boldsymbol{i}_i \\ &+ \left[-\omega_1 \sum_i x_{2_i} x_{1_i} m_i + \omega_2 \sum_i \left(x_{1_i}^2 + x_{3_i}^2\right) m_i - \omega_3 \sum_i x_{2_i} x_{3_i} m_i\right] \boldsymbol{i}_2 \\ &+ \left[-\omega_1 \sum_i x_{3_i} x_{1_i} m_i - \omega_2 \sum_i x_{3_i} x_{2_i} m_i + \omega_3 \sum_i \left(x_{1_i}^2 + x_{2_i}^2\right) m_i\right] \boldsymbol{i}_3\end{aligned} \tag{5.53}$$

The summations in the above equation may be represented concisely as the moments and products of inertia.

$$\begin{aligned} I_{11} &= \sum_i m_i \left(x_{2_i}^2 + x_{3_i}^2\right) \\ I_{22} &= \sum_i m_i \left(x_{1_i}^2 + x_{3_i}^2\right) \\ I_{33} &= \sum_i m_i \left(x_{1_i}^2 + x_{2_i}^2\right) \end{aligned} \tag{5.54}$$

$$I_{kl} = \sum_i m_i x_{k_i} x_{l_i} \qquad \text{for} \qquad k \neq l$$

The above terms permit Eq. (5.53) to be rewritten as

$$\begin{aligned} \boldsymbol{H} = &\ \boldsymbol{i}_1 (\omega_1 I_{11} - \omega_2 I_{12} - \omega_3 I_{13}) \\ &+ \boldsymbol{i}_2 (-\omega_1 I_{21} + \omega_2 I_{22} - \omega_3 I_{23}) \\ &+ \boldsymbol{i}_3 (-\omega_1 I_{31} - \omega_2 I_{32} + \omega_3 I_{33}) \end{aligned} \tag{5.55a}$$

or in matrix form as

$$\boldsymbol{H}^m = \begin{bmatrix} H_1 \\ H_2 \\ H_3 \end{bmatrix} = \begin{bmatrix} I_{11} & -I_{12} & -I_{13} \\ -I_{21} & I_{22} & -I_{23} \\ -I_{31} & -I_{32} & I_{33} \end{bmatrix} \begin{bmatrix} \omega_1 \\ \omega_2 \\ \omega_3 \end{bmatrix} = \boldsymbol{I} \boldsymbol{\omega}_{m/f}^m \tag{5.55b}$$

where the 3×3 matrix is sometimes called the *inertia tensor*. The moment equation may be written from Eq. (5.41) as

$$\boldsymbol{M} = \left.\frac{\mathrm{d}\boldsymbol{H}}{\mathrm{d}t}\right|_m + \boldsymbol{\omega}_{m/f} \times \boldsymbol{H} \tag{5.56a}$$

Equation (5.55a) may be inserted into (5.56a) to give a vector formulation of the moment equations. For the matrix form, we have replaced the vector cross product with the matrix cross product from Eq. (4.29) as follows:

$$M^m = \begin{bmatrix} M_1 \\ M_2 \\ M_3 \end{bmatrix} = I\left(\frac{d\omega^m_{m/f}}{dt}\right) + \Omega^m_{m/f} I \omega^m_{m/f} \tag{5.56b}$$

5.5 Calculation of the Moments and Products of Inertia

The inertia tensor was identified in the last section; an alternate tensor representation is the quadratic form. Figure 5.10 illustrates an arbitrary line L, with unit vector $\boldsymbol{u}$ designating the direction. A mass element m_i is located by the vector $\boldsymbol{r}_i$ in the cartesian frame. The moment of inertia of the body about the line L is

$$I_{LL} = \sum_i m_i \rho_i^2 \tag{5.57}$$

where $\boldsymbol{\rho}_i$ is the vector to the mass element m_i in the direction of the normal to the line L. Equation (5.57) follows from the definitions given in Eqs. (5.54). Our goal is to express I_{LL} in terms of the moments and products of inertia. It is clear that $\boldsymbol{\rho}_i$ is formed from the cross product of $\boldsymbol{r}_i$ with $\boldsymbol{u}$, i.e.,

$$\boldsymbol{\rho}_i = \boldsymbol{r}_i \times \boldsymbol{u} \tag{5.58}$$

Let's assume that the vectors $\boldsymbol{r}_i$ and $\boldsymbol{u}$ may be written as components in the reference frame as

$$\boldsymbol{r}_i = x_{1_i}\boldsymbol{i}_1 + x_{2_i}\boldsymbol{i}_2 + x_{3_i}\boldsymbol{i}_3 \tag{5.59a}$$

$$\boldsymbol{u} = u_1\boldsymbol{i}_1 + u_2\boldsymbol{i}_2 + u_3\boldsymbol{i}_3 \tag{5.59b}$$

$$u_1 = \cos(\alpha_1), \qquad u_2 = \cos(\alpha_2), \qquad u_3 = \cos(\alpha_3) \tag{5.59c}$$

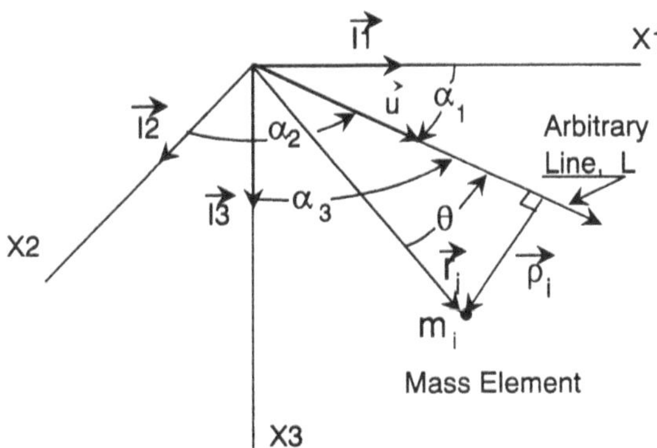

Fig. 5.10 Location of mass element relative to an arbitrary line.

Inserting Eqs. (5.59a) and (5.59b) into Eq. (5.58) and carrying out the extensive manipulations gives the following quadratic form of the inertial tensor:

$$I_{LL} = I_{11}u_1^2 + I_{22}u_2^2 + I_{33}u_3^2 - 2I_{12}u_1u_2 - 2I_{13}u_1u_3 - 2I_{23}u_2u_3 \quad (5.60)$$

We will return to this expression when we discuss torque-free motion in a later chapter.

It should be clear that as the body changes orientation relative to any reference frame, the moments/products of inertia will also change. The components of the unit vector will also change.

For certain orientations of the body relative to the frame, we might expect that the products of inertia will vanish and that the inertia matrix will therefore be diagonalized. The frame orientation under such circumstances is called the *principal axis system*. Equation (5.60) is often used to find the moment of inertia about an arbitrary line. A more general problem might be stated as follows: if the moment-of-inertia matrix is given for one frame and we have a directional cosine matrix relating this frame to a second frame, what is the inertia matrix in this second frame?

Assume we have body-fixed a- and b-frames. Also, we know the DCM between these frames, say, C_a^b. We can relate the angular momentum and angular velocities available in the a-frame to their counterparts in the b-frame as

$$\boldsymbol{H}^a = C_b^a \boldsymbol{H}^b, \qquad \boldsymbol{\omega}_{m/f}^a = C_b^a \boldsymbol{\omega}_{m/f}^b \quad (5.61)$$

but from Eq. (5.55b) we have

$$\boldsymbol{H}^a = I^a \boldsymbol{\omega}_{m/f}^a, \qquad \boldsymbol{H}^b = I^b \boldsymbol{\omega}_{m/f}^b \quad (5.62)$$

By inserting Eqs. (5.61) into the first of Eqs. (5.62) and comparing the results with the second of Eqs. (5.61), we get

$$I^b = C_a^b I^a C_b^a \quad (5.63)$$

This transformation from the a-frame to the b-frame has been encountered before in the transformation of the angular velocity in skew-symmetric form [Eq. (4.34)].

An interesting classical problem might be stated as follows: given an inertia matrix whose elements have been quantized in an arbitrary a-frame, what must the orientation of a b-frame be relative to the a-frame so that the inertia matrix is diagonalized in the b-frame? The orientation of the a-frame relative to the b-frame is given by the directional cosine matrix which transforms vector components in the a-frame into the equivalent components in the b-frame. This requires solving the eigenvalue problem and will not be discussed here.

Application of the Moment Equations—Example 2

Suppose that the re-entry vehicle shown in Fig. 5.11 has both configurational and mass symmetry about the x_1^a-axis. The RV is mounted in a test machine and

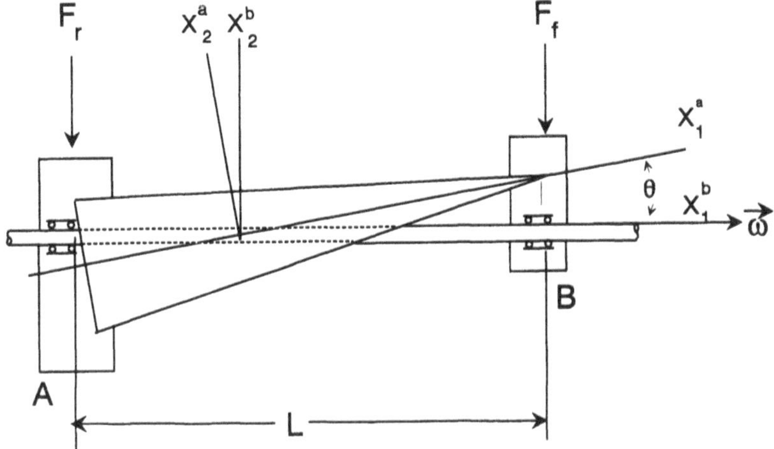

Fig. 5.11 Spin of a re-entry body about nonprincipal axis.

spun about an axis x_1^b at a constant rate of ω rad/s. The x_1^b-axis is offset from the x_1^a-axis by a small angle θ. The goal is to find the forces due to spinning alone at points A and B on the shaft. We wish to express the force in terms of the spin rate ω, the axial and transverse moments of inertia in the a-frame (I_a and I_t), and the offset angle θ.

The inertia tensor of the RV in the a-frame is

$$I^a = \begin{bmatrix} I_a & 0 & 0 \\ 0 & I_t & 0 \\ 0 & 0 & I_t \end{bmatrix} \tag{5.64}$$

The DCM from the a-frame to the b-frame is

$$C_a^b = \begin{bmatrix} \cos(\theta) & -\sin(\theta) & 0 \\ \sin(\theta) & \cos(\theta) & 0 \\ 0 & 0 & 1 \end{bmatrix} \tag{5.65}$$

From Eq. (5.64) we have

$$I^b = C_a^b I^a C_b^a$$

$$I^b = \begin{bmatrix} I_a \cos^2(\theta) + I_t \sin^2(\theta) & (I_a - I_t)\sin(\theta)\cos(\theta) & 0 \\ (I_a - I_t)\sin(\theta)\cos(\theta) & I_a \sin^2(\theta) + I_t \cos^2(\theta) & 0 \\ 0 & 0 & I_t \end{bmatrix} \tag{5.66a}$$

which becomes, upon using the small angle approximation,

$$I^b \approx \begin{vmatrix} I_a & (I_a - I_t)\theta & 0 \\ (I_a - I_t)\theta & I_t & 0 \\ 0 & 0 & I_t \end{vmatrix} \tag{5.66b}$$

Next we apply Eq. (5.56b) to obtain the moments of constraint in the b-frame:

$$M^b = \begin{bmatrix} M_1 \\ M_2 \\ M_3 \end{bmatrix} = I^b \begin{bmatrix} \dot{\omega} = 0 \\ 0 \\ 0 \end{bmatrix} + \begin{bmatrix} 0 & 0 & 0 \\ 0 & 0 & -\omega \\ 0 & \omega & 0 \end{bmatrix} I^b \begin{bmatrix} \omega \\ 0 \\ 0 \end{bmatrix} \tag{5.67a}$$

where $\boldsymbol{\omega} = [\omega, 0, 0]^{\mathrm{T}}$ and is a constant in both magnitude and direction. Inserting Eq (5.66b) gives

$$M^b = \begin{bmatrix} M_1 \\ M_2 \\ M_3 \end{bmatrix} = \begin{bmatrix} 0 & 0 & 0 \\ 0 & 0 & -\omega \\ 0 & \omega & 0 \end{bmatrix} \begin{bmatrix} I_a\omega \\ (I_a - I_t)\theta\omega \\ 0 \end{bmatrix} = \begin{bmatrix} 0 \\ 0 \\ (I_a - I_t)\omega^2\theta \end{bmatrix} \tag{5.67b}$$

Thus, there is a moment about the x_3^b-axis. This moment is a pure couple, so the front and rear forces $\boldsymbol{F}_f$ and $\boldsymbol{F}_r$ are equal in magnitude, i.e.,

$$|\boldsymbol{F}_f| = |\boldsymbol{F}_r| = F = \frac{\omega^2\theta}{L}(I_a - I_t) \tag{5.68}$$

It must be emphasized that the moment M_3 is the moment applied at the collets so that the spin vector will remain along the x_1^b-axis. Hence, the force F_f is in the positive x_2^b-direction, and the force $\boldsymbol{F}_r$ is in the negative x_2^b-direction.

Application of the Moment Equations—Example 3

Earlier we discussed the concept of controlling a re-entry vehicle by offsetting the center of mass. Another option is to use an aerodynamic control system that is very similar to an aircraft elevator. Figure 5.12 shows a split-windward flap system. A cutout at the rear of the RV provides a configurational asymmetry which results in a nonzero trim angle of attack. Flap deflection alters this asymmetry, giving control over the magnitude of this trim angle of attack. Of course, there are dynamics associated with changes in the flap setting, as well as coupling with RV angular velocity and linear acceleration. Our goal here is to obtain the hinge moment equations, or the set of equations that allow load assessment of the moment about the flap hinge.

It is useful to locate the generic mass element m_i three different ways. From the hinge line, its location may be expressed as

$$\boldsymbol{r}_i' = -x_{1_i}\boldsymbol{i}_1 + x_{2_i}\boldsymbol{i}_2 + x_{3_i}\boldsymbol{i}_3 \tag{5.69a}$$

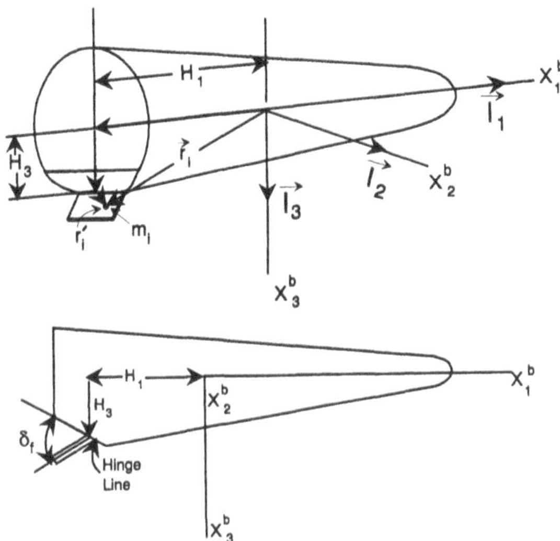

Fig. 5.12 Split-windward flap for pitch/roll control.

From the center of mass of the RV, it is given by

$$\boldsymbol{r}_i = -H_1\boldsymbol{i}_1 + H_3\boldsymbol{i}_3 + \boldsymbol{r}_i' \tag{5.69b}$$

For an inertial observer, it is

$$\boldsymbol{R}_i = \boldsymbol{R}_0 + \boldsymbol{r}_i \tag{5.69c}$$

where $\boldsymbol{R}_0$ locates the center of mass of the re-entry vehicle from the origin of the inertial frame.

The coordinates x_{1_i} and x_{3_i} are related to the flap angle δ_f as follows:

$$\frac{-x_{3_i}}{-x_{1_i}} = \tan(\delta_f) = \frac{x_{3_i}}{x_{1_i}} \tag{5.70}$$

The sign convention in the previous equation indicates that the flap angle is positive for trailing edge "down." The small angle approximation gives

$$x_{3_i} \approx \delta_f x_{1_i} \tag{5.71}$$

H_1 and H_3 are constants which locate the hinge with respect to the center of mass of the RV (H_1 is along the negative i_1-axis, and H_3 is along the positive i_3-axis). Since H_1 and H_3 are constants, we may write

$$\left.\frac{\mathrm{d}\boldsymbol{r}_i}{\mathrm{d}t}\right|_m = \left.\frac{\mathrm{d}\boldsymbol{r}_i'}{\mathrm{d}t}\right|_m \tag{5.72}$$

The value of x_{2_i} does not change with flap deflection, and for a small deflection angle, x_{1_i} is nearly constant. Thus, we may write

$$\left.\frac{\mathrm{d}x_{1_i}}{\mathrm{d}t}\right|_m \approx 0, \qquad \left.\frac{\mathrm{d}x_{2_i}}{\mathrm{d}t}\right|_m \approx 0 \tag{5.73}$$

From Eqs. (5.71), (5.72), and (5.73) we have

$$\left.\frac{\mathrm{d}\boldsymbol{r}_i'}{\mathrm{d}t}\right|_m = \left.\frac{\mathrm{d}\boldsymbol{r}_i}{\mathrm{d}t}\right|_m = \left(x_{1_i}\frac{\mathrm{d}\delta_f}{\mathrm{d}t}\right)\boldsymbol{i}_3 \tag{5.74a}$$

$$\left.\frac{\mathrm{d}^2\boldsymbol{r}_i'}{\mathrm{d}t^2}\right|_m = \left.\frac{\mathrm{d}^2\boldsymbol{r}_i}{\mathrm{d}t^2}\right|_m = \left(x_{1_i}\frac{\mathrm{d}^2\delta_f}{\mathrm{d}t^2}\right)\boldsymbol{i}_3 \tag{5.74b}$$

The moment about the hinge follows from Eq. (5.41) as

$$\boldsymbol{M} = \sum_i \boldsymbol{M}_i = \sum_i \boldsymbol{r}_i' \times m_i\left(\left.\frac{\mathrm{d}^2\boldsymbol{R}_i}{\mathrm{d}t^2}\right|_f\right)$$

or from Eqs. (5.11) and (5.69c) we have

$$\boldsymbol{M} = \sum_i m_i\boldsymbol{r}_i' \times \left[\boldsymbol{a}_0 + \left.\frac{\mathrm{d}^2\boldsymbol{r}_i}{\mathrm{d}t^2}\right|_m + 2\boldsymbol{\omega}_{m/f} \times \left.\frac{\mathrm{d}\boldsymbol{r}_i}{\mathrm{d}t}\right|_m \right.$$
$$\left. + \frac{\mathrm{d}\boldsymbol{\omega}_{m/f}}{\mathrm{d}t} \times \boldsymbol{r}_i + \boldsymbol{\omega}_{m/f} \times (\boldsymbol{\omega}_{m/f} \times \boldsymbol{r}_i)\right] \tag{5.75}$$

Note that we have replaced the derivative $\mathrm{d}^2\boldsymbol{R}_0/\mathrm{d}t^2|_I$ by $\boldsymbol{a}_0$, which is the acceleration of the center of mass of the RV. The derivatives of $\boldsymbol{r}_i'$ have been replaced by the derivatives of $\boldsymbol{r}_i$, as indicated by Eq. (5.74). Although we have considered the time derivatives of x_{3_i} to be significant, the magnitude of x_{3_i} may be ignored if δ_f, the flap deflection, is small. Thus,

$$\begin{aligned}\boldsymbol{r}_i &\approx -H_1\boldsymbol{i}_1 + H_3\boldsymbol{i}_3 - x_{1_i}\boldsymbol{i}_1 + x_{2_i}\boldsymbol{i}_2\\ &\approx -(H_1 + x_{1_i})\boldsymbol{i}_1 + x_{2_i}\boldsymbol{i}_2 + H_3\boldsymbol{i}_3 \\ \boldsymbol{r}_i' &\approx -x_{1_i}\boldsymbol{i}_1 + x_{2_i}\boldsymbol{i}_2\end{aligned} \tag{5.76}$$

We are interested only in the component of the moment about the hinge axis, i.e., in the $\boldsymbol{i}_2$-direction. This moment is given by

$$M_h = \boldsymbol{i}_2 \cdot \boldsymbol{M} \tag{5.77}$$

Next, we sum over all mass elements, assuming that there is symmetry in the variable x_{2_i} and that the center of mass of the flap is a distance e_f from the hinge line (positive aft of the hinge line).

$$\sum_i m_i x_{1_i} = m_f e_f \tag{5.78a}$$

$$\sum_i m_i x_{2_i} = \sum_i m_i x_{3_i} = 0 \tag{5.78b}$$

The moment of inertia of the flap about the hinge axis is

$$\sum_i m_i x_{1_i}^2 = I_f \tag{5.78c}$$

We next apply the condition of Eq. (5.77) to Eq. (5.75). Then we must consider each of the five terms that result; the first term is

$$\boldsymbol{i}_2 \cdot \left[\sum_i m_i \boldsymbol{r}_i' \times \boldsymbol{a}_0\right] = a_{0_3} m_f e_f \tag{5.79a}$$

The second term is

$$\boldsymbol{i}_2 \cdot \left[\sum_i m_i \boldsymbol{r}_i' \times \left.\frac{\mathrm{d}^2 \boldsymbol{r}_i}{\mathrm{d}t^2}\right|_m\right] = I_f \frac{\mathrm{d}^2 \delta_f}{\mathrm{d}t^2} \tag{5.79b}$$

The third term is

$$\boldsymbol{i}_2 \cdot \left[\sum_i m_i \boldsymbol{r}_i' \times \left(2\boldsymbol{\omega}_{m/f} \times \left.\frac{\mathrm{d}\boldsymbol{r}_i}{\mathrm{d}t}\right|_m\right)\right] = 0 \tag{5.79c}$$

The fourth term, which is fairly complex, is

$$\boldsymbol{i}_2 \cdot \left[\sum_i m_i \boldsymbol{r}_i' \times \left(\frac{\mathrm{d}\boldsymbol{\omega}_{m/f}}{\mathrm{d}t} \times \boldsymbol{r}_i\right)\right]$$

$$= \boldsymbol{i}_2 \cdot \left[\sum_i m_i \left(\frac{\mathrm{d}\boldsymbol{\omega}_{m/f}}{\mathrm{d}t}\right)(\boldsymbol{r}_i' \cdot \boldsymbol{r}_i) - \sum_i m_i \boldsymbol{r}_i \left(\frac{\mathrm{d}\boldsymbol{\omega}_{m/f}}{\mathrm{d}t} \cdot \boldsymbol{r}_i'\right)\right]$$

This term becomes

$$\sum_i m_i \frac{\mathrm{d}\omega_2}{\mathrm{d}t}\left[x_{1_i}(H + x_{1_i}) + x_{2_i}^2\right] - \sum_i m_i x_{2_i}\left[-x_{1_i}\frac{\mathrm{d}\omega_1}{\mathrm{d}t} + x_{2_i}\frac{\mathrm{d}\omega_2}{\mathrm{d}t}\right]$$

which simplifies to

$$\frac{\mathrm{d}\omega_2}{\mathrm{d}t}\left(H_1 \sum_i m_i x_{1_i} + \sum_i m_i x_{1_i}^2\right)$$

or

$$\frac{d\omega_2}{dt}(H_1 m_f e_f + I_f) \tag{5.79d}$$

The fifth and final term of Eq. (5.75) is

$$i_2 \cdot \left\{ \sum_i m_i r_i' \times [\omega_{m/f} \times (\omega_{m/f} \times r_i)] \right\}$$

or

$$i_2 \cdot \left\{ \sum_i m_i r_i' \times \left[\omega_{m/f}(\omega_{m/f} \cdot r_i) - r_i(\omega^2_{m/f}) \right] \right\}$$

which becomes, after some manipulation,

$$-\omega_1\omega_3\left(H_1 m_f e_f + I_f\right) - \left(\omega_1^2 + \omega_2^2\right) H_3 e_f m_f \tag{5.79e}$$

Combining all terms of Eqs. (5.79) provides all of the inertial terms associated with the flap; these terms are then set equal to the applied aerodynamic moment (designated in the previous expression as M_h) as follows:

$$\begin{aligned} M_h = {} & a_{0_3}(m_f e_f) + I_f \frac{d^2\delta_f}{dt^2} + \left(\frac{d\omega_2}{dt} - \omega_1\omega_3 \right)(H_1 M_f e_f + I_f) \\ & - \left(\omega_1^2 + \omega_2^2\right) H_3 e_f m_f \end{aligned} \tag{5.80}$$

The applied moment M_h consists of an aerodynamic term M_a and a control term M_c. That is,

$$M_h = M_a + M_c \tag{5.81}$$

There are various ways of expressing the aerodynamic hinge moment M_a, such as

$$M_a = C_h\left(\tfrac{1}{2}\rho V^2 b_f c_f^2\right) \tag{5.82}$$

where b_f and c_f are the span and chord of the flap, respectively, and C_h is the wind-tunnel-derived hinge moment coefficient. Obviously, Eq. (5.80) indicates coupling between the angle of flap deflection δ_f and the parameters of RV motion (i.e., a_{0_3}, ω_1, ω_2, and ω_3).

A simple illustration of the coupling might be identified as a case of the "tail wagging the dog." Suppose that $a_{0_3} = 0 = \omega_3$ and that the control and aerodynamic (i.e., applied) moments are ignored. If the flap is given an angular acceleration, then the RV is also given an angular acceleration, which is given by

$$\frac{d\omega_2}{dt} = -\left(\frac{I_f}{I_f + H_1 m_f e_f}\right)\frac{d^2\delta_f}{dt^2}$$

Thus, imparting angular acceleration to the flap, ("wagging the tail") imparts angular acceleration to the RV ("wagging the dog"). Obviously, in the presence of aerodynamic and control moments as well as body angular rates and acceleration, the coupling between the body and control is much more complex.

This completes the elementary discussion of the inertial part of the force and moment equations. This material will be re-examined in later chapters when the aerodynamic and control forcing functions are further defined.

References

[1]Lindsay, R. L., and Margenau, H., *Foundations of Physics,* Dover, Mineola, NY, 1957.

[2]Britting, K. R., *Inertial Navigation Systems Analysis,* Wiley-Interscience, Wiley, Somerset, NJ, 1971.

Bibliography

Huag, E. J., *Intermediate Dynamics,* Prentice-Hall, Englewood Cliffs, NJ, 1992.

Kaplan, M. H., *Modern Spacecraft Dynamics and Control,* Wiley-Interscience, Wiley, Somerset, NJ, 1976.

6
Keplerian Motion

6.1 Equations of Motion

Figure 6.1 shows a re-entry body (RB) at some point on an exoatmospheric trajectory. (We need not identify the object as a re-entry vehicle because the discussion of this chapter and the following chapter are equally applicable to artificial or natural objects.) We have established an axis system whose origin coincides with the RB. The coordinate frame axes are set as follows: x_1 "up" along the geocentric vector $\boldsymbol{R}_E$; x_2 north, and x_3 to the west, completing the right-handed triad. A fixed, or inertial, axis is located at the Earth's center, with X_3 along the polar axis and X_1 and X_2 in the equatorial plane. Note that when the central angle θ (essentially the latitude) is zero, both axis sets are parallel.

The velocity vector V is confined to the (x_1, x_2) plane of the moving frame and the (X_2, X_3) plane of the inertial frame. The radius vector makes the angle

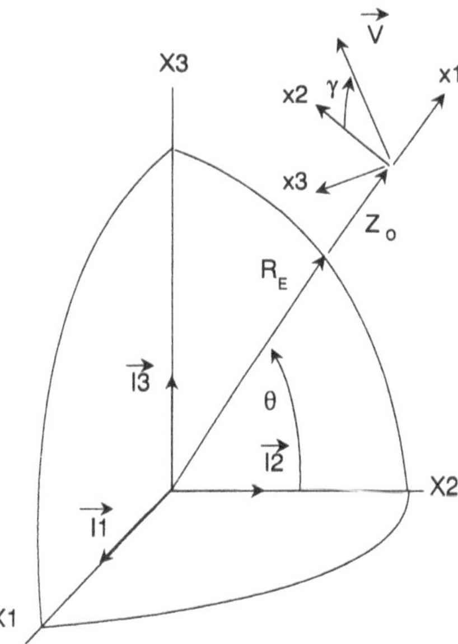

Fig. 6.1 Re-entry body at a typical point on an exoatmospheric trajectory.

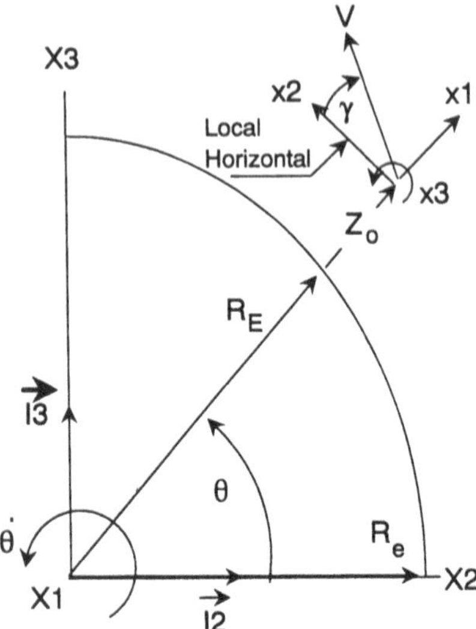

Fig. 6.2 Planar motion of re-entry body.

θ with respect to the equatorial plane. The velocity vector makes an angle γ with respect to the x_2-axis. The angle γ, the flight path angle, is considered positive for a velocity vector above the horizontal. The velocity vector is below the horizontal for nearly all re-entry trajectories studied in the next chapter. However, in this chapter we are considering orbital and suborbital motion, so the velocity vector may often be above the local horizontal.

Since we are ignoring the rotation of the Earth, the geocentric frame may be regarded as inertial. The local frame has its origin at the RB and so cannot be regarded as inertial. We assume that all motion takes place in the (x_1, x_2) plane of the local frame. For convenience, in this analysis we also regard the entire trajectory as being confined to the (X_2, X_3) plane of the geocentric frame. Figure 6.2 shows the planar motion and the appropriate unit vectors.

We can now use the work of the previous chapter to develop the orbital equations. According to Newton's Second Law, the linear acceleration relative to inertial space equals the force per unit mass. Thus, we may write:

$$\frac{\mathrm{d}^2\boldsymbol{r}}{\mathrm{d}t^2} \doteq \ddot{\boldsymbol{r}}^I = \boldsymbol{g}^I \tag{6.1}$$

In the *Keplerian* trajectory, the gravitational force is the only force present. $\boldsymbol{r}^I$ and $\boldsymbol{g}^I$ are the geocentric position vector and the gravity vector, respectively.

We now make use of Eq. (5.14), $\boldsymbol{r} = \boldsymbol{R}_E + \boldsymbol{Z}_0$ replacing f (fixed) with I (inertial) and m (moving) with l (local) to get

$$\ddot{\boldsymbol{r}}^I = \boldsymbol{g}^I = C_l^I\left(\ddot{\boldsymbol{r}}^l + \dot{\Omega}_{l/I}^l\boldsymbol{r}^l + 2\Omega_{l/I}^l\dot{\boldsymbol{r}}^l + \Omega_{l/I}^l\Omega_{l/I}^l\boldsymbol{r}^l\right) \tag{6.2}$$

where

$$r^l = [r,\ 0,\ 0]^{\mathrm{T}}, \qquad \dot{r}^l = [\dot{r},\ 0,\ 0]^{\mathrm{T}}, \qquad \ddot{r}^l = [\ddot{r},\ 0,\ 0]^{\mathrm{T}}$$

$$\Omega^l_{l/I} = \begin{bmatrix} 0 & 0 & 0 \\ 0 & 0 & -\dot{\theta} \\ 0 & \dot{\theta} & 0 \end{bmatrix}, \qquad \Omega^l_{l/I}\Omega^l_{l/I} = -\begin{bmatrix} 0 & 0 & 0 \\ 0 & \dot{\theta}^2 & 0 \\ 0 & 0 & \dot{\theta}^2 \end{bmatrix} \tag{6.3}$$

According to Eq. (4.7) we can transform the gravity vector (gravitational force per unit mass) from the I-frame to the local frame, or l-frame, by

$$g^I = C^I_l g^l \tag{6.4a}$$

where

$$g^l = [-g, 0, 0]^{\mathrm{T}} \tag{6.4b}$$

but $-g$, the component of g along the x_1-axis, is a function of geocentric distance. Thus,

$$g = g_e R_e^2/r^2 = \mu/r^2 \tag{6.4c}$$

where μ is the evaluation of $(g_e R_e^2)$ at the Earth's surface. From Eqs. (6.2) and (6.4a) it is obvious that

$$g^l = \ddot{r}^l + \dot{\Omega}^l_{l/I} r^l + 2\Omega^l_{l/I}\dot{r}^l + \Omega^l_{l/I}\Omega^l_{l/I} r^l \tag{6.5}$$

Inserting Eqs. (6.3) and (6.4) into Eq. (6.5) gives the following for the planar components of acceleration in the local frame:

$$\frac{\mathrm{d}^2 r}{\mathrm{d}t^2} - r\left(\frac{\mathrm{d}\theta}{\mathrm{d}t}\right)^2 + \frac{\mu}{r^2} = 0 \tag{6.6a}$$

$$r\frac{\mathrm{d}^2\theta}{\mathrm{d}t^2} + 2\frac{\mathrm{d}\theta}{\mathrm{d}t}\frac{\mathrm{d}r}{\mathrm{d}t} = 0 \tag{6.6b}$$

If both sides of Eq. (6.6b) are multiplied by r, this equation may be rewritten as

$$\frac{\mathrm{d}}{\mathrm{d}t}\left(r^2\frac{\mathrm{d}\theta}{\mathrm{d}t}\right) = 0$$

or, equivalently,

$$r^2\frac{\mathrm{d}\theta}{\mathrm{d}t} = P \tag{6.7}$$

where P is a constant. The matrix equivalent of Eq. (5.15a) is

$$\boldsymbol{V}^l = \dot{\boldsymbol{r}}^l + \Omega^l_{l/I}\boldsymbol{r}^l \tag{6.8}$$

The vector r^l has been given in Eq. (6.3); the velocity vector in the local frame then becomes

$$\boldsymbol{V}^l = [V\sin(\gamma), V\cos(\gamma), 0]^{\mathrm{T}} \tag{6.9}$$

Inserting Eqs. (6.3) and (6.9) into Eq. (6.8) gives

$$r\frac{\mathrm{d}\theta}{\mathrm{d}t} = V\cos(\gamma)$$

but from Eq. (6.7) we have

$$r^2\frac{\mathrm{d}\theta}{\mathrm{d}t} = rV\cos(\gamma) = r_0V_0\cos(\gamma_0) = P \tag{6.10}$$

Since P is a constant, we can evaluate P anywhere along the trajectory; we choose to evaluate it at trajectory initialization. For a re-entry vehicle this point is given by the location of where the RV off-loads from the final stage of the boost vehicle; for a natural entry body this point of initialization may be taken as the point of first sighting. However chosen, the point where P is evaluated is designated by a zero subscript.

Equation (6.6a) can be rewritten exclusively in terms of r by using Eq. (6.10) to eliminate $\mathrm{d}\theta/\mathrm{d}t$. The result would be a second-order differential equation with the geocentric distance r as the dependent variable and time t as the independent variable. However, the resulting equation is not integrable in closed form, and the periodicity associated with r in orbital motion is not obvious. Therefore we will use Eq. (6.10) to replace time t with central angle θ as the independent variable. Clearly,

$$\frac{\mathrm{d}}{\mathrm{d}t}(\,) = \frac{\mathrm{d}\theta}{\mathrm{d}t}\frac{\mathrm{d}}{\mathrm{d}\theta}(\,) = \frac{P}{r^2}\frac{\mathrm{d}}{\mathrm{d}\theta}(\,) = Pu^2\frac{\mathrm{d}}{\mathrm{d}\theta}(\,) \tag{6.11}$$

where u is the reciprocal of the geocentric distance r, i.e., $u = 1/r$. It then follows that

$$\frac{\mathrm{d}}{\mathrm{d}t}(r) = \frac{\mathrm{d}}{\mathrm{d}t}\left(\frac{1}{u}\right) = Pu^2\frac{\mathrm{d}}{\mathrm{d}\theta}\left(\frac{1}{u}\right) \tag{6.12}$$

which gives

$$\frac{\mathrm{d}r}{\mathrm{d}t} = -P\frac{\mathrm{d}u}{\mathrm{d}\theta} \tag{6.13}$$

From Eqs. (6.11) and (6.13) it follows that

$$\frac{\mathrm{d}^2 r}{\mathrm{d}t^2} = \frac{\mathrm{d}}{\mathrm{d}t}\left(\frac{\mathrm{d}r}{\mathrm{d}t}\right) = Pu^2\frac{\mathrm{d}}{\mathrm{d}\theta}\left(-P\frac{\mathrm{d}u}{\mathrm{d}\theta}\right) = -P^2u^2\frac{\mathrm{d}^2 u}{\mathrm{d}\theta^2} \tag{6.14}$$

and from Eq. (6.10) we have

$$\frac{\mathrm{d}\theta}{\mathrm{d}t} = u^2 P \tag{6.15}$$

By inserting Eqs. (6.13), (6.14), and (6.15) into Eqs. (6.6a) we get

$$\frac{\mathrm{d}^2 u}{\mathrm{d}\theta^2} + u = \frac{\mu}{P^2} = \frac{\mu}{r_0^2 V_0^2 \cos^2(\gamma_0)} \tag{6.16}$$

The above equation indicates that u, the reciprocal of r is harmonic (with the central angle θ). We might have preferred to have the radius vector r vary harmonically with time, but manipulations of Eq. (6.6a) will show that this is not the case.

It is possible to combine some of the orbital parameters into an interesting ratio. Let

$$\lambda = \frac{2\left(\frac{1}{2}V_0^2\right)}{\mu/r_0} \tag{6.17}$$

where V_0^2 is twice the kinetic energy per unit mass at trajectory initialization. It will be shown later that the term μ/r_0 is the square of the linear velocity in a circular orbit of radius r_0. Thus, λ is the ratio of the square of the velocity magnitude at trajectory initialization to the square of the velocity in a circular orbit at the same altitude. Since re-entry trajectories subsequent to boost are always suborbital, we would expect that for the kinds of trajectories of interest λ will be between 0 and 1. Thus, Eq. (6.16) may be written as

$$\frac{\mathrm{d}^2 u}{\mathrm{d}\theta^2} + u = \frac{1}{\lambda r_0 \cos^2(\gamma_0)} \tag{6.18}$$

Our next task is to find both the particular and homogeneous solutions to Eq. (6.18). The particular solution is obvious from Eq. (6.18) and is given by

$$u_p = 1/\lambda r_0 \cos^2(\gamma_0) \tag{6.19a}$$

and the homogeneous solution is of the form

$$u_h = A\sin(\theta) + B\cos(\theta) \tag{6.19b}$$

The constants A and B require the two following initial conditions:

$$u_0 \doteq u|_{\theta=0} = 1/r_0 \tag{6.20a}$$

and, from Eq. (6.13),

$$\left.\frac{du}{d\theta}\right|_0 \doteq \left.\frac{du}{d\theta}\right|_{\theta=0} = -\frac{1}{P}\left.\frac{dr}{dt}\right|_{\theta=0}$$

$$= -\frac{dr/dt}{r(r d\theta/dt)}$$

$$= -\frac{V_0\sin(\gamma_0)}{r_0 V_0\cos(\gamma_0)}$$

$$= -\frac{\tan(\gamma_0)}{r_0} \tag{6.20b}$$

The total solution of Eq. (6.18) can now be written as the sum of the particular and homogeneous solutions as follows:

$$u = u_p + u_h$$

$$= \frac{1}{\lambda r_0\cos^2(\gamma_0)} + A\sin(\theta) + B\cos(\theta) \tag{6.21}$$

Applying the initial conditions of Eqs. (6.20) leads to the following solution:

$$r_0 u(\theta) = \frac{r_0}{r(\theta)} = \frac{1-\cos(\theta)}{\lambda\cos^2(\gamma_0)} + \frac{\cos(\gamma_0+\theta)}{\cos(\gamma_0)} \tag{6.22}$$

An alternate way of writing r_0 is in terms of the Earth's radius R_E and altitude at burnout Z_0 as follows:

$$r_0 = R_E + Z_0 = R_e + Z_0 \tag{6.23}$$

where R_E is the Earth's radius and R_e the radius at the equator. From Fig. 6.3 we see that selecting Z_0, γ_0, and V_0 allows a unique relationship to be written between r (or u) and the central angle θ.

Next, we reformulate Eq. (6.22) in the more or less standard expression for an ellipse. This standard form relates the radius vector r and the central angle θ as follows:

$$r = \frac{a\left(1-e^2\right)}{1-e\cos(\theta-\omega)} \tag{6.24}$$

where a is the semimajor axis, e is the eccentricity, and ω is the argument of the apogee. The numerator $a(1-e^2)$ is known as the *semilatus rectum* of the ellipse. We may manipulate Eq. (6.22) as follows:

$$r = \frac{r_0\lambda\cos^2(\gamma_0)}{1-\cos(\theta)+\lambda\cos(\gamma_0)\cos(\theta+\gamma_0)}$$

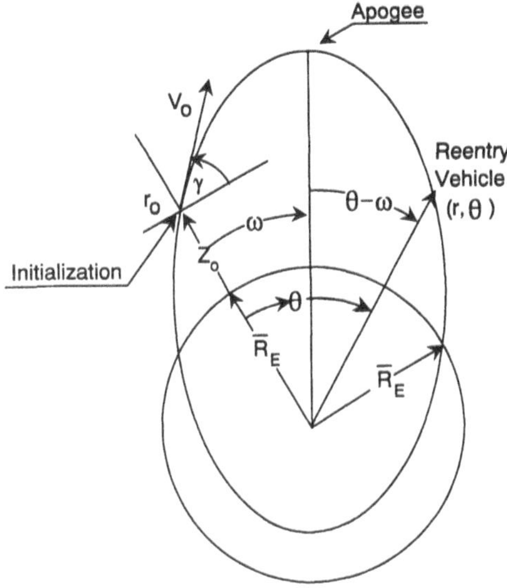

Fig. 6.3 Keplerian trajectory.

or

$$r = \frac{r_0 \lambda \cos^2(\gamma_0)}{1 - e\left\{\cos\theta\left[\frac{1}{e}(1 - \lambda\cos^2\gamma_0)\right] + \sin\theta\left[\frac{1}{e}(\lambda\cos\gamma_0\sin\gamma_0)\right]\right\}} \tag{6.25}$$

Because $\cos(\theta - \omega) = \cos(\theta)\cos(\omega) + \sin(\theta)\sin(\omega)$, we may make the following comparison between Eqs. (6.24) and (6.25):

$$e\sin(\omega) = \lambda\cos(\gamma_0)\sin(\gamma_0) \tag{6.26a}$$

$$e\cos(\omega) = 1 - \lambda\cos^2(\gamma_0) \tag{6.26b}$$

Squaring both sides of Eqs. (6.26) and adding the results gives

$$e = [\lambda(\lambda - 2)\cos^2(\gamma_0) + 1]^{1/2} \tag{6.27a}$$

Next, we divide Eq. (6.26a) by Eq. (6.26b) to give

$$\omega = \tan^{-1}\left[\frac{\lambda\cos(\gamma_0)\sin(\gamma_0)}{1 - \lambda\cos^2(\gamma_0)}\right] \tag{6.27b}$$

Recognizing that,

$$r_0\lambda\cos^2(\gamma_0) = a(1 - e^2)$$

we can use Eq. (6.27a) to get

$$a = r_0/(2 - \lambda) = (R_e + Z_0)/(2 - \lambda) \tag{6.27c}$$

The circular orbit is the simplest of orbits. In a circular orbit the eccentricity is zero. Thus, from Eq. (6.27a) it follows that $\lambda = 1.0$ for $e = 0$, because in a circular orbit γ must always be zero (i.e., the velocity vector must always be normal to the radius vector). From Eq. (6.17) it then follows that the velocity magnitude in a circular orbit must be

$$V = V_0 = \left(\frac{\mu}{r_0}\right)^{1/2} = \left[\frac{g_e R_e^2}{(R_e + Z_0)}\right]^{1/2} \tag{6.28a}$$

However, since $Z_0 \ll R_e$ for most re-entry suborbital trajectories, the above expression yields the following approximation:

$$V_{\text{circular}} = V_0 \approx (g_e R_e)^{1/2} \tag{6.28b}$$

Inserting $R_e = 6.378 \times 10^6$ m and $g_e = 9.81$ m/s^2 results in a circular orbital velocity of 7910 m/s. (The equivalent in English units would be something like $R_e = 2.09 \times 10^7$ ft and $g_e = 32.174$ ft/s^2, which give a circular velocity magnitude of 25,931 ft/s.)

Another transitional orbit is that of a parabola. For a parabola, $e = 1.0$. From Eq. (6.27a) we see that there are three ways to achieve a parabolic orbit: 1) $\lambda = 0$, 2) $\gamma_0 = \pm\pi/2$, and 3) $\lambda = 2$. The $\lambda = 0$ case corresponds to the flat-Earth parabolic trajectory from elementary physics, where the initial speed V_0 is negligibly small in comparison to circular orbital velocity. This particular trajectory is of no interest in the study of problems associated with atmospheric (re-)entry.

The second case, where $\gamma_0 = \pm\pi/2$, gives a sort of degenerate parabola in which the RB trajectory is initiated in a way that the velocity vector is either along the positive or negative radius vector. Such trajectories have little or no practical interest for fairly obvious reasons.

For the third case, where $\lambda = 2$, we have the following from Eq. (6.17):

$$V_{\text{escape}} = V_0 = [2(\mu/r_0)]^{1/2} \tag{6.29a}$$

or, approximately,

$$V_{\text{escape}} = V_0 \approx (2g_e R_e)^{1/2} \tag{6.29b}$$

It should be obvious from Eq. (6.28b) that the parabolic, or escape, velocity is $\sqrt{2}$ times the circular orbital velocity. Equation (6.27c) indicates that as λ approaches 2, the semimajor axis becomes unbounded, which is another representation of the parabola as an escape trajectory.

We can develop a physical sense of the trajectory initiation by rewriting Eq. (6.22) for the special case where $\gamma_0 = 0$ as follows:

$$\frac{r_0}{r(\theta)} = \frac{1 - \cos(\theta) + \lambda\cos(\theta)}{\lambda} \tag{6.30}$$

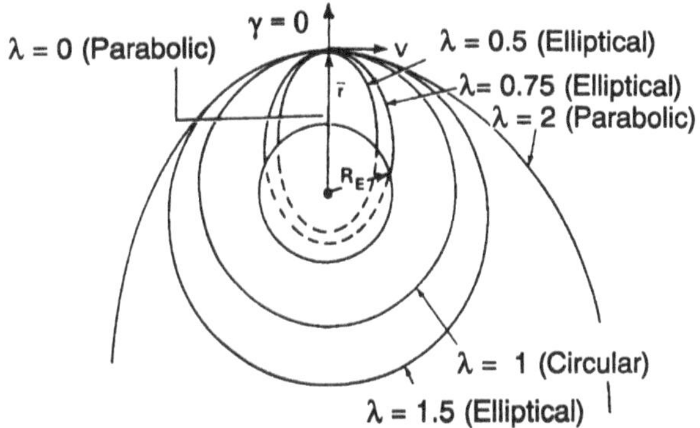

Fig. 6.4 Variation of trajectory geometry with velocity ratio.

Obviously, because Eq. (6.30) is satisfied, when $\lambda = 1$, $r = r_0$ indicating a circular orbit.

For an elliptical orbit, λ lies between 0 and 2; as noted earlier, when $\lambda = 0$ or 2, the orbit is parabolic. A flat-Earth parabola is the result of the first case, and an escape parabola is the result of the second case. For the latter case

$$r(\theta) = \frac{2r_0}{1 + \cos(\theta)} \tag{6.31}$$

As expected, r is unbounded when $\theta = \pi$.

With $\gamma_0 = 0$, we note that Eq. (6.27a) becomes

$$\lambda = 1 \pm e$$

or

$$e = 1 - \lambda \tag{6.32a}$$

$$e = \lambda - 1 \tag{6.32b}$$

As noted for an elliptical trajectory, $0 < \lambda < 2$, and $0 < e < 1$. Thus, according to Eqs. (6.32), there are two values of λ that will result in the same orbital eccentricity. For $0 < \lambda < 1$, the orbit is elliptical and intersects the Earth because the point of orbital initiation is at the orbital apogee. For $1 < \lambda < 2$, the orbit is also elliptical, but the orbit does not intersect the Earth, and the point of orbital initiation is the perigee. Since we are mainly interested in suborbital motion, we consider here values of λ in the range of 0 to 1. The various trajectories are illustrated in Fig. 6.4.

6.2 Impact Equations

Equation (6.22) provides a relationship between r, the geocentric distance to the RB, and θ, the central angle. This equation contains the normalized initial velocity ratio λ and the initial flight path angle γ_0. In this section we use Eq. (6.22) to relate the initial velocity magnitude V_0 and initial flight path angle γ_0 to achieve a fixed value of the central range angle θ. The central range angle is the angle of the subtended arc between trajectory initiation and the point of Earth impact (in the absence of an atmosphere) or the point of atmospheric re-entry.

First, we can replace λ as follows:

$$\lambda = \frac{V_0^2}{\mu/r_0} \tag{6.33a}$$

Also, at impact the dependent/independent variables are, respectively

$$r = R_e, \qquad \theta = \theta_i \tag{6.33b}$$

By inserting Eqs. (6.33) into Eq. (6.22) and solving for the initial velocity V_0 we get

$$V_0 = \left\{\frac{\mu}{r_0}\left[\frac{1-\cos(\theta_i)}{(r_0/R_e)\cos^2(\gamma_0) - \cos(\theta_i+\gamma_0)\cos(\gamma_0)}\right]\right\}^{1/2} \tag{6.34}$$

Equation (6.34) is quite useful for developing a physical sense of how the burnout parameters V_0 and γ_0 control the range. Equation (6.34) provides a relationship between V_0 as a dependent variable and γ_0, θ_i, and Z_0 (as in $r_0 = R_e + Z_0$) as either independent variables or as parameters.

One useful application of Eq. (6.34) is to provide a relationship between trajectory initialization values of V_0 and γ_0 for a fixed value of θ_i. To simplify calculations and to eliminate the initial altitude Z_0 as a variable, we assume that $Z_0 \ll R_e$ and thus let $r_0 = R_e$. A graphical representation of Eq. (6.34) is given in Fig. 6.5 for a number of range angles.

We notice immediately from Fig. 6.5 that for any given range angle there are usually two values of burnout flight path angle γ_0 for a given initial velocity magnitude V_0. For a range angle θ_i of, say, 30 deg and for V_0 fixed at 6000 m/s, the corresponding flight path angles are 7 and 63 deg. The trajectory corresponding to the smaller flight path angle is identified as *underlofted,* and that for the larger flight path angle is *overlofted.* Because of the multivalued nature of the inverse of the function $V_0 = V_0(\gamma_0)$, there are usually two trajectories corresponding to each burnout velocity magnitude for a fixed range.

The radius at trajectory initialization, or boost vehicle burnout, follows from Eq. (6.22) with

$$r(\theta_i) = R_e$$

as

$$\frac{r_0}{R_e} = \frac{R_e + Z_0}{R_e} = \frac{1-\cos(\theta_i)}{\lambda\cos^2(\gamma_0)} + \frac{\cos(\theta_i+\gamma_0)}{\cos(\gamma_0)} \tag{6.35}$$

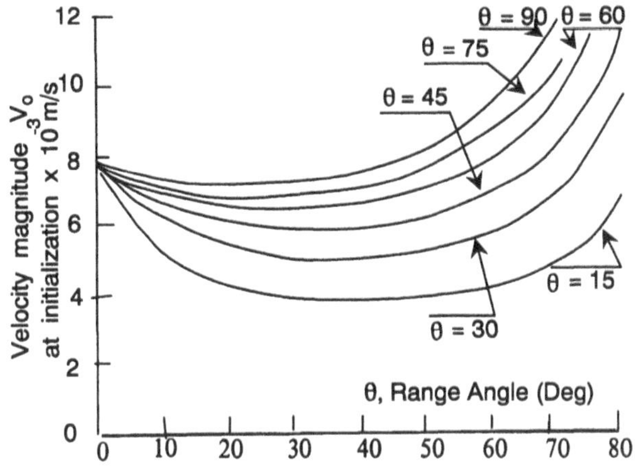

Fig 6.5 Velocity magnitude at burnout vs flight path angle at burnout for several range angles.

If Eq. (6.35) is used to obtain Z_0, the altitude at burnout, then it must be remembered that Z_0 is at least two orders of magnitude less than R_e. Equations (6.34) and (6.35) are of course not independent but are alternate forms of Eq. (6.22).

If we return to Eq. (6.34) with an occasional glance at Fig. 6.5, we note that there seems to be an "optimum" flight path angle at trajectory initialization, which we will designate as $\gamma_0 = \gamma_0^*$. By optimum we mean an angle for which a minimum velocity is required to initiate a trajectory that will subtend a given range angle θ_i. This angle defines the boundary between over- and underlofted trajectories. One can compute γ_0^* by differentiating Eq. (6.34), equating the derivative $dV_0/d\gamma_0$ to zero, and solving for $\gamma_0 = \gamma_0^*$ as follows:

$$\tan(2\gamma_0^*) = \frac{\sin(\theta_i)}{[1 + (Z_0/R_e)] - \cos(\theta_i)} \tag{6.36}$$

In the event that Z_0, the altitude at burnout, may be assumed negligible in comparison to R_e, we may write

$$\gamma_0^* = (\pi/4) - (\theta_i/4) \tag{6.37}$$

It is obvious that as θ_i approaches zero, we have the familiar flat-Earth result which identifies the launch flight path angle for maximum range as $\pi/4$.

In keeping with the flat-Earth approximation, the corresponding velocity V_0 follows from Eq. (6.34) with θ_i approaching zero. [Setting θ_i to zero in Eq. (6.34) will result in an indeterminant form.] Applying the small angle approximation to Eq. (6.34) results in the following:

$$V_0 = \left[\left(\frac{\mu}{2R_e}\right)\left(\frac{\theta_i}{\sin(\gamma_0)\cos(\gamma_0)}\right)\right]^{1/2} = \left[\frac{gR_e\theta_i}{\sin(2\gamma_0)}\right]^{1/2} = \left[\frac{gX}{\sin(2\gamma_0)}\right]^{1/2} \tag{6.38}$$

where X is the range.

6.3 Time of Flight

The size, shape, and planar orientation of a suborbital trajectory of a re-entry body are completely defined by the three parameters a, e, and ω. These parameters are in turn uniquely defined by the initial conditions r_0 (or Z_0), γ_0, and V_0. A unique trajectory must have associated with it a single time of passage over that trajectory. In other words, given the three initial conditions there is a single time from burnout to impact (or from trajectory initiation to a point on the trajectory where the trajectory terminates or the Keplerian assumption is no longer valid).

It was pointed out earlier that time enters into the description of suborbital motion in a very awkward fashion. To make the analysis tractable, we replaced time t with the central angle θ_i as the independent variable. The goal now will be to obtain a relationship between time and this central angle. While it might be preferable to obtain $\theta = \theta(t)$, the only form in which a (t, θ) relationship can be written in elementary functions is as $t = t(\theta)$, i.e., time as a function of central angle θ.

By using Eq. (6.10) we may write

$$\int_0^{\theta(t)} r^2 \mathrm{d}X = \int_0^t [V_0 r_0 \cos(\gamma_0)]\mathrm{d}\tau \tag{6.39}$$

If we make use of Eq. (6.22) (the integrand of the time integral is constant) and replace θ in the integrand with the dummy variable X, we may rewrite Eq. (6.39) as

$$\int_0^t \mathrm{d}\tau = \frac{r_0}{V_0\cos(\gamma_0)} \int_0^{\theta(t)} \left[\frac{1-\cos(X)}{\lambda\cos^2(\gamma_0)} + \frac{\cos(\gamma_0 + X)}{\cos(\gamma_0)}\right]^{-2} \mathrm{d}X \tag{6.40}$$

Integration of the above expression is a formidable task and will not be carried out here. The result for the running variables θ and t is

$$t = \frac{r_0}{V_0\cos(\gamma_0)}\left\{\frac{\tan(\gamma_0)(1-\cos\theta) + (1-\lambda)\sin\theta}{(2-\lambda)\{[(1-\cos\theta)/(\lambda\cos^2\gamma_0)] + [\cos(\gamma_0+\theta)/\cos(\gamma_0)]\}}\right\}$$
$$+\left\{\frac{2r_0}{V_0\lambda[(2/\lambda)-1]^{3/2}}\tan^{-1}\left[\frac{[(2/\lambda)-1]^{1/2}}{\cos(\gamma_0)\cot(\theta/2) - \sin(\gamma_0)}\right]\right\} \tag{6.41a}$$

Let's assume that during the Keplerian phase, the re-entry body subtends an arc of magnitude θ_i. The Keplerian phase is that trajectory segment over which only the gravitational field acts; once the RB has entered the atmosphere, where aerodynamic forces influence and then dominate the motion, Eq. (6.41a) is no longer applicable. However, the time duration of the Keplerian phase is possibly two orders of magnitude greater than that of the re-entry phase; thus, Eq. (6.41a) still can provide a good estimate of the total time of flight. This total time of flight follows from Eq. (6.41a) as

$$T_F = \frac{r_0}{V_0\cos(\gamma_0)}\left\{\frac{\tan\gamma_0(1-\cos\theta_i)+(1-\lambda)\sin\theta_i}{(2-\lambda)\{[(1-\cos\theta_i)/(\lambda\cos^2\gamma_0)]+[\cos(\gamma_0+\theta_i)/\cos\gamma_0]\}}\right\}$$
$$+\left\{\frac{2r_0}{V_0\lambda[(2/\lambda)-1]^{3/2}}\tan^{-1}\left[\frac{[(2/\lambda)-1]^{1/2}}{\cos(\gamma_0)\cot(\theta_i/2)-\sin(\gamma_0)}\right]\right\} \tag{6.41b}$$

Equation (6.41b) expresses the total time of flight T_F as a function of the subtended arc θ_i and the parameters λ, V_0, γ_0, and Z_0; i.e.,

$$T_F = T(\theta_i, \lambda, V_0, \gamma_0, r_0) \tag{6.42}$$

where

$$\lambda = \frac{V_0^2}{\mu/r_0} \tag{6.17}$$

$$V_0 = V_0(\theta_i, r_0, \gamma_0) \tag{6.34}$$

$$R_0 = Z_0 + R_e \tag{6.23}$$

Note that Eq. (6.35) relates r_0 to λ, V_0, and γ_0; however, this equation is just an alternate form of Eq. (6.34) and therefore cannot be included with the above four equations.

One approach to using the above equations is to accept the range angle θ_i as a given. Next, the altitude at trajectory initiation (i.e., Z_0) is selected, which in turn sets the magnitude of the radius vector $\boldsymbol{r}_0$ [Eq. (6.23)]. We then note that with fixed values of θ_i and r_0, Eq. (6.34) provides a continuous relationship between V_0, the initial velocity magnitude, and γ_0, the initial flight path angle. If $\gamma_0 < \gamma_0^*$, then the trajectory is underlofted; if $\gamma_0 > \gamma_0^*$, the trajectory is overlofted.

An overlofted trajectory, as might be expected, has a longer flight time T_F than one which is underlofted. However, it will be shown that an overlofted trajectory is more accurate. In any event we must assume that γ_0 has been set to meet some criterion external to that used in forming the above equations. If r_0, θ_i, and γ_0 are selected, then V_0 is fixed through Eq. (6.34), which in turn fixes λ according to Eq. (6.17). The time of flight is then set by Eq. (6.41).

A variation on the above selection process might be to regard the initial velocity V_0 as the most important parameter, because V_0 is related to the energy delivered during the boost phase. If the attainment of a given velocity is critical, then the greatest range will be reached during the Keplerian phase by selecting the optimum burnout angle γ_0^* [Eq. (6.36)].

We have shown that if the central range angle is given along with the initial velocity magnitude and altitude and that if this velocity is in excess of that set by operating at the optimum burnout angle, then there are two flight path angles—one below γ_0^* (underlofted) and another greater than γ_0^* (overlofted). Thus, given V_0 and γ_0, we can invert Eq. (6.34) and solve for the two flight path angles. A computer program to do this is given in Table 6.1. This program also calculates the time of flight corresponding to the under- and overlofted trajectories. Also calculated by this program is the optimum flight path angle for the given central angle. This angle corresponds to the minimum velocity required to traverse the given central angle. The minimum velocity associated with this optimum angle is calculated according to Eq. (6.34).

Typical output from this program is given in Table 6.2. A velocity magnitude of 7238 m/s results in an underlofted flight path angle of 10.0 deg and an overlofted flight path angle of 42.5 deg. The optimal flight path angle for a central angle of 75 deg is 26.25 deg and from Eq. (6.34) we find that the required initial velocity is 6876 m/s. The corresponding times of flight are also given. It is interesting to note that the overlofted trajectory has a duration of more than twice that of the underlofted trajectory.

It will be shown in the next section that trajectory types (overlofted and underlofted) vary in sensitivity to velocity and positional uncertainties at launch.

Table 6.1 Program to calculate flight path angles

```
!
!
!
! THIS PROGRAM CALCULATES THE FLIGHT PATH ANGLES FOR THE
!  UNDER-LOFTED AND OVER-LOFTED TRAJECTORIES AND THE
!  RESPECTIVE TRAJECTORY TRANSIT TIMES GIVEN THE RANGE
!  CENTRAL ANGLE, THE VELOCITY MAGNITUDE AND THE ALTITUDE
!  AT BOOST TERMINATION. THE PROGRAM ALSO CALCULATES THE
!  FLIGHT PATH ANGLE FOR MINIMUM VELOCITY TO COVER THE
!  SELECTED RANGE ANGLE.
! INPUTS: V0      = VELOCITY AT BOOST TERMINATION (M/S)
!         Z0      = ALTITUDE AT BOOST TERMINATION (M)
!         THETAI = THE RANGE CENTRAL ANGLE (DEG)
! OUTPUTS: GAMM01 = FLIGHT PATH ANGLE UNDERLOFTED (DEG)
!          GAMM02 = FLIGHT PATH ANGLE OVERLOFTED (DEG)
!          GAMM0  = FLIGHT PATH ANGLE MIN VELOCITY (DEG)
!          TIMEUL = TIME FOR UNDERLOFTED TRAJECTORY (SEC)
!          TIMEOL = TIME FOR OVERLOFTED TRAJECTORY (SEC)
!          TIMEOP = TIME FOR MINIMUM TRAJECTORY (SEC)
!
! ENTER V0, Z0, THETAI IN NEXT THREE LINES
```

(continued on next page)

Table 6.1 (continued) Program to calculate flight path angles

```
VO = 7238
ZO = 0
THETAI = 75
M1 = 3.9853E+14
RE = 6.378E+6
R9 = PI/180.0
RO = RE + ZO
TH = THETAI*R9
Z9 = (1-COS(TH))/SIN(TH)
Z9 = Z9*M1/(RO*(VO^2))
Z8 = (COS(TH)-(RO/RE))/SIN(TH)
Z6 = SIN(TH)/((1+(ZO/RE))-COS(TH))
GAMMO = (0.5)*ATN(Z6)
A1 = 1+Z8^2
B1 = 2*Z8*Z9-1
C1 = Z9^2
ROOT = SQR(B1^2-4*A1*C1)
ROOT1 = ROOT
ROOT2 = (-1)*ROOT
X1 = SQR((-B1+ROOT1)/(2*A1))
Z7 = SQR(1-X1^2)/X1
GAMMO1 = ATN(Z7)
X1 = SQR((-B1+ROOT2)/(2*A1))
Z7 = SQR(1-X1^2)/X1
GAMMO2 = ATN(Z7)
CALL TFLT(GAMMO,TH,VO,ZO,TIMEOP)
CALL TFLT(GAMMO1,TH,VO,ZO,TIMEUL)
CALL TFLT(GAMMO2,TH,VO,ZO,TIMEOL)
N1$ = "INITIAL VELOCITY       INITIAL ALTITUDE       CENTRAL
ANGLE"
N2$ = "METERS/SECOND          KILO-METERS            DEGREES"
N3$ = "######.#                    ###.###            ###.##"
N4$ = "FLT. PATH ANG.-UL     FLT. PATH ANG.-OL     FLT. PATH
ANG.-OP"
N5$ = "  DEGREES              DEGREES                DEGREES"
N6$ = "      ###.##                 ###.##
###.##"
N7$ = "TIME - UL             TIME - OL             TIME - OP"
N8$ = "SECONDS               SECONDS               SECONDS"
N9$ = "####.##               ####.##               ####.##"
```

(continued on next page)

Table 6.1 (continued) Program to calculate flight path angles

```
PRINT N1$
PRINT N2$
PRINT USING N3$: V0, Z0/1000, THETAI
PRINT
PRINT N4$
PRINT N5$
PRINT USING N6$: GAMM01/R9, GAMM02/R9, GAMM0/R9
PRINT
PRINT N7$
PRINT N8$
PRINT USING N9$: TIMEUL, TIMEOL, TIMEOP
END
SUB TFLT(G,A,V,Z,T)
! OVER  G: FLIGHT PATH ANGLE (RADIANS)
!       A: SUBTENDED RANGE ANGLE (RADIANS)
!       V: VELOCITY MAGNITUDE (M/S)
!       Z: ALTITUDE (M)
!BACK   T: TIME OF FLIGHT (S)

M1 = 3.9853E+14
RE = 6.378E+6
R = RE+Z
L = (V^2)/(M1/R)
T8 = TAN(G)*(1-COS(A))+(1-L)*SIN(A)
T9 = (1-COS(A))/(L*(COS(G)^2))
T9 = (T9+(COS(G+A)/COS(G)))*(2-L)
T7 = T8/T9
T6 = (COS(G)/TAN(A/2))-SIN(G)
IF ABS(T6)>0.001 THEN
   T8 = (SQR((2/L)-1))/T6
   T9 = ATN(T8)
      IF T8<= 0 THEN
        T9 = T9+PI
      END IF
ELSE
   T9 = PI/2
END IF
   T9 = T9*2*COS(G)/(L*(((2/L)-1)^(3/2)))
   T = (T7+T9)*(R/(V*COS(G)))
END SUB
```

Table 6.2 Typical output from Table 6.1

INITIAL VELOCITY METERS/SECOND	INITIAL ALTITUDE KILO-METERS	CENTRAL ANGLE DEGREES
7238.0	0.000	75.00
FLT. PATH ANG.-U. DEGREES	FLT. PATH ANG. -OL DEGREES	FLT. PATH ANG-OP DEGREES
10.00	42.50	26.25
TIME - UL SECONDS	TIME - OL SECONDS	TIME - OP SECONDS
1271.20	2577.64	1812.41

6.4 Error Analysis

Error analyses take at least two distinct analytical forms. One can separate a dependent or state variable into a nominal term and a perturbation term. If the dependent variable in this separated form is inserted into the differential equations of motion, it is possible to obtain a set of differential equations in the perturbation variable. If the perturbation variable is identified with the uncertainty in the knowledge of the state, then the integration of the perturbation equations indicates how the uncertainty grows with time. In a later chapter we will consider this approach to error analysis when we discuss the Monte-Carlo and covariance propagation approaches.

A second and much simpler approach is considered here. Errors are still equivalent to perturbations of the dependent variables; however, the equations to be perturbed are algebraic rather than differential. These algebraic equations are, in a sense, the integration of the original state differential equations. The algebraic equations of interest here are the impact equations [Eqs. (6.34) and (6.35)].

In these equations the important variables are V_0 (initial velocity magnitude), Z_0 (initial altitude), γ_0 (initial flight path angle), and θ_i (the subtended arc of the Keplerian trajectory). Following the work of Wheelon[1] we obtain the differentials, δ, of the above variables from Eq. (6.34) as follows:

$$\frac{2\delta V_0}{V_0}\left[\frac{1-\cos(\theta_i)}{\lambda\cos^2(\gamma_0)}\right]+\frac{\delta Z_0}{R_e}\left[1+\left(\frac{1-\cos(\theta_i)}{\lambda\cos^2(\gamma_0)}\right)\left(\frac{R_e}{R_e+Z_0}\right)\right]$$
$$-\delta\gamma\left[\frac{\sin(\gamma_0+\theta_i)}{\cos(\gamma_0)}-\left(\frac{\sin(\gamma_0)}{\cos^3(\gamma_0)}\right)\cos(\gamma_0+\theta_i)-\tan(\gamma_0)\left(\frac{1-\cos(\theta_i)}{\lambda\cos^2(\gamma_0)}\right)\right]$$
$$=\delta\theta_i\left[\frac{\sin(\theta_i)}{\lambda\cos^2(\gamma_0)}-\frac{\sin(\gamma_0+\theta_i)}{\cos(\gamma_0)}\right] \quad (6.43)$$

In the above equation $\delta\theta_i$ is used as a measure of the range error; however, a more direct expression might be to write the range error, δR, directly as

$$\delta R = R_e \delta\theta_i \tag{6.44}$$

We can now relate errors in V_0, γ_0, and Z_0 to the range error by means of Eq. (6.43), replacing $\delta\theta_i$ in accordance with Eq. (6.44).

The first error relationship may be written as

$$\frac{\partial R}{\partial V_0} = \frac{2R_e}{V_0}\left[\frac{1-\cos(\theta_i)}{\sin(\theta_i) - \lambda\sin(\gamma_0+\theta_i)\cos(\gamma_0)}\right]$$

If we approximate r_0 as R_e and use Eq. (6.35) to eliminate λ, we can simplify the above expression to get

$$\frac{\partial R}{\partial V_0} = \frac{2R_e}{V_0}\{\sin(\theta_i) + \cot(\gamma_0)[1-\cos(\theta_i)]\} \tag{6.45}$$

The above equation appears to indicate that the range variation with velocity magnitude V_0 depends upon the variables γ_0, V_0, and θ_i. However, if θ_i is considered fixed, then γ_0 and V_0 are not independent but must satisfy Eq. (6.34), one of the impact equations. Figure 6.5 shows the results of this equation graphically.

By way of illustration we may assume that the range angle θ_i is 75 deg. The optimum flight path angle for this range (i.e., γ_0^*) follows from Eq. (6.36) as 26.25 deg. The corresponding velocity magnitude V_0 can be calculated from Eq. (6.34) to give 6876.7 m/s. These values may then be inserted into Eq. (6.45) to obtain the range error sensitivity to the velocity magnitude V_0. Thus, we obtain

$$\frac{\partial R}{\partial V_0} = 4.58 \text{ km}/(\text{m/s})$$

Thus, a velocity magnitude error of 1 m/s at trajectory initiation leads to a 4.58 km error at impact. The range is about 8400 km, so the range error represents 1 part in 1860. Nevertheless, an error of this magnitude is intolerable. The velocity magnitude error must be held to better than 1 m/s (1 part in 7000) to achieve better than, say, a 1 km error at impact.

Equation (6.45) obviously shows that the sensitivity of range error to velocity error increases with increasing range angle θ_i. The relationship between $\partial R/\partial V_0$ and γ_0 is more subtle. Figure 6.5 shows that the required initial speed V_0 will increase whether γ_0 is less than or greater than γ_0^*. Assume that we are considering a trajectory for which the range angle and the initial speed are assigned fixed values which we indicate symbolically as θ_i and V_0. Thus, the only variable in Eq. (6.45) is the flight path angle γ_0, which can take on only two values, corresponding to the underlofted and overlofted trajectories. Obviously, the overlofted trajectory is less sensitive to errors in velocity than the corresponding underlofted trajectory because of the $\cot(\gamma_0)$ term in Eq. (6.45).

Next we must consider range errors that result from errors in the flight path angle. Previously we found a simple expression for the sensitivity of range to errors in the velocity magnitude; in this paragraph and the next, we consider range sensitivity to the direction of the velocity vector. First, we must return to Eq. (6.43) and again make use of Eq. (6.44) to get

$$\frac{\partial R}{\partial \gamma_0} = 2R_e \left[1 - \frac{\sin(\theta_i + 2\gamma_0)}{\sin(2\gamma_0)}\right] \tag{6.46}$$

It should be noted that if $\gamma_0 = \gamma_0^*$ (i.e., if the flight path angle at burnout is set to the value corresponding to the minimum velocity), the range error is insensitive to errors in the flight path angle; i.e.,

$$\left.\frac{\partial R}{\partial \gamma_0}\right|_{\gamma_0 = \gamma_0^*} = 0$$

if we assume that the altitude at burnout is negligible compared to the Earth's radius.

For underlofted trajectories (i.e., $\gamma_0 < \gamma_0^*$) $\partial R/\partial \gamma_0$ is positive, and for overlofted trajectories (i.e., $\gamma_0 > \gamma_0^*$) $\partial R/\partial \gamma_0$ is negative.

If we return to Eq. (6.43) we can determine the effect of an error in altitude Z_0 on range. We obtain

$$\frac{\partial R}{\partial Z_0} = 2\cot(\gamma_0) - \frac{\cos(\gamma_0 + \theta_i)}{\cos(\gamma_0)} \tag{6.47}$$

Consider again the situation for which the optimum burnout flight path angle $\gamma_0 \doteq \gamma_0^* = 26.25$ deg and the central angle $\theta_i = 75$ deg. Under such circumstances,

$$\frac{\partial R}{\partial Z_0} = 4.2731$$

Thus, if the altitude is in error by about 1.0 km, the range error at impact will be 4.27 km.

The plane containing the nominal trajectory may be defined by three points. The most convenient points for our purposes are the center of the Earth, the point of boost vehicle burnout, and the point of impact. The boost guidance system endeavors to place the velocity vector at burnout in this intended, or nominal, trajectory plane. However, as might be expected, a small lateral error in velocity will result in an impact error because the realized trajectory plane will be slightly different than the nominal plane.

The component of the initial velocity vector V_0 normal to the radius vector is $V_0 \cos(\gamma_0)$; if there exists a velocity error δV_n normal to the intended trajectory

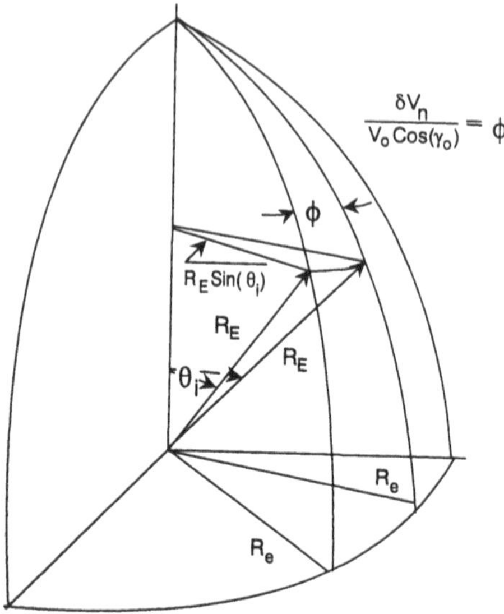

Fig. 6.6 Parameters for the determination of lateral velocity error.

plane, then an angle ϕ develops between the actual and nominal trajectory planes. This angle is easily shown from Fig. 6.6 to be

$$\phi = \frac{\delta V_n}{V_0 \cos(\gamma_0)} \tag{6.48}$$

Although the trajectories may be elliptical, the track of the RV on the Earth will be that of a great circle. The arc length that measures the differences between the two trajectories at impact is designated as δL, which is given by

$$\delta L = (R_e \sin\theta_i)\phi = R_e \sin\theta_i \left[\frac{\delta V_n}{V_0 \cos(\gamma_0)}\right]$$

which simplifies to

$$\frac{\delta L}{\delta V_n} \approx \frac{\partial L}{\partial V_n} = \frac{R_e}{V_0}\frac{\sin(\theta_i)}{\cos(\gamma_0)} \tag{6.49}$$

This important error source is a maximum when the central (or impact) angle θ_i is 90 deg. For a given impact angle, the lateral sensitivity $\delta L/\delta V_n$ is greater for a lofted trajectory than for a shallow trajectory. Using the value $\theta_i = 75$ deg, V_0 is 6876.7 m/s and $\gamma_0 = \gamma_0^*$ is 26.25 deg; the sensitivity of lateral error at impact to the lateral velocity error δV_n is

$$\frac{\partial L}{\partial V_n} = 0.998 \text{ km/(m/s)}$$

Thus, a lateral velocity error of 1.0 m/s results in an impact error of nearly a kilometer.

An error in the displacement of the RV at burnout (or trajectory initiation) causes a positional error at impact. Let's consider positional errors within the nominal trajectory plane and then positional errors normal to the trajectory plane.

A downrange error at booster burnout is designated as δD_0 or $R_e \delta \theta_i$; the corresponding range error at impact is represented as δR or $R_e \delta \theta_i$. Thus, range error sensitivity to downrange position error is simply

$$\frac{\delta R}{\delta D_0} = \frac{R_e \delta \theta}{R_e \delta \theta} = 1 \tag{6.50}$$

A range error of δD_0 translates into an impact error of the same amount.

A displacement error normal to the nominal trajectory of magnitude δN_0 results in an impact range error of δR; the sensitivity of range error at impact to a normal-trajectory positional error can be shown to be

$$\frac{\delta R}{\delta N_0} = \cos(\theta_i) \tag{6.51}$$

The preceding result may be seen in Fig. 6.7 for an RB placed on a longitudinal meridian and launched in a north direction. A displacement from the nominal

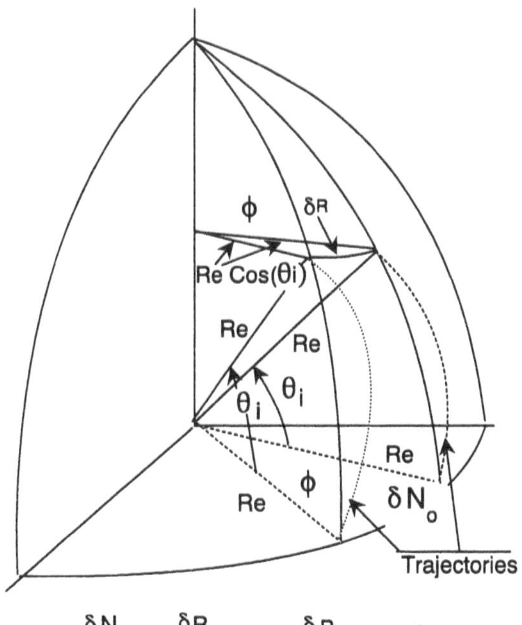

Fig. 6.7 Parameters for the determination of lateral displacement errors.

plane through a distance δN_0 places the RB into another north-directed trajectory. Assume that impact occurs at the subtended impact angle $\theta = \theta_i$. It is easy to show that $\delta R = (R_i \cos \theta_i)\delta\phi$ and $\delta N_0 = R_e \delta\phi$. Taking the ratio of the two expressions gives Eq. (6.51).

We should emphasize at this point that the error sensitivity coefficients developed in this section are equally valid for both a rotating and nonrotating Earth-frame. These coefficients are a measure of the departure of an error-driven trajectory from a nominal trajectory.

The material in this section and the next was taken almost entirely from a paper by Wheelon[1] which was printed in 1959 by the American Rocket Society (now the American Institute of Aeronautics and Astronautics). Certainly, the technology for implementing the ideas expressed by Wheelon has developed enormously over the intervening years. Nevertheless, Wheelon's paper remains a clear and valuable expression of the fundamental physics of ballistic RV's.

Another aspect of RV accuracy examined by Wheelon is concerned with nonspherical contributors to the Earth's gravitational field rather than positional and velocity errors at initiation. This consideration might be expressed as follows: if an impact location were predicted based upon a spherical gravitational model, what would be the displacement of this impact point if a more accurate nonspherical model were used? We consider this question in the next section.

6.5 Oblateness Effects

It is not our intention to prepare a handbook; rather, the ideas in this book are developed from first principles. However, some departure from this approach is necessary in the analysis of the effects of Earth oblateness on the Keplerian trajectory.

Figure 6.8 shows some of the variables that are used in the discussion of trajectories associated with both a spherical gravitational field and a nonspherical oblate field. L_B is the latitude of the re-entry body at trajectory initiation, or the burnout point on the boost trajectory. L_{Ts} and L_{To} are the latitudes of the target, or impact point, for trajectories based upon the spherical and oblate gravitational fields, respectively.

From Eq. (3.70) we may write the gravitational potential, including only the first of the nonspherical terms, as

$$U = \frac{GM_E}{R}\left[1 + \frac{3}{2}J_2\left(\frac{R_e}{R}\right)^2\left(\frac{1}{3} - \cos^2\phi\right)\right] \tag{6.52}$$

In the above equation ϕ is the colatitude of the re-entry body, and the terms R_e (Earth radius at the equator) and J_2 (oblateness term, or second-order Jeffery constant) are given in Table 3.1 as

$$R_e = 6.378 \times 10^6 \text{ m}, \qquad J_2 = 1.0826 \times 10^{-3}$$

The term GM_E in Eq. (6.52) is a bit awkward for our interest in targeting. First, we note that by the definition of U, the gravitational acceleration for a

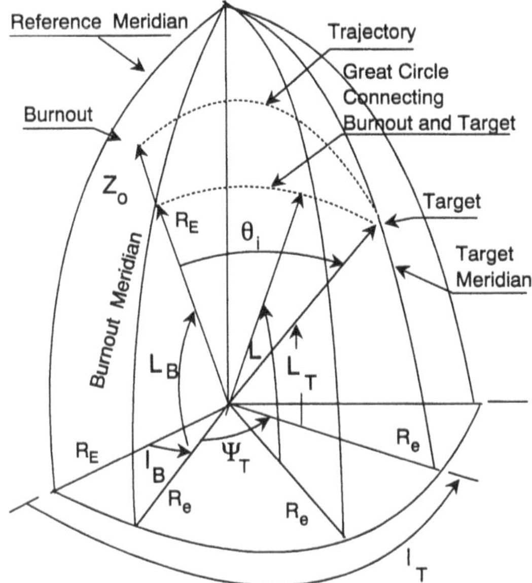

Fig. 6.8 Burnout point and target geometry.

spherical gravitational field may be written as

$$\frac{\partial U}{\partial R} = g = -\frac{GM_E}{R^2}$$

Evaluating the gravitational field at the equator then allows us to replace GM_E as follows:

$$GM_E = g_e R_e^2 = \mu$$

This allows us to rewrite the potential function as

$$U = \frac{\mu}{R_e}\left[\left(\frac{R_e}{R}\right) + \frac{3}{2}J_2\left(\frac{R_e}{R}\right)^3\left(\frac{1}{3} - \cos^2\phi\right)\right] \tag{6.53a}$$

or

$$U = \frac{\mu}{R}\left[1 + \frac{3}{2}J_2\left(\frac{R_e}{R}\right)^2\left(\frac{1}{3} - \cos^2\phi\right)\right] \tag{6.53b}$$

The bracketed term in Eq. (6.53b) is a term which "corrects" the spherical gravity potential.

Our first task is to replace the colatitude ϕ by the central angle θ. From spherical trigonometry it can be shown that

$$\cos(\phi) = \cos(\theta)\sin(L_B) + \sin(\theta)\cos(L_B)\left[1 - \frac{\cos^2(L_T)\sin^2(\psi_T)}{\sin^2(\theta_i)}\right]^{1/2} \tag{6.54}$$

where ψ_T is the difference between the longitude of the target and the longitude of burnout. We can now rewrite Eq. (6.53b) as

$$U = \frac{\mu}{R_e}\left[\frac{R_e}{R} + \frac{3}{2}J_2\left(\frac{R_e}{R}\right)^3 f(\theta)\right] \tag{6.55a}$$

where

$$f(\theta) = \tfrac{1}{3} - [A\cos\theta + B\sin\theta]^2 \tag{6.55b}$$

where A and B are given by

$$A = \sin(L_B) \tag{6.55c}$$

$$B = \cos(L_B)\left[1 - \frac{\cos^2(L_T)\sin^2(\psi_T)}{\sin^2(\theta_i)}\right]^{1/2} \tag{6.55d}$$

The equations of motion given by Eqs. (6.6) were written for a spherical gravitational field. This means that $J_2 = 0$, in which case the derivative of the potential given in Eq. (6.55a) produces Eq. (6.4c). When J_2 is not zero, the gravitational acceleration becomes

$$\begin{aligned} g &= \frac{\partial U}{\partial R}\boldsymbol{I}_R + \frac{1}{R}\frac{\partial U}{\partial \theta}\boldsymbol{I}_\theta \\ &= -\frac{\mu}{R_e}\left[\frac{R_e}{R^2} + \frac{9}{2}\left(\frac{R_e^3}{R^4}\right)f(\theta)\right]\boldsymbol{I}_R + \left[\frac{3}{2}J_2\left(\frac{R_e}{R}\right)^3\frac{\mathrm{d}f}{\mathrm{d}\theta}\right]\boldsymbol{I}_\theta\left(\frac{1}{R}\right) \end{aligned} \tag{6.56}$$

The equations of motion now become

$$\frac{\mathrm{d}^2 r}{\mathrm{d}t^2} - r\left(\frac{\mathrm{d}\theta}{\mathrm{d}t}\right)^2 + \frac{\mu}{R_e}\left[\frac{R_e}{R} + \frac{9}{2}J_2\left(\frac{R_e^3}{R^4}\right)f(\theta)\right] = 0 \tag{6.57a}$$

$$\frac{\mathrm{d}}{\mathrm{d}t}\left(r^2\frac{\mathrm{d}\theta}{\mathrm{d}t}\right) - \frac{3}{2}J_2\left(\frac{R_e}{R}\right)^3\frac{\mathrm{d}f}{R\mathrm{d}\theta} = 0 \tag{6.57b}$$

It is clear that the presence of the oblateness term J_2 means that integrating Eq. (6.57b) will not yield the constant P given in Eq. (6.7) because the presence of the oblateness term introduces a component of gravitational force normal to the radius vector. Wheelon[1] carries out a perturbational solution for Eqs. (6.57), regarding the J_2 terms as higher-order than the spherical terms.

Space does not permit a detailed retracing of Wheelon's analysis. The inclusion of the oblateness effect means that there is a downrange and cross-range displacement of the impact point from that computed using only the spherical gravity model. The downrange and crossrange displacements are indicated by δR and δL, respectively. These displacements of the "oblateness" impact point from the "spherical" impact point are given below in terms of the Jeffery constant J_2, the initial conditions λ and γ_0, and the subtended central angle θ_i.

The downrange error δR is

$$\delta R = -\frac{(3/2)J_2 R_e}{\lambda^2 \cos^3\gamma_0 \sin\gamma_0}\left\{2\left(\frac{1}{3} - A^2\right)(1 - \cos\theta_i) + \left(\lambda\cos^2\gamma_0 - \frac{2}{3}\right)[(1 - \cos\theta_i) + A^2(\cos^2\theta_i + \cos\theta_i - 2) - B^2(1 - \cos\theta_i)^2 - 2AB\sin\theta_i(1 - \cos\theta_i)]\right\} \tag{6.58a}$$

and the crossrange error δL is

$$\delta L = -\left(\frac{2J_2 R_e}{\lambda\cos^2\gamma_0}\right)\left(\frac{\sin\psi_T \cos L_B \cos L_T}{\sin\theta_i}\right)[B(\theta_i - \sin\theta_i) + A(1 - \cos\theta_i)] \tag{6.58b}$$

A program in true basic which implements Eqs. (6.58) is given in Table 6.3, with sample output given in Table 6.4. To provide a representative output, the initial velocity magnitude V_0 and flight path angle γ_0 are set at 7200 m/s and 42.5 deg, respectively. The subtended trajectory angle θ_i is set at 75 deg. It must be emphasized that V_0, γ_0, and θ_i must satisfy Eq. (6.34).

With the above values taken as representative of a Keplerian trajectory, the output shows that the downrange error is about 9 km and the crossrange error is −9 km. The term *error* here means that if targeting were carried out using only the spherical gravity model, then the impact point would be displaced from the targeted point. In this sense, then, there would be an impact error. The term *targeting* here refers to the prediction of the impact point based upon only those conditions which exist at the time of trajectory initiation.

6.6 Earth Rotation Effects

In the error analysis given in the previous section, the rotation of the Earth was ignored. However, we must include the effect of Earth rotation if we are initializing a trajectory which is intended to impact a point on the Earth's surface. Figure 6.9 shows a re-entry body at point B, the point of trajectory initialization. Two other points of interest are $\boldsymbol{r}_{T/B}$, the location of the target T when the RB is at B, and $\boldsymbol{r}_T$, the location of the target at the time of re-entry impact. (Since the target is the intended point of impact, it is tacitly assumed that these two points, target and impact points, will coincide.) As viewed from a nonrotating

Table 6.3 Program to calculate downrange and crossrange positional errors at impact due to Earth oblateness

```
! THIS PROGRAM CALCULATES THE DOWN-RANGE AND CROSS-RANGE
!  POSITIONAL ERRORS AT IMPACT DUE TO THE OBLATENESS OF THE
!  EARTH (NON-SPHERICITY OF THE GRAVITATIONAL FIELD).
!
! INPUT
!      LB0: LATITUDE OF THE RV AT BURNOUT (DEG)
!      LTG: LATITUDE OF TARGET OR INTENDED IMPACT POINT (DEG)
!      V0 : INITIAL (BURNOUT) VELOCITY MAGNITUDE (M/S)
!      GAM0: INITIAL FLIGHT PATH ANGLE (DEG)
!      Z0: INITIAL ALTITUDE
!      PSI: LONGITUDE DIFFERENCE (TARGET - INITIAL) (DEG)
!      THETAI: SUBTENDED CENTRAL ANGLE
! OUTPUT
!      DRAN: DOWN RANGE ERROR (KM)
!      XRAN: CROSS RANGE ERROR (KM)

V0 = 7200.0
Z0 = 0.0
THETAI = 75
GAM0 = 42.5
LB0 = 45.0
LTG = 45.0
PSI = 60.0
MU = 3.9853E+14
OPTION ANGLE DEGREES
N1$ = "+####.##   +####.##     ###.##     #####.#    ###.##"
RE = 6.378E+6
J = 1.0826E-3
R = RE+Z0
LAMD0 = (V0^2)*R/MU
A = SIN(LB0)
B = SQR(1-((COS(LTG)^2)*(SIN(PSI)^2))/(SIN(THETAI)^2))
B = B*COS(LB0)
Z9 = (B^2)*(1-COS(THETAI))^2+2*A*B*SIN(THETAI)*(1-COS(THETAI))
Z9 = (1-COS(THETAI))+(A^2)*(COS(THETAI)^2+COS(THETAI)-2)-Z9
Z9 = (LAMD0*((COS(GAM0)^2)-2/3))*Z9
Z9 = 2*((1/3)-A^2)*(1-COS(THETAI))+Z9
DRAN = ((-1.5)*J*RE/((LAMD0^2)*COS(GAM0)^3*SIN(GAM0)))*Z9
Z8 = A*(1-COS(THETAI))+B*(RAD(THETAI)-SIN(THETAI))
Z8 = SIN(PSI)*COS(LB0)*COS(LTG)*Z8/SIN(THETAI)
XRAN = (-1)*2*J*RE*Z8/(LAMD0*COS(GAM0)^2)
PRINT "DWN-RANGE  X-RANGE  SUB-ARC  INIT VEL  FLT PATH ANG"
PRINT "   KM        KM     DEGREES  M/S       DEGREES"
PRINT
PRINT USING N1$: DRAN/1000,XRAN/1000,THETAI,V0,GAM0
END
```

Table 6.4 Sample output from program in Table 6.3

DWN-RANGE KM	X-RANGE KM	SUB-ARC DEGREES	INIT VEL M/S	FLT PATH ANG DEGREES
+9.11	-9.11	75.00	7200.0	42.5

frame, the target moves along a minor circle through the distance D, which is given by

$$D = (\omega_e r_T \cos L_T) T_F \tag{6.59}$$

where ω_e is the Earth rotation rate, r_T is the geocentric distance to the target, L_T is the latitude of the target, and T_F is the time of flight (the duration of the trajectory from point B to point T). Equation (6.59) shows that knowing the target location in a nonrotating frame requires knowledge of the time of flight when the initial conditions of the trajectory are set at point B. Thus, time of

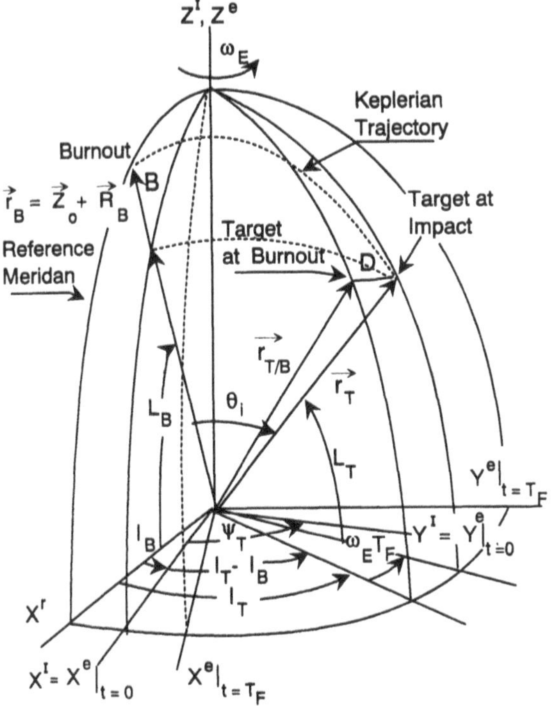

Fig. 6.9 Motion of impact point due to target rotation.

flight is a constraint that must be applied when initializing the trajectory in order to meet the targeting requirements.

The study of Keplerian trajectories subjected to a time-of-flight constraint is known as *Lambert's Problem*. A theorem attributed to Lambert (circa 1761) relates the time of flight T_F to the semimajor axis a, the sum of the radii to the initial and final points, $r_B + r_T$, and the chord c between points B and T. Further discussion of Lambert's problem is given by Kaplan[2] and Battin.[3]

The remainder of this section outlines the equations needed to account for Earth rotation in the initiation of the trajectory. Referring again to Fig. 6.9, we assume that the inertial reference meridian containing the $X^I Z^I$ plane contains the point B (the point of trajectory initiation). The celestial longitude of the target at impact, ψ_T, is given by

$$\psi_T = (l_T - l_B) + \omega_e T_F \tag{6.60}$$

The angle between the Earth-fixed meridians is of course $l_T - l_B$. The burnout and target vectors $\boldsymbol{r}_B$ and $\boldsymbol{r}_T$, respectively, may be conveniently expressed in the inertial frame as

$$\boldsymbol{r}_B^I = [(R_B + Z_0)\cos(L_B), 0, (R_B + Z_0)\sin(L_B)]^{\mathrm{T}} \tag{6.61a}$$

$$\boldsymbol{r}_T^I = [R_T\cos[L_T\cos(\psi_T)], R_T\cos[L_T\sin(\psi_T)], R_T\sin(L_T)]^{\mathrm{T}} \tag{6.61b}$$

The above equations may be simplified by assuming a spherical Earth so that $R_T = R_B = R_E = R_e$, the radius of a spherical Earth.

The range angle θ_i, or the central orbital angle subtended by the re-entry body, follows from Eqs. (6.61) as

$$\theta_i = \cos^{-1}\left[\frac{\boldsymbol{r}_T^{\mathrm{T}}\boldsymbol{r}_B}{R_E(R_E + Z_0)}\right]$$

or from Eq. (6.61),

$$\theta_i = \cos^{-1}[\cos(L_B)\cos(L_T)\cos(\psi_T) + \sin(L_B)\sin(L_T)] \tag{6.62}$$

Fixing the time of flight fixes the angle of inertial or celestial longitude difference ψ_T, which in turns fixes the central angle θ_i of the suborbital arc between points B and T. Equation (6.41b) shows that if the time of flight T_F is fixed, there is an implicit relationship between V_0 and γ_0, the initial or burnout values of velocity magnitude and flight path angle. A minor point to note is that the velocity ratio λ, which also appears in Eq. (6.41b) depends upon both V_0 and $r_B = R_B + Z_0$, where Z_0 is the altitude of the burnout point B. If the vectors $\boldsymbol{r}_B$ and $\boldsymbol{r}_T$ are considered fixed (i.e., $r_B = r_T = R_e$; and L_B, L_T, l_B, l_T are known), then V_0 and λ are merely alternate forms. V_0 and γ_0 must satisfy Eq. (6.34), the impact equation. Thus, there can be at most only a single pair of values of (V_0, γ_0) that will carry the re-entry body from initialization ($\boldsymbol{r}_B$) to target ($\boldsymbol{r}_T$) in a fixed time of flight T_F.

We can now summarize this section by bringing together all relationships relevant to the targeting problem. First, we must emphasize that targeting involves fixing the time of flight T_F, because Earth rotation moves the Earth-fixed target. The celestial longitude difference ψ_T depends upon the time of flight T_F:

$$\psi_T = (l_T - l_B) + \omega_e T_F$$

In the earlier material there was no distinction between velocity relative to the Earth and velocity relative to inertial space because Earth rotation was ignored. At this point we must make the distinction between V_0, the velocity relative to the Earth, and U_0, the velocity relative to inertial space.

The velocity relative to inertial space is

$$\boldsymbol{U}_0 = (l V_0)\boldsymbol{I}^e + [m V_0 + \omega_e (R_B + Z_0)\cos(L_B)]\boldsymbol{J}^e + (n V_0)\boldsymbol{K}^e \tag{6.63a}$$

where $\boldsymbol{I}^e$, $\boldsymbol{J}^e$, $\boldsymbol{K}^e$ are unit vectors along the X^e, Y^e, Z^e axes. Its magnitude is given by

$$U_0 = V_0 \left[l^2 + \left(m + \frac{\omega_e (R_B + Z_0)\cos(L_B)}{V_0} \right)^2 + n^2 \right]^{1/2} \tag{6.63b}$$

where l, m, and n are direction cosines. Note that Eq. (6.63b) may be used to determine V_0, the velocity magnitude relative to the Earth once U_0, the velocity relative to inertial space, is known.

If the trajectory end points ($\boldsymbol{r}_B$ and $\boldsymbol{r}_T$) and time of flight are known, then the subtended angle θ_i is also known.

$$\theta_i = \cos^{-1}[\cos(L_B)\cos(L_T)\cos(\psi_T) + \sin(L_B)\sin(L_T)] \tag{6.62}$$

The time of flight, T_F, and impact equations [Eqs. (6.41b) and (6.34)] may now be used to set unique values of U_0 (rather than V_0) and γ_0.

$$\begin{aligned} T_F = {} & \frac{R_B + Z_0}{U_0 \cos(\gamma_0)} \left\{ \frac{\tan(\gamma_0)(1 - \cos\theta_i) + (1 - \lambda)\sin\theta_i}{(2 - \lambda)\{[(1 - \cos\theta_i)/(\lambda\cos^2\gamma_0)] + [\cos(\gamma_0 + \theta_i)/\cos\gamma_0]\}} \right\} \\ & + \left\{ \frac{2(R_B + Z_0)}{U_0 \lambda [(2/\lambda) - 1]^{3/2}} \tan^{-1} \left[\frac{[(2/\lambda) - 1]^{1/2}}{\cos(\gamma_0)\cot(\theta_i/2) - \sin(\gamma_0)} \right] \right\} \end{aligned} \tag{6.64}$$

and

$$U_0 = \left(\frac{\mu}{R_B + Z_0} \right) \left[\frac{1 - \cos(\theta_i)}{[(R_B + Z_0)/R_B]\cos^2\gamma_0 - \cos(\theta_i + \gamma_0)\cos\gamma_0} \right]^{1/2} \tag{6.65}$$

where the velocity ratio λ given in Eq. (6.17) may be rewritten as

$$\lambda = (R_B + Z_0)U_0^2/\mu \tag{6.66}$$

These equations define unique values of U_0 and γ_0 to bring the re-entry body from $\boldsymbol{r}_B$ to $\boldsymbol{r}_T$ in the time of flight T_F.

With U_0 and γ_0 fixed, we can now write three equations to determine the directional cosines of the velocity vector. First, the flight path angle γ_0 is determined from

$$\begin{aligned}\cos(\pi/2 - \gamma_0) &= \sin(\gamma_0)\\ &= \frac{\boldsymbol{R}_B \cdot \boldsymbol{V}_0}{R_B V_0}\end{aligned}$$

to give

$$\sin(\gamma_0) = l\cos(L_B) + n\sin(L_B) \tag{6.67}$$

Next, we have the requirement that the velocity $\boldsymbol{U}_0$ be in the trajectory plane, i.e.,

$$\frac{\boldsymbol{U}_0}{U_0} \cdot \frac{\boldsymbol{R}_B \times \boldsymbol{R}_T}{R_B R_T} = 0$$

which gives

$$\begin{aligned}l[\cos(L_T)\sin(\psi_T)\sin(L_B)] + m[&\sin(L_T)\cos(L_B)\\ &- \cos(L_T)\sin(L_B)\cos(\psi_T)]\\ &+ n[-\cos(L_B)\cos(L_T)\sin(\psi_T)] = 0\end{aligned} \tag{6.68}$$

Finally, we have the orthogonality requirement

$$l^2 + m^2 + n^2 = 1 \tag{6.69}$$

Setting the initial and final points of the trajectory determines the velocity magnitude U_0 and the angle γ_0 that the velocity vector makes with the local horizontal at trajectory initiation. (The local horizontal plane is that plane whose normal is the vector $\boldsymbol{r}_B$.) Once U_0 and γ_0 are known, it is possible to use the requirement that $\boldsymbol{U}_0$ lie in the trajectory plane and the orthogonality condition to find the directional cosines of the velocity vector $\boldsymbol{U}_0$.

6.7 Deployment Attitudes

The final consideration in this chapter concerns orienting the re-entry body at boost termination in order to meet certain conditions at atmospheric entry or at impact. If the re-entry body encounters the atmosphere at an arbitrary angle of attack, then the side loads will also be arbitrary. Let us define here a zero angle of attack as that angle at which there are no lifting or side loads. Thus, we might be interested in deploying the re-entry vehicle at such an attitude that it will enter the atmosphere at a zero angle of attack. Figure 6.10 illustrates that

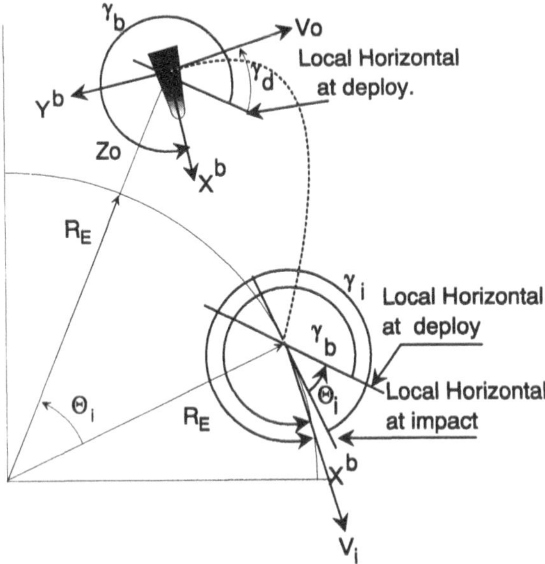

Fig. 6.10 Relationship between body deployment angle, γ_b, and flight path angle at impact, γ_i.

the body deployment angle γ_b, the angle that the X^b-axis, or axis of symmetry, makes with respect to the local horizontal, is

$$\gamma_b = \gamma_i - \theta_i \tag{6.70}$$

where γ_i is the angle that the velocity vector makes with the local horizontal at atmosphere entry and θ_i is the central angle subtended by the trajectory from burnout to impact (or, in this case, atmospheric entry). Note that γ is defined as positive above the local horizontal. Of course, if γ were defined as positive below the local horizontal, then the above equation would become

$$\gamma_b = \gamma_i + \theta_i$$

Another way of setting the deployment angle of the body (the angle which the X^b-axis of symmetry makes with respect to the horizontal) is to meet the *null-miss condition*. This is done as follows.

We expressed the sensitivity of the range R to errors in the velocity magnitude V_0 in Eq. (6.45) and to errors in the flight path angle at deployment γ_0 in Eq. (6.46). These sensitivities were found to be

$$\frac{\partial R}{\partial V_0} = \frac{2R_e}{V_0}\left\{\sin(\theta_i) + \cot(\gamma_0)[1 - \cos(\theta_i)]\right\} \tag{6.45}$$

$$\frac{\partial R}{\partial \gamma_0} = 2R_e\left[1 - \frac{\sin(\theta_i + 2\gamma_0)}{\sin(2\gamma_0)}\right] \tag{6.46}$$

where V_0 is the magnitude of the velocity vector at deployment, γ_0 is the flight path angle at deployment, θ_i is the range angle from deployment to impact, and R_e is the radius of Earth (assumed spherical).

Since $\partial R/\partial V_0$ represents the sensitivity of the range to an error or perturbation in the velocity magnitude, the sensitivity to a velocity vector normal to the velocity is

$$\frac{\partial R}{\partial V_n} = \frac{1}{V_0}\frac{\partial R}{\partial \gamma_0} \tag{6.71}$$

Consequently, a range error due to both a magnitude and directional error in the velocity vector is

$$\Delta R = \left(\frac{\partial R}{\partial V_0}\right)\Delta V + \left(\frac{1}{V}\frac{\partial R}{\partial \gamma_0}\right)\Delta V_n \tag{6.72}$$

Suppose that the axis of symmetry of the re-entry body is along the line X^b as before. The appropriate geometry is illustrated in Fig. 6.11. Let ΔV_x be the velocity error along the line X^b such that this line makes an angle α with respect to the velocity vector. Thus, the two velocity components ΔV and ΔV_n may be written as

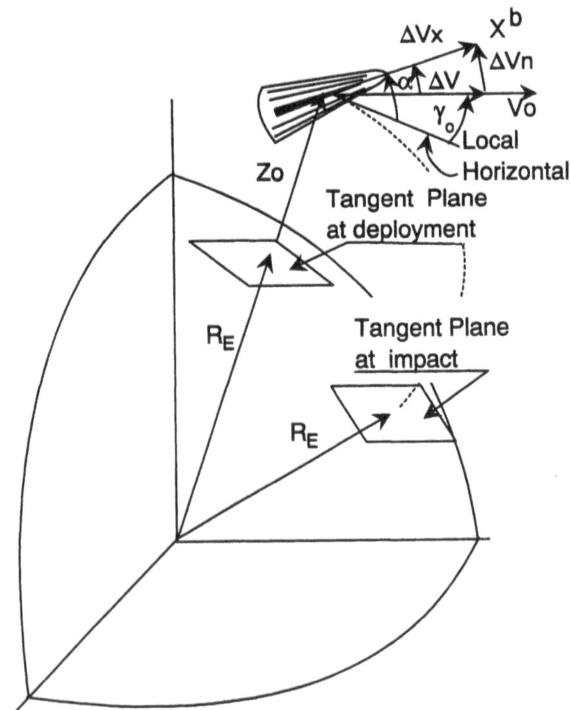

Fig. 6.11 Setting body angle relative to velocity vector for null miss.

$$\Delta V = \Delta V_x \cos(\alpha) \tag{6.73a}$$

$$\Delta V_n = \Delta V_x \sin(\alpha) \tag{6.73b}$$

Inserting Eqs. (6.73) into Eq. (6.72) gives

$$\Delta R = \left[\left(\frac{\partial R}{\partial V_0}\right)\cos(\alpha) + \left(\frac{1}{V_0}\frac{\partial R}{\partial \gamma_0}\right)\sin(\alpha)\right]\Delta V_x \tag{6.74}$$

For zero (or null) error at impact, the bracketed term in Eq. (6.74) must be equal to zero, i.e.,

$$\alpha = \tan^{-1}\left[-V_0\left(\frac{\partial R}{\partial V_0}\right)\bigg/\left(\frac{\partial R}{\partial \gamma_0}\right)\right]\Bigg|_{-\pi/2\leq\alpha\leq\pi/2} \tag{6.75}$$

For a null-miss deployment the re-entry body must be deployed at an angle γ_b, which in turn is related to the flight path angle at deployment γ_0, or simply γ, by the following expression:

$$\gamma_b = \gamma - \alpha \tag{6.76}$$

6.8 Summary

In this chapter we discussed the deployment of the re-entry body as a particle subjected only to the force of a central gravitational field. We confined our attention to this or the Keplerian (gravity-only) phase of the re-entry body trajectory. (The Keplerian phase is sometimes assumed to include orbital motion; however, in the present application the motion is preferably identified as suborbital because the trajectory intersects the Earth's surface.) We then related the orbital parameters such as eccentricity and size of the semimajor axis to the magnitude and direction of the initial velocity vector. We examined briefly the effects of the nonspherical components of the Earth's gravitational field. We then defined targeting to be the selection of trajectory conditions at burnout (which gives the initial velocity vector) that will bring the RB to the intended target point in a preset amount of time. Finally, we considered re-entry vehicle deployment attitudes in order to meet conditions at re-entry such as zero angle of attack at atmospheric entry or Earth impact.

References

[1]Wheelon, A. D., "Free Flight of a Ballistic Missile," *Journal of the American Rocket Society,* Vol. 29, Dec. 1959, pp. 915–926.

[2]Kaplan, M. H., *Modern Spacecraft Dynamics & Control,* Wiley, Somerset, NJ, 1976.

[3]Battin, R. H., *An Introduction to the Mathematics and Methods of Astrodynamics,* AIAA Education Series, AIAA, Washington, DC, 1987.

7
Re-Entry Vehicle Particle Mechanics

7.1 Re-Entry Physics

In Chapter 6 we considered Keplerian motion—motion determined only by the initial conditions and the gravitational field of the central planet. Below an altitude of approximately 60 km, atmospheric contact forces (i.e., aerodynamic loads) become comparable to the gravitational forces. Consequently, aerodynamic forces must be included in trajectory modeling. At much lower altitudes, these aerodynamic forces dominate the motion. (For many re-entry bodies a fairly adequate representation of the essential features of the low-altitude trajectory may be obtained by ignoring the gravitational forces entirely.)

We should note that the sensible atmosphere for re-entry vehicle trajectory analysis has a thickness not much greater than 2% of the Earth's radius. Peak deceleration occurs over a distance that is between 0.2% and 0.5% of the Earth's radius. To put this relative thickness in some perspective, the skin of an apple is usually somewhat greater than 2% of the apple radius. Thus, a body entering the atmosphere at a significant fraction of orbital speed encounters the atmosphere as a near-discontinuity.

In this chapter we develop expressions for the axial and transverse loads which act on a re-entry vehicle. Although an adequate assessment of the inertial loads acting on an atmospheric entry body is essential for the structural design, the management of thermal loads is an even greater challenge. A body entering the atmosphere at a speed of 6000 m/s (19,686 ft/s) has a specific enthalpy of 18 MJ/kg (7745 Btu/lb). To put this quantity in some perspective, we note that carbon vaporizes at 60.3 MJ/kg (28,700 Btu/lb), nickel at 4.23 MJ/kg (1820 Btu/lb), and water at 2.32 MJ/kg (1000 Btu/lb). Obviously, a considerable amount of material would be lost if all of the energy associated with relative atmospheric motion were absorbed through vaporization of the body. Yet many atmospheric entry bodies, both natural and artificial, survive to impact or landing with most of their structural integrity intact. A large portion of the thermal energy must therefore dissipate by means of radiation and in the near flowfield by means of complex irreversible flow processes.

A qualitative relationship can be found between peak inertial and thermal loads. The drag force D may be written as

$$D = \tfrac{1}{2}\rho V^2 S C_D$$

where ρ is the local atmospheric density, S is the reference area (usually the maximum cross-sectional area), V is the velocity magnitude, and C_D is the drag coefficient. If we assume that C_D is constant, then the peak axial loading occurs where ρV^2 is a maximum. The heat flow per unit area, q, at the gas-solid interface is a fraction f of the kinetic energy per unit time per unit area, or power per unit area; i.e.,

$$q = f C_D \left(\rho V^3/2\right)$$

7.2 Equations of Planar Motion

In this section we derive the equations of planar motion of a re-entry vehicle. If the RV is subjected to drag-only, or ballistic flight, the restriction of planar motion results in no loss of generality; however, if the RV is capable of generating lift forces, then restricting it to a planar trajectory does limit the utility of the results. Nevertheless, planar trajectory analysis can lead to closed-form expressions that can be very useful in assessing re-entry vehicle performance.[1,2]

As promised in Chapter 5, Eq. (5.23) will find extensive application in this chapter. Let's consider Eq. (5.23b) (the matrix formulation) and apply it to the planar trajectory shown in Fig. 7.1. This figure shows an RB subjected to the aerodynamic forces of lift and drag as well as a geocentric gravitational field. The following three sets of axes are used in the description:

1) (X^I, Y^I, Z^I)—An inertial frame such that the $X^I Z^I$-plane contains the velocity vector $\boldsymbol{V}$ throughout the motion.

2) (X^l, Y^l, Z^l)—A local frame such that the $X^l Y^l$-plane is that of the local horizontal, i.e., the Z^l-axis is along the local vertical.

3) (X^m, Y^m, Z^m)—A moving frame attached to the RB such that the $X^m Z^m$-plane is coincident with the trajectory plane and the X^m-axis is along the velocity vector at all times.

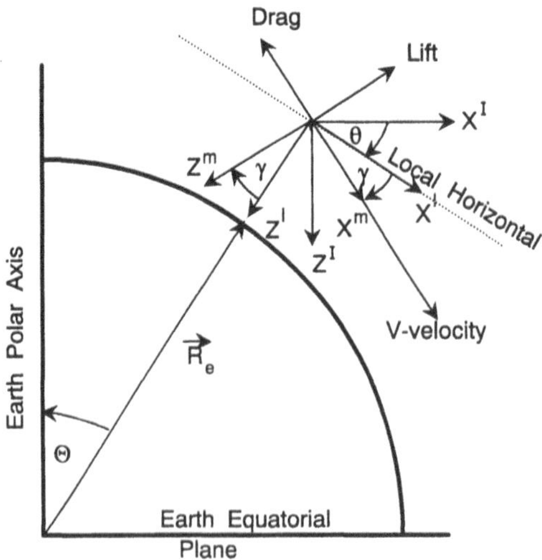

Fig. 7.1 Planar re-entry trajectory.

In the discussion to follow the flight path angle γ is considered positive when the velocity vector is below the local horizontal. Since the Z^l-axis is aligned with the local vertical, the gravitational acceleration vector is coincident with the Z^l-axis. The aerodynamic forces are drag and lift; the drag and lift forces are in the XZ-plane (all axis systems), with drag along the negative X^m-axis and lift normal to the X^m-axis and positive in the direction of the negative Z^m-axis. Because the trajectory is confined to a plane, the Y-axes of all three systems are coincident and positive out of the plane.

First we rewrite Eq. (5.23b) as

$$\begin{aligned}\frac{\boldsymbol{F}^m}{m} &= \frac{\boldsymbol{F}_a^m}{m} + C_l^m \frac{\boldsymbol{F}_g^l}{m} \\ &= \dot{\boldsymbol{V}}^m + \Omega_{m/I}^m \boldsymbol{V}^m\end{aligned} \tag{7.1}$$

In accordance with the definitions of the axis systems, the terms in the above equation may be written as

$$\frac{\boldsymbol{F}_a^m}{m} = \left[-\frac{D}{m},\ 0,\ -\frac{L}{m}\right]^{\mathrm{T}} \tag{7.2a}$$

$$\boldsymbol{F}_g^l/m = \boldsymbol{g}^l = [0, 0, g]^{\mathrm{T}} \tag{7.2b}$$

$$C_l^m = \begin{bmatrix} \cos(\gamma) & 0 & \sin(\gamma) \\ 0 & 1 & 0 \\ -\sin(\gamma) & 0 & \cos(\gamma) \end{bmatrix} \tag{7.2c}$$

$$\Omega_{m/I}^m = \Omega_{l/I}^m + \Omega_{m/l}^m = \begin{bmatrix} 0 & 0 & -(\dot{\theta} + \dot{\gamma}) \\ 0 & 0 & 0 \\ (\dot{\theta} + \dot{\gamma}) & 0 & 0 \end{bmatrix} \tag{7.2d}$$

$$\boldsymbol{V}^m = [V, 0, 0]^{\mathrm{T}} \qquad \dot{\boldsymbol{V}}^m = [\dot{V}, 0, 0]^{\mathrm{T}} \tag{7.2e}$$

Inserting the previous expressions into Eq. (7.1) and separating the components gives the following two equations for planar motion:

$$\frac{\mathrm{d}V}{\mathrm{d}t} = -\frac{D}{m} + g\sin(\gamma) \tag{7.3a}$$

$$V\left(\frac{\mathrm{d}\theta}{\mathrm{d}t} + \frac{\mathrm{d}\gamma}{\mathrm{d}t}\right) = -\frac{L}{m} + g\cos(\gamma) \tag{7.3b}$$

Drag D and lift L may be written in conventional coefficient form as

$$D = C_D \rho V^2 S/2 \qquad L = C_L \rho V^2 S/2 \tag{7.4}$$

The drag coefficient may be written in terms of the ballistic coefficient β, defined as

$$\beta = W/C_D S = mg/C_D S \tag{7.5a}$$

and the lift coefficient may be defined in terms of the lift-to-drag ratio as follows:

$$C_L = \left(\frac{C_L}{C_D}\right) C_D = \left(\frac{C_L}{C_D}\right) \frac{S\beta}{W} \tag{7.5b}$$

Equations (7.3) may be rewritten in terms of the alternate parameters of Eqs. (7.4) and (7.5) as follows:

$$\frac{\mathrm{d}V}{\mathrm{d}t} = -\left(\frac{\rho g}{2\beta}\right) V^2 + g\sin(\gamma) \tag{7.6a}$$

$$V\left(\frac{\mathrm{d}\gamma}{\mathrm{d}t}\right) = -V\left(\frac{\mathrm{d}\theta}{\mathrm{d}t}\right) - \left(\frac{\rho g}{2\beta}\right)\frac{C_L}{C_D} V^2 + g\cos(\gamma) \tag{7.6b}$$

Equations (7.6) are two simultaneous equations which relate three variables: velocity magnitude V, flight path angle (or velocity direction) γ, and central angle θ (or latitude if the trajectory plane is contained in the Earth's meridian plane). These two equations might be identified as *physics equations* in that they result from the application of Newton's Second Law of Motion. In addition, two other equations can also be written based upon the geometry associated with the constrained motion. These equations, called *kinematic equations*, are the horizontal and vertical components of the velocity vector. From Fig. 7.1 it is obvious that

$$\frac{\mathrm{d}h}{\mathrm{d}t} = -V\sin(\gamma) \tag{7.7a}$$

$$\frac{\mathrm{d}\theta}{\mathrm{d}t} = \frac{V\cos(\gamma)}{R_e + h} \tag{7.7b}$$

where h is the altitude of the RB.

We now have four first-order ordinary nonlinear differential equations in terms of the state variables (V, γ, h, θ). We identify these four variables as *states* because they are obtained or evaluated from integration; i.e., a state variable is a variable which is obtained from the process of integration.

In addition, we require two algebraic equations which model the environment: the first gives the atmospheric density, and the second gives the gravitational acceleration. Fairly elaborate models have been developed in Chapters 2 and 3 to account for the altitude variation of density and gravitational acceleration. From Eq. (2.32) we have

$$\rho = \rho_0 e^{-h/H} \tag{7.8a}$$

where $\rho_0 = 1.752\ \mathrm{kg/m^3}$ and $H = 6.7 \times 10^3$ m. From Eq. (3.1) we have

$$g = \frac{R_e^2 g_e}{(R_e + h)^2} \approx g_e\left[1 - \left(\frac{2}{R_e}\right)h\right] \tag{7.8b}$$

According to our definition, neither ρ nor g are states because their evaluation is from an algebraic equation.

We note that the variable θ appears only in derivative form and only in Eq. (7.6b) and (7.7b). Thus, we may eliminate θ by rewriting the state equations as

$$\frac{dV}{dt} = -\left(\frac{\rho g}{2\beta}\right)V^2 + g\sin(\gamma) \tag{7.9a}$$

$$V\frac{d\gamma}{dt} = -\left(\frac{\rho g}{2\beta}\right)V^2\left(\frac{C_L}{C_D}\right) + \cos(\gamma)\left[g - \frac{V^2}{R_e + h}\right] \tag{7.9b}$$

$$\frac{dh}{dt} = -V\sin(\gamma) \tag{7.9c}$$

We can then regard θ to be of interest only in finding the range of the RV, R_a, which may be obtained from the following integration:

$$\begin{aligned} R_a &= R_e\int_{\theta=0}^{\theta} d\theta = R_e\int_0^t\left\{\frac{V\cos(\gamma)}{R_e + h(t)}\right\}dt \\ &\approx \int_0^t [V\cos(\gamma)]dt \end{aligned} \tag{7.10}$$

In the final form above, we have ignored the altitude h in comparison to the Earth's radius R_e.

Further comment must be made regarding the definition of the ballistic coefficient β [Eq. (7.5a)] in SI units and in English units. We must also modify the ballistic coefficient to account for vehicle lift.

According to the definition of β given in Eq. (7.5a), the appropriate units should be Pa, or N/m^2. However, some analysts use units of kg/m^2 in the definition of β. We designate this alternate expression for β by the symbol β_m; the relationship between these two representations is

$$\beta_m = \beta/g \tag{7.11}$$

The consequence of representing the ballistic coefficient in units of kg/m^2 is that $\rho g/2\beta$ [Eqs. (7.9)] becomes $\rho/2\beta_m$.

The proper representation of the ballistic coefficient β when the RV is developing lift is another point of concern. Incremental lift increases the drag during lift production and is identified in Chapter 9 as induced drag. It will be shown that a simple analytical representation of the drag coefficient may be given in the form of the drag polar, which relates induced drag to the square of the instantaneous lift coefficient as follows:

$$C_D = C_{D_0}\left[1 + \left(C_L/C_L^*\right)^2\right] \tag{7.12}$$

The above expression shows that the drag coefficient is formed from two design parameters: C_{D_0}, the zero-lift drag coefficient, and C_L^*, the critical lift coefficient (the lift coefficient at the maximum lift-to-drag ratio). The ballistic coefficient may now be defined in terms of the zero-lift drag coefficient as

$$\beta = W/C_{D_0}S \tag{7.13}$$

Consequently, Eqs. (7.9) should be rewritten as

$$\frac{\mathrm{d}V}{\mathrm{d}t} = -\left(\frac{\rho g}{2\beta}\right)\left[1 + \left(\frac{C_L}{C_L^*}\right)^2\right]V^2 + g\sin(\gamma) \tag{7.14a}$$

$$V\frac{\mathrm{d}\gamma}{\mathrm{d}t} = -\left(\frac{\rho g}{2\beta}\right)\left[1 + \left(\frac{C_L}{C_L^*}\right)^2\right]\left(\frac{C_L}{C_D}\right)V^2 + \cos(\gamma)\left[g - \frac{V^2}{R_e + h}\right] \tag{7.14b}$$

$$\frac{\mathrm{d}h}{\mathrm{d}t} = -V\sin(\gamma) \tag{7.14c}$$

In the following sections we will examine certain restricted solutions to Eqs. (7.14). These closed-form solutions will be used to examine some of the important physical ideas associated with atmospheric entry. Of course, we can also obtain computer solutions to Eqs. (7.14) or even an extension of these equations into three dimensions accompanied by elaborate environmental models. However valuable such work may be in certain contexts such as engagement simulations or targeting, the essential physics of re-entry motion is often obscured.

7.3 Re-Entry Case Studies

Case 1: Horizontal Flight—Drag, No Lift

Horizontal flight requires that the flight path angle γ remain near zero (and that the altitude h remain unchanged). Consequently, we need retain only the first of Eqs. (7.14). Furthermore, since the altitude doesn't change, neither does the density [Eq. (7.8a)]. Thus, we have

$$\frac{\mathrm{d}V}{\mathrm{d}t} = -\left(\frac{\rho g}{2\beta}\right)V^2 \tag{7.15}$$

The above equation integrates to provide the velocity magnitude as a function of time as follows:

$$V = \frac{V_0}{1 + (\rho g/2\beta)V_0 t} \tag{7.16a}$$

Clearly, the velocity magnitude decreases monotonically with time from the initial value of V_0.

An alternate form of horizontal motion follows from Eq. (7.15) by changing the independent variable from time to distance, s, as follows:

$$\frac{dV}{dt} = \frac{dV}{ds}\frac{ds}{dt} = V\frac{dV}{ds} = -\left(\frac{\rho g}{2\beta}\right)V^2$$

The above expression clearly integrates to give

$$V = V_0 \exp[-(\rho g/2\beta)s] \tag{7.16b}$$

Equation (7.16b) is of limited use in the analysis of re-entry vehicle dynamics because horizontal trajectory segments occur only infrequently. However, Eq. (7.16b) might be identified as the *ballistic range equation* because it provides a relationship between velocity magnitude V and downrange distance. In a ballistic range, altitude changes (and hence density changes) are inconsequential.

Example 7.1: Use of the Ballistic Equations

Artillery pieces are usually classified according to bore diameter, often in units of centimeters or inches. The British Army uses an archaic classification system in which the weight of a spherical iron projectile that slightly underfits the bore becomes the identification number. Thus, a "10-pounder" has a bore diameter of 4.50 in. because the bore is loosely fit by a 10-lb iron ball. The diameter of a 10-lb iron ball is actually more like 4.25 in. in order to avoid fit difficulties. A "40-pounder" would have a bore diameter slightly in excess of 6.75 in., the diameter of a 40-lb iron ball.

Let's assume that both iron balls (10- and 40-pounder) have a drag coefficient C_D equal to 0.48.[3] The appropriate parameters for the two projectiles are given in Table 7.1.

Table 7.1 Appropriate parameters for 10-pounder and 40-pounder artillery pieces

10-pounder	40-pounder
$C_D = 0.48$	$C_D = 0.48$
$S = 0.0985$ ft^2	$S = 0.2485$ ft^2
$\beta = 211.51$ lb/ft^2	$\beta = 335.3454$ lb/ft^2
$V_0 = 1000$ ft/s	$V_0 = 1000$ ft/sec
$\rho = 0.00238$ slugs/ft^3	$\rho = 0.00238$ slugs/ft^3

Suppose that we wish to compare the velocity magnitudes of a 10- and 40-pounder at a range of 2500 ft. Using Eq. (7.12) we get

$$V_{10_f} = 1000\exp\{-[(.00238)(32.174)(2500)]/[(2)(211.51)]\} = 636.0 \text{ ft/s}$$

$$V_{40_f} = 1000\exp\{-[(.00238)(32.174)(2500)]/[(2)(335.34)]\} = 751.7 \text{ ft/s}$$

Clearly, the projectile with the larger ballistic coefficient retains the larger velocity at impact. Let's look at the ratio of kinetic energies first at launch:

$$KER = \tfrac{1}{2}W_{10}V_{10_0}^2 / \tfrac{1}{2}W_{40}V_{40_0}^2 = 0.250$$

and then at the 2500-ft range:

$$KER = \tfrac{1}{2}W_{10}V_{10_f}^2 / \tfrac{1}{2}W_{40_f}V_{40_f}^2 = 0.179$$

Clearly, the 10-pounder loses more energy, or places a smaller amount of kinetic energy at a point 2500 ft downrange, than does the 40-pounder.

Case 2: Curved Flight Path over a Flat Earth

In this case, we examine nearly horizontal trajectories in which both lift and gravity cause the trajectory to curve. If we limit the trajectory segment to a small fraction of the Earth's circumference, we may accept the flat Earth assumption (infinite Earth radius R_e). If the flight path angle is kept small enough to justify the small angle approximation, we may simplify Eqs. (7.14a) and (7.14b) as follows:

$$\frac{\mathrm{d}V}{\mathrm{d}t} = -\left(\frac{\rho g}{2\beta}\right)V^2\left[1 + \left(\frac{C_L}{C_L^*}\right)^2\right] \tag{7.17a}$$

$$V\frac{\mathrm{d}\gamma}{\mathrm{d}t} = -\left(\frac{\rho g}{2\beta}\right)\left[1 + \left(\frac{C_L}{C_L^*}\right)^2\right]\frac{C_L}{C_D}V^2 + g \tag{7.17b}$$

The above equations can be rewritten in terms of the arc length s. Equation (7.17a) then integrates to

$$V = V_0\exp\left\{-\left(\frac{\rho g}{2\beta}\right)\left[1 + \left(\frac{C_L}{C_L^*}\right)^2\right]s\right\} \tag{7.18a}$$

The above expression is identical to Eq. (7.16b) except that the drag increase due to lift is included. Of course, both expressions are based upon constant atmospheric density.

Since lift is present we must also develop an expression for the time-dependency of the flight angle γ. First, Eq. (7.17b) must be rewritten in terms of s, the arc-length variable, as follows:

$$V^2\frac{\mathrm{d}\gamma}{\mathrm{d}s} = g - \left(\frac{\rho g}{2\beta}\right)\left[1 + \left(\frac{C_L}{C_L^*}\right)^2\right]\frac{C_L}{C_D}$$

The preceding equation easily integrates to yield

$$\gamma = \gamma_0 + \frac{\beta}{\rho V_0^2}\left[\frac{\exp\left\{\frac{\rho g}{\beta}\left[1+\left(\frac{C_L}{C_L^*}\right)^2\right]s\right\}-1}{1+\left(C_L/C_L^*\right)^2}\right] - \frac{\rho g}{2\beta}\left[1+\left(\frac{C_L}{C_L^*}\right)^2\right]\frac{C_L}{C_D}s \tag{7.18b}$$

We should again point out that Eqs. (7.18) are based upon an assumption of small changes in flight path angle. The consequence of such an assumption is that

$$\sin(\gamma) \approx \gamma, \qquad \cos(\gamma) \approx 1, \qquad \frac{\mathrm{d}h}{\mathrm{d}t} \approx -V\gamma$$

Equations (7.18) are essentially an extension of the ballistic range equation [Eq. (7.16b)] to include lift.

Example 7.2: Ballistic Range Flight with Lift

Ballistic ranges are useful ground-based testing facilities for the simulation of some of the aerodynamic and thermodynamic conditions of re-entry. In such facilities a model is gun-launched nearly along the axis of symmetry of the tube; various optical instruments and photographic equipment are distributed along the walls of the tube to record location, attitude, and time of observation. The trajectory reconstructed from these observations can be used to infer the aerodynamic loads acting on the model.

Our interest here is in a relatively minor operational problem that can occur when a model capable of lift is to be tested. Suppose that we wish to fly a model of a re-entry vehicle which possesses a slight configurational asymmetry. Since the asymmetry will produce a lift force, the trajectory must be initiated so that the model will not strike the walls of the range. Once the size and configuration of the re-entry vehicle have been chosen, the only operational controls that require setting are the magnitude and direction of the initial velocity vector. This initial velocity vector is the velocity at launch termination (muzzle velocity, in the case of a gun launch).

Various conditions will lead to at most a band, or range, of acceptable launch velocity magnitudes: a lower bound is set by testing requirements (Mach or Reynolds numbers), with an upper bound set by gun technology. The goal here is to show how calculations based upon equations derived in this chapter can be used to select (V, γ) pairs which will meet the operational requirement of avoiding a wall strike.

Figure 7.2 presents the essentials of a tubular ballistic testing range along with a trajectory which avoids contact with the range structure. The velocity vector at launch is below the horizontal (positive flight path angle γ at launch). Since the lift is at all times assumed to remain in the vertical plane, the trajectory curves away from the floor. The two critical points are shown: point A at the trajectory nadir and point B at the upper end of the range tube. Given a velocity

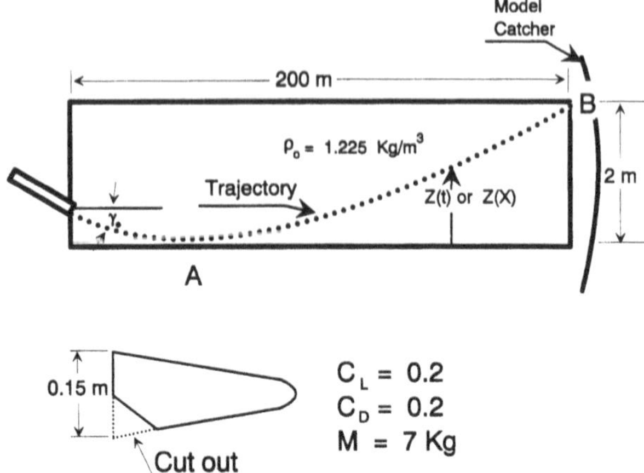

Fig. 7.2 Model trajectory in a ballistic range.

magnitude requirement, we want to find the value or range of values of initial flight path angles that will meet the requirement of avoiding a wall strike for a model with defined mass, drag, and lift characteristics.

In order to show how we might evaluate this operational condition, we will assume that a model with the following characteristics is to be tested in a ballistic range:

$$\begin{aligned} C_D &= 0.2 \\ C_L^* &= 0.2 \\ \text{Mass} &= 7 \text{ kg} \\ \text{Diameter} &= 0.15 \text{ m} \\ \text{Medium density} &= 1.225 \text{ kg/m}^3 \end{aligned}$$

Let's assume that the model is operating at the maximum lift-to-drag ratio, i.e., $C_L = C_L^*$ [see Eq. (7.12)]. Thus, the zero lift drag coefficient is

$$C_{D_0} = \frac{C_D}{1 + (C_L/C_L^*)^2} = \frac{0.2}{2} = 0.1$$

The mass ballistic coefficient is

$$\beta_m = \frac{\beta}{g} = \frac{7}{(0.1)(\pi)(0.15)^2/4} = 3961.2 \text{ kg/m}^2$$

The vertical acceleration, or acceleration in the Z-direction, follows from Eq. (7.17b) as

$$\frac{\mathrm{d}^2 Z}{\mathrm{d}t^2} = -V\frac{\mathrm{d}\gamma}{\mathrm{d}t} = \left(\frac{\rho g}{2\beta}\right)\left[1 + \left(\frac{C_L}{C_L^*}\right)^2\right]\frac{C_L}{C_D}V^2 - g \tag{7.19}$$

Equation (7.19) can be integrated to provide us with the vertical position of the model as a function of time. Since we are primarily interested in the shape of the trajectory, it is more useful to replace time t with downrange distance X as the independent variable. The trajectory can be altered by varying the initial conditions [$V_0 \sin(\gamma_0)$ and Z_0]. However, we are not particularly concerned with the details of the function $Z = Z(X)$, but only with whether or not the trajectory strikes the walls of the range.

First, we combine Eqs. (7.18a) and (7.19) to yield

$$\frac{\mathrm{d}^2 Z}{\mathrm{d}t^2} = K_1 V^2 - g \tag{7.20a}$$

$$V = V_0 e^{-K_2 X} \tag{7.20b}$$

where

$$K_2 = \left(\frac{\rho g}{2\beta}\right)\left[1 + \left(\frac{C_L}{C_L^*}\right)^2\right], \qquad K_1 = K_2\left(\frac{C_L}{C_D}\right)$$

We now may integrate Eq. (7.20b) if we regard the velocity to be the following function of the range distance X:

$$\frac{\mathrm{d}}{\mathrm{d}X}\left(\frac{\mathrm{d}Z}{\mathrm{d}t}\right)\frac{\mathrm{d}X}{\mathrm{d}t} \approx \frac{\mathrm{d}}{\mathrm{d}X}\left(\frac{\mathrm{d}Z}{\mathrm{d}t}\right)V = K_1 V^2 - g$$

The above expression integrates to yield

$$\frac{\mathrm{d}Z}{\mathrm{d}t} = -V_0 \sin(\gamma) - \frac{K_1 V_0}{K_2}(e^{-K_2 X} - 1) - \frac{g}{V_0 K_2}(e^{K_2 X} - 1) \tag{7.20c}$$

Again rewriting the vertical velocity $\mathrm{d}Z/\mathrm{d}t$ in terms of range X as

$$\frac{\mathrm{d}Z}{\mathrm{d}X} = \frac{1}{V}\frac{\mathrm{d}Z}{\mathrm{d}t}$$

Equation (7.20c) integrates to yield the following:

$$Z = Z_0 + \frac{A}{K_2 V_0}(e^{K_2 X} - 1) - \left(\frac{K_1}{K_2}\right)X - \frac{g}{2V_0^2 K_2^2}(e^{2K_2 X} - 1) \tag{7.20d}$$

where

$$A = \left[\frac{K_1}{K_2}V_0 + \frac{g}{V_0 K_2} - V_0 \sin(\gamma_0)\right]$$

Equation (7.20d) may be evaluated for various initial conditions. If the experimentalist can control only Z_0, the height of the launcher above the floor of the range, and γ_0, the angular direction of the launcher, it can be shown that the launcher must be raised above the floor and that the initial velocity vector must be below the horizontal.

For a configuration having the mass and aerodynamic characteristics given in Fig. 7.2, Eq. (7.20d) can be used to compute a trajectory. In order to launch a model at 750 m/s, it is necessary to incline the launcher downward at 1.55 deg. The lowest point in the trajectory is about 12 cm above the range floor; at the end of the range, the model is 1.76 m above the floor (below the 2.0-m requirement). The launcher must be located 1.4 m above the floor. Obviously, there is a range of launcher locations and inclinations that will meet the minimum and maximum trajectory constraints.

Example 7.3: Skip Glider

Another interesting application of Eqs. (7.14) is the description of the path of an entry or re-entry body skipping on the upper boundary of the atmosphere. We noted earlier that the atmosphere is encountered as a near discontinuity by a vehicle possessing high suborbital speeds. The application of lift, mostly, in the direction of the positive Earth radius vector can cause a redirection of the velocity vector. The result is that what was an entry trajectory (with the velocity vector below the local horizontal) becomes an exoatmospheric trajectory (with the velocity vector above the local horizontal). If the initial entry trajectory is suborbital, then the trajectory after the bounce, or skip, will also be suborbital. We might expect to see a succession of skips. Our interest here is in using Eqs. (7.14) to examine motion in the vicinity of the first skip.

It is interesting to note that skipping trajectories have long been considered as a means of extending the range of a suborbital vehicle. The seminal paper on skip gliding is usually identified as that of Sanger and Brent[4]; skip gliding was re-examined at the beginning of the American space commitment in October 1957.[5,6]

Equation (7.14b) may be written as

$$
\begin{aligned}
V\frac{\mathrm{d}\gamma}{\mathrm{d}t} &= V\frac{\mathrm{d}s}{\mathrm{d}t}\frac{\mathrm{d}\gamma}{\mathrm{d}s} \approx V^2\frac{\mathrm{d}\gamma}{\mathrm{d}s} \\
&= -\left(\frac{\rho g}{2\beta}\right)\left[1+\left(\frac{C_L}{C_L^*}\right)^2\right]\left(\frac{C_L}{C_D}\right)V^2 + \cos(\gamma)\left[g - \frac{V^2}{R_e + h}\right]
\end{aligned}
\tag{7.21}
$$

where the atmospheric density is considered constant because altitude changes are small in the vicinity of the skip. Equation (7.21) may be divided through by V^2 to give

$$
\frac{\mathrm{d}\gamma}{\mathrm{d}s} = -\left(\frac{\rho g}{2\beta}\right)\left[1+\left(\frac{C_L}{C_L^*}\right)^2\right]\left(\frac{C_L}{C_D}\right) + \cos\gamma\left[\frac{g}{V^2} - \frac{1}{R_e + h}\right]
\tag{7.22}
$$

One simplification that might be made is to regard the trajectory in the vicinity of the skip as being nearly that of a circular orbit. This assumption is equivalent to regarding the gravitational and centrifugal forces to be in a near balance, i.e.,

$$V \approx V_{\odot} = [g(R_e + h)]^{1/2} = \left[\frac{g_e R_e^2}{R_e + h}\right]^{1/2} \approx (g_e R_e)^{1/2} \tag{7.23a}$$

The integration of Eq. (7.22) then becomes

$$\gamma = \gamma_0 - \left(\frac{\rho g}{2\beta}\right)\left[1 + \left(\frac{C_L}{C_L^*}\right)^2\right]\left(\frac{C_L}{C_D}\right)s \tag{7.23b}$$

An alternate assumption (favored by Vinh; see p. 123 of Ref. 1) is to ignore centrifugal effects in the vicinity of the skip. Such an assumption is equivalent to assuming that the Earth's radius is arbitrarily large. As a consequence, we may simplify Eq. (7.22) to give

$$\frac{d\gamma}{ds} = -\left(\frac{\rho g}{2\beta}\right)\left[1 + \left(\frac{C_L}{C_L^*}\right)^2\right]\frac{C_L}{C_D} + \frac{g}{V^2}\exp\left\{\left(\frac{\rho g}{\beta}\right)\left[1 + \left(\frac{C_L}{C_L^*}\right)^2\right]s\right\} \tag{7.24}$$

We further assume that in the region of vehicle contact with the atmosphere (i.e., the range of the skipping process), the flight path angle is near zero. We also make use of Eq. (7.18a) to express the velocity magnitude as a function of range s. Thus, we may integrate Eq. (7.24) to give

$$\gamma = \gamma_0 - \left\{\left(\frac{\rho g}{2\beta}\right)\left[1 + \left(\frac{C_L}{C_L^*}\right)^2\right]\left(\frac{C_L}{C_D}\right)\right\} s + \frac{\beta}{\rho V_0^2\left(1 + (C_L/C_L^*)^2\right)}\left\{\exp\left[\frac{\rho g}{\beta}\left(1 + \left(\frac{C_L}{C_L^*}\right)^2\right)s\right] - 1\right\} \tag{7.25a}$$

The density, required in both the second and third terms, is a constant; an approximate value may be found from

$$\rho = \rho_0 e^{-h/H} \tag{2.32}$$

which we developed in Chapter 2 and included earlier in this chapter as Eq. (7.8a).

An approximation to the transverse load (in g) due to the skip may be easily found from Eq. (7.14b) to be

$$n_{\perp} = \frac{\rho_0 e^{-h/H}}{2\beta}\left[1 + \left(\frac{C_L}{C_L^*}\right)^2\right]\frac{C_L}{C_D}V_0^2\exp\left\{-\left(\frac{\rho g}{\beta}\right)\left[1 + \left(\frac{C_L}{C_L^*}\right)^2\right]s\right\} \tag{7.25b}$$

where $n_{\perp}$ equals $(V d\gamma/dt)/g$.

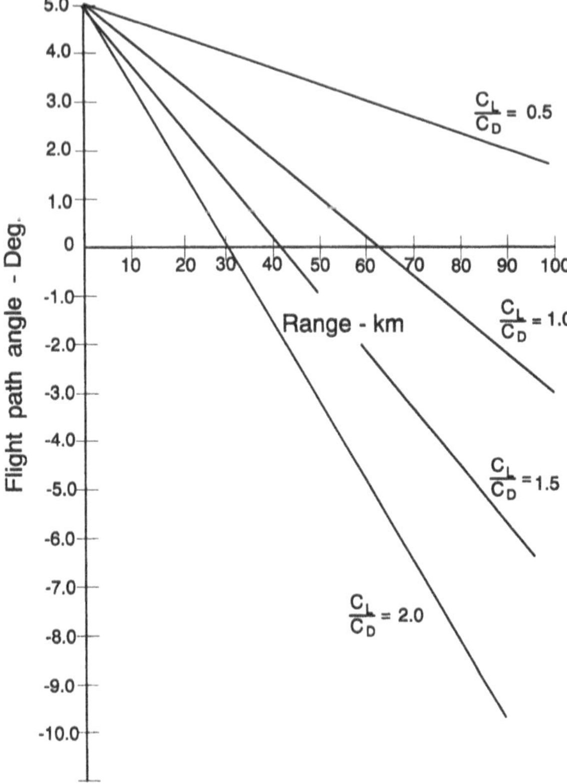

Fig. 7.3 Variation of flight path angle with range for different lift-to-drag ratios.

In the above expression we have assumed that $\cos(\gamma)$ is nearly unity and have ignored the centrifugal acceleration. Note also that we have represented the ballistic coefficient in the "mass form" β_m, or β/g (units of kg/m^2).

Figure 7.3 indicates how the flight path angle might change with range for typical operational and design parameters. It should be obvious that using the skip-gliding technique for range extension of a manned vehicle requires great care in avoiding configurational asymmetries which would result in excessive lift. Even a lift-to-drag ratio that is small by the standards of atmospheric vehicles could be destructive at re-entry speeds.

Case 3: Vertical Re-Entry Without Lift

In cases 1 and 2 the re-entry body's trajectory was limited to essentially horizontal motion. Because of the near constancy of altitude, density variations could be ignored. However, by definition an RB penetrates the atmosphere to a terrestrial impact point. Consequently, we would expect that all re-entry bodies will experience rapid changes in atmospheric density over some segment of the trajectory. The trajectory of case 3 is a simplified trajectory of vertical re-entry without lift, which should demonstrate the essential physics of particle trajectories for which rapid density changes occur.

We start by considering Eqs. (7.14). We can eliminate Eq. (7.14b) because all terms are identically zero: the flight path angle γ remains constant at 90 deg, and lift is not present. Therefore, for these circumstances we may write Eq. (7.14a), the single dynamic equation, Eq. (7.14c), a kinematic equation, and Eq. (7.8a), the atmospheric density simplification, as follows:

$$\frac{\mathrm{d}V}{\mathrm{d}t} = -\left(\frac{\rho g}{2\beta}\right)V^2 + g \tag{7.26a}$$

$$\frac{\mathrm{d}h}{\mathrm{d}t} = -V \tag{7.26b}$$

$$\rho = \rho_0 e^{-h/H} \tag{7.26c}$$

where $H = 6700$ m and $\rho_0 = 1.752$ kg/m^2. Upon inserting Eq. (7.26c) into (7.26a) and using Eq. (7.26b) in a derivative chain rule, we have

$$\frac{\mathrm{d}V}{\mathrm{d}t} = \frac{\mathrm{d}h}{\mathrm{d}t}\frac{\mathrm{d}V}{\mathrm{d}h} = \left[-\frac{1}{2}\right]\frac{\mathrm{d}V^2}{\mathrm{d}h} = -\frac{\rho_0 g\, e^{-h/H}}{2\beta}V^2 + g \tag{7.27}$$

If we make a variable change from V^2 to ξ, Eq. (7.27) becomes

$$\frac{\mathrm{d}\xi}{\mathrm{d}h} - \left(\frac{\rho_0 g\, e^{-h/H}}{\beta}\right)\xi = -2g \tag{7.28}$$

The assumptions associated with the above equation should be re-emphasized: no lift ($C_L = 0$), constant drag coefficient C_D, and a vertical trajectory ($\gamma = \pi/2$).

In Appendix E we have formalized a fairly general method for expressing the nonhomogeneous solution to a first-order differential equation. We now make use of this method in solving Eq. (7.28). First, we must identify the $P(X)$ and $Q(X)$ terms of Eq. (E.1) as

$$P(h) = -\rho_0 e^{-h/H}/\beta_m, \qquad Q(h) = -2g \tag{7.29a}$$

The general solution given in Eq. (E.6) requires the solution of the following indefinite integral:

$$\int^h P(\tau)\mathrm{d}\tau = -\frac{\rho_0}{\beta_m}\int^h e^{-\tau/H}\mathrm{d}\tau = \frac{\rho_0 H}{\beta_m}e^{-\tau/H} \tag{7.29b}$$

where in the third term we have replaced h by the dummy variable, τ. Thus, the solution to Eq. (7.28) becomes

$$\xi = \exp\left[-\left(\frac{\rho_0 H}{\beta_m}\right)e^{-h/H}\right]\int^h\left\{(-2g)\exp\left[\left(\frac{\rho_0 H}{\beta_m}\right)e^{-\tau/H}\right] + C\right\}\mathrm{d}\tau \tag{7.30}$$

We must seek some type of series solution to the above equation because it does not allow for a closed-form solution in terms of elementary functions.

The simplest approach is to bypass the integration entirely by letting $Q(h) = 0$, or, equivalently, by ignoring gravity. Equation (7.30) then becomes

$$\xi = V^2 = C \exp\left[-(\rho_0 H/\beta_m)\, e^{-h/H} \right] \tag{7.31a}$$

or

$$V = C \exp[-(\rho_0 H/2\beta_m)]\, e^{-h/H} \tag{7.31b}$$

To evaluate the constant C we can define the re-entry velocity, V_E as the velocity of the re-entry vehicle at an altitude of 5×10^4 m. The parameters are therefore

$$h = h_E = 5.0 \times 10^4 \text{ m}$$

$$V = V_E$$

$$H = 6700.0 \text{ m}$$

$$\rho_0 = 1.752 \text{ kg/m}^3$$

$$\beta_m = 1.0 \times 10^4 \text{ kg/m}^2$$

It is easy to show that the constant of integration C in the above expression is nearly equal to the velocity magnitude at entry, i.e., $C \approx V_E$. The velocity magnitude V may be expressed as a function of altitude h as follows:

$$V = V_E \exp\left[-(\rho_0 H/2\beta_m)\, e^{-h/H} \right] \tag{7.32a}$$

The axial deceleration (in g) follows from Eq. (7.26a) as

$$n_a = \left(\frac{\rho_0}{2\beta_m g}\right)\left\{ V_E^2 \exp\left[-\left(\frac{\rho_0 H}{\beta_m}\right) e^{-h/H}\right]\right\} e^{-h/H} \tag{7.32b}$$

where $n_a = -\dot{V}/g$ and where V in Eq. (7.26a) has been replaced by the expression in Eq. (7.32a).

Let's return now to Eq. (7.30) and obtain an approximate solution by expanding the exponential in a Taylor series. We must then integrate a sum of terms, (or, equivalently, sum a set of integrals) as follows:

$$\xi = \left\{ -2g \int^h \mathrm{d}\tau - 2g \left[\sum_{n=1}^{\infty} \left(\frac{\rho_0 H}{\beta_m}\right)^n \frac{1}{n!} \int^h e^{-\tau n/H} \mathrm{d}\tau \; + C \right]\right\} \times \exp\left[-\left(\frac{\rho_0 H}{\beta_m}\right) e^{-h/H}\right]$$

Carrying out the integration gives

$$\xi = \left\{ 2gH \sum_{n=1}^{\infty} \left[\left(\frac{\rho_0 H}{\beta_m} \right) \frac{1}{n} \frac{1}{n!} e^{-hn/H} \right] - 2gh + C \right\} \exp\left[-\left(\frac{\rho_0 H}{\beta_m} \right) e^{-h/H} \right]$$

Replacing ξ by V^2 gives

$$V = \left\{ 2gH \sum_{n=1}^{\infty} \left[\left(\frac{\rho_0 H}{\beta_m} \right)^n \frac{1}{n} \frac{1}{n!} e^{-hn/H} \right] - 2gh + C \right\}^{1/2} \exp\left[-\left(\frac{\rho_0 H}{2\beta_m} \right) e^{-h/H} \right] \tag{7.33}$$

If we assume that the series converges, we are then able to evaluate the constant of integration C. We might obtain an approximation to C by using such typical values as

$$\beta_m = 12000 \text{ kg/m}^2$$

$$H = 6700 \text{ m}$$

$$\rho_0 = 1.752 \text{ kg/m}^3$$

$$V_E = 7000 \text{ m/s}$$

These numerical values show that the exponential terms are small, leading to an approximation for C as follows:

$$C \approx 2gH + V_E^2 \tag{7.34}$$

The axial deceleration follows from Eq. (7.26a) as

$$n_a \approx -\left(\frac{\rho_0}{2\beta_m} \right) e^{-h/H} \left\{ 2gH \sum_{n=1}^{\infty} \left[\left(\frac{\rho_0 H}{\beta_m} \right)^n \frac{1}{n!} \frac{1}{n} e^{-hn/H} \right] + V_E^2 \right\} \times \left\{ \exp\left[-\left(\frac{\rho_0 H}{\beta_m} \right) e^{-h/H} \right] \right\} + 1 \tag{7.35}$$

Equations (7.33) and (7.35) may be easily programmed. Such a program is listed in TRUEBASIC™ in Table 7.2, with sample output given in Table 7.3.

Equations (7.33) and (7.35) are restricted to a vertical trajectory; however, in the next case we will see that these results are useful in treating steep, but not necessarily vertical, trajectories as well.

Case 4: Steep Nonvertical Re-Entry Without Lift

In the previous section we considered vertical re-entry without lift. If the results given in Table 7.3 are taken as representative of this class of problem, then it would appear that the contribution of gravity to the total acceleration is

Table 7.2 Program to calculate velocity magnitude and axial deceleration as a function of altitude

```
!THIS PROGRAM CALCULATES THE MAGNITUDE OF THE VELOCITY, V,
! AND THE AXIAL DECELERATION, N, AS A FUNCTION OF ALTITUDE.

! INPUTS:       V0: INITIAL VELOCITY (M/S)
!               B : BALLISTIC COEFFICIENT (KG/M^2)
!               Z0: INITIAL ALTITUDE (M)
! PARAMETERS:   D0: GROUND LEVEL DENSITY (1.752 KG/M^3)
!               H : ATMOSPHERE SPREAD FACTOR (6700 M)
!               G : GRAVITATIONAL ACCELERATION (9.81 M/S^2)
!               STEP: ALTITUDE STEP SIZE (M)
! OUTPUT:       V : VELOCITY (M/S)
!               N : AXIAL ACCELERATION (M/S^2)
!
!
V0 = 6200
Z0 = 1.5E+5
B = 1325.18
STEP = 5000
H = 6700
D0 = 1.752
G = 9.81
OPEN #1: NAME "DECEL.TRU", CREATE NEWOLD
ERASE #1
DIM Z(100),V(100),N(100)
M = Z0/STEP
N1$ = "####.##        -----#.###     ###.#"
N2$ = "VELOCITY       DECELERATION        ALTITUDE"
N3$ = "METERS/SEC         G'S              KILO-METERS"
N4$ = "BALL. COEFF    INITIAL  VELOCITY    INITIAL ALTITUDE"
N5$ = " KG/M^2           METERS/SEC             KILOMETERS"
N6$ = "#####.#        #####.#              ###.#"
Z1 = (D0*H/B)*EXP((-1)*Z0/H)
P = 1
Z3 = 0
T = (V0^2)*EXP(Z1)
FOR I = 1 TO 10
Z2 = (Z1^I)/(I*(I*P))
Z3 = Z3+Z2
P = P*I
NEXT I
C = T-2*G*H*Z3+2*G*Z0
FOR J = 1 TO M+1
Z(J) = Z0-STEP*(J-1)
```

(continued on next page)

Table 7.2 (continued) Program to calculate velocity magnitude and axial deceleration as a function of altitude

```
Z4 = (D0*H/B)*EXP((-1)*Z(J)/H)
P = 1
Z3 = 0
FOR I = 1 TO 10
Z5 = (Z4^I)/(I*(I+P))
Z3 = Z5+Z3
P = P*I
NEXT I
V(J) = SQR(EXP((-1)*Z4)*(2*G*H*Z3-2*G*Z(J)+C))
N(J) = ((D0*(V(J)^2)/(2*B))*EXP((-1)*Z(J)/H))/G-1
NEXT J
PRINT #1: N4$
PRINT #1: N5$
PRINT #1, USING N6$:B,V0,Z0/1000
PRINT #1:
PRINT #1:
PRINT #1: N2$
PRINT #1: N3$
PRINT #1:
FOR J = 1 TO M+1
PRINT #1, USING N1$: V(J), N(J), Z(J)/1000
NEXT J
CLOSE #1
END
```

negligible. Equations (7.32) were developed from the assumption that the gravitational force is negligible; Eqs. (7.33) and (7.35) retained the gravitational force. In this section we again consider the contribution of gravitational acceleration to a steep nonvertical ballistic trajectory.

Let's return to one of the fundamental equations of planar motion, Eq. (7.9b). This equation shows that if the flight path angle γ is to remain constant, then some force must be generated across the trajectory (normal to the velocity vector). Usually we identify such a force as the lift, but for now we need only estimate the magnitude of this force. If this force is negligibly small, then a constant flight path angle might be an acceptable approximation.

Let

$$\frac{d\gamma}{dt} = 0, \qquad \frac{F}{m} = -\left(\frac{\rho g}{2\beta}\right)V^2\frac{C_L}{C_D}$$

Then we have from Eq. (7.9b)

$$\frac{F}{m} = \cos(\gamma)\left(g - \frac{V^2}{R_e + h}\right) \tag{7.36}$$

Table 7.3 Output from program given in Table 7.2

BALL. COEFF KG/M^2	INITIAL VELOCITY METERS/SEC	INITIAL ALTITUDE KILOMETERS
1325.2	6200.0	150.0
VELOCITY METERS/SEC	DECELERATION G'S	ALTITUDE KILO-METERS
6200.00	- 1.000	150.0
6207.91	- 1.000	145.0
6215.80	- 1.000	140.0
6223.69	- 1.000	135.0
6231.56	- 1.000	130.0
6239.43	- 1.000	125.0
6247.29	- 1.000	120.0
6255.13	- 1.000	115.0
6262.97	- 1.000	110.0
6270.79	- 1.000	105.0
6278.61	- .999	100.0
6286.40	- .998	95.0
6294.18	- .996	90.0
6301.92	- .992	85.0
6309.60	- .983	80.0
6317.17	- .963	75.0
6324.50	- .922	70.0
6331.35	- .835	65.0
6337.19	- .651	60.0
6340.90	- .263	55.0
6340.14	.555	50.0
6329.99	2.269	45.0
6300.12	5.831	40.0
6229.22	13.084	35.0
6074.14	27.244	30.0
5751.93	52.418	25.0
5120.61	88.290	20.0
4002.52	114.060	15.0
2382.22	84.965	10.0
838.49	21.462	5.0
307.95	5.390	.0

If the specific force F/m is zero, then

$$V = V_{\odot} = [g(R_e + h)]^{1/2} \tag{7.37}$$

The above equation is essentially a statement that the centrifugal force balances the gravitational force. The velocity given in Eq. (7.37) is the velocity magnitude of circular orbit, or the *first cosmic velocity.*

According to Eq. (7.36), F/m is not zero for the suborbital speeds usually associated with re-entry. However, if we take $V = 6000$ m/s and $h = 60$ km as representative of re-entry conditions, we can calculate the transverse force per unit weight as

$$\begin{aligned} \frac{F}{W} &= \cos(\gamma)\left[1 - \frac{V^2}{(R_e + h)g}\right] \\ &= \cos(\gamma)\left[1 - \frac{(6000)^2}{(9.81)(6.37) \times 10^6 + 6.0 \times 10^4}\right] = 0.4245\cos(\gamma) \end{aligned}$$

Even if $\gamma = 0$, a transverse load of 0.4245 g is negligible in comparison with an axial load which will exceed 10 g for much of the trajectory (see Table 7.3). Thus, for $\mathrm{d}\gamma/\mathrm{d}t = 0$ (and a nonlifting trajectory), we can assume that the centrifugal force is essentially balanced by the gravitational force.

We may now return to Eq. (7.27), emphasizing two conditions: First, since γ is not zero but remains essentially unchanged, we may set $\gamma = \gamma_E$; second, we may neglect the gravitational contribution. Consequently, we may write

$$\int_{V_E}^{V} \frac{\mathrm{d}V}{V} = \int_{h_E}^{h} \left[\frac{\rho_0 e^{-h/H}}{2\beta_m \sin(\gamma_E)}\right] \mathrm{d}h$$

Carrying out the required integrations gives

$$V = V_E \exp\left[\frac{\rho_0 H e^{-h_E/H}}{2\beta_m \sin(\gamma_E)}\right] \exp\left[\frac{\rho_0 H e^{-h/H}}{2\beta_m \sin(\gamma_E)}\right] \tag{7.38}$$

For typical values such as

$$\begin{aligned} \rho_0 &= 1.752 \text{ kg/m}^3 \\ H &= 6700 \text{ m} \\ h_E &= 6.0 \times 10^4 \text{ m} \\ \beta_m &= 3000 \text{ kg/m}^2 \\ \gamma_E &= 60 \text{ deg} \end{aligned}$$

it immediately follows that the first exponential in Eq. (7.38) is near unity; i.e.,

$$e^{2.6 \times 10^{-7}} \approx 1.0$$

Thus, we may simplify Eq. (7.38) in the following form:

$$V = V_E \exp\left[-a_E e^{-h/H}\right] \tag{7.39a}$$

where a_E, the trajectory parameter, is given by

$$a_E = \frac{\rho_0 H}{2\beta_m \sin(\gamma_E)} \tag{7.39b}$$

We could adapt Eqs. (7.33) and (7.35) to a nonvertical descent by replacing $(\rho_0 H/\beta_m)$ by $[\rho_0 H/\beta_m \sin(\gamma_E)]$, or $2a_E$. A more straightforward approach would be to replace β_m by $\beta_m \sin(\gamma_E)$, where we would use the modified ballistic coefficient β_m.

The axial acceleration might be recomposed in terms of the trajectory parameter. From Eqs. (7.32) and (7.39b) we have

$$n_a = \left(\frac{\rho_0 V_E^2}{2\beta_m g}\right)\left[\exp\left(-2a_E e^{-h/H}\right)\right] e^{-h/H} \tag{7.39c}$$

It is of some interest to identify the altitude at which maximum deceleration takes place:

$$\frac{\mathrm{d}n_a}{\mathrm{d}t} = 0 = \left(\frac{\rho_0 V_E^2}{2\beta_m g}\right)\frac{\mathrm{d}}{\mathrm{d}h}\left\{\left[\exp\left(-2a_E e^{-h/H}\right)\right] e^{-h/H}\right\}$$

which gives

$$\tilde{h} = H \ln\left[\frac{\rho_0 H}{\beta_m \sin(\gamma_E)}\right] = H \ln(2a_E) \tag{7.40a}$$

where $\tilde{h}$ is the altitude at which maximum deceleration takes place. Note that $\tilde{h}$ varies inversely with the ballistic coefficient β_m. Massive, low drag-vehicles, (i.e., high β_m) will develop peak axial loads at low altitudes. Of equal interest is the value of the maximum acceleration. Inserting Eq. (7.40a) into Eq. (7.39c) gives

$$\tilde{n}_a = \frac{V_E^2 \sin(\gamma_E)}{2g e H} \tag{7.40b}$$

We can also find the velocity magnitude at which maximum axial deceleration takes place by inserting $\tilde{h}$ from Eq. (7.40a) into Eq. (7.39) to get

$$V = V_E e^{-1/2} = 0.6065 V_E \tag{7.40c}$$

It is interesting to note that the value of maximum axial deceleration is independent of the ballistic coefficient, although, as we have seen, the altitude at which the RV experiences maximum deceleration does depend upon β_m. Figures 7.4 and 7.5 present the results of a computer simulation of re-entry [based upon

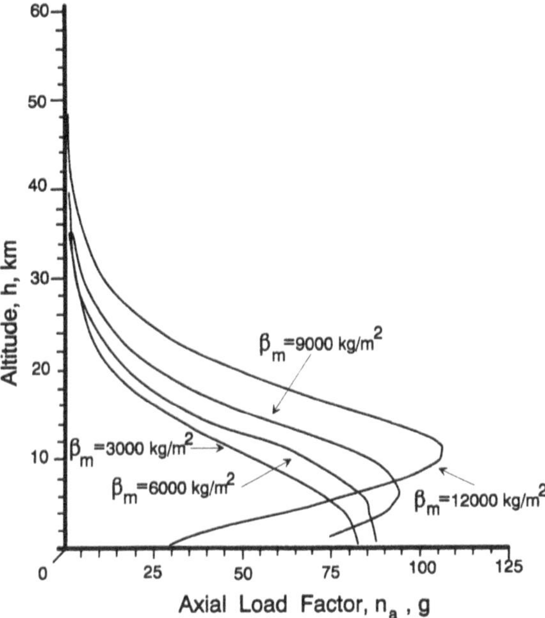

Fig. 7.4a Altitude vs axial load factor for four ballistic coefficients.

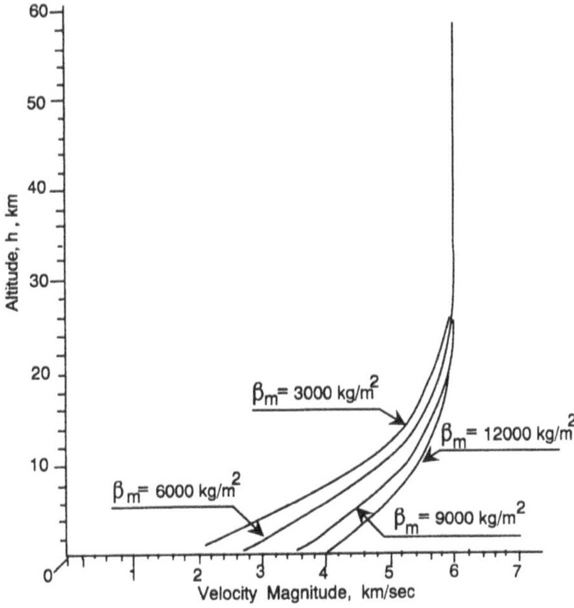

Fig. 7.4b Altitude vs velocity magnitude for four ballistic coefficients.

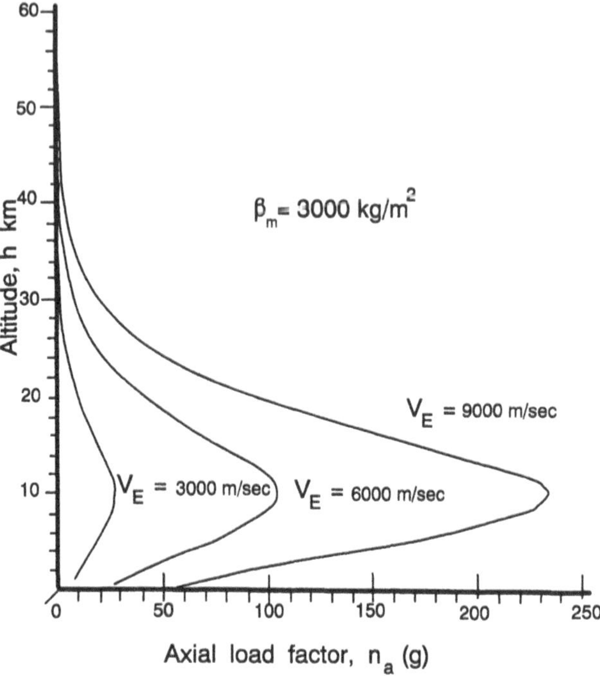

Fig. 7.5a Altitude vs axial load factor for three re-entry velocities.

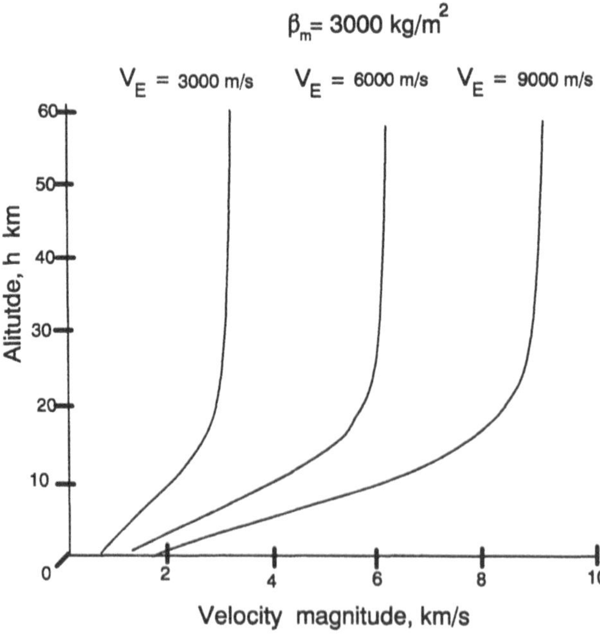

Fig. 7.5b Altitude vs velocity magnitude for three re-entry velocities

Eqs. (7.14)]. We note that in this more detailed model, the maximum deceleration does to some extent depend upon the value of the ballistic coefficient.

Applying Eq. (7.40b) to a Martian entry problem might give misleading results if g and H represent the appropriate Martian values ($g = 3.73$ m/s^2 and $h = 10.6$ km from Table 2.7). If this is done, the units of n_a would be in terms of the Martian gravitational constant, which could of course be changed into Earth units by multiplying by $g_m/g_e = 0.3814$.

Equations (7.40) are of great interest in carrying out "back-of-the-envelope" kinds of preliminary performance estimates of a re-entry vehicle. Therefore, we should assess just how accurate such relationships are by making comparisons to more exact computer simulations.

The computer model used here is that of a particle (drag-only) moving through a 1962 model atmosphere. (The atmospheric model is essentially that given in Appendix A, with break points set as in Table 2.3.) Re-entry is assumed to begin at 60 km, with velocity magnitudes of 3000 and 5000 m/s.

There are two characteristics of Eqs. (7.40) that might be surprising: the altitude of maximum deceleration $\tilde{h}$ is independent of the magnitude of the velocity at entry V_E, and the magnitude of the maximum deceleration $\tilde{n}_a$ is independent of the ballistic coefficient β_m. Figures 7.6 and 7.7 compare the altitude of maximum deceleration and the magnitude of that acceleration from computations based upon Eqs. (7.14). Figures 7.6 and 7.7 summarize the result

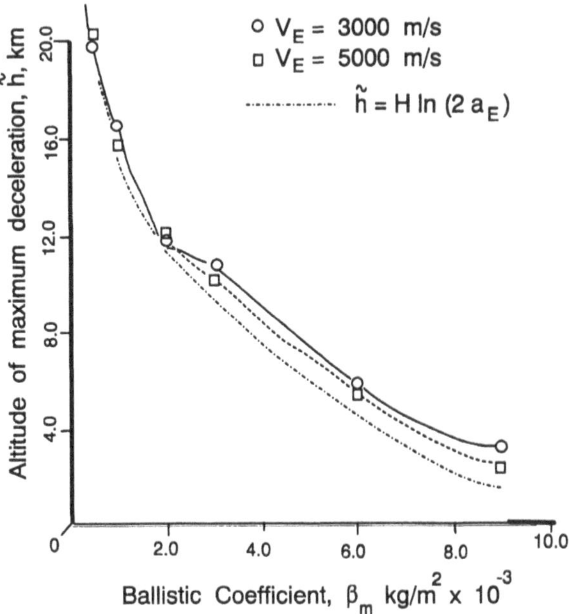

Fig. 7.6 Altitude of maximum deceleration vs ballistic coefficient for two re-entry velocities.

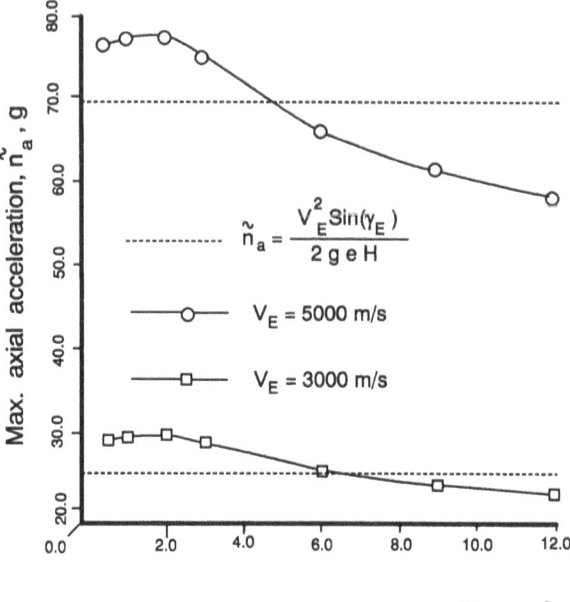

Fig. 7.7 Maximum axial deceleration vs ballistic coefficient for two re-entry velocities.

noted in Figs. 7.4 and 7.5: the altitude of maximum axial deceleration is nearly independent of the magnitude of the entry velocity and follows fairly closely the prediction given by Eq. (7.40a).

However, according to Eq. (7.40b), the magnitude of the maximum deceleration should vary only with initial velocity magnitude V_E and be invariant with the ballistic coefficient. It may be seen clearly from Fig. 7.7 that the value of the maximum deceleration does vary significantly with ballistic coefficient, although Eq. (7.40b) seems to give a sort of average value (at least over the range of ballistic coefficients that were examined). Equation (7.40b) predicts a value of maximum deceleration that is too low for small values of ballistic coefficient and too high for large values.

Case 5: Lifting Re-Entry

In this section we consider a case in which lift is present. Equations (7.14a) and (7.14b) are the fundamental equations we use here, with the modification that the gravitational acceleration is omitted from both equations. It is omitted from Eq. (7.14a) because the aerodynamic loads are at least an order of magnitude greater than the gravitational force and from Eq. (7.14b) because the centrifugal and gravitational forces across the trajectory are essentially in balance. Thus, we have

$$\frac{\mathrm{d}V}{\mathrm{d}t} = -\left(\frac{\rho g}{2\beta}\right)\left[1 + \left(\frac{C_L}{C_L^*}\right)^2\right]V^2 \tag{7.41a}$$

$$V\frac{\mathrm{d}\gamma}{\mathrm{d}t} = -\left(\frac{\rho g}{2\beta}\right)\left[1 + \left(\frac{C_L}{C_L^*}\right)^2\right]V^2\left(\frac{C_L}{C_D}\right) \tag{7.41b}$$

Dividing Eq. (7.41a) by (7.41b) we get

$$\frac{\mathrm{d}V}{V} = \left(\frac{C_L}{C_D}\right)^{-1}\mathrm{d}\gamma \tag{7.42}$$

The above equation may be integrated between the two velocity magnitudes V_1 and V_2 and the corresponding flight path angles γ_1 and γ_2 to give

$$V_2 = V_1 \exp\left[-\left(\frac{\gamma_1 - \gamma_2}{C_L/C_D}\right)\right] \tag{7.43}$$

It must be remembered that C_L/C_D has an associated sign. We note from Eq. (7.41b) that with flight path angle γ positive (velocity vector below the local horizontal), a positive lift diminishes γ. Thus, if γ_2 is less than γ_1, C_L/C_D is positive; if γ_2 is greater than γ_1, C_L/C_D is negative. In either event, changing the flight path angle through the application of lift will diminish the velocity magnitude. A way of avoiding concern for the proper sign of C_L/C_D might be to rewrite Eq. (7.43) as follows:

$$V_2 = V_1 \exp\left[-\mathrm{abs}\left(\frac{\gamma_1 - \gamma_2}{C_L/C_D}\right)\right]$$

$$\Delta V = V_1 - V_2 = V_1\left\{1 - \exp\left[-\mathrm{abs}\left(\frac{\gamma_1 - \gamma_2}{C_L/C_D}\right)\right]\right\} \tag{7.44}$$

where ΔV is the velocity magnitude loss incurred by changing the flight path angle through the application of lift. A special form of Eq. (7.43) might be obtained by replacing $V_1 = V_E$ and $V_2 = V$ to give the velocity magnitude after re-entry as a function of the flight path angle γ as follows:

$$V = V_E \exp\left\{-\frac{\gamma_E - \gamma}{C_L/C_D}\right\} \tag{7.45}$$

Equations (7.43), (7.44), and (7.45), sometimes identified as the *agility equations,* indicate the velocity penalty incurred as a result of changing direction of the velocity vector. Obviously, the larger the lift-to-drag ratio, the lower the velocity magnitude loss in changing the flight path angle $(\gamma_E - \gamma)$.

From Eq. (7.14c) we have

$$\frac{\mathrm{d}h}{\mathrm{d}t} = -V\sin(\gamma) \tag{7.46}$$

Equation (7.41a) may then be rewritten as

$$\frac{\mathrm{d}h}{\mathrm{d}t}\frac{\mathrm{d}V}{\mathrm{d}h} = -V\sin(\gamma)\frac{\mathrm{d}V}{\mathrm{d}h} = -\left(\frac{\rho_0}{2\beta_m}\right)\left[1+\left(\frac{C_L}{C_L^*}\right)^2\right]V^2 e^{-h/H}$$

Using Eq. (7.42), we get

$$\sin(\gamma)\mathrm{d}\gamma = \left(\frac{\rho_0 H}{2\beta_m}\right)\left(\frac{C_L}{C_D}\right)e^{-h/H}\mathrm{d}\left(\frac{h}{H}\right)$$

Integration of the above equation between (γ_1, h_1) and (γ_2, h_2) gives

$$\gamma_2 = \cos^{-1}\left[\left(\frac{\rho_0 H}{2\beta_m}\right)\left(\frac{C_L}{C_D}\right)\left(e^{-h_2/H} - e^{-h_1/H}\right) + \cos(\gamma_1)\right] \tag{7.47}$$

Equation (7.43) provides a relationship between the two dependent variables, V and γ. However, Eq. (7.43) gives no indication of the altitude loss during the change in flight path angle. Equation (7.47) relates the flight path angle to the independent variable h. One approach would be to select an initial flight path angle γ_1 and altitude h_1; the final flight path angle γ_2 would then be computed along with the velocity penalty given by Eq. (7.44).

Equation (7.47) could be rewritten with (γ_1, h_1) = (γ_E, h_E) and with (γ_2, V_2) replaced by (γ, V) as follows:

$$\gamma \approx \cos^{-1}\left[\left(\frac{\rho_0 H}{2\beta_m}\right)\left(\frac{C_L}{C_D}\right)e^{-h/H} + \cos(\gamma_E)\right] \tag{7.48}$$

where we have used the approximation

$$e^{-h_E/H} \approx 0$$

We can combine Eqs. (7.43) and (7.47) to give velocity magnitude as a function of altitude as follows:

$$V_2 = V_1 \exp\left\{-\left(\frac{C_L}{C_D}\right)^{-1}\right.$$

$$\left.\times\left\{\gamma_1 - \cos^{-1}\left[\left(\frac{\rho_0 H}{2\beta_m}\right)^{-1}\left(\frac{C_L}{C_D}\right)\left(e^{-h_2/H} - e^{-h_1/H}\right) + \cos(\gamma_1)\right]\right\}\right\} \tag{7.49}$$

We might now consider the instantaneous value of acceleration. With lift present we must consider two components of acceleration—axial and transverse:

$$-\frac{\boldsymbol{a}}{g} = \boldsymbol{n} = -\frac{\dot{V}}{g}\boldsymbol{i}_a - \frac{V\dot{\gamma}}{g}\boldsymbol{i}_t = n_a\boldsymbol{i}_a + n_t\boldsymbol{i}_t$$

where $\boldsymbol{i}_a$ and $\boldsymbol{i}_t$ are the axial and transverse unit vectors. Again neglecting the gravity terms from Eqs. (7.41) we may write

$$n_a = -\frac{\dot{V}}{g} = \left(\frac{\rho_0 V^2}{2\beta_m g}\right) e^{-h/H} \tag{7.50a}$$

$$n_t = -\frac{V\dot{\gamma}}{g} = \left(\frac{\rho_0 V^2}{2\beta_m g}\right)\left[1 + \left(\frac{C_L}{C_L^*}\right)^2\right]\frac{C_L}{C_D} e^{-h/H} \tag{7.50b}$$

The magnitude of the acceleration is

$$n = \left(n_a^2 + n_t^2\right)^{1/2}$$

$$n = \left(\frac{\rho_0 V^2}{2\beta_m g}\right)\left\{1 + \left[1 + \left(\frac{C_L}{C_L^*}\right)^2\right]^2 \left(\frac{C_L^2}{C_D}\right)\right\}^{1/2} e^{-h/H}$$

We could now replace the velocity V from Eq. (7.49) by letting $V_2 = V$, $h_2 = h$, $V_1 = V_E$, and $h_1 = h_E$ to yield the following:

$$n = \left(\frac{\rho_0 V_E^2}{2\beta_m g}\right)\left\{1 + \left[1 + \left(\frac{C_L}{C_L^*}\right)^2\right]^2 \left(\frac{C_L}{C_D}\right)^2\right\}^{1/2} e^{-h/H}$$
$$\times \exp\left\{-\frac{2}{C_L/C_D}\left\{\gamma_E - \cos^{-1}\left[\left(\frac{\rho_0 H}{2\beta_m}\right)\left(\frac{C_L}{C_D}\right)e^{-h/H} + \cos(\gamma_E)\right]\right\}\right\} \tag{7.50c}$$

7.4 Some Nondimensional Representations

All of the previous analyses have been drawn from the more lengthy discussions by Loh[2] and various extensions and contributions by Vinh.[1] In Section 7.3 we made no attempt to nondimensionalize the fundamental equations of atmospheric entry [Eqs. (7.14)]. "Lumping" the dimensional variables (e.g., velocity magnitude V, altitude h, flight path angle γ) into nondimensional groupings can often isolate combinations which remain more or less invariant during the trajectory or have special properties for classes of trajectories.

Nondimensionalizing may aid in the solution of the fundamental differential equations by recasting these equations into some standard form for which the solutions are readily available. Unfortunately, the introduction of an unfamiliar dependent variable may interfere with an analyst's ability to use intuition.

In order to simplify the results we retain the definition of β as given in Eq. (7.5a), where the drag coefficient C_D contains both the zero-lift drag and the induced drag (drag due to lift). We assume that

$$\beta_m = \beta / g = m / C_D S$$

where the units of β are (N/m^2) and the units of β_m are (kg/m^2).

First, let's rewrite Eq. (7.14a) in accordance with the above modifications to yield

$$\frac{dV}{dt} = -\left(\frac{\rho}{2\beta_m}\right)V^2 + g\sin(\gamma) \tag{7.51}$$

Now, using the chain rule to redefine the independent variable [replacing t by h using Eq. (7.14c)], we get

$$\frac{d(\ )}{dt} = -V\sin(\gamma)\frac{d(\ )}{dh}$$

which alters Eq. (7.17a) to become

$$-V\sin(\gamma)\frac{dV}{dh} = -\left(\frac{\rho V^2}{2\beta_m}\right) + g\sin(\gamma)$$

or

$$\frac{1}{2}\frac{d(V^2)}{dh} - \left[\frac{\rho}{2\beta_m\sin(\gamma)}\right]V^2 = -g$$

Next, we nondimensionalize the velocity magnitude V by the orbital speed at sea level and the altitude by the atmospheric spread parameter H to give

$$\frac{d[V^2/(g_eR_e)]}{d(h/H)} - \left[\frac{\rho H}{\beta_m\sin(\gamma)}\right]\frac{V^2}{g_eR_e} = -\frac{2H}{R_e}$$

It is convenient here to recall the parameter λ of Eq. (6.17), i.e.,

$$\lambda = V^2/g_eR_e$$

The preceding differential equation then becomes

$$\frac{d\lambda}{d(h/H)} - \left[\frac{\rho H}{\beta_m\sin(\gamma)}\right]\lambda = -\frac{2H}{R_e} \tag{7.52a}$$

In like fashion we may rewrite Eq. (7.14b) as

$$-V^2\sin(\gamma)\frac{d\gamma}{d(h/H)} - gH\cos(\gamma)\left(1 - \frac{V^2}{R_eg_e}\right) = -\frac{\rho H}{2\beta_m}\left(\frac{C_L}{C_D}\right)V^2$$

After some manipulation we have

$$\frac{d[\cos(\gamma)]}{d(h/H)} - \frac{H\cos(\gamma)}{R_e}\left(\frac{1}{\lambda} - 1\right) = -\frac{C_L}{C_D}\left(\frac{\rho H}{2\beta_m}\right) \tag{7.52b}$$

Equations (7.52) can be integrated numerically if we introduce the exponential density relationship of Eq. (7.26c). Letting

$$X = \lambda, \qquad Y = \cos(\gamma), \qquad Z = h/H$$

we may rewrite Eqs. (7.52) as

$$\frac{\mathrm{d}X}{\mathrm{d}Z} - \left(\frac{\rho_0 H}{\beta_m}\right)\frac{X e^{-Z}}{(1 - Y^2)^{1/2}} = -\frac{2H}{R_e} \tag{7.53a}$$

$$\frac{\mathrm{d}Y}{\mathrm{d}Z} - \frac{H}{R_e}\left(\frac{1}{X} - 1\right)Y = -\left(\frac{C_L}{C_D}\right)\left(\frac{\rho_0 H}{2\beta_m}\right)e^{-Z} \tag{7.53b}$$

A few comments on Eqs. (7.53) should be made. First, we note that we have two vehicle design/operational parameters: C_L/C_D, the operational lift-to-drag ratio, and β_m, the ballistic coefficient. It must be remembered that β_m as used above is defined as

$$\beta_m = \frac{m}{C_{D_O}\left[1 + \left(C_L/C_L^*\right)^2\right]S} \tag{7.53c}$$

where C_L^* is the lift coefficient at the maximum lift-to-drag ratio. An alternate way of writing Eqs. (7.53) would be to replace the ballistic coefficient as follows:

$$\beta_m \rightarrow \frac{\beta_m}{\left[1 + \left(C_L/C_L^*\right)^2\right]} \tag{7.53d}$$

where β_m in the expression to the right of the arrow is based upon the zero-lift drag coefficient.

Second, we have two planetary parameters: $\rho_0 H$ and H/R_e. Table 7.4 gives these quantities for Earth, Venus, and Mars. There is controversy concerning the Martian and Venusian parameters given in the table; the atmospheric scale factor H depends upon surface temperature and gas molecular weight [see Table 2.7 and Eq. (2.33)], both of which have significant diurnal and annual variations. More importantly, the atmospheric data available for Venus and Mars is extremely limited in comparison to corresponding atmospheric data for Earth.

Table 7.4 Planetary parameters

	Earth	Venus	Mars
$\rho_0 H (\mathrm{kg/m^2})$	1.1738×10^4	1.1×10^6	2.39×10^2
R_e/H	950	415	319

Equations (7.53) are not very useful for applications to re-entry vehicles which enter the atmosphere at shallow flight path angles. Clearly, if the initial or subsequent flight path angle is zero, the singularity in Eq. (7.53a) will cause numerical problems. The origin of this difficulty lies in the selection of $h/H = Z$ as the independent variable. For extremely shallow re-entry, there will be little or no change in $X = V^2/g_eR_e$ with Z; however, X will change with a coordinate variable normal to the radius vector. Ashley[7] points out the difficulties caused by this singularity, particularly when studying the trajectories of skip gliders.

Loh[2] carries out further transformation of Eqs. (7.53) by first differentiating the exponential atmosphere density function as

$$d\rho = -\rho_0 e^{-h/H} d\left(\frac{h}{H}\right) = -\rho d\left(\frac{h}{H}\right) = -\rho \, dZ \tag{7.54}$$

Thus,

$$\frac{d(\ )}{d(h/H)} \equiv \frac{d(\ )}{dZ} = -\rho\frac{d(\ )}{d\rho}$$

Equations (7.52) may then be rewritten as

$$\frac{d\lambda}{d\rho} + \left[\frac{H}{\beta_m \sin(\gamma)}\right]\lambda = \frac{2H}{\rho R_e} \tag{7.55a}$$

$$\frac{d\cos(\gamma)}{d\rho} + \left[\frac{H\cos(\gamma)}{\rho R_e}\right]\left(\frac{1}{\lambda} - 1\right) = \left(\frac{C_L}{C_D}\right)\left(\frac{H}{2\beta_m}\right) \tag{7.55b}$$

where $\lambda = V^2/R_e g_e$. R_e is the radius of the Earth, and λ is the ratio of the kinetic energy in orbit to kinetic energy in a circular orbit at the Earth's surface. H/R_e is a geophysical ratio which for the Earth is about 1/950; for Mars and Venus the appropriate values are given in Table 7.4. For most ballistic re-entry vehicles, λ decreases monotonically from an initial value of slightly less than unity. We will return to Eqs. (7.55) after making note of other ways of representing the equations of planar motion.

A major contributor in the field of atmospheric entry is Vinh.[1] Using the following variable collection

$$u = \frac{1}{2}\left(\frac{V^2}{g_eR_e}\right) = \frac{1}{2}\lambda$$

$$\eta = \left(\frac{SC_D}{2m\beta}\right)\rho = \left(\frac{SC_D\rho_0}{2m\beta}\right)e^{-\beta h}$$

$$\phi = \cos(\gamma)$$

$$\beta = 1/H$$

Vinh obtained the following variation on Eqs. (7.55):

$$\frac{\mathrm{d}u}{\mathrm{d}\eta} = \frac{-2u}{(1-\phi^2)^{1/2}} + \frac{1}{\beta R_e \eta} \tag{7.56a}$$

$$\frac{\mathrm{d}\phi}{\mathrm{d}\eta} = \left(\frac{C_L}{C_D}\right) - \frac{1}{\beta R_e \eta}\left(\frac{1}{2u} - 1\right)\phi \tag{7.56b}$$

It will be left as an exercise for the reader to verify the equivalence of Eqs. (7.56) and (7.55). (Remember that β according to Vinh is the reciprocal of the atmospheric spread factor H and should not be confused with the ballistic coefficient β_m.)

We note that the second term on the left-hand side of Eq. (7.55b) would vanish if $\lambda = 1$; such a condition means that the centrifugal and gravitational forces are in balance, i.e., the RV is in a circular orbit. Recall that in Eq. (7.36) we examined the difference between the gravitational and centrifugal forces across the trajectory. We found that for suborbital re-entry the gravitational and centrifugal forces were not in equilibrium. Nevertheless, we chose to ignore the difference between these two forces in comparison to the axial load. In the absence of lifting loads, this simplification essentially allowed us to ignore the transverse acceleration which is given by Eq. (7.55b).

Loh[2] examined the second term of Eq. (7.55b) for several trajectories (both with and without lift) and came to the conclusion that this term is essentially a constant. Certainly ρ, γ, and λ all vary along any re-entry trajectory; however, Loh's conjecture is that these quantities vary in such a way as to render the second term a constant (at least to some acceptable degree).

Let's examine this assumption, first analytically and then numerically. Let's define the second term of Eq. (7.55b) as *Loh's function,* or $L(\rho)$ as follows:

$$L = L(\rho) = \frac{H}{R_e}\frac{\cos(\gamma)}{\rho}\left(\frac{1}{\lambda} - 1\right) \tag{7.57}$$

Clearly, $L(\rho)$ is a function of ρ, both explicitly and implicitly, through the variation of $L(\rho)$ with λ (or V) and γ with ρ.

According to various workers (e.g., Loh,[2] Ashley,[7] Vinh[1]), the expression given in Eq. (7.57) would be invariant over much of the trajectory, with or without the presence of lift. The qualitative reasoning might go something like this: as the re-entry vehicle descends, the density increases, and the velocity decreases in a compensating fashion. The Loh assumption is best stated by Ashley as follows:

> By careful study of exact numerical results, Loh . . . determined that, for the purposes of integration with respect to ρ . . . $[L(\rho)]$ . . . may be treated as a constant. . . . That is, while the individual quantities ρ, γ, and λ vary along the trajectory, their combined behavior is such as to render the integral insensitive to changes in this term.

Table 7.5 Re-entry body ballistic coefficient bounds

Parameter	Maximum value	Minimum value
$0.0 \leq C_L/C_D \leq 6.6$	Flat-plate	Sphere
$0.001 \leq C_{D_0} \leq 1.0$	C_{D_0} for a sphere	Flat-plate skin friction coefficient
$130 \leq m/A \leq 3000$	Proposed shuttle vehicle[9,10]	Sphere of 0.2-m diameter, density 10^3 kg/m^3
$130 \leq \beta_m \leq 3 \times 10^6$ kg/m^2	(Derived from $\beta_0 = m/C_{D_0}A$)	

Qualitatively at least, the above statement seems reasonable; the decrease in λ along the trajectory would appear to be compensated by an increase in ρ.

Loh's parameter as defined in Eq. (7.57) does not contain the ballistic coefficient β. However, the ballistic coefficient is the important design/operational parameter in determining the variation of λ (or V) over the trajectory (see Fig. 7.4b).

It is interesting to note that the value of the ballistic coefficient used by Loh[2] and repeated by Ashley is very small, falling below the lower limit of most atmospheric entry or re-entry bodies. Loh apparently bases his conclusion on modeling a re-entry body for which β_m is equal to 15.6 kg/m^2 (3.2 lb/ft^2); with a C_D of 1.0 and a body density of 3000 kg/m^3, the re-entry body would be a sphere of 0.8 cm in diameter.

Table 7.5 presents some parametric ranges that may be used to estimate bounds on the ballistic coefficient β_m for nearly all artificial atmospheric entry bodies. Ballistic coefficients of much less than 500 kg/m^2 would appear to be of limited interest.

Is there any justification for regarding Loh's parameter as a constant over a re-entry trajectory for a wide variety of re-entry vehicles? Hough[8] and Freeman[9] indicate that Loh's assumption is valid because of a near balance between centrifugal forces and gravitational forces. Clearly, if λ remains near unity across a significant portion of the trajectory, Loh's parameter will be constant, i.e., zero. However, for a body with an extremely small ballistic coefficient β_m, the velocity ratio λ will decrease rapidly; a balance between centrifugal and gravitational forces cannot exist.

Figure 7.8 presents the variation of the Loh parameter over a range of nondimensional altitudes h/H. Clearly, as β_m decreases, there is a smaller and smaller trajectory segment over which $L(\rho)$ is constant.

An analytical examination may be made of $L(\rho)$. Since H/R_e is about 1/900, the right-hand side of Eq. (7.55a) may be ignored, provided that density is large enough (below an altitude h/H of 6). Equation (7.55a) then integrates from entry (E) to a generic point on the trajectory as follows:

$$\lambda = \lambda_E \exp\left[-\frac{H}{\beta \sin(\gamma)}(\rho - \rho_E)\right] \tag{7.58}$$

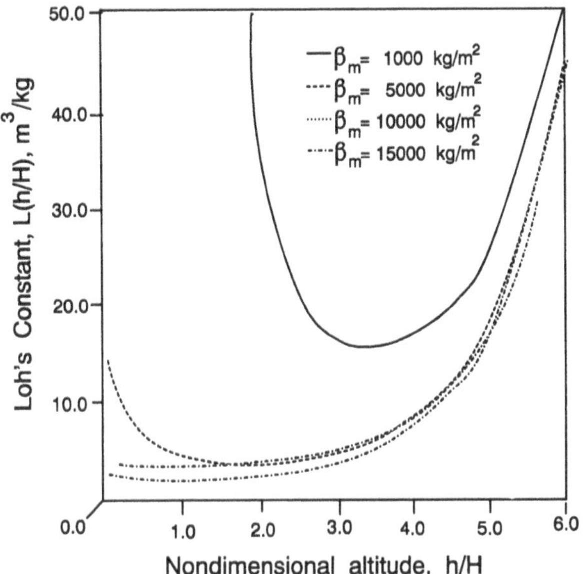

Fig. 7.8 Loh's constant L vs nondimensional altitude h/H.

With the following assumptions

$$\rho_E \approx 0.0, \qquad \gamma = \gamma_E$$

Eq. (7.58) becomes

$$\lambda = \lambda_E \exp\left[-\frac{H\rho}{\beta \sin(\gamma_E)}\right] \tag{7.59}$$

Inserting the above expression into Eq. (7.57) gives

$$L(\rho) \approx \frac{\cos(\gamma_E)}{\rho}\left(\frac{H}{R_e}\right)\left\{\frac{1}{\lambda_E}\exp\left[\frac{H\rho}{\beta \sin(\gamma_E)}\right] - 1\right\} \tag{7.60}$$

If $L(\rho)$ is invariant with density, then the following must hold:

$$\frac{\mathrm{d}L(\rho)}{\mathrm{d}\rho} = 0 \tag{7.61a}$$

which becomes

$$\frac{\left\{\left[\frac{H\rho}{\beta \sin(\gamma_E)}\right] - 1\right\}\exp\left[\frac{H\rho}{\beta \sin(\gamma_E)}\right] + \lambda_E}{\rho^2} = 0 \tag{7.61b}$$

L'Hospital's rule shows that the expression on the left-hand side of Eq. (7.61b) becomes unbounded as the density approaches zero. Since Eq. (7.61b) cannot be valid for arbitrarily small values of density, $L(\rho)$ cannot be assumed constant above some extreme altitude. Subject to such restrictions, Eq. (7.61b) becomes

$$\exp\left\{-\left[\frac{H\rho}{\beta\sin(\gamma_E)}\right]\right\} = \frac{1}{\lambda_E}\left\{1-\left[\frac{H\rho}{\beta\sin(\gamma_E)}\right]\right\} \tag{7.62}$$

Obviously, for small values of the exponent (say, less than 0.5), the right-hand side is an acceptable approximation for the left-hand side. Thus Eq. (7.61b) is approximately true [and $L(\rho)$ is invariant with altitude] if

$$H\rho/\beta\sin(\gamma_E) \le 0.5 \tag{7.63a}$$

and λ_E is near unity. Using the exponential altitude-density relationship, Eq. (7.63a) becomes

$$\frac{h}{H} \ge \ell n[2H\rho_0/\beta\sin(\gamma_E)] \tag{7.63b}$$

Of course, to avoid the singularity associated with zero density [Eq. (7.61b)] and to justify the assumption used in obtaining λ [Eq. (7.59)], an altitude restriction such as the following must apply:

$$h/H \le 6.0 \tag{7.63c}$$

The lower altitude break point [according to Eq. (7.17b)] can be calculated for a re-entry body with a ballistic coefficient β_m of 5000 kg/m^2 and an entry flight path angle γ_E of 45 deg as follows:

$$h/H = \ell n[(2)(6700)(1.752)/(5000)(\sin(45))] = 1.8$$

Figure 7.8 clearly indicates a sudden increase in $L(\rho)$ below a nondimensional altitude of about 1.8.

We might conclude that there is a range of altitudes for which the Loh parameter remains fairly constant. However, this range diminishes as the ballistic coefficient decreases. Loh's assumption is invalid for a re-entry body having a ballistic coefficient comparable to that used in his study. Regardless of the ballistic coefficient (at least for any practical body), the Loh parameter varies with altitude above 5 or 6 scale heights.

If we maintain that Loh's function is essentially a constant for trajectories of interest, then we might integrate Eq. (7.55b) to get

$$\cos(\gamma) = \cos(\gamma_E) + \left\{\left(\frac{C_L}{C_D}\right)\left(\frac{H}{2\beta_m}\right) + \left[\frac{\cos(\gamma)}{\rho}\right]\left(\frac{H}{R_e}\right)\left(1-\frac{1}{\lambda}\right)\right\}(\rho-\rho_E) \tag{7.64}$$

Solving for $\cos(\gamma)$ we obtain

$$\cos(\gamma) = \frac{\cos(\gamma_E) + (C_L/C_D)(H/2\beta_m)(\rho - \rho_E)}{1 + \{(H/R_e)[(1/\lambda) - 1][1 - (\rho_E/\rho)]\}} \tag{7.65}$$

We might now compare Eq. (7.65) to Eq. (7.47). If we let

$$\gamma_E = \gamma_1$$

$$\rho = \rho_0 e^{-h_2/H}$$

$$\rho_E = \rho_0 e^{-h_1/H}$$

$$V^2 = g_e R_e$$

$$\gamma = \gamma_2$$

we see that both equations are equivalent. [It must be remembered that Loh's function is zero when gravitational and centrifugal forces are equal, i.e., when $V^2 = g_e R_e$; the agreement between Eqs. (7.65) and (7.47) says nothing about the validity of the Loh assumption.]

Equation (7.65) is the integral of Eq. (7.55b); we may now consider the integration of Eq. (7.55a). First, we note that this equation is linear in the dependent variable λ in that both λ and $d\lambda/dt$ appear only to the first power. In order to carry out the integration we make use of a simplification suggested by Ashley.[7] We note that the term on the right-hand side of Eq. (7.55a) is

$$\frac{2H}{\rho R_e} = \frac{2}{\rho(R_e/H)} \approx \frac{2e^{h/H}}{\rho_0(R_e/H)}$$

According to Ashley, the term R_e/H is about 900 (950 according to Table 7.4). Since λ is on the order of 1.0, the justification for Ashley's simplification is based upon the following inequality:

$$\frac{H}{\beta_m \sin(\gamma_E)} \gg \frac{2}{\rho(R_e/H)}$$

The reader can explore the limitations of the above inequality; it will be accepted here. As a consequence, we may write Eq. (7.55a) as

$$\sin(\gamma)\frac{d\lambda}{d\rho} = -\left(\frac{H}{\beta_m}\right)\lambda \tag{7.66a}$$

Next, Eq. (7.55b) becomes

$$\sin(\gamma)\frac{d\gamma}{d\rho} = \frac{\cos(\gamma)}{\rho(R_e/H)}\left(\frac{1}{\lambda} - 1\right) - \left(\frac{C_L}{C_D}\right)\left(\frac{H}{2\beta_m}\right) \tag{7.66b}$$

Dividing Eq. (7.66a) by Eq. (7.66b) we get

$$\frac{d\lambda}{d\gamma} = \frac{-(H/\beta_m)\lambda}{\{[\cos(\gamma)/(\rho R_e/H)][(1/\lambda)-1]\} - [(C_L/C_D)(H/2\beta_m)]} \tag{7.67}$$

Integration of the above equation is straightforward if we are once again willing to accept Loh's approximation, i.e., the constancy of the denominator over the trajectory. The integral of Eq. (7.67) is

$$\lambda = \lambda_E \exp\left\{\frac{(H/\beta_m)(\gamma-\gamma_E)}{\{[\cos(\gamma)/(\rho R_e/H)][(1/\lambda)-1]\} - [(C_L/C_D)(H/2\beta_m)]}\right\} \tag{7.68}$$

Equations (7.65) and (7.68) are approximate integrations of Eqs. (7.55), the weak link in these integrations being the Loh assumption.

With the flat-Earth assumption, i.e., $R_e \to \infty$ (which is equivalent to neglecting centrifugal force), Eq. (7.65) reduces to Eq. (7.45). Backing up slightly, we may reconsider integration of Eq. (7.55a) for steep or vertical flight paths and with the term on the right-hand side omitted. (H/R_e is assumed to be negligibly small or zero as it would be for a flat Earth.) The result would be Eq. (7.39a).

Let's re-examine Eqs. (7.65) and (7.68). Ashley[7] points out that these equations express γ and λ in terms of the following three parameters:

1) C_L/C_D: Applied lift-to-drag ratio, which might be expressed in terms of some control variables such as flap deflection angle

2) R_e/H: Planetary size and atmosphere spread parameter

3) H/β_m: Atmosphere and re-entry body interaction parameter

Both equations are functions of flight path angle γ, velocity ratio λ, and atmospheric density ρ (or altitude, once we specify an atmospheric model). If one of these three variables is prescribed, then we may obtain the other two in a fairly straightforward manner.

7.5 Heat Transfer and Dynamics

In this section we describe the interaction between heating loads and vehicle dynamics. This material relies heavily on the analysis of heat transfer/dynamics interaction as presented by Miele,[11] although the seminal paper in this area is without doubt that by Eggers et al.[6]

By definition, the Stanton number C_h may be written as

$$C_h = \frac{q_w}{\rho V c_p(T_r - T_{wl})} \tag{7.69}$$

where q_w is the wall heat flux per unit area in units of J/m^2-s, ρ is the density of the fluid, V is a reference velocity magnitude, c_p is the ratio of specific heats of the gas, T_r is the recovery temperature, and T_{wl} is the local wall temperature.

By Reynolds analogy the Stanton number can be related to the skin-friction coefficient C_f as follows:

$$C_h = \frac{1}{2}\frac{C_f}{Pr} \tag{7.70}$$

where Pr is the Prandtl number.

The fluid mechanics background needed to justify the above relationships cannot be given here. The reader should refer to appropriate texts, such as White's book.[12] As White points out, Eq. (7.70) is strictly valid only for a Pr equal to 1.0 and for a zero pressure gradient through the boundary layer.

We may now rewrite Eq. (7.69), using Eq. (7.70), as follows:

$$q_w = \tfrac{1}{2}C_f \rho V c_p (T_r - T_{wl}) \tag{7.71}$$

Note that in applying Eq. (7.69) we have set the Prandtl number to unity. This approximation is considered acceptable for the semiquantitative discussion to follow.

The recovery temperature T_r may be written in terms of the temperature and velocity in the freestream as follows:

$$T_r = T_\infty + \frac{Pr}{2C_p}V^2$$

T_r is the adiabatic wall temperature, i.e., the temperature that the wall or surface of the RV would attain if no heat were transferred between the flow and the wall.

We can rewrite the above equation in terms of Mach number M by noting that, for a perfect gas,

$$M = \frac{V}{a_\infty}, \qquad a_\infty^2 = \gamma R T_\infty, \qquad \gamma = \frac{c_p}{c_v}, \qquad R = c_p - c_v$$

We may then write the following for the recovery temperature:

$$T_r = T_\infty\left[1 + Pr\left(\frac{\gamma - 1}{2}\right)M^2\right] \tag{7.72}$$

By taking the Prandtl number to be unity and subtracting wall temperature T_{wl} from each side, we get (assuming the RV wall temperature is the same as the ambient temperature T_∞)

$$T_r - T_\infty = T_r - T_{wl} = \tfrac{1}{2}M^2 T_\infty(\gamma - 1) \tag{7.73}$$

where the subscript l refers to local conditions. (Assuming that Pr is unity essentially means that the flow is assumed to be inviscid and that the recovery temperature is identical to the stagnation temperature.)

By using the definition of Mach number given above, we get

$$M^2T_\infty = \frac{V^2}{\gamma R} = \frac{V^2}{(c_p/c_v)(c_p - c_v)} = \frac{V^2}{c_p(\gamma - 1)}$$

Eq. (7.73) now becomes

$$T_r - T_{wl} = V^2/2c_p \tag{7.74}$$

We must emphasize that q_w of Eq. (7.71) is the heat transferred per unit time through a unit area; we designate the heat flow (joules per unit time) through the entire RV as $\dot{Q}$. Clearly, $\dot{Q}$ must be the surface integral of q_w as follows:

$$\dot{Q} = \int_{S_w} q_w \, dS = \int_{S_w} \frac{1}{2} C_f \rho V c_p (T_r - T_{wl}) \, dS$$
$$= \int_{S_w} \frac{1}{2} \frac{C_f}{r} \rho V c_p \left(\frac{V^2}{2c_p}\right) dS$$

Assuming that the reference speed V equals the airspeed at the edge of the boundary layer, we obtain

$$\dot{Q} = \frac{C_f}{4} \rho V^3 S_w \tag{7.75}$$

The above equation states that the heat transfer rate is a maximum when ρV^3 is a maximum. (Recall that the axial loads are a maximum when ρV^2 is a maximum.)

We may now replace the density ρ with the exponential atmosphere [Eq. (2.32)] and express the velocity magnitude V as in Eq. (7.39a) to give

$$\dot{Q} = \frac{C_f}{4} \rho_0 V_E^3 \left[\exp\left(-3a_E e^{-h/H}\right)\right] \tag{7.76}$$

Differentiating $\dot{Q}$ with respect to altitude h gives the altitude of maximum heat transfer, or the altitude at which the maximum heat is flowing into the RV, as follows:

$$h_m = H \ell n(3a_E) = -H \ell n[3\rho_0 H / 2\beta_m \sin(\gamma_E)] \tag{7.77}$$

We may compare Eq. (7.77) with Eq. (7.40a). If we insert Eq. (7.77) into Eq. (7.39a) we get the following velocity magnitude at which the heating rate is a maximum:

$$V_m = V_E e^{-1/3} = 0.7165 \, V_E$$

The previous expression may be contrasted with the velocity magnitude for maximum axial load [Eq. (7.40c)] which is

$$\tilde{V} = V_E e^{-1/2} = 0.6065\ V_E$$

By returning to Eq. (7.26a) and omitting the gravitational effect, we obtain

$$\frac{dV}{dt} = -\left(\frac{g\rho}{2\beta}\right)V^2 \qquad (7.78)$$

Dividing Eq. (7.78) into Eq. (7.75) gives

$$\frac{dQ}{dV} = -\frac{m}{2}\left(\frac{C_f S_w}{C_D S}\right)V \qquad (7.79)$$

where we have replaced β by $m/(C_D S)$.

If we let the initial and final values of heat transferred (Q_E and Q_F, respectively) be

$$Q_E = 0, \qquad Q_F = Q$$

and assume that C_f and C_D are some kind of average values, we get, upon integration of Eq. (7.79),

$$Q = \frac{m}{4}\left(\frac{C_f S_w}{C_D S}\right)\left(V_E^2 - V_F^2\right) \qquad (7.80a)$$

The velocity at impact V_F may be found approximately from Eq. (7.39a) as

$$V_F = V_E \exp\left(-a_E e^{-h/H}\right)_{h=0} = V_E e^{-a_E} \qquad (7.80b)$$

If the law of conservation of energy is applied to the path and we neglect the variation in potential energy, the initial kinetic energy $(m/2)V^2$ must equal the sum of the final kinetic energy plus the energy dissipated through aerodynamic drag. Part of the energy dissipated in drag is contained in the wake; the remainder is transferred to the RV as a heat load.

Let's now define the heat transferred to the RV as a fraction of the total kinetic energy, i.e.,

$$\tilde{Q} = Q \Big/ \left(\frac{m}{2}V_E^2\right)$$

By normalizing Q according to the above expression and replacing V_F according to Eq. (7.80b), we get

$$\tilde{Q} = \frac{1}{2}\left(\frac{C_f S_w}{C_D S}\right)\left(1 - e^{-2a_E}\right) \qquad (7.81)$$

From Eq. (7.39b) we get

$$a_E = \frac{\rho_0 H}{2\beta_m \sin(\gamma_E)} = \frac{\rho_0 C_D S H}{2m \sin(\gamma_E)}$$

We may now look at two extreme cases of Eq. (7.81). Assume first that we have an extremely light RV such that a_E is large (mass m is small). Equation (7.81) then becomes

$$\tilde{Q} = \tilde{Q}_1 = \frac{1}{2}\left(\frac{C_f S_w}{C_D S}\right) \tag{7.82}$$

We may conclude that for the light RV, the fraction of the initial kinetic energy transferred to the RV by conductive heating is one half the ratio of friction drag to total drag. The total drag has as its constituents pressure drag and skin-friction drag. Therefore, in order to minimize the heat conducted to the RV, the friction drag must be made small relative to the pressure drag. The engineering implications are that the light RV must have large pressure drag, i.e., a blunt shape. The graph given in Fig. 7.9 shows how the drag coefficient (based upon Newtonian methods) varies with nose blunting.

It might be of some interest to see how large the ballistic coefficient may be to still allow the approximation of Eq. (7.82) to be acceptable. If we set an upper bound on β_m by requiring that

$$\exp(-2a_E) \leq 0.1$$

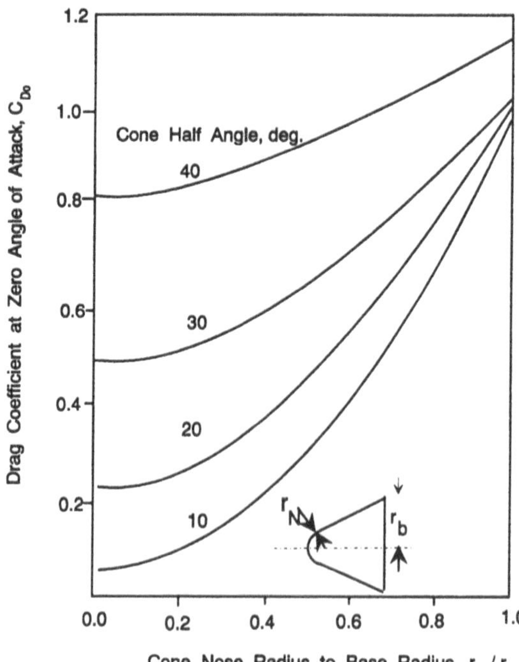

Fig. 7.9 Cone drag coefficient vs bluntness ratio.

then

$$\beta_m \leq \frac{\rho_0 H}{\ell n(.1)\sin(\gamma_E)}$$

By taking the flight path angle γ_E at entry as 75 deg, ρ_0 as 1.172 kg/m^3, and H as 6700 m, we get

$$\beta_m \leq 3530 \text{ kg/m}^2$$

Many RVs have ballistic factors well below this value, so an approximation leading to Eq. (7.82) has more than academic interest.

For a typical practical design, $C_f S_W / C_D S$ is about 1/10. Inserting this value into Eq. (7.82) indicates that the total heat energy transferred to the re-entry vehicle is about 1/20 of the total kinetic energy present at re-entry and is of the same order of magnitude required to vaporize the RV.

Next, consider the engineering requirement that is placed upon a relatively heavy RV. For a heavy RV we may approximate Eq. (7.82) as follows:

$$\begin{aligned}\tilde{Q} &= \frac{1}{2}\left(\frac{C_f S_w}{C_D S}\right)[1 - (1 - 2a_E)] \\ &= \frac{\rho_0 H C_f S_w}{2m \sin(\gamma_E)}\end{aligned} \tag{7.83}$$

Thus, for a heavy RV the fraction of initial kinetic energy transferred to the RV as heat is proportional to the frictional coefficient. Consequently, a shape corresponding to low friction drag is desirable.

We have used the term "relatively heavy" in reference to this type of RV. Obviously, for a vehicle having a value of β_m which is essentially unbounded, Q would be zero or nearly so, because a_E would approach zero. This would mean that only a negligible amount of kinetic energy (in comparison to the total mechanical energy) is converted into thermal energy. For example, one might imagine that a natural entry body several tens of meters in diameter (typically associated with past cataclysmic changes in Earth environment) would carry most of its kinetic energy at entry to the Earth's surface.

References

[1]Vinh, N. S., *Hypersonic and Planetary Entry Flight Mechanics*, University of Michigan Press, 1980.

[2]Loh, W. H. T., *Dynamics and Thermodynamics of Planetary Entry*, Prentice-Hall, Englewood Cliffs, NJ, 1963.

[3]Charters, A. C., and Thomas, R. N., "The Aerodynamic Performance of Small Spheres from Subsonic to High Supersonic Speeds," *Journal of Aeronautical Sciences*, Vol. 12, No. 10, Oct. 1945.

[4]Sanger, E., and Brent, J., "A Rocket Drive for Long Range Bombers," Translation No. CGD-32, Technical Information Branch, Navy Dept., 1944.

[5]Lees, L., Hastwig, F. W., and Cohen, C. B., "Use of Aerodynamic Lift During Entry into the Earth's Atmosphere," *ARS Journal*, Sept. 1959, pp. 633–641.

[6]Eggers, A. J., Allen, H. J., and Neice, S. E., "A Comparative Analysis of the Performance of Long-Range Hypervelocity Vehicles," NACA Technical Report 1382, 1956.

[7]Ashley, H., *Engineering Analysis of Flight Vehicles*, Addison-Wesley, Reading, MA, 1974.

[8]Hough, M. E., "Explicit Guidance along an Optimal Space Curve," *Journal of Guidance, Control, and Dynamics*, Vol. 12, No. 4, July–Aug. 1989, p. 495.

[9]Freeman, D. C., and Powell, R. W., "The Results of Studies to Determine the Impact of Far-Aft Center of Gravity Locations on the Design of a Single-Stage-to-Orbit Vehicle System," AIAA Conference on Advanced Technology for Future Space Systems, AIAA Paper 79-0892, Hampton, VA, May 1979.

[10]Piland, W. M., Talay, T. A., and Stone, H. W., "The Personnel Launch System," *Aerospace America*, AIAA, Nov. 1990.

[11]Miele, A., *Flight Mechanics*, Addison-Wesley, Reading, MA, 1962.

[12]White, F. M., *Viscous Fluid Flows*, McGraw-Hill, Hightstown, NJ, 1974.

8 Decoys and the Identification of Re-Entry Vehicles

8.1 Introduction

Re-entry bodies which are intrinsically benign but which accompany hostile RBs are known as *decoys*. In order for a decoy to be credible it must present to the defense a set of measurable quantities or observables which are sufficiently close numerically to those of a hostile re-entry body to leave considerable doubt as to the threat of the vehicle. Clearly, a defense that has constraints on its response to a perceived threat (such as interceptor inventory limitations) must first identify an observed re-entry body as a threat to some level of confidence before taking defensive action. A response to a decoy means an inventory loss (*false alarm*), and failure to respond to a threat that is not benign means penetration by the offense (*leakage*).

It is assumed here that the observables associated with a decoy and a hostile re-entry body become increasingly distinct with decreasing altitude. Thus, at some altitude after initial detection of a presumed threat and after processing the observables, the defense can discriminate with an acceptable confidence level between the two types of vehicles. The altitude at which this discrimination occurs depends upon the characteristics of the decoy, the accuracy of measurements, and the inventory constraints within which the defense must operate.

In Chapter 7 we identified several parameters that describe the trajectory of a re-entry body. Two parameters that are design-specific are the (zero-lift) ballistic coefficient β and the maximum lift-to-drag ratio. If our interest is confined to ballistic RBs, then the only design parameter of interest is β. (Of course, operational parameters are important also; for example, the angle between the zero-lift line and the velocity vector at entry is a major contributor to the drag and hence the subsequent trajectory.) Thus, it is to be expected that a decoy and a hostile re-entry body will have markedly different ballistic coefficients. However, the defense can measure β only indirectly: a measurement of deceleration may be used to obtain the ballistic coefficient. As altitude decreases we might expect increasing divergence between the measured decelerations of a decoy and a hostile RB. At this point we must acknowledge that there are several observables that are obtained from measurements other than trajectory states (velocity and position). However, since we are concerned in this text primarily with trajectories, we will limit our attention to the ballistic coefficient as the observable that will differentiate between decoys and hostile RVs.

8.2 Estimators

We assume that the defense has radars available to measure the spherical coordinates (range, azimuth, and elevation) of any re-entry body. We are not interested in how such measurements are made, in the errors associated with such measurements, or in the details of any required axis transformations. Let's assume that the coordinates of the re-entry body are available as three Cartesian coordinates. The measurement vector $\boldsymbol{Z}_x$ is given by a time sequence of the X-components of the RV's position, i.e.,

$$\boldsymbol{Z}_x = [X_0, X_1, \ldots, X_{n-1}, X_n]^{\mathrm{T}} \tag{8.1}$$

There are two other measurement vectors associated with the Y- and Z-coordinates, i.e., $\boldsymbol{Z}_y$ and $\boldsymbol{Z}_z$, but we need not concern ourselves with them at this time.

In a batch estimator the measurements are stored as they are obtained. When sufficient data is acquired (according to some criterion), a series of mathematical operations are performed to obtain the estimates. The least-squares estimator is the classic form of a batch estimator. At every data point a fitting function is evaluated as follows:

$$\tilde{X}_k = \hat{X} + \hat{V}_x[T_n - (n-k)\Delta T] + \left(\hat{A}_x/2\right)[T_n - (n-k)\Delta T]^2 \tag{8.2}$$

where ΔT is the data acquisition period, the parameters to be estimated are represented by the vector $\boldsymbol{P}_x = [\hat{X}, \hat{V}_x, \hat{A}_x]^{\mathrm{T}}$, and $(T_n - T_0)$ is the total time duration of the batch sample. The above function and the measurement sequence are illustrated in Fig. 8.1. By minimizing the sum of the squares of the difference between each data point and the corresponding value from the above function [i.e., $(X_k - \tilde{X}_k)$] with respect to each of the parameters, sufficient equations are available to evaluate the parameters.

The parameter vector $\boldsymbol{P}_x$ and the measurement vector $\boldsymbol{Z}_x$ are related by the observation matrix H as follows:

$$\boldsymbol{Z}_x = H\boldsymbol{P}_x \tag{8.3a}$$

where

$$H = \begin{bmatrix} 1 & T_0 & T_0^2/2 \\ 1 & T_1 & T_1^2/2 \\ \vdots & \vdots & \vdots \\ 1 & T_n - (n-k)\Delta T & [T_n - (n-k)\Delta T]^2/2 \\ \vdots & \vdots & \vdots \\ 1 & T_n & T_n^2/2 \end{bmatrix} \tag{8.3b}$$

The least-squares estimation of $\boldsymbol{P}_x$ is given as follows[1]:

$$\boldsymbol{P}_x = (H^{\mathrm{T}}H)^{-1}H^{\mathrm{T}}\boldsymbol{Z}_x \tag{8.4}$$

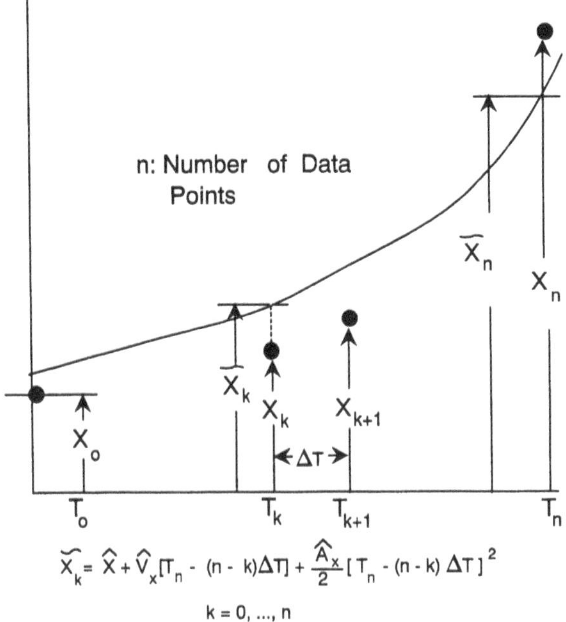

Fig. 8.1 Measurement sequence.

The batch estimator first requires that we make a series of positional measurements, i.e., "fill out" the vector $\boldsymbol{Z}_x$, each element of which is a positional measurement. During the time of data-gathering, no evaluation of the vector $\boldsymbol{P}_x$ is possible. Once sufficient data has been gathered, there is an additional time delay to process the data—processing which requires a matrix transposition, a multiplication, an inverse, and a second set of matrix multiplications. Extensive data-processing time can mean loss of radar track. An additional criticism of the classic least-squares procedure is that it makes no use of the statistical knowledge of measurement accuracy in the formation of the estimates.

A second class of estimators known as recursive estimators makes use of only the most recent measurement and the current estimation, both appropriately weighted according to the relative accuracy of the current state and measurement. The recursive estimator is often known as the Kalman-Bucy, or just Kalman, filter. The storage demands on the recursive estimator are usually significantly less than those required to implement a batch estimator. Most importantly, the recursive filter provides a state estimate at about the same frequency as that with which data becomes available.

In Fig. 8.2 we have indicated a measurement at time $t = T_k$ of the position of a re-entry body (not necessarily a "vehicle" or artifact). If an appropriate sequence of tracking measurements has been made and processed, we should have available sometime after completion of the sequence (at time $t = T_n$) the

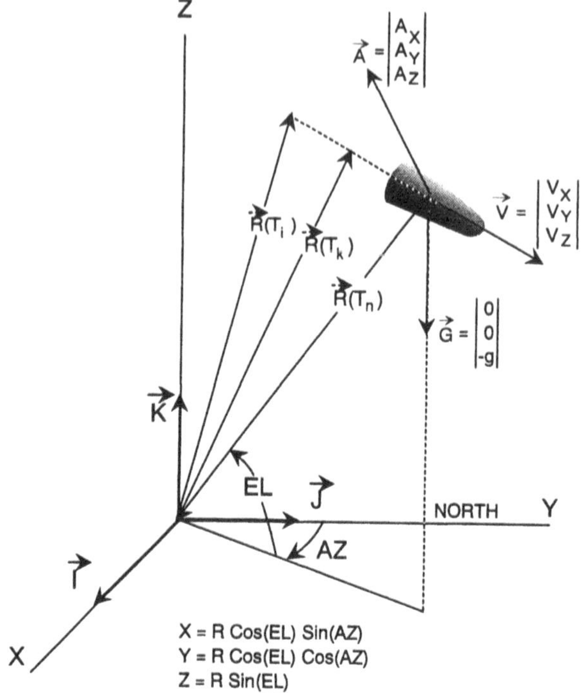

Fig. 8.2 Tracking measurements.

three parameter vectors giving estimates of the position, velocity, and acceleration:

$$\begin{aligned} \boldsymbol{P}_x &= \left[\hat{X}, \hat{V}_x, \hat{A}_x\right]^{\mathrm{T}} \\ \boldsymbol{P}_y &= \left[\hat{Y}, \hat{V}_y, \hat{A}_y\right]^{\mathrm{T}} \\ \boldsymbol{P}_z &= \left[\hat{Z}, \hat{V}_z, \hat{A}_z\right]^{\mathrm{T}} \end{aligned} \tag{8.5}$$

Let us now rewrite Eq. (7.14a) in a three-dimensional form as follows:

$$\frac{\mathrm{d}\boldsymbol{V}}{\mathrm{d}t} \doteq \boldsymbol{A} = \left(\frac{-\rho}{2\beta_m}\right)|\boldsymbol{V}|\boldsymbol{V} + \boldsymbol{g} \tag{8.6}$$

If the re-entry body is not developing any lift, i.e., if the RB appears to be on a ballistic trajectory, the vector $\boldsymbol{A}$ will be nearly antiparallel to the vector $\boldsymbol{V}$. We should note that both density and gravitational acceleration are accurately known functions of altitude Z, i.e.,

$$\rho = \rho(Z), \qquad \boldsymbol{g} = \boldsymbol{g}(Z)$$

We may now rewrite Eq. (8.6) as

$$\hat{A} = \frac{-\rho(\hat{V}^{T}\hat{V})^{1/2}\hat{V}}{2\beta_m} + \boldsymbol{g} \tag{8.7}$$

Next, we can take the inner product of both sides of Eq. (8.7) to get

$$(\hat{A}^{T}A) = (\rho/2\beta_m)^2(\hat{V}^{T}\hat{V})^2 + g^2 - 2(\rho g/2\beta_m)(\hat{V}^{T}\hat{V})^{1/2}\hat{V}_z \tag{8.8}$$

The above equation can be solved for the ballistic coefficient β_m. To simplify the computations we might justifiably ignore the gravitational contribution by setting $\boldsymbol{g} = 0$. Thus,

$$\beta_m = \frac{\rho(Z)}{2}\frac{\hat{V}^{T}\hat{V}}{(\hat{A}^{T}\hat{A})^{1/2}} \tag{8.9}$$

Consequently, an estimate is made of the ballistic coefficient of an observed re-entry body by making estimates of the velocity and acceleration. For our purposes here we identify the ballistic coefficient as an *observable,* although in truth it is inferred from measurements of trajectory states. The observer would presumably have a knowledge of the ballistic coefficient appropriate to the re-entry body of interest. The observed value of ballistic coefficient will undoubtedly differ to some degree from what might be called the standard. The ground-based observer must then decide whether the observed value of β_m is sufficiently close to that of the standard to identify the re-entry body as a threat. The offense endeavors to make the observed ballistic coefficient of a decoy remain within the error bound as long as possible, i.e., to as low an altitude as possible.

The source of confusion between a threatening re-entry body and a decoy becomes clear when we examine Eq. (8.9): the ballistic coefficient is, in one sense, the ratio of the density $\rho(Z)$ and the acceleration magnitude $\boldsymbol{A}^{T}\boldsymbol{A}$. At high altitudes both density and acceleration approach zero, leading to an indeterminate form for the ballistic coefficient. Consequently, at very high altitudes almost any re-entry body will appear about the same as any other re-entry body. As altitude decreases, both density and acceleration become increasingly better defined, with the result that the uncertainty associated with measurement of the ballistic coefficient becomes less. Accuracy in the knowledge of the atmospheric density depends upon the quality of the atmospheric model that the defense has formulated; accuracy in acceleration measurement depends upon the accuracy of the radar system, knowledge of such accuracy (if a statistical estimator is to be used), the estimation algorithm, and the computational speed and precision.

It is the measurable difference between the ballistic coefficients of a re-entry body and a decoy that is used in separating a decoy from a threat. For example, we saw in Chapter 7 that the altitude of maximum deceleration increases with decreasing ballistic coefficient. Depending upon many constraints associated with the entire engagement, there is some altitude at which the defense can have confidence that a distinction can be made between the two types of re-entry bodies so that the appropriate response can be made. If the RB has been tagged as a decoy, then it is removed from radar track; if the RB has been tagged as a threat, then the engagement space is by necessity below this altitude of discrimination, and the appropriate response is then made. In the next section we will consider the decoy credibility problem from a fairly general point of view.

8.3 Decoy Effectiveness

A decoy must have credibility in order to be effective. Because credibility is not an absolute, credibility associated with any decision process must first be assigned some numerical value. In the previous section we briefly discussed ways of quantifying an observation. In this section we will extend the discussion to the decision process that uses measurements as the input to an algorithm that resolves the ambiguity of how to respond to a possible threat.

The decision process is something like the following: if the measurement exceeds some threshold, then the threat is assumed to be credible and defensive action is taken. On the other hand, if the measurement is below such a threshold, then no action is taken. In the present context, "taking action" means launching an interceptor. In Chapter 9 we will show that it is by no means certain that an interceptor launched toward an engagement will always intercept an intended target. However, in this chapter we will assume that a response by the defense is equivalent to nullifying the threat.

Before we try to assign numerical values to the idea of credibility, we must first note that credibility imposes on the defense the possibility of two fundamental decision errors. These errors take the following form:

Leakage: The defense fails to launch an interceptor against a re-entry body (RB) because the RB is assumed to be a decoy. The consequence of this decision is that the RB reaches its intended impact point.

False alarm: The defense launches an interceptor against a decoy because a decoy has been erroneously identified as a re-entry body. The consequence of this decision is that the interceptor inventory has been depleted, with the result that the defense might not be able to respond to a real threat at a later time.

Although leakage and false alarm errors have different consequences for the defense, they are both strongly influenced by how well the decoy duplicates the observables that the defense uses in its decision (discrimination) algorithm. The defense sets the numerical value of the threshold dividing response (launch) from nonresponse (interceptor inventory retention) based on an a priori assessment of a likely engagement construct. Among such considerations might be an estimation of the threat composition (numerical mix of RBs and decoys), the defense's accuracy in measurement, interceptor performance, data processing rate, numerical precision, loss of track of a threat, and saturation of the data acquisition system. In this chapter we will confine ourselves to a binary decision algorithm and draw upon the extensive discussion of the binary decision process by Van Trees.[2]

We begin by defining the observed threat. The term *target* is taken here to mean any observed re-entry body that has any credibility as a threat. Note that an object with credibility can be either a decoy or a re-entry body. In the previous section we identified a specific set of measurements as observables (position, velocity, etc.). Here we identify any observable that is used for discrimination as a signature. A signature might be simply a time-ordered set of observations, say a vector, **Z**:

$$\mathbf{Z} = \left[X_1, X_2, \cdots, X_n\right]^{\mathrm{T}} \tag{8.10}$$

Each component of the signature may be a scalar, for instance,

$$X_i = \text{Peak-to-peak radar cross section (PPRCS)}$$

or a vector, for instance,

$$X_i = [C_1, C_2, C_3] = \text{Three-color sensor} \tag{8.11}$$

The sequence **Z** is used in a discrimination algorithm to give an observation parameter Q as follows:

$$Q = Q(\mathbf{Z}) = \mathbf{Z}^{\mathrm{T}} A \mathbf{Z} \tag{8.12}$$

where A is a square matrix involved in the fitting of a scalar Q to the sequence **Z**.

All of the above is somewhat tangential to the main discussion, except that it is reasonable that a scalar Q may be obtained from the observation of a potential threat. This scalar Q can then be compared with a threshold T as follows:

If $Q \leq T$: Do not launch interceptor at the threat.
If $Q > T$: Launch interceptor at the threat.

If the signature from the decoy and the re-entry body are different, then the decoy would not be mistaken for the re-entry body (and use of the term *decoy* would be meaningless). We therefore make the following assumption: the signature parameter Q is not entirely deterministic. It is a measurement (or it is derived from a set of measurements) and as such is subject to bias and random errors. A stochastic component in the signal vector or sequence **Z** means that there is some uncertainty in assessing the side of the threshold T on which the signal parameter Q lies; consequently, there is a probability that the observed object is a decoy but is assumed to be a re-entry body (false alarm) or that a re-entry body is assumed to be a decoy and is allowed to pass (leakage).

We assume that a Gaussian probability density function describes the uncertainties associated with the signature of both the decoy and the re-entry body. Because the signatures of both the decoy and the re-entry body are obtained using the same equipment, the measurement uncertainties must be identical. Consequently, the standard deviations of both density functions must also be identical. However, because the motion of the decoy and the motion of the re-entry body are different processes, the density functions of decoy and re-entry body must have different means. Without loss of generality we may assume that the mean of the density function of the decoy is zero and that of the re-entry body is nonzero, designated here by m.

First, let's write the two density functions as

$$f_D = \left(\frac{1}{2\pi}\right)^{1/2} \frac{1}{\sigma} \exp\left(-\frac{Q^2}{2\sigma^2}\right), \qquad f_{RB} = \left(\frac{1}{2\pi}\right)^{1/2} \frac{1}{\sigma} \exp\left(-\frac{(Q-m)^2}{2\sigma^2}\right) \tag{8.13}$$

Next, we can normalize the terms in the exponents of Eqs. (8.13) as follows:

$$Y = \frac{Q}{\sigma}, \qquad Y - d = \frac{Q}{\sigma} - \frac{m}{\sigma} \tag{8.14}$$

In Eq. (8.14) the nondimensional mean, d, is the actual mean divided by the standard deviation, σ; or more precisely, d is the nondimensional difference between the means of the re-entry body and the decoy.

$$d = \frac{m}{\sigma} \tag{8.15}$$

We will henceforth identify d as "the credibility parameter." If d is zero, the two density functions of Eqs. (8.13) are equal and the decoy perfectly masks the re-entry body. According to Eq. (8.15) there are two ways that d can be made to approach zero: 1) by matching the signature, Q, between both the RB and the decoy, and 2) by reducing the quality of the measurement, that is, allowing σ to become large. We may also look at Eq. (8.15) in another way: if the offense presents a decoy whose signature closely matches that of the re-entry body, a very accurate measurement (small σ) is necessary to keep d large. The offense controls the numerator and the defense controls the denominator in Eq. (8.15). The offense strives to make d as small as possible (for as long as possible), and the defense wishes to make d as large as possible (as soon as possible). However, engineering limitations, cost, and RB off-loading (substituting decoys for an RB) conspire to make d some finite number. Later we will return to the credibility parameter and how it might vary along a trajectory; for now, we concern ourselves with the calculation of the probabilities of a false alarm (PF), leakage (PL), and detection (PD).

First, we must rewrite Eqs. (8.13) using Eqs. (8.14) as follows:

$$f_D = \left(\frac{1}{2\pi}\right)^{1/2} \exp\left(-\frac{Y^2}{2}\right), \qquad f_{RB} = \left(\frac{1}{2\pi}\right)^{1/2} \exp\left(-\frac{(Y-d)^2}{2}\right) \tag{8.16}$$

In Fig. 8.3 we have illustrated the two error density functions of both the decoy (to the left) and the re-entry body (to the right). Note that we have placed a threshold line, T, somewhere between the maxima of the two functions. The

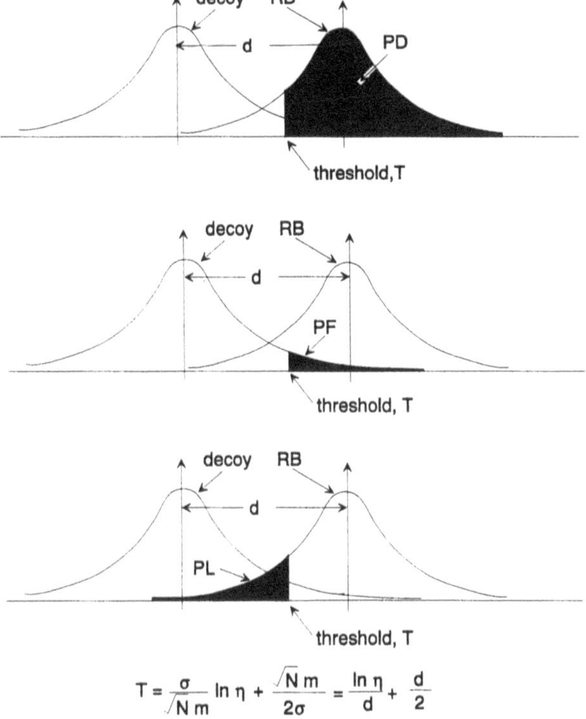

Fig. 8.3 Decoy and re-entry body probability density function.

probability of detection (PD) of a potential threat is the area under the curves to the right of the threshold, T.

$$PD = \left(\frac{1}{2\pi}\right)^{1/2} \int_{T}^{\infty} \exp\left(-\frac{(Y-d)^2}{2}\right) \mathrm{d}Y \tag{8.17}$$

With $Y - d = X$, we have

$$PD = \left(\frac{1}{2\pi}\right)^{1/2} \int_{T-d}^{\infty} \exp\left(-\frac{X^2}{2}\right) \mathrm{d}X \tag{8.18a}$$

Because the error function, erf(X), is defined as

$$\mathrm{erf}(X) = \int_{-\infty}^{X} \left(\frac{1}{2\pi}\right)^{1/2} \exp\left(-\frac{Z^2}{2}\right) \mathrm{d}Z \tag{8.18b}$$

Eq.(8.18a) becomes

$$PD = 1 - \mathrm{erf}(T - d) \tag{8.19}$$

The probability of a false alarm (PF) is the area under the decoy density function to the right of the threshold. From Eqs. (8.16) and (8.18b) we have

$$PF = \int_{T}^{\infty} \left(\frac{1}{2\pi}\right)^{1/2} \mathrm{cxp}\left(-\frac{Y^2}{2}\right) \mathrm{d}Y = 1 - \mathrm{erf}(T) \tag{8.20}$$

The probability of leakage (PL) is the area under the re-entry body density function to the left of the threshold. Again, from Eqs. (8.16) and (8.18b) we have

$$PL = \int_{-\infty}^{T} \left(\frac{1}{2\pi}\right)^{1/2} \exp\left(-\frac{(Y - d)^2}{2}\right) \mathrm{d}Y = \mathrm{erf}(T - d) \tag{8.21}$$

We note that the probability of a false alarm, PF, depends only on the threshold parameter, T, and is independent of the credibility parameter, d. However, the probability of leakage, PL, depends on the difference between the threshold parameter, T, and the credibility, d.

The following point should be emphasized: the defense can only partially control the credibility parameter, d, but it can control the threshold parameter, T. The defense sets T, but it must do so within constraints imposed by a finite interceptor inventory. Equations (8.20) and (8.21) illustrate the dilemma that must be part of the defense construct when there is the possibility of encountering decoys.

A low threshold will reduce the probability of leakage and increase the probability of a false alarm, because a low threshold means that a large percentage of targets are engaged. On the other hand, as the threshold increases, targets are increasingly passed, which in turn increases the probability of leakage. Setting the value of the threshold is a judgment call that must be made by the defense. As pointed out earlier, the driving concern in setting the threshold is the inventory of interceptors. It is not possible to quantify the threshold, T, in a general sense. However, we can pursue a qualitative discussion a bit further using the work of Van Trees.[2]

In a binary hypothesis we assume that either a decoy is present or a re-entry body is present. The presence of a decoy will be designated by H_{d} and the presence of a re-entry body by H_{r}. Obviously, we could have a third possibility—some way to represent "I don't know" or indecision. However, we will confine ourselves to a binary decision algorithm because any indecision concerning a potential threat is equivalent to a decoy assumption or no response. Thus, each time an observation is made (and processed) to give a signature, Q, there are four and only four possible events:

Event	*Decision*	*Consequence*
1. H_d true, choose H_d	Ignore	Retain interceptors
2. H_d true, choose H_r	Launch interceptor	False alarm
3. H_r true, choose H_d	Ignore	Leakage
4. H_r true, choose H_r	Launch interceptor	Eliminate threat

Obviously, the first and fourth decisions correspond to correct choices; the first rightly conserves resources for subsequent deployment, and the fourth activates the defense. The second decision results in a false alarm and the loss of resources without compensation; the third decision is the primary purpose of the decoy, to cause a valid threat to penetrate the defense. We might even go so far as to say that cases (1) and (4) are minor and major victories for the defense, respectively; and cases (2) and (3) are minor and major victories for the offense. We might attempt quantification by assigning numerical costs to the four cases. These cost terms might be defined as follows: C_{dd}, C_{rd}, C_{dr}, C_{rr}, where the first subscript indicates the course of action and the second indicates what was really the case. For example, C_{dr} means the defense believed that a decoy was present, but in reality a re-entry body was present. The result was leakage.

Van Trees[2] defines a risk R as

$$R = C_{dd}P_dP(H_d/H_d) + C_{rd}P_dP(H_r/H_d) + C_{dr}P_rP(H_d/H_r) + C_{rr}P_rP(H_r/H_r) \tag{8.22}$$

The terms P_d and P_r are the a priori probabilities that the defense assigns to its model of the offense construct. Essentially P_d and P_r are the assumed mix of decoys and re-entry bodies. If, for example, there are 25 decoys for every RB, then $P_r = \frac{1}{26}$ and $P_d = \frac{25}{26}$. The typical term, $P(H_r/H_d)$ term in Eq. (8.22), is the Basian probability that H_r would be chosen subject to the condition that H_d is true.

Finally, the cost terms (e.g., C_{rd}) must be assigned by the defense. For example, the defense might set the cost C_{dd} very low (correctly assuming the presence of a decoy) because no resources were expended and no RB leaked. On the other hand, the cost C_{rr} might be higher because an interceptor must be expended. Van Trees[2] suggests that the cost of a wrong decision be higher than the cost of a correct decision; that is,

$$C_{rd} > C_{dd}$$

$$C_{dr} > C_{rr}$$

It would seem that C_{dr}, the cost of assuming a decoy when a re-entry body is present, must be the greatest cost since the defense has been penetrated.

After much manipulation, Van Trees derives a threshold, η, as

$$\eta = \frac{P_d(C_{rd} - C_{dd})}{P_r(C_{dr} - C_{rr})} \tag{8.23}$$

Obviously, if decoys are not present, then $P_d = 0 = \eta$. However, if decoys are present then we would expect that $P_d >> P_r$. Continuing with the idea that a wrong decision should cost more than a correct decision, Eq. (8.23) simplifies to

$$\eta \approx \left(\frac{P_d}{P_r}\right)\left(\frac{C_{rd}}{C_{dr}}\right) \tag{8.24}$$

Because P_d/P_r must be large, then C_{rd}/C_{dr} must be small. This is equivalent to saying that the cost of assuming a re-entry body when presented with a decoy must be much less than the cost of assuming a decoy when presented with a re-entry body. In other words, the defense, when faced with a decoy/re-entry body mix, must regard the cost of a false alarm much less than the cost of leakage.

In adapting the binary decision hypothesis to the re-entry body detection, we seem to be regarding the decoy as a noise process, whereas in reality it is a separate process from that of the re-entry body. We are essentially regarding the signature associated with the re-entry body as a deterministic process corrupted by noise and the signature of the decoy as totally or partially embedded in the noise. If we can separate the signature of the re-entry body beyond some preset threshold, then we assume that we have identified the re-entry body. If the signature is not identifiable, that is, below the threshold, then we are only measuring noise (or a decoy) and the response is to do nothing. Of course, noise will continue to be present in any signature measurement, so there will always be some uncertainty in identification.

The next consideration is how to represent the credibility factor as a function of altitude. This altitude variation will be used to characterize the decoy. Two types of decoys are identified here: 1) black/white and 2) exponential. Figure 8.4 shows that the black/white decoy has perfect credibility (i.e., $d = 0$) above a preset (credibility) altitude; below this altitude the decoy has no credibility ($d = 5$). The credibility parameter varies as a simple discontinuity

$$\begin{aligned} d &= 0 \qquad \text{for} \qquad Z \geq Z_c \\ d &= 5 \qquad \text{for} \qquad Z < Z_c \end{aligned} \tag{8.25a}$$

The exponential decoy may be represented as

$$d = d_c \exp\left[-\left(\frac{Z - Z_c}{Z_c}\right)\right] \tag{8.25b}$$

where Z_c is the credibility altitude of the black/white decoy and d_c is an arbitrary constant (in Fig. 8.4 a value of 2.5 was selected). Equation (8.25b) permits the analyst to vary the decoy credibility with altitude somewhat more realistically than is possible with the black/white decoy.

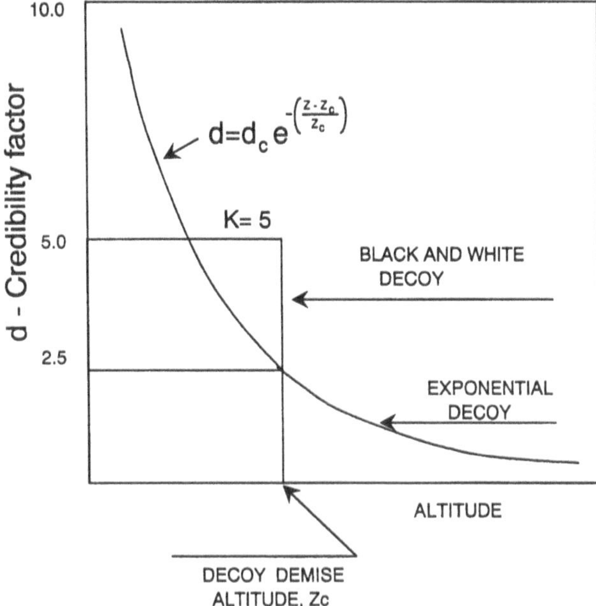

Fig. 8.4 Variation of credibility factor with altitude for exponential and black/white decoys.

Finally, we must consider the setting of the threshold, although any attempt at generalization must be limited to qualitative reasoning; quantitative issues are entirely application dependent. Let's start with the threshold η as defined by Van Trees[2] and given here in Eq. (8.23).

Van Trees[2] assumes that a constant signal is either present or absent in any measurement. Any measurement is accompanied by Gaussian noise that is uncorrelated across successive measurement sequences. For our applications, the presence or absence of a signal is extended to the idea of the presence or absence of a re-entry body; the absence of a re-entry body is equivalent to the presence of a decoy. The probability density functions for decoy (D) and re-entry body (RB) have been given in Eqs. (8.13). Let's assume that the signature Q is simply the sum of the individual measurements, Q_{i}. Thus, for any Q_{i} Eqs. (8.13) become

$$f_{\mathrm{D}}(Q_{\mathrm{i}}/H_{\mathrm{d}}) = \left(\frac{1}{2\pi}\right)^{1/2} \frac{1}{\sigma} \exp\left(-\frac{Q_{\mathrm{i}}^2}{2\sigma^2}\right)$$

$$f_{\mathrm{RB}}(Q_{\mathrm{i}}/H_{\mathrm{r}}) = \left(\frac{1}{2\pi}\right)^{1/2} \frac{1}{\sigma} \exp\left(-\frac{(Q_{\mathrm{i}} - m)^2}{2\sigma^2}\right)$$

Because the noise sequence is statistically uncorrelated with the measurement sequence, the joint probability density is simply the product of the individual density functions across the sequence of N measurements.

$$f_{\mathrm{D}}(Q/H_{\mathrm{d}}) = \prod_{i=1}^{N}\left(\frac{1}{2\pi}\right)^{1/2}\frac{1}{\sigma}\exp\left[-\frac{Q^2}{2\sigma^2}\right] \tag{8.26a}$$

$$f_{\mathrm{RB}}(Q/H_{\mathrm{r}}) = \prod_{i=1}^{N}\left(\frac{1}{2\pi}\right)^{1/2}\frac{1}{\sigma}\exp\left[-\frac{(Q_{\mathrm{i}}-m)^2}{2\sigma^2}\right] \tag{8.26b}$$

Let's take the likelihood ratio as Eq. (8.26b) divided by Eq. (8.26a):

$$\Lambda(Q) = \frac{\prod_{i=1}^{N}\left(\frac{1}{2\pi}\right)^{1/2}\frac{1}{\sigma}\exp\left[-\frac{(Q_i-m)^2}{2\sigma^2}\right]}{\prod_{i=1}^{N}\left(\frac{1}{2\pi}\right)^{1/2}\frac{1}{\sigma}\exp\left[-\frac{Q_i^2}{2\sigma^2}\right]} \tag{8.27}$$

Van Trees cancels the common terms and then takes the logarithm of both sides to get

$$\ell n(\Lambda(Q)) = \frac{m}{\sigma^2}\sum_{i=1}^{N}Q_i - \frac{Nm^2}{2\sigma^2} >^{H_{\mathrm{r}}} \ell n(\eta)$$
$$<_{H_{\mathrm{d}}} \ell n(\eta) \tag{8.28}$$

where we have set the likelihood ratio $\Lambda(Q_i)$ equal to the threshold. After some manipulation we get

$$\sum_{i=1}^{N}Q_i >^{H_{\mathrm{r}}} \frac{\sigma^2}{m}\ell n(\eta) + \frac{Nm}{2}$$
$$<_{H_{\mathrm{d}}} \frac{\sigma^2}{m}\ell n(\eta) + \frac{Nm}{2}$$

or

$$\frac{1}{\sqrt{N}\sigma}\sum_{i=1}^{N}Q_i >^{H_{\mathrm{r}}} \frac{\sigma}{\sqrt{N}m}\ell n(\eta) + \frac{\sqrt{N}m}{2\sigma}$$
$$<_{H_{\mathrm{d}}} \frac{\sigma}{\sqrt{N}m}\ell n(\eta) + \frac{\sqrt{N}m}{2\sigma} \tag{8.29}$$

The symbolism used above, that is,

$$A >^{H_r} B$$
$$A <_{H_d} B$$

means that A must be greater than B for the threat to be identified as a re-entry body and less than B to be identified as a decoy.

Letting

$$d = \frac{\sqrt{N}m}{\sigma} \tag{8.30}$$

the threshold, T, is then

$$T = \frac{\ell n(\eta)}{d} + \frac{d}{2} \tag{8.31}$$

Notice in Eq. (8.30) that σ is a measure of the quality of a single measurement. However, if a large number of measurements $\underline{N}$ are made, then the uncertainty decreases having a standard deviation of $\sigma/\sqrt{N}$.

Both terms on the right side of Eq. (8.31) are based on estimates (on the part of the defense) of the decoy/re-entry body offense construct. There seems to be little point in any discussion of the setting of the threshold quantitatively. We might note in passing that if

$$P_d C_{rd} = P_r C_{dr}$$

of Eq. (8.24), then $\eta = 1[\ell n(\eta) = 0]$ and the threshold $T = d/2$, or the threshold is set midway between the two Gaussian density functions of Fig. 8.3.

We might substitute Eq. (8.31) into Eqs. (8.20) and (8.21) to rewrite probability of false alarm, PF, and the probability of leakage, PL, as

$$PF = 1 - \text{erf}\left(\frac{\ell n \eta}{d} + \frac{d}{2}\right) \tag{8.32a}$$

$$PL = \text{erf}\left(\frac{\ell n \eta}{d} - \frac{d}{2}\right) \tag{8.32b}$$

Figure 8.5 is a plot of PL versus PF for various values of d with η as the varying parameter. For $\eta = 0$, $\ell n(\eta) = -\infty$, the processor always guesses H_r, that is, there is no leakage and the probability of a false alarm is 1.0 (upper right-hand corner). However, when $\eta \to \infty$, the processor always guesses H_d; in this case, the probability of leakage is 1.0 and the probability of a false alarm is 0. As stated earlier, setting numerical values for η is problem dependent.

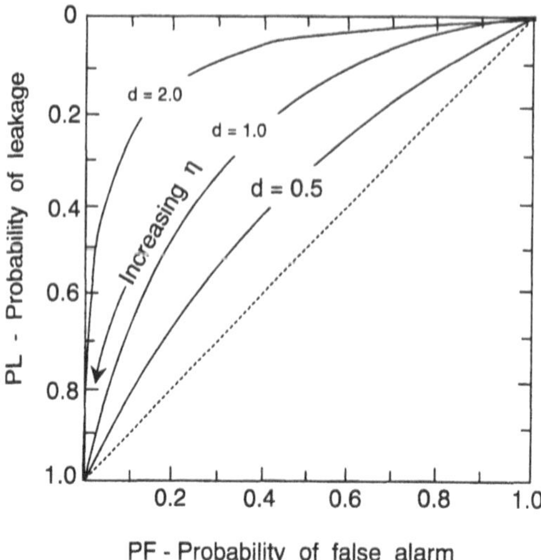

Fig. 8.5 Leakage probability vs. false alarm probability for various credibility parameters.

References

[1]Technical staff, The Analytical Sciences Corporation, *Applied Optimal Estimation*, edited by A. Gelb, M.I.T. Press, Massachusetts Institute of Technology, Cambridge, MA, 1974.

[2]Van Trees, L. L., *Detection, Estimation, and Modulation Theory*, Wiley, Somerset, NJ, 1968.

9
Maneuvering Re-Entry Vehicles: Particle Motion

9.1 Introduction

In Chapter 7 we considered some of the fundamental physics associated with re-entry. A number of closed-form solutions were possible if we made certain simplifying assumptions, such as representing the density-altitude variation by a simple exponential function or ignoring the gravitational force. We also considered the effect of lift in altering the trajectory shape and the velocity penalty exacted by lift production.

In this chapter we consider a re-entry vehicle that is capable of producing lift and how such lift might be used to shape the trajectory. We also model an interceptor directed to terminate the trajectory of such a re-entry vehicle.

We assume that a maneuvering re-entry vehicle (MaRV) alters lift (i.e., shapes the trajectory) according to some preset schedule and is at no time aware of the presence or intent of the interceptor. The MaRV trajectory changes appear to be random to the defense and consequently introduce uncertainty into any interceptor guidance algorithm. The defense will have knowledge of the performance potential of the MaRV. This knowledge might take the form of the numerical value of the maximum lift-to-drag ratio $(L/D)_{\max}$ or the ballistic coefficient β.

Maneuvering is a defensive tactic that a re-entry vehicle uses to confound the guidance algorithms of the interceptor. We assume that during conflict the MaRV has a significant velocity advantage over the interceptor. Consequently, an interceptor following a pure pursuit engagement will usually be unsuccessful in interdiction. Therefore, it is necessary for the defense to "lead" the MaRV. The defense might do this by identifying a projected interception point (for a given MaRV state vector) and then directing the interceptor velocity vector toward such a point. The greater the MaRV's capacity to maneuver, the less time any MaRV state projection will be useful. The interceptor does have a net maneuvering advantage over the MaRV, which can be used to overcome uncertainties introduced by MaRV maneuvering.

In the next section we first consider a simple two-parameter representation of the aerodynamics of both the MaRV and interceptor. We then consider one guidance law for the MaRV to be used in trajectory shaping and two guidance laws for the interceptor.

9.2 Drag Polar

When a re-entry vehicle develops lift (for maneuvering) there is an increase in the aerodynamic drag. This drag rise is identified as *induced drag* because it is induced by the production of lift. Total drag is then the sum of zero-lift drag and induced drag. The zero-lift drag is usually divided into pressure and skin-friction components. For now we are concerned with a simple expression that will indicate the additive nature of the zero-lift and induced drag. Such an expression is designated as a drag polar.

A fairly general drag polar might be written as

$$C_D = C_{D_0} + KC_L^n \tag{9.1}$$

Clearly, Eq. (9.1) represents the drag coefficient as a three-parameter expression. These parameters are C_{D_0}, the zero-lift drag coefficient, K, the induced drag parameter, and n, the drag polar exponent. Later we will assume a value of n equal to 2.0, but for the present n will remain unspecified. Our first consideration is the zero-lift drag coefficient C_{D_0}. Recall from Eq. (7.13) that the ballistic coefficient β contains $C_{D_0}A$, where A is the reference area upon which C_{D_0} is based. The tacit assumption made here is that $C_{D_0}A$ (and hence C_{D_0}) is a constant over much of the trajectory. Of course, no general statement can be made that all re-entry vehicles over all trajectories will have essentially constant values of $C_{D_0}A$. However, Ref. 1 provides some indication of the variation of C_{D_0} with altitude.

In Ref. 1 three different re-entry vehicles were simulated over the entire trajectory. Figure 9.1, taken from this reference, shows the variation of $C_{D_0}A$ with altitude. The vehicles studied were hemispherically blunted cones with half-angles of 12.5 deg (β = 550 lb/ft^2), 6.0 deg (β = 1025 lb/ft^2), and 10.0 deg (β = 1975 lb/ft^2). It should be pointed out that for these vehicles the weight and reference area were chosen to make $C_{D_0}A$ equal to 1.0 ft^2 in the hypersonic region. Examination of Fig. 9.1 indicates that $C_{D_0}A$ remains essentially constant from re-entry (say, at 60 km) to about 10 km. Of course, outside the hypersonic region C_{D_0} is no longer a constant and is usually a function of Mach number and Reynolds number.

We can recast the parameter K in terms of C_{D_0} and the critical lift coefficient C_L^*. The critical lift coefficient is that value of C_L for which the lift-to-drag ratio C_L/C_D is a maximum. First, we rewrite Eq. (9.1) as

$$C_L/C_D = C_L/(C_{D_0} + KC_L^n) \tag{9.2}$$

The derivative of (C_L/C_D) is taken with respect to the lift coefficient C_L and equated to zero. (The value of C_L for which the lift-to-drag ratio is a maximum is by definition the critical lift coefficient C_L^*.) Carrying out this operation gives

$$K = \frac{C_{D_0}}{(n-1)(C_L^*)^n} \tag{9.3}$$

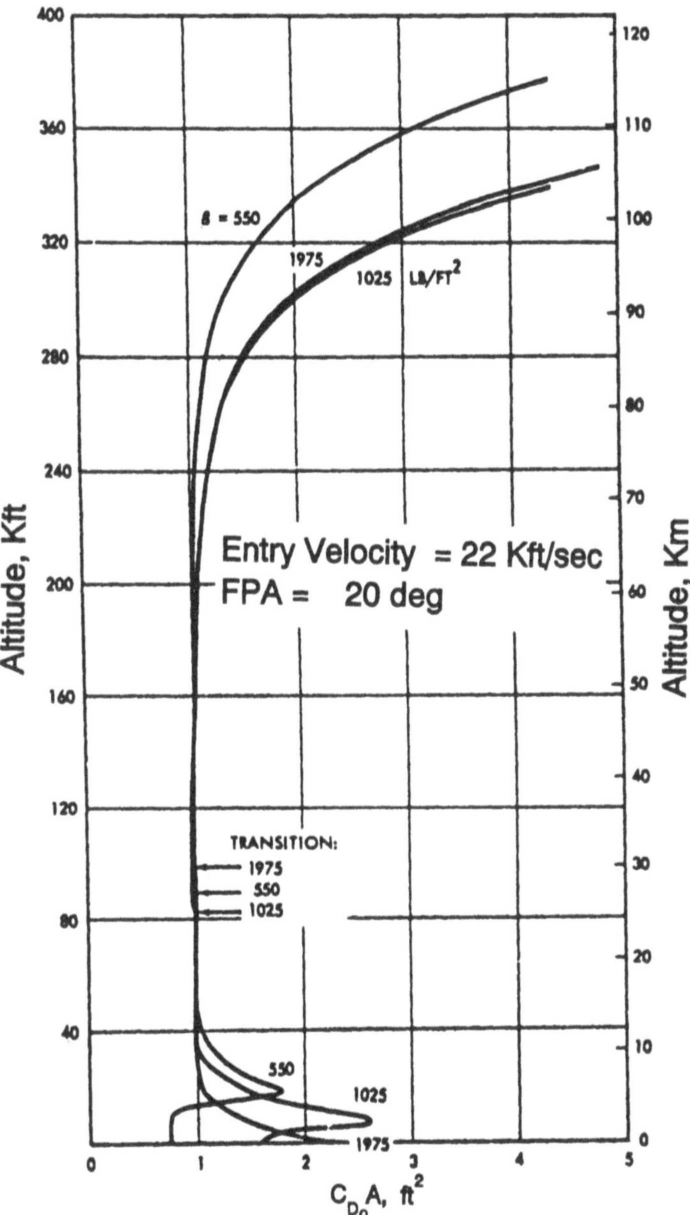

Fig. 9.1 Altitude dependence of drag coefficient of three re-entry vehicles (weight and reference area chosen so that $C_{D_0}A = 1.0$ ft² in hypersonic inviscid regime).

Replacing the induced drag parameter K by the above expression and C_L by C_L^* in Eq. (9.2) gives the maximum lift-to-drag as

$$\frac{C_L}{C_D}_{\max} = \frac{(n-1)C_L^*}{nC_{D_0}} \tag{9.4}$$

Before proceeding further, we should consider the numerical value of the exponent n. For subsonic aircraft, n is generally assigned the value of 2.0. An empirical fit in the hypersonic region for the exponent n is

$$n = -0.09567\,(C_L/C_D)_{\max} + 2.235 \tag{9.5}$$

For a re-entry vehicle having essentially rotational symmetry (conical shape), $(C_L/C_D)_{\max}$ would have a value of about 2.5 at most. The value of n would then be about 1.9958. Obviously, for a vanishing small value of $(C_L/C_D)_{\max}$, n would approach 2.235. Thus, the assumption of $n = 2.0$ might be acceptable.

We may now recast Eqs. (9.1) and (9.2) with the restriction that $n = 2.0$. Equations (9.1) and (9.2) become

$$C_D = C_{D_0} + KC_L^2 \tag{9.6a}$$

$$C_L/C_D = C_L/\left(C_{D_0} + KC_L^2\right) \tag{9.6b}$$

Replacing K according to Eq. (9.3), with $n = 2$, gives

$$C_D = C_{D_0}\left[1.0 + \left(C_L/C_L^*\right)^2\right] \tag{9.7a}$$

$$C_L/C_D = \frac{C_L}{C_{D_0}\left[1.0 + \left(C_L/C_L^*\right)^2\right]} \tag{9.7b}$$

Thus, the drag coefficient and lift-to-drag ratio are specified by two parameters, C_L^* and C_{D_0}. We can use the symbol $\hat{L}$ to represent $(C_L/C_D)_{\max}$. We may now recast Eqs. (9.1) and (9.2) in terms of the parameter pair $(C_L^*, \hat{L})$ using Eq. (9.4) and setting the exponent n to 2.0 as follows:

$$C_D = \frac{1}{2}\left(\frac{C_L^*}{\hat{L}}\right)\left[1.0 + \left(\frac{C_L}{C_L^*}\right)^2\right] \tag{9.8a}$$

$$\frac{C_L}{C_D} = 2\hat{L}\left[\frac{C_L/C_L^*}{1.0 + \left(C_L/C_L^*\right)^2}\right] \tag{9.8b}$$

Finally, we might wish to express C_D and (C_L/C_D) in terms of parameters C_{D_0} and $\hat{L}$ as follows:

$$C_D = C_{D_0}\left[1.0 + \frac{C_L^2}{4\hat{L}^2 C_{D_0}^2}\right] \tag{9.9a}$$

$$\frac{C_L}{C_D} = \frac{(4\hat{L}^2 C_{D_0})C_L}{C_L^2 + 4\hat{L}^2 C_{D_0}^2} \tag{9.9b}$$

Let's now examine some of the characteristics of drag polar representations. We note from Eq. (9.7a) that when the MaRV is operating at the critical lift coefficient or when $(C_L/C_D) = \hat{L}$, that $C_D = 2C_{D_0}$. In other words, when C_L is set at C_L^*, the drag is equally distributed between the zero-lift drag and the induced drag.

If $\hat{L}$ and C_L^* are specified, then the MaRV is considered "designed" and the analyst has available the appropriate lift-to-drag ratio for any value of applied lift coefficient.

Both $\hat{L}$ and C_L^* are regarded as independent parameters. Nevertheless, two interesting relationships may be developed between these two parameters. From Eq. (9.4)

$$C_{D_0} = C_L^*/2\hat{L} \tag{9.10a}$$

and, from Eq. (9.3),

$$K = C_{D_0}/C_L^{*n} = C_{D_0}/C_L^{*2} = 1/2\hat{L}C_L^* \tag{9.10b}$$

Note that a very crude approximation to K might be $1/C_{L_\alpha}$, where C_{L_α} is the change in the lift coefficient with a change in angle of attack. If we examine a class of re-entry vehicles for which C_{D_0} is constant, then $\hat{L}$ is directly proportional to C_L^*. On the other hand, if K is held constant, then there will be an inverse relationship between C_L^* and $\hat{L}$.

A normalization of C_L/C_D by $\hat{L}$ and of C_L by C_L^* such as

$$\bar{L} = (C_L/C_D)/\hat{L}, \qquad \bar{C}_L = C_L/C_L^* \tag{9.11a}$$

gives, from Eq. (9.8b),

$$\bar{L} = 2\bar{C}_L/\left(1 + \bar{C}_L^2\right)$$

A plot of the above function gives a nondimensional form of lift-to-drag ratio vs lift coefficient. In the same spirit, the nondimensional form of the drag polar may be written as

$$C_D/C_{D_0} = \bar{C}_D = 1 + \bar{C}_L^2 \tag{9.11b}$$

Both functions are plotted in Fig. 9.2.

We note that, if $K = 1/C_{L_\alpha}$, then

$$C_{L_\alpha} = 2\hat{L}C_L^* \tag{9.12}$$

As the MaRV becomes more efficient, we would expect $\hat{L}$ to increase; if C_{L_α} remains the same, then an increase in $\hat{L}$ would imply that the critical lift coeffi-

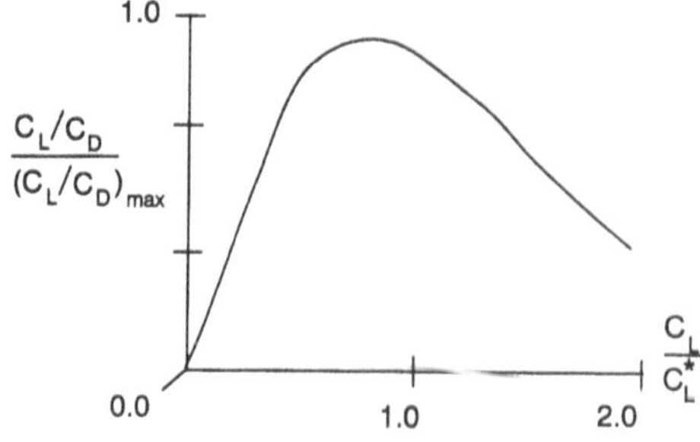

Fig. 9.2a Unitized lift-to-drag ratio vs lift coefficient.

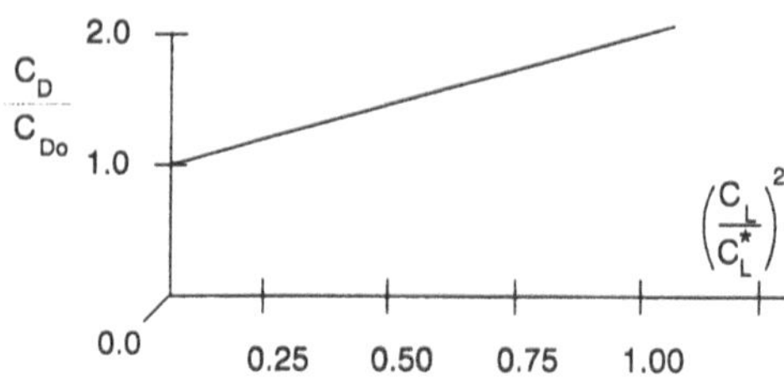

Fig. 9.2b Normalized drag polar.

cient decreases. Consequently, increasing the maximum lift-to-drag ratio means that there is a reduction in the drag associated with the lift, but, as pointed out, there is also a reduction in the value of the critical lift coefficient C_L^*. In order to achieve the necessary lateral acceleration, it might be necessary to operate an aerodynamically efficient re-entry vehicle at a lift coefficient above that corresponding to the maximum lift-to-drag ratio.

The lift-to-drag ratio might take the following form, where the design parameters are the ballistic coefficient β and the maximum lift-to-drag ratio $\hat{L}$:

$$\frac{C_L}{C_D} = 2\hat{L}\left[\frac{(C_LA/W)(\beta/2\hat{L})}{1+(C_LA/W)^2(\beta/2\hat{L})^2}\right] \tag{9.13}$$

where (C_LA/W) is an operational variable, A is the reference area, and W the weight. If we replace β with β_m [see Eq. (7.11)], then W would be replaced by the mass.

There is a range of values for the design parameters that should apply to any practical maneuvering re-entry vehicle. These ranges are given in Table 9.1.

We might now ask how well the drag polar represents the lift and drag characteristics of various aerospace vehicles. A series of charts has been prepared to compare the lift-to-drag ratio vs the lift coefficient using both Newtonian and drag polar representations. These comparisons are given in Figs. 9.3 and 9.4 for both wedges and cones. In Fig. 9.5 similar sets of lift-to-drag characteristics are given for several lifting vehicles. Note that in general the more efficient the vehicle (i.e., the greater the maximum lift-to-drag ratio), the lower the lift coefficient at which $(L/D)_{max}$ occurs.

Table 9.1 Design parameter ranges for maneuvering re-entry vehicles

Parameter	Description of parameter range boundaries
$1.5 \leq (L/D)_{max} < 6.6$	Maximum: Flat plate with skin friction Minimum: Resembles *Apollo* capsule
$0.001 \leq C_{D_0} < 0.160$	Maximum: Wedges with $(L/D)_{max} = 1.5$ Minimum: Flat plate with skin friction alone
$40 \leq W/A \leq 160$ lb/ft^2	Maximum: Payload alone Minimum: 25% of maximum
$250 \leq \beta < 160{,}000$ lb/ft^2	Derived from limits on (W/A) and C_{D_0}

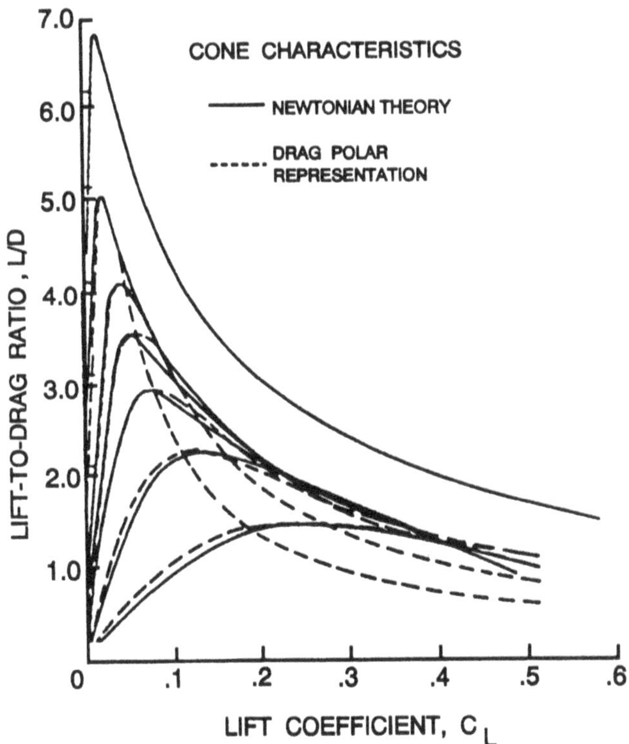

Fig. 9.3 Lift-to-drag ratio vs lift coefficient for cones.

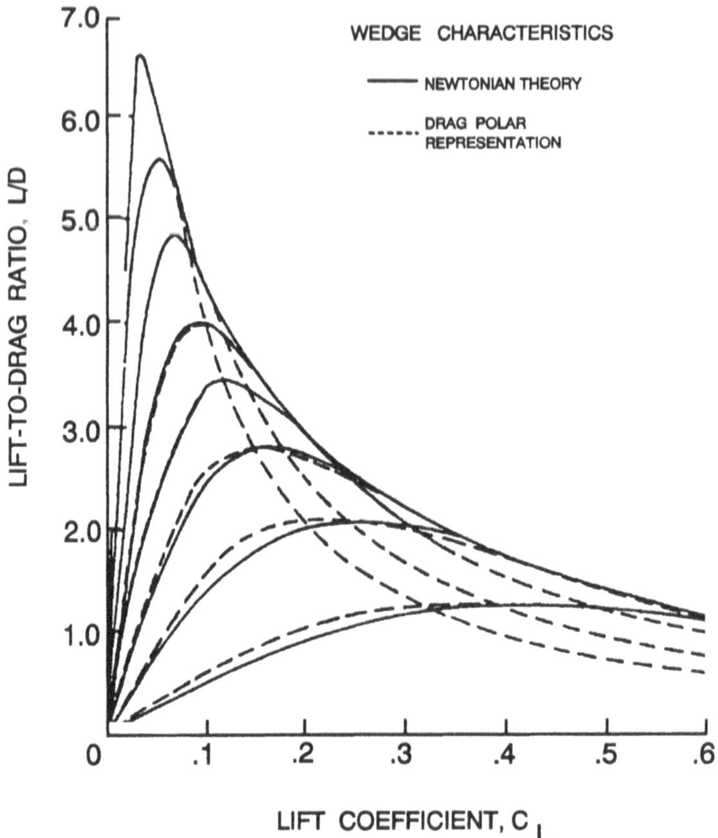

Fig. 9.4 Lift-to-drag ratio vs lift coefficient for wedges.

9.3 MaRV State Equations

The MaRV state equations derived here will be subject to the restriction of a nonrotating flat Earth. These equations will include drag and whatever lift is needed to meet the diveline trajectory shaping requirements. The details of diveline guidance are covered in the next section.

Figure 9.6 shows the positional and velocity states associated with the MaRV, the specific drag and lift forces D/M and L/M, respectively, and the gravity vector $gI3$. In the nonrotating frame, $[X1, X2, X3]^{\mathrm{T}}$, we may write

$$\frac{\mathrm{d}\boldsymbol{R}_m}{\mathrm{d}t} = \boldsymbol{V}_m \tag{9.14a}$$

$$\frac{\mathrm{d}\boldsymbol{V}_m}{\mathrm{d}t} = -\left(\frac{D}{M}\right)\boldsymbol{UV} + \left(\frac{L}{M}\right)\boldsymbol{UL} - g\boldsymbol{I}3 \tag{9.14b}$$

where $\boldsymbol{UV}$ and $\boldsymbol{UL}$ are unit vectors along the velocity lift vectors, respectively.

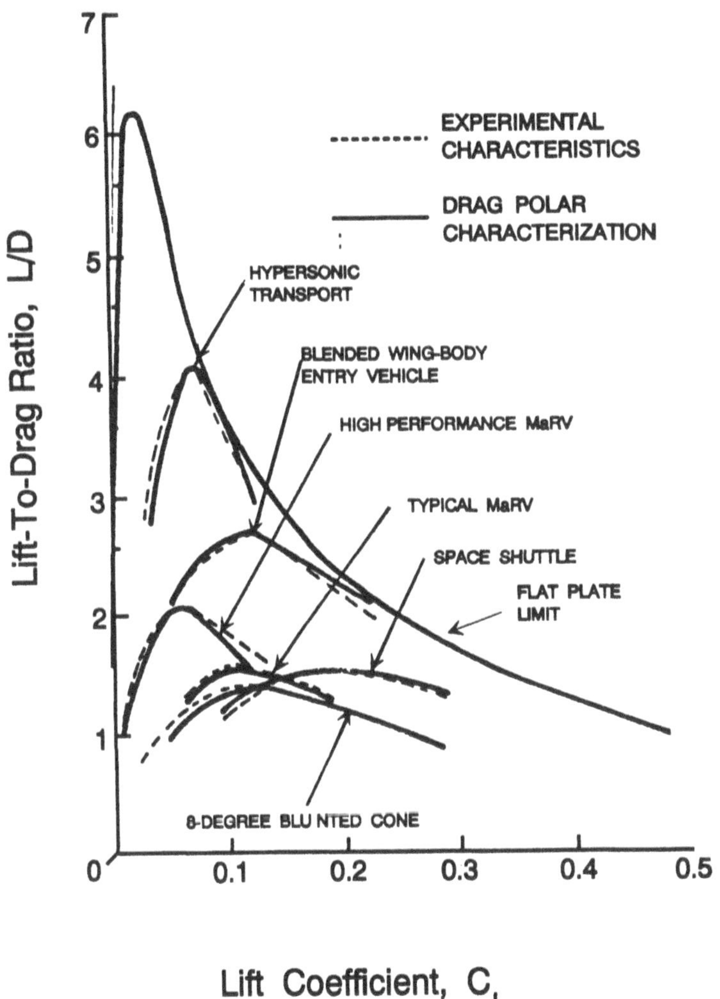

Fig. 9.5 Lift-to-drag ratio vs lift coefficient for various aerospace vehicles.

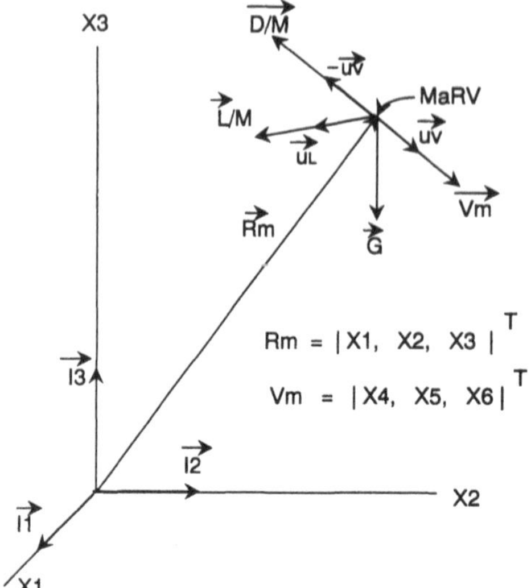

Fig. 9.6 Parameters used to determine MaRV state equations.

The positional and velocity state vectors are given in terms of their components in the inertial frame as

$$R_m^I = [X1, X2, X3]^T \tag{9.15a}$$

$$V_m^I = [X4, X5, X6]^T \tag{9.15b}$$

The magnitude of the lift and drag forces are easily described by the following equations:

$$\begin{aligned}\frac{D}{M} &= \left(\frac{\rho C_D A}{2M}\right) V_m^T V_m \\ &= \left(\frac{\rho}{2\beta_m}\right)\left(X4^2 + X5^2 + X6^2\right)\end{aligned} \tag{9.16a}$$

$$\begin{aligned}\frac{L}{M} &= \left(\frac{\rho C_L A}{2M}\right) V_m^T V_m \\ &= \left(\frac{\rho}{2\beta_m}\right)\frac{C_L}{C_D}\left(X4^2 + X5^2 + X6^2\right)\end{aligned} \tag{9.16b}$$

where β_m is based upon the total (not just zero-lift) drag coefficient. The gravitational acceleration is set at some sea-level value.

Drag is easily defined because it is always along the negative of the velocity vector. Lift, on the other hand, requires only that the lift force be normal to the velocity vector. This requirement does not uniquely set the direction of the lift vector. Let's describe the *lift space* as that plane whose normal is coincident with the velocity vector. We must then span the lift space with two orthogonal

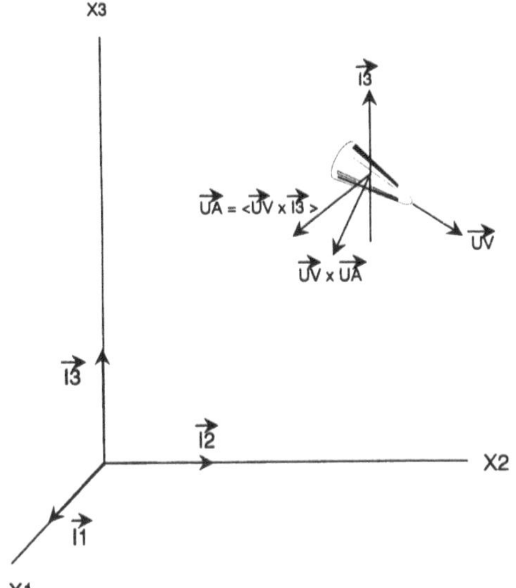

Fig. 9.7 MaRV aerodynamic frame.

unit vectors. Figure 9.7 presents a lift axis system. First, we define a unit vector along the velocity vector as

$$\begin{aligned} \boldsymbol{UV} &= \frac{\boldsymbol{V}_m}{(\boldsymbol{V}_m^{\mathrm{T}}\boldsymbol{V}_m)^{1/2}} \\ \boldsymbol{UV}^I &= \left[\frac{X4}{V_m}, \frac{X5}{V_m}, \frac{X6}{V_m}\right]^{\mathrm{T}} \end{aligned} \tag{9.17}$$

where

$$V_m = \left[X4^2 + X5^2 + X6^2\right]^{1/2}$$

The second axis, designated $\boldsymbol{UA}$, is normal to both the velocity vector $\boldsymbol{UV}$ and the vertical unit vector $\boldsymbol{I3}$.

$$\begin{aligned} \boldsymbol{UA} &= \langle \boldsymbol{UV} \times \boldsymbol{I3} \rangle \\ \boldsymbol{UA}^I &= \left[\frac{X5}{VA}, \frac{-X4}{VA}, 0\right]^{\mathrm{T}} \end{aligned} \tag{9.18}$$

where

$$VA = \left(X4^2 + X5^2\right)^{1/2} = \left[\boldsymbol{V}_m^{\mathrm{T}}\boldsymbol{V}_m - X6^2\right]^{1/2}$$

Consequently, the unit vector $\boldsymbol{UA}$ is always horizontal because it is always normal to $\boldsymbol{I3}$, and, of course, $\boldsymbol{UA}$ is also normal to the velocity vector.

A third unit vector, $\boldsymbol{UB}$, is defined normal to both $\boldsymbol{UA}$ and $\boldsymbol{UV}$ as follows:

$$\begin{aligned} \boldsymbol{UB} &= \langle \boldsymbol{UV} \times \boldsymbol{UA} \rangle = \boldsymbol{UV} \times \boldsymbol{UA} \\ \boldsymbol{UB}^I &= \left[\frac{X4X6}{(VA)V_m}, \frac{X5X6}{(VA)V_m}, -\frac{VA}{V_m} \right]^{\mathrm{T}} \end{aligned} \tag{9.19}$$

Note that although $\boldsymbol{UB}$ is gencrally not vertical, $\boldsymbol{UB}$ always lies in the plane defined by $\boldsymbol{I3}$ and $\boldsymbol{UV}$. Since lift must always be normal to the velocity vector, the lift vector must always lie in the plane defined by unit vectors $\boldsymbol{UA}$ and $\boldsymbol{UB}$. The frame defined by [$\boldsymbol{UV}$, $\boldsymbol{UA}$, $\boldsymbol{UB}$] is called the *aerodynamic* frame in that the specific drag force, D/M, is always along the negative of $\boldsymbol{UV}$ and the specific lift force, L/M, is spanned by the vectors $\boldsymbol{UA}$ and $\boldsymbol{UB}$. The directional cosine matrix from the aerodynamic frame (designated by A) to the inertial frame (designated by I) is

$$C_A^I = \begin{bmatrix} \uparrow & \uparrow & \uparrow \\ \boldsymbol{UV}^I & \boldsymbol{UA}^I & \boldsymbol{UB}^I \\ \downarrow & \downarrow & \downarrow \end{bmatrix} = \begin{bmatrix} \frac{X4}{V_m} & \frac{X5}{VA} & \frac{X4X6}{(VA)V_m} \\ \frac{X5}{V_m} & -\frac{X4}{VA} & \frac{X5X6}{(VA)V_m} \\ \frac{X6}{V_m} & 0 & -\frac{VA}{V_m} \end{bmatrix} \tag{9.20}$$

The components of the aerodynamic force vectors in the A-frame or aerodynamic frame are

$$\boldsymbol{f}^A = \left[-\frac{D}{M}, \frac{L}{M}\cos(\phi), \frac{L}{M}\sin(\phi) \right]^{\mathrm{T}} \tag{9.21}$$

where ϕ is the angle that the specific lift L/M makes with the $\boldsymbol{UA}$-axis. Thus, we might rewrite the state equations [Eqs. (9.14)] in matrix form as

$$\frac{\mathrm{d}\boldsymbol{R}_m^I}{\mathrm{d}t} = \boldsymbol{V}_m^I \tag{9.22a}$$

$$\frac{\mathrm{d}\boldsymbol{V}_m^I}{\mathrm{d}t} = C_A^I \boldsymbol{f}^A - \boldsymbol{G}^I \tag{9.22b}$$

where

$$\boldsymbol{G}^I = [0, 0, -g]^{\mathrm{T}}$$

The vector form of Eqs. (9.22) is

$$\frac{\mathrm{d}\boldsymbol{R}_m}{\mathrm{d}t} = \boldsymbol{V}_m \tag{9.23a}$$

$$\frac{dV_m}{dt} = -\left(\frac{D}{M}\right)UV + \left(\frac{L \cdot UA}{M}\right)UA + \left(\frac{L \cdot UB}{M}\right)UB - gI_3 \qquad (9.23b)$$

where the lift vector $L^A = [0, L\cos(\phi), L\sin(\phi)]^T$. In the state equations [either Eq. (9.22) or Eq. (9.23)] we note that lift can be controlled by setting the level of lift L (or C_L) and controlling the direction of the lift through the angle ϕ.

The trajectory of the MaRV will respond to changes in the settings of C_L and ϕ. However, what is lacking is a guidance algorithm which will relate ϕ and C_L to meet trajectory requirements. Fairly simple maneuvers are readily carried out using the state formulation given by Eqs. (9.23). For example, to accomplish a pull-down, the angle ϕ of Eq. (9.21) is set equal to 90 deg, placing the lift vector along **UB**, which in turn lies in a vertical plane. Lift of some magnitude (say, that corresponding to the critical lift coefficient) is applied, and the maneuver is carried out. Horizontal turns follow similarly by setting ϕ equal to 0 or 180 deg.

The unit lift vector, with components in the A-frame, is

$$UL^A = [0, \cos(\phi), \sin(\phi)]^T$$

The elements are sequentially the components of UL^A along the **UV**, **UA**, and **UB** unit vectors. The components of the same vector in the inertial frame are as follows:

$$UL^I = [UL1, UL2, UL3]^T = C_A^I[0, \cos(\phi), \sin(\phi)]^T \qquad (9.24)$$

We will not make further use of the A-frame; rather, we will develop an algorithm which will directly calculate the unit lift vector **UL** in the inertial, or I-frame. Before doing this we can rewrite the MaRV state equations once again in a form which is an expansion of Eqs. (9.14) as

$$\frac{dX1}{dt} = X4, \qquad \frac{dX2}{dt} = X5, \qquad \frac{dX3}{dt} = X6$$

$$\frac{dX4}{dt} = -\left(\frac{\rho}{2\beta_m}\right)\left[1 + \left(\frac{C_L}{C_L^*}\right)^2\right]V_m X4 + \left(\frac{\rho}{2\beta_m}\right)\left(\frac{C_L}{C_D}\right)\left[1 + \left(\frac{C_L}{C_L^*}\right)^2\right]V_m^2 UL1$$

$$\frac{dX5}{dt} = -\left(\frac{\rho}{2\beta_m}\right)\left[1 + \left(\frac{C_L}{C_L^*}\right)^2\right]V_m X5 + \left(\frac{\rho}{2\beta_m}\right)\left(\frac{C_L}{C_D}\right)\left[1 + \left(\frac{C_L}{C_L^*}\right)^2\right]V_m^2 UL2$$

$$\frac{dX6}{dt} = -\left(\frac{\rho}{2\beta_m}\right)\left[1 + \left(\frac{C_L}{C_L^*}\right)^2\right]V_m X6 + \left(\frac{\rho}{2\beta_m}\right)\left(\frac{C_L}{C_D}\right)\left[1 + \left(\frac{C_L}{C_L^*}\right)^2\right]V_m^2 UL3 - g$$

$$(9.25)$$

where β_m is based upon only the zero-lift drag.

In the next section we consider a way of specifying the elements of the unit lift vector **UL**.

9.4 Diveline Guidance

The purpose of diveline guidance (DG) is to provide both a trajectory-shaping and a trajectory-targeting algorithm. Prior to using the diveline guidance algorithm, one or more divelines are established. Such lines intersect the Earth in the vicinity of the target, with the final diveline intersecting the target. The DG algorithm then sets a direction to the lift vector such that the velocity vector of the MaRV is placed along the diveline. A diveline is specified by four numbers: two numbers are the coordinates of the origin of the diveline on the surface of the Earth; the remaining two numbers are directional numbers that define the slope, or attitude, of the line (i.e., the azimuth of the vertical plane containing the diveline and the elevation angle of the diveline in that vertical plane). Most of the development in this section is based upon the work of Gracey[2] and Cameron.[3]

The geometry of a diveline is illustrated in Fig. 9.8. Here we note that the diveline is described by direction only. The diveline unit vector is given by

$$\boldsymbol{UD}^I = [UD1, UD2, UD3]^{\mathrm{T}} \tag{9.26}$$

where

$$UD1 = \cos(D1)\sin(D2)$$

$$UD2 = \cos(D1)\cos(D2)$$

$$UD3 = \sin(D1)$$

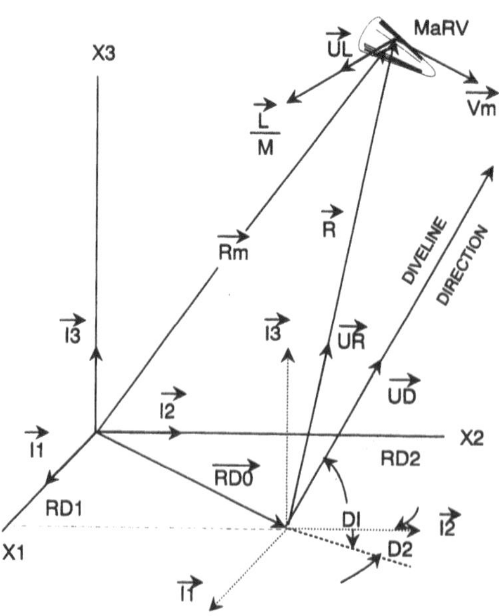

Fig. 9.8 Diveline geometry.

Clearly, only two quantities, i.e., the angles $D1$ and $D2$, are needed to define the direction of the diveline, and two more are needed to locate the origin of the diveline on the Earth's surface. The origin of only the final diveline is at the target. The diveline's point of origin is located by the vector

$$\boldsymbol{RD0}^I = [RD1, RD2, 0]^{\mathrm{T}} \tag{9.27}$$

The position of the MaRV is located from the origin of the coordinate system by the vector $\boldsymbol{R}_m$ and from the origin of the diveline by the vector $\boldsymbol{R}$.

The vector $\boldsymbol{R}$ is obtained from the following vectors

$$\boldsymbol{R} = \boldsymbol{R}_m - \boldsymbol{RD0} \tag{9.28}$$

The DG algorithm appropriately sets the direction of the lift vector to direct the MaRV velocity along the diveline. However the maneuver need not be completed before the MaRV is directed toward a new diveline. Only in the case of the final diveline must the maneuver be completed, i.e., the MaRV velocity must be brought into coincidence with the diveline before impact, to meet targeting requirements.

The vector $\boldsymbol{R}$ given in Eq. (9.28) may be unitized to give

$$\boldsymbol{UR}^I \doteq \langle \boldsymbol{R}^I \rangle = [UR1, UR2, UR3]^{\mathrm{T}} \tag{9.29}$$

To implement diveline guidance we must first construct a vector $\boldsymbol{W}$, defined as

$$\boldsymbol{W} = \boldsymbol{UR}(\boldsymbol{UR} \cdot \boldsymbol{UD}) - (G)\boldsymbol{UD} \tag{9.30}$$

where G is the diveline guidance gain. (Note that in this chapter we are also using $\boldsymbol{G}$ to represent the gravitational acceleration vector; since the DG gain G never appears directly in the MaRV state equations, there should be no cause for confusion.)

To give geometric significance to the above equation, let's define the vector $\boldsymbol{Q}$ as

$$\begin{aligned} \boldsymbol{Q} &= (\boldsymbol{UD} \times \boldsymbol{UR}) \times \boldsymbol{UR} \\ &= \boldsymbol{UR} \times (\boldsymbol{UR} \times \boldsymbol{UD}) \end{aligned} \tag{9.31a}$$

The vector $\boldsymbol{Q}$ lies in a plane defined by the vectors $\boldsymbol{UD}$ and $\boldsymbol{UR}$. Using a familiar cross-product theorem, Eq. (9.31a) may be written as

$$\boldsymbol{Q} = \boldsymbol{UR}(\boldsymbol{UD} \cdot \boldsymbol{UR}) - \boldsymbol{UD} \tag{9.31b}$$

If we compare Eqs. (9.30) and (9.31b), we note that $\boldsymbol{Q} = \boldsymbol{W}$ if the gain term G equals 1.0. On the other hand, if $G = 0.0$, $\boldsymbol{W}$ is coincident with $\boldsymbol{UR}$ and of magnitude $\cos(\theta)$. Thus, by varying the gain from 0.0 to 1.0, the direction of vector $\boldsymbol{W}$ varies from being along $\boldsymbol{UR}$ to being normal to $\boldsymbol{UR}$.

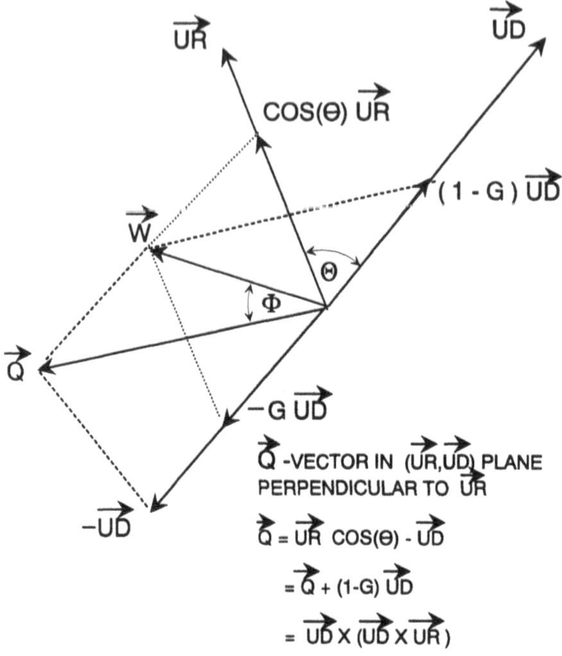

Fig. 9.9 Definition of vector *W*.

Figure 9.9 illustrates the relevant geometry of vectors $\boldsymbol{Q}$, $\boldsymbol{UR}$, $\boldsymbol{UD}$, and $\boldsymbol{W}$. We may summarize such relationships as follows:

$$\begin{aligned}\boldsymbol{Q} &= (\boldsymbol{UD} \times \boldsymbol{UR}) \times \boldsymbol{UR} \\ &= \boldsymbol{UR} \times (\boldsymbol{UR} \times \boldsymbol{UD}) \\ &= \boldsymbol{UR}(\boldsymbol{UD} \cdot \boldsymbol{UR}) - \boldsymbol{UD} \\ &= \boldsymbol{UR}\cos\theta - \boldsymbol{UD}\end{aligned}$$

$$|\boldsymbol{Q}| = \sin(\theta) \tag{9.32}$$

The vector $\boldsymbol{W}$ may be written as

$$\begin{aligned}\boldsymbol{W} &= \boldsymbol{UR}(\boldsymbol{UR} \cdot \boldsymbol{UD}) - (G)\boldsymbol{UD} \\ &= \boldsymbol{UR}\cos(\theta) - (G)\boldsymbol{UD} \\ &= \boldsymbol{Q} + (1 - G)\boldsymbol{UD}\end{aligned}$$

$$|\boldsymbol{W}| = [G^2 + (1 - 2G)\cos^2(\theta)]^{1/2} \tag{9.33}$$

The reader can now check that when $G = 0.0$ and 1.0, the respective magnitudes of $\boldsymbol{W}$ are $\cos(\theta)$ and $\sin(\theta)$.

We might ask what the gain G is when the vector $\boldsymbol{W}$ is normal to $\boldsymbol{UD}$, the diveline pointing vector. For $\boldsymbol{W}$ to be normal to $\boldsymbol{UD}$ we must have

$$\boldsymbol{W} \cdot \boldsymbol{UD} = 0.0 = \boldsymbol{UD} \cdot \boldsymbol{Q} + (1 - G)\boldsymbol{UD} \cdot \boldsymbol{UD}$$

Upon solving for G we get

$$G = \cos^2(\theta)$$

Inserting this value into Eq. (9.33) gives

$$|\boldsymbol{W}| = \sin(\theta)\cos(\theta) = \sin(2\theta)/2$$

Table 9.2 summarizes the values of the gain G and the corresponding magnitudes and directions of the vector $\boldsymbol{W}$. We leave it as an exercise for the reader to show that

$$\cos(\phi) = \frac{G\sin(\theta)}{[(1 - 2G)\cos^2(\theta) + G^2]^{1/2}}$$

which provides a relationship between ϕ, the angle between $\boldsymbol{W}$ and $\boldsymbol{Q}$, and θ, the angle that $\boldsymbol{UR}$ makes with respect to $\boldsymbol{UD}$.

The next goal is to set the direction of the lift vector, i.e, to assign numerical values to the components of the unit vector $\boldsymbol{UL}$ in the inertial frame. Lift must at all times be perpendicular to the velocity vector $\boldsymbol{UV}$. We need another vector to establish a unique direction to the lift vector: the vector $\boldsymbol{W}$. The vector $\boldsymbol{W}$ lies in a plane defined by $\boldsymbol{UR}$ and $\boldsymbol{UD}$, the positional and diveline unit vectors, specifying the MaRV position and diveline direction from the diveline origin. We define the unit lift vector as

$$\boldsymbol{UL} = < (\boldsymbol{W} \times \boldsymbol{UV}) \times \boldsymbol{UV} > \tag{9.34}$$

To gain some insight into the above triple product, we may use Eq. (9.30) to rewrite the above expression as

$$\begin{aligned} \boldsymbol{UL} &= < \boldsymbol{UV} \times (\boldsymbol{UV} \times \boldsymbol{W}) > \\ &= < \cos(\theta)[(\boldsymbol{UR} \times \boldsymbol{UV}) \times \boldsymbol{UV}] - G(\boldsymbol{UD} \times \boldsymbol{UV}) \times \boldsymbol{UV} > \\ &= < \cos(\theta)[\boldsymbol{UV} \times (\boldsymbol{UV} \times \boldsymbol{UR})] - G[\boldsymbol{UV} \times (\boldsymbol{UV} \times \boldsymbol{UD})] > \end{aligned} \tag{9.35}$$

Table 9.2 Relationship between the gain G and the magnitude of the vector W

G (diveline gain)	$\lvert \boldsymbol{W} \rvert$	Direction
0.0	$\cos(\theta)$	Along $\boldsymbol{UR}$
$\cos^2(\theta)$	$\sin(2\theta)/2$	Normal to $\boldsymbol{UD}$
1.0	$\sin(\theta)$	Normal to $\boldsymbol{UR}$

We see that the lift vector is composed of the sum of two vectors, both of which are normal to the velocity vector as required. The first vector lies in a plane defined by the vectors **UV** and **UR**, and the second lies in a plane defined by vectors **UV** and **UD**. The relative contribution of each of the vectors is controlled by the multipliers $\cos(\theta)$ and G; $\cos(\theta)$ measures the angular separation of **UR** and **UD**, and G is the gain associated with the diveline guidance algorithm.

First, let's write the unit vectors that will be needed later:

$$\begin{aligned} \boldsymbol{UV}^I &= \left[\frac{X4}{V_m}, \frac{X5}{V_m}, \frac{X6}{V_m}\right]^{\mathrm{T}} \\ \boldsymbol{UR}^I &= \left[\frac{X1}{R}, \frac{X2}{R}, \frac{X3}{R}\right]^{\mathrm{T}} \\ \boldsymbol{UD}^I &= [\cos(D1)\sin(D2), \cos(D1)\cos(D2), \sin(D2)]^{\mathrm{T}} \end{aligned} \tag{9.36}$$

where

$$V_m = \left(X4^2 + X5^2 + X6^2\right)^{1/2}$$

$$R = \left(X1^2 + X2^2 + X3^2\right)^{1/2}$$

Let's rewrite Eq. (9.34) in the following form:

$$\begin{aligned} \boldsymbol{UL} &= < \boldsymbol{UV} \times (\boldsymbol{UV} \times \boldsymbol{W}) > \\ &= < \boldsymbol{UV}(\boldsymbol{UV} \cdot \boldsymbol{W}) - \boldsymbol{W} > \end{aligned} \tag{9.37a}$$

The magnitude M of the vector enclosed in the brackets is:

$$\begin{aligned} M &= \left[|\boldsymbol{W}|^2 - (\boldsymbol{UV} \cdot \boldsymbol{W})^2\right]^{1/2} \\ &= \left[(G^2 + (1 - 2G)\cos^2(\theta)) \right. \\ &\quad \left. - (UV1W1 + UV2W2 + UV3W3)^2\right]^{1/2} \end{aligned} \tag{9.37b}$$

The diveline gain may be set at some value between 0.0 and 1.0 as indicated in Table 9.2. The gain can be varied throughout the maneuver as the angle between **UR** and **UD** varies or simply set at a constant value, say, 0.5. In either case, the above vector **UL** may be calculated from Eqs. (9.36) (using the diveline direction unit vector **UD** and the state unit vectors **UR** and **UV**). **UD** is a constant until a new diveline is selected, and **UR** and **UV** are available at the end of each integration step. Thus, the unit lift vector **UL** may be written as

$$\begin{aligned} \boldsymbol{UL}^I &= [UL1, UL2, UL3]^{\mathrm{T}} \\ \boldsymbol{UL} &= \tfrac{1}{M}[\boldsymbol{UV}(\boldsymbol{UV} \cdot \boldsymbol{W}) - \boldsymbol{W}] \end{aligned} \tag{9.38}$$

All operations in the above equations are easily carried out from the development of the three unit vectors given in Eqs. (9.36). Therefore, we have means of calculating the "right-hand side" of the integrand of the state equations [Eq. (9.25)]. Our next consideration is to determine the interception point.

Again, we must emphasize that the development in Section 9.4 is based upon the work of Gracey[2] and Cameron.[3] A computer program of the trajectory of a MaRV using diveline guidance is given in Appendix D.

9.5 Determining the Projected Interception Point

We have provided the MaRV state equations, which include an algorithm for directing the MaRV lift vector for trajectory shaping and targeting. At this point we assume that the interceptor state equations have also been developed. Consequently, we are describing a 12-state system which may be separated into four sets—MaRV positional and velocity states and interceptor positional and velocity states—as follows:

$$\begin{aligned} \boldsymbol{R}_m &= [X1, X2, X3]^{\mathrm{T}} \qquad & \boldsymbol{V}_m &= [X4, X5, X6]^{\mathrm{T}} \\ \boldsymbol{R}_I &= [X7, X8, X9]^{\mathrm{T}} \qquad & \boldsymbol{V}_I &= [X10, X11, X12]^{\mathrm{T}} \end{aligned} \tag{9.39}$$

The reason that we are deferring the development of the interceptor state equations is that guidance concerns are an important part of these equations. It must be remembered that the interceptor is aware of at least the positional MaRV states and uses this information to continually construct a projected interception point (PIP).

In considering the interception problem we must remember that although the interceptor has a maneuvering advantage over the MaRV, the MaRV has something like a two-to-one advantage in speed. Consequently, interceptor algorithms based upon pursuit navigation, i.e., placing the interceptor velocity vector along the interceptor-MaRV line of sight, will not generally be successful. Deviated pursuit, in which the interceptor's velocity vector direction is biased ahead of the line of sight, might enjoy more success. The algorithm that is to follow is essentially a form of deviated pursuit.

First, we ballistically extend the velocity vector of the MaRV. In other words, we assume that the MaRV will progress along an extension of its instantaneous velocity vector. The position of the MaRV along this velocity extension depends upon how rapidly the velocity magnitude diminishes with distance. It was shown in Eq. (7.17b) that velocity magnitude decreases exponentially with drag coefficient. To be consistent, only the zero-lift drag coefficient should be used. However, the defense knows that the MaRV is likely to maneuver; although the direction of lift is not known, it may be assumed that the drag coefficient will increase due to lift-induced drag. Therefore, the defense will likely increase the drag coefficient to some degree over the zero-lift value. The projected interception point, which must lie along this line defined by the velocity projection, is set by a weighting process using an estimate of the interceptor's velocity magnitude.

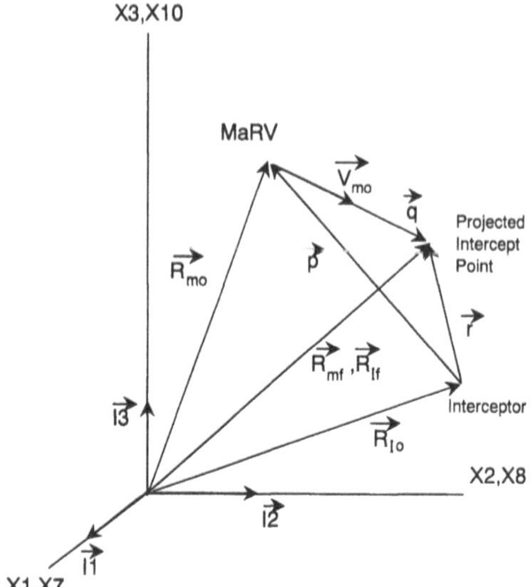

Fig. 9.10 Interceptor geometry.

Figure 9.10 illustrates the application of an interception algorithm. The vector $\boldsymbol{p}$ is the line-of-sight vector from the interceptor to the MaRV at the beginning of the computations to determine the projected interception point. The vector $\boldsymbol{q}$ is the directed distance from the present MaRV position to the PIP. Thus, we have

$$\boldsymbol{p} = \boldsymbol{R}_{m_0} - \boldsymbol{R}_{I_0} = [(X1 - X7), (X2 - X8), (X3 - X9)]^{\mathrm{T}}$$

$$\boldsymbol{q} = \boldsymbol{R}_{m_f} - \boldsymbol{R}_{m_0} = \boldsymbol{R}_{I_f} - \boldsymbol{R}_{m_0} \tag{9.40}$$

$$\boldsymbol{r} = \boldsymbol{R}_{m_f} - \boldsymbol{R}_{I_0} = \boldsymbol{R}_{I_f} - \boldsymbol{R}_{I_0}$$

where the subscripts 0 and f refer to the initial and final states, respectively. From the above relationship we have

$$\boldsymbol{r} = \boldsymbol{p} + \boldsymbol{q} \tag{9.41}$$

The magnitude of the vector $\boldsymbol{r}$ is the directed distance from the present position of the interceptor to the PIP. The vector $\boldsymbol{r}$ is the interceptor's distance-to-go vector. The square of the magnitude of $\boldsymbol{r}$ is

$$|\boldsymbol{r}|^2 = \boldsymbol{r} \cdot \boldsymbol{r} = (\boldsymbol{p} + \boldsymbol{q}) \cdot (\boldsymbol{p} + \boldsymbol{q}) = \boldsymbol{p} \cdot \boldsymbol{p} + \boldsymbol{q} \cdot \boldsymbol{q} + 2\boldsymbol{p} \cdot \boldsymbol{q}$$

Let V_{I_a} be the estimate of the average interceptor speed over the distance and T be the estimated time of passage; we then have

$$V_{I_a}^2 T^2 = \boldsymbol{r} \cdot \boldsymbol{r} = \boldsymbol{p} \cdot \boldsymbol{p} + \boldsymbol{q} \cdot \boldsymbol{q} + 2\boldsymbol{p} \cdot \boldsymbol{q} \tag{9.42}$$

Now we return to Eq. (9.40), where we make use of the idea of ballistic projection as follows:

$$q = \begin{bmatrix} X4(T) + A1\frac{T^2}{2} \\ X5(T) + A2\frac{T^2}{2} \\ X6(T) + A3\frac{T^2}{2} \end{bmatrix} \tag{9.43}$$

where ($A1$, $A2$, $A3$) are the estimates of the MaRV acceleration, which for our purposes will be taken as the right-hand side of the last three MaRV state equations [Eqs. (9.25)], i.e.,

$$A1 = \frac{dX4}{dt}, \qquad A2 = \frac{dX5}{dt}, \qquad A3 = \frac{dX6}{dt}$$

The accelerations for the zero-lift case are given as

$$A1 = -\left(\frac{\rho}{2\beta_m}\right)V_m X4, \qquad A2 = -\left(\frac{\rho}{2\beta_m}\right)V_m X5, \qquad A3 = -\left(\frac{\rho}{2\beta_m}\right)V_m X6$$

where β_m is the ballistic coefficient based upon zero-lift drag.

There is some obvious inconsistency here in that the right-hand side of the velocity state equations can contain some lift as well as drag contributions; when lift is present, drag is increased over its ballistic value because of induced drag. If such concerns present some difficulty then we could write

$$A1 = -\left(\frac{\rho}{\beta_m}\right)V_m X4, \qquad A2 = -\left(\frac{\rho}{\beta_m}\right)V_m X5, \qquad A3 = -\left(\frac{\rho}{\beta_m}\right)V_m X6$$

where we have (arbitrarily) assumed a MaRV operating at critical lift coefficient that results in twice the zero-lift drag.

If the vector $\boldsymbol{q}$ is defined according to Eq. (9.43) and the vector $\boldsymbol{p}$ according to Eq. (9.40), we will obtain from Eq. (9.42) a quartic in T, the time-to-go, as follows:

$$\begin{aligned} AT^4 + BT^3 + (C1 + C2)T^2 + DT + E &= 0 \\ AT^4 + BT^3 + CT^2 + DT + E &= 0 \end{aligned} \tag{9.44a}$$

The coefficients of the above equation are given in Table 9.3. Shorthand notation for the above coefficients is

$$\begin{aligned} &A = \boldsymbol{A}_m^{\mathrm{T}}\boldsymbol{A}_m, \qquad B = \boldsymbol{V}_m^{\mathrm{T}}\boldsymbol{A}_m, \qquad C1 = \boldsymbol{V}_m^{\mathrm{T}}\boldsymbol{V}_m - V_{I_a} \\ &C2 = \boldsymbol{A}_m^{\mathrm{T}}\boldsymbol{p}, \qquad C = C1 + C2, \qquad D = 2\boldsymbol{V}_m^{\mathrm{T}}\boldsymbol{p}, \qquad E = \boldsymbol{p}^{\mathrm{T}}\boldsymbol{p} \end{aligned} \tag{9.44b}$$

where $\boldsymbol{A}_m$ is the MaRV acceleration.

Table 9.3 Coefficients for Eq. (9.44a)

Coefficient	Description
$A = (A1^2 + A2^2 + A3^2)/4$	MaRV acceleration
$B = (X4)(A1) + (X5)(A2) + (X6)(A3)$	MaRV velocity/acceleration
$C1 = X4^2 + X5^2 + X6^2 - V_{I_a}^2$	MaRV-interceptor velocity coupling
$C2 = A1(X1 - X7) + A2(X2 - X8) + A3(X3 - X9)$	MaRV acceleration/MaRV-interceptor separation
$D = 2[X4(X1 - X7) + X5(X2 - X8) + X6(X3 - X9)]$	MaRV velocity/MaRV-interceptor separation
$E = (X1 - X4)^2 + (X2 - X8)^2 + (X3 - X9)^2$	MaRV-interceptor separation

The goal now is to obtain the roots of Eq. (9.44a). Only roots that are real positive numbers are of any interest. All six coefficients must be calculated from information available in the MaRV and interceptor state equations. MaRV position and velocity states are available from the integration of the state equations; the MaRV acceleration is available directly from the state equations. We also require the interceptor positional states; the quantity V_{I_a} is of course an estimate of the interceptor's speed over the range-to-go vector.

A fairly straightforward method of solving the quartic of Eq. (9.44) is to use a variation of the Newton-Raphson method of obtaining real roots of a polynomial. Here, the coefficients A and B are ignored at first. Then the first estimate of the time-to-go T is available from the real, positive solution of the remaining quadratic:

$$T1 = \left[-D \pm (D^2 - 4\,CE)^{1/2} \right] / 2C \tag{9.45}$$

By ignoring A and B we are essentially assuming that the MaRV acceleration is zero. To be consistent we should also ignore $C2$, because it contains MaRV acceleration. However, this will not be done here because the real motive for ignoring A and B is to "jump-start" a solution.

Equation (9.45) gives us a first estimate of the solution. We then set the second estimate as

$$T2 = T1 + \varepsilon \tag{9.46}$$

where ε is obviously a corrective term. Inserting $T2$ into Eq. (9.44) allows us to solve for ε as follows:

$$\varepsilon \approx -\left[\frac{A(T1)^4 + B(T1)^3 + C(T1)^2 + D(T1) + E}{4A(T1)^3 + 3B(T1)^2 + 2C(T1) + D} \right] \tag{9.47}$$

Once ε is obtained, the improved estimate of T is obtained from Eq. (9.46). An iterative estimation procedure may then be carried out to obtain subsequent improvements in the process.

A problem arises when the argument of the square root in Eq. (9.45) becomes negative, i.e.,

$$(D^2 - 4CE) < 0 \tag{9.48}$$

The physical meaning of the above inequality is that there exists no point in space at which the interceptor can reach the MaRV. As pointed out earlier, the MaRV has a significant velocity advantage over the interceptor; for an engagement geometry where the interceptor is attempting to overtake the MaRV (interceptor and MaRV velocity vectors nearly collinear), there can be no defined PIP. Note from Table 9.3 that if V_{I_a} is increased, $C1$ and hence C is decreased. Under such circumstances an unrealistically large value of V_{I_a} is used to effect a solution; then the interceptor must wait until the MaRV turns to target, and a fire control solution can be obtained using a more realistic value of interceptor speed. There are other computational anomalies associated with obtaining the real roots of the quartic. These can usually be handled with some embellishments to the Newton-Raphson method. Next, we consider how to use the value of the time-to-go T to obtain the projected intercept point.

Assuming that T, $\boldsymbol{V}_m$, and $\boldsymbol{A}_m$ are available, we can now calculate the interceptor distance-to-go vector, ***RIP***, as the sum of the $\boldsymbol{p}$ and $\boldsymbol{q}$ vectors.

$$\boldsymbol{r} = \boldsymbol{p} + \boldsymbol{q} = \begin{bmatrix} RIP1 \\ RIP2 \\ RIP3 \end{bmatrix} = \begin{bmatrix} (X1 - X7) + X4(T) + A1(T^2/2) \\ (X2 - X8) + X5(T) + A2(T^2/2) \\ (X3 - X9) + X5(T) + A3(T^2/2) \end{bmatrix} \tag{9.49}$$

The vector $\boldsymbol{r} = \boldsymbol{RIP}$ will be used in the next section to guide the interceptor to the PIP. To reiterate: the MaRV states $[X1, \ldots X6]^{\mathrm{T}}$ are available from a solution to the MaRV state equations; the MaRV acceleration vector $[A1, A2, A3]^{\mathrm{T}}$ is taken directly from the MaRV state equation; the interceptor positional vector $[X7, X8, X9]^{\mathrm{T}}$ is found from the interceptor state equations. V_{I_a} is just an estimate of the average velocity of the interceptor over the remaining distance-to-go vector $\boldsymbol{r} = \boldsymbol{RIP}$.

9.6 Interceptor Guidance Equations

The goal of an interceptor guidance algorithm is to align the interceptor velocity vector ***VI*** with the range-to-go vector $\boldsymbol{r} = \boldsymbol{RIP}$. Figure 9.11 shows misalignment between the range-to-go vector and the interceptor velocity vector. Both vectors are presented in unit form as

$$\begin{aligned} \boldsymbol{UR} &= \frac{\boldsymbol{RIP}}{|\boldsymbol{RIP}|} = \left[\frac{RIP1}{|\boldsymbol{RIP}|}, \frac{RIP2}{|\boldsymbol{RIP}|}, \frac{RIP3}{|\boldsymbol{RIP}|}\right]^{\mathrm{T}} \\ &= [UR1, UR2, UR3]^{\mathrm{T}} \end{aligned} \tag{9.50}$$

where

$$|\boldsymbol{RIP}| = \left(RIP1^2 + RIP2^2 + RIP3^2\right)^{1/2}$$

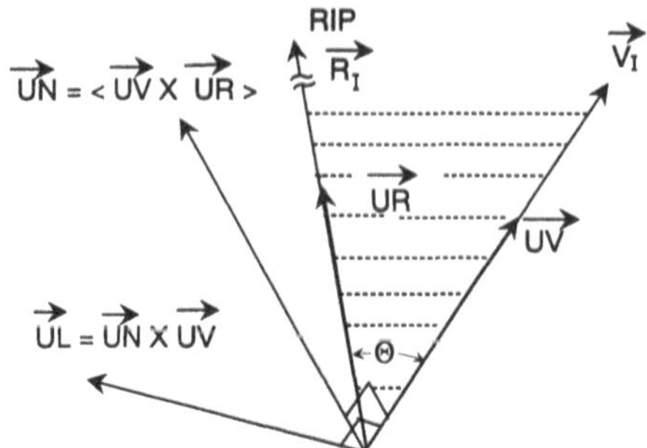

Fig. 9.11 Geometry of cross-product steering.

and

$$\boldsymbol{UV} = \left[\frac{X10}{VI}, \frac{X11}{VI}, \frac{X12}{VI}\right]^{\mathrm{T}}$$
$$= [UV1, UV2, UV3]^{\mathrm{T}} \tag{9.51}$$

where

$$VI = \left(X10^2 + X11^2 + X12^2\right)^{1/2}$$

We first construct a unit vector normal to the plane defined by ***UR*** and ***UV*** as

$$\boldsymbol{UN} = < \boldsymbol{UV} \times \boldsymbol{UR} >$$

and a unit vector in the plane defined by ***UR*** and ***UV*** and at the same time normal to ***UV*** as

$$\boldsymbol{UL} = \boldsymbol{UN} \times \boldsymbol{UV} = < (\boldsymbol{UV} \times \boldsymbol{UR}) \times \boldsymbol{UV} > \tag{9.52}$$

Let θ be the angle between ***UR*** and ***UV***. Thus, we can show that

$$\boldsymbol{UL} = (\boldsymbol{UV} \times \boldsymbol{UR}) \times \boldsymbol{UV} / \sin(\theta) \tag{9.53}$$

Thus, the interceptor lift must be directed along the vector ***UL***. Obviously, there is a singularity in Eq. (9.53) when θ equals 0. Under these circumstances the vectors ***RIP*** and ***VI*** would be collinear, and there would be no need to apply lift.

To avoid the singularity a threshold value is set for the angle θ; if θ is below the threshold, no lift is applied to the interceptor. The angle θ is easily found as

$$\cos(\theta) = \boldsymbol{UR} \cdot \boldsymbol{UV}$$

$$\theta = \cos^{-1}[(UR1)(UV1) + (UR2)\,(UV2) + (UR3)\,(UV3)] \tag{9.54}$$

Since the interceptor is capable of generating continuous lift, we can command a lift coefficient that is proportional to the angle θ. A second problem of sorts is introduced when considering lift demand. Referring to Fig. 9.11 we note that only enough lift should be applied to the interceptor to bring the interceptor velocity vector ***VI*** into collinearity with the vector ***RIP*** at the PIP. The generation of lift greater than the minimum will bring the velocity and range-to-go vectors into coincidence sooner, i.e., before the vector ***RIP*** goes to zero. However, generating more lift than needed will result in an induced drag penalty, because the induced drag varies with the square of the lift coefficient.

One approach that might be taken to minimize drag losses is to bring the interceptor along a circular path that will pass through the PIP. It is easy to show that the required lift coefficient is something like

$$C_L = \frac{4W\sin(\theta/4)}{g\rho A(RIP)} \tag{9.55}$$

where W is the weight of the interceptor, A is the reference area, ρ is the atmospheric density, and θ is the turning angle. If the interceptor generates a lift coefficient lower than that dictated by Eq. (9.55), there will likely be an overshoot of the designated PIP. The lift control can be "tightened" by replacing the "4" of Eq. (9.55) with K_L as shown where:

$$4 \leq K_L \leq 20$$

with the upper limit arbitrarily taken to be 20.

We can now return to Eq. (9.55); by using the small angle approximation we get

$$C_L = K_L W\theta / [4g\rho A(RIP)] \tag{9.56}$$

Complications arise in applying the simple relationship given above. First, we note that the required lift coefficient varies inversely with density and hence directly with altitude. The applied lift coefficient varies with weight, which might make the guidance task in the boost phase more difficult because the weight is decreasing rapidly. The moderate-to-high altitude interceptions occur in the coast phase, where weight is constant after burnout.

A modification of Eq. (9.56) might also be made in that the constant K_L can be varied, say, from 1.0 to 20. Let T be the time-to-go [available from the solution of the quartic of Eq. (9.44)].

$$1.0 < K_L = 4(T_s/T) < 20 \tag{9.57}$$

Thus, when the time-to-go is equal to the set time T_s, only sufficient lift will be generated to bring the interceptor along a circular arc into the PIP. If T is greater than this set time, the interceptor will move in the general direction of the PIP but will not expend its velocity in an attempt to pass through a PIP that may change drastically with an MaRV maneuver. On the other hand, as the time-to-go becomes small (less than T_s), then the interceptor generates increasing lift to bring it to the PIP. As time-to-go diminishes, the MaRV is unable to move the PIP significantly through maneuvering.

Up to this point we have developed a rather basic form of guidance called cross-product steering (to which we have added a few embellishments). Another type of guidance algorithm that is more generally accepted is *proportional guidance*. Like any guidance algorithm introduced into a particular environment, there are many variations of the basic idea of proportional guidance.

In its simplest form as applied to MaRV interception, proportional guidance requires that the linear velocity vector of the interceptor be rotated at an angular rate relative to inertial spacc that is proportional to the rotational rate of the line of sight to the MaRV. Figure 9.12 illustrates the essential geometry. If γ_I is the flight path angle of the interceptor, proportional navigation requires that

$$\frac{d\gamma_I}{dt} = K\omega_{m/I} \tag{9.58}$$

where $\omega_{m/I}$ is the magnitude of the angular rotation of the line-of-sight vector from the interceptor to the target, or MaRV, and K is the guidance constant.

The required interceptor acceleration is provided by lift and is given by

$$\begin{aligned} a_L &= V_I \frac{d\gamma_I}{dt} \\ &= K\omega_{m/I} V_I \end{aligned} \tag{9.59}$$

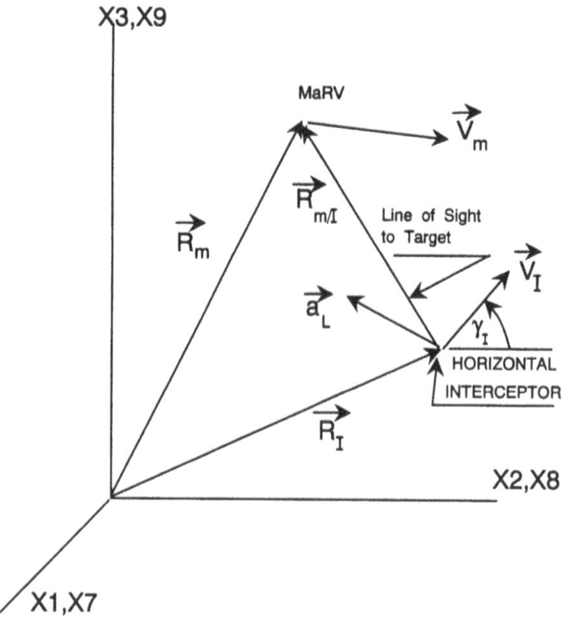

Fig. 9.12 Proportional guidance.

The magnitude of the angular velocity is given as follows:

$$|\boldsymbol{\omega}_{m/I}| = \frac{|\boldsymbol{V}_{m/I}|}{|\boldsymbol{R}_{m/I}|}\sin(\boldsymbol{V}_{m/I}, \boldsymbol{R}_{m/I}) \tag{9.60}$$

where $\boldsymbol{V}_{m/I} = \boldsymbol{V}_m - \boldsymbol{V}_I$ and the argument of the sine function is the angle between $\boldsymbol{V}_{m/I}$ and $\boldsymbol{R}_{m/I}$.

The vector $\boldsymbol{\omega}_{m/I}$ is normal to the plane defined by the MaRV relative velocity vector $\boldsymbol{V}_{m/I}$ and the line-of-sight vector $\boldsymbol{R}_{m/I}$. Consequently, the unit vector indicating the direction of the angular velocity vector is

$$\boldsymbol{e}_w = \frac{\boldsymbol{R}_{m/I} \times \boldsymbol{V}_{m/I}}{|\boldsymbol{R}_{m/I}||\boldsymbol{V}_{m/I}|\sin(\boldsymbol{V}_{m/I}, \boldsymbol{R}_{m/I})}$$

Thus,

$$\boldsymbol{\omega}_{m/I} = |\boldsymbol{\omega}_{m/I}|\boldsymbol{e}_w = \frac{|\boldsymbol{V}_{m/I}|\sin(\boldsymbol{V}_{m/I}, \boldsymbol{R}_{m/I})}{|\boldsymbol{R}_{m/I}|\sin(\boldsymbol{V}_{m/I}, \boldsymbol{R}_{m/I})}\frac{(\boldsymbol{R}_{m/I} \times \boldsymbol{V}_{m/I})}{|\boldsymbol{R}_{m/I}||\boldsymbol{V}_{m/I}|}$$

or

$$\boldsymbol{\omega}_{m/I} = \frac{\boldsymbol{R}_{m/I} \times \boldsymbol{V}_{m/I}}{\boldsymbol{R}_{m/I} \cdot \boldsymbol{R}_{m/I}} \tag{9.61a}$$

The vector form of Eq. (9.59) is

$$\boldsymbol{a}_L = K\boldsymbol{\omega}_{m/I} \times \boldsymbol{V}_I \tag{9.61b}$$

The time-to-go T is found as

$$T = -\frac{1}{|\boldsymbol{V}_{m/I}|}\left(\boldsymbol{R}_{m/I} \cdot \frac{\boldsymbol{V}_{m/I}}{|\boldsymbol{V}_{m/I}|}\right)$$

Obviously, T is influenced only by the component of $\boldsymbol{V}_{m/I}$ that is along the separation (or line-of-sight) vector $\boldsymbol{R}_{m/I}$; the negative sign accounts for the fact that the component of $\boldsymbol{V}_{m/I}$ along $\boldsymbol{R}_{m/I}$ must be in the negative direction, i.e., tending to decrease the magnitude of $\boldsymbol{R}_{m/I}$. Thus, we have for T the following expression:

$$T = -\frac{\boldsymbol{R}_{m/I} \cdot \boldsymbol{V}_{m/I}}{\boldsymbol{V}_{m/I} \cdot \boldsymbol{V}_{m/I}} \tag{9.62}$$

We might modify the guidance constant K, sometimes called the *navigation ratio,* by multiplying by the ratio of the relative velocity component and the interceptor velocity component along the line-of-sight vector as follows:

$$\boldsymbol{a}_L = K\left(\frac{\boldsymbol{V}_{m/I} \cdot \boldsymbol{R}_{m/I}}{\boldsymbol{V}_I \cdot \boldsymbol{R}_{m/I}}\right)\boldsymbol{\omega}_{m/I} \times \boldsymbol{V}_I \tag{9.63}$$

Extensive simulation has shown that proportional navigation gives penetration results comparable to the PIP/cross-product steering discussed previously. The obvious advantage of proportional navigation is that there is no need to go through the process of identifying the projected intercept point with the attending difficulties of factoring a quartic to obtain the time of flight. Proportional navigation sets the required acceleration magnitude as well as direction; a check then must be made to determine if the demanded side load exceeds the maximum allowable and if the required lift coefficient is within the capability of the interceptor. Cross-product steering, on the other hand, provides only a steering direction; the lift coefficient must be set to meet various terminal conditions [see, for example, Eq. (9.55)]. Here, too, checks must be carried out to ensure that maximum side loads are not exceeded.

In the next section we present a derivation of the interceptor state equations. This formulation is given in terms of cross-product steering.

9.7 Interceptor State Equations

The interceptor state equations may be written along the lines of the MaRV state equations. We have selected the interceptor positional state vector to be $\boldsymbol{R}_I = [X7, X8, X9]^{\mathrm{T}}$ and the velocity state vector to be $\boldsymbol{V}_I = [X10, X11, X12]^{\mathrm{T}}$. We have

$$\frac{\mathrm{d}\boldsymbol{R}_I}{\mathrm{d}t} = \boldsymbol{V}_I \tag{9.64a}$$

$$\frac{\mathrm{d}\boldsymbol{V}_I}{\mathrm{d}t} = \left(\frac{TH}{M} - \frac{D}{M}\right)\boldsymbol{UV} + \left(\frac{L}{M}\right)\boldsymbol{UL} - g\boldsymbol{I}_3 \tag{9.64b}$$

where the thrust and drag per unit mass, respectively, are given as follows:

$$\frac{TH}{M} = \frac{I_{\mathrm{sp}}}{M_p + M_b}\frac{\mathrm{d}M_p}{\mathrm{d}t} + \frac{(P_e - P_a)A_e}{M_p + M_b} \tag{9.65a}$$

$$\frac{D}{M} = \frac{\rho A V_I^2}{2(M_p + M_b)}\left(C_{D_0} + KC_L^2\right) \tag{9.65b}$$

Equation (9.65a) is a fairly standard form of the rocket equation, where I_{sp} is the specific impulse of the interceptor rocket propellant, $\mathrm{d}M_p/\mathrm{d}t$ is the mass flow of the rocket, A_e is the exhaust area of the rocket nozzle, and P_e and P_a are the rocket exhaust and atmospheric pressures, respectively. The terms M_p and M_b are the mass of the propellant and the burnout, or structural, mass of the rocket.

In the drag equation it is not practical to make use of the ballistic coefficient. Unlike the MaRV, which has constant mass (except for small losses due to ablation), the interceptor undergoes rapid mass reduction during boost. Note that in both the thrust and drag equations, there is an implied dependence upon state $X9$, the interceptor altitude, through the atmospheric pressure in Eq. (9.65a) and atmospheric density in Eq. (9.65b).

We may now rewrite the interceptor state equations in scalar form as follows:

$$\frac{dX7}{dt} = X10, \qquad \frac{dX8}{dt} = X11, \qquad \frac{dX9}{dt} = X12$$

$$\begin{aligned}
\frac{dX10}{dt} &= \left(\frac{TH}{M} - \frac{D}{M}\right)UV1 + \left(\frac{L}{M}\right)\left(\frac{UR1}{\sin(\theta)} - \frac{UV1}{\cos(\theta)}\right) \\
\frac{dX11}{dt} &= \left(\frac{TH}{M} - \frac{D}{M}\right)UV2 + \left(\frac{L}{M}\right)\left(\frac{UR2}{\sin(\theta)} - \frac{UV2}{\cos(\theta)}\right) \\
\frac{dX12}{dt} &= \left(\frac{TH}{M} - \frac{D}{M}\right)UV3 + \left(\frac{L}{M}\right)\left(\frac{UR3}{\sin(\theta)} - \frac{UV3}{\cos(\theta)}\right) - g
\end{aligned} \tag{9.66}$$

The unit vectors UV and UR were discussed at length in the preceding section. The angle θ was calculated in Eq. (9.54). The lift per unit mass L/M is

$$\frac{L}{M} = \frac{\rho A(V_I)^2}{2(M_p + M_b)} C_L \tag{9.67}$$

We also discussed in the preceding section ways for selecting the lift coefficient C_L. In Eqs. (9.56), for example, we related C_L to the angular separation between the interceptor velocity $\boldsymbol{V}_I$, and the range-to-go vector $\boldsymbol{R}_{m/I}$. Certain conditions should be placed upon Eq. (9.67). First, the lateral maneuvering loads must be set at some upper limit due to structural considerations. Thus, L/M must have an upper bound. Secondly, there is a limitation in the magnitude of C_L, which is set by the configuration. Third, a lower limit must be set upon the separation angle θ to avoid the obvious singularities in Eqs. (9.66).

Finally, there are auxiliary equations or tables which must be included with the above state equations. We have already alluded to the atmospheric density ρ [Eqs. (9.65b) and (9.67)] and atmospheric pressure P_a [Eq. (9.65a)]. There must also be an equation to account for rocket mass flow, i.e.,

$$M_p = M_{p_0} - \left(\frac{dM_p}{dt}\right)t \tag{9.68}$$

where M_p is the propellant mass at time t, M_{p_0} is the propellant mass at the beginning of boost, and dM_p/dt is the propellant mass flow rate.

In the next section we present some results taken from various simulations of a MaRV being engaged by an interceptor using projected intercept point (PIP) and cross-product steering.

9.8 Simulation Results

In this section we present two illustrations of trajectories of maneuvering re-entry vehicles. In Fig. 9.13 a MaRV trajectory is given for which the lift

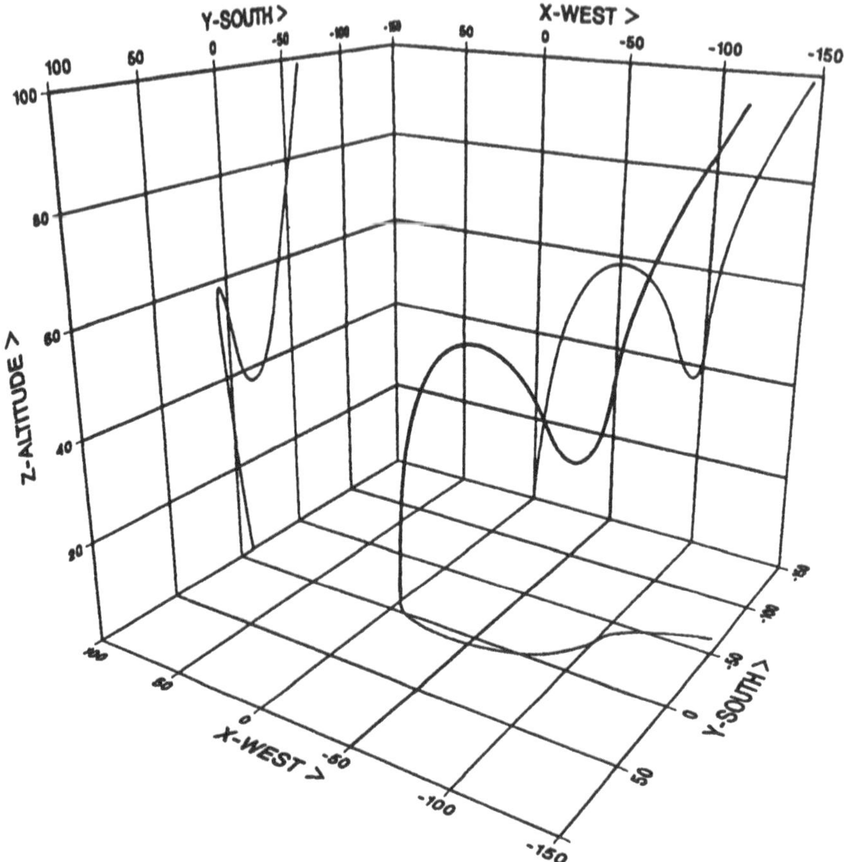

Fig. 9.13 Representative MaRV trajectory—random lift generation.

vector was set to produce a pull-up to be followed by a tuck, or pull-down. Both maneuvers occur essentially in a vertical plane. Diveline guidance was used only to meet targeting requirements.

In Fig. 9.14 the MaRV is subjected to four sequential divelines. It is not convenient to illustrate the divelines, but it may be seen that great deviation from a ballistic trajectory is possible using diveline guidance including significant out-of-plane maneuvering.

9.9 Other Guidance Laws and Summary

By way of definition we should first consider the distinction between guidance and control, terms which are often used synonymously. According to Bryson,[4] guidance concerns the strategy of maintaining a prescribed or nominal flight path in the presence of off-nominal conditions. Control concerns the strategy of maintaining the angular orientation of the vehicle subject to guidance, environmental, and payload constraints. In this section the re-entry vehicle is still

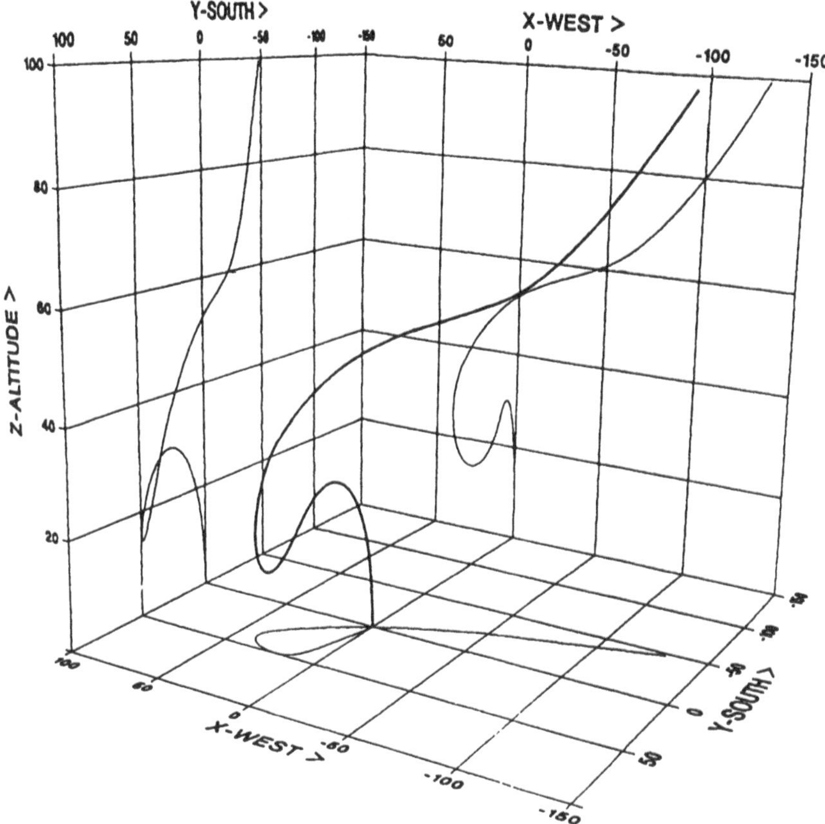

Fig. 9.14 Representative MaRV trajectory—diveline guidance.

limited to a point mass. In ignoring the moment equations, the tacit assumption is made that the vehicle is always in trim and that the control strategy is being implemented perfectly.

In Section 9.4 we considered a guidance law that defined a lift direction that would place the MaRV's velocity vector along a preset line (diveline). As pointed out, diveline guidance could be used to shape the MaRV trajectories for evasion purposes and as a means to meet the targeting requirements. In Section 9.5 we considered an elementary method for determining a projected interception point (PIP). This fire control algorithm could then be used to identify a projected interception point. In Section 9.6 we examined a fundamental guidance law to bring the interceptor to the projected interception point.

In this book we are primarily concerned with the dynamics of a re-entry vehicle and only to a limited extent with the guidance of such vehicles for targeting and evading an active defense or for guiding an interceptor toward an engagement. Most of the activity associated with guidance law development and refinement has found application in short-range tactical missiles. An interesting survey paper by Pastrick et al.[5] should be consulted for both its excellent overview and extensive references to the literature (up to 1981). In addition, a fundamental resource in guidance law design and implementation is Zarchan's comprehensive effort.[6]

We now outline several guidance laws which might be applicable to both interceptor and MaRV modeling.

The most fundamental guidance law for interception must be the line of sight (LOS), or beam rider. In essence the vehicle is directed from a remote command station to remain on a line of sight to the target. This method is obviously an extension of an aimed kinetic energy weapon where the projectile is directed along the initial line of sight. Adding guidance during interceptor transit can correct for off-nominal conditions (wind, density, etc.) or target motion.

LOS guidance may be divided into the command line-of-sight guidance (CLOS) and the beam rider (BR). In CLOS, the controller maintains the LOS and, through an uplink transmission, directs the interceptor to remain on the LOS. In the BR method an electromagnetic beam is sensed by the interceptor. There is then some indication of an off-beam position. An autopilot using a control algorithm then redirects the interceptor into the beam. Only in the case of a non-maneuvering target are both the interceptor's position vector and velocity vector in the beam; in the general case the velocity vector is not within the beam because the interceptor's velocity vector must continually be rotated in order to maintain the interceptor within the beam.

Figure 9.15 shows the engagement of a MaRV on a constant-altitude, constant-speed path. Clearly,

$$
\begin{aligned}
\mathrm{d}s^2 &= \mathrm{d}r^2 + r^2\mathrm{d}\theta^2 \\
\left(\frac{\mathrm{d}r}{\mathrm{d}\theta}\right)^2 + r^2 &= \left(\frac{\mathrm{d}s}{\mathrm{d}\theta}\right)^2
\end{aligned}
\tag{9.69}
$$

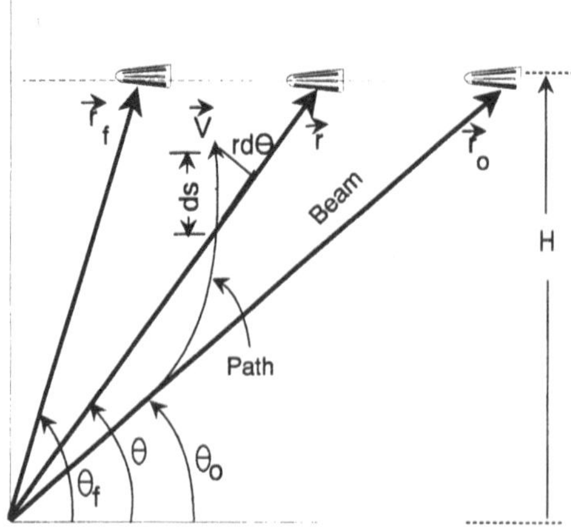

Fig. 9.15 Beam rider guidance.

The horizontal distance traveled by the target since the launch of the interceptor is given by

$$V_T t = H[\cot(\theta_0) - \cot(\theta)]$$

where V_T is the target speed (assumed constant), t is the time of flight of the inteceptor, and H is the target altitude; θ is the beam angle with respect to the horizontal. It is easily shown that

$$\frac{d\theta}{dt} = \frac{V_T}{r}\sin^2(\theta) \tag{9.70a}$$

and, by definition,

$$V_I = \frac{ds}{dt} \tag{9.70b}$$

After some manipulation we may rewrite Eq. (9.69) to give

$$\left(\frac{dr}{d\theta}\right)^2 + r^2 = \tilde{K}^2\csc^4(\theta) \tag{9.71a}$$

where r is the range to the interceptor; the constant $\tilde{K}$ may be expressed in terms of the altitude H and the interceptor-to-target velocity magnitude ratio $V_I/V_T = P$ as

$$\begin{aligned}\tilde{K} &= \left(\frac{V_I}{V_T}\right)H \\ &= PH\end{aligned} \tag{9.71b}$$

Within the limitations of this simplified example of a constant-speed, constant-altitude MaRV, we must find the solution of Eq. (9.71a) to obtain $r = r(\theta)$; θ is in turn set by the geometry of the moving target. Locke[7] presents a series solution to Eq. (9.71a). Such procedures are dated by 40 years; however, the analysis does provide some insight into the transverse acceleration demands.

If Eq. (9.71a) were solved, impact would take place when

$$r_f \sin(\theta_f) = H \tag{9.71c}$$

Again, θ is a single-valued function of the target position,and r follows from the solution to Eq. (9.71a). When the condition of Eq. (9.71c) is satisfied, engagement has occurred. For a modern simulation, Eq. (9.71c) is of interest in evaluating the bounds of the problem; however, a computer simulation would not restrict modeling to a target of constant altitude and velocity magnitude.

To remain within the beam, the interceptor is required to turn at a rate $\mathrm{d}\phi/\mathrm{d}t$ given by

$$\frac{\mathrm{d}\phi}{\mathrm{d}t} = \frac{2V_T}{H}\sin^2(\theta)\left\{1 + \frac{r\cot(\theta)}{[\tilde{K}^2\csc^4(\theta) - r^2]^{1/2}}\right\} \tag{9.72a}$$

The corresponding lateral acceleration (normal to the velocity vector) is

$$a_N = V_I\frac{\mathrm{d}\phi}{\mathrm{d}t} \tag{9.72b}$$

The derivation of Eq. (9.72a) is straightforward, but space does not permit the details to be presented here. The reader is referred to Locke.[6] Obviously, Eqs. (9.72) permit an estimate of interceptor maneuvering requirements, at least within the geometric restrictions set forth in Fig. 9.15.

Pursuit guidance is another useful interceptor guidance strategy. The most instinctive of all guidance schemes, pursuit guidance involves placing the velocity vector of the interceptor on a line of sight to the target. The interceptor must have velocity superiority over the target [i.e., the parameter P of Eq. (9.71b) must be greater than 1]. Pursuit guidance and many of its derivatives are discussed by Locke[7] and Howe,[8] and in a review by Teng and Phillips.[9] Three variations of pursuit guidance are as follows: 1) velocity pursuit, already discussed, where the interceptor velocity is always directed toward the target; 2) attitude pursuit, where the interceptor center line is pointed toward the target; 3) deviated pursuit, where a constant nonzero angle is maintained between the velocity vector and the line of sight to the target.

As an example we consider deviated pursuit. Deviated pursuit is a strategy in which the angle between the interceptor velocity vector and the line of sight is a fixed constant. The geometry of deviated pursuit is shown in Fig. 9.16. The angle δ is the deviation, or offset angle. In pure pursuit the angle δ is equal to zero. As the target maneuvers, the inertial angle of the line of sight will change. Lift must be generated at the interceptor to maintain the offset angle. γ_m and γ_I are the flight path angles of the MaRV and the interceptor, respectively.

Because of limited space here we restrict engagement to a nonmaneuvering target. For simplicity we further assume that the target is moving horizontally at constant velocity (i.e., $\gamma_m = 0.0$). The equations of motion along and normal to the LOS, respectively, are

$$\frac{\mathrm{d}R}{\mathrm{d}t} = V_T\cos(\phi) - V_I\cos(\delta) \tag{9.73a}$$

$$R\frac{\mathrm{d}\phi}{\mathrm{d}t} = -V_T\sin(\phi) + V_I\sin(\delta) \tag{9.73b}$$

where $\mathrm{d}\phi/\mathrm{d}t$ is the rotation rate of the line of sight since $\mathrm{d}\gamma_m/\mathrm{d}t$ is 0.0.

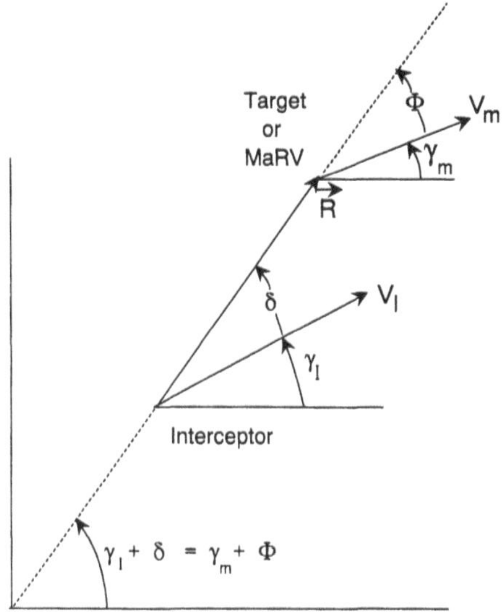

Fig. 9.16 Deviated pursuit.

Dividing Eq. (9.73a) by Eq. (9.73b) gives

$$\frac{1}{R}\frac{\mathrm{d}R}{\mathrm{d}t} = \left[\frac{\cos(\phi) - P\cos(\delta)}{-\sin(\phi) + P\sin(\delta)}\right]\left(\frac{\mathrm{d}\phi}{\mathrm{d}t}\right) \tag{9.74}$$

where P is the interceptor-target velocity ratio as given in Eq. (9.71b). Integration of Eq. (9.74) results in the following expression:

$$R = R_0\left[\frac{\sin(\phi) - P\sin(\delta)}{\sin(\phi_0) - P\sin(\delta)}\right]^{[P\cos(\delta)/K]-1} \times \left[\frac{1 - P\sin(\delta)\sin(\phi_0) + K\cos(\phi_0)}{1 - P\sin(\delta)\sin(\phi) + K\cos(\phi)}\right]^{P\cos(\delta)/K} \tag{9.75a}$$

where

$$K = \left[1 - P^2\sin^2(\delta)\right]^{1/2} \tag{9.75b}$$

and ϕ_0 and R_0 are the intial values of ϕ and R, respectively, and where,

$$P^2\sin^2(\delta) < 1 \tag{9.75c}$$

It should be emphasized that the derivation of Eq. (9.75a) is subject to the restriction of Eq. (9.75c). The trajectory where

$$P^2\sin^2(\delta) > 1$$

spirals around the target and fails to achieve an interception. Thus, Eq. (9.75c) places a restriction on the deviation, or lead angle, δ for a given velocity ratio P. Note the inverse relationship between P and δ. If

$$P \sin(\delta) = 1$$

then

$$R = R_0 \left[\frac{1 - \sin(\phi_0)}{1 - \sin(\phi)}\right] \exp\left\{P \cos(\delta) \left[\tan\left(\frac{\pi}{4} + \frac{\phi_0}{2}\right) - \tan\left(\frac{\pi}{4} + \frac{\phi}{2}\right)\right]\right\} \tag{9.76}$$

The time of flight T_f is given by

$$T_f = \frac{R_0[P + \cos(\phi_0 + \delta)]\sec(\delta)}{V_T(P^2 - 1)} \tag{9.77}$$

The time of flight is finite and positive provided that

$$P > 1 \qquad \text{and} \qquad \delta < \pi/2$$

The turning rate $d\phi/dt$ follows from Eq. (9.73b) as

$$\frac{d\phi}{dt} = V_T \left[\frac{P \sin(\delta) - \sin(\phi)}{R}\right] \tag{9.78}$$

The range-to-go R may be substituted from Eq. (9.75a), again subject to the condition of Eq. (9.75c), to give

$$\frac{d\phi}{dt} = F(\phi_0)\left\{\frac{[1 - P \sin(\delta) \sin(\phi) + K \cos(\phi)]^{P \cos(\delta)/K}}{[\sin(\phi) - P \sin(\delta)]^{[P \cos(\delta)/K]-2}}\right\} \tag{9.79a}$$

where

$$F(\phi_0) = -\frac{V_T}{R_0}\left\{\frac{[\sin(\phi_0) - P \sin(\delta)]^{[P \cos(\delta)/K]-1}}{[1 - P \sin(\delta) \sin(\phi_0) + K \cos(\phi_0)]}\right\}\left(\frac{P \cos(\delta)}{K}\right) \tag{9.79b}$$

The required lateral acceleration a_N is given by Eq. (9.72b).

In Section 9.6 we discussed cross-product steering and proportional guidance. The latter is probably the most popular or pervasive guidance algorithm in use, particularly for radar-controlled guided missiles. Locke[7] defines proportional guidance as a strategy in which the rate of change of interceptor heading is directly proportional to the rate of rotation of the line of sight. This statement is consistent with Eq. (9.58).

Confining ourselves to a two-dimensional case, Eq. (9.58) leads to the following expression for the transverse acceleration a_N:

$$a_N = V_I \frac{d\gamma}{dt} = K V_I \omega_{m/I} \tag{9.80}$$

Zarchan,[6] who has written extensively on guidance strategies (particularly proportional guidance), rewrites Eq. (9.80) as

$$a_N = K\omega_{m/I}\frac{\mathrm{d}R}{\mathrm{d}t} \tag{9.81}$$

where the interceptor velocity magnitude V_I of Eq. (9.80) has been replaced by the closing rate $\mathrm{d}R/\mathrm{d}t$.

We might consider a form of biased proportional guidance suggested by Grogg[10] as a strategy to be used against a maneuvering re-entry vehicle where a target point might be identified. In Fig. 9.17 a MaRV is shown in a diving turn toward an intended or presumed intended target. First we let $\boldsymbol{a}_e$ be the acceleration the interceptor would possess in a straight-line flight. Straight-line flight is taken here to mean the condition where the total acceleration $\boldsymbol{a}_I$ is collinear with the velocity vector as follows:

$$\boldsymbol{a}_e = (\boldsymbol{a}_I \cdot \boldsymbol{V}_I)\frac{\boldsymbol{V}_I}{V_I^2} \tag{9.82}$$

We now define the relative acceleration as

$$\boldsymbol{a}_r = \frac{2(\boldsymbol{R}_m \times \boldsymbol{V}_m) \times \boldsymbol{V}_m}{\boldsymbol{R}_m \cdot \boldsymbol{R}_m} - \boldsymbol{a}_e \tag{9.83}$$

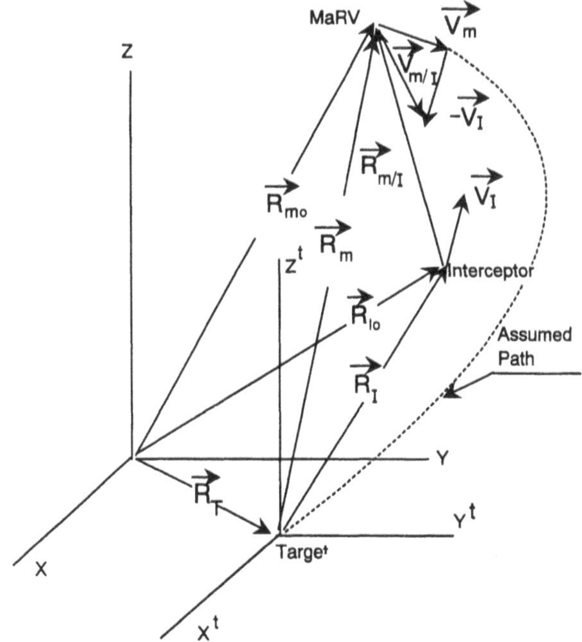

Fig. 9.17 Biased proportional guidance.

The line-of-sight rate for nonbiased guidance was given in Eq. (9.61a). A biased line-of-sight rate may be given as

$$\boldsymbol{\omega}_b = \frac{\boldsymbol{R}_{m/I} \times \boldsymbol{a}_r}{2\boldsymbol{R}_{m/I} \cdot \boldsymbol{V}_{m/I}} \tag{9.84}$$

where, as in Eqs. (9.61), the relative positional and velocity vectors are expressed as

$$\boldsymbol{R}_{m/I} = \boldsymbol{R}_m - \boldsymbol{R}_I, \qquad \boldsymbol{V}_{m/I} = \boldsymbol{V}_m - \boldsymbol{V}_I$$

The desired lateral acceleration $\boldsymbol{a}_L$ then becomes

$$\boldsymbol{a}_L = K\left(\frac{\boldsymbol{V}_{m/I} \cdot \boldsymbol{R}_{m/I}}{\boldsymbol{V}_I \cdot \boldsymbol{R}_{m/I}}\right)(\boldsymbol{\omega}_{m/I} - K_b\boldsymbol{\omega}_b) \times \boldsymbol{V}_I \tag{9.85}$$

where $\boldsymbol{\omega}_b$ is given in Eq. (9.84), $\boldsymbol{\omega}_{m/I}$ is given in Eq. (9.61a), and K_b is a weighting constant for the biased line-of-sight rate. Equations (9.63) and (9.85) become identical if K_b is set equal to zero.

We next consider an algorithm that might be useful in guiding a re-entry vehicle to an intended impact point. This algorithm is identified here as the *tangent cubic guidance law.*[11] This guidance law fits a cubic between the current vehicle position and its desired final position subject to two tangent constraints.

In Fig. 9.18 the re-entry vehicle is being given a lift component to follow a cubic path to the impact point. To simplify the analysis, assume that a reference

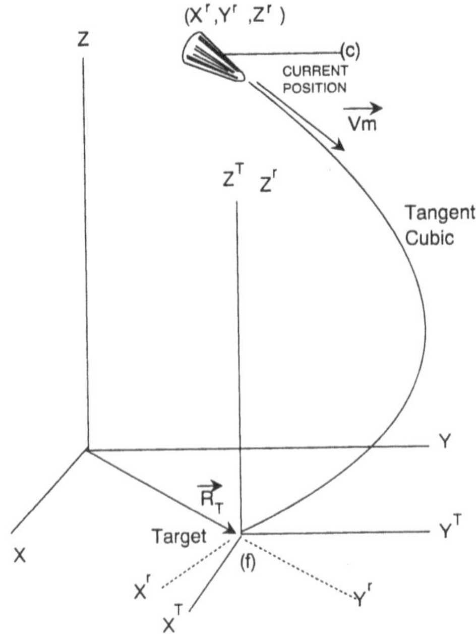

Fig. 9.18 Tangent cubic guidance.

axis system (γ-frame) is located at the intended impact point (i.e., the target). The orientation of the γ-frame is such that the X-Z plane contains the velocity vector of the MaRV. The equation of the cubic is

$$Z = Z(X) = AX^3 + BX^2 + CX + D \tag{9.86}$$

Obviously, we have four constants that must be determined by constraints placed on the trajectory. Here we follow the work of Page and Rogers[11] and use two initial conditions and two terminal conditions. These might be as follows

1) Current position

$$Z = Z_{\text{current}} = Z_c, \qquad X = X_{\text{current}} = X_c$$

2) Current flight path angle

$$\frac{dZ/dt}{dX/dt} = \left.\frac{dZ}{dX}\right|_{\text{current}} = \tan(\gamma)$$

3) Final position

$$Z_f = 0.0, \qquad X_f = 0.0$$

4) Final flight path angle

$$\left.\frac{dZ}{dX}\right|_{\text{final}} = \tan(\gamma_f)$$

The acceleration command required to implement this guidance strategy is the normal acceleration a_N, i.e,

$$a_N = \frac{\boldsymbol{V}_m \cdot \boldsymbol{V}_m}{R} = \frac{V_m^2}{R}$$

where V_m is the MaRV velocity magnitude; R is the radius of curvature, which can be written in terms of $Z(X)$ as

$$\begin{aligned} R &= \frac{[1 + (dZ/dX)^2]^{3/2}}{(d^2Z/dX^2)} \\ &= \frac{[1 + \tan^2(\gamma)]^{3/2}}{(d^2Z/dX^2)} \end{aligned} \tag{9.87}$$

Using the constraints previously specified, we get the following expression for the commanded acceleration a_N:

$$a_N = \frac{V_m^2 \cos^2(\gamma)}{X_c}\left[4\tan(\gamma) + 2\tan(\gamma_f) - 6\frac{Z_c}{X_c}\right] \tag{9.88}$$

Thus, knowing the current MaRV velocity magnitude V_m, position (X_c, Z_c), and current and final flight path angles γ and γ_f, it is possible to specify a required command acceleration.

Page and Rogers[11] suggest changing the final constraint to one of commanded acceleration at the target, i.e.,

$$\left.\frac{\mathrm{d}^2 Z}{\mathrm{d}X^2}\right|_{\text{final}} = 0.0$$

The above constraint gives the following for the command acceleration:

$$a_N = \frac{3V_{m_x}}{XV_m}\left[\tan(\gamma) - \frac{Z_c}{X_c}\right] \tag{9.89}$$

where $V_{m_x} = V_m \cos(\gamma)$.

For a parabolic curve, the final acceleration cannot be zero because

$$Z = AX^2 + BX + C \tag{9.90a}$$

$$\left.\frac{\mathrm{d}^2 Z}{\mathrm{d}X^2}\right|_{X = X_f = 0} = 2A \tag{9.90b}$$

Such a requirement would set $A = 0.0$ and reduce the parabola to a straight line. Applying the first three of the previously given constraints results in the following expression for the lateral acceleration:

$$a_N = \frac{2V_{m_x}^3}{V_m X}\left[\tan(\gamma) - \frac{Z_c}{X_c}\right] \tag{9.91}$$

Many other guidance algorithms might be investigated. The work of Locke[7] is interesting as a source of guidance strategies but is rather dated. A modern and comprehensive treatment of guidance is contained in the excellent work by Zarchan.[6]

References

[1]Finke, R.G., "Reentry Vehicle Dispersion Due to Atmospheric Variations," Institute for Defense Analysis, Science and Technology Division, Alexandria, VA, Research Paper P-506, Aug. 1969.

[2]Gracey, C., Cliff, E. M., Lutz, F. H., and Kelley, H. J., "Fixed-Trim Guidance Analysis," Final Report Contract N60921-80-0090, Naval Surface Warfare Center, Dahlgren, VA, 1981; Gracey, C., et al., "Fixed-Trim Reentry Guidance Analysis," *Journal of Guidance and Control,* Vol. 5, No. 6, 1982.

[3]Cameron, J. D. M., "Explicit Guidance Equations for Maneuvering Reentry Vehicles," General Electric Company, Philadelphia, PA, 1973.

[4]Bryson, A. E., "New Concepts in Control Theory—1959–1984," *Journal of Guidance,* Vol. 8, No. 4, 1985, pp. 417–425.

[5]Pastrick, H. L., Seltzer, S. M., and Warren, M. E., "Guidance Laws for Short-Range Tactical Missiles," *Journal of Guidance and Control,* March/April 1981.

[6]Zarchan, P., *Tactical and Strategic Missile Guidance,* Vol. 124, Progress in Astronautics and Aeronautics, AIAA, Washington, DC, 1990.

[7]Locke, A. S., *Guidance,* D. Van Nostrand (now Van Nostrand-Reinhold-Wadworth), Florence, KY, 1955, Chap. 12.

[8]Howe, R. M., "Guidance," *Systems Engineering Handbook,* McGraw-Hill, Hightstown, NJ, 1965.

[9]Teng, L., and Phillips, P. L., "Application of Non-linear Filter to Short-Range Missile Guidance," *Proceedings Journal of Astronautical Sciences,* June 1968, pp. 138–147.

[10]Grogg, B. B., *The WINSOME Manual, Vol. I—Program Description,* General Research Corp., Santa Barbara, CA, 1976.

[11]Page, J. A., and Rogers, R. O., "Guidance and Control of Maneuvering Reentry Vehicles," *Conference on Decision and Control, and Symposium on Adaptive Processes Proceedings,* Vol. 1, Institute of Electrical and Electronic Engineers, Piscataway, NJ, Dec. 7–9, 1977, pp. 659–664.

Bibliography

Abzug, M. J., "Vector Methods in Homing Guidance," *Journal of Guidance and Control,* May/June 1979, pp. 253–255.

Clemow, J., *Missile Guidance,* Temple Press, London, 1960.

Cunningham, P. P., "Lambda Matrix Terminal Control for Missile Guidance," *Journal of Spacecraft and Rockets,* July 1968, pp. 119–121.

Garnell, P., *Guided Weapon Control Systems,* 2nd ed., Brassey's Defence Publications, London, 1980.

Guelman, M., "The Closed Form Solution of True Proportional Navigation," *Transactions on Aerospace and Electronic Systems,* July 1976, pp. 472–482.

Guelman, M., "A Qualitative Study of Proportional Navigation," IEEE Transactions on Aerospace and Electronic Systems, July 1971, pp. 337–343.

Meyer, J. A., and Bland, J. G., "Final Value Missile Homing Guidance," *Journal of Spacecraft and Rockets,* Feb. 1967, pp. 270–280.

York, R. J., and Pastrick, H. L., "Optimal Terminal Guidance with Constraints at Final Time," *Journal of Spacecraft and Rockets,* July 1977, pp. 381, 382.

10
Angular Motion During the Exoatmospheric (Keplerian) Phase

10.1 Introduction

In Chapter 6 we discussed the exoatmospheric motion of a re-entry vehicle approximated by a particle. Representing the re-entry body as a particle means that angular motion is ignored. However, angular motion during the exoatmospheric, or Keplerian, phase is important inasmuch as the attitude of the body during re-entry is set during the initiation of this phase. Body attitude at re-entry sets the aerodynamic loads during the early portion of the endoatmospheric trajectory segment.

The only force acting during the Keplerian phase is due to the planetary gravity field; there is no associated gravitational moment if effects due to field gradient are ignored. In the absence of moments the angular momentum vector remains constant in magnitude and direction. Therefore, during the Keplerian phase, the RV has associated with it a constant angular momentum vector; as we shall show, the orientation of this angular momentum vector can have a significant influence on the attitude of the re-entry body at entry.

Before investigating torque-free motion, we will briefly discuss one way a re-entry vehicle is given angular momentum. (We make here the distinction between re-entry body and re-entry vehicle; *body* refers to any object, natural or artificial, which enters the atmosphere, whereas *vehicle* refers to an artifact only.)

10.2 Re-Entry Vehicle Deployment

Figure 10.1 shows a typical deployment trajectory. We assume that any events preceding the deployment of the final stage are tangential to the discussion of re-entry vehicle deployment. The final stage (FS) trajectory segment begins before final stage ignition. When flight control receives a command to commence final stage, it initiates final stage thrust vector control (TVC). After the passage of an interval of time set by the guidance, flight control issues commands for previous stage separation and final stage ignition. During the final stage, flight guidance performs acceleration and velocity computations to meet targeting requirements.

After final stage burnout, flight control initiates separation and ejection of the final stage motor. The ejection motor, mounted in the forward dome of the final stage, is ignited. The result is that the final stage motor backs out of the

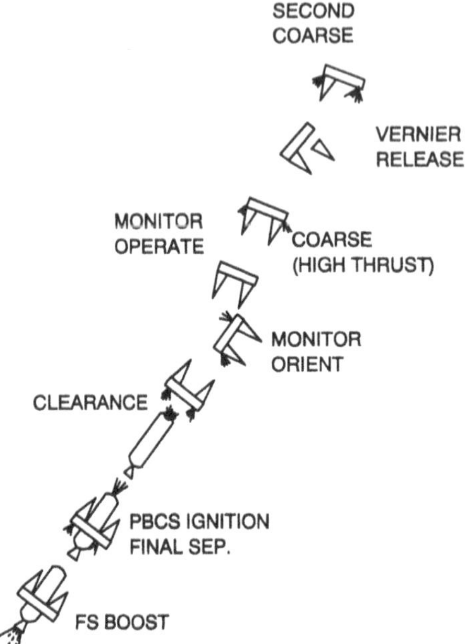

Fig. 10.1 Representative deployment trajectory.

equipment section. The thrusters on the equipment section then move this section away from the final stage motor.

Once the final stage is cleared, low-level thrusters in the equipment section position and orient the equipment section in space. During the maneuvering associated with the post-boost control system (PBCS), it is necessary to avoid re-entry vehicle disturbance caused by the near-presence of exhaust gases from the PBCS nozzle. A maneuver called the plume avoidance maneuver (PAM) involves shutting down the offending jet and operating the other jets to rotate and back away the equipment section from the released re-entry vehicle. It is during this PAM that the re-entry vehicle gets a variety of disturbances. Some of the details of the PAM are shown in Fig. 10.2.

For the purposes of the remaining discussion in this chapter, we assume that the re-entry vehicle is placed in an intended orbit, is influenced only by a uniform gravitation field, and has been spun about an axis that is close, but not necessarily coincident, to the axis of symmetry. In the next section we "rediscover" some nineteenth-century geometric and analytic concepts that provide an understanding of the torque-free motion that is characteristic of the Keplerian trajectory.

10.3 Analytical Treatment of Torque-Free Motion

Background

When a re-entry vehicle is placed into its suborbital trajectory, it is usually spun about one of its principal axes of inertia. In this section we examine

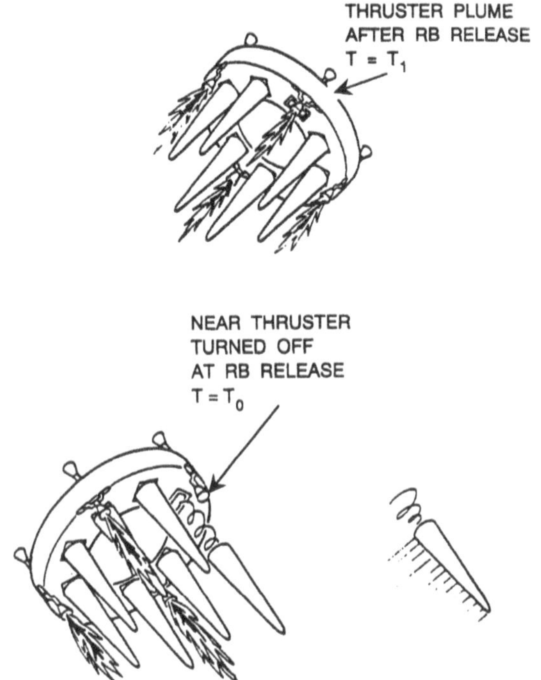

Fig. 10.2 Re-entry body ejection and plume avoidance maneuver (PAM).

the purpose of imparting spin to the vehicle. We also consider the effects of perturbations to the intended spin direction.

The analytical and geometric treatment of torque-free motion was brought to maturity during the nineteenth century. A principal contributor was L. Poinsot, who in 1834 provided a geometrical treatment of torque-free angular motion. We will consider this work later.

Torque-free motion is difficult to reproduce under laboratory conditions at the Earth's surface. The presence of an atmosphere limits the assumption of no torque, and gravitational force limits the duration of the observable motion. Wertz[1] provides a photographic record (albeit of poor quality) of the oscillations of a lunar satellite just after launch. The photographic sequence of the oscillatory motion was obtained by a crew member using a 12-frame-per-second hand-held camera. Analytical treatments that follow rely heavily on similar presentations by Regan,[2] Greenwood,[3] Wiesel,[4] and Kaplan.[5]

Recall Eq. (5.56a), which provides a general expression for the time rate of change of angular momentum. First, we replace m (moving frame) with b (body frame). We change the reference frame from f (fixed) to I (inertial). We get

$$\boldsymbol{M}^b = \left.\frac{\mathrm{d}\boldsymbol{H}^b}{\mathrm{d}t}\right|_b + \Omega^b_{b/I}\boldsymbol{H}^b \tag{10.1a}$$

or

$$\begin{aligned} \boldsymbol{M}^b &= [M_1, M_2, M_3]^{\mathrm{T}} \\ &= I^b \frac{\mathrm{d}\boldsymbol{\omega}^b_{b/I}}{\mathrm{d}t} + \Omega^b_{b/I} I^b \boldsymbol{\omega}^b_{b/I} \end{aligned} \tag{10.1b}$$

where I^b is the inertia tensor [Eq. (5.55b)] evaluated in a body-fixed frame and $\boldsymbol{\omega}^b_{b/I}$ is the angular velocity vector of the body relative to inertial space with components also in the body frame. $\Omega^b_{b/I}$ is, of course, the skew-symmetric form of $\boldsymbol{\omega}^b_{b/I}$.

We now assume that the re-entry vehicle has mass symmetry about each of the orthogonal body-fixed axes; i.e., the 1-, 2-, and 3-axes are principal axes. Furthermore, we assume that the angular velocity is nominally along the 1-axis and equal to ω_0. It is for this reason that the 1-axis (in this case) will be referred to as the *spin axis* even though there may be components of angular velocity normal to this axis. However, these normal components of angular velocity will be considered as higher-order or as velocity perturbations in comparison to $\boldsymbol{\omega}_0$. Thus, the angular velocity may be written as

$$\boldsymbol{\omega}^b_{b/I} = [\omega_0, 0, 0]^{\mathrm{T}} + [\omega_1, \omega_2, \omega_3]^{\mathrm{T}} = [\omega_0 + \omega_1, \omega_2, \omega_3]^{\mathrm{T}}$$

$$\frac{\mathrm{d}\boldsymbol{\omega}^b_{b/I}}{\mathrm{d}t} = [\dot{\omega}_1, \dot{\omega}_2, \dot{\omega}_3]^{\mathrm{T}}$$

where

$$\omega_1, \omega_2, \omega_3 \ll \omega_0 \tag{10.2}$$

We assume torque-free motion in the present use of Eqs. (10.1). Thus, the scalar form of these equations is

$$I_1 \frac{\mathrm{d}\omega_1}{\mathrm{d}t} + \omega_2\omega_3(I_3 - I_2) = 0 \tag{10.3a}$$

$$I_2 \frac{\mathrm{d}\omega_2}{\mathrm{d}t} + (\omega_0 + \omega_1)\omega_3(I_1 - I_3) = 0 \tag{10.3b}$$

$$I_3 \frac{\mathrm{d}\omega_3}{\mathrm{d}t} + (\omega_0 + \omega_1)\omega_2(I_2 - I_1) = 0 \tag{10.3c}$$

If we continue to assume that the components of angular velocity normal to the spin axis are perturbational, we have

$$\boldsymbol{\omega}^b_{b/I} = [\omega_0 + \omega_1, \omega_2, \omega_3]^{\mathrm{T}} \approx [\omega_0, \omega_2, \omega_3]^{\mathrm{T}}$$

which gives the following simplified forms of Eqs. (10.3):

$$\frac{d\omega_1}{dt} = \omega_2\omega_3\left(\frac{I_2 - I_3}{I_1}\right) \tag{10.4a}$$

$$\frac{d\omega_2}{dt} = \omega_0\omega_3\left(\frac{I_3 - I_1}{I_2}\right) \tag{10.4b}$$

$$\frac{d\omega_3}{dt} = \omega_0\omega_2\left(\frac{I_1 - I_2}{I_3}\right) \tag{10.4c}$$

If the 1-axis is coincident with an axis of rotational symmetry ($I_2 = I_3$), then Eq. (10.4a) would clearly show that

$$\frac{d\omega_1}{dt} = 0 = \frac{d}{dt}(\omega_0 + \omega_1) \tag{10.5}$$

or

$$(\omega_0 + \omega_1) = \text{constant}$$

Even if $I_2 \neq I_3$, we might still invoke Eq. (10.5) because the perturbational assumption might justify neglecting the product $\omega_2\omega_3$ of Eq. (10.4a). Assuming that Eq. (10.5) is justified, we have the following after manipulation of Eqs. (10.4b) and (10.4c):

$$\frac{d^2\omega_2}{dt^2} + \omega_0^2\left[\frac{(I_1 - I_2)(I_1 - I_3)}{I_2 I_3}\right]\omega_2 = 0 \tag{10.6a}$$

$$\frac{d^2\omega_3}{dt^2} + \omega_0^2\left[\frac{(I_1 - I_2)(I_1 - I_3)}{I_2 I_3}\right]\omega_3 = 0 \tag{10.6b}$$

Both of the above equations are uncoupled second-order linear equations with either a harmonic (sinusoidal) or exponential solution. If the motion is sinusoidal, then the natural undamped frequency is given by

$$\alpha = \omega_0\left[(I_1 - I_2)(I_1 - I_3)/I_2 I_3\right]^{1/2} \tag{10.7}$$

The solution to Eqs. (10.6) then becomes

$$\omega_2 = A_2 e^{i\alpha t} + B_2 e^{-i\alpha t} \tag{10.8a}$$

$$\omega_3 = A_3 e^{i\alpha t} + B_3 e^{-i\alpha t} \tag{10.8b}$$

If α is real, these equations are easily rewritten in the more familiar form of harmonic motion as

$$\omega_2 = \Omega_2 \sin[\alpha(t - t_0)] \tag{10.9a}$$

$$\omega_3 = \Omega_3 \sin[\alpha(t - t_0)] \tag{10.9b}$$

If, on the other hand, α is imaginary, then the exponents of Eq. (10.8) are real. One exponent will be positive, with the result that the perturbational angular velocity will increase without bound.

In our example we chose the spin axis to be the 1-axis. For harmonic motion, α of Eq. (10.7) must be real, which can occur only if one of the following conditions is satisfied:

$$I_1 > I_2 \qquad \text{and} \qquad I_1 > I_3 \tag{10.10a}$$

or

$$I_1 < I_2 \qquad \text{and} \qquad I_1 < I_3 \tag{10.10b}$$

Harmonic (bounded) motion is assured only if the angular velocity is placed initially along a major or minor principal axis.

Figure 10.3 shows the unperturbed angular velocity vector $[\omega_0, 0, 0]^T$ and the perturbational angular velocity vector $\boldsymbol{\delta\omega}_p = [\omega_1, \omega_2, \omega_3]^T$. Provided that the unperturbed angular velocity vector is along a major or minor principal axis, the motion of the angular velocity vector as "seen" in the body frame

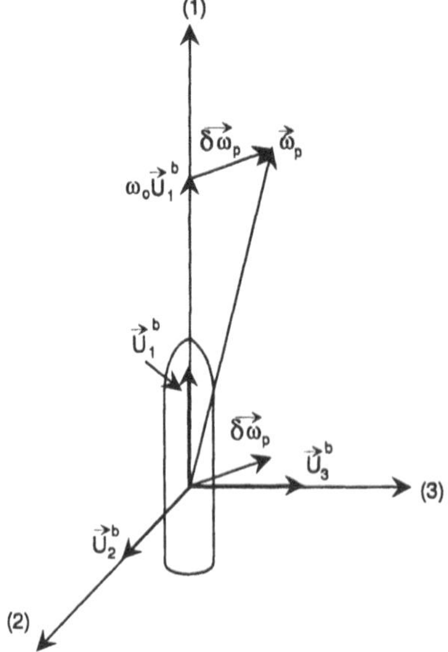

Fig. 10.3 Angular velocity components.

will be that of an undamped sinusoid. On the other hand, if the angular velocity vector is in the vicinity of the intermediate principal axis, the angular velocity vector will diverge exponentially.

Let's now introduce a further simplification by assuming that the 1-axis is a minor principal axis and an axis of rotational mass symmetry. Under these circumstances we have

$$\begin{aligned} I_2 &= I_3 = I_T \\ I_1 &= I < I_T \end{aligned} \tag{10.11a}$$

Equation (10.7) then becomes

$$\alpha = \omega_0\,[(I_T - I)/I_T] \tag{10.11b}$$

There might be some logical difficulties with our assumption that the product $\omega_2\omega_3$ in Eq. (10.4a) is of higher order and therefore negligible. With the restriction to a body of axial symmetry ($I_2 = I_3$), the right side of Eq. (10.4a) is equal to zero.

The harmonic solution for ω_2 given in Eq. (10.9a) is

$$\omega_2 = \Omega_2 \sin[\alpha(t - t_0)] \tag{10.9a}$$

We may solve for ω_3 by first rewriting Eq. (10.4b) as

$$\omega_3 = \frac{I_2}{\omega_0(I_3 - I_1)}\frac{\mathrm{d}\omega_2}{\mathrm{d}t} = \frac{I_T}{\omega_0(I_T - I)}\frac{\mathrm{d}}{\mathrm{d}t}\{\Omega_2 \sin[\alpha(t - t_0)]\}$$

or

$$\begin{aligned} \omega_3 &= \Omega\cos[\alpha(t - t_0)] \\ &= \Omega \sin\,[\alpha(t - t_0) + (\pi/2)] \end{aligned} \tag{10.12}$$

The constant Ω is the magnitude of the oscillation, and the constant t_0 allows for an arbitrary phase angle.

We should emphasize at this point that the oscillation that we are considering is that of the angular velocity as observed in a body frame with the 1-axis coincident to an axis of rotational mass symmetry. Also, the changing direction of the angular velocity as observed in a body-fixed frame and expressed by Eq. (10.12) is only for torque-free motion. Consequently, the direction and magnitude of the angular momentum vector $\boldsymbol{H}$, although fixed in inertial space, will appear to move in a body-fixed frame. We examine next the motion (if any) of $\boldsymbol{H}$, $\boldsymbol{\omega}$, and $\boldsymbol{U}_1^b$ (the unit vector along the body axis of symmetry), first in a body frame and then in an inertial frame.

Figure 10.4 illustrates the behavior of the angular velocity vector $\boldsymbol{\omega}$ as seen in a body-fixed frame. The angular velocity has a constant component ω_0 along the axis-of-symmetry vector $\boldsymbol{U}_1^b$; the projection of the angular velocity vector

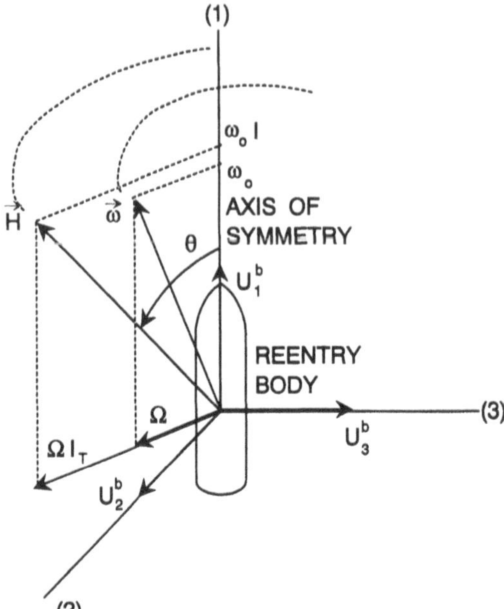

Fig. 10.4 Angular velocity and angular momentum in a body frame.

on the plane of symmetry (spanned by the U_2^b and U_3^b unit vectors) undergoes circular motion at a frequency of α rad/s. We might conclude from Eq. (10.11b) that the frequency α is less than the component of $\boldsymbol{\omega}$ along the 1-axis; however, for very slender configurations where $I \ll I_T$, α does approach ω_0.

Next, consider the angular momentum vector $\boldsymbol{H}$. The components of $\boldsymbol{H}$ in a body frame may be expressed as follows:

$$\boldsymbol{H}^b = I^b \boldsymbol{\omega}_{b/I}^b = \begin{bmatrix} I & 0 & 0 \\ 0 & I_T & 0 \\ 0 & 0 & I_T \end{bmatrix} \begin{bmatrix} \omega_0 \\ \Omega \sin[\alpha(t - t_0)] \\ \Omega \cos[\alpha(t - t_0)] \end{bmatrix}$$

which becomes

$$\boldsymbol{H}^b = \begin{bmatrix} H_1 \\ H_2 \\ H_3 \end{bmatrix} = \begin{vmatrix} I\omega_0 \\ I_T\Omega \sin[\alpha(t - t_0)] \\ I_T\Omega \cos[\alpha(t - t_0)] \end{vmatrix} \tag{10.13}$$

We note that although the angular momentum remains constant in inertial space, it must appear to rotate when observed from the body-fixed frame. To see that this must be the case, we note that the angular momentum is formed by premultiplying the angular velocity $\boldsymbol{\omega}_{b/I}^b$ by a matrix (the inertia tensor) whose elements are all constants. Thus, the inertia tensor changes the direction and magnitude of the angular velocity vector to produce the angular momentum vector. Although the angular momentum and angular velocity vectors are not coincident (except in the uninteresting case of spherical symmetry), both appear to trace out a cone whose axis of symmetry is the body's axis of symmetry and whose vertex is at the body's center of mass. The angle between the angular momentum vector and the axis of symmetry is found by taking the inner product of $\boldsymbol{H}$ with the unit vector along the axis of mass symmetry $\boldsymbol{U}_1^b$.

$$\cos\theta = \frac{\boldsymbol{H}\cdot\boldsymbol{U}_1^b}{|\boldsymbol{H}|} = \frac{(\boldsymbol{H}^b)^{\mathrm{T}}\boldsymbol{U}_1^b}{[(\boldsymbol{H}^b)^{\mathrm{T}}\boldsymbol{H}^b]^{1/2}}$$

From Eq. (10.13) we can easily show that

$$\cos\theta = I\omega_0/\left(I^2\omega_0^2 + I_T^2\Omega^2\right)^{1/2} \tag{10.14}$$

Figure 10.4 presents the two important vectors, angular momentum $\boldsymbol{H}$ and angular velocity $\boldsymbol{\omega}_{b/I}^b = \boldsymbol{\omega}$, from the perspective of a body-fixed reference frame. During atmospheric flight the direction of the vertical (normal to the geoid) is usually taken as a reference direction. However, during the Keplerian, or exoatmospheric, segment, the concept of a vertical is meaningless. Keplerian motion is as close to torque-free motion as has ever been realized. Under such circumstances the angular momentum vector is invariant in inertial space and therefore serves as a meaningful reference direction.

We will now in effect change our perspective from a body-fixed frame to an inertial-frame. First, take the 1-axis (axis of mass symmetry) of the body as coincident with the angular momentum vector. Rotate about the 1-axis through the angle ψ, then about the new 2-axis through the angle θ, and finally about the new 1-axis through the angle ϕ. If we refer to Appendix C, we note that the above E_{121} sequence corresponds to an Euler rotation scheme as follows:

$$C_1(\theta_1)C_2(\theta_2)C_1(\theta_3) \tag{10.15}$$

In order to be consistent with the notation of Chapter 4 and Appendix C, we always identify θ_3 as the first angle of rotation; in this case it is a rotation about the 1-axis. Our second rotation is through the angle θ_2; in this case it is a rotation about the new 2-axis. Finally, our third rotation is through the angle θ_1 and about the new 1-axis. Obviously, after the first and second rotations the 1- and 2-axes are no longer aligned with the original 1- and 2-axes. The rotation sequence is shown in Fig. 10.5 and defines a direction cosine matrix (DCM) from the inertial frame to the body frame. In our particular application we identify the three sequential angles as follows:

$$\theta_1 = \phi, \qquad \theta_2 = \theta, \qquad \theta_3 = \psi \tag{10.16}$$

Our next task is to relate the components of the angular velocity in the body frame to the derivatives of the Euler angles given above. Reference should again be made to the E_{121} entry in Table C.3 of Appendix C.

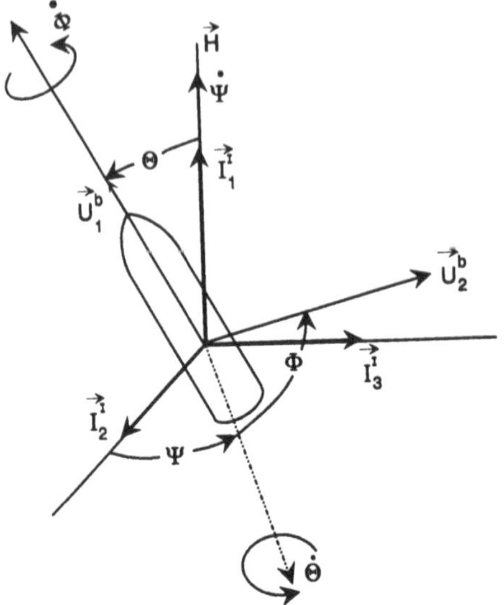

Fig. 10.5 Angular motion in the inertial frame.

$$\omega_1 = \dot{\phi} + \dot{\psi}\cos\theta = \omega_0 \tag{10.17a}$$

$$\omega_2 = \dot{\theta}\cos\phi + \dot{\psi}\sin\theta\sin\phi = \dot{\psi}\sin\theta\sin\phi = \Omega\sin\alpha t \tag{10.17b}$$

$$\omega_3 = -\dot{\theta}\sin\phi + \dot{\psi}\sin\theta\cos\phi = \dot{\psi}\sin\theta\cos\phi = \Omega\cos\alpha t \tag{10.17c}$$

Note that we have made use of the fact that θ must be a constant since all terms on the right-hand side of Eq. (10.14) are constants.

We can now square both sides of Eqs. (10.17b) and (10.17c) and add them together to get

$$(\dot{\psi})^2 \sin^2\theta(\sin^2\phi + \cos^2\phi) = \Omega^2(\sin^2\alpha t + \cos^2\alpha t)$$

which reduces to

$$\dot{\psi} = \Omega/\sin\theta \tag{10.18}$$

From Eq. (10.14) we get

$$\sin\theta = I_T\Omega/\left(I^2\omega_0^2 + I_T^2\Omega^2\right)^{1/2} \tag{10.19}$$

so Eq. (10.18) becomes

$$\begin{aligned}\dot{\psi} &= \left(I^2\omega_0^2 + I_T^2\Omega^2\right)^{1/2}/I_T \\ &= \omega_0\left[(I/I_T)^2 + (\Omega/\omega_0)^2\right]^{1/2}\end{aligned} \tag{10.20}$$

Up to now we have acquired four angular rates: ω_0, Ω, $\dot{\phi}$, and $\dot{\psi}$. Before proceeding, let's define each of these angular rates:

ω_0: Total spin rate; the projection of the angular rotation vector $\boldsymbol{\omega}^b_{b/I}$ along the axis of symmetry of the body.

Ω: The magnitude of the vector representing the projection of the angular velocity into a plane normal to the spin axis, or axis of symmetry. This vector rotates about the axis of symmetry at a rate of α rad/s.

$\dot{\psi}$: A measure of the rate at which the axis of symmetry precesses, or rotates about, the angular momentum vector. (Some engineers refer to $\dot{\psi}$ as the *nutation* rate. However, nutation is better referred to as a higher-frequency "nodding" motion.)

$\dot{\phi}$: The relative spin rate, or the spin rate of the body-fixed axis system relative to a frame that is precessing about $\boldsymbol{H}$.

We can now relate the precessional and relative spin rates by first dividing Eq. (10.17c) into Eq. (10.17b) to get

$$\tan\phi = \tan\alpha t$$

or

$$\phi = \alpha t, \qquad \frac{\mathrm{d}\phi}{\mathrm{d}t} = \alpha$$

From this relationship and from Eq. (10.11b) we have

$$\frac{\mathrm{d}\phi}{\mathrm{d}t} = \omega_0\left[1 - \left(\frac{I}{I_T}\right)\right]$$

We can write

$$\omega_0 = \left(\frac{I_T}{I_T - I}\right)\frac{\mathrm{d}\phi}{\mathrm{d}t}$$

and use this relationship to replace ω_0 in Eq. (10.17a) to get

$$\frac{\mathrm{d}\psi}{\mathrm{d}t} = \left(\frac{I}{I_T - I}\right)\left(\frac{1}{\cos\theta}\right)\frac{\mathrm{d}\phi}{\mathrm{d}t}$$

$$= \left[\frac{(I/I_T)}{1-(I/I_T)}\right]\left(\frac{1}{\cos\theta}\right)\frac{\mathrm{d}\phi}{\mathrm{d}t} \tag{10.21}$$

Typically, re-entry vehicles are slender configurations so as to meet the demands of re-entry thermal and pressure loads. Thus, we would nearly always expect I/I_T to be less than unity. Equation (10.21) indicates that under such circumstances the precession and relative spin rates are in the same direction; i.e., precession is "prograde." Many nonentry orbital vehicles are designed so that I/I_T is greater than unity; in this case the precession is retrograde, or in a direction opposite to that of relative spin.

Figure 10.6 illustrates the vector relationship between the angular momentum vector $\boldsymbol{H}$, the angular velocity vector $\boldsymbol{\omega}$, and the axis of symmetry for the case where $I_T > I$. From this figure and Eq. (10.18) we see that δ, the angle between the angular velocity vector and the axis of symmetry, is given by

$$\tan\delta = \frac{(\mathrm{d}\psi/\mathrm{d}t)\sin\theta}{\omega_0} = \frac{\Omega}{\omega_0} \tag{10.22}$$

Next, using Eq. (10.19) we get

$$\tan\theta = I_T\Omega/I\omega_0$$

or

$$\tan\delta = (I/I_T)\tan\theta \tag{10.23}$$

Obviously, for a slender body (i.e., $I/I_T < 1$) δ must be less than θ. We leave it as an exercise for the reader to show that

$$\frac{\mathrm{d}\psi}{\mathrm{d}t} = |\boldsymbol{\omega}|\left[\sin^2\delta + \left(\frac{I}{I_T}\right)^2\cos^2\delta\right]^{1/2} \tag{10.24}$$

We must point out that the angular momentum vector $\boldsymbol{H}$, the angular velocity vector $\boldsymbol{\omega}$, and the axis of symmetry remain in the same plane throughout the

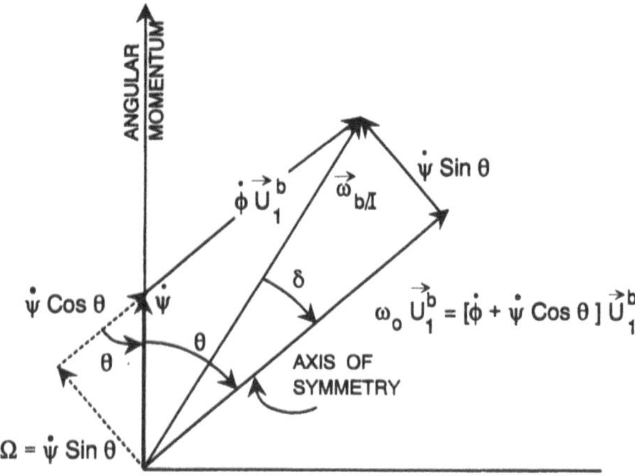

Fig. 10.6 Vector relationships for torque-free motion of an axisymmetric body.

precessional motion. The reader can demonstrate this coplanar requirement by showing that the following is true:

$$(\boldsymbol{H} \times \boldsymbol{U}_1) \cdot \boldsymbol{\omega}_{b/I} = 0 \tag{10.25}$$

(Note that the direction of the axis of symmetry is designated by the unit vector $\boldsymbol{U}_1$.)

Thus, we may summarize the motion of an axially symmetric re-entry vehicle in torque-free angular motion after release from the boost vehicle. It is assumed that the re-entry vehicle is spun about an axis that is nearly coincident with the minor axis of mass symmetry at a rate ω_0. If the transverse component of the angular velocity, Ω, is of order 0.1, i.e., O(0.1), the corresponding transverse component of angular momentum is O(1) because the transverse moment of inertia is usually an order of magnitude greater than that of the axial moment of inertia. The plane defined by the angular momentum vector, the angular velocity vector, and the axis of symmetry rotates about the angular momentum vector. This rotation is called *precession*; this precessional rate may be computed using Eq. (10.20). The angle between the axis of symmetry and the angular momentum vector, designated here as θ, is computed from Eq. (10.14); the angle δ between the angular velocity vector and the axis of symmetry is given by Eq. (10.23).

In the limiting case for which the angular velocity is entirely along the axis of symmetry $\boldsymbol{U}_1^b$, i.e., $\Omega = 0$, both θ and δ would be zero; the precessional rate would then be equal to the angular velocity rate multiplied by the ratio of the axial to transverse moments of inertia.

In the interesting case for which both θ, and hence δ, are nonzero, an observer traveling with the spinning re-entry vehicle during the Keplerian phase would observe the axis of symmetry of the re-entry vehicle rotating about some invisible line in space. This line would define the direction of the angular momentum vector. The observer would note the body spinning about its axis of symmetry at a rate maybe 5 to 10 times that of the precessional rate. [The relationship between the spin rate and the precessional rate is given in Eq. (10.20).]

If both the angular velocity and angular momentum vectors were visible, the angular motion of an axisymmetric prolate re-entry vehicle might best be described by two cones of coincident vertices, one fixed in space (the space cone) and the other, known as the body cone, rolling upon the space cone. The line of contact of these two cones would define the direction of the angular velocity vector. Figure 10.7 illustrates the three vectors and their relationship to the space and body cones. The angular velocity vector defines the axis of instantaneous rotation of the re-entry vehicle in inertial space. This axis of rotation precesses about the angular momentum vector, which is along the axis of symmetry of the space cone. Since the body cone rolls on the space cone without slipping, we can easily show that

$$\frac{\mathrm{d}\phi}{\mathrm{d}t} = \frac{\mathrm{d}\psi}{\mathrm{d}t}\left[\frac{\sin(\theta - \delta)}{\sin\delta}\right] \tag{10.26}$$

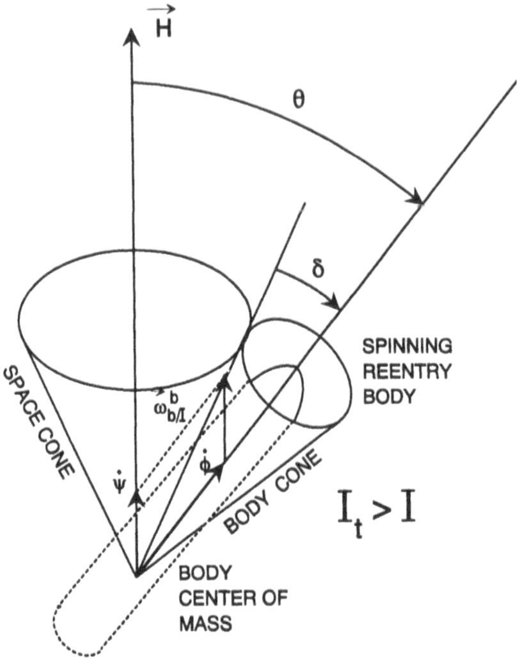

Fig. 10.7 Geometric representation of the prograde precession of a rigid body.

From Eqs. (10.19) and (10.23) we have

$$\tan\theta = (\Omega/\omega_0)(I/I_T)^{-1}, \qquad \tan\delta = (\Omega/\omega_0) \tag{10.27}$$

Since Ω/ω_0 and I/I_T are O(0.1), $\tan\delta$ is also O(0.1), but $\tan\theta$ is O(1). A small transverse angular velocity applied to the re-entry vehicle at the end of boost can result in an order of magnitude larger misalignment of the re-entry vehicle with respect to the angular momentum vector.

Obviously, it is desirable to control the angle of attack at entry for two separate but related reasons. Since the axis of symmetry is precessing about the vector $\boldsymbol{H}$, fixing the angular momentum in a direction parallel to the velocity vector at entry will essentially set the initial angle of attack to the angle θ. However, since the precessional rate depends to some extent upon the transverse components of the angular velocity vector, there is no way of controlling the orientation of the angle-of-attack plane at entry. Thus, uncertainty of the direction of the initial aerodynamic loads at entry could easily translate into unacceptably large impact point dispersion. The second concern is with the magnitude of the initial transverse loads associated with large initial angles of attack.

We identify two ways the re-entry vehicle might acquire angular velocity components transverse to the principal axis. One is by imparting spin to an axis which is slightly misaligned with respect to the principal axis. A second way, discussed in the following example, is for the re-entry vehicle to be subjected to a moment impulse during separation; the physical origins of this moment impulse are unimportant here but might arise, for instance, from the gas plume of an adjacent thruster.

Example

We wish to find a relationship between an unintentional moment impulse and the angle between the angular momentum and the axis of symmetry. We also wish to find the corresponding precessional period.

We assume that the re-entry vehicle has been ejected from the boost vehicle without transverse angular velocity. Consequently, the angular momentum vector $\boldsymbol{H}_0$ is assumed to be aligned with the re-entry vehicle's axis of symmetry. The re-entry vehicle is then given a transverse force $\boldsymbol{F}$ as shown in Fig. 10.8. Thus,

$$\boldsymbol{M} = \boldsymbol{F}l$$

where l is the couple separation. It is not important whether $\boldsymbol{F}$ is a concentrated force or is obtained from the integration of a pressure distribution. The net result is a couple of magnitude M. Let's assume that the force $\boldsymbol{F}$ and hence couple $\boldsymbol{M}$ act over a small interval of time, producing a moment impulse $\hat{\boldsymbol{M}}$ as follows:

$$\hat{\boldsymbol{M}} = \int_0^t \boldsymbol{M}\mathrm{d}t, \qquad \boldsymbol{M} = \frac{\mathrm{d}\hat{\boldsymbol{M}}}{\mathrm{d}t} \tag{10.28}$$

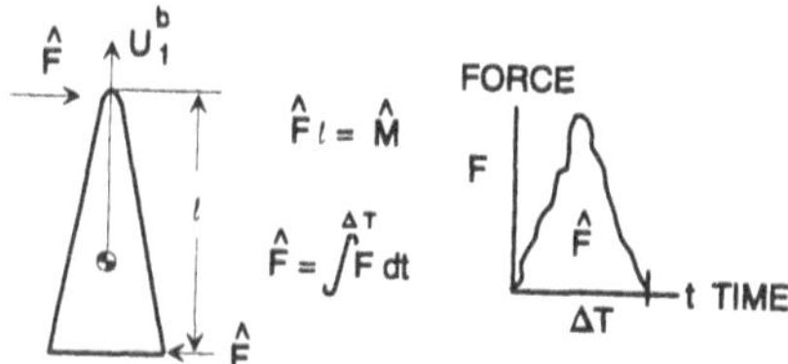

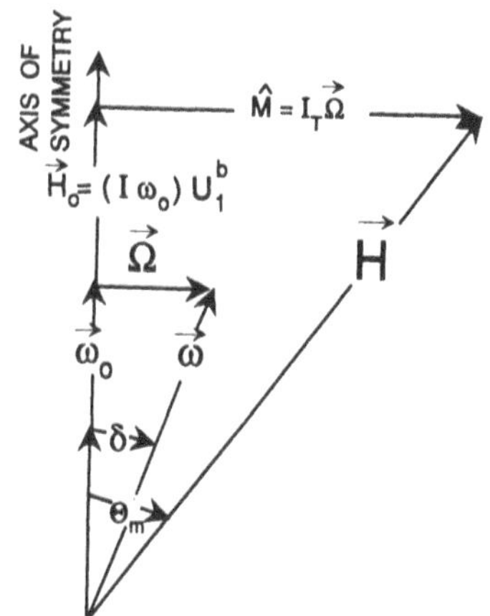

Fig. 10.8 Components of vectors *H* and *ω* before and after the application of a moment impulse.

Since

$$\frac{\mathrm{d}\boldsymbol{H}}{\mathrm{d}t} = \boldsymbol{M} = \frac{\mathrm{d}\hat{\boldsymbol{M}}}{\mathrm{d}t}$$

we have

$$\Delta\boldsymbol{H} = \hat{\boldsymbol{M}}$$

Thus, the angular momentum vector after the impulse is applied changes from its initial value of $\boldsymbol{H}_0$ as follows:

$$\boldsymbol{H} = \boldsymbol{H}_0 + \Delta\boldsymbol{H} = I\omega_0 + \hat{\boldsymbol{M}} \tag{10.29}$$

where we have assumed that ω_0, the angular velocity imparted to the re-entry vehicle at release, is entirely along the $\boldsymbol{U}_1^b$-axis and $\hat{\boldsymbol{M}}$ is normal to $\boldsymbol{U}_1^b$. From Fig. 10.8 it is clear that

$$\tan\theta = \frac{\hat{M}}{I\omega_0} = \tan\theta_m \tag{10.30}$$

where we have restricted the angle θ to a maximum of θ_m. Since the axial moment of inertia I and the angular velocity ω_0 are fixed, restricting the angle θ to an upper limit θ_m sets a tolerance level on the moment impulse $\boldsymbol{M}$.

The unintentional angular moment $\Delta\boldsymbol{H}$ can also be written as follows:

$$|\Delta\boldsymbol{H}| = \Omega I_T = (\omega_0\tan\delta)I_T = \hat{M} \tag{10.31}$$

where Ω is the magnitude of the transverse angular velocity. The precessional rate $\dot{\psi}$ follows from Eq. (10.20) as

$$\frac{\mathrm{d}\psi}{\mathrm{d}t} = \frac{[(I\omega_0)^2 + (I_T\Omega)^2]^{1/2}}{I_T} \tag{10.20}$$

Using Eq. (10.31) allows us to write

$$\frac{\mathrm{d}\psi}{\mathrm{d}t} = \left[\left(\frac{I}{I_T}\right)^2\omega_0^2 + \left(\frac{\hat{M}}{I_T}\right)^2\right]^{1/2} = \omega_0\left[\left(\frac{I}{I_T}\right)^2 + \left(\frac{I}{I_T}\right)^2\left(\frac{\hat{M}}{I\omega_0}\right)^2\right]^{1/2}$$

or, from Eq. (10.30),

$$\begin{aligned}\frac{\mathrm{d}\psi}{\mathrm{d}t} &= \omega_0\left(\frac{I}{I_T}\right)(1 + \tan^2\theta_m)^{1/2} \\ &= \omega_0\left(\frac{I}{I_T}\right)\sec\theta_m \end{aligned} \tag{10.32}$$

The precessional period T_p follows as

$$T_p = \frac{2\pi}{|\dot{\psi}|} = \left(\frac{2\pi}{\omega_0}\right)\left(\frac{I_T}{I}\right)\cos\theta_m \tag{10.33}$$

where

$$\theta_m = \tan^{-1}\left(\frac{\hat{M}}{I\omega_0}\right) \tag{10.34}$$

10.4 Torque-Free Motion—Poinsot Construction

As pointed out in the previous section, angular momentum is conserved in torque-free motion. Because there are no nonconservative forces acting, kinetic energy due to angular rotation is also conserved. First, let's examine the magnitude of the angular momentum vector. Angular momentum was discussed in Chapter 5. In matrix form the angular momentum and angular velocity are related as follows:

$$\boldsymbol{H}^b = I^b \boldsymbol{\omega}_{b/I}^b \tag{10.35}$$

The square of the magnitude of $\boldsymbol{H}^b$ is given by the inner product of the vector with itself, i.e.,

$$\begin{aligned} H^2 &= (\boldsymbol{H}^b)^{\mathrm{T}}\boldsymbol{H}^b \\ &= (I^b\boldsymbol{\omega}_{b/I}^b)^{\mathrm{T}}(I^b\boldsymbol{\omega}_{b/I}^b) \\ &= (\boldsymbol{\omega}_{b/I}^b)^{\mathrm{T}}(I^b)^{\mathrm{T}}I^b\boldsymbol{\omega}_{b/I}^b \end{aligned} \tag{10.36}$$

With

$$I^b = \begin{bmatrix} I_1 & 0 & 0 \\ 0 & I_2 & 0 \\ 0 & 0 & I_3 \end{bmatrix} \qquad \text{and} \qquad \boldsymbol{\omega}_{b/I}^b = \begin{bmatrix} \omega_1 \\ \omega_2 \\ \omega_3 \end{bmatrix}$$

Equation (10.36) becomes

$$H^2 = I_1^2\omega_1^2 + I_2^2\omega_2^2 + I_3^2\omega_3^2 \tag{10.37}$$

As we have pointed out, kinetic energy due to angular rotation must also be conserved. The kinetic energy of a rotating body is in principle equal to the sum of the kinetic energy E_i of its constituent particles, i.e.,

$$E_i \doteq \frac{1}{2}m_i \left.\frac{\mathrm{d}\boldsymbol{r}_i}{\mathrm{d}t}\right|_I \cdot \left.\frac{\mathrm{d}\boldsymbol{r}_i}{\mathrm{d}t}\right|_I \tag{10.38}$$

where $\mathrm{d}\boldsymbol{r}_i/\mathrm{d}t\,|_I$ is the velocity relative to inertial space of the vector from the center of mass of the body to the constituent particle (see Fig. 5.3). From Eq. (5.10), assuming $\boldsymbol{R}_0 = 0$, we have

$$\left.\frac{\mathrm{d}\boldsymbol{r}_i}{\mathrm{d}t}\right|_I = \left.\frac{\mathrm{d}\boldsymbol{r}_i}{\mathrm{d}t}\right|_m + \boldsymbol{\omega}_{b/I} \times \boldsymbol{r}_i = \boldsymbol{\omega}_{b/I} \times \boldsymbol{r}_i \tag{10.39}$$

where the first term on the right is identically zero because of the assumed body rigidity.

We may insert Eq. (10.39) into Eq. (10.38) and sum over all mass particles as follows:

$$\begin{aligned} E = \sum_i E_i &= \frac{1}{2}\sum_i m_i(\boldsymbol{\omega}_{b/I} \times \boldsymbol{r}_i) \cdot \left.\frac{\mathrm{d}\boldsymbol{r}_i}{\mathrm{d}t}\right|_I \\ &= \frac{1}{2}\sum_i \boldsymbol{\omega}_{b/I} \cdot \left(\boldsymbol{r}_i \times m_i \left.\frac{\mathrm{d}\boldsymbol{r}_i}{\mathrm{d}t}\right|_I\right) \\ &= \frac{1}{2}\boldsymbol{\omega}_{b/I} \cdot \sum_i \left(\boldsymbol{r}_i \times m_i \left.\frac{\mathrm{d}\boldsymbol{r}_i}{\mathrm{d}t}\right|_I\right) \end{aligned} \tag{10.40}$$

The bracketed term in Eq. (10.40) is the angular momentum $\boldsymbol{H}$, so

$$E = \tfrac{1}{2}\left(\boldsymbol{\omega}^b_{b/I}\right)^{\mathrm{T}} \boldsymbol{H}^b = \tfrac{1}{2}\boldsymbol{\omega}_{b/I} \cdot \boldsymbol{H} \tag{10.41}$$

The above expression becomes, after applying Eq. (10.35),

$$E = \tfrac{1}{2}\left(I_1\omega_1^2 + I_2\omega_2^2 + I_3\omega_3^2\right) \tag{10.42}$$

Equations (10.37) and (10.42) define ellipsoidal surfaces. Let's rewrite both equations to produce the H-ellipsoid, or the angular momentum ellipsoid, and the E-ellipsoid, or the energy ellipsoid.

$$\frac{\omega_1^2}{(H/I_1)^2} + \frac{\omega_2^2}{(H/I_2)^2} + \frac{\omega_3^2}{(H/I_3)^2} = 1 \tag{10.43a}$$

$$\frac{\omega_1^2}{(2E/I_1)} + \frac{\omega_2^2}{(2E/I_2)} + \frac{\omega_3^2}{(2E/I_3)} = 1 \tag{10.43b}$$

Since the vector $\boldsymbol{\omega}_{b/I}$ has been expressed in terms of body-frame components, the above ellipsoids are fixed in the body frame. The angular velocity vector must, for any motion for which H and E are fixed, lie simultaneously on the surfaces of both ellipsoids.

Thus, Eqs. (10.37) and (10.42) or Eqs. (10.43a) and (10.43b) provide two conditions that must be satisfied by the angular velocity: for any arbitrary torque-free motion, the angular velocity vector must lie on the surface of both ellipsoids at the same time.

Note that, if the ellipsoids were imagined fixed to the body, the major axes of both would correspond to the minor principal inertia axes of the body because of the inverse and inverse square-root relationships. In other words, the major axes of the H- and E-ellipsoids are given respectively for the 1, 2 , and 3 body axes as follows:

$$\frac{H}{I_1}, \frac{H}{I_2}, \frac{H}{I_3} \qquad \text{and} \qquad \left(\frac{2E}{I_1}\right)^{1/2}, \left(\frac{2E}{I_2}\right)^{1/2}, \left(\frac{2E}{I_3}\right)^{1/2}$$

The second observation must be that the H- and E-ellipsoids are not geometrically similar. The length of the major axes of each ellipsoid are set by the initial values of angular momentum H and energy E. Therefore, let us assume that the E-ellipsoid is interior to the H-ellipsoid but that the E-ellipsoid cuts through the H-ellipsoid, making a closed curve of intersection. This curve, known as the *polhode*, must be the locus of the angular velocity vector as it is observed from the body frame (see Fig. 10.9). To incorporate some of the work from the previous section, suppose we set the U_1^b-axis as that of nominal spin but acquire some transverse components of angular velocity. Thus, the angular velocity will not lie entirely along the U_1^b-axis, but would appear, in a body-fixed frame, to rotate about the U_1^b-axis.

The following discussion closely parallels the excellent treatment of torque-free motion given by Wiesel.[4] Suppose throughout the discussion the magnitude of the angular momentum vector is constant. This means that the H-ellipsoid will remain fixed in the body-frame. However, we will assume that the kinetic energy is decreased in fixed increments due to some unspecified mechanism.

Figure 10.10 shows the entire sequence of energy reduction moving from left to right, top to bottom. The maximum energy condition shows the polhode as a point located on the U_1^b-axis. Diminishing the rotational kinetic energy allows the H-ellipsoid to become "exposed," and the polhode curve then appears as a closed

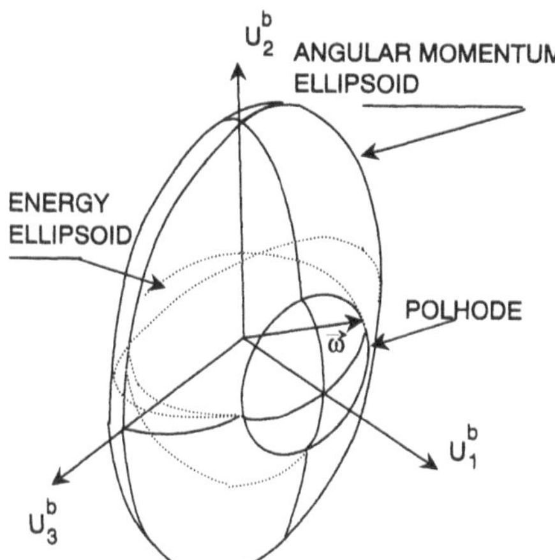

Fig. 10.9 Intersection of angular momentum and kinetic energy momentum ellipsoids. (From *Space Flight Dynamics* by W. E. Wiesel, McGraw-Hill Publisher.)

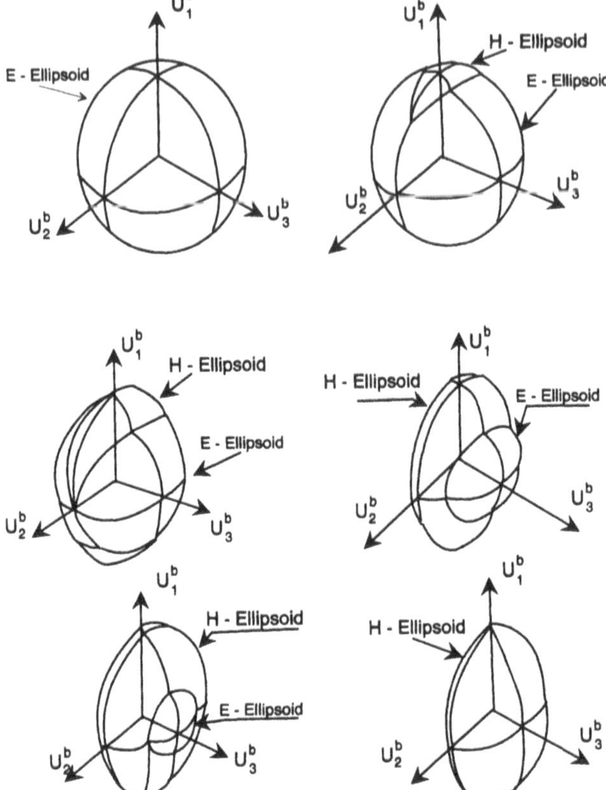

Fig. 10.10 Fixed angular momentum ellipsoid in the presence of a diminishing energy ellipsoid.

curve surrounding the U_1^b-axis. Of course, there are two polhode curves, since a reversal of the angular velocity would result in a second curve surrounding the U_1^b-axis in the negative direction.

Diminishing the energy still further enlarges the polhode curves until they meet at the U_2^b-axis. During this entire energy reduction process, more and more of the H-ellipsoid is exposed. Continual reduction of the rotational energy separates the two polhodes into two separate branches. Finally, at the minimum-energy condition the E-ellipsoid is entirely within the H-ellipsoid. The minimum energy condition shows the polhode as a point located on the U_3^b-axis.

Figure 10.11 is a summary of the various paths that are taken by the angular velocity vector, first for motion in the vicinity of the U_1^b-axis, then in the vicinity of the U_2^b-axis, and finally in the vicinity of the U_3^b-axis. Note that for angular rotation that is nearly along either the U_1^b and U_3^b (minor and major, respectively) the angular velocity is bounded. The term *bounded,* rather than *stable,* is more precise since the angular velocity vector remains in the vicinity of, rather than converging on, the spin axis. When the intermediate inertia axis is selected as

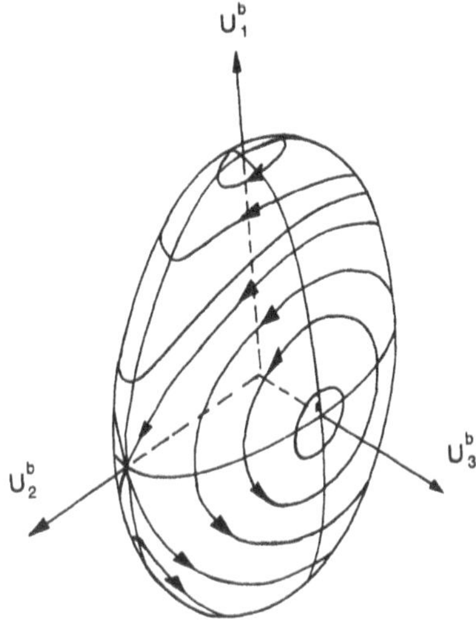

Fig. 10.11 Polhode path on the angular momentum ellipsoid.

the spin axis (in this case, the U_2^b-axis), there is a departure of the angular velocity vector from that axis.

We showed earlier that for the axisymmetric case, the angular velocity vector undergoes a coning motion about either of the stable axes. Thus, we would expect that the polhodes surrounding the two stable points are circular.

The preceding discussion centering around Figs. 10.10 and 10.11 was concerned with observations from a body-frame. Although we tracked the path of the angular velocity vector on the surface of the energy ellipsoid, we did not locate the angular momentum except for requiring it to remain on the H-ellipsoid.

Figure 10.12 shows an inertial frame with the angular momentum located vertically. Again, our discussion parallels the description given by Wiesel[4] and

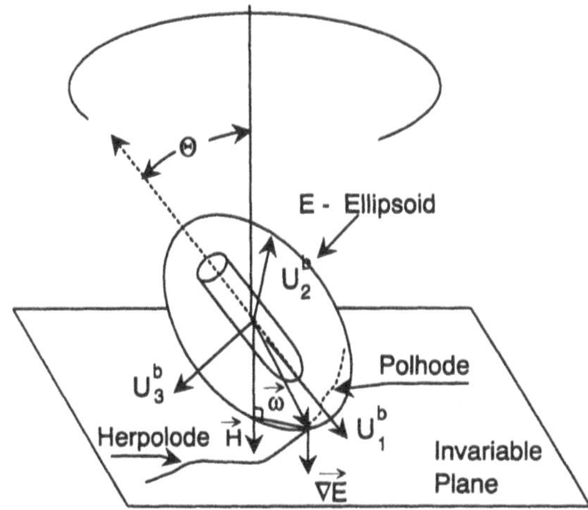

Fig. 10.12 Poinsot's construction.

to some degree Regan.[2] Kaplan[5] and Hughes[6] treat the Poinsot construction as well, with the most extensive treatment given by Hughes.

The component of the angular velocity vector along $\boldsymbol{H}$ (inertially fixed) follows from Eqs. (10.37) and (10.42) as

$$\frac{\boldsymbol{\omega}_{b/I} \cdot \boldsymbol{H}}{|\boldsymbol{H}|} = \frac{1}{|\boldsymbol{H}|}\left[I_1\omega_1^2 + I_2\omega_2^2 + I_3\omega_3^2\right] \tag{10.44}$$

Obviously, the component of $\boldsymbol{\omega}_{b/I}$ along $\boldsymbol{H}$ must be constant since both $\boldsymbol{E}$ and $\boldsymbol{H}$ are constant. Therefore, the tip of the angular velocity vector must at all times lie in a plane that is perpendicular to the angular momentum vector. This plane is called the invariable plane. The center of the E-ellipsoid is coincident with the center of mass of the re-entry vehicle. The normal to the ellipsoid is along the gradient of the E-ellipsoid, i.e.,

$$\begin{aligned}\nabla E &\doteq \left[\boldsymbol{U}_1^b \frac{\partial()}{\partial\omega_1} + \boldsymbol{U}_2^b \frac{\partial()}{\partial\omega_2} + \boldsymbol{U}_3^b \frac{\partial()}{\partial\omega_3}\right]\left(\frac{1}{2}I_1\omega_1^2 + \frac{1}{2}I_2\omega_2^2 + \frac{1}{2}I_3\omega_3^2\right) \\ &= (I_1\omega_1)\boldsymbol{U}_1^b + (I_2\omega_2)\boldsymbol{U}_2^b + (I_3\omega_3)\boldsymbol{U}_3^b \end{aligned} \tag{10.45}$$

Clearly, the gradient of the E-ellipsoid is the angular momentum vector $\boldsymbol{H}$. The E-ellipsoid must be tangent to the invariable plane at the point of contact; the point of contact locates the angular velocity vector, which in turn defines the instantaneous axis of body rotation.

Since the E-ellipsoid is fixed to the re-entry vehicle, any motion of this ellipsoid must be equivalent to body motion. Therefore, the motion of the re-entry vehicle as seen in an inertial frame is identical to that of the E-ellipsoid (with its origin at the center of mass of the re-entry vehicle) rolling without slipping on the invariable plane.

For an axisymmetric re-entry vehicle, the spin axis of the body generates a cone in inertial space. For a body with no axis of symmetry, the motion is more complex, and the E-ellipsoid does not have a circular cross section normal to the spin axis. Therefore, the rotation of the ellipsoid on the invariable plane causes the spin axis to undergo a "nodding," or nutational motion, in addition to the coning motion. This nutational motion takes place at twice the frequency of the precessional motion. This nutational motion may be understood by rolling an ellipse on a flat surface (invariable plane) in such a way that the path taken by the contact point on the plane is more or less circular. Because of the elliptical shape, the radius from the centroid of the ellipse to the contact points shrinks and grows twice while the ellipse traces out the circular (i.e., precessional) motion. Of course, if the ellipse reduces to a circle, the nutational motion vanishes.

10.5 Relative Motion

Wiesel[4] provides an interesting development of the Clohessy-Wiltshire equations, which allow us to analyze the relative motion of two re-entry vehicles which are launched in the same vicinity. Although Wiesel places more emphasis on orbital docking maneuvers, the same analysis can be applied to two re-entry

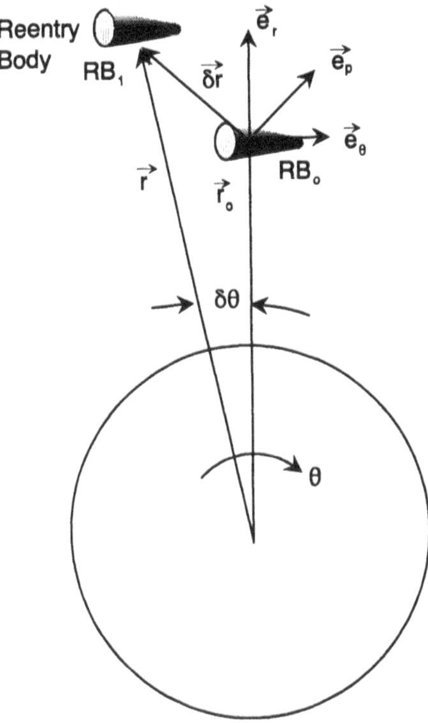

Fig. 10.13 Dynamics of two neighboring re-entry bodies.

bodies which are physically separated but traveling along nearly identical trajectories.

Consider two neighboring re-entry vehicles as depicted in Fig. 10.13. Let's take as a reference one re-entry vehicle at a radius of $\boldsymbol{r}_0$. We consider the second re-entry vehicle to have relative coordinates $(\delta r,\ r_0\delta\theta,\ \delta y)$ as follows:

$$\begin{aligned} \boldsymbol{r} &= (r_0 + \delta r)\boldsymbol{e}_r + (r_0\delta\theta)\boldsymbol{e}_\theta + (\delta y)\boldsymbol{e}_p \\ &= \boldsymbol{r}_0 + \delta\boldsymbol{r} \end{aligned} \tag{10.46a}$$

with relative velocity and acceleration given by

$$\left.\frac{\mathrm{d}\boldsymbol{r}}{\mathrm{d}t}\right|_e \doteq \delta\boldsymbol{v} = (\delta\dot{r})\boldsymbol{e}_r + (r_0\delta\dot{\theta})\boldsymbol{e}_\theta + (\delta\dot{y})\boldsymbol{e}_p \tag{10.46b}$$

$$\left.\frac{\mathrm{d}^2\boldsymbol{r}}{\mathrm{d}t^2}\right|_e \doteq \delta\boldsymbol{a} = (\delta\ddot{r})\boldsymbol{e}_r + (r_0\delta\ddot{\theta})\boldsymbol{e}_\theta + (\delta\ddot{y})\boldsymbol{e}_p \tag{10.46c}$$

The e-frame rotates about the Earth at an angular rate $|\boldsymbol{\omega}_{e/I}| = \mathrm{d}\theta/\mathrm{d}t$. From Eq. (6.10) we have

$$\boldsymbol{\omega}_{e/I} = \left(r_0 V_0 \cos\gamma_0 / r_0^2\right)\boldsymbol{e}_p \tag{10.47}$$

Thus, we may write the following for the acceleration relative to inertial space:

$$\left.\frac{d^2\boldsymbol{r}}{dt^2}\right|_I \doteq \boldsymbol{a}_g = \left.\frac{d^2\boldsymbol{r}}{dt^2}\right|_e + 2\boldsymbol{\omega}_{e/I} \times \left.\frac{d\boldsymbol{r}}{dt}\right|_e$$

$$+ \frac{d\boldsymbol{\omega}_{e/I}}{dt} \times \boldsymbol{r} + \boldsymbol{\omega}_{e/I} \times (\boldsymbol{\omega}_{e/I} \times \boldsymbol{r}) \tag{10.48}$$

For convenience, let's ignore the angular acceleration term, although for a noncircular orbit the angular velocity of the reference frame will not remain constant. Next, we note that

$$\boldsymbol{a}_g = -\frac{\mu \boldsymbol{r}}{r^3} = -\frac{\mu[(r_0 + \delta r)\boldsymbol{e}_r + r_0\delta\theta \boldsymbol{e}_\theta + \delta y \boldsymbol{e}_p]}{(r_0^2 + \delta r^2 + 2r_0\delta r + r_0^2\delta\theta^2 + \delta y^2)^{3/2}}$$

At this point we will define the term *neighborhood*. Only perturbational variables are retained: products of such variables (higher-order terms) are ignored. So two re-entry vehicles are in the same neighborhood if we can get useful results after we make the above higher-order simplifications. Thus, the above equation becomes, after expanding the denominator using the binomial theorem,

$$\boldsymbol{a}_g = -\left(\frac{\mu \boldsymbol{r}_0}{r_0^3}\right) - \frac{\mu}{r_0^3}[(-2\delta r)\boldsymbol{e}_r + (r_0\delta\theta)\boldsymbol{e}_\theta + \delta y \boldsymbol{e}_p] \tag{10.49}$$

Inserting Eqs. (10.46) and (10.49) into Eq. (10.48) and collecting terms gives

$$\delta\ddot{y} + \omega^2\delta y = 0 \tag{10.50a}$$

$$r_0\delta\ddot{\theta} + 2\omega\delta\dot{r} = 0 \tag{10.50b}$$

$$\delta\ddot{r} - 2\omega r_0\delta\dot{\theta} - 3\omega^2\delta r = 0 \tag{10.50c}$$

where $\omega^2 = \mu/r_0^3 = V_0^2\cos^2\gamma_0/r_0^2$. It is interesting that the out-of-plane coordinate δy appears in only one equation and is uncoupled from $r_0\delta\theta$ and δr.

Let's assume that the initial conditions for Eq. (10.50a) are given as

$$\delta y\,|_{t=0} \doteq \delta y_0, \qquad \delta\dot{y}\,|_{t=0} \doteq \delta\dot{y}_0$$

Then the solution of Eq. (10.50a) is that of harmonic motion, i.e.,

$$\delta y = \delta y_0 \cos\omega t + \frac{\delta\dot{y}_0}{\omega}\sin\omega t \tag{10.51}$$

Equation (10.50b) may be rewritten as

$$r\delta\ddot{\theta} + 2\omega\delta\dot{r} = 0 = \frac{d}{dt}(r_0\delta\dot{\theta} + 2\omega\delta r)$$

which integrates to give

$$\delta\dot{\theta} = \delta\dot{\theta}_0 + (2\omega/r_0)(\delta r_0 - \delta r) \tag{10.52}$$

For the initial conditions

$$\delta\dot{\theta}|_{t=0} \doteq \delta\dot{\theta}_0, \qquad \delta r|_{t=0} \doteq \delta r_0$$

The above equation, like Eq. (10.50a), represents harmonic motion, except this time with a constant term on the right. Thus, we must solve for the homogeneous and particular solutions (represented here by subscripts h and p, respectively), i.e.,

$$\delta r_h = A\cos\omega t + B\sin\omega t \qquad \delta r_p = 4\delta r_0 + (2r_0/\omega)\delta\dot{\theta}_0$$

Upon evaluating the constants A and B, the complete solution for the radial variable δr is

$$\delta r(t) = -\left(\frac{2}{\omega}r_0\delta\dot{\theta}_0 + 3\delta r_0\right)\cos\omega t + \frac{\delta\dot{r}_0}{\omega}\sin\omega t + \left(4\delta r_0 + \frac{2}{\omega}r_0\delta\dot{\theta}_0\right) \tag{10.53}$$

Next, we insert Eq. (10.53) into Eq. (10.52) to get

$$\delta\dot{\theta}(t) = -\left(3\delta\dot{\theta}_0 + \frac{6\omega}{r_0}\delta r_0\right) + \left(4\delta\dot{\theta}_0 + \frac{6\omega}{r_0}\delta r_0\right)\cos\omega t - \left(\frac{2}{r_0}\delta\dot{r}_0\right)\sin\omega t \tag{10.54a}$$

The above expression may then be integrated to give

$$\delta\theta(t) = -\delta\theta_0 - \left(3\delta\dot{\theta}_0 + \frac{6\omega}{r_0}\delta r_0\right)t + \left(\frac{4}{\omega}\delta\dot{\theta}_0 + \frac{6}{r_0}\delta r_0\right)\sin\omega t + \left(\frac{2}{\omega r_0}\delta\dot{r}_0\right)(\cos\omega t - 1) \tag{10.54b}$$

The relative position solutions are given by Eqs. (10.51), (10.53), and (10.54b). The relative velocity solutions follow from differentiating these equations. From Eqs. (10.50) we notice that six integrations must be performed to obtain six states. These states are separated into velocity states and positional states as follows:

$$\frac{\mathrm{d}\delta\boldsymbol{r}}{\mathrm{d}t} \doteq \delta\boldsymbol{v} = [\delta\dot{r}, r_0\delta\dot{\theta}, \delta\dot{y}]^{\mathrm{T}} \tag{10.55a}$$

$$\delta\boldsymbol{r} = [\delta r, r_0\delta\theta, \delta y]^{\mathrm{T}} \tag{10.55b}$$

In referring to the above states we have dropped the word *relative*.

Wiesel then writes the solution in the form of a matrix which transitions the states from their initial values forward in time as follows:

$$\begin{bmatrix} \delta r \\ r_0\delta\theta \\ \delta y \\ \delta\dot{r} \\ r_0\delta\dot{\theta} \\ \delta\dot{y} \end{bmatrix} = \begin{bmatrix} [\phi_{rr}]_{3\times 3} & [\phi_{rv}]_{3\times 3} \\ [\phi_{vr}]_{3\times 3} & [\phi_{vv}]_{3\times 3} \end{bmatrix} \begin{bmatrix} \delta r_0 \\ r_0\delta\theta_0 \\ \delta y_0 \\ \delta\dot{r}_0 \\ r_0\delta\dot{\theta}_0 \\ \delta\dot{y}_0 \end{bmatrix} \tag{10.56}$$

where the partitions of the transition matrix are as follows:

$$|\phi_{rr}|_{3\times 3} = \begin{bmatrix} 4 - 3\cos\omega t & 0 & 0 \\ 6(\sin\omega t - \omega t) & -1 & 0 \\ 0 & 0 & \cos\omega t \end{bmatrix} \tag{10.57a}$$

$$|\phi_{rv}|_{3\times 3} = \begin{bmatrix} \frac{1}{\omega}\sin\omega t & \frac{2}{\omega}(1 - \cos\omega t) & 0 \\ \frac{2}{\omega}(\cos\omega t - 1) & \frac{4}{\omega}\sin\omega t - 3t & 0 \\ 0 & 0 & \frac{1}{\omega}\sin\omega t \end{bmatrix} \tag{10.57b}$$

$$|\phi_{vr}|_{3\times 3} = \begin{bmatrix} 3\omega\sin\omega t & 0 & 0 \\ 6\omega(\cos\omega t - 1) & 0 & 0 \\ 0 & 0 & -\omega\sin\omega t \end{bmatrix} \tag{10.57c}$$

$$|\phi_{vv}|_{3\times 3} = \begin{bmatrix} \cos\omega t & 2\sin\omega t & 0 \\ -2\sin\omega t & -3 + 4\cos\omega t & 0 \\ 0 & 0 & \cos\omega t \end{bmatrix} \tag{10.57d}$$

We may now rewrite Eq. (10.56) in terms of positional and velocity states as

$$\delta \boldsymbol{r}(t) = \phi_{rr}\delta\boldsymbol{r}_0 + \phi_{rv}\delta\boldsymbol{v}_0 \tag{10.58a}$$

$$\delta \boldsymbol{v}(t) = \phi_{vr}\delta\boldsymbol{r}_0 + \phi_{vv}\delta\boldsymbol{v}_0 \tag{10.58b}$$

Wiesel indicates that the above equations may be applied to docking maneuvers between two orbiting spacecraft. However, we are concerned here primarily with suborbital re-entry vehicles for which the angle θ will have a maximum of about $\pi/2$. We should also note the presence of the time-proportional terms in

Eq. (10.54). This equation shows that the state $\boldsymbol{r}$ will increase in time unless the initial conditions are chosen such that

$$\delta\dot{\theta}_0 + (2\omega/r_0)\delta r_0 = 0 \tag{10.59}$$

Due to the short duration of the Keplerian phase, the drifting will not be a major concern. However, Eqs. (10.58) might find application to the following types of studies related to the Keplerian trajectory of a re-entry vehicle: 1) investigation of collisions between re-entry vehicles having nearly the same initial conditions; 2) setting initial conditions on an interceptor to impact a re-entry vehicle during the Keplerian phase; and (3) error analysis or the study of the effects of off-nominal initial conditions. Although the motivations behind types 1 and 2 are at opposite poles, the methods are certainly related: the first deals with avoiding a collision, and the second deals with causing one.

We might set $t = T$ as the time of interception, or collision, and require that

$$\delta\boldsymbol{r}(T) = 0$$

Thus, from Eq. (10.58a) we have

$$\delta\boldsymbol{v}_0 = -\phi_{rv}^{-1}(T)\phi_{rr}(T)\delta\boldsymbol{r}_0 \tag{10.60}$$

Thus, by knowing the separation between the two vehicles, i.e., $\delta\boldsymbol{r}_0 = \delta\boldsymbol{r}(t = 0)$, we may calculate the incremental velocity state required to effect a collision.

Equations (10.58) might have some use in performing an error analysis. Assume that a nominal trajectory is available. The relative positional/velocity state vector would then be the deviations from the nominal or intended trajectory acquired at boost termination (the initiation of the Keplerian trajectory). Consequently, the re-entry vehicle follows a trajectory somewhat different from that of the nominal trajectory. The analyst might then be able to ascertain the error state at re-entry or to carry the Keplerian trajectory to impact to assess the contribution of errors at boost termination on total error.

References

[1]Wertz, J. R. (ed.), *Astrophysics and Space Science Library–Vol. 73—Space Craft Attitude Determination and Control,* Computer Sciences Corporation, D. Reidel Publishing, Klumer-Boston, Hingham, MA, 1985, Chap. 16.

[2]Regan, F. J., *Re-Entry Vehicle Dynamics,* AIAA Education Series, AIAA, New York, 1984, Chap. 7.

[3]Greenwood, D. T., *Principles of Dynamics,* 2nd ed., Prentice-Hall, Englewood Cliffs, NJ, 1988.

[4]Wiesel, W. E., *Space Flight Dynamics,* McGraw-Hill, New York, 1989, Chap. 5.

[5]Kaplan, M. H., *Modern Space Craft Dynamics and Control,* Wiley, Somerset, NY, 1976, Chap. 2.

[6]Hughes, P. C., *Spacecraft Attitude Dynamics,* Wiley, Somerset, NY, 1986.

11
Flowfield Description

11.1 Introduction

In Chapter 7 it was seen that a re-entry body encounters severe inertial and thermal loads created by its motion through the atmosphere. In fact, at low altitudes aerodynamic forces dominate the trajectory to the extent that gravitational forces may be neglected in the early analysis and design phases.

The forces (integrated pressure and shear stresses) and moments acting on the re-entry body govern its linear and angular motion. While one is accustomed to treating the atmosphere as a continuum, the rarefied upper atmosphere cannot be treated as such, and the effect of individual molecules on the re-entry body must be considered. At lower altitudes, even though the atmosphere may be considered a continuum, the temperature distribution in the flowfield is closely coupled to the pressure and viscous stress distributions, significantly complicating an aerothermodynamicist's problem. In addition, at high temperatures air can no longer be considered a perfect gas, and chemical reactions between the constituents of the atmosphere and the re-entry body surface material must be taken into account.

Figure 11.1 shows some of the key events in a typical re-entry trajectory. Above the *initial pitch over* altitude, the low atmospheric density results in the aerodynamic forces being dominated by the gyrodynamic forces. But even though the aerodynamic forces are small, we have seen in Chapter 5 that even small initial errors grow to become quite significant at impact. Below the *first resonance* altitude, the aerodynamic forces, discussed in Sections 11.7 and 11.8, overwhelm the gyrodynamic forces.

Therefore, to accurately predict a re-entry body's trajectory, it is crucial to understand its aerodynamics throughout the flight envelope. Also, the severe thermal environment significantly alters the structural, material, and dielectric properties of the substructure, heatshield, and radome/sensor windows.

Any attempt to cover in detail the various aspects of hypersonics and high-temperature gasdynamics will completely overwhelm the discussion of dynamics—the main focus of this book. In this chapter we will first briefly discuss the environment a re-entry body can expect to encounter. Then an introduction to the effects of the environment is given. Finally, simplified force and moment determination methods are provided. For a more in-depth and detailed understanding of re-entry aerothermodynamics, classic[1–3] and modern[4,5] texts are available.

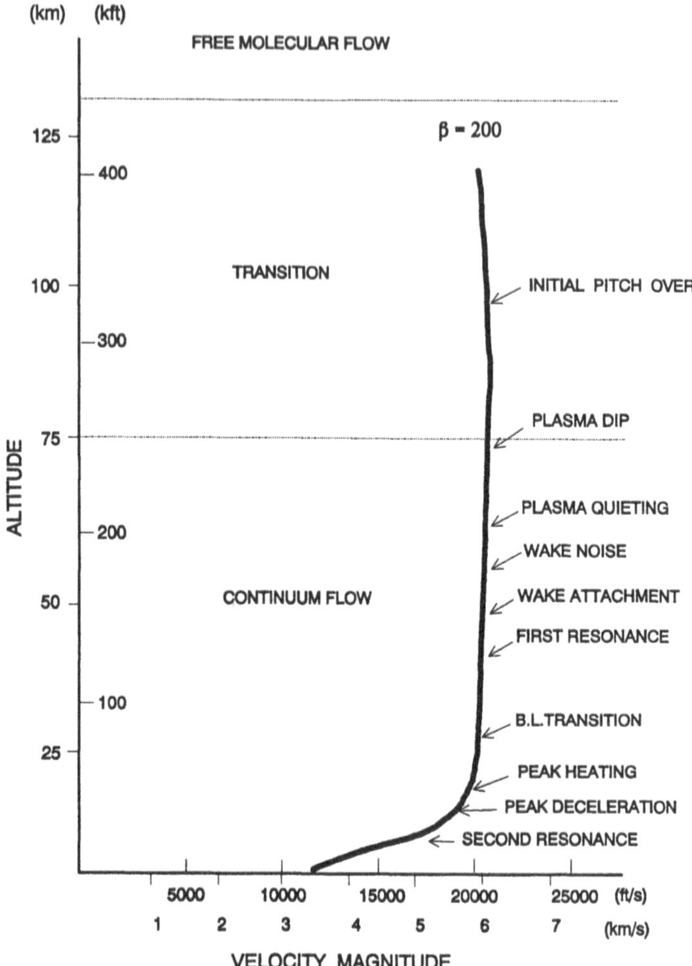

Fig. 11.1 Flow regimes with typical re-entry events.

11.2 Flowfield Determination

To determine the trajectory and attitude of a re-entry body, a flight dynamicist needs to know the forces and moments acting on the re-entry body. Three techniques are available to determine the aerothermodynamics of a re-entry body: 1) flight testing, 2) wind-tunnel testing, and 3) analytic or numerical solution of equations governing fluid flow.

Flight testing is the most accurate, but obviously also the most expensive, method of determining the aerodynamics. Since a boost and deployment vehicle would be required to flight-test a re-entry body, flight testing is not utilized until the late stages of development of the entire system.

Wind-tunnel testing, on the other hand, is less expensive and less restrictive than flight testing, but still quite expensive, labor-intensive, and time-consuming. Since few wind tunnels can accept full-scale models for testing, scale models are tested by matching nondimensional parameters (Mach number, Reynolds number, etc.) expected during flight. Unfortunately, the facilities that can accurately simulate the important re-entry conditions are limited in number and cover only a small portion of the re-entry flight envelope.

The most attractive option in the initial analysis and design phase is to determine the flowfield by solving the governing equations. Unfortunately, the governing equations are nonlinear partial differential equations, which have frustrated most attempts to obtain analytical solutions except in the simplest of cases. With the advent of digital computers, it is possible to solve the governing equations numerically. This has spawned a new field called *computational fluid dynamics* (CFD), but even with state-of-the-art computers and numerical schemes, CFD solutions are not sophisticated enough to render wind tunnels obsolete.

11.3 Fluid Flow Governing Equations

All fluid flowfields are governed by the following three conservation laws:

1) Conservation of mass: Continuity equation
2) Conservation of momentum: Newton's Second Law of Motion
3) Conservation of energy: First Law of Thermodynamics

These conservation laws in conjunction with a *definition of the fluid* provide the system of equations required to determine the flowfield for a given set of *boundary conditions*.

If the atmosphere around the re-entry body is modeled as a continuum, the conservation equations lead to a system of equations based on the bulk properties of air known as the *Navier-Stokes equations*. (Strictly speaking, the Navier-Stokes equations refer only to the momentum conservation equation but frequently, and here, include the continuity and energy equations.) On the other hand, a rarefied flowfield is described by the *Boltzmann equation*, which is based on molecular mechanics.

11.4 Definition of Fluid: Microscopic and Macroscopic Structure of Gases

At the macroscopic level, various thermodynamic properties of a *real gas* may be related by Van der Waals' equation of state

$$\left(p + a/v^2\right)(v - b) = RT \tag{11.1}$$

where p is the pressure, T the temperature, v the specific volume, and R the gas constant. The constants a and b are particular to a gas and account for intermolecular forces (significant only at very high pressures and low temperatures) and the volume of void between the particles in the gas, respectively. But for most aerodynamic applications, a and b are negligible and Eq. (11.1) reduces to the more familiar *perfect gas* equation of state

$$p = \rho RT \tag{11.2}$$

where the density $\rho = 1/v$. Equation (11.2) was first introduced in Chapter 2 as Eq. (2.8) in the modeling of the properties of the atmosphere.

Gases at a microscopic level consist of a large number of particles (molecules, atoms, electrons, etc.) that are in constant random motion, colliding with each other and with the confining boundaries. Macroscopic properties such as temperature, velocity, etc., are actually averages of all particles in a certain volume or over an interval of time. The particles of a flowing gas can be considered to have a random velocity distribution superimposed on the macroscopic flow velocity.

Owing to the random motion and collisions, the state of the particles changes discontinuously and frequently. Therefore, a particle introduced into a volume of gas at equilibrium will be indistinguishable from the original particles after a sufficient number of collisions. The time required to render the introduced particle indistinguishable from the original particles is called the *relaxation time* τ.

If a flowfield is not in thermodynamic equilibrium, i.e., the macroscopic properties vary spatially, the *intermolecular* and *molecule-surface* collisions result in momentum, energy, and mass transport within the gas and at the bounding surfaces. The associated transport properties are described by coefficients of viscosity, thermal conductivity, and mass diffusivity and are related to gradients in velocity, temperature, and mass concentration, respectively.

For a gas at rest, by relating classical thermodynamics to the kinetic theory of gases (accounting for the particle nature of gases), it can be shown that

$$\frac{p}{\rho} = \frac{1}{3}\frac{\sum mC^2}{\sum m} = \frac{2}{3}e_{\text{tr}} = RT \tag{11.3}$$

where m is the particle mass, C the particle velocity, and e_{tr} the specific translational energy of the gas. This equation shows that temperature is a measure of the molecular kinetic energy. The secondary thermodynamic properties may also be determined from kinetic theory. By definition, the specific heats at constant volume c_v and at constant pressure c_p are

$$\begin{aligned} c_v &\equiv (\partial e/\partial T)_v \\ c_p &\equiv (\partial h/\partial T)_p \end{aligned} \tag{11.4}$$

where e is the specific internal energy and h is the specific enthalpy.

The total specific internal energy of the gas is the sum of the translational, rotational, vibrational, electronic, and the zero-state energies. At moderate temperatures only the translational and rotational modes, if any, of a molecule are excited. The specific heats under such conditions are constant, and the gas is considered to be *calorically* perfect. Classical statistical mechanics relates the specific heats to the number of active degrees of freedom n_f as follows:

$$c_p = \frac{n_f + 2}{2}R \qquad \text{and} \qquad c_p - c_v = R \tag{11.5}$$

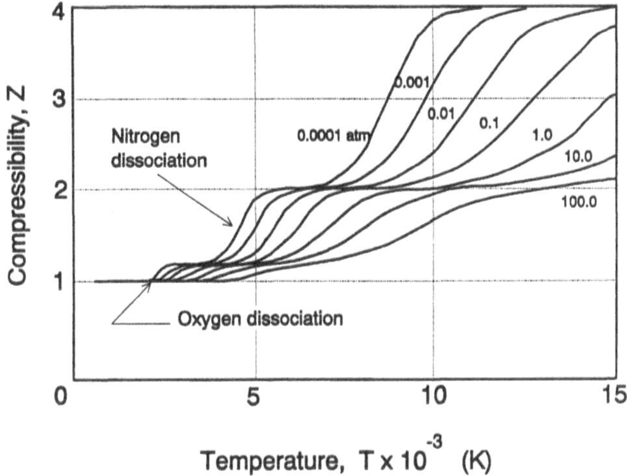

Fig. 11.2 Compressibility of air as a function of temperature.

For a diatomic gas $n_f = 5$ (three translational and two rotational degrees of freedom), and near room temperature, air (consisting primarily of N_2 and O_2) may be considered a diatomic gas such that

$$c_v = \frac{5}{2}R, \qquad c_p = \frac{7}{2}R, \qquad \gamma \equiv \frac{c_p}{c_v} = \frac{7}{5} \tag{11.6}$$

At temperatures where the vibrational modes of the molecules are excited (above 600 K for air), the specific heats are no longer constant. If c_p, c_v, and γ vary with temperature alone, the gas is considered to be *thermally* perfect (the term *perfect gas* will be used to refer to a calorically *and* thermally perfect gas).

As the temperature is increased further, the constituent gases begin to dissociate (~ 3000 K) and then ionize (~ 10,000 K). The concentration of resultant monatomic species, ions and electrons will depend upon temperature as well as pressure (or density) and so may not be considered to behave even as a thermally perfect gas.

This deviation from the perfect gas law [Eq. (11.2)] can be quantified by the *compressibility factor* $Z = P/(\rho RT)$ and is variously referred to as *real-gas effects* or *high-temperature effects*. Figure 11.2 illustrates the variability of Z (for air) as a function of temperature and pressure as calculated by Hansen.[6]

11.5 Flow Regimes

Consider a flowfield where the number density of particles and hence the intermolecular collision frequency is high. The relaxation time for a particle moving through such a flowfield will be short, and the associated relaxation distance small. If this relaxation distance is small compared to the *characteristic* dimension of the flowfield, the macroscopic properties can be considered to vary continuously, and the gas can be treated as a *continuum*.

In contrast, a *rarefied* flowfield will have a long relaxation time and hence a longer relaxation distance. If this relaxation distance is not small compared to the characteristic flowfield dimension, the state of the individual particles and their interaction with other particles and the boundaries must be considered.

One measure of the relaxation distance in a gas is the *mean free path* between collisions, λ, defined in Chapter 2 as follows:

$$\lambda = 1/\sqrt{2}\pi\sigma^2 n \tag{11.7}$$

The mean free path depends only upon the number density of the gas, n, which is a function of the re-entry body's altitude. Though the effective diameter of the gas particles, σ, depends upon the temperature of the gas, the variation of σ is not as significant a factor as the density. Note that the mean free path is defined in a reference frame moving with the flow, with a velocity equal to the macroscopic velocity of the gas.

To help us quantify the relative importance of treating the fluid as a collection of particles vis-à-vis a continuum, we use a nondimensional parameter called the *Knudsen number*, *Kn*, defined as

$$Kn = \frac{\lambda}{L} = \frac{\text{mean free path}}{\text{characteristic flowfield dimension}} \tag{11.8}$$

A high *Kn* indicates the importance of the particulate nature of the fluid and that the Boltzmann equation must be employed, whereas a low *Kn* permits treatment of the fluid as a continuum and the use of the Navier-Stokes equations. Interestingly, it has been shown that for flows for which $Kn \rightarrow \infty$, the Boltzmann equation solution converges asymptotically to the Navier-Stokes equation solution.

When the density is low, i.e., at high altitudes, the mean free path is large (Knudsen number is high), and the flowfield cannot be treated as a continuum. Even though it is of less importance to a flight dynamicist, the Knudsen number can be high when the characteristic dimension is small, e.g., flow through a shock wave where the characteristic dimension would be the thickness of the shock wave.

At this point it may appear that *Kn* is a function of only the re-entry body's altitude; however the characteristic dimension L is not the same at all altitudes. In fact, under certain conditions, the characteristic dimension, or length, is a function of the flow velocity.

The characteristic length could be a dimension of a flowfield characteristic or of the body. For a blunt cone the base radius may be appropriate. For a lifting body the mean aerodynamic chord of the lifting surface may be the correct choice. On the other hand, for flows past flat plates, the boundary layer thickness or the distance from the leading edge may be the appropriate characteristic dimension.

The Knudsen number can also be defined in terms of two nondimensional parameters, namely the *Mach number* and *Reynolds number*, which quantify macroscopic properties of the flow. The Mach number M is a normalized flow speed where the normalizing factor is the speed of sound, a measure of the compressibility of the fluid, i.e.,

$$M = \frac{V}{a} = \frac{\text{macroscopic speed of gas}}{\text{speed of sound in the gas}} \tag{11.9}$$

where $a = \sqrt{\gamma RT}$ (for a perfect gas). The Reynolds number is the ratio of inertia to viscous forces in the fluid, i.e.,

$$Re = \rho VL/\mu \tag{11.10}$$

where μ is the coefficient of viscosity of the fluid and L is the characteristic dimension of the flowfield. From kinetic theory,[7] μ may be defined in terms of the mean free path λ as follows:

$$\mu = \tfrac{1}{2}\rho\lambda a\sqrt{8/\pi\gamma} \tag{11.11}$$

By substituting Eqs. (11.9), (11.10), and (11.11) into Eq. (11.8) we get

$$Kn = 1.25\sqrt{\gamma}M/(Re) \tag{11.12}$$

In high Reynolds number ($Re \gg 1$) flows, the characteristic dimension of importance to viscous forces is the boundary layer thickness δ, which is proportional to $1/(Re)^{1/2}$ for laminar boundary layers. Thus, Eq. (11.12) becomes

$$Kn \propto \left(M/\sqrt{Re}\right) \tag{11.13}$$

Both Eqs. (11.12) and (11.13) indicate that the particulate nature of the gas is important if the Mach number is high and/or the Reynolds number is low.

To get an appreciation for the magnitude of the Knudsen number as a function of altitude, we shall consider a blunt-nosed cone re-entering the atmosphere. Since the cone base area is often used to define the force and moment coefficients, we choose the base radius r_b as the characteristic dimension in the Knudsen number. The number density n can be expressed in terms of ρ and the gas particle mass m. Using the exponential atmosphere model for ρ [Eq. (2.35)], Eq. (11.8) becomes

$$Kn = \frac{m/\left(\sqrt{2}\pi\sigma^2\rho_0 e^{-h/H}\right)}{r_b} \tag{11.14}$$

The particle mass can be determined from the molecular weight $\hat{M}$ and Avogadro's number Na

$$m = \hat{M}/Na \tag{11.15}$$

Assuming a constant value of $\hat{M} = 28.9$ kg/kmole (Fig. 2.3 shows this assumption of constant $\hat{M}$ to be valid only to a 100 km altitude) and $\sigma = 3.7 \times 10^{-10}$ m, the Knudsen number for $r_b = 1$ m is shown as a function of altitude in Fig. 11.3a. Figure 11.3b shows lines of constant Knudsen number as a function of Mach number and Reynolds number, defined in Eqs. (11.12) and (11.13).

Based on the relative importance of the particulate nature of air, quantified by the Knudsen number, a re-entry trajectory can be divided into three major regimes:

1) Free molecular flow ($Kn \gg 1$): $M \,/\, Re > 3$

2) Transition (from free molecular to continuum flow) regime: $3 > M/Re$; and $M/(Re)^{1/2} > 0.01$ ($Re \gg 1$)

3) Continuum flow ($Kn \ll 1$): $0.01 > M/(Re)^{1/2}$

Free molecular flow is dominated by molecule-surface interaction with negligible interaction between incident and reflected particles. Continuum flow, on the other hand, is dominated by intermolecular collisions. In the transition regime between these two extremes, both intermolecular and molecule-surface collisions are important.

As a re-entry body moves from one flow regime to another, flow characteristics change continuously and therefore there are no discrete Knudsen numbers demarcating the various flow regimes; experience and broad rules of thumb guide us in defining the regimes. It is generally accepted that for $M/Re > 3$ the flow is in the free molecular regime. $M/\sqrt{Re} < .01$ means that the continuum fluid dynamic phenomena are applicable.

Another ambiguity is introduced by the fact that the density and hence the mean free path in a flowfield around a re-entry body can vary quite widely. Therefore, the freestream Knudsen number, based on the freestream mean free path λ_∞, may not be the most judicious choice for determining the flow regime. A more

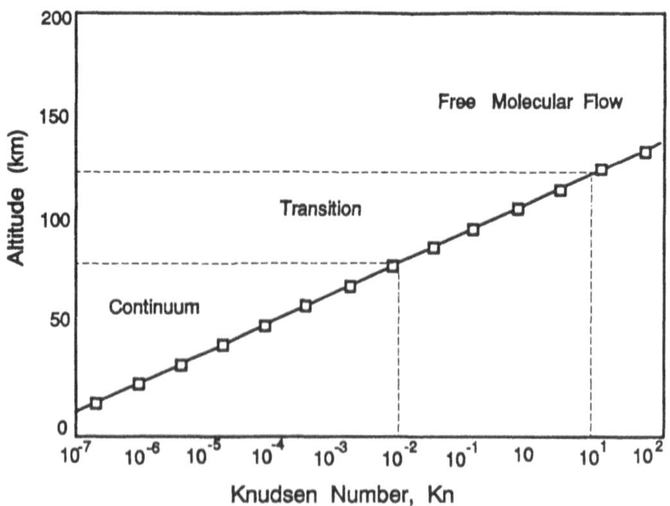

Fig. 11.3a Knudsen number vs altitude (flow regimes based on base area).

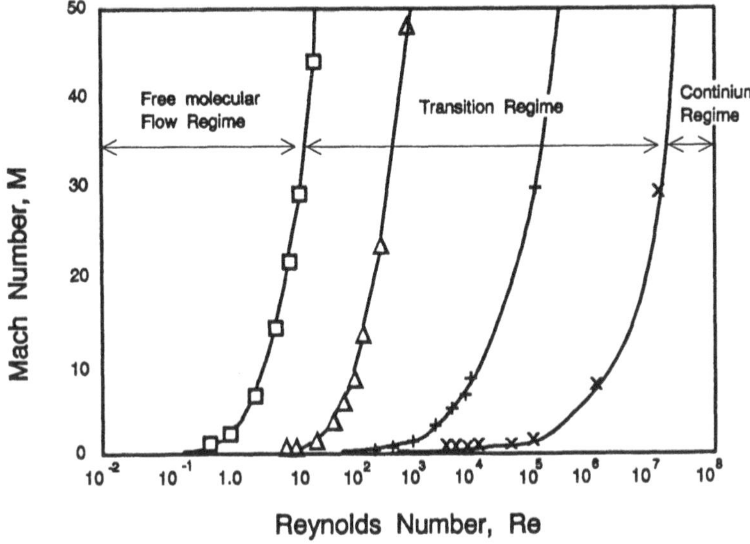

Fig. 11.3b Lines of constant Knudsen number (flow regimes based on Mach number and Reynolds number).

appropriate Knudsen number is based on a corrected mean free path λ_w that accounts for the state of particles emitted from the body,[8] i.e.,

$$\lambda_w = \frac{4}{\sqrt{\pi\gamma}}\left(\frac{T_w}{T_\infty}\right)^{1/2}\frac{\lambda_\infty}{M_\infty} \tag{11.16}$$

where T_w is the wall temperature, T_∞ is the free stream temperature, and M_∞ is the freestream Mach number.

Bird[9] proposes a broad rule of thumb (see Fig. 11.4) for determining the flow regime based on the *local* (as opposed to the freestream) Knudsen number.

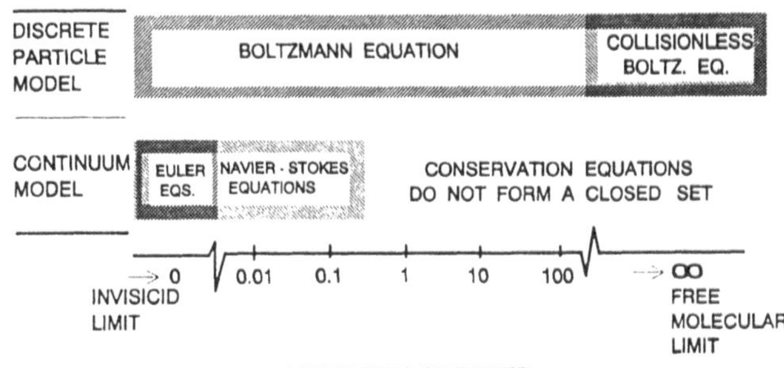

Fig. 11.4 Validity of conventional mathematical models as a function of local Knudsen number (from Bird[9]).

11.6 Free Molecular Flow Regime: $Kn \gg 1$

As a body re-enters the upper atmosphere, the mean free path will be much larger than the characteristic dimension of the body, and most particles reflecting from its surface will be far downstream before interacting with another particle. Therefore, the incident flow is undisturbed by the re-entry body and is often referred to as a *collisionless flow*. In other words, the flowfield in the vicinity of the re-entry body will be dominated by particle-surface collisions, and the number of collisions between incident and reflected particles will be negligible.

Since particle-surface interactions are the mechanism for momentum and energy transfer to the body, an undisturbed incident stream of particles should simplify the determination of the forces, moments, and energy transfer. Unfortunately, molecule-surface interactions are not very well understood. In spite of extensive study, the most useful models are still the *specular* and *diffuse* reflection models postulated by Maxwell in 1879.

The specular reflection model assumes that the particles are perfectly elastic and that they reflect off a solid surface with the tangential velocity unchanged while the normal velocity is reversed (see Fig. 11.5). This model predicts only normal momentum transfer and no tangential momentum or energy transfer and is therefore greatly oversimplified, but it is still a useful reference point.

The diffuse reflection model, by contrast, assumes that the velocity of the reflected particles is independent of the incident velocity. The particle is reflected with a Maxwellian distribution corresponding to the temperature of the reflected molecule. In other words, the reflected particles would have the same velocity distribution as if emanating from an imaginary gas behind the reflecting surface (the imaginary gas having no relative macroscopic velocity with respect to the surface).

The temperature T_r of the reflected particle may be different from either the wall temperature T_w or the incident gas temperature T_i. The energy transfer

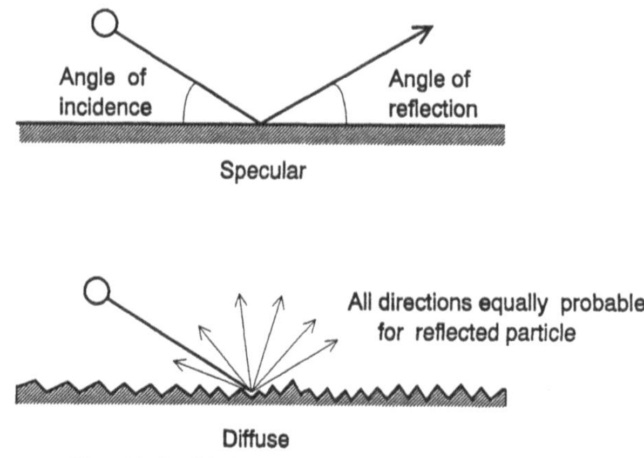

Fig. 11.5 Molecule-surface interaction models.

due to particle-surface interaction is defined by the *thermal accommodation coefficient* a_c, in terms of the energy fluxes dE as follows:

$$a_c = \frac{\mathrm{d}E_i - \mathrm{d}E_r}{\mathrm{d}E_i - \mathrm{d}E_w} \tag{11.17}$$

where dE_i and dE_r are the energy fluxes incident on and reflected from the surface. The term dE_W represents the energy flux if the emitted particles had a Maxwellian distribution corresponding to T_w.

Along the lines of a_c, a *normal momentum accommodation coefficient* σ_N and a *tangential momentum accommodation coefficient* σ_T can also be defined as follows:

$$\begin{aligned} \sigma_N &\equiv (p_i - p_r)/(p_i - p_w) \\ \sigma_T &\equiv (\tau_i - \tau_r)/\tau_i \end{aligned} \tag{11.18}$$

For diffuse reflection $a_c = \sigma_N = \sigma_T = 1$, and for specular reflection $a_c = \sigma_N = \sigma_T = 0$. Experiments with carefully prepared surfaces have shown a_c and σ_N to be significantly lower than 1. For interaction of air with most engineering surfaces, experimental data, indicate that $a_c \approx \sigma_N \approx 1$. It is expected that $\sigma_T \approx 1$ as well.

If one assumes a Maxwellian velocity distribution, the momentum and energy flux at a surface may be determined. It can be shown[10] that the pressure and shear stress acting on an elemental area inclined to the freestream at an angle θ are

$$\begin{aligned} \frac{\Delta p}{q_\infty}s^2 = &\left[\frac{(2-\sigma_N)}{\sqrt{\pi}}s\sin\theta + \frac{\sigma_N}{2}\left(\frac{T_w}{T_\infty}\right)^{1/2}\right]\exp(-s^2\sin^2\theta) \\ &+\left[(2-\sigma_N)\left(\tfrac{1}{2}+s^2\sin^2\theta\right)\right. \\ &\left.+\frac{\sigma_N}{2}\left(\frac{T_w}{T_\infty}\right)^{1/2}\sqrt{\pi}s\sin\theta\right][1+\operatorname{erf}(s\sin\theta)] \end{aligned} \tag{11.19}$$

$$\frac{\Delta\tau}{q_\infty}s^2 = \frac{\sigma_T s\cos\theta}{\sqrt{\pi}}\left\{\exp(-s^2\sin^2\theta) + \sqrt{\pi}s\sin\theta\,[1+\operatorname{erf}(s\sin\theta)]\right\} \tag{11.20}$$

The heat transfer q at the surface is

$$\begin{aligned} \frac{\beta_\infty^3 q}{\rho_\infty} = &\left[\left(s^2 + \frac{\gamma}{\gamma-1} - \frac{\gamma+1}{2(\gamma-1)}\frac{T_w}{T_\infty}\right)\left\{\exp(-s^2\sin^2\theta)\right.\right. \\ &\left.+\sqrt{\pi}s\sin\theta\,[1+\operatorname{erf}(s\sin\theta)]\right\} \\ &\left.-\tfrac{1}{2}\exp(-s^2\sin^2\theta)\right]\left[a_c/\left(4\sqrt{\pi}\right)\right] \end{aligned} \tag{11.21}$$

where the freestream *molecular speed ratio* s_∞ is analogous to a Mach number and the parameter β_∞ is analogous to an inverse continuum speed of sound, i.e.,

$$s_\infty = V_\infty \beta_\infty = V_\infty / \sqrt{2RT_\infty} \tag{11.22}$$

where the subscript ∞ refers to freestream conditions.

It is apparent from Eqs. (11.19), (11.20), and (11.21) that the skin friction depends only on the incident stream and the reflection characteristics, whereas the pressure depends on the wall temperature, which in turn is a function of a_c.

At this time, let us make note of a limiting case that will be of interest when the continuum regime is discussed—a cold surface ($T_w/T_\infty \approx 1$) with a very high speed ratio ($V_\infty \rightarrow \infty$). The pressure coefficient C_p and the coefficient of shear C_f for windward surfaces ($\sin\theta > 0$) are

$$C_p \equiv \frac{\Delta p}{q_\infty} \rightarrow 2(2 - \sigma_N)\sin^2\theta$$
$$C_f \equiv \frac{\Delta \tau}{q_\infty} \rightarrow 2\sigma_T \cos\theta \sin\theta \tag{11.23a}$$

and for leeward surfaces ($\sin\theta < 0$) are

$$C_p \rightarrow 0 \qquad \text{and} \qquad C_f \rightarrow 0 \tag{11.23b}$$

Note that $q_\infty \equiv \frac{1}{2}\rho_\infty V_\infty^2 = \frac{1}{2}\gamma M_\infty^2 p_\infty = s_\infty^2 p_\infty$.

For simple shapes, Eqs. (11.19) and (11.20) can be integrated over the surface to determine the forces acting on the body. Schaaf and Chambre[10] list references where aerodynamic forces have been calculated. Table 11.1 lists the coefficient of drag C_D in terms of the normal and tangential momentum accommodation coefficients (σ_N and σ_T, respectively) for some simple shapes.[11]

For shapes where closed-form integration is not feasible, the surface may be approximated by a series of flat panels. The sum of the forces and moments on the individual panels will determine the total force and moment on the body:

$$\frac{F}{q_\infty S} = \sum_{i=1}^{m}\left\{C_{p_i}(-\hat{\boldsymbol{n}}_i) + C_{T_i}\left[\hat{\boldsymbol{n}}_i \times (\hat{\boldsymbol{V}}_\infty \times \hat{\boldsymbol{n}}_i)\right]\right\} \mathrm{d}s$$
$$\frac{M}{q_\infty S c} = \sum_{i=1}^{m}\left[r_i \times \left\{C_{p_i}(-\hat{\boldsymbol{n}}_i) + C_{T_i}\left[\hat{\boldsymbol{n}}_i \times (\hat{\boldsymbol{V}}_\infty \times \hat{\boldsymbol{n}}_i)\right]\right\}\right] \mathrm{d}s \tag{11.24}$$

where $\hat{\boldsymbol{n}}_i$ is the panel outward unit normal and $\hat{\boldsymbol{V}}_\infty$ is the unit freestream velocity vector.

Since we are assuming collisionless flow, the use of Eqs. (11.18), (11.19), and (11.20) in Eq. (11.24) is, strictly speaking, valid only for convex bodies.

Table 11.1 C_D in terms of σ_N and σ_T

$C_{D_{\text{Free Molecular}}}$	=	$(C_{D_{\text{Pressure}}})$	+	$(C_{D_{\text{Shear}}})$
Disk ($\perp$ to flow)		$\left(2(2-\sigma_N)+\frac{\sigma_N}{s}\sqrt{\pi\frac{T_w}{T_\infty}}\right)$		—
Disk ($\parallel$ to flow)		—		$\left(\frac{\sigma_T}{s\sqrt{\pi}}\right)$
Sphere		$\left((2-\sigma_N)+\frac{2}{3}\frac{\sigma_N}{s}\sqrt{\pi\frac{T_w}{T_\infty}}\right)$	+	(σ_T)
Sharp cone		$\left((2-\sigma_N)\sin^2\theta_c+\frac{\sigma_N}{s}\sqrt{\pi\frac{T_w}{T_\infty}}\sin\theta_c\right)$	+	$(2\sigma_T\cos^2\theta_c)$
Circular cylinder		$\left(\frac{4}{3}(2-\sigma_N)+\frac{\sigma_N}{4s}\sqrt{\pi^3\frac{T_w}{T_\infty}}\right)$	+	$\left(\frac{2}{3}\sigma_T\right)$

Table 11.2 lists a FORTRAN program to compute the forces and moments acting on an arbitrary body in free molecular flow.

For complex shapes, where particles reflected from one surface may impinge on another surface before escaping downstream, Eqs. (11.18), (11.19), and (11.20) must be modified considerably. An alternative is a probabilistic numerical approach that simulates a large sample of particle trajectories. Absence of intermolecular collision in free molecular flow allows each of the particle trajectories to be generated independently. The statistics of the particles impinging on and emitted from the surface will determine the forces and moments. This approach is called the *test-particle Monte Carlo* method.

Table 11.2 FORTRAN program to calculate the forces and moments in the free molecular flow regime

```
C
      PROGRAM IMPACT
C------THIS PROGRAM CALCULATES FORCES AND MOMENTS ON AN ARBITRARY BODY
C------APPROXIMATED BY FLAT FOUR CORNERED PANELS
      DIMENSION X(100,4),Y(100,4),Z(100,4),XCRN(4),YCRN(4),ZCRN(4)
      DIMENSION UNITV(3), NORM(3), SHEAR(3), CENTR(3)
      DIMENSION FORCE(3), MOM(3), FORI(3), MOMI(3)
      REAL LREF, MINF, MOM, MOMI, NORM
      PI = ACOS(-1.0)
      DTR = PI/180.
      RTD = 1./DTR
C------AREF - Ref. Area; LREF - Ref. Length; MINF - Free stream Mach No.
```

(continued on next page)

Table 11.2 (continued) FORTRAN program to calculate the forces and moments in the free molecular flow regime

```
C------SIGMAN - Normal Accommodation Coeff.; SIGMAT - Tangential A. C.
C------GAMMA - Ratio of Specific heats ;
C------ALFA - Angle of attack (deg.); BETA - Sideslip angle (deg.)
      READ (11,101) AREF,LREF,MINF,SIGMAN,SIGMAT,TWTINF,GAMMA,ALFA,BETA
      write(6,101) AREF,LREF,MINF,SIGMAN,SIGMAT,TWTINF,GAMMA,ALFA,BETA
      SINF = MINF * SQRT(GAMMA/2.)
      ALFAR = ALFA * DTR
      BETAR = BETA * DTR
C
C------Read panel no., and x,y,z for each corner (1,4) of panel.
C------Input corners in Anti-Clockwise direction (looking at panel from
C------outside body). Note: Three corners should not be co-linear unless
C------corners 4 & 1 are coincident (Triangular panel: corner4=corner1).
C
      DO 400 IPNL = 1, 100
      READ (11,102,END=100) IPANL
      write (6,102) IPANL
      do 400 j = 1, 4
      read (11,*) ICORN, X(IPNL,J),Y(IPNL,J),Z(IPNL,J)
      write (6,*) ICORN, X(IPNL,J),Y(IPNL,J),Z(IPNL,J)
400   CONTINUE
C
101   FORMAT (10X,9(F10.3,/,10X)/)
102   FORMAT (7X,I3)
103     FORMAT (4X,I1,5X,3(F10.4))
C
1000  IPNLMX = IPNL - 1
C     WRITE (12,*) 'IPNLMX=', IPNLMX
C------Unit velocity vector in Body-fixed coordinate frame.
      UNITV(1) = COS(BETAR) * COS(ALFAR)
      UNITV(2) = SIN(BETAR)
      UNITV(3) = COS(BETAR) * SIN(ALFAR)
C
      DO 1250 I = 1, 3
            FORCE(I) = 0.0
            MOM(I) = 0.0
1250  CONTINUE
C
      DO 2000 IPANL = 1, IPNLMX
C
            DO 1500 ICONR = 1, 4
            XCRN(ICORN) = X(IPANL,ICORN)
            YCRN(ICORN) = Y(IPANL,ICORN)
            ZCRN(ICORN) = Z(IPANL,ICORN)
1500        CONTINUE
```

(continued on next page)

Table 11.2 (continued) FORTRAN program to calculate the forces and moments in the free molecular flow regime

```
C------Calculate Area, Deflection angle, Normal, and Shear vectors panel
      CALL GEOM(XCRN,YCRN,ZCRN,CENTR,UNITV,NORM,SHEAR,AREA,DELTAR)
      THETA = 90. - DELTA*RTD
c     WRITE (12,*) ('Panel no. ',IPANL,';  CNTR=',CENTR)
c     WRITE (12,*) ('CTNR=',CENTR,'; NRM=',NORM,'; SHR=',SHEAR)
c------Calculate Pressure and Shear forces based on Free Molecular Theory
      CALL PRESSFM (SINF,SIGMAN,SIGMAT,TWTINF,DELTAR,CP,CT)
c     WRITE (12,*) 'AREA=', AREA, '; THETA=',THETA
c     WRITE (12,*) 'CP=', CP,'; CT=', CT
      DO 1750 I = 1, 3
      FORI(I) = -(CP*NORM(I) -CT*SHEAR(I)) * AREA
      FORCE(I) = FORCE(I) + FORI(I)
1750  CONTINUE
      WRITE (6,*) ipanl, 'Delta Force=', FORI
c     write (12,*) 'total Force =', Force
C------Calculate Moment contribution of each panel.
      CALL CROSS (CENTR,FORI,MOMI)
      DO 1800 I = 1, 3
      MOM(I) = MOM(I) + MOMI(I)
1800  CONTINUE
C------Nondimensionalize Forces and Moments
2000  CONTINUE
      CFA = FORCE(1)/AREF
      CFY = FORCE(2)/AREF
      CFN = FORCE(3)/AREF
C
      CML =-MOM(1)/(AREF*LREF)
      CMM = MOM(2)/(AREF*LREF)
      CMN =-MOM(3)/(AREF*LREF)
C
      write (12,*)
      write (12,*) 'Mach No.=',minf,'; Speed ratio, s=', sinf
      write (12,*) 'Sigma Norm.=',sigman,'; Sigma Tang.=',sigmat,
     +  '; Tw/Tinf =',twtinf
      write (12,*) 'Ref. Area=', aref,'; Ref. Length=',lref
      write (12,*)
      write (12,*) 'Angle of attack=',alfa,'; sideslip=',beta
      WRITE (12,*) 'Force Coeff.:  CFA,CFY,CFN=',CFA,',','CFY',',',CFN
      WRITE (12,*) 'Moment Coeff.: CML,CMM,CMN=',CML,',',CMM,',',CMN
C
      STOP
      END
C     ***********************************************************
      SUBROUTINE PRESSFM (SINF,SIGMAN,SIGMAT,TWTINF,DELTAR,CP,CT)
C
C     Purpose:     Free Molecular Flow Pressure and Shear Stress
```

(continued on next page)

Table 11.2 (continued) FORTRAN program to calculate the forces and moments in the free molecular flow regime

```
C     Comments:    For Newtonian Pressure
C                     SINF very large, SIGMAN = 1.0,
C                     SIGMAT = 0.0, TWTINF = 1.0.
C
            SN2M = 2. - SIGMAN
            SPI = SQRT (ACOS(-1.0))
            SCOSD = SINF * COS(DELTAR)
            SSIND = SINF * SIN(DELTAR)
            S2C2D -= SCOSD * SCOSD
            EXPS2 = EXP(-S2C2D)
            ERFSCD1 = 1.0 + ERF(SCOSD)
            TERM1 = sn2m * SCOSD/SPI
            TERM2 = sigman * SQRT(TWTINF)/2.0
            TERM12 = (TERM1 + TERM2) * EXPS2
            TERM3 = sn2m * (0.5 + S2C2D)
            TERM4 = TERM2 * SPI * SCOSD
      CP = (TERM12 + (TERM3 + TERM4)*ERFSCD1)/(SINF * SINF)
      CT = sigmat*SSIND*(EXPS2 + (SPI*SCOSD*ERFSCD1))/(SPI*SINF*SINF)
C
      RETURN
      END
C     ***********************************************************
      SUBROUTINE GEOM (X,Y,Z,CENTR,UNITV,NORM,SHEAR,AREA,DELTAR)
C
      DIMENSION X(4),Y(4),Z(4),UNITV(3),NORM(3),TANG(3),SHEAR(3)
      DIMENSION VEC12(3),VEC13(3),VEC14(3),VEC24(3),CENTR(3)
      DIMENSION VECA1(3),VECA2(3)
      REAL MAGNA1, MAGNA2, MNORM, MSHEAR, NORM
        CALL DIFF (X(2),Y(2),Z(2),X(1),Y(1),Z(1),VEC12)
        CALL DIFF (X(3),Y(3),Z(3),X(1),Y(1),Z(1),VEC13)
        CALL DIFF (X(4),Y(4),Z(4),X(1),Y(1),Z(1),VEC14)
        CALL DIFF (X(4),Y(4),Z(4),X(2),Y(2),Z(2),VEC24)
      CALL CROSS (VEC13,VEC24,NORM)
      CALL MAGNTD (NORM, MNORM)
      DO 10 I = 1, 3
            NORM(I) = NORM(I)/MNORM
10    CONTINUE
      CALL CROSS (NORM, UNITV, TANG)
c     write (12,*) ('norm=', norm, 'tang=',tang)
      CALL CROSS (TANG, NORM, SHEAR)
      CALL MAGNTD (SHEAR, MSHEAR)
c     write (12,*) ('shear=',shear, 'mshear=',mshear)
      DO 20 I = 1, 3
            SHEAR(I) = SHEAR(I)/MSHEAR
20    CONTINUE
c     write (12,*) ('shear=',shear, 'mshear=',mshear)
      CALL CROSS (VEC12, VEC13, VECA1)
```

(continued on next page)

Table 11.2 (continued) FORTRAN program to calculate the forces and moments in the free molecular flow regime

```
      CALL CROSS (VEC13, VEC14, VECA2)
      CALL MAGNTD (VECA1, MAGNA1)
        AREA1 = 0.5 * MAGNA1
      CALL MAGNTD (VEGA2, MAGNA2)
        AREA2 = 0.5 * MAGNA2
        AREA = AREA1 + AREA2
      CALL CENTROID (X,Y,Z,AREA,AREA1,AREA2,CENTR)
      CALL ANGLE (UNITV, NORM, DELTAR)
      RETURN
      END
C     ***********************************************************
      SUBROUTINE CENTROID (X,Y,Z,AREA,AREA1,AREA2,CENTR)
C     Purpose:    Determine Centroid of each panel
      DIMENSION CENTR(3),X(4),Y(4),Z(4),V1H23(3),V1H43(3)
        XH23 = 0.5 * (X(2) + X(3))
        YH23 = 0.5 * (Y(2) + Y(3))
        ZH23 = 0.5 * (Z(2) + Z(3))
      CALL DIFF (XH23,YH23,ZH23,X(1),Y(1),Z(1),V1H23)
        XBAR1 = X(1) + (2.* V1H23(1)/3.)
        YBAR1 = Y(1) + (2.* V1H23(2)/3.)
        ZBAR1 = Z(1) + (2.* V1H23(3)/3.)
        XH43 = 0.5 * (X(4) + X(3))
        YH43 = 0.5 * (Y(4) + Y(3))
        ZH43 = 0.5 * (Z(4) + Z(3))
      CALL DIFF (XH43,YH43,ZH43,X(1),Y(1),Z(1),V1H43)
        XBAR2 = X(1) + (2.* V1H43(1)/3.)
        YBAR2 = Y(1) + (2.* V1H43(2)/3.)
        ZBAR2 = Z(1) + (2.* V1H43(3)/3.)
        CENTR(1) = (AREA1 * XBAR1 + AREA2 * XBAR2)/AREA
        CENTR(2) = (AREA1 * YBAR1 + AREA2 * YBAR2)/AREA
        CENTR(3) = (AREA1 * ZBAR1 + AREA2 * ZBAR2)/AREA
      RETURN
      END
C     ************************************************************
      SUBROUTINE ANGLE (VEC1, VEC2, DELTAR)
C     Purpose:    Determine angle between two vectors
      DIMENSION VEC1(3), VEC2(3)
      REAL MVEC1, MVEC2, NUMER
        NUMER = VEC1(1)*VEC2(1) + VEC1(2)*VEC2(2) + VEC1(3)*VEC2(3)
      CALL MAGNTD (VEC1, MVEC1)
      CALL MAGNTD (VEC2, MVEC2)
        CSTHT =-NUMER/(MVEC1*MVEC2)
        DELTAR = ACOS(CSTHT)
      RETURN
      END
```

(continued on next page)

Table 11.2 (continued) FORTRAN program to calculate the forces and moments in the free molecular flow regime

```
C     ************************************************************
      SUBROUTINE MAGNTD (VEC, MAGN)
C     Purpose:    Determine Magnitude of a vector
      DIMENSION VEC(3)
      REAL MAGN
        MAGN = SQRT(VEC(1)*VEC(1) + VEC(2)*VEC(2) + VEC(3)*VEC(3))
      RETURN
      END
C     ************************************************************
      SUBROUTINE CROSS (VEC1, VEC2, VEC1X2)
C     Purpose:    Determine Vector Cross Product
      DIMENSION VEC1(3), VEC2(3), VEC1X2(3)
        VEC1X2(1) = VEC1(2) * VEC2(3) - VEC1(3) * VEC2(2)
        VEC1X2(2) = VEC1(3) * VEC2(1) - VEC1(1) * VEC2(3)
        VEC1X2(3) = VEC1(1) * VEC2(2) - VEC1(2) * VEC2(1)
      RETURN
      END
C     ************************************************************
      SUBROUTINE DIFF (XTO,YTO,ZTO,XFROM,YFROM,ZFROM,VFTO)
C     Purpose:    Calculate Difference of Two Vectors
      DIMENSION VFTO(3)
        VFTO(1) = XTO - XFROM
        VFTO(2) = YTO - YFROM
        VFTO(3) = ZTO - ZFROM
      RETURN
      END
C     ************************************************************
      REAL FUNCTION ERF (XX)
C     NAME:         FUNCTION ERF
C     PURPOSE:      TO DETERMINE VALUE FOR ERROR FUNCTION
C                   (ERF)
C     INPUT
C     PARAMETERS:   NAME      DESCRIPTION
C                   X         COORDINATE FOR ERROR FUNCTION
C     OUTPUT
C     PARAMETERS:   NAME      DESCRIPTION
C                   ERF       FUNCTION ERF
C     COMMENTS:     ERROR FUNCTION IS DEFINED AS FOLLOWS:
C                       ERF = 2.0 * F / SQRT (PI)
C                   WHERE F IS THE INTEGRAL FROM 0 TO X OF
C                   EXP (- U * U)
C     DEFINE DISTRIBUTION FUNCTION
      F(U) = EXP (- U * U)
      X = ABS (XX)
      PI = 3.14159
      N = INT (X / 0.2) * 2 + 1
```

(continued on next page)

Table 11.2 (continued) FORTRAN program to calculate the forces and moments in the free molecular flow regime

```
      IF (N .LT. 5) N = 5
      DELX = X / FLOAT (N - 1)
      ERF = 0,0
      DO 10 I = 1, N
         XF = FLOAT (I - 1) * DELX
         A = 2.0
         ITEST = (-1) ** I
         IF (ITEST .EQ. 1) A = 4.0
         IF (I .EQ. 1 .OR. I .EQ. N) A = 1.0
         ERF = ERF + A * F (XF)
   10 CONTINUE
         ERF = ERF * X / (3.0 * FLOAT (N - 1))
         ERF = 2.0 * ERF / SQRT (PI)
         IF (XX .LT. 0.0) ERF = - ERF
         RETURN
         END
                        Input file for program IMPACT
 AREF =   1.
 LREF =   4.
 MINF =   20.00
 sigman = 1.0
 sigmat = 1.0
 TRTINF = 2.0
 GAMMA  = 1.4
 ALFA   = 5.
 BETA   = 0.
     "Panel corners read in anti-clockwise direction
         looking at panel from outside the body."
ipanel=  1
   1     0.0        0.0       0.0
   2     2.0        1.0       0.0
   3     2.0       -1.0       0.0
   4     0.0       -0.0       0.0
ipanel=  2
   1     0.0        0.0       0.0
   2     2.0        0.0       1.0
   3     2.0        1.0       0.0
   4     0.0       -0.0       0.0
ipanel=  3
   1     0.0        0.0       0.0
   2     2.0       -1.0       0.0
   3     2.0        0.0       1.0
   4     0.0       -0.0       0.0
ipanel=  4
   1     2.0        1.0       0.0
   2     4.0        1.0       0.0
   3     4.0       -1.0       0.0
   4     2.0       -1.0       0.0
```

(continued on next page)

Table 11.2 (continued) FORTRAN program to calculate the forces and moments in the free molecular flow regime

```
ipanel=  5
    1     2.0        -1.0       0.0
    2     4.0        -1.0       0.0
    3     4.0         0.0       1.0
    4     2.0         0.0       1.0
ipanel=  6
    1     2.0         0.0       1.0
    2     4.0         0.0       1.0
    3     4.0         1.0       0.0
    4     2.0         1.0       0.0
ipanel=  7
    1     4.0        -1.0       0.0
    2     4.0         1.0       0.0
    3     4.0         0.0       1.0
    4     4.0        -1.0       0.0
```

11.7 Continuum Flow Regime: $Kn \ll 1$

Before we discuss the transition from free molecular flow to continuum flow, let us examine the continuum regime. It is in this regime that the re-entry body will experience the maximum inertial and thermal loads, which will drive most aspects of vehicle design. Figure 11.6 illustrates a flowfield about a typical re-entry body. Before going into any details, let us first examine a very general flowfield. Then we can compare and contrast this general flow problem to re-entry aerothermodynamics.

As we mentioned earlier in conjunction with a definition of the gas and a set of boundary conditions, the Navier-Stokes equations will determine the flowfield

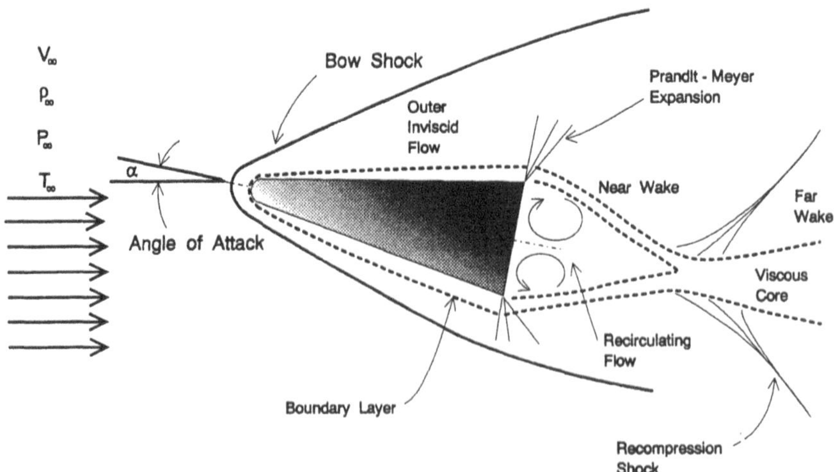

Fig. 11.6 Typical re-entry body flowfield.

around the body. For a very wide range of problems, air may be considered a perfect gas as defined by Eqs. (11.2) and (11.5). To define the problem completely, boundary conditions are also required. The farfield boundary condition is that any disturbance caused by the presence of the body should remain finite.

At the body surface, in general, the velocity normal to a surface (solid, impermeable surface) must be zero, and viscosity will cause the tangential velocity also to be zero (see Fig. 11.7a). Thus,

$$\begin{aligned} \boldsymbol{V}\cdot\boldsymbol{n} \equiv v &= 0 \\ V|_w &= 0 \end{aligned} \tag{11.25}$$

In addition, the gas temperature at the body surface must match the specified wall temperature. Since the wall temperature will depend upon the heat conduction into the interior of the re-entry body, it will generally not be known a priori. When the wall temperature is specified, the flow problem is decoupled from the internal heat conduction problem (see Fig. 11.7b). The condition where there is no heat transfer through the wall is called the *adiabatic* wall condition. The specified wall temperature is given by

$$T(x) = T_w(x) \tag{11.26a}$$

and the adiabatic wall condition is expressed as

$$(\partial T/\partial n)_w = 0 \tag{11.26b}$$

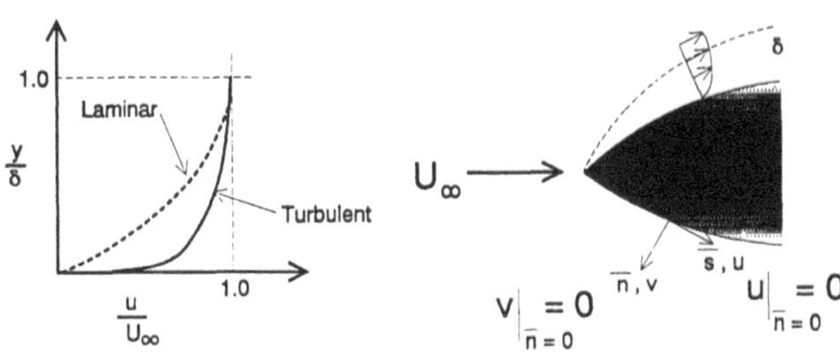

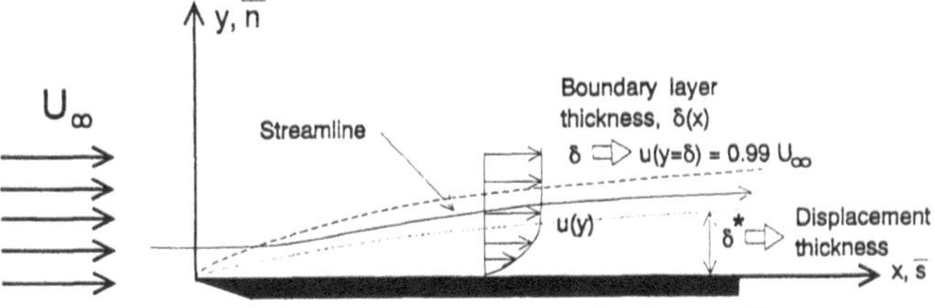

Fig. 11.7a Boundary-layer velocity profiles.

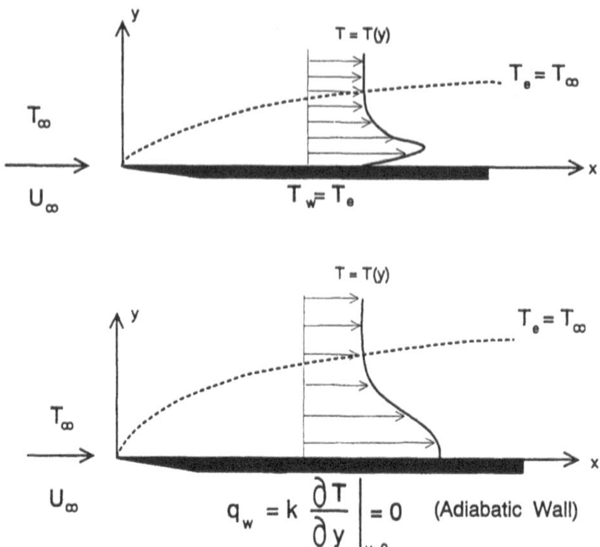

Fig. 11.7b Boundary-layer temperature profiles.

The resulting equations are a set of coupled, nonlinear, partial differential equations. The Navier-Stokes equations are a mixed set of elliptic-parabolic equations for steady flow and are hyperbolic-parabolic when the unsteady term is retained. In addition to being analytically intractable, the Navier-Stokes equations do not appear to lend themselves to a numerical solution of complete configurations even on computers of the foreseeable future.[12]

Based on the local features, the flowfield around the re-entry body may be divided into three regions (see Fig. 11.6): 1) the boundary layer, 2) the outer inviscid flow region, and 3) the wake region.

The boundary layer is the region adjacent to the body surface, dominated by viscous effects and heat conduction. The shearing effect of viscosity causes the velocity to vary from zero at the wall, satisfying the wall boundary conditions, to the outer flow velocity at the edge of the boundary-layer (see Fig. 11.7a). The boundary-layer thickness $\delta_{99\%}$ (or δ) is defined as the distance normal to the wall where the local velocity is 99% of the "freestream" (or edge) velocity. To satisfy the mass conservation equation, the streamlines will be deflected away from the wall because of the slower-moving fluid accumulating near the wall. The amount of deflection of the streamlines outside the boundary layer is called the displacement thickness δ^* and is a function of the distance from the leading edge and the velocity profile in the boundary layer. The velocity profile in the boundary layer determines the viscous forces acting on the body. The wall shear stress τ_w is given by

$$\tau_w = \mu \left(\partial u / \partial n\right)_w \tag{11.27a}$$

and the local skin friction coefficient $C_f(x)$ is given by

$$C_f(x) = \tau_w / \tfrac{1}{2}\rho_e V_e^2 \tag{11.27b}$$

Similarly, the temperature must vary from the wall to the outer edge to satisfy the boundary conditions and to account for the temperature rise due to kinetic energy changes (adiabatic) and friction (see Fig. 11.7b). The heat transfer rate at the wall may be expressed by Fourier's law of heat conduction. The wall heat transfer rate q_w and Stanton number C_h are given by

$$\begin{aligned} q_w &= -k\,(\partial T/\partial y)_w \\ C_h &= q_w / \rho_e u_e (h_{aw} - h_w) \end{aligned} \tag{11.28}$$

where k is the thermal conductivity and h_{aw} and h_w are the adiabatic wall enthalpy (enthalpy for $q_w = 0$) and the actual wall enthalpy, respectively. The subscribt e refers to the conditions at the edge of the boundary layer.

When the speed of sound is much greater than the freestream flow velocity ($U \ll a$), the density may be considered to be constant (incompressible flow). In addition, if the coefficient of viscosity and heat conduction are also constant, the energy equation is decoupled from the momentum equation, significantly simplifying the problem. However, this luxury is not afforded in high-speed flow, where the density may no longer be considered a constant.

Vorticity (a measure of angular velocity of a fluid element), a product of the velocity gradient caused by viscous forces, is convected downstream much faster than it diffuses normal to the surface. Therefore, for high Reynolds number flows, the boundary layer is usually quite thin. Hence, the viscous effects can be assumed to be confined to the boundary layer. For thin boundary layers, the pressure is constant normal to the surface and is assumed to be imposed by the outer flow at the outer edge of the boundary layer. An order-of-magnitude analysis of a two-dimensional flow with these boundary layer approximations allows us to reduce the number of governing equations by one; these equations are called the boundary layer equations.[13] Furthermore, similarity solutions allow us to transform these still complex, coupled, partial differential equations to ordinary differential equations.[13] Although similarity solutions are restricted to a single geometry, they provide extraordinary insight into the problem.

Using similarity techniques, Van Driest[14] calculated the flat plate boundary-layer profiles for a range of freestream Mach numbers and wall temperatures. Note that the boundary-layer thickness and temperature increase dramatically with Mach number (see Fig. 11.8a). The coupling of the energy equation and the momentum equation is illustrated when we compare the results of the adiabatic and nonadiabatic wall flows (see Fig. 11.8b). The boundary layer is much thicker for an adiabatic wall than for a cold wall. Since the shear stress and heat transfer rate are proportional to the velocity and temperature gradients [Eqs. (11.27) and (11.28) respectively], a thicker boundary layer results in lower shear stress and heat transfer rates at the wall.

The outer flow region, assumed to be inviscid (i.e., viscosity is confined to the boundary layer) and nonconducting, governs the pressure forces acting on the body. The outer flow may be determined independently of the boundary

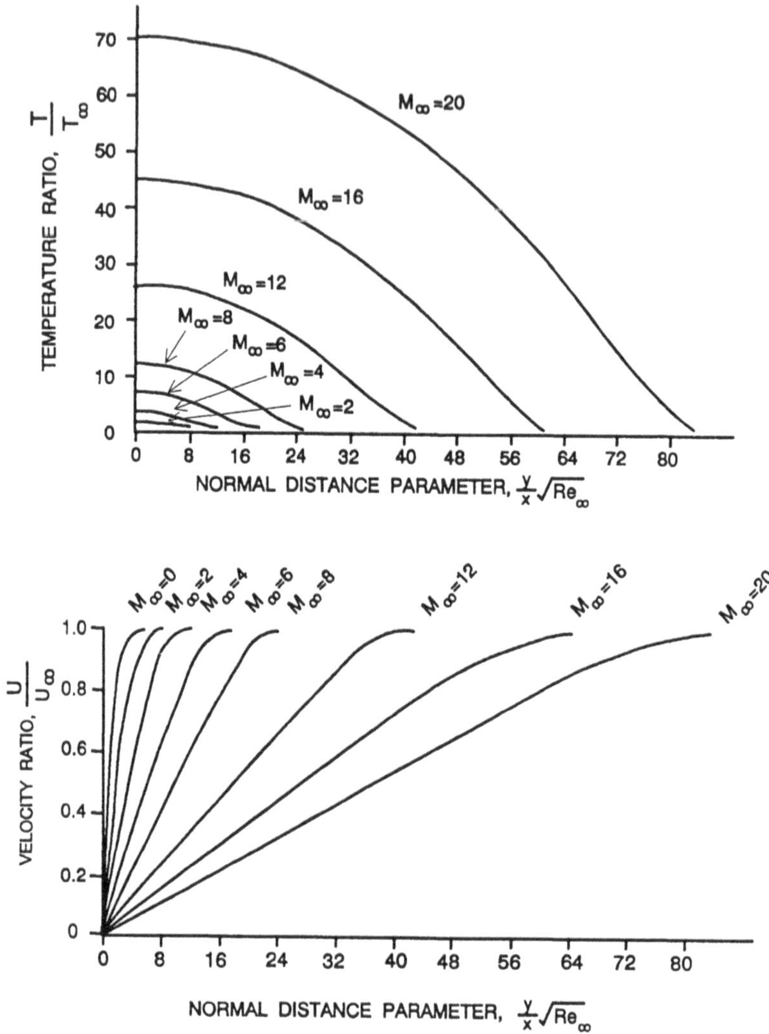

Fig. 11.8a Temperature and velocity distribution across laminar boundary layers on an insulated flat plate at various Mach numbers (from Van Driest[14]).

layer by using a fictitious body that is the original body with the displacement thickness superimposed. The inviscid pressure and velocity field determined in the outer region will provide the boundary conditions for the outer edge of the boundary layer.

For the inviscid outer flow region, we neglect the viscous and the heat transfer terms from the Navier-Stokes equations to obtain a simplified set of equations called the *Euler equations*. Even though the Euler equations are also coupled, nonlinear partial differential equations, they are of a lower order than the Navier-Stokes equations. The unsteady Euler equations are hyperbolic, but the steady

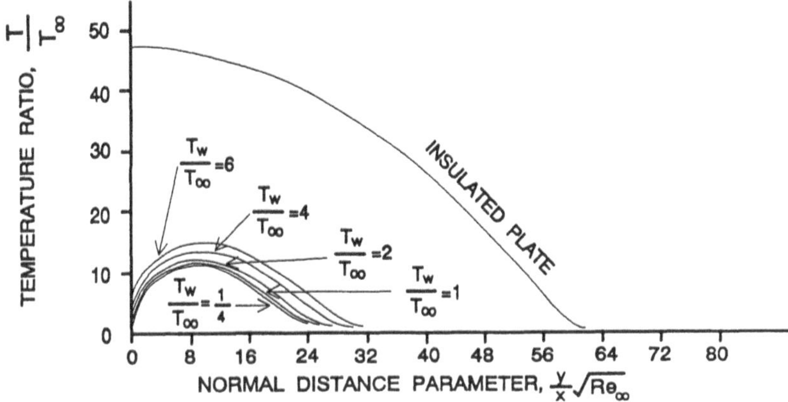

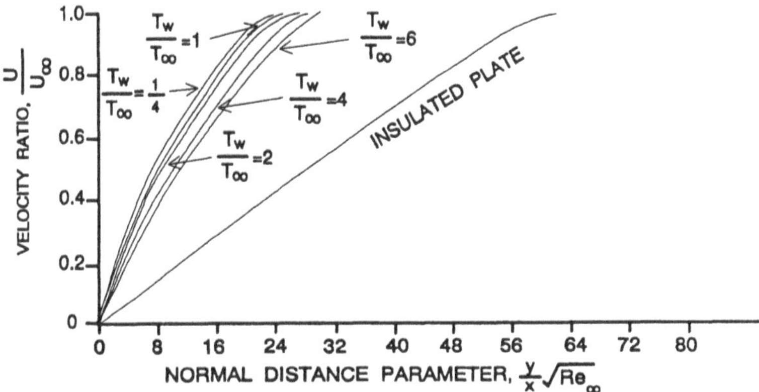

Fig. 11.8b Temperature and velocity distribution across laminar boundary layer for $M_\infty = 16$ and various wall-to-freestream temperature ratios (from Van Driest[14]).

state equations change their character from elliptic in subsonic flow ($M_\infty < 1$) to hyperbolic in *supersonic* flow ($M_\infty > 1$). Let us now try to get a physical understanding of the reasons for and effects of the change in character of the flow with a change in M_∞.

A disturbance (point source) in a fluid creates acoustic waves (infinitesimal pressure pulses) that radiate spherically with the speed of sound, a, "telegraphing" the presence of the source to the flowfield. Figure 11.9 illustrates the propagation of the acoustic waves from a moving source. When the speed of the source, U, is less than a, the acoustic waves will "warn" the surrounding fluid of the approach of the source. When the source moves faster than a, the "warnings" cannot propagate ahead of the source and are confined to a cone of influence, called the Mach cone. In a source-fixed frame of reference, with a uniform flow, U, past the source, for subsonic flow ($U < a$) the disturbance

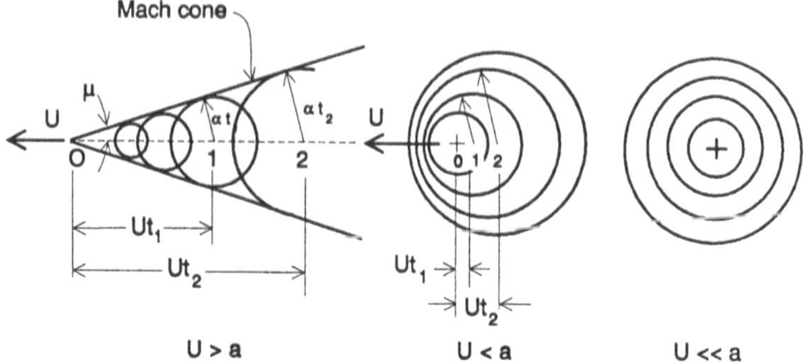

Fig. 11.9 Wave propagation from a moving source.

influences the entire flowfield, while in a supersonic flow ($U > a$) the influence is limited to a zone demarcated by the Mach lines. The Mach lines are inclined to the local flow by the Mach angle, μ, a characteristic angle associated with the Mach number:

$$\mu = \sin^{-1}(1/M) \tag{11.29}$$

Hence the subsonic problem will require the boundary conditions to be specified over the entire outer boundary and is called a *boundary value problem*. The supersonic problem, on the other hand, is an initial value problem, requiring boundary conditions only at an initial stage with the solution obtained by marching downstream.

The Euler equations may be further simplified for bodies that create only *small perturbations* to the freestream. The resultant small perturbations equations[15] are linear, except when $M_\infty \to 1$ (*transonic*) or $M_\infty \gg 1$ (*hypersonic*). A flow that contains both subsonic and supersonic regions is referred to as transonic flow (see Fig. 11.10). Since the character of the governing equations changes at

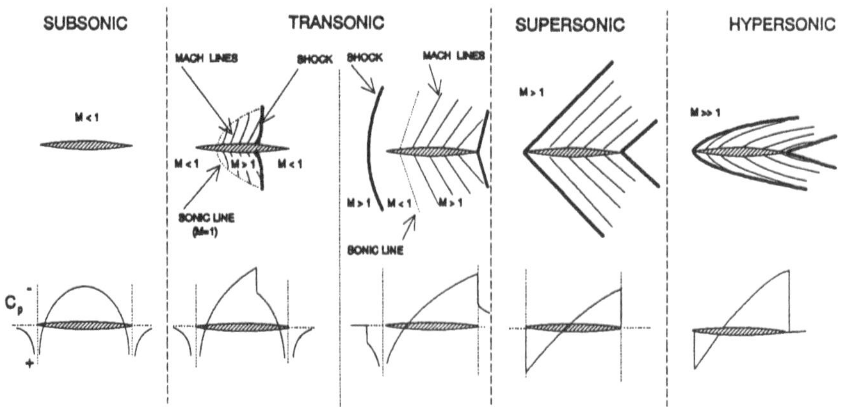

Fig. 11.10 Continuum flow regimes.

$M = 1$, the solution technique must also be modified accordingly. In addition, a strong interaction between the boundary layer and shock waves in transonic flow further complicates any solution process. Hypersonic flow will be discussed in more detail in the next section.

The wake region flow is quite complex and is dominated by a recirculating *dead-air* zone behind the separation point (see Fig. 11.6). In addition, the vorticity diffuses from the boundary layer merging with the outer flow. The wake flow determines the base pressure and hence the base drag. In hypersonic separated base flow, a good approximation is $p_{\text{base}} = p_\infty$. If the flow does not separate, the wake region may be treated as a combination of the boundary layer and outer inviscid flow.

Based on the character of the governing equations and the accuracy of some of the approximations, the continuum flow regime may be subdivided into

a) incompressible flow—$M_\infty \leq 0.3$ (constant ρ)
b) compressible subsonic flow—$0.3 < M_\infty < 1$
c) transonic flow—$0.8 < M_\infty < 1.2$
d) supersonic flow—$M_\infty > 1$
e) hypersonic flow—$M_\infty > 5$

Figure 11.10 illustrates some of the salient features of the various flow categories. In the continuum regime, with the possible exception of the space shuttle and a few drawing board concepts, like the NASP, all re-entry system designs are driven by hypersonic flow. Therefore, after a brief discussion of the other flow categories, we will direct our discussions toward hypersonic flows.

First, let us consider incompressible subsonic flow past a body. We know that the constant density medium in incompressible flow decouples the momentum and energy equations. If we simplify the problem further and consider an inviscid flowfield, the Euler equation (momentum equation) integrates to yield *Bernoulli's* equation:

$$\tfrac{1}{2}\rho V^2 + p = \text{constant (along a streamline)} \tag{11.30}$$

If we now increase M_∞ and density can no longer be considered a constant, the energy equation for the adiabatic flow of a perfect gas may be written as

$$c_p T + \tfrac{1}{2}V^2 = c_p T_0 \equiv h_0 = \text{constant} \tag{11.31}$$

and the compressible flow momentum equation (sometimes referred to as the compressible flow Bernoulli's equation) becomes

$$\frac{V^2}{2} + \frac{\gamma}{\gamma - 1}\frac{p}{\rho} = \text{constant (along a streamline)} \tag{11.32}$$

It should be noted that Bernoulli's equations are valid throughout the flowfield if the flow is *irrotational* (vorticity, $\Omega = \nabla \times V = 0$).

For an *isentropic* (adiabatic, inviscid, nonconducting) flow, the relationships between the stagnation conditions and local flow properties are

$$\frac{a_0^2}{a^2} = \frac{T_0}{T} = 1 + \frac{\gamma - 1}{2} M^2$$

$$\frac{p_0}{p} = \left(1 + \frac{\gamma - 1}{2} M^2\right)^{\gamma/(\gamma-1)} \tag{11.33}$$

$$\frac{\rho_0}{\rho} = \left(1 + \frac{\gamma - 1}{2} M^2\right)^{1/(\gamma-1)}$$

where the subscript 0 refers to stagnation conditions.

Equations (11.33) are valid only in isentropic flows. However when a supersonic flow is compressed through a shock wave, the process is no longer isentropic. A shock wave is quite thin (to the order of 10^{-7} m) and from a macroscopic perspective may be considered to be a discontinuity in the flow. The flow behind a shock that is normal to the flow is always subsonic. Associated with the increase in entropy and static pressure across a shock is a decrease in the total pressure while the total temperature remains unchanged. The flow-field properties behind a normal shock (Fig. 11.11a) may be calculated using the *Rankine-Hugoniot* relations:

$$\frac{p_2}{p_1} = 1 + \frac{2\gamma}{\gamma + 1}\left(M_1^2 - 1\right)$$

$$\frac{\rho_2}{\rho_1} = \frac{u_1}{u_2} = \frac{(\gamma + 1)M_1^2}{(\gamma - 1)M_1^2 + 2}$$

$$\frac{T_2}{T_1} = 1 + \frac{2(\gamma - 1)}{(\gamma + 1)^2}\frac{\gamma M_1^2 + 1}{M_1^2}\left(M_1^2 - 1\right) \tag{11.34}$$

$$M_2^2 = \left(1 + \frac{\gamma - 1}{2} M_1^2\right) \Big/ \left(\gamma M_1^2 - \frac{\gamma - 1}{2}\right)$$

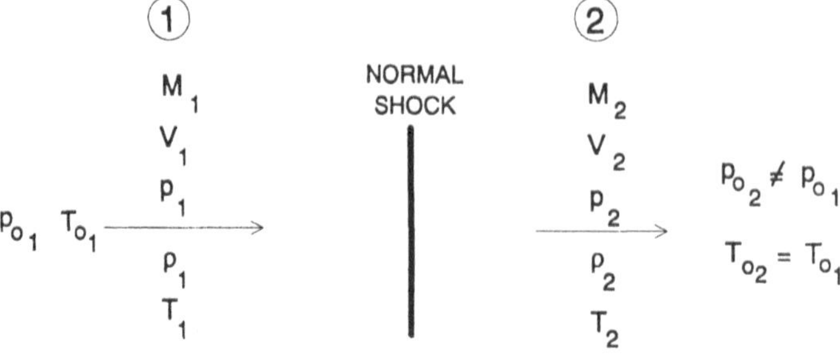

Fig. 11.11a Flow across a normal shock.

Now consider a two-dimensional flowfield. When the flow is not normal to the shock, the velocity component normal to the shock is modified as if going through a normal shock, and the component parallel to the shock must be equal on either side of the shock to satisfy the continuity equation. Therefore, the flow behind the shock will be deflected (see Fig. 11.11b). Conversely, if a supersonic flow is compressed by deflecting the flow, an oblique shock may be formed. Furthermore, the flow in the region between the shock and the wedge will be uniform.

Since an oblique shock acts as a normal shock to the component of flow normal to it, the normal shock equations [Eqs. (11.34)] given above may be applied to the oblique shock problem (see Fig. 11.11b). M_1 and M_2 are replaced by their normal components $M_1 \sin\beta$ and $M_2 \sin(\beta - \theta)$, where β is the shock angle and θ is the flow deflection angle. The relationship among the shock angle, deflection angle, and properties across an oblique shock are given in Eqs. (11.35),

$$
\begin{aligned}
\frac{p_2}{p_1} &= 1 + \frac{2\gamma}{\gamma+1}\left(M_1^2 \sin^2\beta - 1\right) \\
\frac{\rho_2}{\rho_1} &= \frac{(\gamma+1)M_1^2 \sin^2\beta}{(\gamma-1)M_1^2 \sin^2\beta + 2} \\
\frac{T_2}{T_1} &= 1 + \frac{2(\gamma-1)}{(\gamma+1)^2}\,\frac{\gamma M_1^2 \sin^2\beta + 1}{M_1^2 \sin^2\beta}\left(M_1^2 \sin^2\beta - 1\right) \\
\tan\theta &= 2\cot\beta\,\frac{M_1^2 \sin^2\beta - 1}{M_1^2(\gamma + \cos 2\beta) + 2}
\end{aligned}
\tag{11.35}
$$

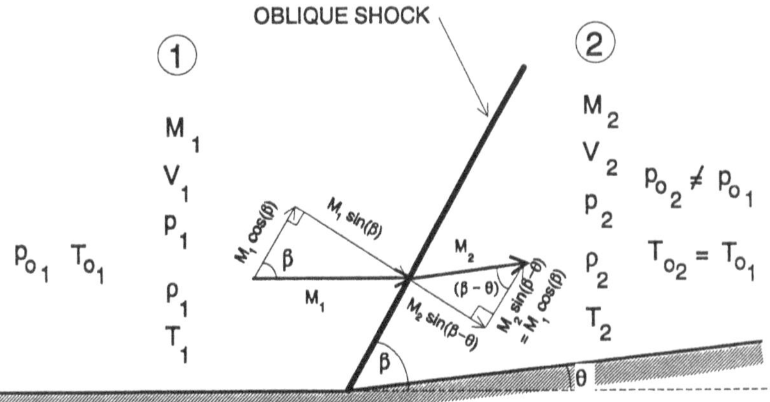

Fig. 11.11b Flow across an oblique shock.

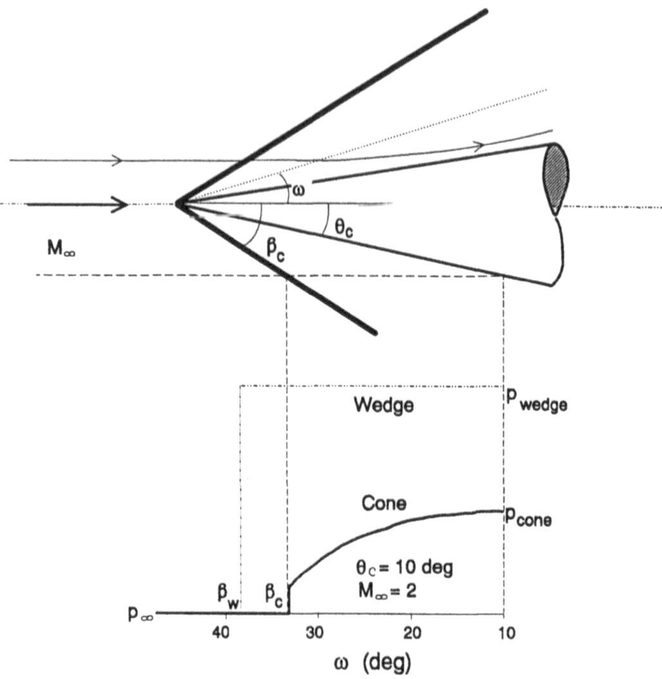

Fig. 11.12 Supersonic flow over a sharp cone.

Extensive graphical and tabular solutions to Eqs. (11.35) are available from NASA.[16]

In the three-dimensional case of a flow past a cone, determining the properties behind the shock is not as simple as the two-dimensional case. Unlike a wedge flow, the properties behind a shock in *conical flow* are constant along any ray emanating from the vertex of the cone (see Fig. 11.12). The conical shock angle β_c and the properties behind the shock are tabulated for a range of Mach numbers M_∞, cone angles θ_c, and ray angles ω, by Kopal[17] and Sims.[18]

For both two- and three-dimensional flow, if the flow deflection angle is too large, an oblique shock solution does not exist and a curved shock, detached from and normal at the leading edge (see Fig. 11.13), is formed. Since the flow behind the normal part of the curved shock will be subsonic, the shape and standoff distance of the curved shock depends upon the geometry of the body and must be calculated along with the entire flowfield. An additional problem associated with a curved shock is that the entropy increase across the shock varies with the shock angle. Hence, even with a constant entropy freestream, the flow behind the shock will have an entropy gradient, resulting in complications that will be discussed in the next section.

When a supersonic flow is expanded by a convex surface (see Fig. 11.14), the flow remains isentropic throughout. The resulting change in the flow direction is related to the Mach number by the *Prandtl-Meyer* function, i.e.,

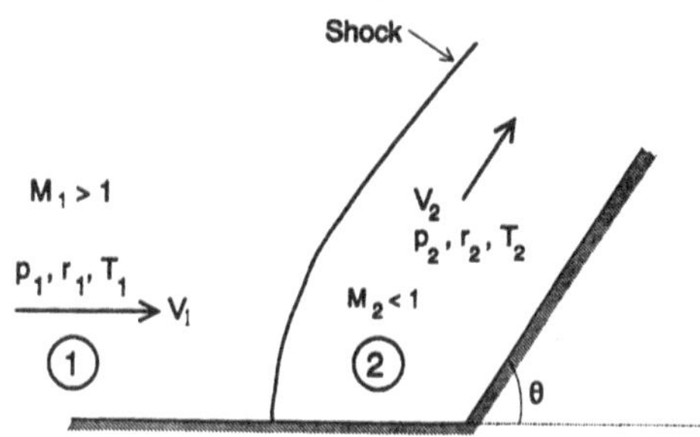

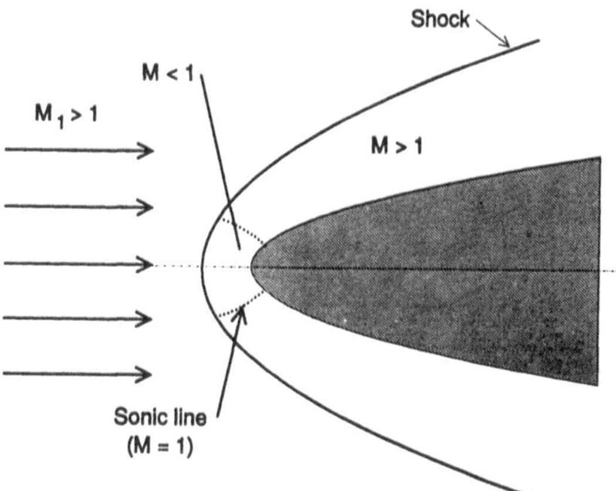

Fig. 11.13 Detached shock.

$$\nu(M) = \sqrt{\frac{\gamma+1}{\gamma-1}} \tan^{-1} \sqrt{\frac{\gamma-1}{\gamma+1}(M^2-1)} - \tan^{-1} \sqrt{M^2-1} \tag{11.36}$$

so that $\nu = 0$ corresponds to $M = 1$. It is also possible to compress a flow isentropically (without creating a shock) by a smooth concave turn. The relationship between the initial Mach number, turning angle θ (in radians), and the final Mach number is given by

$$\nu(M_2) = \nu(M_1) - |\theta| \tag{11.37a}$$

for compression and

$$\nu(M_2) = \nu(M_1) + |\theta| \tag{11.37b}$$

for expansion. Since the expansion process is isentropic and we are considering only isentropic compression here, the stagnation conditions are unchanged.

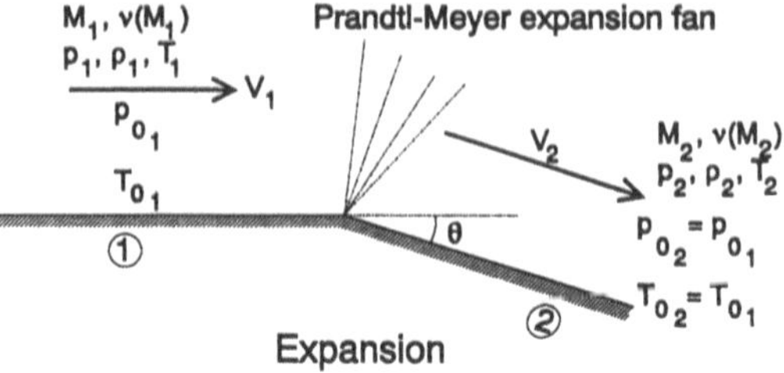

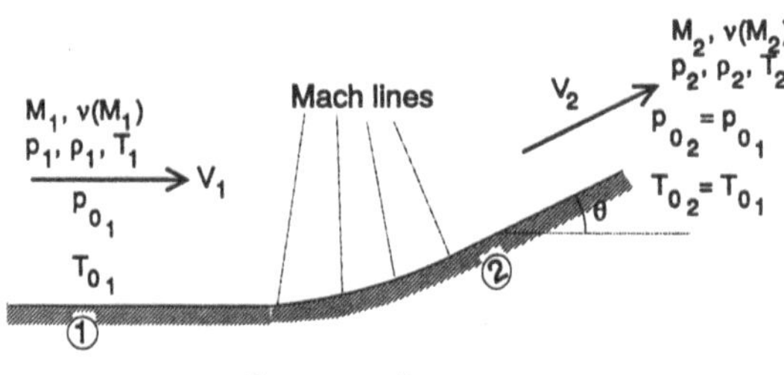

Fig. 11.14 Isentropic flow processes.

Hence, the local flow properties can be determined from Eqs. (11.33). Equations (11.33), (11.34), and (11.36) have been tabulated for a range of Mach numbers.[16]

We have already stated that the steady-state Euler equations in supersonic flow are hyperbolic. Therefore, if we know the flow properties at some initial plane (or line for a two-dimensional case), the method of characteristics (MOC) is ideally suited for determining the flowfield downstream of the initial plane (or line). Refer to Fig. 11.15. Let line AB be the initial condition line, i.e., the line along which all of the flow properties are known. We know that the influence of a generic point on the flow downstream is limited to a cone defined by the Mach lines [Eq. (11.29)]. These Mach lines are, in fact, the characteristic lines, along which the compatibility parameters K are constant. Along a right-running characteristic,

$$K_{-} = \text{constant} = \theta - \nu(M) \tag{11.38a}$$

and along a left-running characteristic,

$$K_{+} = \text{constant} = \theta + \nu(M) \tag{11.38b}$$

where θ is the flow direction and ν is the Prandtl-Meyer function [Eq. (11.35)]. We can discretize the problem and consider two neighboring points, labeled 1 and 2 in Fig. 11.15, on AB. The characteristic lines, approximated by straight line segments, emanating from these points intersect at point 3. Hence, point 3

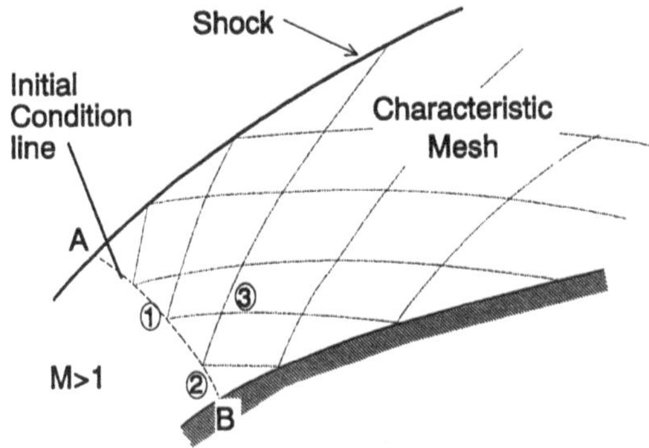

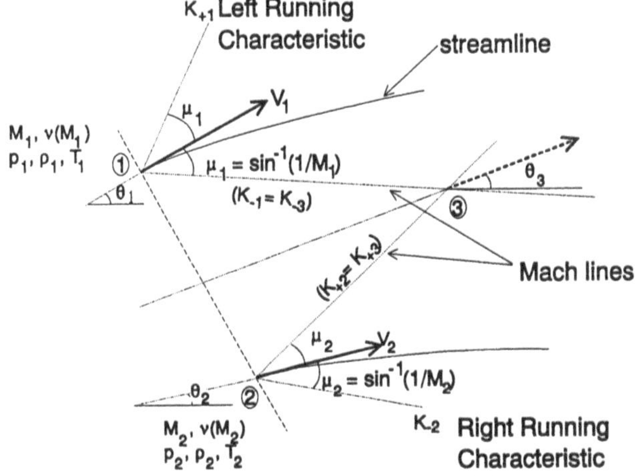

Fig. 11.15 Method of characteristics.

is influenced by both 1 and 2. Since K must be constant along a characteristic, $(K_-)_1 = (K_-)_3$ and $(K_+)_2 = (K_+)_3$, and we can calculate θ_3 and ν_3. Proceeding in a like manner from the other points on AB and giving special treatment to the boundaries, the entire flowfield may be charted. A more detailed discussion may be found in Zucrow's text.[19]

In three-dimensional flow, the Mach lines and characteristic lines would be replaced with characteristic surfaces. Here the geometry gets quite complicated and other numerical techniques are preferred. Anderson and coworkers[12] provide a detailed discussion of various computational methods.

11.8 Hypersonic Flow: $M_\infty \gg 1$

As discussed earlier, the Navier-Stokes equations govern hypersonic flow also. What distinguishes hypersonic flow from supersonic flow is the magnitude of the inaccuracies of most of the supersonic flow approximations. With increasing Mach number, these inaccuracies become progressively larger and finally render the approximations invalid in hypersonic flow. With this fact in mind, it is generally accepted that as $M_\infty \to 5$ the hypersonic effects cannot be neglected.

To be more specific, there are four main features of hypersonic flows that deny us the comfort of supersonic flow approximations as $M_\infty \to 5$: 1) high-temperature effects, 2) viscous interaction between outer flow and the boundary-layer region, 3) entropy gradients, and 4) thin shock layer.

If one were to follow the trend set by supersonic aerodynamics, a typical hypersonic body would be an extremely slender body with a sharp leading edge. The sharp leading edge would no doubt reduce the drag. However, in hypersonic flow, one cannot ignore the heat transfer to the body. Using similarity techniques, Van Driest[14] showed that the stagnation region wall heating rate $(q_w)_s$ is proportional to the square root of the edge velocity gradient $(\mathrm{d}u_e/\mathrm{d}s)_s$. DeJarnette et al.[20] show the relationship between the velocity gradient and the stagnation point radius R to be as follows:

$$\left(\frac{\mathrm{d}u_e}{\mathrm{d}s}\right)_s = \frac{V_\infty}{R}\sqrt{1.85\frac{\rho_\infty}{\rho_s}}$$

$$\therefore (q_w)_s \propto \sqrt{\left(\frac{\mathrm{d}u_e}{\mathrm{d}s}\right)_s} \propto \frac{1}{\sqrt{R}} \tag{11.39}$$

It is obvious from the preceding equation that a sharp nose ($R \to 0$) would never survive the stagnation point heating; therefore, all hypersonic vehicles require blunt leading edges. In fact, all manned vehicles present an almost flat leading edge ($R \to \infty$) in order to minimize the heat transfer rate [Eq. (11.39)], thereby maintaining a survivable atmosphere inside the vehicle; also the high drag helps reduce the velocity to enable a soft touchdown.

Now consider a re-entry body, say a blunt wedge (10-deg semivertex angle), traveling at $M_\infty = 15$. Using Eq. (11.34) and assuming $\gamma = 1.4$, the Mach number behind the detached normal shock is 0.38. For an ambient temperature of 288.1 K, the temperature behind the normal shock is 12,871 K, and the stagnation temperature on the nose (assuming only isentropic processes between shock and body) is 13,248 K [from Eq. (11.33)]; compare this to the Sun's surface temperature of about 6500 K! (Remember that perfect gas assumptions are not valid at these temperatures; for a real gas the stagnation temperature would be lower.)

Because of viscous dissipation (friction between adjacent fluid layers) caused by the no-slip boundary condition [Eq. (11.25)], the boundary layer is also a region of very high temperatures (see Fig. 11.8). Therefore, the high-temperature fluid in the boundary layer will transfer heat to the body until the temperature gradient at the wall is zero (adiabatic wall). White[13] has shown (for $q_w = 0$) that for a perfect gas with constant specific heats

$$T_w = T_{aw} = T_e\left(1 + \sqrt{Pr}\,\frac{\gamma - 1}{2}M_e^2\right)$$

where

$$Pr = \frac{\mu c_p}{k} = \frac{\text{Viscous dissipation}}{\text{Heat conduction}} = 0.72 \tag{11.40}$$

for air. The subscript e refers to the conditions at the edge of the boundary layer. Going back to our example and using Eq. (11.34) or flow tables,[16] we get a wedge shock angle of 13.5 deg and a temperature behind this shock of 955 K. The Mach number behind the oblique shock is 7.37, and the adiabatic wall temperature as calculated from Eq. (11.40) is 9758 K.

High-Temperature Effects

We have already seen that the approximation of air as a perfect gas breaks down at temperatures much lower than the stagnation temperature calculated above. The definition of the fluid must include the fact that the vibrational modes of the molecules have been excited at around 600 K and that at higher temperatures dissociation and ionization products will lead to chemical reactions. Therefore, the gas will consist of diatomic and monatomic particles as well as ions and electrons. We have already seen that monatomic gases like atomic oxygen and nitrogen, exhibiting fewer degrees of freedom, have thermodynamic properties different from those of diatomic gases like O_2 and N_2 [Eq. (11.5)]. The specific enthalpy h, the specific internal energy e, and even the gas constant R for the mixture will depend on the concentrations of the species x_i (e.g., x_{O_2}, x_O, x_{N_2}, x_{NO}, x_e). And since the concentrations depend on both temperature and pressure, the gas cannot be approximated even as a thermally perfect gas. We have

$$\begin{aligned} h &= h(T, x_i) = h(T, p) \\ e &= e(T, x_i) = e(T, p) \end{aligned} \tag{11.41}$$

The preceding equations assume that the equilibrium concentrations are reached instantaneously and result in *local equilibrium flow*. Local equilibrium flow is applicable when the characteristic chemical reaction time τ_c is much shorter than the fluid residence time τ_f. This situation usually occurs at lower altitudes where a higher density and shorter mean free path result in smaller reaction times.

Now consider an aero-assisted orbital transfer vehicle (AOTV). Such a vehicle would spend long periods of time in the low-density region of the Earth's atmosphere where aerodynamic forces, rather than retrorockets, are used to decelerate from a higher orbit to a lower one. In this low-density region, the shock layer may be in chemical and thermodynamic nonequilibrium. In *nonequilibrium flow*, when τ_c is comparable to τ_f, the concentrations of the various species will be functions of time in addition to being functions of temperature and pressure, i.e.,

$$
\begin{aligned}
h &= h[T, x_i(t)] = h[T, p, t] \\
e &= e[T, x_i(t)] = e[T, p, t]
\end{aligned}
\tag{11.42}
$$

Park[21] discusses nonequilibrium flow in detail.

In contrast to local equilibrium flow, when $\tau_c \gg \tau_f$ (i.e., reactions are assumed to proceed extremely slowly), the species concentrations may be assumed to be frozen in time; this condition is referred to as *frozen flow*. An example of frozen flow is the process in a shock. Even though the flow properties change dramatically in a shock, the thermodynamic and chemical properties of the fluid are essentially unchanged since the width of the shock is so small that the residence time is negligibly small, i.e., $\tau_f \approx 0$.

The high temperatures are important not only to the fluid properties, but also the material properties of the re-entry body. We must keep in mind that the temperatures reached in the region of the stagnation point are much higher than the melting point of most materials; unless some form of active cooling is provided, the stagnation region will ablate. In fact, controlled ablation is a common method of thermal management for re-entry bodies. The rate of ablation depends on the thermodynamic properties of the heat shield. Not only will the presence of the ablated materials in the flowfield complicate the chemical reaction process, but also the shape of the body itself will be altered, especially in the stagnation regions (see Fig. 11.16). Hence, the heat transfer rate not only influences the flowfield properties but also governs the rate of change of the shape of the nose tip. Furthermore, as the heat shield vaporizes, or *outgasses*, the process of the gases leaving the surface with a finite velocity alters the boundary conditions and hence the flowfield. The effects of the outgassing on the flowfield pressure are relatively unimportant, but a complication is introduced when the re-entry body is spinning.

A typical re-entry body will be assumed here to be a blunt nose cone. At an angle of attack, the heating rate will reach a maximum on the windward side and a minimum on the leeward side. Though there will be correspondingly high outgassing on the windward side compared to the leeward side (see Fig. 11.17a), the flowfield will be symmetric about the angle-of-attack plane. When the re-entry body is spinning about its axis of symmetry, the heating rate distribution will still be symmetric about the angle-of-attack plane. However, since the ablation process has a finite rate, there will be a phase lag between

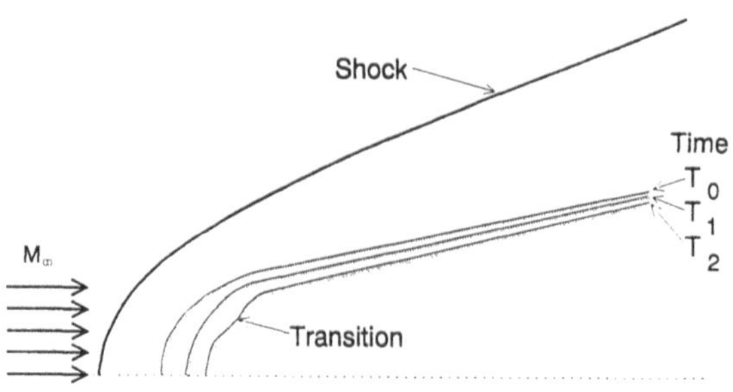

Fig. 11.16 Nosetip ablation history.

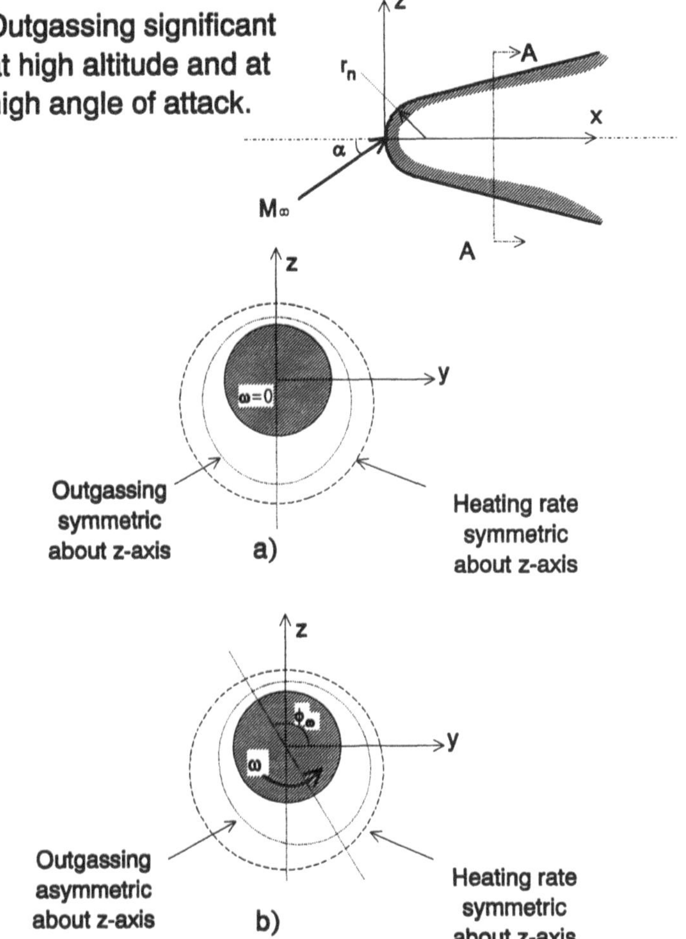

Fig. 11.17 Interaction of outgassing and spin: a) $\omega = 0$, b) $\omega \neq 0$.

the ablation distribution and the heating rate distribution (see Fig. 11.17b). This results in side forces and yawing moments that complicate the dynamics of the body and must be taken into account.

Finally, we must remember that in most applications of the energy equation, conduction is assumed to be the primary source of heat transfer. But for certain re-entry conditions, such as for AOTVs and planetary probes, the high temperatures ($\approx$ 10,000 K) result in radiation (as opposed to conduction) becoming the primary mechanism of heat transfer.

Viscous Interaction

We know that, for a constant-pressure process, an increase in temperature results in a lower density. Therefore, to maintain the same mass flow in the higher-temperature boundary layer associated with supersonic flows, the boundary-layer thickness will have to increase. It has been shown using similarity techniques that

$$\frac{\delta}{x} \propto \frac{M_\infty^2}{\sqrt{Re_x}} \tag{11.43}$$

Hence, for a given Reynolds number, a hypersonic ($M_\infty \gg 1$) boundary layer will be significantly thicker than a supersonic boundary layer.

Let us now consider supersonic flow past a flat plate with a sharp leading edge at zero angle of attack (see Fig. 11.18). We have seen that the outer inviscid flow may be determined by considering inviscid flow past a fictitious body with a shape of the original body augmented by the boundary-layer displacement thickness δ^*. In supersonic flow δ^* is negligible, and the pressure at the edge of the boundary layer $p_e(x)$ is the same as the freestream pressure p_∞.

However, in hypersonic flow, where δ is much larger [Eq. (11.43)] than in supersonic flow, the flow deflection caused by δ^* induces a leading edge shock. Associated with this boundary layer induced shock is an increase in $p_e(x)$ near the leading edge. This result should not be unexpected since we know that an increase in pressure is associated with an increase in temperature. This induced shock alters the outer region flowfield, which in turn affects the boundary layer, causing higher skin friction and heat transfer rates. This interaction between the boundary layer and the outer inviscid region is called *viscous interaction*; the two regions can no longer be determined independently. Viscous interaction is more pronounced near the leading edge, where $d\delta^*/dx$, and hence the flow deflection, is the greatest. A viscous interaction parameter χ allows us to evaluate the relative importance of the interaction between the inviscid and viscous regions as follows:

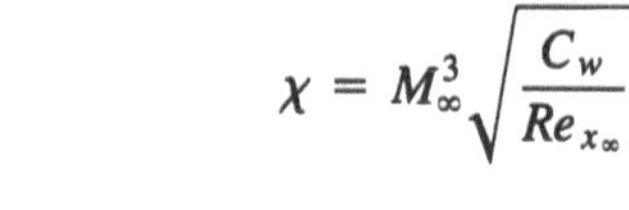

$$\chi = M_\infty^3 \sqrt{\frac{C_w}{Re_{x_\infty}}}$$

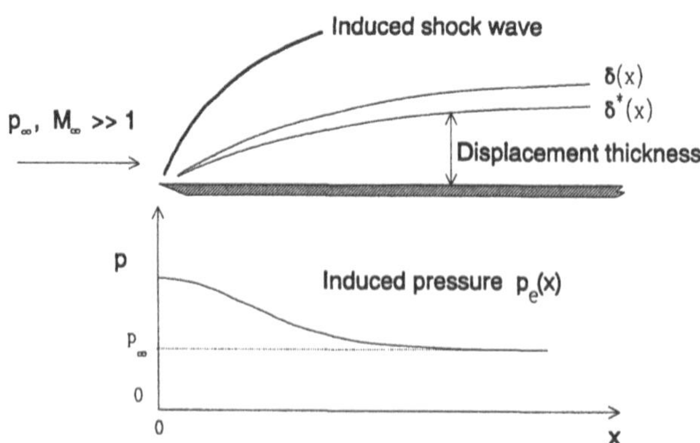

Fig. 11.18 Boundary layer induced shock wave and pressure distribution over flat plate.

where

$$C_w = \frac{\rho_w \mu_w}{\rho_\infty \mu_\infty} \tag{11.44}$$

C_w is the Chapman-Rubesin parameter, Re_{x_∞} is the Reynolds number based on the distance x from the leading edge and the freestream conditions. In regions of strong viscous interaction ($\chi \gg 1$), the outer inviscid and boundary-layer flows must be solved simultaneously.

If one assumes that the boundary layer acts as slender wedge with an angle equal to the displacement thickness gradient, it can be shown that the ratio of the induced pressure to the freestream pressure p_i/p_∞ is proportional to the viscous interaction parameter χ for strong interaction ($\chi > 3$). The correlation of induced pressure for an insulated flat plate in air, taken from Hayes and Probstein's classic work, is illustrated in Fig. 11.19.

Let us define a nondimensional coefficient of pressure C_p as

$$C_p \equiv \frac{p - p_\infty}{q_\infty} = \frac{2}{\gamma M^2}\left(\frac{p}{p_\infty} - 1\right) \approx \frac{2}{\gamma M^2}\frac{p}{p_\infty} \tag{11.45}$$

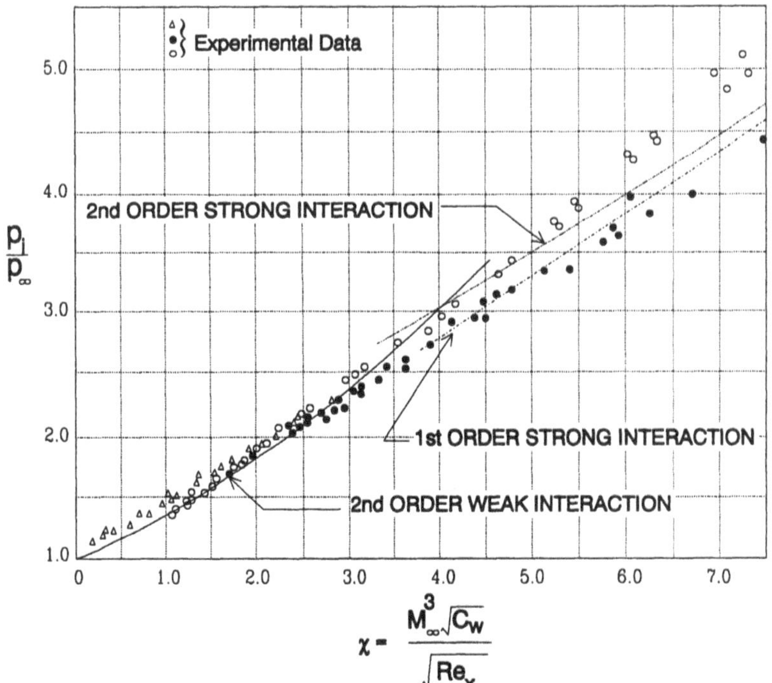

Fig. 11.19 Weak interaction and strong interaction correlations (from Hayes and Probstein[1]).

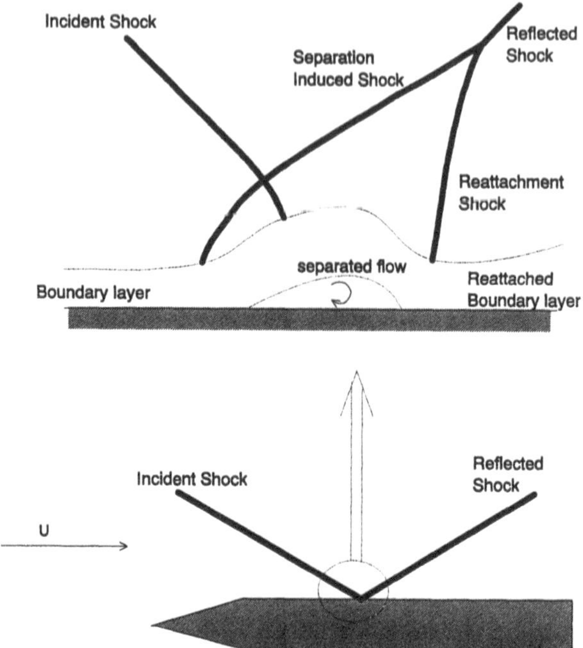

Fig. 11.20 Boundary layer/shock wave interaction.

Since for hypersonic flow $p/p_\infty \gg 1$.

From Fig. 11.19 we see that $p_i/p_\infty \propto \chi$, so we can define a new viscous interaction parameter $\bar{V}$ as

$$C_p \propto \frac{M_\infty}{\sqrt{Re_{x_\infty}}} \sqrt{C_w} \equiv \bar{V} \tag{11.46}$$

If we look back to Eq. (11.13), we see that $\bar{V}$ is similar to the Knudsen number for viscous flow.

Another form of viscous-inviscid interaction occurs when a shock wave impinges on a surface and interacts with the boundary layer. The flow deflection caused by the impinging shock causes local flow separation, and the increased pressure causes the boundary layer to reattach. The reattachment of the boundary layer creates a shock that merges with the reflected shock (see Fig. 11.20). This region of viscous-inviscid interaction is a region of very high heat transfer rates due to the thinner boundary layers caused by the interaction; it has the potential to cause catastrophic failure if not adequately accounted for.

Entropy Gradients

We have seen that to reduce the heat transfer rate, a re-entry body must have blunt, rather than sharp, leading edges. This results in a normal shock at the leading edge, which curves around to an oblique (or conical) shock downstream.

The entropy increase of a streamline passing through the normal shock at the leading edge will be much higher than that of a streamline passing through the oblique shock. This entropy gradient results in a rotational ($\Omega = \nabla \times V \neq 0$) flowfield.

In the outer flow, the Euler equations can be solved by the method of characteristics with some modifications to account for the entropy gradient normal to the streamlines. (Entropy is still constant along a streamline.) In addition to compatibility equations along the Mach lines, two more compatibility equations must be satisfied along the streamline passing through point 3 (see Fig. 11.15). The rotational method of characteristics is discussed by Zucrow and Hoffman.[19]

The boundary layer, itself a region of rotational flow, has imposed on it an entropy layer, which is also a region of rotational flow. Therefore, in addition to specifying $p_e(x)$ and $u_e(x)$, we need to specify the edge entropy $s_e(x)$ also. In irrotational outer flow, the vorticity created in the boundary layer is convected downstream much more quickly than it can diffuse, thereby keeping the boundary layer thin. When the outer flow is rotational, the vorticity from the outer flow interacts with the boundary layer vorticity, creating computational challenges.

Thin Shock Layer

Consider a 10-deg wedge re-entering the atmosphere at $M_\infty = 20$. From the θ-β-M map,[16] we see that the shock angle $\beta = 12.8$ deg. We also know from Eq. (11.35) that the rise in density across a shock increases with M_∞, i.e.,

$$\lim_{\substack{M_\infty \to \infty \\ \gamma \to 1}} \frac{\rho_2}{\rho_\infty} \to \frac{\gamma + 1}{\gamma - 1} \to \infty \tag{11.47}$$

thereby allowing the mass flow to "squeeze" through a narrower region, resulting in a thin *shock layer*—the region between the shock and the body. Similarly, from Eq. (11.35), in the hypersonic limit ($M_\infty \to \infty$) for a slender body (small θ and hence small β), we have

$$\lim_{\gamma \to 1} \frac{\beta}{\theta} \to \frac{\gamma + 1}{2} \to 1 \tag{11.48}$$

With a thick boundary layer in a thin shock layer, it seems logical that there would be significant interaction between the viscous boundary layer and the inviscid outer layer. This interaction violates many of the assumptions made in supersonic flow, complicating the solution process. Fortunately, the thin shock layer allows other approximations that greatly simplify the solution process; these will be discussed in the next sections.

Since the solution of the full Navier-Stokes equations is very computationally intensive and the boundary-layer equations do not model the interaction between the inviscid outer layer and the viscous boundary layer, a compromise set of equations known as the *thin-layer*, or *parabolized*, Navier-Stokes equations is usually used. The parabolized equations are a mixed set of hyperbolic-parabolic equations in the streamwise direction and are easier to solve than the elliptic Navier-Stokes equations. Unlike the boundary-layer equations, the parabolized equations are valid in both the viscous and inviscid regions of the flowfield; Anderson discusses these at length.[12]

Having noted the complexities raised by the interactions between the viscous boundary layer and the inviscid outer layer, let us for the moment concentrate on the boundary layer alone. We have seen that the wall shear stress and heat transfer depend on the velocity and temperature gradients at the wall. Therefore, when the velocity and temperature profiles in the boundary layer change, the new gradient at the wall will determine the shear stress and heat transfer rate.

We know that the governing equations are the Navier-Stokes equations. If the solution to this complicated set of equations could be determined for specified boundary conditions, the pressure, velocity, and temperature would be known. Hence, the forces and moments acting on the body and the heat transfer could be calculated. If the boundary conditions are steady, we would expect the solution also to be steady, but we know that at the microscopic level the flow is quite random. When the flow properties are randomly unsteady, we can resolve any property into a mean value (represented by an over bar) plus a fluctuating value (represented by a prime), i.e.,

$$Q = \bar{Q} + Q' \tag{11.49}$$

where

$$\bar{Q} = \frac{1}{T}\int_{t_0}^{t_0+T} Q \, \mathrm{d}t$$

and the integration period T is large compared to the fluctuation period. All of the flow properties in the Navier-Stokes equations can be resolved into mean and randomly fluctuating components.

If the time averages are taken, the resulting equations are similar to the Navier-Stokes equations. The differences are that the flow properties are replaced by their mean values and that there is an additional term called the Reynolds stress term, which is a function of the turbulent inertia tensor. The nine components of this tensor are additional unknowns introduced into the problem without accompanying equations to determine them.

$$\begin{aligned} \text{Reynolds stress term} &\equiv \rho \frac{\partial}{\partial x_j}\left(\overline{u_i' u_j'}\right) \\ \text{Turbulent inertia tensor} &\equiv \overline{u_i' u_j'} \qquad i, j = 1, 2, 3 \end{aligned} \tag{11.50}$$

where $u_i = \bar{u}_i + u_i'$ is the component of velocity along the x_i direction. To define the inertia tensor, detailed knowledge of turbulent structure is required.

When the fluctuations about the mean are small, the Reynolds stress term is negligible, and the flow is considered *laminar*. For laminar flow the time-averaged equations are identical to the Navier-Stokes equations. In a stable laminar boundary layer, any disturbance (e.g., induced by freestream turbulence, wall roughness, etc.) diminishes rapidly and becomes negligible. But when the laminar boundary layer becomes unstable, the disturbances start growing. The instability initially manifests itself as unstable two-dimensional waves, called *Tollmien-Schlichting* waves. The Tollmien-Schlichting waves develop into three-dimensional eddies, which break down into regions of high local shear with three-dimensional fluctuations. These regions of high shear coalesce and in this intensely fluctuating flow, called *turbulent flow*, the Reynolds stress term is never negligible. Since there are no physical laws currently available to determine the turbulent inertia tensor, it must be empirically determined.

Unfortunately, the transition process is one of the least understood aspects of fluid dynamics. The transition from laminar to turbulent flow occurs over a finite distance. However, for most engineering purposes, transition is approximated to occur at a point, and the process between the stable laminar flow and the fully turbulent flow is ignored. The flow upstream of this point is assumed to be laminar, and downstream of this point the flow is considered turbulent. The transition point is usually empirically correlated to a Reynolds number based on local flow conditions and either the distance from the leading edge or the local momentum thickness.

Refer again to Fig. 11.1. The boundary-layer on a body above the boundary-layer transition altitude is by definition completely laminar. At the transition altitude, which in turn depends upon several factors, the boundary layer near the base of the body becomes turbulent, and the transition point moves forward as the flow conditions change. For most ballistic applications, where the altitude change is rapid, the transition location virtually jumps to the nose of the body. However, for advanced concepts such as the powered aerospace plane, the movement of the transition location may be more gradual, and in applications such as SCRamjets, the location of transition, is critical to successful operation.

The high level of mixing due to the intense fluctuations in turbulent flow causes the high-energy fluid near the boundary layer edge to interact with the low-engery wall flow and the fluid outside the boundary layer. This results in higher velocity and temperature gradients near the wall and a much thicker boundary layer. The higher temperature and velocity gradients near the wall (see Fig. 11.21), result in significantly higher shear stress and heat transfer rates at the wall in turbulent flows.

The higher shear stress means higher drag forces. In hypersonic flow, viscous forces are on the order of 10–15% of the total forces, and the accuracy of transition prediction may not appear important. Actually, the forces and moments are affected more by the higher ablation rate associated with the higher heat transfer rate. The change in forces due to ablation-induced shape change can be significant. If part of the body has a laminar boundary layer and the rest a turbulent boundary layer, the ablation rates and hence the body shape changes will be discontinuous. In addition, asymmetric transition can result in asymmetric ablation of the nosetip and hence significant moments.

It is therefore imperative that the transition point be known accurately. The location of the transition point is influenced by numerous major and minor factors, often interrelated and difficult to interpret. The most valuable information that can be obtained from the great mass of available transition data is the

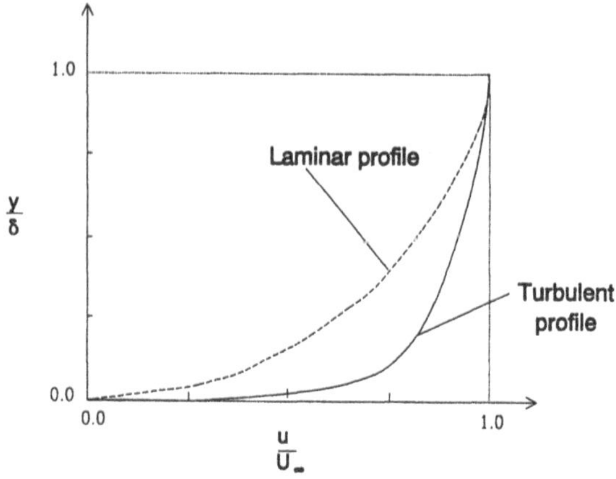

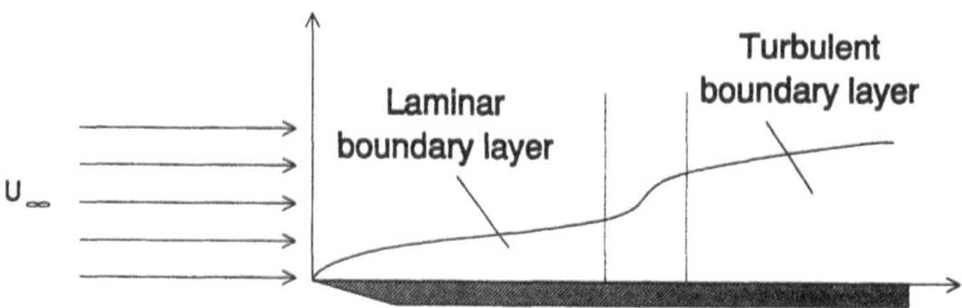

Fig. 11.21 Laminar and turbulent boundary-layer velocity profiles.

trend of the data, not the absolute magnitude of the transition Reynolds number. According to Stetson,[22] the factors that are thought to have a major effect on the transition Reynolds number include the Mach number, nosetip bluntness, angle of attack (crossflow), freestream Reynolds number (unit Reynolds number), freestream turbulence (environment), wall temperature, surface roughness, pressure gradient, and outgassing (mass transfer).

Having waded through the complexities of rigorously determining the flowfield and the forces and moments on bodies in hypersonic flow, let us move on to some approximate methods.

11.9 Impact Methods

In the early vehicle design stage, it is impractical to use the Navier-Stokes equations, or even the parabolized Navier-Stokes equations, because of attending computational complexities. Instead, numerically simple techniques known as *impact methods* are used for approximating the forces and moments acting on the body. Impact methods assume that the pressure at any point depends primarily upon the shape of the body at that point and the freestream conditions.

In the seventeenth century, Sir Issac Newton postulated a theory to determine the fluid dynamic forces on a body in incompressible flow. According to *New-*

tonian impact theory, a fluid particle impinging on the surface of a body would lose all momentum normal to the surface, whereas the momentum tangential to the surface would be unaffected. This model assumes that the fluid particles have no random velocity component usually associated with a microscopic particle of gas. Let us use this theory to determine the forces acting on a flat plate of area A inclined to the freestream flow V_∞ at an angle θ (see Fig. 11.22a). The volume of fluid $\mathrm{d}\upsilon$ impinging on the flat plate in time $\mathrm{d}t$ is

$$\mathrm{d}\upsilon = (V_\infty\,\mathrm{d}t)(A\sin\theta) \tag{11.51}$$

and the mass flux is

$$\dot{m} \equiv \frac{\mathrm{d}m}{\mathrm{d}t} = \rho_\infty V_\infty A \sin\theta \tag{11.52}$$

where ρ_∞ is the freestream density.

Newton's Second Law states that the force exerted by the fluid is equal to the time rate of change of momentum of the fluid. We can resolve the fluid velocity

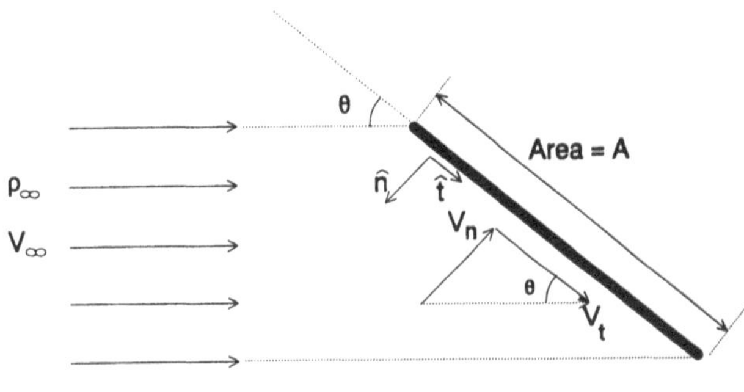

a) Flat plate in uniform flow

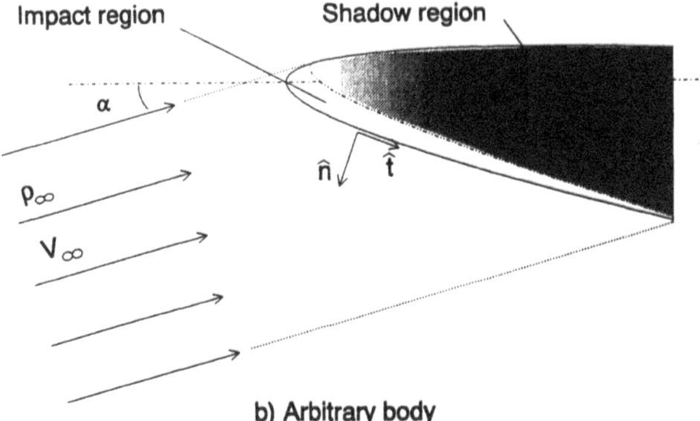

b) Arbitrary body

Fig. 11.22 Newtonian aerodynamics.

into components that are normal, $V_n = V_\infty \sin\theta$, and tangential, $V_t = V_\infty \cos\theta$, to the flat plate. Since the tangential momentum is unchanged, only the normal momentum exerts a force, i.e.,

$$F_n \equiv -\dot{m}\Delta V_n = \rho_\infty V_\infty^2 A \sin^2\theta$$
$$F_t \equiv -\dot{m}\Delta V_t = 0 \tag{11.53}$$

and, therefore, Newtonian C_p and C_f for the windward or *impact* side are

$$C_p \equiv \frac{\Delta p}{q_\infty} = 2\sin^2\theta \qquad C_f \equiv \frac{\Delta\tau}{q_\infty} = 0 \tag{11.54}$$

Newtonian theory assumes that the leeward or *shadow* side does not interact with the flow (see Fig. 11.22b) and hence there is no change in the leeward pressure and shear (leeward $C_p = C_f = 0$).

For a given shape, C_p may be integrated over the surface of the body to determine the Newtonian forces and moments. One must be careful in distinguishing between the windward side and the leeward (or shadow) regions of the body.

For simple shapes it is possible to integrate Eq. (11.54) analytically to obtain the forces and moments. It is relatively easy to show that for a sphere the drag coefficient $C_D = 1$. The results for a sharp cone[23] are

$$C_A = 2\sin^2\theta_c + \sin^2\alpha(1 - 3\sin^2\theta_c)$$
$$C_N = \cos^2\theta_c \sin 2\alpha \tag{11.55}$$
$$C_M = -\tfrac{2}{3}\left(C_N/\tan\theta_c \cos^2\theta_c\right)$$

where θ_c is the cone angle, α is the angle of attack (see Fig. 11.23a), C_A is the axial force coefficient, C_N is the normal force coefficient, and C_M is the pitching moment coefficient. Equation (11.55) assumes that there are no shadow surfaces, i.e., $\theta_c > \alpha$. For a sphere cone, the sphere and the sharp cone can be merged such that the slope is continuous (see Fig. 11.23b). The drag coefficient of the truncated sphere that forms the nose can be shown to be

$$C_{D_{\text{nose}}} = 1 - \sin^4\theta_c \tag{11.56}$$

Subtracting the drag contribution of the front of the sharp cone from the complete sharp cone drag at $\alpha = 0$ and adding the spherical nose contribution gives

$$\begin{aligned} D_{\text{total}} &= D_{\text{nose}} + D_{\text{frustum}} \\ &= (1 - \sin^4\theta_c)q_\infty\pi r_n^2 + 2\sin^2\theta_c q_\infty \pi(r_b^2 - r_n^2\cos^2\theta_c) \end{aligned} \tag{11.57}$$

where r_n is the radius of the spherical nose and r_b is the cone base radius.

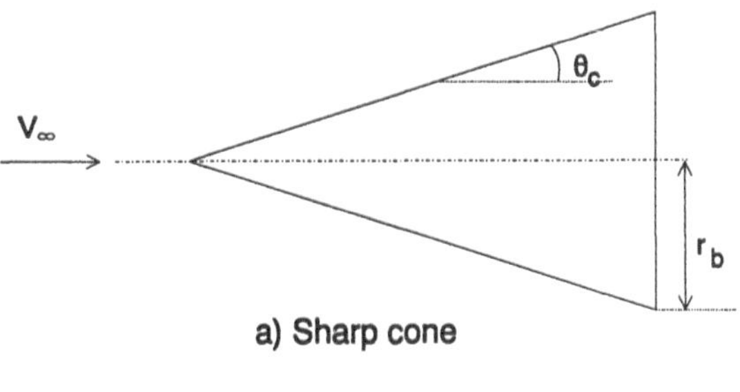

a) Sharp cone

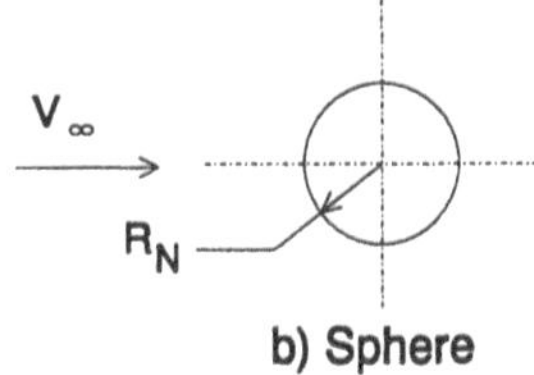

b) Sphere

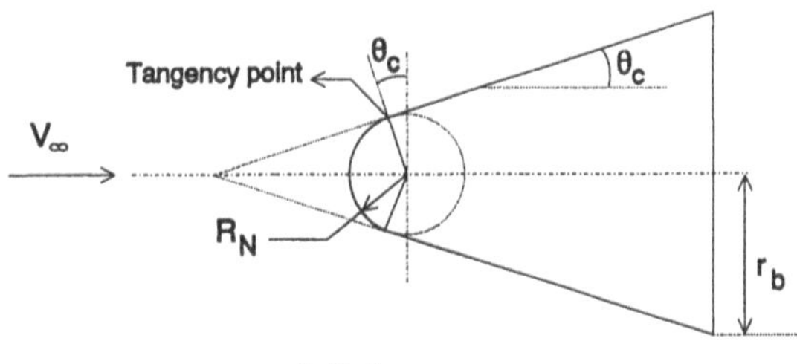

c) Sphere cone

Fig. 11.23 Simple bodies for Newtonian aerodynamics.

If one defines the bluntness ratio, or simply the bluntness, as $B = r_n/r_b$, the drag coefficient based on r_b of the complete sphere cone at $\alpha = 0$ becomes

$$C_{D_{\text{sphere cone}}} = 2\sin^2\theta_c + B^2(1 - 2\sin^2\theta_c\cos^2\theta_c - \sin^4\theta_c) \tag{11.58}$$

The forces and moments for any body of revolution can be approximated by treating the body as a series of truncated sharp cones (see Fig. 11.24) and using Eq. (11.55) to give

$$C_A = C_{A_1}\frac{r_1^2}{r_N^2} + \sum_{i=2}^{N} C_{A_i}\frac{r_i^2 - r_{i-1}^2}{r_N^2} \qquad C_N = C_{N_1}\frac{r_1^2}{r_N^2} + \sum_{i=2}^{N} C_{N_i}\frac{r_i^2 - r_{i-1}^2}{r_N^2}$$

$$C_M = C_{M_1}\frac{r_1^3}{r_N^3} + \sum_{i=2}^{N}\left[C_{M_i}\frac{r_i^3 - r_{i-1}^3}{r_N^3} - C_{N_i}\frac{(r_i^2 - r_{i-1}^2)}{r_N^2}\frac{(s_i - x_i)}{r_N}\right] \tag{11.59}$$

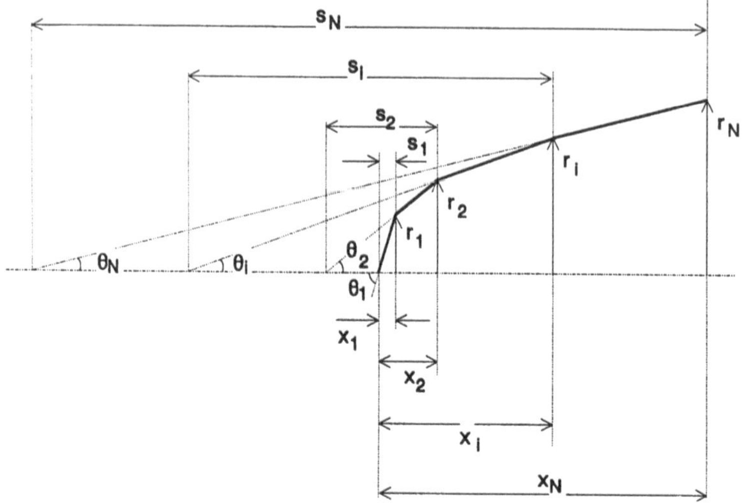

Fig. 11.24 Geometry for aerodynamics of body of revolution.

The program to calculate the forces and moments for a body of revolution is listed in Table 11.3.

Upon comparing Newtonian theory to the limiting free molecular flow case of Eq. (11.23), we see that the results are identical if $\sigma_N = 1$ (incident normal momentum absorbed by surface) and $\sigma_T = 0$ (tangential momentum unchanged). It is therefore possible to use the program written for free molecular flow (see Table 11.2) to determine the continuum (Newtonian) aerodynamics by setting $s_\infty \gg 1$, $\sigma_N = 1$, and $\sigma_T = 0$. The Newtonian C_p on each panel depends only on the slope of that panel relative to the freestream. Using $s_\infty \gg 1$ automatically accounts for whether or not the panel is shadowed because of the characteristics of the error function in Eq. (11.19) [$\text{erf}(-\infty) = -1$ and $\text{erf}(\infty) = 1$].

From Eq. (11.34), we have the following for the pressure ratio across an oblique shock in the hypersonic limit:

$$\lim_{M_\infty \to \infty} C_p \to \frac{4}{\gamma + 1} \sin^2 \beta \tag{11.60}$$

We saw from Eq. (11.48) that in hypersonic flow $\beta \approx \theta$ for $\gamma = 1$. Again setting $\gamma = 1$, Eq. (11.60) gives a result identical to Eq. (11.54), lending further credence to the soundness and applicability of Newtonian theory.

We can see from Fig. 11.25 that the Newtonian pressure distribution over a sphere cone compares favorably, until the tangency point is reached, to that calculated using the Euler equations. (Morrison et al.[24] have tabulated the pressure distribution over a range of sphere cones for a range of Mach numbers using an Euler solver.) Since the Newtonian result depends upon only the local surface

Table 11.3 FORTRAN program to calculate Newtonian forces and moments

```
      PROGRAM ARBCONE
C *** NEWTONIAN AERODYNAMICS FOR ARBITRARY BODIES OF REVOLUTION
C *** This program does not account for shadowing. Therefore,
C *** angle of attack should always be less than the cone angle.
C *** PROGRAMMER: CARLOS A. LOPEZ
      REAL X(500), R(500), T(500), S(500)
C *** DEFINE PROGRAM CONSTANTS
      PI = ACOS (-1.0)
      RD = PI / 180.0
C *** READ GEOMETRY (500 POINT LIMIT)
C *** Caution: X(i) should never be .GE. X(i+1)
      J = 1
5       READ (5, *, END = 10) X(J), R(J)
        J = J + 1
        GO TO 5
10    CONTINUE
      N = J - 1
C *** LOCAL CONE ANGLES AND VIRTUAL VERTEX DISTANCES
      S(1) = X(1)
      T(1) = ATAN (R(1) / X(1))
      DO I = 2, N
      T(I) = ATAN ((R(I) - R(I-1)) / (X(I) - X(I-1)))
      S(I) = R(I) / ((R(I) - R(I-1)) / (X(I) - X(I-1)))
      END DO
C *** NEWTONIAN AERODYNAMICS BASED ON A SHARP CONE (CONESUB)
      WRITE (6, *) ' '
      WRITE (6, '(15X, A)') 'NEWTONIAN ARBITRARY BODY AERODYNAMICS'
      WRITE (6, *) ' '
      DO I = 1, 91
        ALPHA = FLOAT (I - 1) * RD
        IF (T(1) .GT. 0.0) THEN
          CALL CONESUB (T(1), ALPHA, CAI, CNI, CMI)
          CA = CAI * R(1) ** 2 / R(N) ** 2
          CN = CNI * R(1) ** 2 / R(N) ** 2
          CM = CMI * R(1) ** 3 / R(N) ** 3
        END IF
        DO K = 2, N
          IF (T(K) .GT. 0.0) THEN
            CALL CONESUB (T(K), ALPHA, CAI, CNI, CMI)
            CA = CA + CAI * (R(K) ** 2 - R(K-1) ** 2) / R(N) ** 2
            CN = CN + CNI * (R(K) ** 2 - R(K-1) ** 2) / R(N) ** 2
            CM = CM + CMI * (R(K) ** 3 - R(K-1) ** 3) / R(N) ** 3
                 +(CNI * (R(K) ** 2 - R(K-1) ** 2) / R(N) ** 3)
               *(S(K) - X(K))
          END IF
        END DO
```

(continued on next page)

Table 11.3 (continued) FORTRAN program to calculate Newtonian forces and moments

```
C  ***   PRINT RESULTS
         AA = ALPHA / RD
         WRITE (6, 15) 'AA = ',AA, 'CA = ',CA, 'CN = ',CN, 'CM = ',CM
15       FORMAT (4(2X, A, F10.4))
      END DO
      STOP
      END
C
      SUBROUTINE CONESUB (THETA, ALPHA, CA, CN, CM)
C  ***   NEWTONIAN AERODYNAMICS OF A SHARP CONE
      PI = ACOS(-1.0)
      SINT = SIN (THETA)
      COST = COS (THETA)
      TANT = TAN (THETA)
      SINA = SIN (ALPHA)
      COSA = COS (ALPHA)
      TANA = TAN (ALPHA)
      T1 = 1.0 - 3.0 * SINT ** 2
      CA = 2.0 * SINT ** 2 + SINA ** 2 * T1
      CN = COST ** 2 * SIN (2.0 * ALPHA)
      IF (ALPHA .GT. THETA) THEN
         BETA = ASIN (TANT / TANA)
         COSB = COS (BETA)
         T2 = (BETA + PI / 2.0) / PI
         T3 = TANT / TANA
         T4 = COSB * SIN (2.0 * ALPHA) * SIN (2.0 * THETA)
         T5 = COSB / (3.0 * PI)
         CA = CA * T2 + 3.0 * T4 / (4.0 * PI)
         CN = CN * (T2 + T5 * (T3 + 2.0 / T3))
      END IF
      CM = -CN * 2.0 / (3.0 * TANT * COST * COST)
      RETURN
      END
```

angle, effects of complex phenomena such as overexpansion and recompression associated with the tangency point are absent. The forces determined by Newtonian theory closely match experimental results (see Fig. 11.26). However, the discrepancies in the pressure distribution near the nose can lead to significant inaccuracies in the moments.

To understand why Newtonian theory does so well qualitatively, picture the thin shock layer in hypersonic flow: the shock and body are almost coincident, and the freestream impinging on the body/shock appears to lose its normal momentum and move parallel to the body/shock. The Newtonian theory models this phenomenon and therefore proves to be quite a good approximation for hypersonic flow (but fails in the incompressible flow regime for which it was postulated).

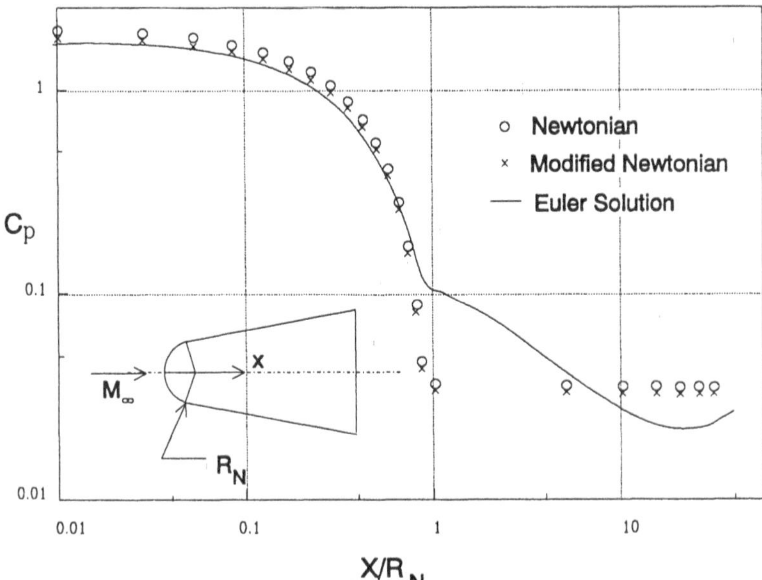

Fig. 11.25 Coefficient of pressure comparisons (α = 0 deg, M_∞ = 30).

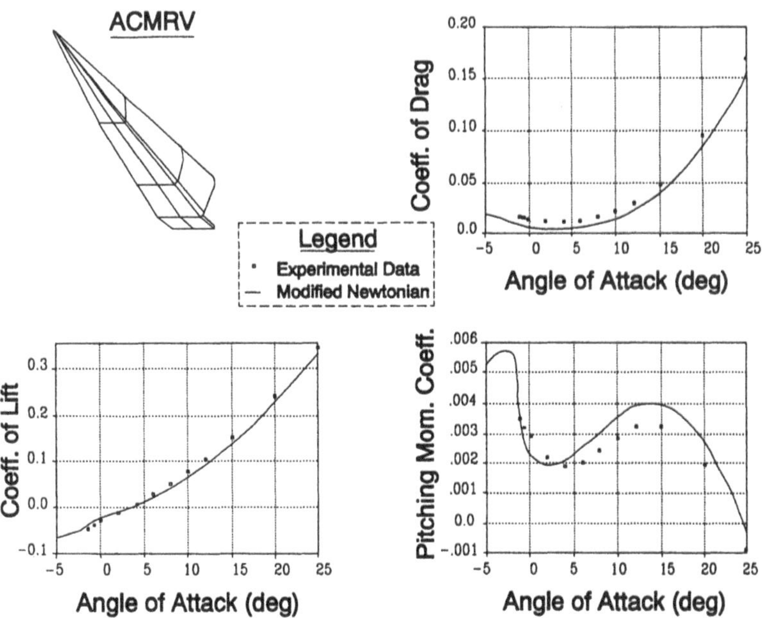

Fig. 11.26 Comparison of modified Newtonian force and moment predictions with experimental data.

Lees[25] improved upon Newtonian theory (see Fig. 11.25) by modifying the constant 2.0 in Eq. (11.54) by using $C_{p_{\max}}$, the stagnation pressure coefficient behind the normal shock at the leading edge of the body, i.e.,

$$C_{p_{\max}} = \frac{p_{0_2} - p_\infty}{q_\infty} = \frac{(p_{0_2}/p_\infty) - 1}{(\gamma/2)M_\infty^2} \tag{11.61}$$

where the subscript 0_2 refers to the stagnation value behind the shock [Eqs. (11.33) and (11.34)]; these values are tabulated in Ref. 16. The *modified Newtonian* theory is a function of the local surface inclination and the freestream Mach number. Other improvements that have been made to Newtonian theory are discussed in some detail by Hayes and Probstein[1] and Truitt.[2]

Unlike Newtonian theory, most of the other impact methods are not supported by the physics of the flow. Instead they are based upon intuition and/or empiricism. Two of the more frequently used methods are the *tangent cone* and *tangent wedge* methods. These methods predict the pressure at any point on an arbitrary body by assuming that the point is on an imaginary sharp cone/wedge and by using the cone/oblique shock tables (from Refs. 17 and 16, respectively). The imaginary cone/wedge angle is equal to the local deflection angle of the body.

A FORTRAN program in the public domain known as the supersonic/hypersonic arbitrary body program (S/HAB)[26] includes 15 different impact methods and an integral boundary layer solution for approximate but efficient calculation of pressure and shear forces on arbitrary bodies. This includes empirical methods that work well for specific bodies; some methods are applicable even down to the high supersonic regime.

11.10 Transition Flow Regime

As mentioned earlier, both intermolecular and molecule-surface interactions are important in the transition flow regime. This regime is not as well understood as the free molecular and continuum regimes.

The transition regime may be broadly subdivided into three regions based on the generally predominant phenomena in that region: 1) nearly free molecular flow, 2) merged viscous region, and 3) slip flow.

In free molecular flow (discussed in Section 11.6) the mean free path is so large that particles emitted from the surface of the body do not interact with the freestream particles until they are far downstream of the body. As the re-entry body descends into the denser atmosphere, the flowfield is still dominated by molecule-surface interactions, but shorter mean free path results in some (not negligible) interaction between the incident and the emitted particles. This part of the transition regime is referred to as the *nearly free molecular flow* region ($0.1 < M/Re < 3$).

To compute the nearly free molecular flowfield, the density, velocity, and temperature of the emitted particle cloud, assuming purely free molecular flow, are first computed. Then a small percentage of the freestream particles are allowed to be scattered by the emitted particle cloud. This requires an intermolecular collision model in addition to molecule-surface interactions. (Two very simple molecule-surface interaction models were discussed in Section 11.6.)

As the transition regime is entered from the continuum flow regime, it is assumed that the first failure of continuum theory occurs at the fluid-surface interface. The continuum boundary condition of continuity of tangential velocity and temperature [Eqs. (11.25) and (11.26)] at the surface break down, leading to slip and temperature-jump boundary conditions. This region adjacent to the continuum regime is called the *slip flow* region ($0.1 > M/(Re)^{1/2} > 0.01$). In the boundary layer, a region near the surface known as the *Knudsen layer* exists where molecule-surface interactions are also important. In continuum flow the Knudsen layer is so thin (on the order of mean free paths) that its effects are negligible, and the flow is dominated by intermolecular collision effects. However, the lower density in the slip flow region allows the presence of Knudsen layer to be felt. Although the idea of "continuum (Navier-Stokes equation)-plus-slip" is not supported by the Boltzmann equation, it has yielded a substantial body of practical and satisfactory results.[27]

Upon moving to a higher altitude from the slip flow region, the continuum model of a shock breaks down, i.e., the shock can no longer be treated as a discontinuity. The lower density causes the shock to "smear" out and merge with the boundary layer. Here the intermolecular and molecule-surface interactions take on equal importance. When this region is approached from the nearly free molecular side, the formation of a shock and boundary layer starts to become apparent. Most of the incident particles will interact several times with emitted particles before impinging on the body, creating the predecessors to a shock and boundary layer.

Recent attempts to solve this part of the transition regime with the direct-simulation Monte Carlo (DSMC) method have been quite fruitful. In this method, a very large number (10^4) of simulation particles, each representing significantly more (10^{12}) real particles, are followed as they progress through the flowfield. The simulation particles interact with other particles and the surface of the body. Again, as in nearly free molecular flow, intermolecular and surface-molecule interaction models are required. Figure 11.27 shows excellent agreement between DSMC and experiment.[28]

Rather than solving the computer-intensive Boltzmann equations or the Navier-Stokes-plus-slip equations, a more practical but approximate solution is a smooth fit, or bridging function, between the more easily obtainable continuum and free molecular solutions. In this technique, the normalized force coefficient $\bar{C}_X$ is assumed to be a function of the Knudsen number. $\bar{C}_x$ is given by

$$\frac{C_x - C_{x_{\text{cont}}}}{C_{x_{FM}} - C_{x_{\text{cont}}}} = \bar{C}_x = F(Kn) \tag{11.62}$$

Experimental data are then correlated to determine this function empirically. The best possible solution would be one correlation function that would be applicable to all bodies at any angle of attack and temperature for all forces and moments. It would be more realistic to expect a different correlation for each class of bodies to provide more accurate results. However, lack of experimental data in the early design stages requires using either a correlation that attempts to encompass different classes of bodies or a correlation derived for a similar body.

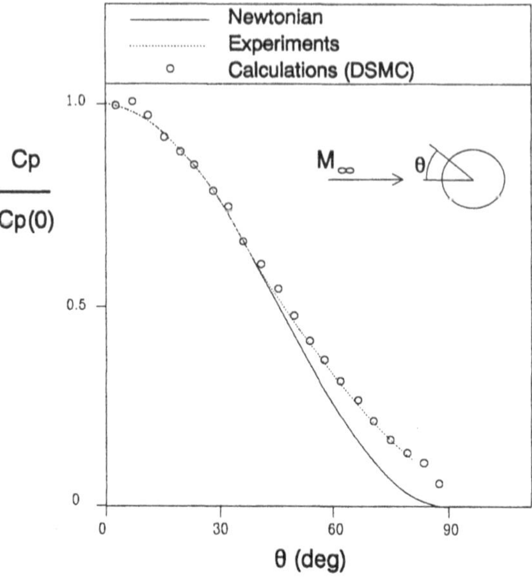

Fig. 11.27 Pressure distribution along the body surface (M_∞ = 22.4, Kn = 0.1).

Koppenwallner and Legge[11] found that the normalized drag coefficient correlates well with $Kn_0/\sin\theta_c$ for a sharp cone, where the Knudsen number Kn_0 is based upon the stagnation viscosity (see Fig. 11.28). To accommodate experimental data for spheres, cylinders, and cones, Gorenbukh[28] uses the following error function correlation:

$$F = \phi(x) = \frac{1}{\sqrt{2\pi}} \int_{-\infty}^{x} e^{-y^2/2} dy \tag{11.63}$$

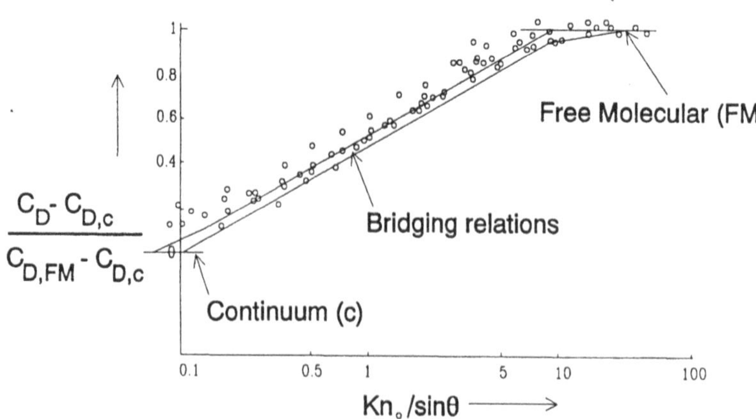

Fig. 11.28 Cone drag in the transition regime.

where

$$x = \log_{10}(Kn) + a/\sigma$$

The Knudsen number is based upon a diameter of the body (base diameter for cones). Statistical manipulation of the data resulted in $a = 0.878$ and $\sigma = 0.77$. Although this correlation falls into the middle of the data, the scatter is significant (see Fig. 11.29). Instead of using Kn as the correlation parameter, a Reynolds number function based on either stagnation viscosity[29] or reference conditions[30] improves the correlation. Potter[30] uses the same correlation for both drag and lift coefficients with a simple modification. The drag coefficient uses a function of the projected frontal area, whereas the lift coefficient uses both the projected frontal area and the projected planform area.

For the space shuttle data book, a sine function of Kn is used to correlate both the normal and axial force coefficients in the transition regime,[31] i.e.,

$$\bar{C}_A = \sin^2 \omega$$
$$\bar{C}_N = \sin^2 \omega \tag{11.64}$$

where

$$\omega = \pi \frac{3 + \log_{10} Kn}{8}$$

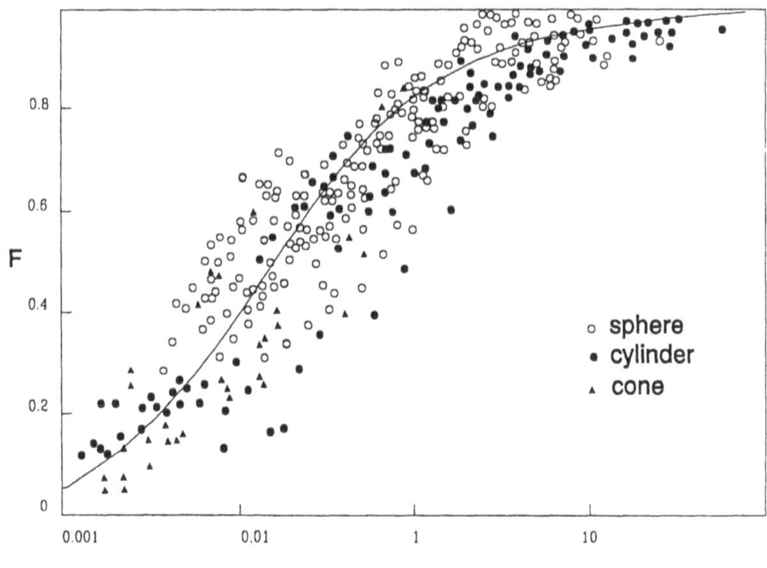

Fig. 11.29 Transition regime correlation.

Here Kn is based on the freestream conditions and the mean aerodynamic chord. The correlation was modified to account for the difference between the axial and normal force coefficients. The modification term is a function of the projected normal-to-axial area ratio. The correlation is continually modified as flight data are accumulated.

11.11 Numerical Example

Having briefly addressed a wide range of complexities associated with determining the exact forces and moments acting on a re-entry body, let us use some of the simplifications and approximations discussed in the preceding sections to determine approximate axial and normal coefficients for a simple sphere cone. We will consider a 10-deg cone with base radius of 1 meter and a bluntness of 0.1 re-entering at Mach 20. If one assumes a perfect gas, the specific heat ratio γ is 1.4 and the freestream speed ratio $s_\infty = M_\infty(\gamma/2)^{1/2}$ is 16.733.

First we need to determine the flow regime as a function of the re-entry body's altitude. From Fig. 11.3a we see that above an altitude of 125 km ($Kn \approx 10$) the re-entry body is in the free molecular regime and below 80 km ($Kn \approx 0.01$) in the continuum regime.

The force coefficients in the free molecular regime may be determined using Eqs. (11.19) and (11.20). As we mentioned earlier, a reasonable assumption is $a_c \approx \sigma_N \approx \sigma_T \approx 1$. Assuming the re-entry body traverses the free molecular regime in a short period of time, we would not expect any significant surface heating, and so $T_w \approx T_\infty$. Employing the program listed in Table 11.2, we see that for a zero angle of attack $C_A = 2.0141$. At an angle of attack of 5 deg, the free molecular $C_A = 2.0$ and $C_N = 0.1986$.

For the same re-entry body in the continuum regime, Eq. (11.58) shows us that at zero angle of attack $C_A = 0.0697$. To determine the force coefficients at an angle of attack, we use the program listed in Table 11.2. We saw earlier that with $s_\infty \gg 1$ (i.e., $M_\infty \gg 1$), $\sigma_N = 1$, $\sigma_T = 0$, and $T_W/T_\infty = 0$ the free molecular flow results from Eq. (11.23) are equivalent to the Newtonian results. At a 5-deg angle of attack we see that the continuum $C_A = 0.0758$ and $C_N = 0.170$.

We may now complete the picture by using Eqs. (11.62) and (11.63) to determine the drag coefficient in the transitional regime ($0.01 < Kn < 10$). Figure 11.30 shows the variation of C_A at 0-deg and 5-deg angles of attack and C_N at a 5-deg angle of attack as the re-entry body descends through the atmosphere.

11.12 Summary

We have seen that any stationary volume of gas is a collection of particles in constant random motion. Associated with this random motion are intermolecular collisions and collisions between the particles and adjacent surfaces. If the mean free path between collisions λ is small compared to the characteristic dimension L of the flow, the gas may be treated as a continuum. But if λ is much larger than L, the gas must be treated as a collection of discrete but interacting particles.

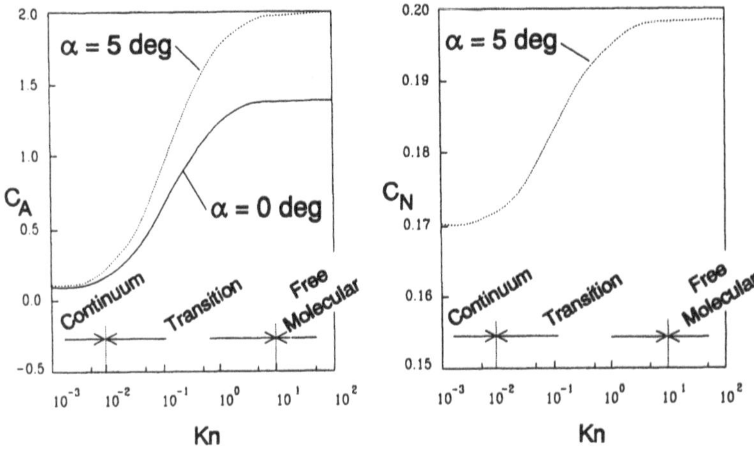

Fig. 11.30 Force coefficients on a 10-deg sphere cone (R_b = 1 m, B = 0.1) in continuum, transition, and free molecular flow regimes.

A dimensionless parameter called the Knudsen number ($Kn = \lambda/L$) was defined. When $Kn \gg 1$ molecule-surface interactions dominate the flow and the regime is called the free molecular flow. At the other extreme, where $Kn \ll 1$, intermolecular collisions dominate the flow and the atmosphere in this regime may be treated as a continuum. In between lies the transition regime, where neither intermolecular nor molecule-surface interactions are negligible.

Although the equations governing fluid flow have been known for a long time, the complexity of the equations has made analytical solution all but impossible except for a small number of simple cases. The governing equation based on the particulate nature of air is called the Boltzmann equation. In the continuum regime the governing equations are the Navier-Stokes equations. The Navier-Stokes equations are actually a special case of the Boltzmann equation.

In Section 11.6 we discussed the free molecular flow regime. Based on a Maxwellian incident velocity distribution, the pressure, shear stress, and heat transfer to a simple convex body are given in Eqs. (11.19) through (11.21). An ambiguity arises in determining the accommodation coefficients for the molecule-surface interaction models. A computer program for calculating the forces and moments on an arbitrary convex body is provided. The case of concave bodies is not as straightforward since the incident particles on some parts of the body may have been emitted from other parts of the body and do not conform to the Maxwellian distribution of the freestream particles.

Next the continuum regime was discussed. If the flow is not hypersonic, air may be treated as a perfect gas. In this regime the flowfield could be separated in the outer inviscid region, the boundary layer adjacent to the body and the wake.

In the outer region, the fluid is assumed to be inviscid and nonconducting whereas viscosity and heat conduction are dominant only in a thin region adjacent to the body. We also saw that the outer inviscid region could be treated independently from the thin boundary layer.

In hypersonic flow there are four problems to contend with: 1) high temperature effect—perfect gas assumptions invalid, 2) viscous interaction—outer flow and boundary layer coupled, 3) entropy gradients—viscous interaction and MOC is more complicated, and 4) thin shock layer—viscous interaction, but allows for impact methods.

For flows without separation, the wake may be treated as a combination of an outer inviscid flow and an inner viscous core. A separated flow wake will have a region of recirculating air. A good approximation in hypersonic flow is to assume the base pressure to be equal to the freestream pressure ($C_{p_{base}} = 0$).

Although viscous forces are not a dominant part of the total force, the heat-transfer-induced outgassing and ablation processes can play a significant role. As a re-entry body descends through the atmosphere, at a critical Reynolds number the boundary layer near the base of the body transitions from laminar to turbulent. With increasing Reynolds number, this transition point moves forward on the body. This transition process depends upon many factors that are not very well understood.

Associated with a turbulent boundary layer is a higher shear stress, but more importantly, a higher heat transfer rate. The higher heat transfer rate results in higher ablation rates. The unpredictability of transition means unpredictable shape changes due to ablation and hence unreliable force predictions. Finally, asymmetric shape changes, which are not uncommon, can cause significant moments.

Having discussed the difficulties faced in trying to determine the hypersonic forces and moments, we see that simple impact methods provide acceptable predictions. Impact methods predict the pressure based only upon the local flow deflection angle and freestream conditions.

The most widely applicable impact method is the modified Newtonian method. We show that the computer program for free molecular flow is applicable for Newtonian flow when the speed ratio is high and the accommodation coefficients are judiciously chosen.

Less rigorously derived methods that include empirical methods are available. While they may be more accurate than Newtonian predictions for certain classes of bodies, these empirical methods are severely limited in their applicability.

The transition regime from free molecular to continuum is subdivided into three regions. Adjacent to the free molecular regime is the nearly free molecular region. Interactions between the incident particles and particles emitted from the surface, which were negligible in the free molecular regime, begin to play an appreciable role. As the density increases, the beginnings of a shock and boundary layer become apparent in the merged viscous region. In this region, both intermolecular and molecule-surface interactions are important. As the re-entry body descends further into the denser atmosphere, the intermolecular collisions start to dominate the flow and this region is called the slip flow region. The shock and boundary layers are well defined. But in a thin layer near the surface the molecule-surface interactions are not yet negligible and macroscopic velocity and temperature at the wall do not match the wall velocity and temperature. The slip flow region is adjacent to the continuum regime and differs from continuum flow only near the body surface.

A simple and practical method of determining forces and moments in the transition regime is to fit a smooth curve between the continuum regime ($Kn \rightarrow 0$) and the free molecular regime ($Kn \rightarrow \infty$) as a function of Kn, based on correlation of available data for similar bodies.

Finally, some of the approximate methods are used to determine the force coefficients for a typical re-entry body in all three regimes.

References

[1]Hayes, W. D., and Probstein, R. F., *Hypersonic Flow Theory*, Academic Press, New York, 1959.

[2]Truitt, R. W., *Hypersonic Aerodynamics*, Roland, New York, 1959.

[3]Vincenti, W. G., and Kruger, C. H., Jr., *Introduction to Physical Gas Dynamics*, Robert E. Krieger Publishing, Malabar, FL, 1965.

[4]Anderson, J. D., *Hypersonic and High Temperature Gas Dynamics*, McGraw-Hill, New York, 1989.

[5]Bird, G. A., *Molecular Gas Dynamics*, Clarendon Press, Oxford, 1976.

[6]Hansen, C. F., "Approximations for the Thermodynamic and Transport Properties of High-Temperature Air," NACA-TN-4150, Washington, DC, 1958.

[7]John, J. E. A., *Gas Dynamics*, Allyn & Bacon, Boston, MA, 1969.

[8]Cox, R. N., and Crabtree, L. F., *Elements of Hypersonic Aerodynamics*, English University Press, London, 1965.

[9]Bird, G. A., "Low-Density Aerothermodynamics," *Thermophysical Aspects of Re-Entry Flows*, edited by James N. Moss and Carl D. Scott, Vol. 103, Progress in Astronautics and Aeronautics, AIAA, New York, 1986.

[10]Schaaf, S. A., and Chambre, P. L., "Flow of Rarefied Gases," *High-Speed Aerodynamics and Jet Propulsion*, Vol. III, Fundamentals of Gas Dynamics, edited by H. W. Emmons, Princeton University Press, Princeton, NJ, 1958.

[11]Koppenwallner, G., and Legge, H., "Drag of Bodies in Rarefied Hypersonic Flow," *Thermophysical Aspects of Re-Entry Flows*, edited by James N. Moss and Carl D. Scott, Vol. 103, Progress in Astronautics and Aeronautics, AIAA, New York, 1986.

[12]Anderson, D. A., Tannehill, J. C., and Pletcher, R. H., *Computational Fluid Mechanics and Heat Transfer*, Hemisphere, Washington, DC, 1984.

[13]White, F. M., *Viscous Fluid Flow*, McGraw-Hill, New York, 1974.

[14]Van Driest, E. R., "Investigation of Laminar Boundary Layer in Compressible Fluids Using the Crocco Method," NACA-TN-2597, Washington, DC, 1952.

[15]Liepmann, H. W., and Roshko, A., *Elements of Gasdynamics*, Wiley, New York, 1957.

[16]Ames Research Staff, "Equations, Tables, and Charts for Compressible Flow," NACA Rept. 1135, 1953.

[17]Kopal, Z., "Tables of Supersonic Flow around Cones," Massachusetts Institute of Technology, Tech. Rept. No. 1, 1947.

[18]Sims, J. L., "Tables for Supersonic Flow around Right Circular Cones at Zero Angle of Attack," NASA SP-3004, 1964.

[19]Zucrow, M. J., and Hoffman, J. D., *Gas Dynamics*, Wiley, New York, 1976.

[20]DeJarnette, F. R., Hamilton, H. H., and Weilmuenster, K. J., "A Review of Some Approximate Methods Used in Aerodynamic Heating Analyses," *Journal of Thermophysics*, Vol. 1, No. 1, Jan. 1987.

[21]Park, C., *Nonequilibrium Hypersonic Aerothermodynamics*, Wiley, New York, 1990.

[22]Stetson, K. F., "Comments on Hypersonic Boundary-layer Transition," Wright Research and Development Center, WRDC-TR-90-3057, Wright-Patterson AFB, OH, 1990.

[23]Regan, F. J., *Re-Entry Vehicle Dynamics*, AIAA Education Series, American Institute of Aeronautics and Astronautics, Inc., New York, 1984.

[24]Morrison, A. M., Solomon, J. M., Ciment, M., and Ferguson, R. E., *Handbook of Inviscid Sphere-Cone Flow Fields and Pressure Distributions—Volume II*, NSWC/WOL/TR-75-45, 1975.

[25]Lees, L., "Hypersonic Flow," Proceedings of the 5th International Aerospace Conference, Los Angeles, Institute of Aeronautical Sciences, New York, 1955.

[26]Gentry, A. E., Smyth, D., and Oliver, W., "The Mark IV Supersonic-Hypersonic Arbitrary-Body Program, Technical Report," AFFDL-TR-73-159, Wright-Patterson AFB, OH, 1973.

[27]Sherman, F. S., "The Transition from Continuum to Molecular Flow," Vol. 1, *Annual Review of Fluid Mechanics*, edited by W. R. Sears and M. Van Dyke, Annual Reviews, Palo Alto, CA, 1969.

[28]Wetzel, W., and Oertel, H., "Monte Carlo Simulations of Hypersonic Flows Past Blunt Bodies," *Journal of Thermophysics*, Vol. 4, No. 2, April 1990, pp. 157–161.

[29]Gorenbukh, P. I., "A Correlation for Drag Coefficients of Bodies in Hypersonic Flow of Rarefied Gases," *Fluid Mechanics—Soviet Research*, Vol. 15, No. 4, July–Aug. 1986, pp. 56–64; translated from Uchenyye Zapiski TsAGI, *17*, 2, pp. 99–105.

[30]Potter, J. L., "Procedure for Estimating Aerodynamics of Three-Dimensional Bodies in Transitional Flow," *Rarefied Gas Dynamics: Theoretical and Computational Techniques*, edited by E. P. Muntz, D. P. Weaver, and D. H. Campbell, Vol. 118, Progress in Astronautics and Aeronautics, AIAA, Washington, DC, 1989.

[31]Blanchard, R. C., and Buck, G. M., "Rarefied-Flow Aerodynamics and Thermosphere Structure for Shuttle Flight Measurements," *Journal of Spacecraft and Rockets*, Vol. 23, No. 1, Feb. 1986. pp. 18–24.

12
Angular Motion During Re-Entry

12.1 Introduction

In Chapter 7 we discussed particle motion during re-entry. For those results to have any utility we must be able to ignore the finite dimensions of the re-entry vehicle (RV). The tacit assumption that justifies approximating the re-entry vehicle by a particle is that the vehicle remains in *trim* at all times. *Trim* is the condition for which the sum of all the moments acting on the vehicle is identically zero. Thus, should any moment unbalance occur due to changes in the aerodynamic environment, there must be ample control moment available to reset the moment sum to zero instantaneously. In Chapter 9 we continued with the particle approximation for maneuvering re-entry vehicles. Under these circumstances the control system must vary the trim points of the vehicle in order to control the amount of transverse load. The control system then is regarded as a regulator although the trajectory-shaping algorithm will continually reset the trim points (angle of attack) in order to vary the lift force.

The particle assumption is actually an assumption that the vehicle possesses a control system that has infinite band width, i.e., the system can completely suppress any vehicle dynamics as it changes operating points and can maintain a zero moment sum at all times. Furthermore, the control system can move between trim points with arbitrary speed. Thus, to the control system the vehicle appears to have zero impedance. In this chapter we will consider some closed form treatments of the angular motion of the vehicle, acknowledging that the vehicle has a finite, non-zero impedence.

12.2 Planar Motion

In this section we will derive the equations of planar motion. Although we should be able to arrive at the two force equations and one moment equation almost by inspection, we will use this presentation as an opportunity to apply the methods of analysis developed in Chapters 4 and 5.

In Fig 12.1 we note four distinct axis systems: 1) inertial, 2) local, 3) body, and 4) velocity. The (X^I,Y^I,Z^I); (X^l,Y^l,Z^l); (X^b,Y^b,Z^b); (X^v,Y^v,Z^v) axes are defined, respectively, as follows: the I-frame has the X- and Y-axes in the equatorial plane with the Y-axis out of the paper and the Z-axis along the polar axis; the l-frame has the X-axis along the local horizontal, Z down and Y also out of the paper; the b-frame is fixed to the vehicle with the X-axis along the direction of thrust, if any, and the Y-axis parallel to the Y-axis of the local

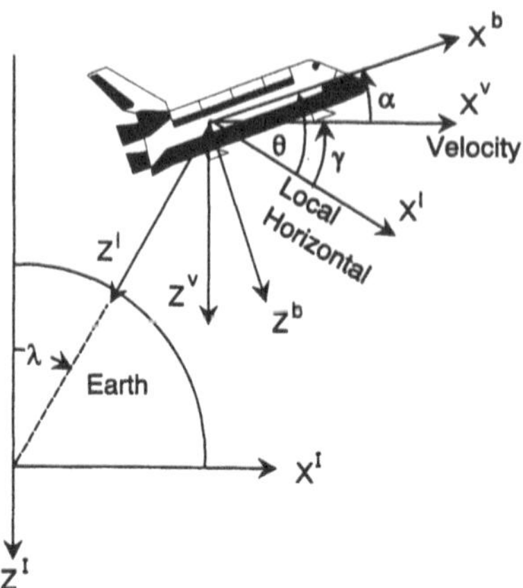

Fig. 12.1 Axis systems.

frame; the v-frame has the X-axis along the velocity vector, and the Y-axis parallel to the Y-axis of the local frame; the Y-axes of all frames are positive out of the paper.

The angles λ, α, γ, and θ are, respectively, the co-latitude or the angle between X^l and X^I axes; the angle of attack or the angle between the X^b and the X^v axes; the flight path angle or the angle between the X^v and the X^l axes; the pitch angle or the angle between the X^b and X^l axes.

By inspection we may write the following for the direction cosine matrices (DCM):

$$
C_l^b = \begin{bmatrix} \cos\theta & 0 & -\sin\theta \\ 0 & 1 & 0 \\ \sin\theta & 0 & \cos\theta \end{bmatrix} \qquad C_l^v = \begin{bmatrix} \cos\gamma & 0 & -\sin\gamma \\ 0 & 1 & 0 \\ \sin\gamma & 0 & \cos\gamma \end{bmatrix}
$$

$$
C_l^I = \begin{bmatrix} \cos\lambda & 0 & -\sin\lambda \\ 0 & 1 & 0 \\ \sin\lambda & 0 & \cos\lambda \end{bmatrix} \qquad C_v^b = \begin{bmatrix} \cos\alpha & 0 & -\sin\alpha \\ 0 & 1 & 0 \\ \sin\alpha & 0 & \cos\alpha \end{bmatrix} \tag{12.1}
$$

Angular velocities are entirely along the Y-axis so they are expressed as follows:

$$
\begin{aligned}
\boldsymbol{\omega}_{b/l}^b &= [0, \dot{\theta}, 0]^{\mathrm{T}} & \boldsymbol{\omega}_{v/l}^b &= [0, \dot{\gamma}, 0]^{\mathrm{T}} \\
\boldsymbol{\omega}_{l/I}^b &= [0, -\dot{\lambda}, 0]^{\mathrm{T}} & \boldsymbol{\omega}_{b/v}^b &= [0, \dot{\alpha}, 0]^{\mathrm{T}} \\
& & \boldsymbol{\omega}_{b/I}^b &= [p, q, r]^{\mathrm{T}}
\end{aligned} \tag{12.2}
$$

Note that the superscript b is used with the angular velocity vectors. Since all angular velocities are along the Y-axis and all Y-axes are collinear, any superscript, e.g., I, v, l would also be satisfactory.

Clearly,

$$\boldsymbol{\omega}^b_{b/I} = \boldsymbol{\omega}^b_{l/I} + \boldsymbol{\omega}^b_{b/l} = [0, (\dot{\theta} - \dot{\lambda}), 0]^{\mathrm{T}} = [p, q, r]^{\mathrm{T}} \tag{12.3}$$

where we have designated the b-frame components of the angular velocity of the body relative to inertial space as $[p, q, r]^{\mathrm{T}}$.

The linear velocity vector may be expressed alternatively in the b- and v-frames as follows:

$$\boldsymbol{V}^b = [U_b, 0, W_b]^{\mathrm{T}} \qquad \boldsymbol{V}^v = [V, 0, 0]^{\mathrm{T}} \tag{12.4}$$

The forces acting on the re-entry vehicle are separated into aerodynamic, thrust, and gravitational. The aerodynamic forces F_a may also be given in either a body or velocity frame, that is,

$$\boldsymbol{F}^b_a = [X, 0, Z]^{\mathrm{T}} \qquad \boldsymbol{F}^v_a = [-D, 0, -L]^{\mathrm{T}} \tag{12.5}$$

where D and L are the drag and lift forces. Thrust is most readily expressed in the b-frame. We will assume that the thrust T acts in the direction of the positive X^b-axis, although in a more detailed analysis we would surely have to include thrust misalignment with respect to X^b. Thus, we may write

$$\boldsymbol{T}^b = [T, 0, 0]^{\mathrm{T}} \tag{12.6}$$

The gravitational force per unit mass g acts in the positive Z^l direction, or

$$\boldsymbol{g}^l = [0, 0, g]^{\mathrm{T}} \tag{12.7}$$

The acceleration of the RV, or equivalently the force per unit mass, is defined as follows:

$$\boldsymbol{a}^I \equiv \frac{\mathrm{d}}{\mathrm{d}t}[\boldsymbol{V}^I]$$

which may be written as

$$\boldsymbol{a}^I = \frac{\mathrm{d}}{\mathrm{d}t}[C^I_v \boldsymbol{V}^v] = C^I_v[\dot{\boldsymbol{V}}^v + \Omega^v_{v/I}\boldsymbol{V}^v] \tag{12.8}$$

or

$$\boldsymbol{a}^v = \left(C^I_v\right)^{\mathrm{T}} \boldsymbol{a}^I = [\dot{\boldsymbol{V}}^v + \left(\Omega^v_{l/I} + \Omega^v_{v/l}\right)\boldsymbol{V}^v] \tag{12.8}$$

From Newton's Second Law we may write

$$\boldsymbol{a}^v = \frac{\boldsymbol{F}^v_a}{m} + \frac{\boldsymbol{T}^v}{m} + \boldsymbol{g}^v = \frac{\boldsymbol{F}^v_a}{m} + \frac{1}{m}C^v_b\boldsymbol{T}^b + C^v_l\boldsymbol{g}^l \tag{12.9}$$

where m is the mass of the RV. Equating Eqs. (12.8) and (12.9) gives

$$\dot{\boldsymbol{V}}^v + \left(\Omega^v_{l/I} + \Omega^v_{v/l}\right)\boldsymbol{V}^v = (1/m)\left(\boldsymbol{F}^v_a + C^v_b\boldsymbol{T}^b\right) + C^v_l\boldsymbol{g}^l \tag{12.10a}$$

The preceding equation then becomes

$$\begin{bmatrix} \dot{V} \\ 0 \\ 0 \end{bmatrix} + \begin{bmatrix} 0 & 0 & (\dot{\gamma} - \dot{\lambda}) \\ 0 & 0 & 0 \\ -(\dot{\gamma} - \dot{\lambda}) & 0 & 0 \end{bmatrix} \begin{bmatrix} V \\ 0 \\ 0 \end{bmatrix} = \frac{1}{m} \begin{bmatrix} -D + T\cos(\alpha) \\ 0 \\ -L - T\sin(\alpha) \end{bmatrix} + \begin{bmatrix} -g\sin(\gamma) \\ 0 \\ g\cos(\gamma) \end{bmatrix} \tag{12.10b}$$

The aerodynamic and thrust forces—sometimes identified as contact forces—may be resolved into two components, one, F_T, tangential to the flight path, the other, F_N, normal to the flight path. These components are expressed in the following manner:

$$F_T = -D + T\cos(\alpha) \qquad F_N = L + T\sin(\alpha) \tag{12.11}$$

Equation (12.10b) then may be written in the following component form accounting for the changes in velocity magnitude and in velocity direction:

$$\frac{\mathrm{d}V}{\mathrm{d}t} = \frac{F_T}{m} - g\sin(\gamma) \tag{12.12}$$

$$V\frac{\mathrm{d}\gamma}{\mathrm{d}t} = \frac{F_N}{m} - g\cos(\gamma) + V\frac{\mathrm{d}\lambda}{\mathrm{d}t} \tag{12.13}$$

We may now use a kinematic relationship between velocity $\boldsymbol{V}$ and the position vector $\boldsymbol{R}$ to introduce the altitude $\boldsymbol{h}$; we then use a second relationship to eliminate the latitude rate $\mathrm{d}\lambda/\mathrm{d}t$ in Eq. (12.13). By way of definition we have

$$\boldsymbol{V}^I \equiv \frac{\mathrm{d}\boldsymbol{R}^I}{\mathrm{d}t} = \frac{\mathrm{d}}{\mathrm{d}t}\left[C_l^I \boldsymbol{R}^l\right] = C_l^I\left[\frac{\mathrm{d}R^l}{\mathrm{d}t} + \Omega_{l/I}^l R^l\right]$$

where

$$\boldsymbol{R}^l = [0, 0, -(R_e + h)]^{\mathrm{T}}$$

Multiplying both sides of the preceding equation by $(C_l^I)^{\mathrm{T}}$, we get

$$(C_l^I)^{\mathrm{T}}\boldsymbol{V}^I = \boldsymbol{V}^l = C_v^l \boldsymbol{V}^v = (C_l^v)^{\mathrm{T}}\boldsymbol{V}^v = \frac{\mathrm{d}\boldsymbol{R}^l}{\mathrm{d}t} + \Omega_{l/I}^l \boldsymbol{R}^l \tag{12.14a}$$

The matrices of the preceding equation may be filled in to give

$$\begin{bmatrix} \cos(\gamma) & 0 & \sin(\gamma) \\ 0 & 1 & 0 \\ -\sin(\gamma) & 0 & \cos(\gamma) \end{bmatrix} \begin{bmatrix} V \\ 0 \\ 0 \end{bmatrix} = \begin{bmatrix} 0 \\ 0 \\ -\dot{h} \end{bmatrix} + \begin{bmatrix} 0 & 0 & -\dot{\lambda} \\ 0 & 0 & 0 \\ +\dot{\lambda} & 0 & 0 \end{bmatrix} \begin{bmatrix} 0 \\ 0 \\ -(R_e + h) \end{bmatrix} \tag{12.14b}$$

These become

$$\frac{dh}{dt} = V\sin(\gamma) \tag{12.15a}$$

and

$$\frac{d\lambda}{dt} = \frac{V\cos(\gamma)}{(R_e + h)} \tag{12.15b}$$

Using Eq. (12.15a), we may rewrite Eq. (12.13) as follows:

$$V\frac{d\gamma}{dt} = \frac{F_N}{m} - \cos(\gamma)\left[g - \frac{V^2}{(R_e + h)}\right] \tag{12.16}$$

Solving equation (12.15a) yields the altitude of the re-entry vehicle as a function of time. Altitude is of course measured along the local vertical; Eq. (12.15b) provides us with the path length of the RV trajectory along the Earth's surface as

$$S = \int^{\lambda} R_e d\lambda = R_e \int^{\lambda} d\lambda$$

which may be rewritten as

$$S = R_e \int_0^t \frac{V(t)\cos[\gamma(t)]}{[R_e + h(t)]} dt \tag{12.17}$$

The required time functions $V(t)$, $\gamma(t)$, and $h(t)$ are all available from the solutions of Eqs. (12.12), (12.15), and (12.16).

The moment equations may now be introduced; however, for planar motion, only the moment about the Y-axis is nonzero. Let us assume that the body-fixed axes are principal axes. Thus, from the definition of the moment $\boldsymbol{M}^I$, or

$$\boldsymbol{M}^I = \frac{d}{dt}[\boldsymbol{H}^I]$$

we have

$$\boldsymbol{M}^I = \frac{d}{dt}\left[C_b^I \boldsymbol{H}^b\right] = C_b^I\left[\dot{\boldsymbol{H}}^b + \Omega_{b/I}^b \boldsymbol{H}^b\right]$$

This equation becomes

$$\boldsymbol{M}^b = \frac{d\boldsymbol{H}^b}{dt} + \Omega_{b/I}^b \boldsymbol{H}^b = I^b \frac{d\boldsymbol{\omega}_{b/I}^b}{dt} + \Omega_{b/I}^b I^b \boldsymbol{\omega}_{b/I}^b \tag{12.18}$$

where I^b is the inertia tensor and the applied moment (components in the body frame) is as follows:

$$\boldsymbol{M}^b = [M_x, M_y, M_z]^{\mathrm{T}}$$

The inertia tensor for a principal axis system is diagonal, that is,

$$I^b = \begin{bmatrix} I_{xx} & 0 & 0 \\ 0 & I_{yy} & 0 \\ 0 & 0 & I_{zz} \end{bmatrix}$$

The angular velocity may be expressed in either a column or skew-symmetric form as follows:

$$\boldsymbol{\omega}^b_{b/I} = [p, q, r]^{\mathrm{T}}$$

$$\Omega^b_{b/I} = \left(\boldsymbol{\omega}^b_{b/I}\right)^* = \begin{bmatrix} 0 & -r & q \\ r & 0 & -p \\ -q & p & 0 \end{bmatrix}$$

For planar motion, inserting the preceding equations into Eq. (12.18) gives

$$\begin{bmatrix} M_x \\ M_y \\ M_z \end{bmatrix} = \begin{bmatrix} I_{xx} & 0 & 0 \\ 0 & I_{yy} & 0 \\ 0 & 0 & I_{zz} \end{bmatrix} \begin{bmatrix} 0 \\ \dot{q} \\ 0 \end{bmatrix} + \begin{bmatrix} 0 & 0 & q \\ 0 & 0 & 0 \\ -q & 0 & 0 \end{bmatrix} \begin{bmatrix} I_{xx} & 0 & 0 \\ 0 & I_{yy} & 0 \\ 0 & 0 & I_{zz} \end{bmatrix} \begin{bmatrix} 0 \\ q \\ 0 \end{bmatrix}$$

or

$$I_{yy} \frac{\mathrm{d}q}{\mathrm{d}t} = M_y \tag{12.19}$$

The applied moment M_y will now be separated into four constituent parts. These are M_a (aerodynamic), M_T (thrust), M_c (control), and M_g(gravitational).

Let us examine the gravitational moment first. The gravitational moment has its origin in the gradient of the gravity field across the re-entry vehicle. Identifying the (2)-axis with the Y- or pitch axis and the (1)- and (3)-axes with the X- and Z-axes, we have, from Eq. (F.8a) of Appendix F,

$$M_g = \frac{3\mu X_0 Z_0}{R_0^5}(I_{xx} - I_{zz}) \tag{12.20}$$

R_0 is the geocentric distance of the center of mass of the vehicle. The coordinates (X_0, Z_0) are the coordinates of the center of mass of the vehicle in the local or l-frame. It is easily shown that

$$[X_0, 0, Z_0]^{\mathrm{T}} = C^b_l\,[0, 0, -R_0]^{\mathrm{T}}$$

From Eq. (12.1) we obtain the DCM C_l^b

$$X_0 = R_0 \sin(\theta) \qquad Z_0 = -R_0 \cos(\theta) \tag{12.21}$$

With $\mu = gR_e^2 \approx gR_0^2$ [see Eq. (6.4c)] we obtain

$$M_g = -(3g/2R_0)(I_{xx} - I_{zz})\sin(2\theta) \tag{12.22}$$

If the roll moment of inertia (I_{xx}) is less than the yaw moment of inertia (I_{zz}), which is usually the case for aerodynamic vehicles, then the gravitational (gradient) moment will result in angular motion that is exponentially divergent. During nearly all of the re-entry trajectory, the aerodynamic moments dominate the angular motion and the gravitational moment may be neglected. However, when the vehicle is in orbital motion prior to re-entry, the gravitational moment may be a major contributor to vehicle attitude and therefore must be retained.

The thrust moment, of course, arises in the attitude-control thruster. Figure 12.2 depicts a single stage to orbit (SSTO) vehicle in which a thruster is displaced from the X^b-axis through a distance ℓ_T. Included in this figure is a thrust history. We next average this thrust and multiply this average by the thrust axis offset to obtain an average moment. Note that ℓ_T is taken as

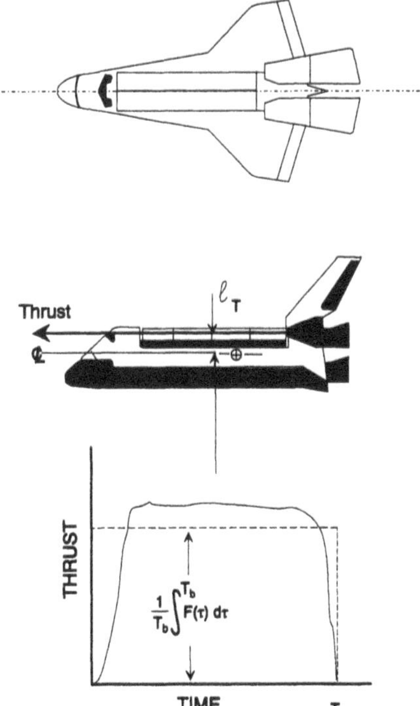

Fig. 12.2 Thrust moment.

negative when the offset is along the negative Z^b-axis. A positive offset will result in a positive pitching moment, or

$$M_T = \frac{\ell_T}{T_b}\int_0^{T_b} F(\tau)\mathrm{d}\tau \qquad (12.23)$$

Let us now turn our attention to the aerodynamic and control moments. First we will nondimensionalize the moment M and normal force N to define the pitching moment and normal force coefficients. This step is accomplished as follows:

$$C_m = M/\tfrac{1}{2}\rho V^2 S\bar{c} \qquad C_N = N/\tfrac{1}{2}\rho V^2 S \qquad (12.24)$$

where ρ is the density, V the velocity magnitude, S the reference area (wing area in case of a SSTO vehicle), and $\bar{c}$ the reference length, taken here to be a representative wing chord length called the *mean aerodynamic chord.* We will not discuss how this chord is numerically evaluated for a given wing planform. For further information the reader may consult References 1 and 2. Table 12.1 presents some important mass and geometric properties of a representative shuttle.

Before discussing any analytical forms, we should identify the typical control system that might be present on a manned lifting re-entry vehicle. Unlike most atmospheric vehicles, the shuttle is a tailless configuration (see Fig. 12.3). In a wing-only vehicle, the wing must provide not only the main lifting force but also the appropriate control moments to keep the vehicle in trim. (*Trim* is the condition of zero net moment.)

Roll control is provided by differentially activated elevons; deflected together, the elevons provide pitch control. Elevons combine the functions that are usually associated with elevators (pitch control) and ailerons (roll control) on typical atmospheric vehicles. Pitch trim is provided primarily by body flap. Directional control is provided by a conventional rudder, although it is usually coordinated with elevon setting. In addition the rudder can be split to provide additional drag as required during the terminal phase of the re-entry trajectory.

Table 12.1 Shuttle geometric/mass data

Symbol	Definition	Data
b	Wing span	23.79 m
l	Body length	32.77 m
S	Wing area	249.77 m^2
h	Tail height	10.55 m
$\bar{c}$	Mean aerodynamic chord	12.06 m
I_{xx}	Roll moment of inertia	1.0141×10^6 kg-m^2
I_{yy}	Pitch moment of inertia	7.7104×10^6 kg-m^2
I_{zz}	Yaw moment of inertia	7.8704×10^6 kg-m^2
I_{xz}	Plane of symmetry product of inertia	1.99×10^5 kg-m^2

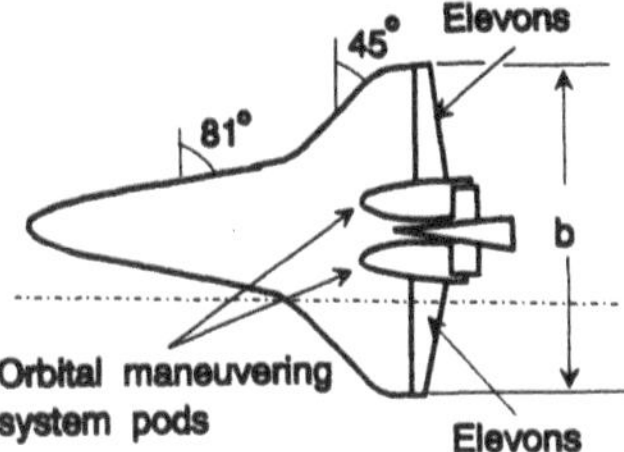

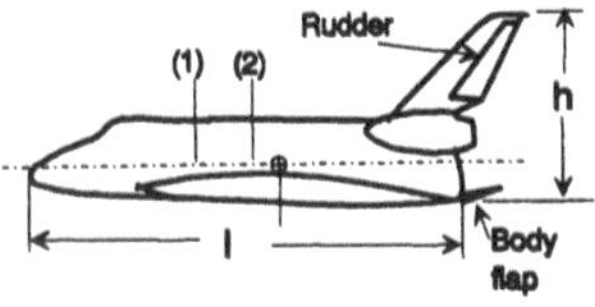

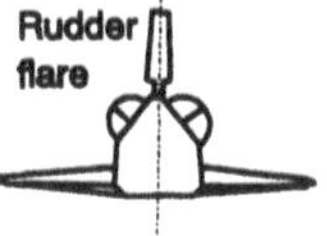

Fig. 12.3 Typical shuttle configuration.

12.3 Static Stability

We must now investigate the analytical form of the aerodynamic pitching moment. The coefficient C_m, as used below, is restricted to the aerodynamic contribution. Note in Fig. 12.3 that there are three aerodynamic controls (rudder, body flap, and elevons). In addition there are the passive aerodynamic loads of the medium on the fuselage and wings. Thus, with, δ_e and δ_F as the elevon and body flap deflection angles respectively (positive when trailing edge is down) we may write

$$C_m = C_m(\alpha, \delta_e, \delta_F) \tag{12.25}$$

Let us assume that in the preceding equation we can separate the passive aerodynamic effects, i.e., α-dependent, from the control effects, i.e., (δ_e, δ_F)-dependent, to write the following:

$$C_m = C_m(\alpha) + C_{m_0}(\delta_e, \delta_F)$$

or, linearizing in angle of attack α,

$$C_m = C_{m_\alpha}\alpha + C_{m_0}(\delta_e, \delta_F) \tag{12.26}$$

We are implying that the moment coefficient C_m is evaluated at some longitudinal station. If this station coincides with the center of mass and if the control settings are selected such that this moment is zero, then the vehicle is said to be *trimmed*. According to Eq. (12.26), trimming the vehicle at a given center of mass location implies a unique relationship between the angle of attack at trim α_{trim} and the control variables (δ_e, δ_F) that is described in the following manner:

$$\alpha_{\text{trim}} = -C_{m_0}(\delta_e, \delta_F)/C_{m_\alpha} \tag{12.27}$$

Figure 12.4 is an example of static aerodynamic data for the Shuttle.[5] Note that the pitching moment has a negative slope. Although it is not obvious from this figure, we can shift the pitching moment curve vertically (without changing slope) by appropriately setting the control variables (δ_e, δ_F) and thereby alter the trim angle of attack α_{trim}.

We may now introduce two concepts that are useful in describing aerodynamic controls and how such controls are used to set trim conditions. These concepts are the *center of pressure* and the *aerodynamic center* (or *ac*).

The center of pressure is identified for our purposes as that longitudinal station at which the component of the aerodynamic moment vector normal to the vehicular plane of symmetry is zero. We can easily show that if the controls are locked or set to some fixed position, the center of pressure would move longitudinally with changes in angle of attack. However, we will defer this demonstration until we have derived Eq. (12.30). This movement of the center of pressure with angle of attack (fixed control settings) means that either the vehicle could operate at only one trimmed angle of attack or the center of mass

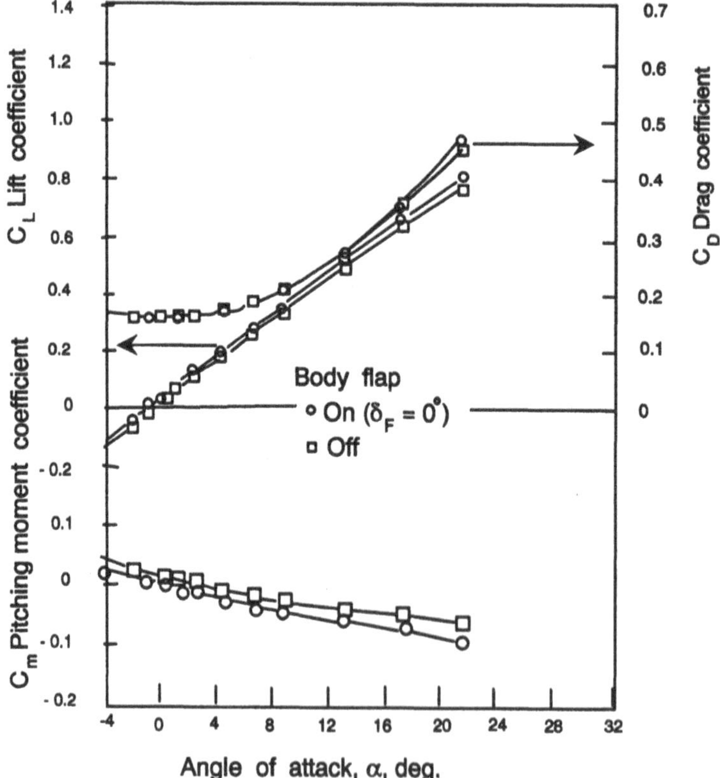

Fig. 12.4 Lift, drag, and pitching moment vs angle of attack for a Mach number of 1.90 and a center of gravity of 0.65 body lengths. Courtesy of NASA, Ref. 5.

would have to be moved to change the trim angle of attack. Obviously neither of these conditions is acceptable for most practical re-entry vehicles, certainly not for a manned orbiter. Allowing the controls a range of movement produces a corresponding range of trim angles of attack for a fixed center of mass location. A range of control movement permits a corresponding longitudinal range of center of mass locations over which the vehicle is trimmable.

In carrying out the longitudinal moment sum, we find that a concept with more utility than the center of pressure is the aerodynamic center. Subsonic wing theory identifies a longitudinal station on the wing, the aerodynamic center, at which the moment sum will be invariant with angle of attack. The aerodynamic moment at the aerodynamic center will not necessarily be zero but the derivative of the pitching moment, $\mathrm{d}C_{m_{ac}}/\mathrm{d}\alpha$, will be essentially zero. As compressibility effects become significant with increasing Mach number, the concept of an aerodynamic center remains useful although the location of the aerodynamic center will move rearward.

For a fixed Mach number there will be some variation of pitching moment with angle of attack at extreme angles of attack. Under such circumstances we might consider the idea of the movement of the aerodynamic center with angle of attack or just abandon the whole concept of a point of moment invariance with angle of attack. For our purposes we will avoid such compromises by restricting the angle of attack range to those values for which $\mathrm{d}C_{m_{ac}}/\mathrm{d}\alpha = 0$.

Figure 12.5 is the profile view of an orbiter. The exact vehicle shape is irrelevant except that we are limiting our consideration to a so-called tailless configuration. In the tailless vehicle, the wing must provide both the sustaining lift loads and the control moments to trim the vehicle. Note in this figure that the unit of distance is given as a fractional part of the mean aerodynamic chord $\bar{c}$.

If we take moments about the center of mass, we obtain

$$\begin{aligned} M &= M_{ac} + [L\cos(\alpha) + D\sin(\alpha)]\,(h - h_{ac})\,\bar{c} \\ &\quad + [L\sin(\alpha) - D\cos(\alpha)]\,z\,\bar{c} \end{aligned} \tag{12.28}$$

We can reproduce the preceding equation in coefficient form using

$$C_{m_{ac}} = \frac{M_{ac}}{\frac{1}{2}\rho V^2 S\bar{c}} \qquad C_L = \frac{L}{\frac{1}{2}\rho V^2 S} \qquad C_D = \frac{D}{\frac{1}{2}\rho V^2 S} \tag{12.29}$$

Equation (12.28) then becomes

$$C_m = C_{m_{ac}} + (h - h_{ac})C_N \tag{12.30}$$

Note that we have ignored the center of mass offset term, or $z\bar{c}$, and have made use of the relationship for the normal force coefficient C_N in the following manner:

$$C_N = C_L\,\cos(\alpha) + C_D\,\sin(\alpha)$$

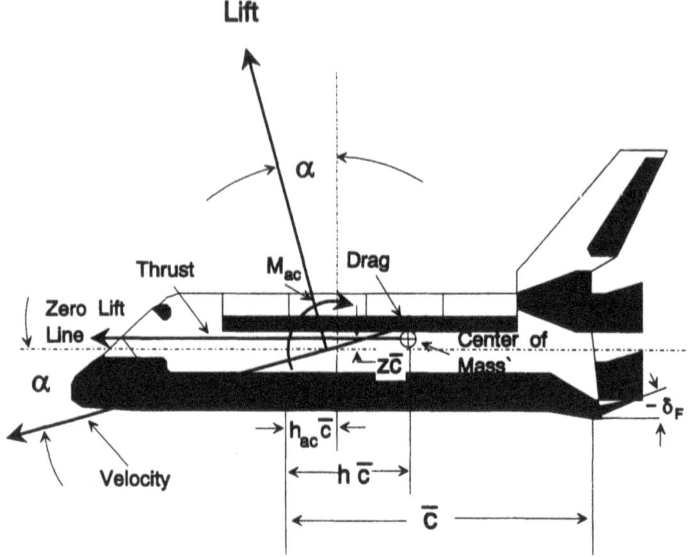

α: angle of attack
$\bar{c}$: mean aerodynamic chord
h : fraction of mean aerodynamic chord

Fig. 12.5 Aerodynamic forces and moments.

For small angles of attack, C_L and C_N are nearly equal; for our analytical studies such interchanges are convenient, although for large angles of attack, C_N and C_L are markedly different. A useful approximation to Eq. (12.30) for a small angle of attack is this:

$$C_m = C_{m_{ac}} + (h - h_{ac})C_L \tag{12.31}$$

As we indicated, the utility of the aerodynamic center idea is that there is a point at which the moment is invariant with changes in the angle of attack. Thus, we have

$$\frac{\mathrm{d}C_m}{\mathrm{d}\alpha} \equiv C_{m_\alpha} = (h - h_{ac})C_{L_\alpha} \tag{12.32}$$

The term $(h - h_{ac})$ is sometimes identified as the *static margin* since it is the remaining distance (in mean aerodynamic chords) through which the center of mass must move in order to make the pitching moment derivative with respect to angle of attack equal to zero.

We have pointed out for the vehicle to be in trim (and hence controllable) the moment sum must be zero at the center of mass. Thus, from Eq. (12.30) we find that

$$C_{m_{ac}}(\delta_e, \delta_F) = -(h - h_{ac})C_N \tag{12.33a}$$

We will later set δ_F and/or δ_e to meet the above trim requirement. For the vehicle to be stable it must return to a trim condition (trim angle of attack) with the appropriate restoring moments supplied by the passive aerodynamics. In other words, setting the controls is equivalent to setting the trim angle of attack but only the passive aerodynamics (C_L) together with the static margin $(h - h_{ac})$ affect vehicle stability. For the standard axis system a moment about the positive Y-axis, i.e., a moment that tends to increase angle of attack, will incur a negative moment from the passive aerodynamics tending to decrease the angle of attack provided the vehicle is statically stable. Thus the condition for static stability becomes

$$C_{m_\alpha} \equiv \frac{dC_m}{d\alpha} < 0 \tag{12.33b}$$

According to Eq. (12.32), the pitching moment derivative C_{m_α} is set by the aerodynamic load (essentially, lift), the location of the aerodynamic center, and the location of the center of mass. C_L and h_{ac} depend upon vehicle shape and of course h locates the center of mass. Since vehicle shape is a settled issue when operational analysis is considered, we might think of C_{m_α} (and hence static stability) solely as a function of center of mass location.

Let us briefly return to Eq. (12.30) to demonstrate that the center of pressure must move with angle of attack. Consider two longitudinal stations (1) and (2), located from the reference point by $h_{(1)}$ and $h_{(2)}$. The moment coefficient at station (2) is related to the moment at (1) from statics as:

$$C_{m(2)} = C_{m(1)} + C_N(h_{(2)} - h_{(1)}) \tag{12.34}$$

Let us assume that the vehicle is trimmed at station (2). Hence we may designate $h_{(2)} = h_{cp}$. We may also linearize the pitching moment at station (1) as

$$C_{m(1)} = C_m = C_{m_0}(\delta_F) + C_{m_\alpha}\alpha \tag{12.35}$$

where we have separated the moment into control (body flap) and angle of attack contributions and linearized the angle of attack term. Since the vehicle is trimmed at station (2), by definition $C_{m(2)} = 0$. So we obtain from Eq. (12.34)

$$h_{cp} = h_1 - [C_{m_\alpha}\alpha + C_{m_0}(\delta_F)]/C_N \tag{12.36}$$

Now if we take the derivative of h_{cp} with respect to angle of attack and set this to zero, we find the condition for the invariance of center of pressure with angle of attack to be

$$C_{m_\alpha}C_N - C_{N_\alpha}[C_{m_\alpha}\alpha + C_{m_0}(\delta_F)] = 0$$

or

$$C_{m_0}(\delta_F) = 0 \tag{12.37}$$

since C_{N_α} cannot be zero. Equation (12.37) requires that the trim angle of attack [Eq. (12.27)] be always zero, clearly an impractical requirement for a lifting vehicle. Therefore the center of pressure must move with angle of attack.

Let us now complete the linearization of Eq. (12.26). If we assume that the elevons and flaps cause only linear changes in the pitching moment, we may write

$$C_m = C_{m_{0_{ac}}} + C_{m_{\delta_e}}\delta_e + C_{m_{\delta_F}}\delta_F + C_{m_\alpha}\alpha \tag{12.38a}$$

where $C_{m_{0_{ac}}}$ is the pitching moment coefficient at $\delta_e = 0$ and $\delta_F = 0$. In Eq. (12.33a) we noted that the pitching moment at the aerodynamic center could be varied by altering the control settings. However, when the moment at the aerodynamic center is linearized, there is a pure couple that is present regardless of the control settings.

Let us write the linear equations of the pitching moment and lift coefficients. We will assume that the angle of attack is defined to be zero when the vehicle experiences no net lifting force. Furthermore, we will assume that the lift force depends upon angle of attack and control settings. Thus, we have

$$\begin{aligned} C_L(\alpha, \delta_e, \delta_F) &= \frac{\partial C_L}{\partial \alpha}\alpha + \frac{\partial C_L}{\partial \delta_e}\delta_e + \frac{\partial C_L}{\partial \delta_F}\delta_F \\ &= C_{L_\alpha}\alpha + C_{L_{\delta_e}}\delta_e + C_{L_{\delta_F}}\delta_F \end{aligned} \tag{12.38b}$$

For the remainder of this section, let us assume that only the elevon is controlling pitch. Thus, we will ignore δ_F, the flap setting. The trim condition is defined as that condition for which the moment sum is zero. We are primarily interested in the angle of attack and elevon setting when the vehicle is in trim. Thus, we may simultaneously solve Eqs. (12.38a) and (12.38b) to give

$$\delta_{e_{\text{trim}}} = \frac{C_{m_{0_{ac}}} C_{L_\alpha} + C_{L_{\text{trim}}} C_{m_\alpha}}{C_{L_{\delta_e}} C_{m_\alpha} - C_{L_\alpha} C_{m_{\delta_e}}} \tag{12.39a}$$

$$\alpha_{\text{trim}} = -\frac{C_{m_{\delta_e}} C_{L_{\text{trim}}} + C_{L_{\delta_e}} C_{m_{0_{ac}}}}{C_{L_{\delta_e}} C_{m_\alpha} - C_{m_{\delta_e}} C_{L_\alpha}} \tag{12.39b}$$

Alternative forms that have some utility easily follow from the preceding equations, including

$$\alpha_{\text{trim}} = (C_{L_{\text{trim}}} - C_{L_{\delta_e}}\delta_{e_{\text{trim}}})/C_{L_\alpha} \tag{12.39a}$$

$$\delta_{e_{\text{trim}}} = -(C_{m_{0_{ac}}} + C_{m_\alpha}\alpha_{\text{trim}})/C_{m_{\delta_e}} \tag{12.39b}$$

In a conventional atmospheric vehicle (e.g., an airplane), the elevator controls pitch angle; the elevator is a hinged portion of a small horizontal stabilizing surface several wing chords behind the wing. Thus, the trimming force (and resulting moment) reside in one surface (horizontal stabilizer), whereas the lifting force resides in another surface, the wing. However, in a tailless vehicle (such as a manned re-entry orbiter), the wing camber must be changed in order to change trim angle of attack. The elevon, being at the trailing edge of the wing or the main lifting surface, changes the camber of the wing and hence alters C_{m_0}; the elevon also changes the lift on the wing. This change in lift makes an additional contribution to the change in the pitching moment. This increase in lift, or

$$\Delta C_L = \frac{\partial C_L}{\partial \delta_e} \delta_e$$

must be transferred to the center of mass from the aerodynamic center to provide a second contribution to moment change from elevon deflection, or

$$\frac{\partial C_L}{\partial \delta_e} \delta_e (h - h_{\text{ac}})$$

We may now obtain the following expression for the total change in pitching moment due to elevon deflection:

$$\Delta C_m = \frac{\partial C_{m0}}{\partial \delta_e} \delta_e + \frac{\partial C_L}{\partial \delta_e} \delta_e (h - h_{\text{ac}}) \tag{12.40a}$$

In this expression the first term on the right is the change in pitching moment due to the elevon changing of wing camber; the second term is the transfer of the change in lift to the center of mass.

Next we take the derivative of the preceding expression with respect to the elevon angle to obtain the total moment change due to the elevon. Obviously if the center of mass were at the aerodynamic center then the second term in the above expression would vanish, that is,

$$\frac{\partial C_m}{\partial \delta_e} = \left[\frac{\partial C_{m0}}{\partial \delta_e} + \frac{\partial C_L}{\partial \delta_e} (h - h_{\text{ac}}) \right] \rightarrow \left. \frac{\partial C_{m0}}{\partial \delta_e} \right|_{h = h_{\text{ac}}} \tag{12.40b}$$

To reiterate: the elevon (as a pitch controller) creates a change in the pure couple caused by wing camber and a change in the lift on the wing [as seen in Eq. (12.40a)]. This change in lift causes an additional moment about the center of mass. The incremental couple may be moved to the center of mass unchanged.

In Eqs. (12.39a) and (12.39b) we note that the following seven terms must be available before we can determine if the vehicle may be trimmed at a given center of mass location:

C_{m_0} the pitching couple, often associated with wing camber or flow curvature; the couple that exists when the normal force is zero

$C_{m_{0_{ac}}}$ the pitching couple (C_{m_0}) when $\delta_e = 0$ and $\delta_F = 0$

$C_{m_{\delta_e}}$ the change in pitching moment with change in elevon angle [see Eq. (12.40b)]

C_{m_α} the static pitching moment derivative or the change in pitching moment with change in angle of attack

$C_{L_{\text{trim}}}$ the lift coefficient at trim conditions

$C_{L_{\delta_e}}$ the change in lift with a change in elevon angle

C_{L_α} the change in lift coefficient with a change in angle of attack; the lift curve slope

For example, at a Mach number of 5.0, the derivative C_{L_α} varies from 0.02870/radian around zero degrees angle of attack to 0.2006/radian at 35 degrees angle of attack.[5] Over the same range for angle of attack and at the same Mach number, C_{m_α} varies from -0.0287 to -0.3438. All derivatives are available from wind tunnel measurements. Equations (12.39) may have only limited value in the analysis of manned re-entry vehicles in that the range of angle of attack over which the vehicle is operated (particularly at high Mach numbers) often limits the applicability of linear analysis. Of course linear methods might have utility in the vicinity of the trim angle of attack. We would expect significant nonlinearities in the variation of the lift and pitching moment with the control variables.

We may get some idea of the large variation of angle of attack over the flight envelope during re-entry by considering a typical profile[4] given in Fig. 12.6. This figure is only meant to be representative.

Let us now consider an approach to analyzing trim conditions of a typical vehicle where aerodynamic nonlinearities are present. For this work we might plot the pitching moment coefficient C_m versus angle of attack. Figure 12.7 represents a typical set of curves. Note that curves are moved downward by altering the control settings. Remember that positive control deflections for both the elevon and flap are trailing edge downward: trailing edge downward increases the lift on the wing and decreases (makes more negative) the pitching moment. The trim condition occurs when the pitching moment is zero. If one sets the body flap δ_F to 10 degrees and the elevon angle δ_e to zero degrees, the moment is zero at about 20 and again at about 40 degrees angle of attack. Also note that at the lower trim angle the vehicle is unstable since the derivative

$$C_{m_\alpha} = \frac{\mathrm{d}C_m}{\mathrm{d}\alpha}$$

is positive whereas at the upper trim angle of attack the vehicle is stable since the derivative is negative.

We may also present essentially the same information in a more convenient form by replacing the angle of attack with the normal force coefficient. The normal force coefficient varies linearly (or nearly so) with angle of attack. Thus

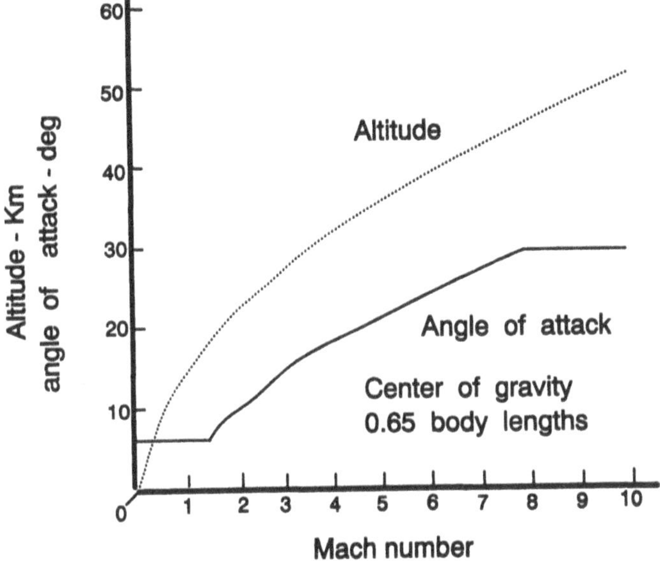

Fig. 12.6 Shuttle re-entry profile—altitude and angle of attack vs Mach number (forward center of gravity). Courtesy of NASA Langley Research Center, Ref. 5.

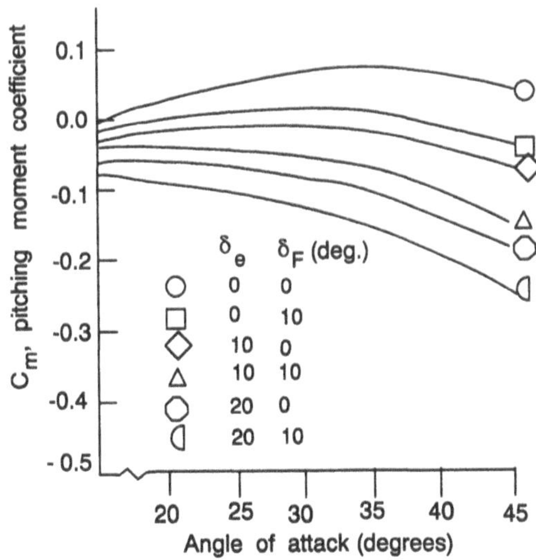

Fig. 12.7 Pitching moment coefficient vs angle of attack at a Mach number of 20.3 and a center of mass at 0.715 body lengths. Courtesy of NASA Langley Research Center, Ref. 5.

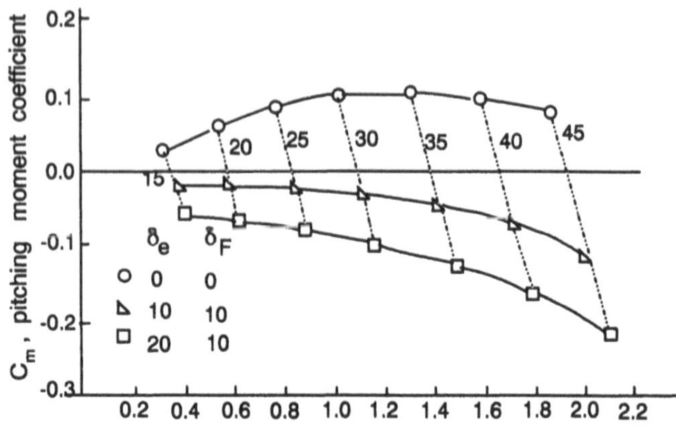

Fig. 12.8 Pitching moment coefficient vs normal force coefficient for angle attack range of 15 to 45 degrees (data from Ref. 3).

in Fig. 12.8 we have the corresponding plot of C_m versus C_N. Again we note that the application of positive signed control inputs reduces or makes more negative the pitching moment. Although not identified as such, there is a range of control settings that can trim the vehicle over the angle of attack range from 15 to 45 degrees. In other words the pilot has available the appropriate control input that can set a lift coefficient for trajectory shaping and at the same time keep the vehicle in trim.

Both Figs. 12.7 and 12.8 were derived based on a forward center of mass location of 0.715 body lengths. Now let us examine what happens if the center of mass is moved rearward through a distance of ΔX meters. First, take the forward center of mass as a reference location and designate the pitching moment at this point by the superscript r (for reference). As we move the center of mass rearward the pitching moment coefficient at the new center of mass location is given as

$$C_m = C_m^r + (\Delta X/\bar{c})C_N$$

We may now replace the term $(\Delta X/\bar{c})$ by ζ to give

$$C_m = C_m^r + \zeta C_N \tag{12.41}$$

This equation indicates that as the center of mass is moved rearward the pitching moment coefficient becomes less negative or more positive. Thus, all of the C_m versus C_N curves given in Figs. 12.7 and 12.8 move *upward* as ζ increases. We note that the lowest curve in these figures correspond to the maximum control settings. In considering vehicle trim with rearward movement of the center of mass, we need only examine the C_m-C_N function (curve) for which the controls (body flap, δ_F, and elevon, δ_e) are at their maximum values.

The problem we have set then becomes the following: how far aft may the center of mass be positioned for the vehicle to be trimmable?

First we must define the range for angle of attack over which we are interested in trimming the vehicle. Since lift varies almost linearly with angle of attack, trajectory requirements that set the lift range are equivalently setting the angle of attack range. If α^* is within the closed interval (α_1, α_2) that defines the range of acceptable angles of attack, then we may define the critical center of mass location as that value for which $C_m = 0$ for $\alpha = \alpha^*$ and the C_m-α function has a local maximum at $\alpha = \alpha^*$. The first condition requires that

$$C_m(\alpha^*) = 0 = C_m^r(\alpha^*) + \zeta C_N(\alpha^*)$$

or

$$\zeta = -C_m^r(\alpha^*)/C_N(\alpha^*) \tag{12.42a}$$

The second condition occurs when the C_m-α or C_m-C_N curve has a local maximum. We then seek to find the center of mass location for which this maximum occurs at $C_m = 0$. From Eq. (12.41) we have

$$\frac{\mathrm{d}C_m^r}{\mathrm{d}C_N} + \zeta = \frac{\mathrm{d}C_m}{\mathrm{d}C_N} = \frac{\mathrm{d}C_m}{\mathrm{d}\alpha}\left(\frac{\mathrm{d}C_N}{\mathrm{d}\alpha}\right)^{-1} = 0$$

or

$$\zeta = -\frac{\mathrm{d}C_m^r}{\mathrm{d}C_N} = \frac{-\mathrm{d}C_m^r/\mathrm{d}\alpha}{\mathrm{d}C_N/\mathrm{d}\alpha} \tag{12.42b}$$

In Fig. 12.9 are indicated the C_m-C_N curves for various center of mass locations and maximum control deflections. Note that the vehicle is trimmable at a value of C_N of about 0.8 (corresponding to an angle of attack of about 21 degrees) for the most rearward location of the center of mass at 0.745 body lengths. The vehicle also appears to be trimmable at an angle of attack of about 47 degrees. Since such curves move upward as the control settings are reduced (see Fig. 12.8), this vehicle would not be trimmable for any angle of attack within the range of 21 to 45 degrees.

If the center of mass is moved back to 0.735 body lengths, a reduction of the control settings would apparently permit this vehicle to be trimmed over a 15 to 45 degree angle of attack range. For a center of mass somewhere between 0.735 and 0.745 body lengths, there would be a local maximum in the C_m-C_N curve permitting one trim point. The exact position of the center of mass and the applicable normal force coefficient might be calculated from Eq. (12.42b).

Vinh and Lin[3] provide a very interesting graphical technique for assessing the rear-most location of the center of mass. Aerodynamic trim occurs when the pitching moment is zero. The two conditions are given in Eqs. (12.42), the first when $C_m = 0$ for a given angle of attack α^* and the second when the C_m-C_N curve has a local maximum at $\alpha = \alpha^*$.

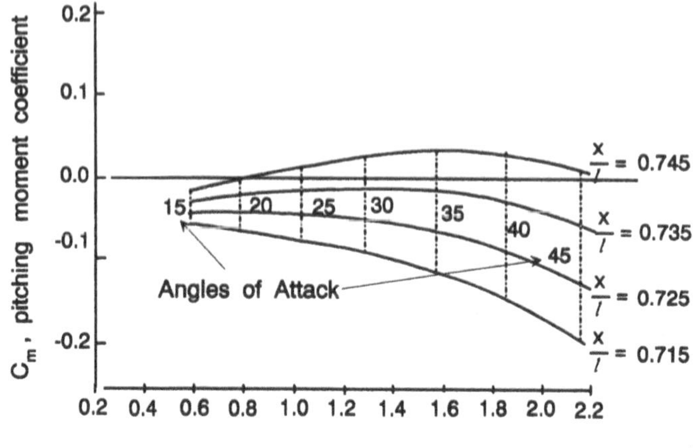

Fig. 12.9 Pitching moment coefficient vs normal force coefficient at maximum control deflection for various center of mass locations (data from Ref. 3).

Rewriting Eq. (12.42a), we have

$$C_m^r = -\zeta C_N \tag{12.43}$$

The preceding equation may be depicted on a C_m-C_N graph as a straight line of slope $-\zeta$ through the origin. This line might intersect the C_m-C_N curve at one or two points (first condition) or be tangent to the curve (second condition). We will *tag* each line with the appropriate center of mass location, $[\zeta(\bar{c}/l) + (X/l)^r]$, where $(X/l)^r$ is the location of the reference center of mass (in this case, 0.715). In Fig. 12.10 we have presented the C_m-C_N curve at a center of mass location of 0.715 body lengths and maximum control deflection. Note that the straight lines corresponding to $\zeta = 0.742$ and 0.745 have two points of intersection. For the $\zeta = 0.745$ line, the intersections are at an angle of attack of about 23 and 43 degrees.

The vehicle is *not* trimmable for any angle of attack between these intersection points. For a center of mass in which $\zeta = 0.739$ there is only one angle of attack (for maximum control deflection) at which the vehicle can be trimmed. As the control settings are diminished the curve moves upward; depending upon the control settings there is a range of angles of attack for which the vehicle is trimmable.

For any center of mass location aft of 0.739 body lengths, there would be a range of angles of attack for which the vehicle is not trimmable. For a center of mass location of, say, 0.742 the vehicle would be trimmable for angles of attack somewhat lower than 25 degrees and higher than 40 degrees, but within the interval 25 to 40 degrees the vehicle is not trimmable for any control setting. The C_m-C_N curve is associated with the maximum control settings.

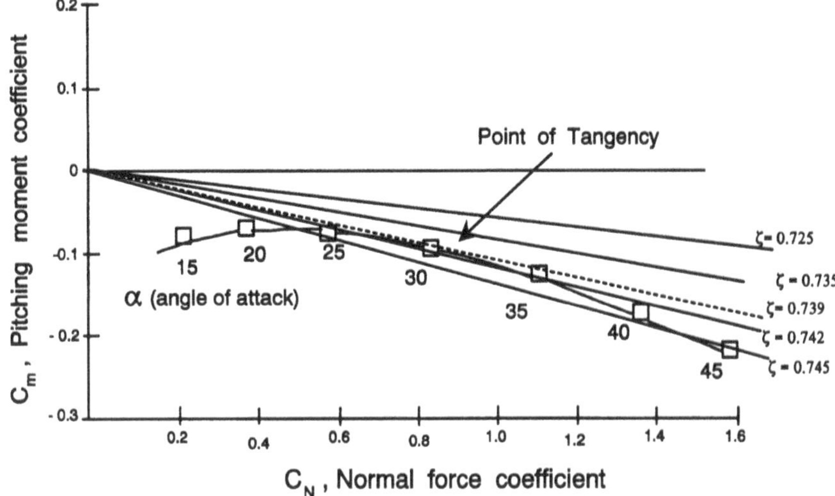

Fig. 12.10 Conditions for trim.

12.4 Phugoid and Spiral Motion

An atmospheric vehicle at low speeds undergoes an oscillatory motion of its center of mass as the total energy periodically changes between kinetic and potential energy. During phugoid motion the angle of attack remains essentially constant while the center of mass oscillates about some reference altitude. For example, if the vehicle is subjected to an increase in lift, it will rise above the reference condition, increasing potential energy; however, the vehicle's speed will diminish as it climbs, reducing kinetic energy. When the lift falls below the weight, the vehicle will descend, gathering speed as potential energy decreases. If we assume that the process is energy conservative, the gains and losses in one form of energy (potential or kinetic) appear in the other form: the total energy sum remains constant.

For a manned orbiter re-entering the atmosphere there is also a phugoid-like motion. However, two important effects present in the phugoid motion of an orbiter are ignored in the analysis of the phugoid motion of the low-speed atmospheric vehicle. These effects are significant density changes during motion and centrifugal effects caused by the trajectory following a curved Earth.

First, let us rewrite Eqs. (12.15a), (12.12), and (12.16). Note that we have replaced dh by dR in Eq. (12.15a), F_T by $(T - D)$ in Eq. (12.12) and F_N by L and $(R_e + h)$ by R in Eq. (12.16).

$$\frac{\mathrm{d}R}{\mathrm{d}t} = V\sin(\gamma) \tag{12.44a}$$

$$m\frac{\mathrm{d}V}{\mathrm{d}t} = T - \frac{1}{2}\rho V^2 S C_D - mg\sin(\gamma) \tag{12.44b}$$

$$mV\frac{d\gamma}{dt} = \frac{1}{2}\rho V^2 S C_L - m\left(g - \frac{V^2}{R}\right)\cos(\gamma) \tag{12.44c}$$

Next, we will define a nondimensional arc length $\bar{S}$ as

$$\bar{S} = \frac{1}{R_0}\int_0^t V(t)\mathrm{d}t \qquad \frac{\mathrm{d}\bar{S}}{\mathrm{d}t} = \frac{V}{R_0} \tag{12.45a}$$

We will now nondimensionalize Eqs. (12.44a), (12.44b), and (12.44c) and change the independent variable from time t to arc length $\bar{S}$ in this manner:

$$\frac{\mathrm{d}}{\mathrm{d}t}(\,) = \frac{\mathrm{d}}{\mathrm{d}\bar{S}}(\,)\frac{\mathrm{d}\bar{S}}{\mathrm{d}t} = \frac{V}{R_0}\frac{\mathrm{d}}{\mathrm{d}\bar{S}}(\,) = \frac{V}{R_0}(\,)' \tag{12.45b}$$

We may now take the following as reference conditions:

$$R = R_0 \qquad V = V_0 \qquad \gamma = 0$$

$$T = \tfrac{1}{2}\rho_0 S C_{D_0} V_0^2 \tag{12.46}$$

$$\tfrac{1}{2}\rho_0 S C_{L_0} V_0^2 = m\left[g_0 - \left(V_0^2/R_0\right)\right]$$

where the subscript "0" refers to the undisturbed or reference trajectory. Note that the settings of the lift and drag coefficients remain unchanged for vehicle departure from the reference conditions, i.e., $C_D = C_{D_0}$ (note that C_{D_0} is the reference trajectory drag coefficient and should not be confused with zero-lift drag coefficient) and $C_L = C_{L_0}$. We should note in passing that the subscript to the density, i.e., ρ_0, in the foregoing usage indicates nominal conditions—not sea level conditions (as was the practice in Chapter 2).

We may now rewrite Eqs. (12.44a), (12.44b), and (12.44c) using the preceding reference conditions and changing the independent variable from time t to nondimensional arc length $\bar{S}$. That is,

$$R'/R_0 = \sin(\gamma) \tag{12.47a}$$

$$\frac{V'}{V} = \frac{\rho_0 R_0}{2\beta}\left[\left(\frac{V_0}{V}\right)^2 - \left(\frac{\rho}{\rho_0}\right)\right] - \frac{gR_0}{V^2}\sin(\gamma) \tag{12.47b}$$

$$\gamma' = \left(\frac{C_{L_0}}{C_{D_0}}\right)\left(\frac{\rho_0 R_0}{2\beta}\right)\left(\frac{\rho_0}{\rho}\right) - \left[\frac{gR_0}{V^2} - \frac{R_0}{R}\right]\cos(\gamma) \tag{12.47c}$$

where we have taken β as the ballistic coefficient, $m/C_{D_0}S$. Next we introduce the perturbational variables ΔV, ΔR, and γ, or

$$V = V_0 + \Delta V \qquad R = R_0 + \Delta R \qquad \gamma = 0 + \gamma \tag{12.48}$$

We must account for variations in the gravitational acceleration g and the density ρ. We will assume that the gravitational field is adequately represented as spherical and the atmospheric density by an exponential variation with altitude; that is,

$$\frac{g}{g_0} = \frac{R_0^2}{(R_0 + \Delta R)^2} \approx \left(1 - \frac{2\Delta R}{R_0}\right) \tag{12.49a}$$

$$\frac{\rho}{\rho_0} = e^{-\Delta R/H} \approx \left[1 - \frac{\Delta R}{R_0}\left(\frac{R_0}{H}\right)\right] \tag{12.49b}$$

Next we insert the perturbational variables in Eqs. (12.47a), (12.47b), and (12.47c). Before doing this, let us define the following lumped parameters to ease the complexity of the expressions:

$$B = \rho_0 R_0/2\beta \qquad \lambda = V_0^2/g_0 R_0 \qquad A = R_0/H \tag{12.50a}$$

We must also eliminate C_{L_0} through the use of the third expression in Eq. (12.46). This equation may be rewritten as

$$\left(\frac{\rho_0 S R_0 C_{D_0}}{2m}\right)\frac{C_{L_0}}{C_{D_0}} = \frac{g_0 R_0}{V_0^2} - 1 = \frac{1}{\lambda} - 1$$
$$B\frac{C_{L_0}}{C_{D_0}} = \frac{1-\lambda}{\lambda} \tag{12.50b}$$

With the parameter groupings of Eqs. (12.50a) and (12.50b), we have the perturbational forms of Eqs. (12.47a), (12.47b), and (12.47c).

$$\frac{\Delta R'}{R_0} = \gamma \tag{12.51a}$$

$$\frac{\Delta V'}{V_0} = AB\left(\frac{\Delta R}{R_0}\right) - 2B\left(\frac{\Delta V}{V_0}\right) - \frac{1}{\lambda}\gamma \tag{12.51b}$$

$$\gamma' = \left[\frac{2}{\lambda} - 1 - \left(\frac{1-\lambda}{\lambda}\right)A\right]\frac{\Delta R}{R_0} + \frac{2}{\lambda}\left(\frac{\Delta V}{V_0}\right) \tag{12.51c}$$

The parameter λ follows from Eq. (6.17). The parameter B is the ratio of $\rho_0 R_0$ to β and A is a measure of R_0 in terms of the atmospheric scale factor H. In Table 7.4 we provided these data:

$$\rho_e R_e \approx \rho_0 R_0 = 1.1738 \times 10^7 \text{ kg/m}^2$$

$$\frac{R_e}{H} \approx \frac{R_0}{H} \approx 950$$

Equations (12.51a), (12.51b), and (12.51c) are simultaneous in the states, that is,

$$X = \left[\frac{\Delta R}{R_0}, \frac{\Delta V}{V_0}, \gamma\right]^{\mathrm{T}} \tag{12.52a}$$

The system matrix F may be written from Eqs. (12.51a,), (12.51b), and (12.51c) as

$$\frac{\mathrm{d}X}{\mathrm{d}\bar{S}} = FX \tag{12.52b}$$

The roots of the determinant of the resolvent matrix provide L, the eigenvalues of the system, as

$$\Delta = \det\left\{[F - LI]\right\} = 0 \tag{12.52c}$$

where I is the identity matrix.

The determinant following from Eqs. (12.51a), (12.51b), and (12.51c) is written in this manner:

$$\Delta = \begin{vmatrix} -L & 0 & 1 \\ AB & -(2B+L) & -(1/\lambda) \\ (2/\lambda) - 1 - A[(1-\lambda)/\lambda] & 2/\lambda & -L \end{vmatrix} = 0$$

Expanding by first row minors gives

$$-L\begin{vmatrix} -(2B+L) & -(1/\lambda) \\ 2/\lambda & -L \end{vmatrix} + \begin{vmatrix} AB & -(2B+L) \\ (2/\lambda) - 1 - A[(1-\lambda)/\lambda] & 2/\lambda \end{vmatrix} = 0$$

or, after Ref. 3,

$$L^3 + 2BL^2 + \omega^2 L - 2B\xi^2 = 0 \tag{12.53}$$

where

$$\omega^2 = \left[(1-\lambda)(2 + A\lambda) + \lambda^2\right]/\lambda^2 \tag{12.54a}$$

$$\xi^2 = (2 - \lambda + A\lambda)/\lambda \tag{12.54b}$$

If $0 < \lambda < 1$, it is easily shown that both ω and ξ must be real.

Equation (12.53) has three roots, one real and two complex conjugates. Let us assume that the real root is small compared to the magnitude of the complex roots as it is for atmospheric vehicles. Therefore, we may write

$$L_1 = 2B\,(\xi/\omega)^2 \tag{12.55}$$

The solution of the state equation will be characterized by one positive exponent and a sinusoidal or oscillatory motion. For now let us concentrate on the real root, which Vinh and Lin[3] identify as the *spiral mode*. In aircraft analysis, the spiral mode is a divergent motion usually encountered as a positive real root in the roll-yaw characteristic equation. Here we are analyzing only the force equations.

In general, taking $\boldsymbol{X}$ as representative of the state vector of Eq. (12.52a) we have a solution in terms of the nondimensional arc length $\bar{S}$, or

$$\boldsymbol{X}(\bar{S}) = \boldsymbol{C}e^{L_1\bar{S}} = \boldsymbol{C}e^{[2B(\xi/\omega)^2\bar{S}]} \tag{12.56a}$$

where $\boldsymbol{C}$ is a vector constant. Since $\bar{S} = (V/R_0)t$ we may rewrite the preceding expression as:

$$\boldsymbol{X}(t) = \boldsymbol{C}\exp\left[2B\left(\frac{\xi}{\omega}\right)^2\frac{V}{R_0}t\right] \tag{12.56b}$$

Obviously the time constant of the spiral mode is

$$T_s \approx \frac{R_0}{V_0}\left(\frac{1}{2B}\right)\left(\frac{\omega}{\xi}\right)^2 = \left(\frac{2m}{\rho_0 C_{D_0} S V_0}\right)\left(\frac{\omega}{\xi}\right)^2$$

To get an idea of how each of the states will change with time, let us write the solution of the radial state as

$$\Delta R'/R_0 = \gamma = C_1 L_1 e^{L_1\bar{S}} \tag{12.57a}$$

From Eq. (12.51c) and Eq. (12.57a) we have:

$$\frac{\Delta V}{V_0} = \frac{C_1}{2}\left[L_1^2\lambda - 2 + \lambda + (1-\lambda)A\right]e^{L_1\bar{S}} \tag{12.57b}$$

Remember that A is about 950 and that L_1 varies with ρ_0. Thus for high altitude where ρ_0 is low and λ is approaching unity, the bracketed term in Eq. (12.57b) is negative; at lower altitudes this term is positive. If we rewrite Eq. (12.57b) using Eq. (12.57a) we obtain

$$\frac{\Delta V}{V_0} = \frac{1}{2}\left[L_1^2\lambda - 2 + \lambda + (1-\lambda)A\right]\frac{\Delta R}{R_0} \tag{12.58}$$

With the bracketed term negative at high altitudes, the velocity will decrease in an outward spiral (ΔR positive) and increase in an inward spiral (ΔR negative). At lower altitudes where the bracketed term is positive, the reverse occurs: an outward spiral increases the speed and an inward spiral decreases the speed.

It might be of some interest to calculate the bracketed term to gain some appreciation for magnitudes. Let us rewrite this term as

$$\left[L_1^2\lambda + (1-\lambda)A + \lambda - 2\right] = F \tag{12.59}$$

where from Eq. (12.55)

$$L_1 = \frac{(\rho_{sl} e^{-h/H})(R_e + h)}{\beta} \left[\frac{(2 - \lambda + A\lambda)\lambda}{(1 - \lambda)(2 + A\lambda) + \lambda^2} \right]$$

and typical values for the constants are

$$\rho_{sl} = 1.752 \text{ kg/m}^3$$

$$R_e = 6.39 \times 10^6 \text{ m}$$

$$A = 950$$

$$H = 6700 \text{ m}$$

Note that we are using the subscript *sl* for sea level rather than 0 as was done in earlier work; the change is necessary since 0 has been used in this chapter to represent the nominal conditions of the trajectory.

The calculation of the ballistic coefficient for a representative vehicle is obtained from data contained in Ref. 5. For example,

$$C_D = 0.6 \qquad m = 1.9 \times 10^6 \text{ kg} \qquad S = 577.0 \text{ m}^2$$

$$\beta = m/C_D S = 5488 \text{ kg/m}^2$$

In Fig. 12.11 we have plotted the function F from Eq. (12.59) as a function of λ for $h = 40$ km and 60 km. Note that F decreases for small values of

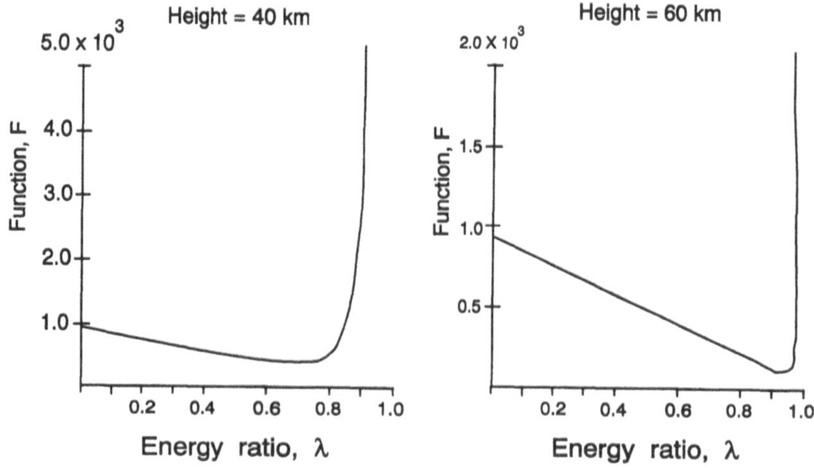

Fig. 12.11 Function vs lambda for altitudes of 40 and 60 kilometers.

λ but eventually becomes singular at $\lambda = 1$. In any event, F remains positive, which means that for an increase in altitude, i.e., ΔR positive, the velocity also increases.

We might now examine the phugoid or oscillatory motion. First recall Eq. (12.53). This equation has one real root, which we have associated with the spiral mode; the complex pair describes damped oscillatory motion.

If we assume initially that the phugoid frequency is ω (i.e., we are ignoring the constant term, $2B\xi^2$ and dividing through by L_1) we get

$$\omega(\text{rad/s}) = \frac{\left[(1-\lambda)(2+A\lambda)+\lambda^2\right]^{1/2}}{\lambda}\left(\frac{V_0}{R_0}\right) \tag{12.60}$$

(Remember that ω as given in Eq. (12.54a) is in radians per nondimensional arc length.) Let us examine the preceding equation at two extremes: when λ is small or nearly zero, we have the condition of the atmospheric vehicle; when λ is slightly less than unity, we have the orbiter just after re-entry.

For an airplane we might ignore the density gradient, i.e., set $A = R_0/H = 0$ since the density gradient is proportional to $1/H$. As $\lambda \to 0$ and $A = 0$, Eq. (12.60) simplifies to the following:

$$\omega = \frac{\left[(1-\lambda)(2+A\lambda)+\lambda^2\right]^{1/2}}{(V_0^2/g_0R_0)(R_0/V_0)}\left(\frac{V_0}{R_0}\right) = \frac{\sqrt{2}g}{V_0} \tag{12.60}$$

The result is a phugoid period of

$$P_p = \sqrt{2}\pi(V_0/g) \tag{12.61a}$$

If the density gradient is included, i.e.,

$$A \approx \frac{R_e}{H} = -\frac{R_e}{\rho}\frac{\mathrm{d}\rho}{\mathrm{d}h}$$

then the phugoid period for an airplane becomes

$$P_p = \frac{2\pi(V_0/g)}{\left[2 - \dfrac{1}{\rho}\dfrac{\mathrm{d}\rho}{\mathrm{d}h}\left(\dfrac{V_0^2}{g}\right)\right]^{1/2}} \tag{12.61b}$$

At the other extreme for a near-orbital vehicle, we have, from Eq. (12.60),

$$\omega = \lim_{\lambda\to 1}\frac{\left[(1-\lambda)(2+A\lambda)+\lambda^2\right]^{1/2}}{V_0/g} = \frac{g}{V_0}$$

where the period becomes

$$P_0 = 2\pi g/V_0 \tag{12.61c}$$

which is the period for orbital motion.

A slight improvement in accuracy is possible by *dividing out* the real root of the cubic [Eq. (12.53)]. In other words, since $2B(\xi/\omega)^2$ is the real root we may examine the quadratic after dividing the cubic through by the real root. That is,

$$\frac{L^3 + 2BL^2 + \omega^2 L - 2B\xi^2}{L - 2B\,(\xi/\omega)^2}$$

gives as a quadratic approximation

$$L^2 + 2B\left[1 + (\xi/\omega)^2\right]L + \left\{\omega^2 + 4B^2\,(\xi/\omega)^2\left[1 + (\xi/\omega)^2\right]\right\} = 0 \qquad (12.62)$$

The frequency of the oscillation of the phugoid motion can then be written as

$$\omega_p = \frac{V_0}{R_0}\left\{\omega^2 + 4B^2\left(\frac{\xi}{\omega}\right)^2\left[1 + \left(\frac{\xi}{\omega}\right)^2\right]\right\}^{1/2} \qquad (12.63\text{a})$$

where ω is given in Eq. (12.54a).

The damping ratio is given by the following:

$$\zeta = \frac{B\left[1 - (\xi/\omega)^2\right]}{\omega_p\,(R_0/V_0)} \qquad (12.63\text{b})$$

As a simple check on Eq. (12.63b) we can ignore $(\xi/\omega)^2$ and replace ω_p with the airplane value given earlier. In such a case we can easily show that

$$\zeta = \left(\sqrt{2}/2\right)\left(C_L/C_D\right)^{-1} \qquad (12.63\text{c})$$

Since the lift-to-drag ratio of an airplane usually exceeds 10, the damping ratio ζ for an airplane is around 0.05, indicating that the phugoid motion is lightly damped.

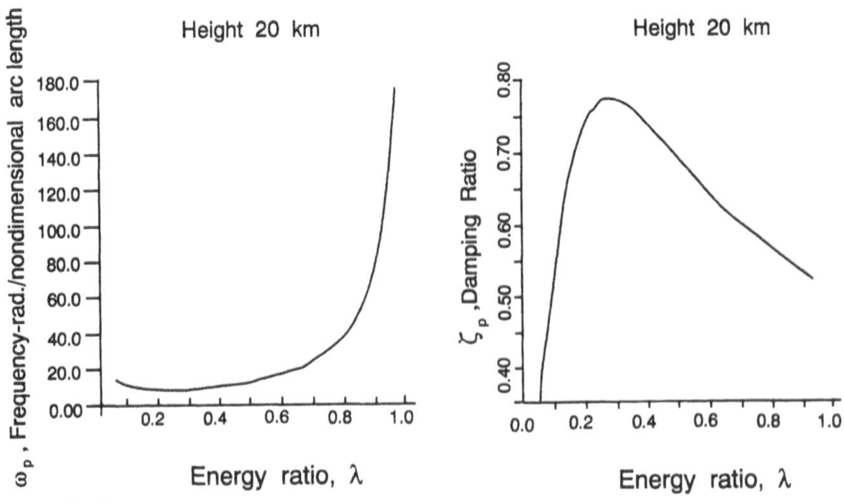

Fig. 12.12a Nondimensional frequency and damping ratio vs energy ratio for an altitude of 20 km.

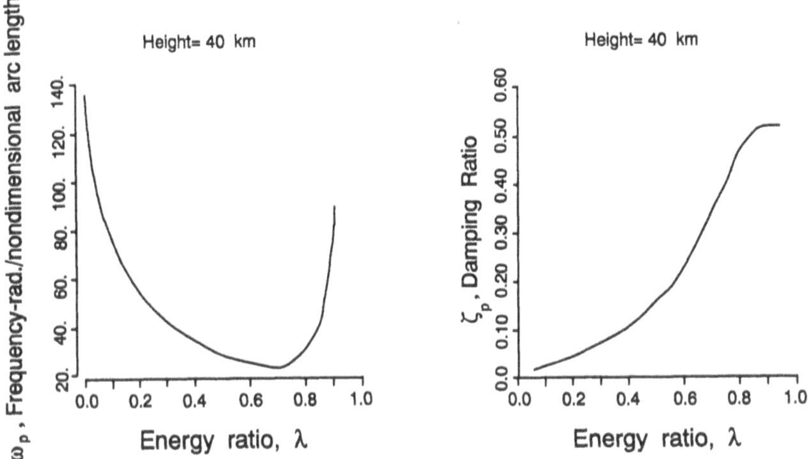

Fig. 12.12b Nondimensional frequency and damping ratio vs energy ratio for an altitude of 40 km.

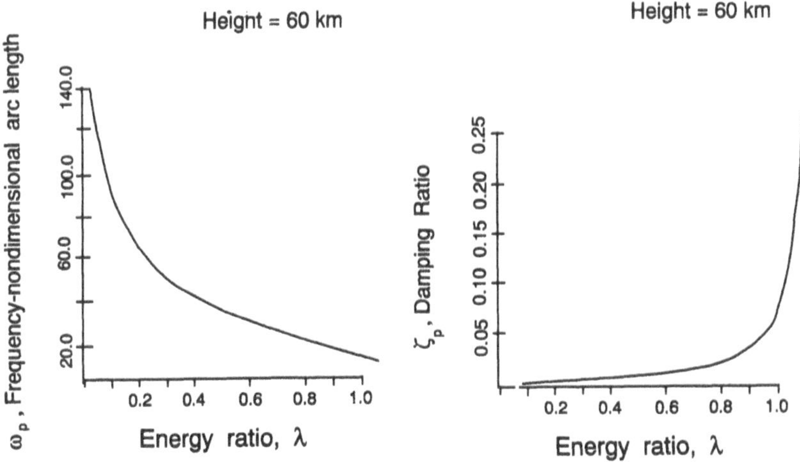

Fig. 12.12c Nondimensional frequency and damping ratio vs energy ratio for an altitude of 60 km.

Next we consider the roots of the phugoid quadratic, [see Eq. (12.62)]. The two roots for the oscillatory case are given as

$$L_{1,2} = -\zeta\omega_n \pm i\omega_n\sqrt{1-\zeta^2} \tag{12.64}$$

Oscillatory motion occurs when the damping ratio is between 0 and 1.0; if ζ is greater than unity, then the motion is overdamped and nonoscillatory. We have calculated $\omega_n = \omega_p$ [see Eq. (12.63a)] and ζ [see Eq. (12.63b)] versus λ for three altitudes (20, 40, and 60 km). The results are given in Figs. 12.12a, 12.12b, and 12.12c. Note that 20 km, the damping is high at a λ value of about

0.25; at higher altitudes the damping ratio remains low, increasing only at the higher values of λ. In these figures the frequence ω_n is in units of radians per nondimensional arc length $\bar{S}$ [see Eqs. (12.45)].

12.5 Aerodynamic Force and Moments in a Body Frame

The force and moment equations for a rigid body were developed in Chapter 5. We may use forms presented in Eq. (5.53) for the force and Eq. (5.56) for the moment. These are repeated here with f for fixed replaced by b for body, that is,

$$\frac{\mathrm{d}\boldsymbol{V}^b}{\mathrm{d}t} + \boldsymbol{\Omega}^b_{b/I}\boldsymbol{V}^b = \left(\frac{\boldsymbol{F}^b_a}{m}\right) + C^b_l \boldsymbol{g}^l \tag{12.65a}$$

$$I^b \frac{\mathrm{d}\boldsymbol{\omega}^b_{b/I}}{\mathrm{d}t} + \boldsymbol{\Omega}^b_{b/I} I^b \boldsymbol{\omega}^b_{b/I} = \boldsymbol{M}^b \tag{12.65b}$$

where $\boldsymbol{V}^b$ and $\boldsymbol{\omega}^b_{b/I}$ are the linear and angular velocities of the b-frame relative to inertial space; I^b is the inertia tensor and $\boldsymbol{F}^b$ and $\boldsymbol{M}^b$ are the force and moment vectors; $\boldsymbol{g}^l$ is the specific gravitational force in a local frame and m is the vehicle mass.

The linear velocity vector will be divided into a steady or unperturbed value $\boldsymbol{V}_\infty$ and a small variable value $\boldsymbol{v}$, that is,

$$\boldsymbol{V}^b = \boldsymbol{V}^b_\infty + \boldsymbol{v}^b = [U_\infty, 0, W_\infty]^{\mathrm{T}} + [u, v, w]^{\mathrm{T}} \tag{12.66a}$$

$$\boldsymbol{V}^b = \left[\left(1 + \frac{u}{U_\infty}\right), \frac{v}{U_\infty}, \left(\frac{w - W_\infty}{U_\infty}\right)\right]^{\mathrm{T}} U_\infty \tag{12.66b}$$

Now let

$$\tan(\alpha_T) \approx \alpha_T = \frac{W_\infty}{U_\infty} \approx \frac{W_\infty}{|\boldsymbol{V}_\infty|}$$

$$\tan(\alpha) \approx \alpha = \frac{w}{U_\infty} + \frac{W}{U_\infty} \approx \frac{w + W_\infty}{|\boldsymbol{V}_\infty|}$$

$$\tan(\beta) \approx \beta = \frac{v}{U_\infty} \approx \frac{v}{|\boldsymbol{V}_\infty|}$$

$$\alpha - \alpha_T \approx \frac{w}{U_\infty}$$

α_T is the trim angle of attack and $(\alpha - \alpha_T)$ is the perturbation about the trim angle of attack. Note that we have assumed that since α_T is small, we can replace U_∞ with the magnitude of V_∞. If this is done then Eq. (12.66b) can be rewritten as

$$V^b = V_\infty [1, \beta, (\alpha - \alpha_T)]^{\mathrm{T}} \tag{12.66c}$$

The variables describing magnitude and direction of the velocity vector are V_∞, $(\alpha - \alpha_T)$, and β.

We may now express the force and moment vectors that originate in the aerodynamic pressure distribution in the following manner:

$$F^b = [X, Y, Z]^{\mathrm{T}}$$

$$M^b = [L, M, N]^{\mathrm{T}}$$

where

$$X = q_\infty S C_x \tag{12.67a}$$

$$Y = q_\infty S\left[-\beta C_{y_\beta} + \dot{\beta}(c/2V_\infty)C_{y_{\dot{\beta}}}\right] \tag{12.67b}$$

$$Z = q_\infty S\left[\alpha C_{z_\alpha} + \dot{\alpha}(c/2V_\infty)C_{z_{\dot{\alpha}}}\right] \tag{12.67c}$$

$$L = q_\infty S c\left[p(c/2V_\infty)C_{l_p} + C_{l_0}\right] \tag{12.67d}$$

$$M = q_\infty S c\left[(\alpha - \alpha_T)C_{m_\alpha} + p(c/2V_\infty)\beta C_{m_{p\beta}} + q(c/2V_\infty)(C_{m_q} + C_{m_{\dot{\alpha}}})\right] \tag{12.67e}$$

$$N = q_\infty S c\left[-\beta C_{n_\beta} + p(c/2V_\infty)\alpha C_{n_{p\alpha}} + r(c/2V_\infty)(C_{n_{\dot{\beta}}} - C_{n_r})\right] \tag{12.67f}$$

A few comments are in order concerning the preceding expressions. Note that the reference area S is common to all six loads. For bodies of revolution, S is taken as the base or maximum cross-sectional area. For vehicles having extremely small wings or strakes, S continues to be the base area. However, for lifting vehicles, S is usually taken as the exposed or wetted wing area. The reference length c is common to all six loads; for bodies of revolution c is taken as either the body length, body radius, or body diameter. For winged vehicles, c can be the mean aerodynamic chord for the X, Z, M loads and the wing semi-span for the Y, L, N loads. The overriding criterion is, of course, that the reference area and length be the same area and length that were used in defining the nondimensional force and moment coefficients. Such variations, important as they are in setting loads are tangential to our main concerns.

The various derivatives are in general usage. For example,

$$C_{y_\beta} = \frac{\partial C_y}{\partial \beta} = \frac{\partial C_y}{\partial (v/V_\infty)} = \frac{1}{\frac{1}{2}\rho V_\infty^2 S}\frac{\partial Y}{\partial \beta}$$

Note also that for the derivatives to be nondimensional the variables for differentiation must also be nondimensional, or

$$C_{m_q} \equiv \frac{\partial C_m}{\partial(qc/2V_\infty)} \qquad C_{m_{p\beta}} \equiv \frac{\partial^2 C_m}{\partial\beta\,\partial(pc/2V_\infty)}$$

We may now write out the force and moment equations as follows:

$$\dot{u} + qW_\infty - rU = (X/m) - g\sin(\theta) \tag{12.68a}$$

$$\dot{v} + rU_\infty - pW_\infty = (Y/m) + g\cos(\theta)\sin(\phi) \tag{12.68b}$$

$$\dot{w} + pv - qU_\infty = (Z/m) + g\cos(\theta)\cos(\phi) \tag{12.68c}$$

$$I_x\dot{p} + (I_z - I_y)rq = L \tag{12.68d}$$

$$I_y\dot{q} + (I_x - I_z)pr = M \tag{12.68e}$$

$$I_z\dot{r} + (I_y - I_x)pq = N \tag{12.68f}$$

In the preceding equations we assumed that the body frame is also a principal frame; for this reason we have used a single subscript to denote the appropriate axis in defining the moments/products of inertia.

12.6 Rolling Moment

In this section we will consider the simplest of all angular motion, roll about the axis of symmetry. Let us first rewrite the rolling moment equation as

$$I_x\frac{\mathrm{d}p}{\mathrm{d}t} + (I_z - I_y)rq = L \tag{12.69a}$$

For a rotationally symmetric vehicle, $I_z = I_y$ and so roll is uncoupled from yaw and pitch. Replacing the rolling moment L by its equivalent in Eq. (12.67d), we have

$$\frac{\mathrm{d}p}{\mathrm{d}t} = \frac{q_\infty S c}{I_x}\left[C_{l_0} + p\left(\frac{c}{2V_\infty}\right)C_{l_p}\right] \tag{12.69b}$$

which we may rewrite as

$$\frac{\mathrm{d}}{\mathrm{d}t}\left[p + \frac{C_{l_0}}{C_{l_p}}\left(\frac{2V_\infty}{c}\right)\right] - \left(\frac{q_\infty S c^2 C_{l_p}}{2V_\infty I_x}\right)\left[p + \left(\frac{C_{l_0}}{C_{l_p}}\right)\left(\frac{2V_\infty}{c}\right)\right] = 0$$

since $\left(C_{l_0}/C_{l_p}\right)(2V_\infty/c)$ is assumed invariant with time.

Integration of the preceding expression may take the following form:

$$p(t) = p_{ss} + [p(0) - p_{ss}]\,e^{t/T} \tag{12.70}$$

where the steady state spin rate p_{ss} and the roll time constant are defined as

$$p_{ss} = -\frac{C_{l_0}}{C_{l_p}}\left(\frac{2V_\infty}{c}\right) \qquad T = \frac{2V_\infty I_x}{q_\infty S c^2 C_{l_p}}$$

From the preceding equation we see that the roll rate p decays from an initial value $p(0)$ to a final value p_{ss}. The time constant T must be negative, which means in turn that C_{l_p} must be negative. If there is a roll driving moment, i.e., C_{l_0} is nonzero, the roll rate will settle on some steady value, which we identified as p_{ss}. Without a roll driving moment the roll rate will approach zero as a final value.

It might be of some interest to calculate C_{l_p} and C_{l_0}. Although there are no general methods for calculating C_{l_0}, the roll driving torque for a body of revolution, we might surmise that such torque is caused by the ablative pattern set up in the heat shield.

One contributor to roll torque can arise from the offset of the center of mass from the axis of configurational symmetry. Mass offset is illustrated in Fig. 12.13. If we re-examine Eq. (5.40) we see that the first term is the mass offset contribution to angular momentum, i.e.,

$$\boldsymbol{h}_0 = \sum_i^\infty (m_i \boldsymbol{r}_i) \times \boldsymbol{V}_\infty = m\left[R_y \boldsymbol{j}^b + R_z \boldsymbol{k}^b\right] \times \left[U\boldsymbol{i}^b + V\boldsymbol{j}^b + W\boldsymbol{k}^b\right]$$

Carrying out the manipulations, we get

$$\boldsymbol{h}_0 = \left[(R_y W - R_z V)\boldsymbol{i}^b + (R_z U)\boldsymbol{j}^b + (-R_y U)\boldsymbol{k}^b\right]$$

If we confine our interest to roll rate, then we have

$$\boldsymbol{\omega}_{b/I}^b = p\boldsymbol{i}^b$$

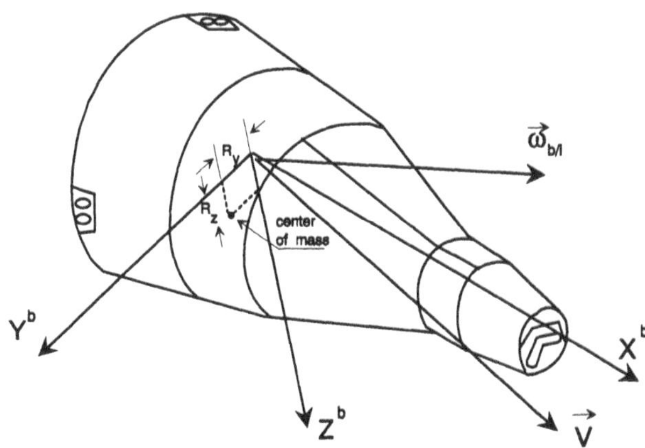

Fig. 12.13 Definition of mass-offset.

so that

$$
\begin{aligned}
l_0 &= \left[\left.\frac{\mathrm{d}\boldsymbol{h}_0}{\mathrm{d}t}\right|_b + p\boldsymbol{i}^b \times \boldsymbol{h}_0 \right] \cdot \boldsymbol{i}^b \\
&= m\left[R_y \dot{W} - R_z \dot{V}\right] = mR_y a_z - mR_z a_y
\end{aligned} \tag{12.71}
$$

The total roll moment L consists of a driving moment L_0, a damping moment $L_p p$, and a mass offset moment l_0, and is expressed as

$$
\begin{aligned}
L &= L_0 + l_0 + \frac{\partial L}{\partial p} p \\
&= C_{l_0}(q_\infty S c) + C_{l_p}\left(q_\infty S c^2 / 2V_\infty\right) p - mR_z a_y + mR_y a_z
\end{aligned} \tag{12.72}
$$

With mass offset, the rolling moment equation is now coupled with the force equation through the acceleration terms a_y and a_z.

Next let us calculate the rolling moment. In Fig. 12.14a we see a portion of a vehicle rotating about the X^b-axis, i.e., the axis of symmetry. Figure 12.14b is a plan view containing the axis of symmetry. We shall assume that shear stress is developed at the surface with a component τ_c along the conical generator and an orthogonal component τ_ϕ; let ψ be the angle between the resultant stress and the concial generator, as

$$\psi = \tan^{-1}(\tau_\phi / \tau_c)$$

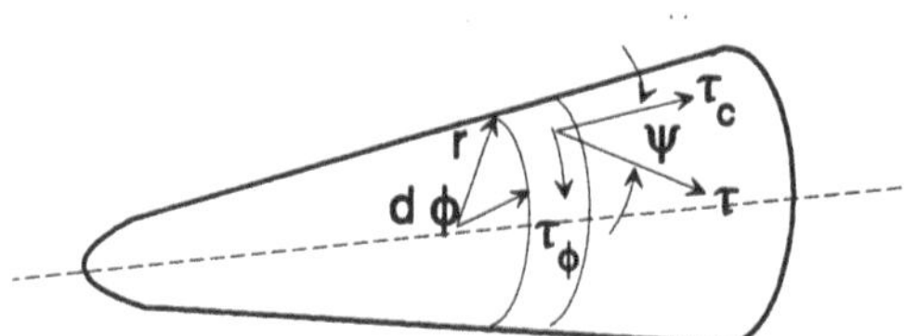

Fig. 12.14a Representative frictional element.

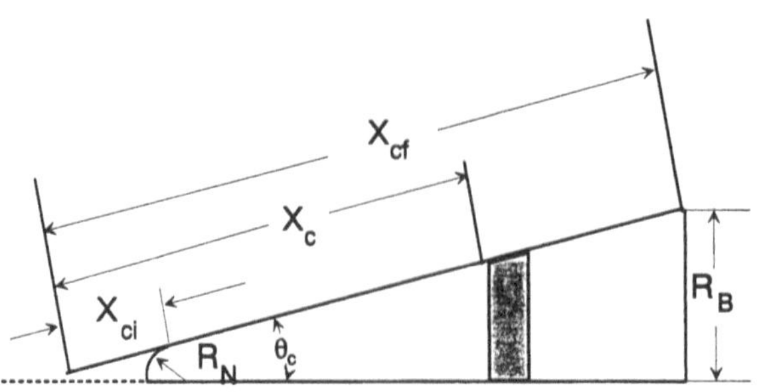

Fig. 12.14b Plan view.

The elemental area is

$$dA = 2\pi r \, dX_c$$

and the elemental roll torque is

$$dl = -2\pi r^2 \tau_\phi dX_c$$

Integration of the previous expression provides the total roll torque, or

$$l = -2\pi \int_{X_{ci}}^{X_{cf}} \tau_\phi r^2 \, dX_c \tag{12.73}$$

where X_c is a coordinate along the conical generator of the vehicle. The limits of integration are defined as follows: X_{ci} is the point on the generator where the nose radius has its point of tangency with the conical generator; X_{cf} is the rearward-most point on the generator. These points are indicated in Fig. 12.14b.

The roll damping derivative C_{l_p} may now be expressed as

$$C_{l_p} = \frac{\partial C_l}{\partial (pc/2V_\infty)} = \frac{l}{q_\infty S c\,(pc/2V_\infty)} = \frac{l}{q_\infty (\pi R^2{}_B)(2R_B)(pR_B/V_\infty)} \tag{12.74a}$$

where we have replaced the generic reference length c by the body base diameter, $2R_B$, where R_B is the body base radius.

Replacing l in Eq. (12.73) by the previous expression, we have

$$C_{l_p} = -2\pi \left[\frac{\int_{X_{ci}}^{X_{cf}} \tau_\phi r^2 dX_c}{q_\infty(\pi R_B^2)(2R_B)(pR_B/V_\infty)} \right] \tag{12.74b}$$

Now we let the angle ψ be equal to the helix angle (and make the reasonable assumption that ψ is small), we get

$$\psi = pr/U_e$$

where U_e is the flow velocity just outside of the boundary layer. Let us write the shear stress τ in coefficient form, indicating that the coefficient is a function of the helix angle. That is,

$$\tau_\phi = \tau\psi = q_e C_f\,(pr/U_e) \tag{12.75}$$

where q_e is the dynamic pressure at the edge of the boundary layer.

Now r in Eq. (12.74b) is related to X_c as

$$r = X_c \sin(\theta_c)$$

We can rewrite Eq. (12.74b) to get

$$C_{l_p} = -\left(\frac{q_e}{q_\infty}\right)\sin^3(\theta_c)\int_{(R_N/R_B)\csc(\theta_c)}^{\csc(\theta_c)} C_f\left(\frac{V_\infty}{U_e}\right)\left(\frac{X_c}{R_B}\right)^3 \mathrm{d}\left(\frac{X_c}{R_B}\right) \tag{12.76}$$

Obviously once C_f has been selected as a function of X_c/R_B, integration of Eq. (12.76) can be carried out.

Let us calculate the damping-in-roll derivative C_{l_p} and make comparisons to measurements. At a Mach number of 7.7 and a Reynolds number of 6.23×10^6 (based upon body length), a cone half-angle of 7.25 degrees and a bluntness of 7.5%, the measured value of C_{l_p} is 0.00250. First we must calculate the skin friction coefficient C_f. From the Blasius relationship the skin friction of a flat plate with laminar boundary layer is

$$C_f = 1.328/Re^{1/2} = 5.32 \times 10^{-4}$$

The skin friction coefficient should vary with $X_c^{1/2}$, but in the interest of simplicity we will keep C_f a constant. We must have values for the ratios q_e/q_∞ and U_e/V_∞, the ratio of local to free stream dynamic pressure and the airspeed to the free stream value, respectively. Here we will rely on the work of Linnell and Bailey[7] and use of the following relationships for P_e/P_∞, T_e/T_∞, and U_e/V_∞, local edge pressure, temperature, and velocity ratios with free stream conditions.

For $M_\infty \sin(\theta_c) > 1$

$$\frac{T_e}{T_\infty} = \left[1 + \exp\left\{-1 - 1.52\,[M_\infty \sin(\theta_c)]\right\}\right]\left\{1 + \frac{[M_\infty \sin(\theta_c)]^2}{4}\right\} \tag{12.77a}$$

and for $M_\infty \sin(\theta_c) < 1$

$$\frac{T_e}{T_\infty} = 1 + 0.35\,[M_\infty \sin(\theta_c)]^{3/2} \tag{12.77b}$$

$$\frac{P_e}{P_\infty} = 1 + \frac{0.7M_\infty^2\left[4\sin^2(\theta_c)\right]\left[2.5 + 8.0\left(\sqrt{M_\infty^2 - 1}\right)\sin(\theta_c)\right]}{1 + 16.0\left(\sqrt{M_\infty^2 - 1}\right)\sin(\theta_c)} \tag{12.77c}$$

$$\frac{U_e}{V_\infty} = \cos(\theta_c)\left[1 - \frac{\sin(\theta_c)}{M_\infty}\right]^{1/2} \tag{12.77d}$$

Since $M_\infty \sin(\theta_c) = 0.972$, we will use Eq. (12.77b) for T_e/T_∞. It is easily shown from the equation of state of an ideal gas that

$$\frac{q_e}{q_\infty} = \left(\frac{U_e}{V_\infty}\right)^2\left(\frac{P_e}{P_\infty}\right)\left(\frac{T_e}{T_\infty}\right)^{-1}$$

Using $\theta_c = 7.25$ degrees and $M_\infty = 7.7$ we can easily calculate

$$U_e/V_\infty = 0.984 \qquad P_e/P_\infty = 2.644 \qquad T_c/T_\infty = 1.335$$

Now carrying out the integration indicated in Eq. (12.76) we get

$$C_{l_p} = \tfrac{1}{4}(q_e/q_\infty)\csc(\theta_c)C_f\,[V_\infty/U_e]\left[1-(R_N/R_B)^4\right] \tag{12.77e}$$

Solving Eq. (12.77e) yields

$$C_{l_p} = -0.00206$$

This result is in fair agreement with the measured value of -0.00250.

Let us calculate the roll time constant T of Eq. (12.70). For an altitude h of 15 km, a velocity of 4500 m/s, a roll moment of inertia of 0.972 kg-m^2, a base diameter of 0.44 meters and a roll damping derivative C_{l_p} of -0.00206, we get

$$T = 2V_\infty I_x/q_\infty S(2R_B)^2 C_{l_p} = 19.076 \text{ s}$$

or the time to half amplitude is 13.22 seconds.

An interesting paper by Glover shows that we can relate C_{lp}, the damping-in-roll derivative, to C_A, the axial shear force coefficient.[8] The axial shear force coefficient is the component of the frictional force acting in the downstream direction as shown in Fig. 12.15. That is,

$$C_A = \frac{\cos(\theta_c)}{q_\infty \pi R_B^2}\int_{X_{ci}}^{X_{cf}} dF$$

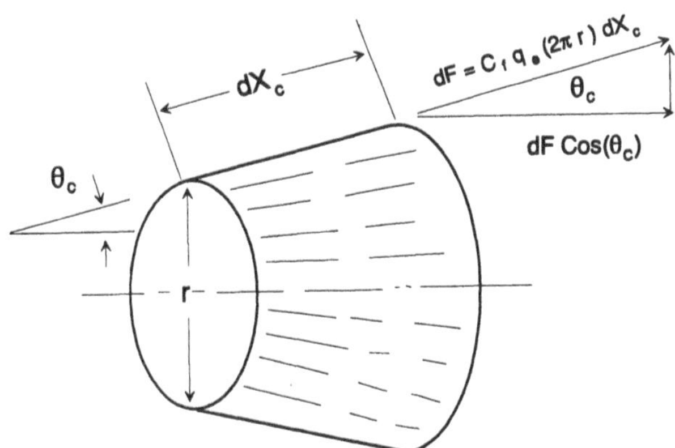

Fig. 12.15 Resolution of the differential shear force into a downstream component.

where

$$\mathrm{d}F = C_f q_e (2\pi r)\,\mathrm{d}X_c$$

or

$$C_A = (q_e/q_\infty)\cos(\theta_c)\int_{(R_N/R_B)\csc(\theta_c)}^{\csc(\theta_c)} 2C_f q_e (2\pi r)\,\mathrm{d}(X_c/R_B)$$

But since $r/X_c = \sin(\theta_c)$, we have

$$C_A = (q_e/q_\infty)\sin(2\theta_c)\left[\int_{(R_N/R_B)\csc(\theta_c)}^{\csc(\theta_c)} C_f (X_c/R_B)\,\mathrm{d}(X_c/R_B)\right] \tag{12.78}$$

According to Glover, C_{l_p} and C_A are related in this manner:

$$C_{l_p}/C_A = C/\cos^2(\theta_c) \tag{12.79}$$

where the parameter C is defined over the range

$$-0.25 < C < -0.20$$

The exact value of C is set by the cone bluntness R_N/R_B. For a nose bluntness of 0.22, C equals 0.225. Obviously C_f is related to C_A [see Eq. (12.78)] as is C_{lp} [see Eq. (12.76)]. If we were to assume that C_f is constant in the integrations of Eqs. (12.76) and (12.78) then for large Mach numbers we would have the following for

$$C = \frac{\left[1 - (R_N/R_B)^4\right]}{4\left[1 - (R_N/R_B)^2\right]} = \frac{1}{4}\left[1 + (R_N/R_B)^2\right]$$

Since C_f must vary with (X_c/R_N), the preceding expression is only approximate.

12.7 Pitching Moment Equations in an Exponential Atmosphere

Another closed-form solution of interest is that of planar pitching motion, i.e., angular velocity components roll p and yaw r are identically zero. Let us divide Eq. (12.68c) through by U_∞ and then replace velocity components by angles from Eq. (12.66b). Thus,

$$\dot{\alpha} - q = \frac{\rho_\infty V_\infty S}{2m}\left[C_{z_\alpha} + C_{z_{\dot{\alpha}}}\dot{\alpha}\left(\frac{c}{2V_\infty}\right)\right] \tag{12.80}$$

where we have taken V_∞ as an acceptable approximation to U_∞. The derivative $C_{z_{\dot{\alpha}}}$ (and later $C_{m_{\dot{\alpha}}}$) are measures of flow lag. In hypersonic flows the translational speed of the flow is so great relative to the speed of cross flow induced by angular motion that the flow field adjusts nearly instantaneously to changes in cross flow. We shall also replace C_{z_α} (which is always negative) with the normal force derivative C_{N_α}. Consequently, Eq. (12.80) becomes (after neglecting $C_{z_{\dot\alpha}}$),

$$\dot{\alpha} - q = -\left(\frac{\rho_\infty V_\infty S}{2m}\right)(C_{N_\alpha}\alpha) \tag{12.81a}$$

The pitching moment equation follows from Eq. (12.67e) and (12.68e) as

$$\dot{q} = \frac{\rho_\infty V_\infty^2 S c}{2 I_{yy}}\left[C_{m_\alpha}\alpha + \left(\frac{qc}{2V_\infty}\right)C_{m_q}\right] \tag{12.81b}$$

where we have assumed that the trim angle of attack α_T is zero.

In formulating the preceding equation we have neglected the flow lag moment $C_{m_{\dot\alpha}}$; since the roll rate is also zero we have also omitted the Magnus derivative $C_{m_{p\alpha}}$. We can now differentiate Eq. (12.81a) to get

$$\dot{q} = \ddot{\alpha} + \left(\frac{\rho_\infty V_\infty S}{2m}\right)C_{N_\alpha}\dot{\alpha} + \left(\frac{S C_{N_\alpha}\alpha}{2m}\right)\frac{\mathrm{d}}{\mathrm{d}t}(\rho_\infty V_\infty) \tag{12.81c}$$

The third term on the right allows changes in free stream conditions of both density and velocity. In some applications (wind tunnel testing, say), $\rho_\infty V_\infty$ may be regarded as a constant. However, for most re-entry trajectories, $\rho_\infty V_\infty$ will change a few orders of magnitude from entry to impact.

We may eliminate q from Eq. (12.81b) by substituting in Eq. (12.81a) and then use the resulting expression to remove $\dot{q}$ from Eq. (12.81c) to get

$$\ddot{\alpha} + \frac{\rho_\infty V_\infty}{2}\left[\left(\frac{S}{m}\right)C_{N_\alpha} - 2\left(\frac{S}{m}\right)\left(\frac{c}{2K_y}\right)^2 C_{m_q}\right]\dot{\alpha} + \frac{\rho_\infty V_\infty}{2}\left\{\left[-V_\infty\left(\frac{S}{m}\right)\left(\frac{c}{K_y^2}\right)C_{m_\alpha} - 2\left(\frac{\rho_\infty V_\infty}{2}\right)\left(\frac{S}{m}\right)^2\left(\frac{c}{2K_y}\right)^2 C_{m_q}C_{N_\alpha}\right] + \frac{\mathrm{d}}{\mathrm{d}t}\left[\ln\left(\frac{\rho_\infty V_\infty}{2}\right)\left(\frac{S}{m}\right)C_{N_\alpha}\right]\right\}\alpha = 0 \tag{12.82}$$

Note that we have used the radius of gyration $K_y^2 = I_{yy}/m$ as a substitute for the pitch moment of inertia. We will take Eq. (12.82) as the general equation for planar pitching motion. We now examine some interesting closed-form solutions to this equation.

Equation (12.82) is a second-order variable coefficient equation in that $\rho_\infty V_\infty$ and V_∞ are functions of the independent variable time. For a wind tunnel test where these quantities are constant, the solution is that of a linear, constant coefficient second-order equation. However, for most re-entry trajectories the solution is more complicated.

Let us look at the last term in the second brackets, that is,

$$\frac{\mathrm{d}}{\mathrm{d}t}[\ln(\rho_\infty V_\infty)]$$

First, from Eq. (7.26a) let us write

$$\rho \frac{dV}{dt} = -\frac{(\rho V)^2}{2\beta} + g \approx \frac{(\rho V)^2}{2\beta} \tag{12.83a}$$

We have omitted for convenience the ∞ subscript from ρ and V and have neglected the gravitational acceleration. Next, we note that

$$\begin{aligned} \frac{d}{dt}(\rho V) &= \rho \frac{dV}{dt} + V \frac{d\rho}{dt} \\ &= -\frac{(\rho V)^2}{2\beta} - V^2 \sin(\gamma_E) \frac{d\rho}{dZ} \end{aligned} \tag{12.83b}$$

where we have made use of the identity

$$\frac{d}{dt}(\,) = -V \sin(\gamma_E) \frac{d}{dZ}(\,)$$

If we assume an exponential atmosphere from Eq. (2.32), then

$$\rho = \rho_0 e^{-Z/H} \qquad \frac{d\rho}{dZ} = -\frac{\rho}{H}$$

Note that we are again representing the density at $Z = 0$ by ρ_0. We may write

$$\frac{d}{dt} \ln(\rho V) = \frac{1}{\rho V} \frac{d(\rho V)}{dt} = -\frac{\rho V}{2\beta} + \frac{V}{H} \sin(\gamma_E) \tag{12.83c}$$

This expression may be substituted into Eq. (12.82) if we are willing to accept the approximation that $V \sin(\gamma_E)$ is the vertical velocity throughout the trajectory.

We may check on the constancy of $\rho_\infty V_\infty$ with altitude Z by writing, from Eqs. (2.32) and (7.39),

$$\rho_\infty = \rho_0 e^{-Z/H} \tag{2.32}$$

$$V_\infty = V_E \exp\left(-a_E e^{-Z/H}\right) \tag{7.39a}$$

so that we may form the ρV product on the left of Eq. (12.83c):

$$\rho_\infty V_\infty = (\rho_0 V_E) \exp\left\{-\left[\frac{Z}{H} + a_E e^{-Z/H}\right]\right\} \tag{12.84a}$$

Recall from Eq. (7.39b) that the term a_E is given as

$$a_E = \rho_0 H / [2\beta \sin(\gamma_E)]$$

Taking the derivative of Eq. (12.84a) with respect to Z, we get

$$\frac{\mathrm{d}}{\mathrm{d}Z}(\rho_\infty V_\infty) = \rho_0 V_E \exp\left\{-\left[\frac{Z}{H} + a_E e^{-Z/H}\right]\right\} \frac{\mathrm{d}}{\mathrm{d}Z}\left\{-\left[\frac{Z}{H} + a_E e^{-Z/H}\right]\right\} \tag{12.84b}$$

The stationary value of $\rho_\infty V_\infty$ would require that

$$\frac{\mathrm{d}}{\mathrm{d}Z}\left(-\frac{Z}{H} - a_E e^{-Z/H}\right) = -\frac{1}{H} + \frac{a_E}{H} e^{-\tilde{Z}/H} = 0$$

or the constant-$\rho_\infty V_\infty$ altitude $\tilde{Z}$ is

$$\tilde{Z} = H \ln a_E \tag{12.84c}$$

Clearly the rate of change of $\rho_\infty V_\infty$ with altitude is zero only at a certain altitude $\tilde{Z}$, which varies with a_E [see Eq. (7.39b)]. [Note from Eq. (7.40a) that $\tilde{h} = H \ln(2)$, where $\tilde{h}$ is the altitude of maximum axial deceleration.] Therefore, $\rho_\infty V_\infty$ cannot be constant over the entire trajectory.

We may compare $\rho_\infty V_\infty$ at two altitudes using Eq. (12.84a), that is,

$$\frac{(\rho_\infty V_\infty)_2}{(\rho_\infty V_\infty)_1} = \left\{\exp\left[-\left(\frac{Z_2}{H} - \frac{Z_1}{H}\right)\right]\right\} \exp\left[a_E\left(e^{-Z_1/H} - e^{-Z_2/H}\right)\right] \tag{12.85}$$

Let us consider the following values:

$$Z_2 = 15\ \text{km} \qquad Z_1 = 45\ \text{km}$$

$$\beta = 6000\ \text{kg/m}^2$$

$$a_E = 0.1$$

Consequently the ratio of (2) to (1) in Eq. (12.85) is about 0.012, which means that $\rho_\infty V_\infty$ at altitude (1) is about 90 times the value at altitude (2)—a change of about two orders of magnitude.

Despite the demonstration that $\rho_\infty V_\infty$ changes greatly during re-entry, we first assume that $\rho_\infty V_\infty$ is a constant over a segment of the trajectory. We return to Eq. (12.82), making the further simplification of ignoring the force derivative, C_{N_α}. As a result, we have

$$\ddot{\alpha} + \frac{\rho_\infty V_\infty}{2}\left[-2\frac{S}{m}\left(\frac{c}{2K_y}\right)^2 C_{m_q}\right]\dot{\alpha} + \frac{\rho_\infty V_\infty}{2}\left[-V_\infty\left(\frac{S}{m}\right)\left(\frac{c}{K_y^2}\right)C_{m_\alpha}\right]\alpha = 0 \tag{12.86a}$$

The undamped natural frequency ω_n and the damping ratio ζ are given as

$$\omega_n = \left[-\frac{\rho_\infty V_\infty^2}{2}\left(\frac{S}{m}\right)\left(\frac{c}{K_y^2}\right)C_{m_\alpha}\right]^{1/2} \tag{12.86b}$$

$$\zeta = \frac{1}{2\omega_n}\left[-2\left(\frac{S}{m}\right)\left(\frac{c}{2K_y}\right)^2 C_{m_q}\right]\frac{\rho_\infty V_\infty}{2} \tag{12.86c}$$

Let us concentrate on the frequency. If we use Eqs. (2.32) and (7.39a), we have

$$\omega_n = \left[V_E \exp\left(-a_E e^{-Z/H}\right)\right]\left[-\frac{\rho_0}{2}e^{-Z/H}\left(\frac{S}{m}\right)\left(\frac{c}{K_y^2}\right)C_{m_\alpha}\right]^{1/2} \tag{12.87}$$

We of course recognize that Eq. (12.87) cannot be considered as part of the solution of Eq. (12.86a). Because we first assumed that $\rho_\infty V_\infty$ was invariant with changing altitude, we cannot, after obtaining the frequency, reintroduce an expression for variation in ρ_∞ and V_∞.

Table 12.2 presents some metrology and aerodynamic characteristics of a typical re-entry vehicle. If we use the following values,

$\gamma_E = +90.0$ deg $V_E = 5000$ m/s $Z = 25$ km

$\rho_{sl} = 1.752$ kg/m^3 $H = 6700$ m

Table 12.2 Physical characteristics of a generic re-entry vehicle

Parameter	Value	Units
Nose radius, R_N	1.98 (0.77)	cm (in)
Base radius, R_B	22 (8.67)	cm (in)
Bluntness, R_N/R_B	0.09	
Reference length, $c = R_B$	22 (8.76)	cm (in)
Length, L	1.52 (5)	m (ft)
Mass, m	92.0 (6.3)	kg (slugs)
Reference area, $S = \pi R_B^2$	0.152 (1.64)	m^2 (ft^2)
Roll moment of inertia, I_{xx}	0.972 (0.72)	kg-m^2 (slug-ft^2)
Pitch moment of inertia, I_{yy}	9.32 (6.88)	kg-m^2 (slug-ft^2)
Yaw moment of inertia, I_{zz}	9.32 (6.88)	kg-m^2 (slug-ft^2)
C.G. from nose, X_0/L	0.61	
Static margin	0.071	
Pitching moment derivative, C_{m_α}	−0.520	
Normal force derivative, C_{N_α}	2.15	
Pitch damping moment, C_{m_q}	−8.0	
Pitch damping force, C_{N_q}	−0.35	
Zero-lift drag coefficient, C_{D_0}	0.100	
Ballistic factor, β_m	6050 (1239.1)	kg/m^2 (lb/ft^2)
Pitch-yaw radius of gyration, K_y	0.318 (1.045)	m (ft)

and the data contained in Table 12.2 we obtain

$$\omega_n = 30.6 \text{ rad/s} = 4.87 \text{ Hz}$$

or about 1 km per oscillation.

Let us now return to Eq. (12.82); although we will make some simplifications, we will retain the altitude or time variation of $\rho_\infty V_\infty$. To make the analysis tractable, we will omit the damping derivative C_{m_q}, and, as before, the force derivative C_{N_α}. Thus, Eq. (12.82) may be written

$$\ddot{\alpha} - \left[\frac{\rho_\infty V_\infty^2}{2}\left(\frac{C_{m_\alpha} cS}{I_{yy}}\right)\right]\alpha = 0 \tag{12.88a}$$

We now define the *static stability parameter* P_s, defined by Martin[9] as

$$P_s = -C_{m_\alpha} Sc/I_{yy} \tag{12.88b}$$

For a statically stable vehicle, P_s must be positive. The (undamped) natural frequency can be written as

$$\omega_n = P_s^{1/2}\left(\rho_\infty V_\infty^2/2\right)^{1/2} = q_\infty^{1/2} P_s^{1/2} \tag{12.88c}$$

We may now transform Eq. (12.88a) into a more tractable form by introducing the variable ξ as:

$$\xi = \frac{2\omega_n H}{V_\infty \sin(\gamma_E)} = \left[\frac{2H}{V_\infty \sin(\gamma_E)}\right]\left(\frac{P_s \rho_\infty}{2}\right)^{1/2} V_\infty$$

$$= \left[\frac{2\rho_0 H^2 P_s}{\sin^2(\gamma_E)}\right]^{1/2} e^{-Z/2H} = \xi_E e^{-Z/2H} \tag{12.88d}$$

In the preceding equation we have introduced once again the exponential atmospheric model. All terms in the parentheses are constant, and their values depend upon size, mass distribution, static stability, and entry angle. These terms are lumped in the parameter ξ_E as:

$$\xi_E = \left[\frac{2\rho_E H^2 P_s}{\sin^2(\gamma_E)}\right]^{1/2} \tag{12.88e}$$

Equation (12.88a) may now be rewritten as:

$$\ddot{\alpha} + \xi^2\left[\frac{V_E \sin(\gamma_E)}{2H}\right]^2 \alpha = 0 \tag{12.89}$$

We may now transform the independent variable from time to the parameter ξ. The first derivative becomes

$$\frac{\mathrm{d}\alpha}{\mathrm{d}t} = \frac{\mathrm{d}\alpha}{\mathrm{d}\xi}\frac{\mathrm{d}\xi}{\mathrm{d}Z}\frac{\mathrm{d}Z}{\mathrm{d}t} \approx \frac{\mathrm{d}\alpha}{\mathrm{d}\xi}\left\{\left(-\frac{1}{2H}\xi_E e^{-Z/2H}\right)[-V_E\sin(\gamma_E)]\right\}$$

$$= \left[\frac{V_E\sin(\gamma_E)}{2H}\right]\xi\frac{\mathrm{d}\alpha}{\mathrm{d}\xi} \tag{12.90a}$$

The second derivative then follows as

$$\frac{\mathrm{d}^2\alpha}{\mathrm{d}t^2} = \left[\frac{V_E\sin(\gamma_E)}{2H}\right]\frac{\mathrm{d}}{\mathrm{d}\xi}\left(\xi\frac{\mathrm{d}\alpha}{\mathrm{d}\xi}\right)\frac{\mathrm{d}\xi}{\mathrm{d}Z}\frac{\mathrm{d}Z}{\mathrm{d}t}$$

$$= \left[\frac{V_E\sin(\gamma_E)}{2H}\right]\left(\frac{\mathrm{d}\alpha}{\mathrm{d}\xi} + \xi\frac{\mathrm{d}^2\alpha}{\mathrm{d}\xi^2}\right)\left(\frac{-\xi}{2H}\right)\left[-V_E\sin(\gamma_E)\right]$$

$$= \left[\frac{V_E\sin(\gamma_E)}{2H}\right]^2\left(\xi\frac{\mathrm{d}\alpha}{\mathrm{d}\xi} + \xi^2\frac{\mathrm{d}^2\alpha}{\mathrm{d}\xi^2}\right) \tag{12.90b}$$

Inserting Eq. (12.90b) into Eq. (12.89) gives

$$\frac{\mathrm{d}^2\alpha}{\mathrm{d}\xi^2} + \frac{1}{\xi}\frac{\mathrm{d}\alpha}{\mathrm{d}\xi} + \alpha = 0 \tag{12.91}$$

Recall that Bessel's equation in standard form is

$$\frac{\mathrm{d}^2Y}{\mathrm{d}X^2} + \frac{1}{X}\frac{\mathrm{d}Y}{\mathrm{d}X} + \left(1 - \frac{\nu^2}{X^2}\right)Y = 0 \tag{12.92}$$

Note the similarity between Eqs. (12.91) and (12.92) if Y is replaced by α and X by ξ, and ν is set to zero. Many texts on advanced calculus note that the solution to Eq. (12.92) can be expressed as:

$$Y = C_1J_\nu(X) + C_2J_{-\nu}(X)$$

where ν is not equal to zero or an integer, and

$$Y = C_1J_0(X) + C_2Y_0(X)$$

where ν equals zero. In general J_ν, $J_{-\nu}$, and Y_ν, $Y_{-\nu}$ are Bessel functions of the first and second kind, respectively, of order ν. Graphical representations of these functions are given in Figs. 12.16a and 12.16b.

In terms of (α, ξ) the solution of Eq. (12.91) may be written as:

$$\alpha = C_1J_0(\xi) + C_2Y_0(\xi) \tag{12.93a}$$

since $\xi = 0$ in Eq. (12.91).

Now with some manipulation and the use of Eq. (12.88d), we may write:

$$\frac{\mathrm{d}\alpha}{\mathrm{d}t} = \omega_n\frac{\mathrm{d}\alpha}{\mathrm{d}\xi}$$

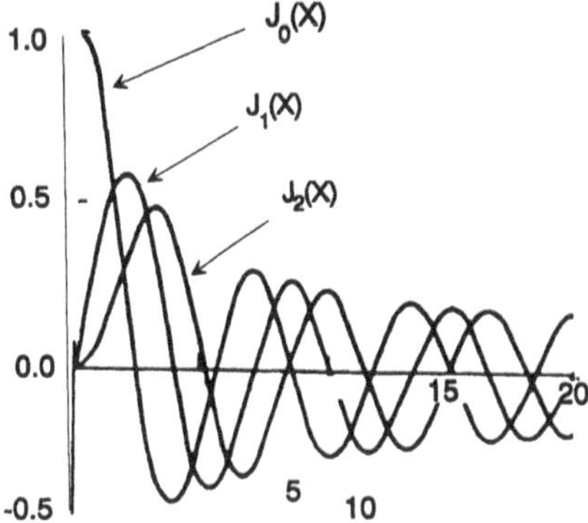

Fig. 12.16a Bessel functions of the first kind of order 0, 1, 2.

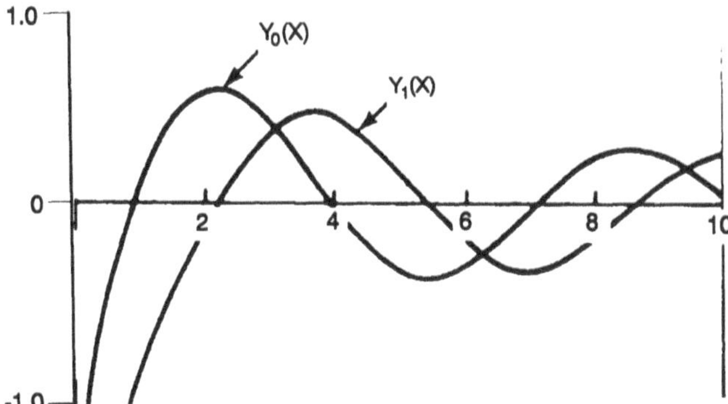

Fig. 12.16b Bessel functions of the second kind of order 0, 1.

Thus,

$$\frac{d\alpha}{dt} = \omega_n \left[C_1 J_0'(\xi) + C_2 Y_0'(\xi) \right] \tag{12.93b}$$

If we now take α_E and $\dot{\alpha}_E$ as the initial conditions or conditions at entry, we may solve for constants C_1 and C_2 of Eq. (12.93b) as

$$C_1 = \frac{\alpha_E Y_0'(\xi_E) - (\dot{\alpha}/\omega_n)_E \, Y_0(\xi_E)}{J_0(\xi_E) Y_0'(\xi_E) - J_0'(\xi_E) Y_0(\xi_E)} \tag{12.94a}$$

$$C_2 = \frac{(\dot{\alpha}/\omega_n)_E \, J_0(\xi_E) - \alpha_E J_0'(\xi_E)}{J_0(\xi_E) Y_0'(\xi_E) - J_0'(\xi_E) Y_0(\xi_E)} \tag{12.94b}$$

where the subscript E refers to conditions at entry. Next we must evaluate ξ_E. From Eq. (12.88d) we have

$$\xi_E \equiv \xi|_E = \left[2\rho_0 H^2 P_s / \sin^2(\gamma_E)\right]^{1/2} e^{-Z_E/2H}$$

From the data in Table 12.2 we get

$$P_s = -\frac{C_{m_\alpha} S c}{I_{yy}} = -\frac{(-0.520)(0.152)(0.22)}{(9.32)} = 1.9 \times 10^{-3}\text{m/kg}$$

so

$$\xi_E = \left[\frac{2(1.752)(6700)^2(1.9 \times 10^{-3})}{\sin^2(15.0)}\right]^{1/2} e^{-150.0/2(6.7)} = 0.0291$$

where we have taken altitude at entry as 150 km and entry angle a shallow 15 degrees.

For small values of the argument, the Bessel functions $J_0(\xi)$ and $Y_0(\xi)$ and their derivatives may be approximated as

$$J_0(\xi_E) \approx 1.0$$

$$Y_0(\xi_E) \approx (2/\pi)\ln(0.891\xi_E)$$

$$J_0'(\xi_E) \approx -\xi_E/2$$

$$Y_0'(\xi_E) \approx \frac{2}{\pi\xi_E}\left[1 + \frac{\xi_E^2}{2}\ln\left(\frac{1.1236}{\xi_E}\right)\right]$$

The constants of integration may be written from Eqs. (12.94a) and (12.94b) as

$$C_1 \approx \alpha_E + [2\dot{\alpha}_E H / V_E \sin(\gamma_E)] \ln(1.1236/\xi_E) \tag{12.95a}$$

$$C_2 \approx \alpha_E \omega_n J_0'(\xi_E) = -\alpha_E \omega_n J_1(\xi_E) \tag{12.95b}$$

We now investigate the effect that an exponential atmosphere has on the envelope of the angle of attack. Let us assume that $C_1 \neq 0$ but $C_2 = 0$. From Eqs. (12.93a) and (12.93b) we get

$$\alpha = \alpha_E J_0(\xi) \tag{12.96a}$$

$$\dot{\alpha} = \alpha_E \omega_n J_0'(\xi) = -\alpha_E \omega_n J_1(\xi) \tag{12.96b}$$

Note that in Eq. (12.96b) we have made use of the following identity:

$$-J_1(\xi) = J_0'(\xi) \equiv \frac{\mathrm{d}J_0(\xi)}{\mathrm{d}\xi}$$

The general form of this identity is

$$\xi J_n'(\xi) = nJ_n(\xi) - \xi J_{n+1}(\xi)$$

(see page 95, Ref. 13).

The function $J_1(\xi)$ is a Bessel function of the first kind of order (1). When the order n is a positive or negative integer or zero, the Bessel function can be calculated from the following series:

$$J_n(\xi) = \sum_{s=0}^{\infty} \frac{(-1)^s}{s!(n+s)!}\left(\frac{\xi}{2}\right)^{n+2s} \tag{12.97}$$

The definition of $Y_n(\xi)$—Bessel function of the second kind, order n [see Eqs. (12.93a) and (12.93b)]—is too complicated to be given here. However, one may consult Ref. 13 for further details.

Note that $J_0(\xi)$ resembles a cosine except that the envelope to the amplitude decreases as the argument ξ increases; the separation of successive peaks increases also with ξ.

Since we are interested in the amplitude of the angle of attack at altitudes below 30 km, we can look for an approximation to the Bessel functions $J_0(\xi)$ and $J_1(\xi)$ for larger values of ξ. Table 12.3 gives the value of ξ for various altitudes below 100 km. We note from this table that ξ increases rapidly with decreasing altitude. At atmospheric entry the value of ξ is small, allowing us to use the small value approximation to the Bessel function in setting the initial conditions [Eqs. (12.93)]. However, we are mainly interested in motion well down into the atmosphere, say, below 60 km. We can obtain the results in terms of the more familiar trigonometric functions by using the asymptotic approximation to the Bessel functions because of the large value of ξ at low altitude, that is,

$$J_0(\xi) = (2/\pi\xi)^{1/2}\cos[\xi - (\pi/4)] \tag{12.98a}$$

$$J_1(\xi) = (2/\pi\xi)^{1/2}\cos[\xi - (3\pi/4)] \tag{12.98b}$$

Table 12.3 Variation of trajectory variable ξ with altitude

Z (km)	ξ
100	1.21
60	24.0
30	225.0
15	683.0
10	992.0
5	1441.0

Thus, below 60 km we can write

$$\alpha = \alpha_E \, (2/\pi\xi)^{1/2} \cos[\xi - (\pi/4)] \tag{12.99a}$$

$$\dot{\alpha} = \alpha_E \, (2/\pi\xi)^{1/2} \, \omega_n \cos[\xi - (3\pi/4)] \tag{12.99b}$$

From Eq. (12.99a) we write for the envelope to the angle of attack the following:

$$\alpha_{\text{env}} = \alpha_E \left(\frac{2}{\pi\xi}\right)^{1/2} = \alpha_E \left[\frac{2\sin^2(\gamma_E)}{\pi^2 \rho_0 H^2 P_s}\right]^{1/4} e^{Z/4H} \tag{12.100a}$$

At the beginning of this section we assumed the aerodynamic damping was zero, i.e., $C_{m_q} = 0$. In spite of the absence of aerodynamic damping, we note that the angle of attack envelope decreases as the altitude decreases and the density increases.

Next we can go to Eq. (12.99b) and consider the altitude variation of the envelope of the angle of attack rate. The result is

$$\dot{\alpha}_{\text{env}} = \alpha_E \, \omega_n \left[\frac{2\sin^2(\gamma_E)}{\pi^2 \rho_0 H^2 P_s}\right]^{1/4} e^{Z/4H} \tag{12.100b}$$

We may rewrite the preceding equation by using Eqs. (2.32), (7.39a), and (12.88) to give

$$\dot{\alpha}_{\text{env}} = \alpha_E \, V_E \left[\frac{\rho_0 P_s \sin^2(\gamma_E)}{2\pi^2 H^2}\right]^{1/4} \exp\left[-\left(a_E \, e^{-Z/H} - \frac{Z}{4H}\right)\right] \tag{12.100c}$$

where, of course,

$$a_E = \rho_0 H / 2\beta \sin(\gamma_E)$$

From Eqs. (12.88a), (12.88b), and (12.88c) we may write

$$\ddot{\alpha} = -\omega_n^2 \alpha = -\alpha_E \, \omega_n^2 J_0(\xi)$$

According to the approximations of Eqs. (12.98a) and (12.98b) we get

$$\ddot{\alpha} = -\alpha_E \, \omega_n^2 \, (2/\pi\xi)^{1/2} \cos[\xi - (\pi/4)]$$

The maximum acceleration, or the acceleration of the angle of attack envelope, is

$$\ddot{\alpha}_{\text{env}} = \alpha_E \, \omega_n^2 \left[\frac{2}{\pi\xi}\right]^{1/2} = \alpha_E \, \omega_n^2 \left[\frac{2\sin^2(\gamma_E)}{\pi^2 \rho_0 H^2 P_s}\right]^{1/4} e^{Z/4H} \tag{12.100d}$$

or, equivalently,

$$\ddot{\alpha}_{\text{env}} = \alpha_E \, V_E^2 \left[\frac{P_s^3 \rho_0^3 \sin^2(\gamma_E)}{8\pi^2 H^2}\right]^{1/4} \exp\left[-\left(2a_E \, e^{-Z/H} - \frac{3Z}{4H}\right)\right] \tag{12.100e}$$

Table 12.4 contains a listing of a program that provides the pitch frequency and the angle of attack acceleration envelopes versus altitude. Graphical output from the program is presented in Figs. 12.17 and 12.18. Note that the peak values of pitch frequency increase with increasing β; however, the altitude at maximum pitch frequency is higher for the lower values of β. The reason is of course that a low value of β means a large velocity decrement and hence low dynamic pressure; a large value of β, on the other hand, means that the vehicle reaches a lower altitude while retaining a large fraction of its initial velocity. Angular velocity and angular acceleration maxima increase with increasing values of β. Large angular acceleration will result in large tangential accelerations. For example, if we assume α_E is 30 degrees, a vehicle having a β of 10,000 kg/m^2 would have a peak angular acceleration as follows:

$$\ddot{\alpha} = (\pi/6)(588.0) = 307.0 \text{ rad/s}^2$$

For a point on the vehicle 0.5 meters from the axis of rotation, the peak acceleration in g's would be

$$n = (307.0)(0.5)/(9.81) = 15.7\ g$$

A final consideration is to examine the variation in angle of attack, rather than the angle of attack envelope, with altitude. Our starting point again will be Eq. (12.82), which we will reformulate in nondimensional altitude, Z/H, rather than time. Since time appears in Eq. (12.82) only as the variable of differentiation, we may write

$$\frac{\mathrm{d}\alpha}{\mathrm{d}t} = \frac{\mathrm{d}\alpha}{\mathrm{d}(Z/H)}\frac{\mathrm{d}(Z/H)}{\mathrm{d}t} \approx -\frac{V\sin(\gamma_E)}{H}\frac{\mathrm{d}\alpha}{\mathrm{d}(Z/H)} \tag{12.101a}$$

and

$$\frac{\mathrm{d}^2\alpha}{\mathrm{d}t^2} = \frac{\mathrm{d}\dot{\alpha}}{\mathrm{d}(Z/H)}\frac{\mathrm{d}(Z/H)}{\mathrm{d}t} \approx \left(\frac{V\sin(\gamma_E)}{H}\right)^2\left[\frac{\mathrm{d}^2\alpha}{\mathrm{d}(Z/H)^2} + \frac{1}{V}\frac{\mathrm{d}V}{\mathrm{d}(Z/H)}\frac{\mathrm{d}\alpha}{\mathrm{d}(Z/H)}\right] \tag{12.101b}$$

but from Eq. (7.39a) we have

$$\frac{1}{V}\frac{\mathrm{d}V}{\mathrm{d}(Z/H)} = a_E e^{-Z/H} \tag{12.101c}$$

Substituting Eqs. (12.101a), (12.101b), and (12.101c) into Eq. (12.82) gives us

$$\frac{\mathrm{d}^2\alpha}{\mathrm{d}(Z/H)^2} + \tilde{a}_E e^{-Z/H}\frac{\mathrm{d}\alpha}{\mathrm{d}(Z/H)} + \frac{1}{4}\left(\xi_E^2 e^{-Z/H} + \tilde{b}_E^2 e^{-2Z/H}\right)\alpha = 0 \tag{12.102a}$$

Table 12.4 TRUE BASIC program to calculate the envelope of the angle of attack, the angle of attack rate and the angle of attack acceleration, and the pitch frequency versus altitude

```
! INPUTS:       ZE: INITIAL ALTITUDE (METERS)
!               VE: INITIAL VELOCITY MAGNITUDE (METERS/SECOND)
!               BETA: BALLISTIC COEFFICIENT (KG./SQ. METER)
!               GAMMA: INITIAL ENTRY ANGLE (DEGREES)
!               H: ATMOSPHERE SPREAD FACTOR (METERS)
!               RHO: MODIFIED SEA LEVEL DENSITY (KG/CU-METERS)
!               DH: STEP SIZE (METERS)
!               CMA: STATIC PITCHING MOMENT DERIVATIVE
!               C: REFERENCE LENGTH (METERS)
!               IYY: PITCH MOMENT OF INERTIA
ZE = 60000
VE = 6000
BETA = 6050.52
GAMMA = 70.0
RHO = 1.752
H = 6.7E3
DH = 2.0E3
CMA = (-1.0)*0.520
C = 0.44
IYY = 9.32
RTD = 180.0/PI
GAMMA = GAMMA/RTD
S = PI*(C^2)/4.0
PS = (-1.0)*CMA*S*C/(2.0*IYY)

DIM Z(100),A(100),ALPHA(100),ALDOT(100),ALDDOT(100),W(100)
Z(1) = ZE
AE = RHO*H/(2.0*BETA*SIN(GAMMA))
K1 = (2.0*(SIN(GAMMA))^2/(((PI*H)^2)*RHO*PS))^(0.25)

FOR I = 1 TO 60
 W(I) = VE*SQR(PS*RHO/2)*EXP(-AE*EXP(-Z(I)/H))*EXP(-Z(I)/(2*H))
 ALPHA(I) = K1*EXP(Z(I)/(4.0*H))
 ALDOT(I) = W(I)*K1*EXP(Z(I)/(4.0*H))
 ALDDOT(I) = (W(I)^2)*K1*EXP(Z(I)/(4.0*H))
 Z(I+1) = Z(I)-DH
 Z9 = I
 IF Z(I)<0 THEN EXIT FOR
NEXT I
N1$ = "###.#      #####.#        ####.#       ###.#"
N2$ = "####.#     #####.#       #.###^^^^     ##.####     ####.####"
N3$ = "ALTITUDE PITCH-FREQ. ANG-ATT ENV ANG-VEL ENV ANG-ACC ENV."
PRINT "ALTITUDE","VELOCITY","BETA","GAMMA"
PRINT "AT ENTRY","AT ENTRY","------","AT ENTRY"
PRINT "KILO-METERS","METERS/SEC.","KG/SQ-METERS","DEGREES"
```

(continued on next page)

Table 12.4 (continued) TRUE BASIC program to calculate the envelope of the angle of attack, the angle of attack rate and the angle of attack acceleration, and the pitch frequency versus altitude

```
PRINT
PRINT USING N1$:ZE/1000,VE,BETA,GAMMA*RTD
PRINT
PRINT N3$
PRINT
FOR I = 1 TO (Z9-1)
PRINT USING N2$:Z(I)/1000,W(I),ALPHA(I),ALDOT(I),ALDDOT(I)
NEXT I
END
```

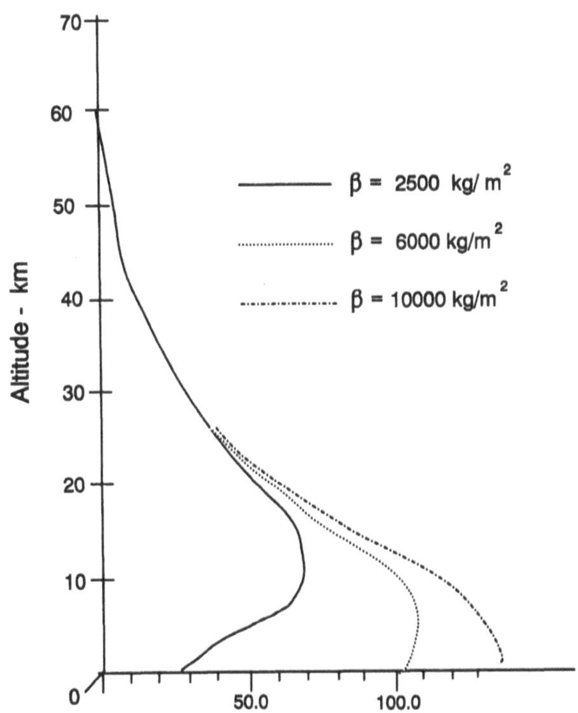

Fig. 12.17 Pitch frequency vs altitude.

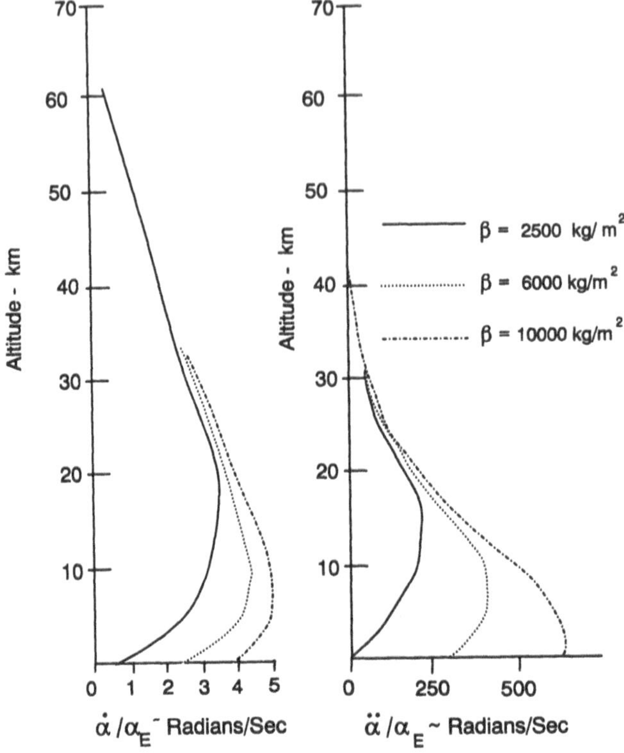

Fig. 12.18 Angular velocity envelope and angular acceleration envelope vs altitude.

where

$$\tilde{a}_E = a_E\left[1 - \frac{C_{N_\alpha}}{C_D} + 2\left(\frac{c}{2K_y}\right)^2 \frac{C_{m_q}}{C_D}\right]^{1/2} \tag{12.102b}$$

$$\tilde{\xi}_E = 2a_E^{1/2}\left[-\left(\frac{H}{\sin(\gamma_E)}\right)\left(\frac{c}{K_y^2}\right)\frac{C_{m_\alpha}}{C_D} + \frac{C_{N_\alpha}}{C_D}\right]^{1/2} \tag{12.102c}$$

$$\tilde{b}_E = 2a_E\left[-\frac{C_{N_\alpha}}{C_D} - 2\left(\frac{c}{K_y^2}\right)\frac{C_{m_q}C_{N_\alpha}}{C_D^2}\right]^{1/2} \tag{12.102d}$$

Martin identifies $\tilde{a}_E$ as the dynamic stability parameter[9]; if $C_{N_\alpha} = C_{m_q} = 0$, as was our earlier premise, then $\tilde{a}_E = a_E$. The additional term $\tilde{b}_E$ would be zero, if we ignored lift and damping contributions.

Martin (originally Allen[14]) transformed the angle of attack in the following manner,

$$\alpha = \tilde{\alpha} \exp\left[(\tilde{a}_E/2)e^{-Z/H}\right] \tag{12.102e}$$

with the result that Eq. (12.102a) may be rewritten as

$$\frac{d^2\tilde{\alpha}}{d(Z/H)^2} + \frac{1}{4}\left[\left(\tilde{\xi}_E^2 + 2\tilde{a}_E\right)e^{-Z/H} + \left(\tilde{b}_E^2 - \tilde{a}_E^2\right)e^{-2Z/H}\right]\tilde{\alpha} = 0$$

or

$$\frac{d^2\tilde{\alpha}}{d(Z/H)^2} + \frac{1}{4}\left(\tilde{\xi}_E^2 + 2\tilde{a}_E\right)e^{-Z/H}\left[1 + \left(\frac{\tilde{b}_E^2 - \tilde{a}_E^2}{\tilde{\xi}_E^2 + 2\tilde{a}_E}\right)e^{-Z/H}\right]\tilde{\alpha} = 0 \quad (12.102f)$$

The second term in the brackets is negligible for a slender re-entry vehicle, especially at high altitudes. Thus, Eq. (12.102f) may be written as

$$\frac{d^2\tilde{\alpha}}{d(Z/H)^2} + \left[\left(\frac{\tilde{\xi}_E^2 + 2\tilde{a}_E}{4}\right)e^{-Z/H}\right]\tilde{\alpha} = 0 \qquad (12.103)$$

The solution to Eq. (12.103) may be written as

$$\tilde{\alpha} = C_1 J_0\left[\left(\tilde{\xi}_E^2 + 2\tilde{a}_E\right)^{1/2} e^{-Z/2H}\right] + C_1 Y_0\left[\left(\tilde{\xi}_E^2 + 2\tilde{a}_E\right)^{1/2} e^{-Z/2H}\right]$$

or using Eq. (12.102e),

$$\alpha = \exp\left(\frac{\tilde{a}_E}{2}e^{-Z/H}\right)\left\{C_1 J_0\left[\left(\tilde{\xi}_E^2 + 2\tilde{a}_E\right)^{1/2} e^{-Z/2H}\right]\right.$$
$$\left. + C_2 Y_0\left[\left(\tilde{\xi}_E^2 + 2\tilde{a}_E\right)^{1/2} e^{-Z/2H}\right]\right\} \qquad (12.104)$$

where C_1 and C_2 are given in Eq. (12.94a) or approximately in Eq. (12.95).

Note that the appearance of $\tilde{a}_E$ in the exponent is the reason why we have associated dynamic stability with this parameter: in Figs. 12.16 we noted that the Bessel functions of the first and second kind, order zero, damp out for increasing values of the argument. However, for a sufficiently large value of $\tilde{a}_E$ [Eq. (12.102b)] the angle of attack, although oscillatory in Z/H, may grow with decreasing altitude.

Martin simplifies Eq. (12.104) by first ignoring $\tilde{a}_E$ in comparison to $\tilde{\xi}_E$ in the argument of the Bessel function.[9] Then, using Eq. (12.88d), we obtain

$$\alpha = \exp\left\{\tfrac{1}{2}\tilde{a}_E e^{-Z/H}\right\}\left[C_1 J_0(\tilde{\xi}) + C_2 Y_0(\tilde{\xi})\right] \qquad (12.105)$$

where $\tilde{\xi} = \tilde{\xi}_E e^{-Z/2H}$. Let us repeat Eq. (12.99a) for comparison, that is,

$$\alpha = \alpha_E\,(2/\pi\xi)^{1/2}\cos[\xi - (\pi/4)] \qquad (12.99a)$$

As we showed in Table 12.3, ξ increases rapidly with decreasing altitude. At altitudes below 60 km, ξ is in excess of 24.0. Thus, we may combine the preceding equations for α to get an analytical expression for the envelope, or

$$\alpha_{env} = \alpha_E \left(2/\pi\tilde{\xi}_E\right)^{1/2} e^{Z/4H} \exp\left(\tilde{a}_E e^{-Z/H}/2\right) \tag{12.106}$$

An alternate form of Eq. (12.106) is easily shown to be

$$\alpha_{env} = \alpha_E \left(2/\pi\tilde{\xi}_E\right)^{1/2} (V_\infty/V_E)^{-\tilde{a}_E/2a_E} (\rho_\infty/\rho_0)^{-1/4}$$

Clearly (V_∞/V_E) will decrease and ρ_∞/ρ_0 will increase with decreasing altitude.

Returning to Eq. (12.106), we see that the exponential, exp $(Z/4H)$, will decrease with decreasing altitude; however, the second exponential will increase under the same circumstances. We will leave as an exercise for the reader to show that the altitude for minimum value of the envelope or the altitude for angle of attack divergence is

$$Z_{(\alpha_{env})\min} = H \ln(2\tilde{a}_E) = H \ln[2a_E(\tilde{a}_E/a_E)] \tag{12.107}$$

Apparently, for values of $\tilde{a}_E > 0.5$ the angle of attack envelope can diverge. When we examine Eq. (12.102b), we note that $\tilde{a}_E$ can be either positive or negative; a positive value requires small damping and large drag. If $\tilde{a}_E$ is negative then there will be no divergence in the angle of attack envelope.

The pitch frequency, in cycles per atmospheric scale height H, is, from Eq. (12.103),

$$\omega_{\tilde{\alpha}} = \frac{(\tilde{\xi}_E^2 + 2\tilde{a}_E)^{1/2}}{2} e^{-Z/2H} \approx \frac{\tilde{\xi}_E}{2} e^{-Z/H}$$

where we have neglected $\tilde{a}_E$ in comparison to $\tilde{\xi}_E$.

The pitch frequency in radians per second is obtained by multiplying the preceding expression by

$$\frac{V\ \sin(\gamma_E)}{H} = \left[\frac{V_E}{H} \exp\left(-a_E e^{-Z/H}\right)\right] \sin(\gamma_E)$$

where we have made use of Eq. (7.39a). The result is the following expression:

$$\omega_\alpha = \tfrac{1}{2}\tilde{\xi}_E\,[V_E\ \sin(\gamma_E)/H]\,e^{-Z/2H} \exp\left(-a_E e^{-Z/H}\right) \tag{12.108a}$$

The altitude of maximum aerodynamic frequency may be shown to be

$$Z_{(\omega_\alpha)\max} = H \ln(2a_E) \tag{12.108b}$$

Without further derivation, we give Martin's results for the envelopes of the angle of attack velocity and acceleration and their corresponding altitudes of maximum values,[9] as follows:

$$\dot{\alpha}_{env} = \left[\alpha_E \left(\frac{2}{\pi}\right)^{1/2}\right] \tilde{\xi}_E^{1/2} \left[\frac{V_E \sin(\gamma_E)}{H}\right] e^{-Z/4H} \exp\left[\left(\frac{1}{2}\tilde{a}_E - a_E\right) e^{-Z/H}\right] \tag{12.109a}$$

$$Z_{(\dot{\alpha})\max} = H\ln\left[2a_E\left(2 - \frac{\tilde{a}_E}{a_E}\right)\right] \tag{12.109b}$$

$$\ddot{\alpha}_{\text{env}} = \left[\frac{\alpha_E}{(8\pi)^{1/2}}\right](\tilde{\xi}_E)^{3/2}\left(\frac{V_E\sin(\gamma_E)}{H}\right)^2 e^{-3Z/4H}\exp\left[\left(\frac{1}{2}a_E - 2\tilde{a}_E\right)e^{-Z/H}\right] \tag{12.110a}$$

$$Z_{(\ddot{\alpha})\max} = H\ln\left[2a_E\left(\frac{4}{3} - \frac{\tilde{a}_E}{3a_E}\right)\right] \tag{12.110b}$$

A program for calculating the first and second derivatives of the angle of attack envelopes is given in the accompanying program (Table 12.4) written in TRUE BASIC.

References

[1]Etkin, B., *Dynamics of Flight, Stability and Control*, 2nd ed., John Wiley & Sons, Somerset, NJ, 1982.

[2]Irving, F. G., *Introduction to the Static Stability of Low Speed Aircraft*, Pergammon Press Inc., New York, 1966.

[3]Vinh, N. S., and Lin, C.- F., "Longitudinal Control Effectiveness and Entry Dynamics of a Single-Stage-to-Orbit Vehicle," N82-27351, Univ. Michigan, Ann Arbor, MI, 1982.

[4]Boyden, R. P., and Freeman, D. C., "Subsonic and Transonic Dynamic Characteristics of a Space Shuttle Orbiter," NASA-TN-D-8042, NASA Langley Research Center, Hampton, VA, Nov. 1975.

[5]Freeman, D. C., Boyden, R. A., and Davenport, E. E., "Supersonic Dynamic Stability Characteristics of a Space Shuttle Orbiter," NASA-TN-D-8043, NASA Langley Research Center, Hampton, VA, Jan. 1976.

[6]Freeman, D. C., and Powell, R. W., "Impact of Far Aft Center of Gravity for a Single-Stage-to-Orbit Vehicle," *Journal of Spacecraft and Rockets*, Vol. 17, No. 4, July–Aug. 1980, pp. 311–315.

[7]Linnell, R. D., and Bailey, J. Z., "Similarity Rule Estimation Methods for Cones and Parabolic Noses," *Journal of Aeronautical Sciences*, Vol. 23, Aug. 1956, pp. 796, 797.

[8]Glover, L., "Effects on Roll Rate of Mass and Aerodynamic Asymmetries for Ballistic Reentry Bodies," *Journal of Spacecraft and Rockets*, Vol. 2, March–April 1965, pp. 220–225.

[9]Martin, J. J., *Atmospheric Entry—An Introduction to Its Science and Engineering*, Prentice-Hall, Old Tappan, NJ, 1966.

[10]Fredrick, H. R., and Dore, F. J., "The Dynamic Motion of a Missile Descending through the Atmosphere," *Journal of Aeronautical Sciences*, Vol. 22, 1955, pp. 628–638.

[11]Norling, R. A., "Altitude Stabilization for Slowly Tumbling Reentry Vehicle," *Journal of American Rocket Society*, Vol. 32, 1962, pp. 1867–1870.

[12]Leon, H. I., "Angle of Attack Convergence of a Spinning Missile Descending through the Atmosphere," *Journal of Aerospace Sciences*, Vol. 28, 1958, pp. 480–484.

[13]Sneddon, I. N., *Special Functions of Mathematical Physics and Chemistry*, Oliver and Boyd, Interscience, New York, 1956.

[14]Allen, H. J., "Motion of a Ballistic Missile Angularly Misaligned with the Flight Path upon Entering the Atmosphere and Its Effect upon Aerodynamic Heating, Aerodynamic Loads and Miss Distance," NACA-TN-4048, NASA, Washington, DC, 1959.

13
Error Analysis

13.1 Introduction

In this chapter we will consider two related but distinct concepts that form the basis for the evaluation of the trajectory of a re-entry vehicle. To clarify these concepts, let us note that in a conflict we might first consider an issue that is primarily a concern of an offense and then that of concern to a defense.

Assume that an offense wishes to characterize the value of a typical trajectory by the proximity of its impact point to an intended impact point (i.e., a target). Qualitatively, it would seem, the closer the better. However, uncertainties in the initial conditions of the trajectory, along with uncertainties in the physical properties of the atmosphere through which the RV must traverse, contribute to uncertainties in the impact point. In this chapter we will consider ways of relating uncertainties in the initial and environmental conditions to uncertainties in position (including impact point).

Continuing with the conflict example, we might indicate the second issue as a defense problem. The defense must examine the trajectory of a re-entry vehicle to identify it as a threat, identify the likely threatened area and perhaps use several measurables in an algorithm to expend and direct an active defense. The question here is what such a defense can do to minimize the effects of errors embedded in the measurements.

Of course both of the preceding concerns need not be confined to a conflict. For a re-entry vehicle of any type, impact point uncertainties are of fundamental importance for issues such as safety and vehicle recovery. One of the goals of a testing program might be the determination of the ballistic coefficient. Obviously the confidence in such a parameter reflects uncertainties in radar and optical tracking measurements.

13.2 Circular Error Probable (CEP)

A variety of error sources contribute to a discrepancy between the position of the intended target and the position of the re-entry vehicle at impact. Although these sources of error have some identifiable physical origin, it is not possible to assign numerical values of arbitrary precision to these errors. However, it is usually feasible to establish a mean value of a given error source and a distribution of likely values about this mean. For example, atmospheric properties such as density and wind velocities can be given mean values and a Gaussian distribution characterized by a standard deviation of likely values about this mean.

The procedure that we will use is to choose a nominal trajectory based upon the means of all error sources. Errors are then treated as a zero-mean process about this nominal trajectory.

Let the nominal state or the state vector associated with the nominal trajectory be designated by $\boldsymbol{X}_n$ and the actual state vector by $\boldsymbol{X}$. Let us define the error in the state or the state error vector $\boldsymbol{e}$ as

$$\boldsymbol{e} \doteq \boldsymbol{X} - \boldsymbol{X}_n \tag{13.1}$$

A precise statistical description of the error state is given by the state covariance matrix P, defined[1] as the expectation of all possible pairs of error vector components, or

$$P \doteq E\left[\boldsymbol{e}\,\boldsymbol{e}^{\mathrm{T}}\right] = E\left[(\boldsymbol{X} - \boldsymbol{X}_n)(\boldsymbol{X} - \boldsymbol{X}_n)^{\mathrm{T}}\right] \tag{13.2}$$

Equation (13.2) is a very useful form for assessing the size of the error vector and the degree of coupling among the components of the error vector. First, note that

$$\boldsymbol{X}, \boldsymbol{X}_n, \boldsymbol{e} = \mathbb{R}^{nx1}$$

$$P = \mathbb{R}^{nxn}$$

For a rigid body representation n is equal to 12, and for a particle representation n is equal to 6. To keep the discussion tractable, we will restrict the error vector to positional states only, in which case n is equal to 3. Consequently, P is a 3×3 matrix. Our goal here is to describe the impact errors in terms of a single entity that we will call the circular error probable, or the CEP. However, before doing this we will first consider a more general form of the description of impact errors.

A companion to the error covariance matrix (ECM), P is the error quadratic form (EQF) of the covariance matrix[1,2]:

$$Q = E[\boldsymbol{e}^{\mathrm{T}} \boldsymbol{A} \boldsymbol{e}] \tag{13.3}$$

The EQF (limited here to positional errors) can be represented in three-space as an ellipsoid (see Fig. 13.1). Note that the ellipsoid is of finite size at the beginning of the re-entry trajectory. Thus, the re-entry trajectory begins with uncertainties (confined to positional errors only in this case).

These errors describe positional uncertainties of the re-entry vehicle due to errors acquired during the boost, deployment and Keplerian phases. In addition we note that the error ellipsoid grows in size and rotates throughout the re-entry segment, indicating that uncertainties attributable to the environment will make further contributions to positional errors. The ellipsoid will grow also because velocity errors translate into positional errors over time.

In Fig. 13.1 we note that there are three reference frames: an inertial or I-frame, a local or l-frame, and a trajectory or t-frame. The orientation of the

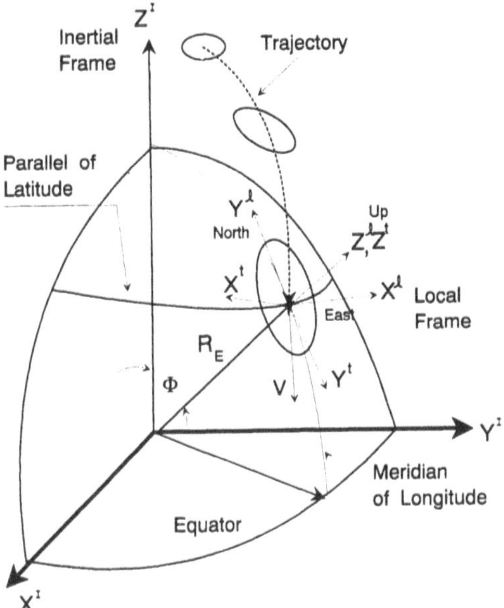

Fig. 13.1 Propagation of the error ellipsoid.

inertial frame is arbitrary, although for the present we will assume that the X^I, Y^I axes are parallel to the Earth's equatorial plane. The local frame originates at the impact point of the nominal trajectory; the Z^l-axis is along the local vertical, the Y^l is to the north, and the X^l-axis is to the east.The third axis system translates with the re-entry body; the Y^t-axis is horizontal and down range (acute angle between Y^t and the velocity vector), Z^t along the local vertical, and X^t cross range to the right. A fourth axis system, not shown in Fig. 13.1, is introduced to describe the error ellipsoid. This axis system, designated as the b-frame, is aligned with the principal axes of the error ellipsoid and translates with the re-entry body. Obviously when the error covariance matrix P is expressed in the b-frame, the matrix is diagonalized. From Eq. (13.3) we indicate the appropriate frame by the superscript. Thus,

$$Q = E[(\boldsymbol{e}^I)^{\mathrm{T}} A^I (\boldsymbol{e}^I)]$$

or

$$\begin{aligned} Q &= E[(C_b^I \boldsymbol{e}^b)^{\mathrm{T}} A^I (C_b^I \boldsymbol{e}^b)] \\ &= E[(\boldsymbol{e}^b)^{\mathrm{T}} (C_I^b A^I C_b^I) \boldsymbol{e}^b] \\ &= E[(\boldsymbol{e}^b)^{\mathrm{T}} A^b \boldsymbol{e}^b] \end{aligned} \tag{13.4}$$

where

$$A^b = C_I^b A^I C_b^I$$

In other words, the A matrix of the error quadratic form (EQF) is transformed from the inertial to the b-, or principal axis, frame by the preceding operation. The eigenvalues of the P^I matrix are the diagonal elements of the P^b matrix. The rows (columns) of the $C_I^b(C_b^I)$ matrix are the eigenvectors. The eigenvectors are the directional cosines of the unit vectors $(\boldsymbol{i}^b, \boldsymbol{j}^b, \boldsymbol{k}^b)$ in the I-, or inertial, frame.

Next we will assume that the transformation, i.e., the DCM C_I^l, between the I-frame and the l-frame is available; since the nominal trajectory will locate the impact point, the appropriate transformation can easily be determined. If we ignore Earth's rotation (usual practice in error analysis), then the l-frame is also inertial; otherwise, the elements of the C_I^l are time dependent.

Figure 13.2 shows the relationship that might exist between the l-, b-, and t-frames. We notice that γ is the flight path angle, or the angle that the velocity vector $\boldsymbol{V}$ makes with the horizontal at impact. Note furthermore that the error ellipsoid describes an ellipse of expanding and then contracting size as it passes

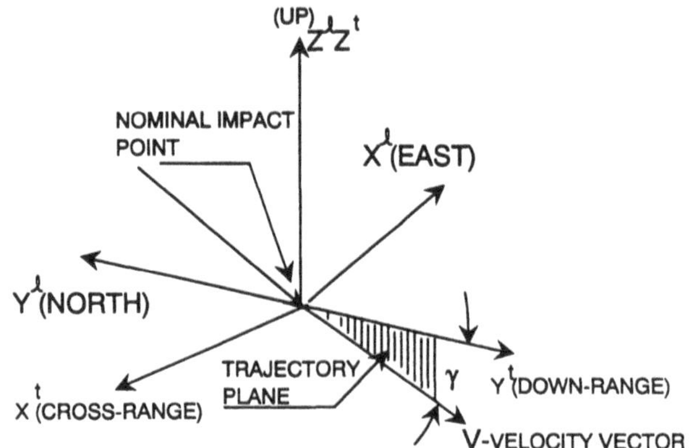

Fig. 13.2a Relationship among the local and trajectory frames.

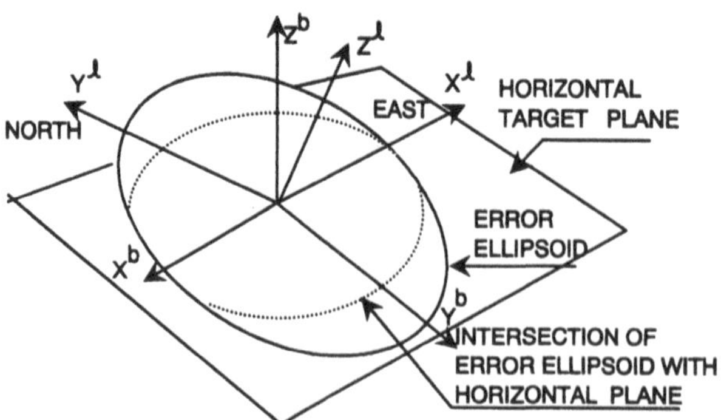

Fig. 13.2b Intersection of error ellipsoid with target plane.

through the target plane (horizontal plane tangent to the earth at the nominal impact point).

The error vector $\boldsymbol{e}$ may be written in component form as

$$\boldsymbol{e} = [e_c, e_d, e_v]^{\mathrm{T}} \tag{13.5}$$

where e_c, e_d, and e_v are the cross-range, down-range, and vertical resolution of the error vector. However, we are interested only in the down-range and cross-range components at impact. Thus, the error in the vertical direction must be converted into an equivalent down-range and cross-range contribution. From Fig. 13.3 we see that the error in the vertical direction can be converted into a down-range error because of the uncertainty δt in the time of impact. The impact time uncertainty is given as

$$\delta t = \frac{e_v}{|\boldsymbol{V}|\sin(\gamma)} \tag{13.6a}$$

with the corresponding contribution to down-range error δD of

$$\begin{aligned} \delta D &= \delta t|\boldsymbol{V}|\cos(\gamma) \\ &= \cot(\gamma)e_v \end{aligned} \tag{13.6b}$$

The cross-range error δC is not affected by an error in the vertical direction. Thus, we may write for the down-range and cross-range total displacement error

$$\delta D = 0e_c + 1e_d + \cot(\gamma)e_v \tag{13.7a}$$

$$\delta C = 1e_c + 0e_d + 0e_v \tag{13.7b}$$

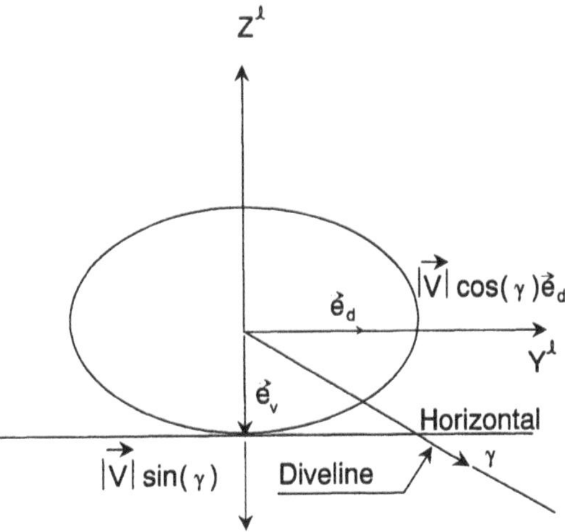

Fig. 13.3 Conversion of vertical error into down-range error.

The preceding equations may be rewritten in matrix form as

$$\begin{bmatrix} \delta D \\ \delta C \end{bmatrix} = \begin{bmatrix} 0 & 1 & \cot(\gamma) \\ 1 & 0 & 0 \end{bmatrix} \begin{bmatrix} e_c \\ e_d \\ e_v \end{bmatrix} = B\boldsymbol{e} \tag{13.8}$$

where $\boldsymbol{e}$, the error vector, has been given in Eq. (13.5). Thus, the error covariance matrix (2 × 2) in the target plane is given by the matrix I^t as

$$I^t = BP^tB^{\mathrm{T}}$$

$$I^t = \begin{bmatrix} I_{11} & I_{12} \\ I_{21} & I_{22} \end{bmatrix} \tag{13.9}$$

$$I^t = \begin{bmatrix} E(\delta D \delta D) & E(\delta D \delta C) \\ E(\delta C \delta D) & E(\delta C \delta C) \end{bmatrix} = \begin{bmatrix} \sigma_d^2 & \sigma_{dc} \\ \sigma_{cd} & \sigma_c^2 \end{bmatrix}$$

The I matrix, or the impact covariance matrix, is obviously symmetric.

The error covariance matrix P in the l-, or local, frame follows, for example, from Eq. (4.34), that is,

$$P^l = C_I^l P^I C_l^I \tag{13.10}$$

We mentioned earlier that in most error analyses, the rotation of the Earth is ignored. Consequently, the elements of the DCM C_I^l are constants.

Equation (13.9) requires the error covariance matrix in the target or the t-frame. The P matrix may be transformed from the general local frame to the t-frame by a similar transformation, in the following manner:

$$P^t = C_l^t P^l C_t^l \tag{13.11}$$

The DCM C_l^t requires knowledge of the direction of the velocity vector at impact in the l-frame; both Z^t and Z^l are collinear.

The EQF, or Q, of Eq. (13.4) corresponds to an ellipsoid whose centroid is at the impact point of the nominal trajectory. The ellipse in the horizontal plane is designated the error ellipse and the corresponding EQF, q is given as

$$\begin{aligned} q &= E\left[[\delta D \; \delta C] \begin{bmatrix} a_{11} & a_{12} \\ a_{21} & a_{22} \end{bmatrix} \begin{bmatrix} \delta D \\ \delta C \end{bmatrix}\right] \\ &= I_{11}a_{11} + I_{12}(a_{12} + a_{21}) + I_{22}a_{22} \\ &= \sigma_d^2 a_{11} + \sigma_{dc}(a_{12} + a_{21}) + \sigma_c^2 a_{22} \end{aligned} \tag{13.12}$$

The principal axes of the error ellipse are obtained from the eigenvalues of the impact covariance matrix I^t, as follows:

$$\det\begin{bmatrix} I_{11}-\lambda & I_{12} \\ I_{21} & I_{22}-\lambda \end{bmatrix} = 0$$

or

$$\lambda^2 - (I_{11}+I_{22})\lambda + (I_{11}I_{22} - I_{12}^2) = 0 \tag{13.13}$$

where we have made use of the symmetry of the impact covariance matrix, i.e., $I_{12} = I_{21}$. After some manipulation we get for the principal values

$$\sigma_{\max}^2 = \tfrac{1}{2}\left\{(I_{11}+I_{22}) + [(I_{11}-I_{22})^2 + 4I_{12}^2]^{1/2}\right\} \tag{13.14a}$$

$$\sigma_{\min}^2 = \tfrac{1}{2}\left\{(I_{11}+I_{22}) - [(I_{11}-I_{22})^2 + 4I_{12}^2]^{1/2}\right\} \tag{13.14b}$$

where, of course,

$$I_{11} = \sigma_d^2 \qquad I_{22} = \sigma_c^2 \qquad I_{12} = I_{21} = \sigma_{cd}$$

The error ellipse is given in Fig. 13.4. Note that the principal axes (X^b, Y^b) are not necessarily aligned with the trajectory axes. We now have enough information to evaluate this angle of misalignment, or ψ.

First we note that

$$I^b = \begin{bmatrix} \sigma_{\max}^2 & 0 \\ 0 & \sigma_{\min}^2 \end{bmatrix} \tag{13.15}$$

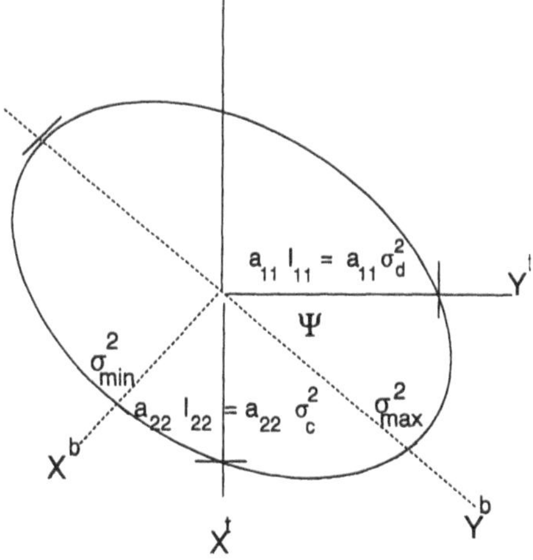

Fig. 13.4 Impact error ellipse.

Next we transform from the t- to the b-frame, or

$$\begin{bmatrix} e_y \\ e_x \end{bmatrix}^b = \begin{bmatrix} \cos(\psi) & \sin(\psi) \\ -\sin(\psi) & \cos(\psi) \end{bmatrix} \begin{bmatrix} \delta D \\ \delta C \end{bmatrix}^t \tag{13.16}$$

But we have

$$\begin{aligned} I_{12}^b &= I_{21}^b = E[e_x^b e_y^b] = 0 \\ &= E\left\{ [\delta D \cos(\psi) + \delta C \sin(\psi)][-\delta D \sin(\psi) + \delta C \cos(\psi)] \right\} = 0 \end{aligned}$$

which becomes

$$\tfrac{1}{2}[\sigma_c^2 - \sigma_d^2] \sin(2\psi) + \sigma_{dc} \cos(2\psi) = 0$$

or finally, to give an expression for the misalignment angle, ψ,

$$\psi = \frac{1}{2} \tan^{-1} \left[\frac{2\sigma_{dc}}{\sigma_d^2 - \sigma_c^2} \right] = \frac{1}{2} \tan^{-1} \left[\frac{2I_{12}}{I_{11} - I_{22}} \right] \tag{13.17}$$

We may now review the discussion so far. If we know the covariance matrix P in the I-frame [see Eq. (13.2)], we can get P in the local or trajectory frame using Eqs. (13.10) and (13.11). Knowing the trajectory flight path angle at impact γ, we can obtain the impact covariance matrix I from Eq. (13.9). Then using the elements of the I matrix (in the t-frame) we can obtain the maximum and minimum variances from Eq. (13.14) and the orientation of the major axis of symmetry relative to the down-range direction Y^t.

The next consideration is the computation of a single statistical parameter to use as a measure of accuracy at impact. We have shown that having the ECM, or P, available at impact is sufficient for calculating the maximum and minimum variances, σ_{max}^2 and σ_{min}^2, respectively. We will now show how both variances can be combined to form a single statistical performance measure, the circular error probable, or, for short, the CEP.

The CEP is defined as the radius of the circle centered at the target (or the impact point of the nominal trajectory) within which 50 percent of the trajectories impact. Fundamental to the concept of the CEP is the tacit assumption that the statistics of the initial conditions and the forcing functions remain the same from trajectory to trajectory. Let us define X^b as the axis of minimum dispersion. Thus,

$$\sigma_x = \sigma_{min} \qquad \sigma_y = \sigma_{max} \tag{13.18}$$

Obviously, if

$$\sigma_x = \sigma_y = \sigma_{min} = \sigma_{max}$$

then the impact pattern would be truly circular; in this special case the CEP would simply be some constant times σ_x.

When the cross-range and down-range statistical parameters are different, then we need three parameters to describe the statistics; these can be either

$$\sigma_d^2 \qquad \sigma_c^2 \qquad \sigma_{dc}$$

or

$$\sigma_{\max}^2 \qquad \sigma_{\min}^2 \qquad \psi$$

However, even with different cross- and down-range statistical parameters we can still give meaning to the idea of the CEP, although some ambiguity must be present when three independent parameters are replaced by a single parameter. The definition of CEP remains the same, namely, the circle, centered at the nominal impact point, whose area is equal to the area bounded by the curve enclosing 50 percent of the impact points and whose centroid is at the impact point of the nominal trajectory.

Assume that two horizontal and orthogonal directions are used for locating the impact point of a typical trajectory (from the impact point of the nominal trajectory), say X and Y, perhaps a cross-range and down-range coordinate. We can then write the following bivariate probability density function:

$$f(x, y) = \frac{1}{2\pi\sigma_x\sigma_y} e^{(-1/2)[(X^2/\sigma_x^2)+(Y^2/\sigma_y^2)]} \tag{13.19}$$

where we have assumed that there is no statistical correlation between the two channels. The probability that an impact will occur within the region R is

$$\begin{aligned} P[(X^2 + Y^2)^{1/2} < R] &= \int\int_R f(X, Y)\mathrm{d}X\,\mathrm{d}Y \\ &= \int_0^X \int_0^Y \frac{1}{2\pi\sigma_x\sigma_y} e^{(-1/2)[(X^2/\sigma_x^2)+(Y^2/\sigma_y^2)]}\mathrm{d}Y\,\mathrm{d}X \end{aligned} \tag{13.20}$$

Making use of the following variable transformations,

$$X = r\cos(\theta)$$

$$Y = r\sin(\theta)$$

allows the integral of Eq. (13.20) to be rewritten as

$$P(r \leq R) = \frac{1}{2\pi\sigma_x\sigma_y} \int_0^{2\pi} \int_0^R e^{(-r^2/2)\{[\cos^2(\theta)/\sigma_x^2]+[\sin^2(\theta)/\sigma_y^2]\}} r\,\mathrm{d}r\,\mathrm{d}\theta \tag{13.21}$$

Now for the special case where

$$\sigma_x = \sigma_y = \sigma \qquad \text{and} \qquad P(r \leq R) = 0.5$$

we have

$$0.5 = \frac{1}{2\pi\sigma^2}\int_0^{2\pi}\int_0^R e^{-r^2/2\sigma^2} r \, dr \, d\theta$$

With some manipulation we get the following:

$$R = [2\ell n(2)]^{1/2}\sigma$$

or

$$\text{CEP} = R = 1.1774\sigma \tag{13.22}$$

If the variances are not equal then the integration of Eq. (13.21) becomes much more difficult. The use of the idea of a CEP is so extensive that we will go through the procedure. First, let us define the correlation coefficient as

$$\rho = \frac{\sigma_x}{\sigma_y} = \frac{\sigma_{\min}}{\sigma_{\max}} \tag{13.23}$$

Then the integral in Eq. (13.21) may be rewritten as follows:

$$P(r \leq R) = \frac{1}{2\pi\sigma_y^2\rho}\int_0^{2\pi}\int_0^R e^{-r^2/2\sigma_y^2} e^{\{[-r^2/(2\sigma_y^2\rho^2)][1-\rho^2][\cos^2(\theta)]\}} r \, dr \, d\theta$$

Now let us examine the inner integral, which contains θ. We write

$$I = \int_0^{2\pi} e^{-[r^2/(2\sigma_y^2\rho^2)][1-\rho^2][\cos^2(\theta)]} d\theta$$

If we define A as follows,

$$A = \frac{r^2(1-\rho^2)}{2\sigma_y^2\rho^2}$$

then the previous equation may be rewritten as

$$I = e^{-A/2}\int_0^{2\pi} e^{[(-A/2)\cos(2\theta)]} d\theta$$

Then we have

$$P(r \leq R) = \frac{1}{2\pi\sigma_y^2\rho}\int_0^R e^{-r^2/2\sigma_y^2} e^{-A/2}\left[\int_0^{2\pi} e^{(-A/2)\cos(2\theta)} d\theta\right] r \, dr$$

which may be manipulated further to give

$$P(r \leq R) = \frac{1}{2\pi\sigma_y^2\rho}\int_0^R e^{\{-[r^2(1+\rho^2)]/4\sigma_y^2\rho^2\}}\left[\int_0^{2\pi} e^{(-A/2)\cos(2\theta)}\mathrm{d}\theta\right] r\,\mathrm{d}r$$

Now if we make the following variable definition,

$$\bar{r} = r/\sigma_y$$

the previous integral becomes

$$P(\bar{r} \leq \frac{R}{\sigma_y}) = \frac{1}{2\pi\rho}\int_0^{R/\sigma_y} e^{-[\bar{r}^2(1+\rho^2)/(4\rho^2)]}\left[\int_0^{2\pi} e^{(-A/2)\cos(2\theta)}\mathrm{d}\theta\right]\bar{r}\,\mathrm{d}\bar{r} \quad (13.24)$$

Note that the inner integral of Eq. (13.24) is definite, but that the outer integral has a variable upper limit. To be consistent with our definition of CEP, the function on the left must be

$$P(\bar{r} \leq \frac{R}{\sigma_y}) = 0.5 \quad (13.25)$$

Thus, when the integration is carried out, we have a function of two variables, that is,

$$R/\sigma_y \qquad \text{and} \qquad \rho$$

Then the variable R/σ_y may be manipulated so that Eq. (13.25) is satisfied.

Let us consider the inner integral first, that is,

$$I = \int_0^{2\pi} e^{(-A/2)\cos(2\theta)}\mathrm{d}\theta$$

Next we may replace the exponential by the appropriate series in order to approximate a closed form solution as follows:

$$I = \int_0^{2\pi}\left\{1 + \left[-\frac{A}{2}\cos(2\theta)\right] + \frac{1}{2!}\left[-\frac{A}{2}\cos(2\theta)\right]^2 + \cdots + \frac{1}{n!}\left[-\frac{A}{2}\cos(2\theta)\right]^n + \cdots\right\}\mathrm{d}\theta$$

It is easy to show that all the odd powers vanish so that the integral appears as

$$I = 2\pi F(A) \quad (13.26a)$$

where

$$F(A) = 1 + \frac{1}{2!}\frac{1}{2}\left(\frac{A}{2}\right)^2 + \frac{1}{4!}\left(\frac{1\cdot 3}{2\cdot 4}\right)\left(\frac{A}{2}\right)^4 + \frac{1}{6!}\left(\frac{1\cdot 3\cdot 5}{2\cdot 4\cdot 6}\right)\left(\frac{A}{2}\right)^6$$
$$+ \frac{1}{8!}\left(\frac{1\cdot 3\cdot 5\cdot 7}{2\cdot 4\cdot 6\cdot 8}\right)\left(\frac{A}{2}\right)^8 + \frac{1}{10!}\left(\frac{1\cdot 3\cdot 5\cdot 7\cdot 9}{2\cdot 4\cdot 6\cdot 8\cdot 10}\right)\left(\frac{A}{2}\right)^{10} \cdots \quad (13.26b)$$

If Eqs. (13.26) are inserted into Eq. (13.24) we have

$$P\left(\bar{r} \le \frac{R}{\sigma_y}\right) = 0.5 = \frac{1}{\rho}\int_0^{R/\sigma_y} F\left[\frac{\bar{r}^2(1-\rho^2)}{2\rho^2}\right] e^{-[\bar{r}^2(1+\rho^2)]/(2\rho^2)}\bar{r}\,\mathrm{d}\bar{r} \quad (13.27)$$

Equation (13.27) can be integrated using Simpson's rule for an estimated value of the upper limit, R/σ_y. If the integral is not sufficiently close to 0.5, then R/σ_y is adjusted until it is. This process of trial and error can be carried out manually or with some searching algorithm in a program. In either event, once the integral is acceptably close to 0.5, the correlation coefficient ρ is then changed and the process repeated. The results are plotted in Fig. 13.5. Because of the awkwardness of this procedure, two analytical relationships were fitted to the resulting curve, one for a value of ρ less than 0.35, and a second for a value of ρ greater than 0.35 (and of course less than or equal to unity). These relationships are taken from Ref 3. That is,

$$\mathrm{CEP} = \sigma_{\max}\left\{0.674 + \left(\frac{\sigma_{\min}}{\sigma_{\max}}\right)\left[0.0786 + 0.2573\left(\frac{\sigma_{\min}}{\sigma_{\max}}\right) + 1.1108\left(\frac{\sigma_{\min}}{\sigma_{\max}}\right)^2\right]\right\}$$
$$0.0 < \frac{\sigma_{\min}}{\sigma_{\max}} \le 0.35 \quad (13.28)$$

and

$$\mathrm{CEP} = 0.562\sigma_{\max} + 0.615\sigma_{\min} \qquad 0.35 < \left(\frac{\sigma_{\min}}{\sigma_{\max}}\right) \le 1.0 \quad (13.29)$$

Assuming that we have the error covariance matrix P in an inertial frame, we can now review the steps necessary to obtain the CEP or the circular error probable:

1) Transform the P matrix into the local and then into the trajectory frame.
2) Obtain the projection matrix B using the flight path angle γ [see Eq. (13.8)].
3) Obtain the impact covariance matrix I' [see Eq. (13.9)].
4) Obtain the maximum and minimum variances [see Eq. (13.14)].
5) Obtain the misalignment angle ψ [see Eq. 13.17)].
6) Calculate the correlation coefficient ρ [see Eq. (13.23)].

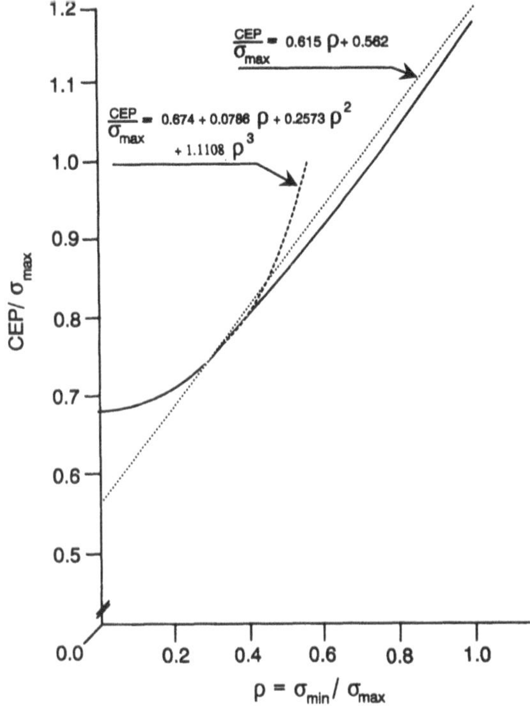

Fig. 13.5 CEP/σ_{max} vs ρ.

7) If ρ equals 1.0, calculate the CEP from Eq. (13.22).
8) If ρ is less than 0.35, use Eq. (13.28).
9) If ρ is less than 1.0 but greater than 0.35, use Eq. (13.29).

13.3 Covariance Propagation and Monte Carlo Simulations

In the previous section we introduced the error ellipsoid, which corresponds to the error covariance matrix P. It was shown that the eigenvalues of P may be reduced to a single parameter called the CEP, or the circular error probable. In this section we will continue with the concept of the error covariance matrix and show two different ways to use the ECM to assess system accuracy.

First let us assume that the re-entry vehicle has just re-entered the atmosphere. From Newton's Second Law, we can relate the three positional coordinates—$X1, X2, X3$—to the applied forces of aerodynamics and gravity. The resulting set of three second-order differential equations have been derived and analyzed in several other parts of this book. Rather than repeating this work, we will write a set of six first-order equations. These equations will require six initial conditions: three positional and three velocity. To simplify the procedure, we will define the $X1$-axis to be cross range (positive to the right from the direction of the trajectory), the $X2$-axis down range, and the $X3$-axis up. The velocity components in the $X1, X2, X3$ direction are $X4, X5, X6$. The six equations of motion are as follows:

$$\frac{dX1}{dt} = X4 \qquad \frac{dX2}{dt} = X5 \qquad \frac{dX3}{dt} = X6$$

$$\frac{dX4}{dt} = -\left(\frac{\rho C_D S g}{2W}\right) V\ X4$$

$$\frac{dX5}{dt} = -\left(\frac{\rho C_D S g}{2W}\right) V\ X5 \tag{13.30}$$

$$\frac{dX6}{dt} = -\left(\frac{\rho C_D S g}{2W}\right) V\ X6 - g$$

where the velocity magnitude V is

$$V = [(X4)^2 + (X5)^2 + (X6)^2]^{1/2}$$

and ρ is the density at altitude $X3$, S is the reference area (usually the maximum cross-sectional area), C_D is the drag coefficient, and W is the weight (or W/g, the mass).

A compact way to represent the previous set of equations is as a matrix, or

$$\frac{d\boldsymbol{X}}{dt} = \boldsymbol{F}(\boldsymbol{X})$$

where

$$\frac{d}{dt}\begin{bmatrix} X1 \\ X2 \\ X3 \\ X4 \\ X5 \\ X6 \end{bmatrix} = \begin{bmatrix} f_1(\boldsymbol{X}) \\ f_2(\boldsymbol{X}) \\ f_3(\boldsymbol{X}) \\ f_4(\boldsymbol{X}) \\ f_5(\boldsymbol{X}) \\ f_6(\boldsymbol{X}) \end{bmatrix} \tag{13.31}$$

The appropriate functions are easily identified from Eqs. (13.30). For example,

$$f_3(\boldsymbol{X}) = X6$$

$$f_5(\boldsymbol{X}) = -(\rho_0 C_D S g/2W)\, e^{-X3/H}\, V\ X5$$

where we have assumed an exponential atmosphere [see Eq. (2.32)].

In assessing errors a useful procedure is to define one trajectory—a nominal trajectory—assumed error free. The impact point of this trajectory is assumed to be the intended impact point or simply is taken as the reference point; the remaining trajectories then are perturbations (by some error source) about this nominal trajectory. We are not interested in characterizing the error source, i.e., defining the physical origin of this error source. We simply wish to describe quantitatively the deviations from a nominal trajectory that are a consequence of this error source.

The state vector $\boldsymbol{X}$ may then be separated into two parts: a nominal part, $\boldsymbol{X}_0$ and a perturbed part, $\tilde{\boldsymbol{X}}$, as

$$\boldsymbol{X} = \boldsymbol{X}_0 + \tilde{\boldsymbol{X}} \tag{13.32a}$$

Figure 13.6 shows both the nominal and the perturbed trajectory, with the three variables of Eq. (13.32a) indicated; the error ellipsoid may grow and rotate throughout the nominal trajectory.

Similarly, the nonlinear vector function $\boldsymbol{F}$ can be linearized and written as follows:

$$\boldsymbol{F} \approx \boldsymbol{F}_0 + \frac{\mathrm{d}\boldsymbol{F}}{\mathrm{d}\boldsymbol{X}}(\boldsymbol{X} - \boldsymbol{X}_0) = \boldsymbol{F}_0 + \frac{\mathrm{d}\boldsymbol{F}}{\mathrm{d}\boldsymbol{X}}\tilde{\boldsymbol{X}} \tag{13.32b}$$

where a subscript 0 indicates a nominal trajectory. Next we insert Eqs. (13.32a) and (13.32b) into Eq. (13.31) to give

$$\frac{\mathrm{d}\boldsymbol{X}_0}{\mathrm{d}t} + \frac{\mathrm{d}\tilde{\boldsymbol{X}}}{\mathrm{d}t} \approx \boldsymbol{F}(\boldsymbol{X})|_{\boldsymbol{X}=\boldsymbol{X}_0} + \frac{\mathrm{d}\boldsymbol{F}}{\mathrm{d}\boldsymbol{X}}(\boldsymbol{X} - \boldsymbol{X}_0) = \boldsymbol{F}(\boldsymbol{X}_0) + \left.\frac{\mathrm{d}\boldsymbol{F}}{\mathrm{d}\mathbf{X}}\right|_{\boldsymbol{X}_0}\tilde{\boldsymbol{X}}$$

But from Eq. (13.31) we obtain

$$\frac{\mathrm{d}\boldsymbol{X}_0}{\mathrm{d}t} = \boldsymbol{F}(\boldsymbol{X}_0) \tag{13.33}$$

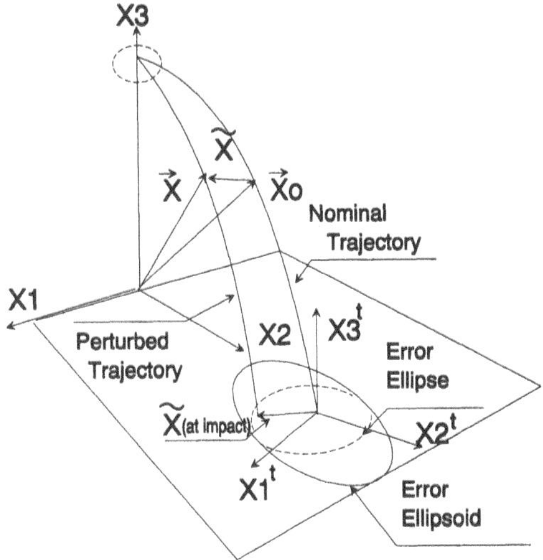

Fig. 13.6 Nominal and perturbed trajectories.

Consequently,

$$\frac{d\tilde{X}}{dt} = \left.\frac{dF}{dX}\right|_{X_0} \tilde{X} = A\tilde{X} \tag{13.34}$$

Equation (13.33) is the nonlinear differential equation set that describes the nominal trajectory. Equation (13.34), on the other hand, is a linear differential equation that describes perturbation about this nominal trajectory. The matrix A is the Jacobian of the perturbational variables, whose elements are as follows:

$$A = [a_{ij}] = \left[\frac{\partial f_i}{\partial X_j}\right]_{X=X_0} \tag{13.35a}$$

For the six-state system under consideration, A is of dimension 6×6. Matrix A can be written as follows:

$$A = \begin{bmatrix} 0 & 0 & 0 & 1 & 0 & 0 \\ 0 & 0 & 0 & 0 & 1 & 0 \\ 0 & 0 & 0 & 0 & 0 & 1 \\ 0 & 0 & a_{43} & a_{44} & a_{45} & a_{46} \\ 0 & 0 & a_{53} & a_{54} & a_{55} & a_{56} \\ 0 & 0 & a_{63} & a_{64} & a_{65} & a_{66} \end{bmatrix} \tag{13.35b}$$

Now if we set

$$K_D = \frac{\rho_0 S C_D g}{2W} = \frac{\rho_0 S C_D}{2m} \tag{13.36}$$

where W is the weight and m is the mass, then we can easily show, for example, that

$$a_{53} = \frac{\partial f_5}{\partial X3} = \frac{K_D}{H} e^{-X3_0/H} V_0 \, X5_0 \tag{13.37a}$$

Also,

$$a_{55} = -K_D e^{-X3_0/H} \left(V_0 + \frac{X5_0^2}{V_0}\right) \tag{13.37b}$$

$$a_{64} = -K_D e^{-X3_0/H} \left(\frac{X4_0 \, X6_0}{V_0}\right) \tag{13.37c}$$

We must emphasize again that the elements of matrix A are evaluated using the current states of the nominal trajectory (denoted by subscript 0). Even though the elements of A are time varying, reflecting the time-varying states of the nominal trajectory, the elements of A are treated as time-constant for integration of the perturbational variable $\tilde{X}$. Thus, Eq. (13.34) is treated as a homogeneous matrix equation (or set of homogeneous scalar equations).

A convenient way of representing the integration of Eq. (13.34) or equivalently a way of indicating the relationship between the perturbational state vector $\tilde{X}$ at time t_i and t_{i+1} is the use of the transition matrix, ϕ. That is,

$$\tilde{X}_{i+1} = \phi(t_{i+1}, t_i)\tilde{X}_i \tag{13.38}$$

If ϕ is assigned some fixed value by the nominal states, then according to Eq. (13.38) ϕ may be accepted as a matrix that transitions the perturbation or error state vector $\tilde{X}$ over the time interval t_i to t_{i+1}. Since we are assuming that, however evaluated, the elements of A are constant, then the system represented by Eq. (13.34) is defined as *stationary*. Gelb identifies the transition matrix ϕ formally with the matrix exponent and, more usefully, with the matrix series,[4] or

$$\phi(t_{i+1}, t_i) = e^{A(t_{i+1}-t_i)} = e^{A\Delta t}$$

$$\phi = I + A\Delta t + \frac{1}{2!}A^2\Delta t^2 + \frac{1}{3!}A^3\Delta t^3 + \cdots + \frac{1}{n!}A^n\Delta t^n \tag{13.39}$$

where

$$A^n = A \cdot A \cdot A \cdot A \cdot \cdots \cdot A$$

In the previous section we indicated that the error ellipsoid is the quadratic form of the covariance matrix. We define the error covariance matrix at time t_i, P_i, as

$$P_i = E[\tilde{X}_i\tilde{X}_i^{\mathrm{T}}] \tag{13.40a}$$

where $\tilde{X}_i$ is of course the perturbational or error state vector [see Eq. (13.32a)] at time t_i; X is a 6×1 matrix and therefore P is a 6×6 matrix. Since

$$P_{i+1} = E[\tilde{X}_{i+1}\tilde{X}_{i+1}^{\mathrm{T}}] \tag{13.40b}$$

insertion of Eq. (13.38) into (13.40b) gives

$$P_{i+1} = \phi_i P_i \phi_i^{\mathrm{T}} \tag{13.41}$$

The preceding equation provides for a discrete propagation of the covariance matrix. Alternatively, we can use the first two terms of Eq. (13.39) to get

$$\begin{aligned} P_{i+1} &\approx [(I + A\Delta t)P_i(I + A\Delta t)^{\mathrm{T}}] \\ \frac{P_{i+1} - P_i}{\Delta t} &= AP_i + P_iA^{\mathrm{T}} \end{aligned} \tag{13.42}$$

Note that we have not subscripted the system matrix A. It is appropriate to use subscripts on the error covariance matrix P since the subscripts indicate the value of P at the beginning and end of the interval. However, the elements of matrix A are varying continuously and can be considered constant over the interval i to $i+1$ only as an approximation. The left-hand side of the second expression in Eq. (13.42) when taken to a limit as the time interval becomes arbitrarily small is the definition of the derivative. That is,

$$\frac{\mathrm{d}P}{\mathrm{d}t} = A(t)P(t) + P(t)A(t)^{\mathrm{T}} \tag{13.43}$$

The preceding equation may be identified as a form of the Riccati equation. Briefly, our goal is to integrate this equation from the point of atmospheric entry to impact of the nominal trajectory.

Developing the appropriate algorithms for the integration of the Riccati equation is an imposing task. For the ensuing discussion the Riccati integration was done in the ACSL™, a higher level simulation language.[5]

To simplify the results, only a velocity error is included at trajectory initiation. The initial conditions of the nominal trajectory are as follows:

$$X = X1 = 0.0 \text{ m}$$

$$Y = X2 = 0.0 \text{ m}$$

$$Z = X3 = 2.5 \times 10^4 \text{ m}$$

$$\frac{\mathrm{d}X}{\mathrm{d}t} = X4 = 0.0 \text{ m/s}$$

$$\frac{\mathrm{d}Y}{\mathrm{d}t} = X5 = 3.535 \times 10^3 \text{ m/s}$$

$$\frac{\mathrm{d}Z}{\mathrm{d}t} = X6 = -3.535 \times 10^3 \text{ m/s}$$

Again note that only the velocity $(X4, X5, X6)$ will be given an error; the position vector $(X1, X2, X3)$ is assumed to be exact. Such a restriction is meaningful only in the context of a demonstration: in any practical situation the positional errors must also be present since the error covariance matrix at re-entry would reflect velocity errors at the beginning of boost, resulting in both positional and velocity errors at re-entry.

With the preceding restriction let us set the velocity variance at 100 $(\text{m/s})^2$. To parallel the covariance propagation idea, i.e., the integration of Eq. (13.43), to impact, we will first examine a 100-trajectory Monte Carlo simulation.

Based on a standard deviation of 10 m/s for each velocity component, the initial error covariance matrix was used to initiate the 100 trajectories (i.e., provide the initial conditions). The 100 trajectories were determined by integrating the first order ordinary differential equations in matrix form [Eq. (13.31)]. All trajectories were terminated at the time of impact of the nominal trajectory. The error covariance matrix at impact was then calculated. In Table 13.1 only the impact positional error sub-matrix appears, since our interest is confined to impact errors. From this matrix we can then calculate the eigenvalues and eigenvectors.

One way of looking at the eigenvalues/vectors is to consider the error covariance matrix in an arbitrary reference frame. The numerical values appearing in the matrix depend upon the orientation of the reference frame. The eigenvectors then establish a directional cosine matrix, which locates a frame such that the P matrix is diagonalized. These diagonal elements are the eigenvalues; the directional cosines of each of the axes of the frame in which P is diagonalized are the eigenvectors. Since P is always diagonally symmetric the eigenvalues and the eigenvector components are real. In Fig. 13.7 one can see the error ellipsoid and the relationship between the principal axes of the ellipsoid and the target frame.

Next we propagate the covariance matrix from penetration to impact, based upon Eq. (13.43). At first glance it would appear that covariance propagation is more *efficient* (than the Monte Carlo method), since only one trajectory equation (the nominal trajectory) needs to be integrated whereas an adequate

Table 13.1 Use of Monte Carlo method to compute error covariance matrices

Initial velocity error covariance matrix

$$\begin{bmatrix} 91.96 & 15.14 & -14.20 \\ 15.14 & 81.75 & -11.95 \\ -14.20 & -11.95 & 78.68 \end{bmatrix}$$

Impact positional error covariance matrix

$$\begin{bmatrix} 4477.9 & 321.2 & -268.5 \\ 321.2 & 3768.2 & 995.6 \\ -268.5 & 995.6 & 897.22 \end{bmatrix}$$

Eigenvalues

$L1 = 554.0\ \text{m}^2 \qquad \sigma_1 = 23.54\ \text{m}$

$L2 = 3983.1\ \text{m}^2 \qquad \sigma_2 = 63.11\ \text{m}$

$L3 = 4606.2\ \text{m}^2 \qquad \sigma_3 = 67.87\ \text{m}$

Eigenvectors

$U1$	$U2$	$U3$
0.0897	0.3917	0.9157
−0.3029	−0.8651	0.3998
0.9488	−0.3132	0.0410

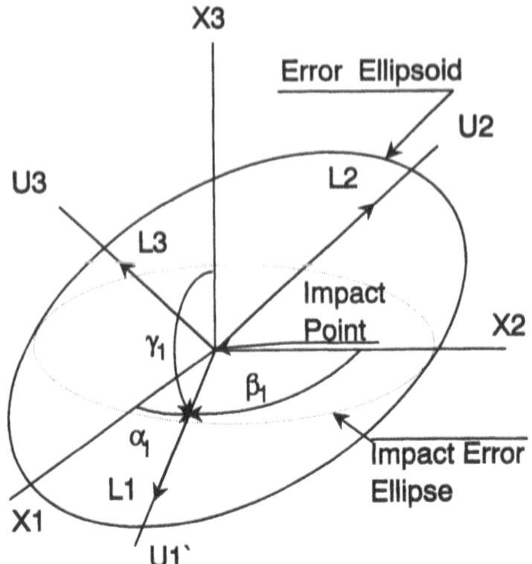

Fig. 13.7 Error covariance matrix and orientation with target frame.

Monte Carlo simulation might require 100 or so trajectories. The not-so-obvious complexities of the integration of the Riccati equation mean both methods are computationally intensive. Depending upon the software available (and algorithm chosen), the covariance propagation takes less effort on the part of the analyst than does the setup surrounding the Monte Carlo method. Of course the most important consideration concerns the adequacy of the covariance propagation. Table 13.2 presents the results obtained using the covariance propagation method. The initial covariance matrix is omitted since it is the same in both methods.

By examining the root sum square of the trace of the error covariance matrix obtained from both the Monte Carlo and the covariance propagation methods, it is possible to make a comparison between both of these techniques. (The trace of any square matrix is always equal to the sum of the eigenvalues.)

MONTE CARLO	COVARIANCE PROPAGATION
$\sigma_{\text{rms}} = 95.62$ m	$\sigma_{\text{rms}} = 95.48$ m

Before proceeding to other topics, let us obtain the size and orientation of the error ellipse. The assumption here is that we have available the eigenvalues and the eigenvectors of the error covariance matrix.

As we pointed out earlier, the components of the eigenvectors are the directional cosines of each of the principal axes of the error ellipsoid. If these vector components (i.e., the directional cosines) are used to form the rows of a matrix, that matrix is a directional cosine matrix (DCM) from the trajectory axes, say, (X^t, Y^t, Z^t) to the body (principal axes) frame, (X^b, Y^b, Z^b). In terms of the DCM symbolism used earlier we have

$$X^b = C_t^b X^t \tag{13.44a}$$

Table 13.2 Covariance propagation

Impact positional error covariance matrix

$$\begin{bmatrix} 4471.2 & 319.5 & -267.9 \\ 319.5 & 3751.2 & 995.4 \\ -267.9 & 995.4 & 893.6 \end{bmatrix}$$

Eigenvalues of impact error covariance matrix (positional)

$L1 = 547.1\ \text{m}^2$	$\sigma_1 = 23.39$ m
$L2 = 3972.2\ \text{m}^2$	$\sigma_2 = 63.03$ m
$L3 = 4596.8\ \text{m}^2$	$\sigma_3 = 67.80$ m

Eigenvectors

$U1$	$U2$	$U3$
0.0896	0.3860	0.9482
−0.3047	−0.8670	−0.3151
0.9482	−0.3151	0.3999

Since we are interested in the maximum ellipse that is formed by the error ellipsoid as it passes through the horizontal plane (tangent to the nominal impact point), we may write Eq. (13.44a) as

$$\begin{bmatrix} X \\ Y \\ Z \end{bmatrix}^b = C_t^b \begin{bmatrix} X \\ Y \\ 0 \end{bmatrix}^t = (C_b^t)^{\mathrm{T}} \begin{bmatrix} X \\ Y \\ 0 \end{bmatrix}^t \tag{13.44b}$$

where the Z^t coordinate is zero in the impact plane. The error ellipsoid is given as

$$\frac{(X^b)^2}{\sigma_1^2} + \frac{(Y^b)^2}{\sigma_2^2} + \frac{(Z^b)^2}{\sigma_3^2} = 1 \tag{13.45}$$

where $(\sigma_1, \sigma_2, \sigma_3)$ are the eigenvalues of the error ellipsoid. By replacing (X^b, Y^b, Z^b) with (X^t, Y^t), we have the equation of an ellipse. This ellipse will lie in the X^t-Y^t plane. The major and minor axes of this ellipse will not necessarily be coincident with the X^t, Y^t axes [X^t, cross range (positive to the right of the direction of the velocity vector) and Y^t, down range]. However, a simple coordinate transformation will determine the orientation of the major axis of this ellipse with respect to the Y^t or the down-range direction. The transformation will also determine the major and minor axes of the error ellipse.

Table 13.3 provides a program, written in TRUEBASIC™, which computes the eigenvalues and the eigenvector components of an error covariance matrix. The program also computes the major and minor axes of the error ellipse with centroid at the nominal target and the CEP according to Eqs. (13.28) and (13.29).

Table 13.3 Program to calculate eigenvalues and eigenvectors and CEP for an error covariance matrix

```
 THIS PROGRAM PERFORMS THE FOLLOWING FUNCTIONS:
! THE INPUT IS THE ERROR COVARIANCE MATRIX IN A TARGET FRAME
! THE OUTPUT IS THE FOLLOWING:
! THE EIGENVALUES AND EIGENVECTORS OF THE ERROR COV. MATRIX
! THE MAJOR AND MINOR AXES OF THE ERROR ELLIPSE AT THE TARGET
! THE ORIENTATION OF THE MAJOR AXIS OF THE ERROR ELLIPSE
!     RELATIVE TO THE DOWN-RANGE DIRECTION
! TARGET FRAME: X: CROSS RANGE; Y: DOWN RANGE; Z UP
! DCM: CP2T-FROM PRINCIPAL FRAME OF ERROR ELLIPSE TO TARGET FRAME
! DCM: CT2P-FROM TARGET FRAME TO PRINCIPLE FRAME OF ERROR ELLIPSE

DIM P(3,3), Z(3), P9(3,3), P8(3,3), I(3,3), CP2T(3,3), CT2P(3,3)
DIM Z1(3,3), P1(3,3)

OPEN #1: NAME "EIG.TRU", CREATE NEWOLD
ERASE #1

! READ IN THE ELEMENTS OF THE ERROR COVARIANCE MATRIX
PRINT #1: "      COVARIANCE MATRIX      "
MAT READ P
DATA 4.4779E3, 3.21255E2, -2.6851E2
DATA 3.21255E2, 3.76816E3, 9.9562E2
DATA -2.6851E2, 9.9562E2, 8.9722E2
! READ OUT THE COVARIANCE MATRIX AS A CHECK
PRINT
MAT PRINT #1: P
PRINT
! CALCULATE THE COEFFICIENTS OF THE CHARACTERISTIC EQUATION
A0 = P(1,1)*(P(2,3)^2)+P(2,2)*(P(1,3)^2)+P(3,3)*(P(1,2)^2)
A0 = A0-P(1,1)*P(2,2)*P(3,3)-2*P(1,2)*P(1,3)*P(2,3)
A1 = P(1,1)*P(2,2)+P(1,1)*P(3,3)+P(2,2)*P(3,3)
A1 = A1-((P(1,2)^2)+(P(1,3)^2)+(P(2,3)^2))
A2 = (-1.0)*(P(1,1)+P(2,2)+P(3,3))
A3 = 1
! DEFINE THE CHARACTERISTIC EQUATION
DEF FNX(Z) = A0+A1*Z+A2*(Z^2)+A3*(Z^3)
DEF FND(Z) = A1+2.0*A2*Z+3.0*A3*(Z^2)

! SET COUNTER FOR EIGENVALUES AND BOUNDS FOR SEARCH
J = 1
XHIGH = ABS(P(1,1))+ABS(P(2,2))+ABS(P(3,3))
XLOW = (-1.0)*XHIGH
```

(continued on next page)

Table 13.3 (continued) Program to calculate eigenvalues and eigenvectors and CEP for an error covariance matrix

```
! BEGIN SEARCH FOR EIGENVALUES
S = 1.0
FOR X = 0.0 TO XHIGH STEP S
   Y1 = FNX(X)
   Y2 = FNX(X+S)
   IF SGN(Y1) <> SGN(Y2) THEN
       X0 = X
       Y1 = FNX(X0)
       Y1D = FND(X0)
       DO WHILE ABS(Y1)> 1.0E-6
           D = (-1.0)*Y1/Y1D
           X0 = X0+D
           Y1 = FNX(X0)
           Y1D = FND(X0)
           IF Y1D < 1.0E-12 THEN EXIT DO
       LOOP
       Z(J) = X0
       J = J+1
   END IF
NEXT X

! OBTAIN THE EIGENVECTORS
MAT P9 = IDN
FOR L = 1 TO 3
   MAT P8 = (Z(L))*P9
   MAT I = P-P8
   W1 = 1.0
   W2 = I(2,2)*I(1,3)-I(1,2)*I(2,3)
   W3 = (I(2,1)*I(1,2)-I(1,1)*I(2,2))/W2
   W2 = (I(1,1)*I(2,3)-I(2,1)*(I(1,3)))/W2
   W = SQR(W1^2+W2^2+W3^2)
   CP2T(1,L) = W1/W
   CP2T(2,L) = W2/W
   CP2T(3,L) = W3/W
NEXT L

MAT CT2P = TRN(CP2T)
MAT Z1 = P*CP2T
MAT P1 = CT2P*Z1

PRINT #1: "    EIGENVALUES    "
MAT PRINT #1: Z
PRINT
```

(continued on next page)

Table 13.3 (continued) Program to calculate eigenvalues and eigenvectors and CEP for an error covariance matrix

```
PRINT #1: "    EIGENVECTORS    "
MAT PRINT #1: CP2T
PRINT

! CALCULATE THE MAJOR AND MINOR AXES OF THE ERROR ELLIPSE
! AND THE ORIENTATION OF THE MAJOR AXIS WRT TO DOWN RANGE

A = (CP2T(1,1)^2)/Z(1)+(CP2T(2,1)^2)/Z(2)+(CP2T(3,1)^2)/Z(3)
B = (CP2T(1,2)^2)/Z(1)+(CP2T(2,2)^2)/Z(2)+(CP2T(3,2)^2)/Z(3)
C = CP2T(1,1)*CP2T(1,2)/Z(1)
C = C+(CP2T(2,1)*CP2T(2,2)/Z(2))
C = C+(CP2T(3,1)*CP2T(3,2)/Z(3))
C = 2.0*C
PSI = 0.5*ATN(2.0*C/(A-B))
D = SQR(1.0/(A*(COS(PSI)^2)+B*(SIN(PSI)^2)+C*SIN(2.0*PSI)))
E = SQR(1.0/(A*(SIN(PSI)^2)+B*(COS(PSI)^2)-C*SIN(2.0*PSI)))
IF D>E THEN
   SIGMAX = D
   SIGMIN = E
ELSE
   SIGMAX = E
   SIGMIN = D
END IF

PRINT #1: "MAX. AND MIN. STANDARD DEVIATION OF ERROR ELLIPSE"
PRINT #1: SIGMAX, SIGMIN
PRINT

PRINT #1: "ANGLE BETWEEN MAJOR AXIS AND DOWN RANGE",
PSI*180.0/PI,"DEGREES"
PRINT
! CALCULATE THE CEP
RAT = SIGMIN/SIGMAX
IF RAT > 0.0 AND RAT <= 0.35 THEN
   CEP = SIGMAX*(0.674+RAT*(0.0786+RAT*0.2573+1.1108*(RAT^2)))
ELSE
   CEP = 0.562*SIGMAX+0.615*SIGMIN
END IF

PRINT #1: "CIRCULAR ERROR PROBABLE EQUALS";"    "; CEP
END
```

(continued on next page)

Table 13.3 (continued) Program to calculate eigenvalues and eigenvectors and CEP for an error covariance matrix

```
                              OUTPUT
                         COVARIANCE MATRIX
          4477.9              321.255         -268.51
           321.255           3768.16           995.62
          -268.51             995.62           897.22
                            EIGENVALUES
        554.01319            3983.1351        4606.1317
                           EIGENVECTORS
      8.9722373e-2           .39175331         .91568512
      -.30286956            -.86512005         .39979663
       .9487992             -.31320383        4.1029531e-2
  MAX. AND MIN. STANDARD DEVIATION OF ERROR ELLIPSE
    74.836344          43.006992
ANGLE BETWEEN MAJOR AXIS AND DOWN RANGE: -22.814908 DEGREES
CIRCULAR ERROR PROBABLE: 68.507326
```

We will now compare the Monte Carlo and the covariance propagation methods where not only the initial conditions (velocity only here) but also the density ρ and the drag coefficient C_D vary. The nominal or mean values and the standard deviation are also given, that is,

	VX	VY	VZ	ρ_0	C_D
Mean	0.0	3535.53	3535.53	1.752	0.10
σ	10.0	10.0	10.0	0.07	0.004

(Note that we are using the subscript 0 to the density to indicate the mean value of the sea level value of density; recall that the exponential atmosphere model has two parameters, ρ_0 and the atmospheric spread factor H.)

For a given Monte Carlo trajectory the Gaussian random number generator is entered with the appropriate standard deviation, the output is added algebraically to the mean value, and the trajectory is run. The error covariance matrix P about the nominal trajectory impact point is then manipulated to provide the eigenvalues and eigenvectors. The random numbers used in the Monte Carlo process are then combined to form an error covariance matrix by correlating each state taken two at a time. The resulting matrix then initializes a covariance propagation procedure that is terminated at the time of nominal trajectory impact.

Let us discuss the results. However, before doing this we must first write the system matrix A with the states *enhanced* to account for two additional variables, ρ_0 and C_D. Unlike the positional/velocity states, neither ρ_0 nor C_D has associated differential equations. Both ρ_0 and C_D appear in the velocity state equations. We will now write the system matrix as an enhancement of Eq. (13.35b), or

$$\frac{dt}{dt}\begin{bmatrix}\widetilde{X1}\\\widetilde{X2}\\\widetilde{X3}\\\widetilde{X4}\\\widetilde{X5}\\\widetilde{X6}\\\tilde{\rho}_0\\\tilde{C}_D\end{bmatrix}=\begin{bmatrix}0&0&0&0&0&1&0&0\\0&0&0&0&0&0&1&0\\0&0&0&0&0&0&0&1\\0&0&a_{43}&a_{44}&a_{45}&a_{46}&a_{47}&a_{48}\\0&0&a_{53}&a_{54}&a_{55}&a_{56}&a_{57}&a_{58}\\0&0&a_{63}&a_{64}&a_{65}&a_{66}&a_{67}&a_{68}\\0&0&0&0&0&0&0&0\\0&0&0&0&0&0&0&0\end{bmatrix}\begin{bmatrix}\widetilde{X1}\\\widetilde{X2}\\\widetilde{X3}\\\widetilde{X4}\\\widetilde{X5}\\\widetilde{X6}\\\tilde{\rho}_0\\\tilde{C}_D\end{bmatrix} \tag{13.46}$$

where the mean values of the sea level density and drag coefficient are identified with overbars, i.e.,

$$\rho_0 = \bar{\rho}_0 + \tilde{\rho}_0$$

and

$$C_D = \bar{C}_D + \tilde{C}_D$$

We may rewrite the preceding matrix using the A matrix of Eq. (13.35b) as

$$\frac{d}{dt}\begin{bmatrix}\tilde{X}\\\tilde{\rho}_0\\\tilde{C}_D\end{bmatrix}=\left[\begin{array}{c:c}\begin{array}{ccc}&&\\&&\\&&\\&A&\\&&\\&&\\\cdots&0&\cdots\\\cdots&0&\cdots\end{array}&\begin{array}{cc}0&0\\0&0\\0&0\\a_{47}&a_{48}\\a_{57}&a_{58}\\a_{67}&a_{68}\\0&0\\0&0\end{array}\end{array}\right]\begin{bmatrix}\tilde{X}\\\tilde{\rho}_0\\\tilde{C}_D\end{bmatrix} \tag{13.47}$$

The elements a_{i7} and a_{i8} involve differentiation with respect to the density parameter ρ_0 and the drag coefficient C_D. For example,

$$a_{47} = \frac{\partial f_4}{\partial \rho_0} = -\frac{\bar{C}_D S g}{2W} e^{-X3_0/H} V_0 X4_0 \tag{13.48a}$$

$$a_{58} = \frac{\partial f_5}{\partial C_D} = -\frac{\bar{\rho}_0 S g}{2W} e^{-X3_0/H} V_0 X5_0 \tag{13.48b}$$

Note that we are assuming that $C_D = C_{D_0}$ since lift is not present, i.e.,

$$C_D = C_{D_0}\left[1 + \left(\frac{C_L}{C_L^*}\right)^2\right]\Bigg|_{C_L=0} = C_{D_0}$$

A comparison of eigenvalues/vectors, first for the Monte Carlo and then the covariance propagation procedures, is provided in Tables 13.4 and 13.5.

We note immediately that the eigenvalues (principal values of the error ellipse) are in close agreement between the two methods. It might appear that the eigenvectors are not comparable. For example, let us compare the $U1$ vector components between the Monte Carlo and covariance propagation methods: -0.7084 to -0.6736. The first is the Monte Carlo value, the second the covariance propagation value. The first represents an angle of 135.10 deg, whereas the second is an angle of 132.34 deg—less than a 3.0-deg difference. The angle between the eigenvectors from the Monte Carlo and covariance matrix propagation methods is shown to be

$$U1, U1 = 5.83 \text{ deg} \qquad U2, U2 = 0 \text{ deg} \qquad U3, U3 = 2.3 \text{ deg}$$

In this section we have presented two common methods for carrying out error analyses of re-entry vehicles. Which method is better? First we note that both methods give essentially the same result. Which is the more intensive computationally? Note the following: if we restrict our consideration to a six-state system, the 100 trajectories of the Monte Carlo approach require in a six-state system the integration of 600 first-order differential equations. The covariance propagation method requires that we obtain the six states of the nominal

Table 13.4 Monte Carlo—enhanced state

Impact positional error covariance matrix

$$\begin{bmatrix} 4456.18 & 374.06 & -333.89 \\ 374.06 & 59857.2 & -56752.3 \\ -333.89 & -56752.3 & 60322.8 \end{bmatrix}$$

Eigenvalue of impact error covariance matrix (positional only)

$L1 = 3336.49\ \text{m}^2$	$\sigma_1 = 57.76$ m
$L2 = 4454.73\ \text{m}^2$	$\sigma_2 = 66.74$ m
$L3 = 116845.00\ \text{m}^2$	$\sigma_3 = 341.83$ m

Eigenvectors

$U1$	$U2$	$U3$
−0.263	0.9996	−0.00445
−0.7084	0.01551	−0.7056
−0.7053	0.2173	0.7085

Table 13.5 Covariance propagation—extended state

Impact positional covariance matrix

$$\begin{bmatrix} 4471.2 & 319055 & -267092 \\ 319055 & 64721.1 & -56066.7 \\ -267092 & -56066.7 & 54306.1 \end{bmatrix}$$

Eigenvalue of impact error covariance matrix (positional)

$L1 = 3205.36\ \text{m}^2$	$\sigma_1 = 56.61$ m
$L2 = 4469.90\ \text{m}^2$	$\sigma_2 = 66.86$ m
$L3 = 445823.00\ \text{m}^2$	$\sigma_3 = 340.33$ m

Eigenvectors

$U1$	$U2$	$U3$
0.01363	0.9999	−0.00374
−0.6736	0.00642	−0.7391
−0.7390	0.01259	0.6736

trajectory, i.e., six integrations. In addition we must propagate a 6×6 matrix. However, since the matrix is symmetrical this would require the integration of 21 first-order equations, bringing the total to 27 equations. It appears that the covariance propagation method is more efficient than the Monte Carlo approach. We should point out that the algorithms for integrating the Monte Carlo equations are well established; Riccati integrators in certain applications might be open to some question. Further information on the solution of the Riccati equation is given in an excellent summary in Ref. 7; additional useful information may also be found in Refs. 8 and 9. Finally, remember that the Riccati equation uses a *linearized* transition matrix that could result in significant errors for a highly nonlinear system.

13.4 State Estimation

In this section we will treat the re-entry trajectory as a phenomenon that is observed from a fixed ground-based location. Since the observed (and implied) states of this system are generally changing with time, we will call the re-entry trajectory a *dynamic process*. Further, we will assume that some, but not all, of the states that define this dynamic process are observable at regular time intervals. Such observations or measurements are made with limited accuracy. More precisely, embedded in these measurements is a noise process independent of the measurements themselves. This noise process can only be described in some statistical fashion and cannot be separated from the measurements.

In addition to the measurement noise, there is the additional uncertainty associated with the process itself, which we will identify as process noise. As we will see, measurements are combined according to some algorithm with the most recent estimate to provide a new state vector estimate; this new state vector estimate must be *transitioned* over the time interval between measurements. The equations used in this transition procedure are the system model or the set of differential equations that represent the dynamics of the system. The nonlinear transition equations are represented in Eq. (13.31), and a linearized version appears in Eq. (13.34). An incomplete model of the system (as any mathematical model of a physical system must be) will only imperfectly bring the last state vector estimate over time to the next measurement time. Thus the algorithm that provides the (new) estimate of the state vector must perforce combine measurements corrupted by a nonremovable noise process with a state vector estimate that has been imperfectly transitioned over the time interval between measurements. A few examples of process noise follow.

An environmental variable such as atmospheric density is present in the $\boldsymbol{F}$-vector of Eq. (13.31). Density has both a deterministic and a noise content. Even though the actual value of atmospheric density could in principal be evaluated from measurements and thermodynamic relationships, the physics cannot be entirely determined by the observer. Thus, from the point of view of such a ground-based observer, atmospheric density must contain a stochastic component that can only be bounded in a statistical sense.

Probably the most significant contributor to process uncertainty is the ability of the re-entry vehicle to maneuver. An object observed from the ground might be suspected capable of maneuvering, that is, of developing an aerodynamic force (lift) normal to the velocity vector. For example, space debris, possessing both an irregular shape and angular velocity, might develop lift forces that can only be described as random by a ground observer. Similarly a re-entry vehicle capable of maneuvering according to some pre-set schedule must also appear as a random process.

We will lump as process noise all those contributors to the system model that cannot be determined by the observer, even though such contributors might be preset (vehicle maneuvering) or regarded as deterministic in principle (atmospheric density).

We will now discuss the use of the Kalman (or Kalman-Bucy) filter as an algorithm that will in a sense balance the effects of process and measurement noise to provide an estimate of the state vector. First we will limit the system to six states associated with particle motion—three positions ($X1, X2, X3$) and three velocities ($X4, X5, X6$). The measurement vector $\mathbf{Z}$ will be limited to the three positional states only. The remaining three components of the velocity vector are estimated. Thus in a sense we have an underdetermined system—three measurements are available at each measurement time, but six distinct quantities are to be estimated. We will then *enhance* the state vector by adding two more states to be estimated, that is, two orthogonal components of the lift vector.

The Kalman algorithm has been extensively utilized in a wide variety of applications (arguably some of which are only marginally appropriate). The algorithm has been called a "recursive solution to Gauss' original least-squares problem."[3] While the Kalman algorithm provides an estimation based upon the idea of minimum least squares, it differs from the classical least squares approach in significant ways.

First, the method of least squares is a batch operation, in that all the measurements must be available before processing. The Kalman algorithm is recursive in that only the present or most recent measurement is needed: past measurements affect the results by setting a weighting matrix (Kalman gain), although the influence of past measurements diminishes exponentially with time.

Second, least squares is nonstatistical: the statistical characteristics of the noise are not used in the fitting. Kalman filtering, on the other hand, uses the statistical properties of the measurement and process noise in setting the gain matrices.

Third, and most important, in least squares estimation the parameters being estimated are usually the coefficients of algebraic polynomials (an iterative procedure may be used when the parameters to be estimated appear in the argument of various functions). Kalman filtering, on the other hand, estimates *states* or entities that are obtained from the integration of differential equations. Thus, Kalman filtering is best applied to a dynamic process in which the states are continuously changing according to a model that gives an acceptable approximation to the process.

In summary, let us define characteristics of the Kalman estimator:

1) Recursive: It stores and manipulates only the current or most recent measurement.

2) Linear: All manipulations require linear algebra.

3) Optimum: In the least squares sense it provides the best estimation of states in the presence of measurement and process noise by minimizing the trace of the error covariance matrix.

4) Dynamic: The estimation is of state variables that are constantly changing with time.

5) Statistical: It incorporates the statistical characteristics of the noise; both model and measurement noise are either an infinite bandwidth white noise or a limited bandwidth Markov process that is reducible to white noise.

The essential characteristics of the recursive filter may be seen from the following relationship, which is used to estimate the current average $\hat{\boldsymbol{X}}_k$ of a sequence of scalar measurements. $\boldsymbol{Z}_i$ is an element in such a measurement sequence, from the first measurement up to and including the kth measurement, so that

$$\boldsymbol{Z} = [\boldsymbol{Z}_1, \boldsymbol{Z}_2, \ldots, \boldsymbol{Z}_K, \ldots, \boldsymbol{Z}_{K-1}, \boldsymbol{Z}_K]^{\mathrm{T}} \tag{13.49}$$

The average $\hat{\boldsymbol{X}}_k$ may be obtained in two ways: batch, in which all measurements are processed one at a time, or recursive, in which only the current measurement is used. The following example provides batch and recursive estimates of the average. The batch estimator is given in Eq. (13.50a) and the recursive estimator is given in both Eqs. (13.50b) and (13.50c).

$$\hat{\boldsymbol{X}}_k = \frac{1}{k}\sum_{i=1}^{k} Z_i \tag{13.50a}$$

The preceding relationship may be rewritten as

$$\hat{X}_k = \frac{1}{k}\left[\sum_{i=1}^{k-1} \mathbf{Z}_i + \mathbf{Z}_k\right]$$

$$= [(k-1)/k]\hat{X}_{k-1} + (1/k)\mathbf{Z}_k \tag{13.50b}$$

or in an alternative form,

$$\hat{X}_k = \hat{X}_{k-1} + (1/k)[\mathbf{Z}_k - \hat{X}_{k-1}] \tag{13.50c}$$

Obviously, the recursive relationship of Eqs. (13.50b) and (13.50c) requires only the previous average (estimate) and the current measurement. The batch algorithm of Eq. (13.50a), on the other hand, requires that all previous measurements be stored: when a new measurement becomes available, the averaging (data processing) must be repeated.

The recursive algorithm of Eq. (13.50b) clearly forms the present estimation by first multiplying the last estimate by $(k-1)/k$ and then the current measurement by $(1/k)$ and then algebraically combining both to form the new estimate. The multipliers or *gains* depend upon the number of measurements. Note that the following is true as k increases without bound:

$$\lim_{k\to\infty}\left(\frac{k-1}{k}\right) \to 1 \qquad \lim_{k\to\infty}\left(\frac{1}{k}\right) \to 0$$

The implication is that as more measurements, Z_k, become available the new estimate is indistinguishable from the previous estimate. The result is that measurements are ignored unless they are drastically different from the average. Even so, as time goes on and more measurements are made, even a noticeably anomalistic measurement is essentially ignored—the filter has "gone to sleep." Future measurements are essentially ignored.

The recursive averaging procedure has some of the characteristics of the Kalman filter: the new estimate is the weighted sum of the previous estimate and the current measurement. However, there are some very important differences between the Kalman algorithm and the averaging algorithm just given. In the averaging algorithm, for example, X, the quantity being estimated (the average in this case), cannot change with time, or at least there is no provision in the algorithm to account for an average that might change somewhat with time. In addition, although the procedure produces the average of the measurement stream, there is no justification for using the average as some kind of estimate of the measurements themselves. Implicit in the average in the first place is that the measurements Z and the average X must be the same physical quantity. Finally, even if the average is "felt" to be representative of the measurement set, no use is made of any uncertainty in the measurements themselves in weighing the last estimate (average) and the current measurement.

The Kalman filter addresses the foregoing limitations while retaining the following form of Eq. (13.50b):

$$\begin{pmatrix}\text{New}\\ \text{estimate}\end{pmatrix} = \text{Gain1}\begin{pmatrix}\text{Old}\\ \text{estimate}\end{pmatrix} + \text{Gain2 (Measurement)}$$

We will change the subscripting notation used in Eq. (13.50) to rewrite the procedure as follows:

$$\hat{X}_k(+) = \hat{X}_k(-) + K_k\,[Z_k - H\hat{X}_k(-)] \qquad (13.51)$$

Notice that this relationship is identical to Eq. (13.50c) except that there is a transformation between the measurement and the estimate: the entity being estimated need not necessarily be the same physical quantity as the measurement. The filter is quite useful in estimating position and velocity even though only position has been measured. The transformation between measurement and state is contained in the *observation matrix* H. Note also the use of the terms $\hat{X}_k(+)$ and $\hat{X}_k(-)$. The (+) in the first term means the current estimate or the state estimate after processing by the filter algorithm has been completed; the (−) in the second term means the previous estimate transitioned to the time of the current or kth measurement.

An excellent derivation of the Kalman algorithm may be found in Chapter 4 of Ref. 4. Another readable discussion (albeit less rigorous) is contained in Chapter 5 of Ref. 6. Here we will merely state the appropriate equations for the algorithm, with the intention of using the generally accepted symbolic forms. In the development leading to Eq. (13.41) we introduced the transition matrix and used this matrix to transition the error covariance matrix P. In Eq. (13.52a), which follows, the transition matrix is used to transition the state X over the discrete interval $(k-1)$ to k (at which time the next measurement is made), or

$$X_k = \phi_k X_{k-1} + W_k \qquad (13.52a)$$

Note also the presence of the process noise vector W_k. Noise accounts for uncertainty in modeling the phenomenon under study. As noted earlier, this uncertainty may have its source in environmental causes or in the maneuvers of a re-entry vehicle, maneuvers that are entirely unpredictable to a ground observer.

The measurements are related to the state as

$$Z_k = H_k X_k + V_k \qquad (13.52b)$$

where H is the measurement matrix and V_k accounts for the presence of noise in the measurement. The estimation of the state (which is a balance between the measurement and the transitioned previous estimate) follows from Eq. (13.50b) as

$$\hat{X}_k(+) = K1_k\hat{X}_k(-) + K2_k Z_k \qquad (13.52c)$$

The error e_k in the state vector before and then after the measurement-estimation process is

$$e_k(-) = \hat{X}_k(-) - X_k \tag{13.52d}$$

$$e_k(+) = \hat{X}_k(+) - X_k \tag{13.52e}$$

We can now relate the state error after measurement to the state error before measurement using Eqs. (13.52) in the following manner:

$$e_k(+) = [K1_k + K2_k H_k - I]X_k + K1_k e_k(-) + K2_k V_k \tag{13.53}$$

The Kalman algorithm is based upon the assumption that the error vector e_k and the measurement noise V_k have zero mean, i.e.,

$$E[e_k(+)] = 0 = E[e_k(-)] \tag{13.54a}$$

$$E[V_k] = 0 \tag{13.54b}$$

However, the state mean is not necessarily zero, i.e.,

$$E[X_k] \neq 0 \tag{13.55}$$

Consequently, from the estimate of Eq. (13.53) and using Eqs. (13.54) and (13.55) we get

$$(K1_k + K2_k H_k - I) = 0 \tag{13.56a}$$

Letting $K2_k = K_k$, we get

$$K1_k = I - K2_k H_k = I - K_k H_k \tag{13.56b}$$

Thus Eq. (13.53) becomes

$$e_k(+) = (I - K_k H_k)\, e_k(-) + K_k V_k \tag{13.56c}$$

Equation (13.52c) then is

$$\hat{X}_k(+) = \hat{X}_k(-) + K_k[Z_k - H_k \hat{X}_k(-)] \tag{13.57a}$$

or, equivalently,

$$\hat{X}_k(+) = (I - K_k H_k)\hat{X}_k(-) + K_k Z_k \tag{13.57b}$$

At this point we emphasize a fundamental assumption: the state error and measurement noise are independent zero-mean processes. The argument goes that if there is a known bias, say, in the measurement noise, then this bias can be removed. It is the responsibility of the user to remove any bias in the measurements. Otherwise the Kalman algorithm is being used inappropriately. (In using a modification of the method of least squares for parameter estimation, the measurement bias can be assumed to be present and such bias is represented simply as another parameter to be estimated.)

By way of review there are four covariance matrices that enter the formulation, or

$$E[\boldsymbol{W}_i \boldsymbol{W}_k^{\mathrm{T}}] = Q, i = k$$
$$= 0, i \neq k \tag{13.58}$$

$$E[\boldsymbol{V}_i \boldsymbol{V}_k^{\mathrm{T}}] = R, i = k$$
$$= 0, i \neq k \tag{13.59}$$

$$E[\boldsymbol{e}_k(-)\boldsymbol{e}_k^{\mathrm{T}}(-)] = P_k(-) \qquad E[\boldsymbol{e}_k(+)\boldsymbol{e}_k^{\mathrm{T}}(+)] = P_k(+) \tag{13.60}$$

Equations (13.58) and (13.59) require that both the process noise $\boldsymbol{W}_k$ and the measurement noise $\boldsymbol{V}_k$ be *white* (infinite band width) noise. If we insert Eq. (13.56c) into Eq. (13.60) and apply the requirement that the error vector be independent of the noise vector, i.e.,

$$E[\boldsymbol{e}_k(-)\boldsymbol{V}_k^{\mathrm{T}}] = E[\boldsymbol{V}_k \boldsymbol{e}_k^{\mathrm{T}}(-)] = 0 \tag{13.61}$$

we get

$$P_k(+) = (I - K_k H_k)P_k(-)(I - K_k H_k)^{\mathrm{T}} + K_k R_k K_k^{\mathrm{T}} \tag{13.62}$$

Equation (13.62) relates the covariance matrix after measurement to the covariance matrix before measurement. The error covariance matrix $P_k(+)$ becomes a measure of how well the algorithm is estimating the state vector.

The remaining problem is the determination of the K-matrix. The other terms in Eqs. (13.62) and (13.57b) can be assigned numerical values. As Gelb points out, the gain matrix K is determined from the requirement that the trace of the error covariance matrix be minimized.[4] Let

$$J_k \doteq \mathrm{trace}[P_k(+)]$$

First we note the following identity:

$$\frac{\partial}{\partial A}[\mathrm{trace}(ABA^{\mathrm{T}})] = 2AB \tag{13.63}$$

where A is a symmetric matrix and B is conformable. If we now use the identity given by Eq. (13.63) in Eq. (13.62) and set the resulting expression to zero, we get after some manipulation

$$K_k = P_k(-)H_K^{\mathrm{T}}[H_k P_k(-)H_k^{\mathrm{T}} + R_k]^{-1} \tag{13.64}$$

Inserting this value for K_k into Eq. (13.62) gives

$$P_k(+) = (I - K_k H_k) P_k(-) \tag{13.65}$$

In the following example for the scalar case, let

$$P_k \to p^2 \qquad R_k \to \sigma_N^2 \qquad \text{and} \qquad H_k \to 1$$

The Kalman gain matrix follows from Eq. (13.64) as

$$K_k = p^2(p^2 + \sigma_N^2)^{-1} = \frac{p^2}{p^2 + \sigma_N^2}$$

Inserting the foregoing expression into Eq. (13.57b) gives after some manipulation

$$\hat{X}_K(+) = \left(\frac{\sigma_N^2}{p^2 + \sigma_N^2}\right)\hat{X}(-) + \left(\frac{p^2}{p^2 + \sigma_N^2}\right) Z_k$$

The new estimation of the state $\hat{X}_k(+)$ is thus shown to be the weighted sum of the previous estimation and the most recent measurement, Z_k. There are two extremes: the noise is very large, i.e., σ_N^2 is very large relative to the state uncertainty, or the noise is extremely small relative to the state uncertainty. These two conditions are given below:

$$\sigma_N^2 \gg p^2 \qquad \left(\frac{\sigma_N^2}{p^2 + \sigma_N^2}\right) \to 1 \qquad \left(\frac{p^2}{p^2 + \sigma_N^2}\right) \to 0$$

$$\sigma_N^2 \ll p^2 \qquad \left(\frac{\sigma_N^2}{p^2 + \sigma_N^2}\right) \to 0 \qquad \left(\frac{p^2}{(p^2 + \sigma_N^2)}\right) \to 1$$

In the first case (large measurement noise—small process noise) the best estimate is to use the last best estimate and to ignore the measurement. Conversely, in the second case (small measurement noise—large system uncertainty) only the measurement is used to estimate the state and the last best estimate is ignored. For most useful systems we can expect that operation is not near either of these extremes and that both the current measurement and the last best estimate would be appropriately weighted to provide the current best estimate.

Returning to the vector representation, we note that the state vector $\boldsymbol{X}$ is propagated between measurements through the transition matrix ϕ_k. We assume that $\hat{\boldsymbol{X}}_k(+)$, the current state estimate, must also be propagated between measurements (and estimations).

Equations (13.52d) and (13.52e) give the error vector before and after measurement-estimation as

$$\boldsymbol{e}_k(+) = \hat{\boldsymbol{X}}_k(+) - \boldsymbol{X}_k \qquad \boldsymbol{e}_{k+1}(-) = \hat{\boldsymbol{X}}_{k+1}(-) - \boldsymbol{X}_{k+1}$$

and from Eq. (13.52a)

$$X_{k+1} = \phi_k X_k + W_{k+1}$$

$$\hat{X}_{k+1}(-) = \phi_k \hat{X}_k(+)$$

so that

$$e_{k+1}(-) = \phi_k \hat{X}_k(+) - (\phi_k X_k + W_k)$$

Therefore,

$$P_{k+1}(-) = \phi_k P_k(+) \phi_k^{\mathrm{T}} + Q_k \tag{13.66}$$

where Q_k is the matrix given in Eq. (13.58). We can now summarize the Kalman algorithm:

1) Have available the measurement vector $\boldsymbol{Z}_k$ and evaluate the measurement matrix H_k and the system matrix A_k. That is,

$$Z_k = H_k X_k$$

$$\phi_k = I + \sum_{n=1}^{\infty} \frac{1}{(n!)} A_k^n \Delta t^n$$

2) Have available from previous estimation the state estimation vector and error covariance matrix, transitioned from the previous measurement, that is,

$$\hat{X}_k(-), \qquad P_k(-)$$

3) Calculate the Kalman gain matrix as

$$K_k = P_k(-) H_k^{\mathrm{T}} [H_k P_k(-) H_k^{\mathrm{T}} + R_k]^{-1}$$

4) Update the state estimation vector as

$$\hat{X}_k(+) = \hat{X}_k(-) + K_k [Z_k - H_k \hat{X}_k(-)]$$

or

$$\hat{X}_k(+) = (I - K_k H_k) \hat{X}_k(-) + K_k Z_k$$

5) Update the error covariance matrix, that is,

$$P_k(+) = (I - K_k H_k) P_k(-)$$

6) Transition state estimation to next measurement time, or

$$\hat{X}_{k+1}(-) = \phi_k \hat{X}_k(+)$$

7) Transition error covariance matrix to next measurement time as

$$P_{k+1}(-) = \phi_k P_k(+)\phi_k^{\mathrm{T}} + Q_k$$

8) Go to step 1 and repeat the process.
9) Print out

$$\hat{X}_{k+1}(+), \qquad P_{k+1}(+)$$

Now the state estimate is available, which is the purpose of the whole exercise. Before leaving this subject, we should make a point concerning step 7. In applying this step to transition the error covariance matrix, we must include the Q matrix representing the process noise. If this matrix is omitted, then the tacit assumption is made that the model of the process is perfect. Consequently, the matrix P gets smaller (say, the trace approaches zero). From step 3, the gain matrix also diminishes. Likewise, the second term in step 4 diminishes and the first term in step 4 approaches the identity matrix. The result is that measurements are ignored and the new estimate is equal to the old estimate transitioned forward. In essence the filter has *gone to sleep*. This is of course what should happen: if the model is perfect then there is no need for measurements and consequently no need for an estimation algorithm.

We must emphasize that the Kalman filter is linear, in that the transition matrix is derived from a set of first-order linear differential equations. However, many, if not most, processes are nonlinear. To extend the utility of such estimation algorithms to the nonlinear system, we must linearize the equations of motion about a nonlinear nominal or reference trajectory as was done in Eqs. (13.33) and (13.34). Two frequently encountered terms are *Linearized Filter* and *Extended Filter*. In Eq. (13.34) the system matrix A is the Jacobian given in Eq. (13.35a). These derivatives, which constitute the elements of matrix A are evaluated at the state variables of the nominal trajectory for the linear filter and at the state variables of the nominal trajectory corrected by the estimates for the extended filter. For the Linearized filter

$$\left.\frac{\partial f_i}{\partial X_j}\right|_{X=X_0}$$

and for the Extended filter

$$\left.\frac{\partial f_i}{\partial X_j}\right|_{X=X_0+\hat{X}}$$

Finally we should emphasize that when the state equations are linear, the Kalman filter algorithm is an optimal estimator in the least-squares sense. The extensions of the algorithm to a nonlinear process through either the Linearized or Extended adaptations are not necessarily optimum.

The effort discussed in the next section uses the Linearized filter. A discussion of the relative merits of each approach is given by Brown in Chapter 9 of Ref. 2. A summary of Kalman extensions is given in Ref. 10.

13.5 State Acquisition

In Section 13.3 we discussed briefly the propagation of the covariance matrix as associated with a re-entry vehicle subjected to initial velocity errors together with environmental uncertainties. In this section we will assume that the re-entry vehicle is capable of maneuvering. The model will require eight states for adequate representation. Although the capacity of the RV to maneuver is acknowledged, we assume that the actual maneuvers cannot be predicted; maneuvering therefore contributes to process uncertainty. We will assume that the re-entry vehicle lift is sequenced according to some preset schedule. This means that re-entry vehicle maneuvering does not reflect a response to actions of the defense. Nevertheless, whereas the lift program is preset, the ensuing motion of the RV appears to the ground observer to be a random process, a process in which successive observations are correlated.

Figure 13.8 indicates the essential geometry associated with the radar observation. Only the positional coordinates of the RV are available to the ground-based tracking system, even though complete description of the ballistic motion requires six states. Also, in this figure the lift space is spanned by two orthogonal unit vectors, ***UA*** and ***UB***.

The unit vector ***UA*** is defined as normal to the unit velocity vector ***UV*** and the unit vertical vector ***I***3.

$$UA = \left[\frac{X5}{(X4^2 + X5^2)^{1/2}}, \frac{-X4}{(X4^2 + X5^2)^{1/2}}, 0\right]^T \tag{13.67a}$$

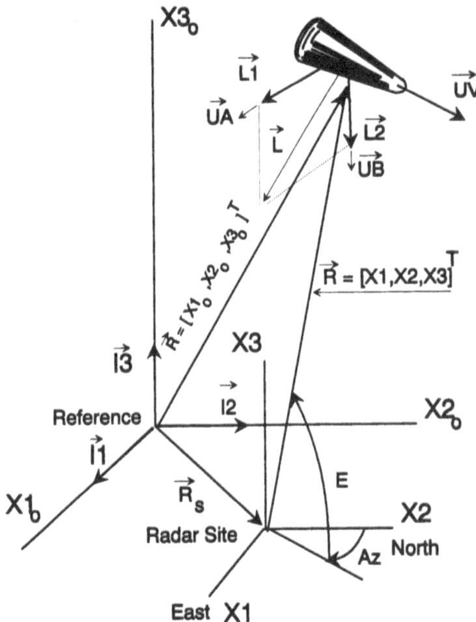

Fig. 13.8 Radar and tracking measurements.

The second unit vector, ***UB***, is defined as normal to both ***UA*** and ***UV***, that is,

$$\boldsymbol{UB} = \left[\frac{X4X6}{(X4^2+X5^2)^{1/2}V}, \frac{X5X6}{(X4^2+X5^2)^{1/2}V}, -\frac{(X4^2+X5^2)^{1/2}}{V}\right]^{\mathrm{T}} \tag{13.67b}$$

The components of both unit vectors are evaluated by the nominal trajectory at each measurement. In an engagement simulation, the components are evaluated at each integration step. Notice also in Fig. 13.8 that there are two additional states: the lift components $L1$ and $L2$, or the coefficients of lift in the ***UA*** and the ***UB*** directions, respectively.

It is not appropriate to treat noise associated with lift as a white (infinite band width) process. The infinite band width means that if we were to examine a system at a succession of intervals, the noise present at two successive intervals, regardless of how closely spaced the intervals, would be statistically independent. (By examining a system we mean making measurements of at least some of the states. In the modeling or simulation of a system, examination means having available the output from the state integrator, although only some such states may be used in a smoothing, predicting or estimating algorithm.)

We will use the term *time constant* as a measure of the time duration required by the re-entry body to change the lift setting, say, the time required for the maximum lift to reverse direction. We will assume that this time constant is large compared to the sampling period of a sequence of ground-based observations.

One way of analytically representing the Gauss-Markov (GM) process is with the autocorrelation function. For a GM process this function is

$$\phi = \sigma_M^2 e^{-\alpha\tau} \tag{13.68}$$

where τ is the correlation time and α is the reciprocal of the time constant, which, as we have pointed out, is a measure of how fast the RV can change lift, say, between the maximum value and the no-lift condition. We will return to this equation after we introduce some additional concepts.

First we must recognize that the variance of a zero-mean process is the square of the standard deviation σ and is equal to the second moment of the probability density function $f(a)$ as

$$\sigma^2 = \int_{-\infty}^{+\infty} a^2 f(a)\mathrm{d}a \tag{13.69}$$

For a maneuvering vehicle we will make use of the discrete/continuous probability density function $f(a)$, as suggested by Singer.[11] The application of lift is obviously a two-component problem: the lift coefficient must be assigned a value and a direction in the lift-space defined by unit vectors ***UA*** and ***UB*** of Eqs. (13.67).

Whereas lift is applied in a preset (deterministic) sequence, from the point of view of the ground-based observer lift is a stochastic process. Since lift cannot be adequately treated as a white noise process, it will be regarded as a first-order Gauss-Markov process. (White noise is the limit of a Gauss-Markov process with the covariance becoming very large and correlation time going to zero.)

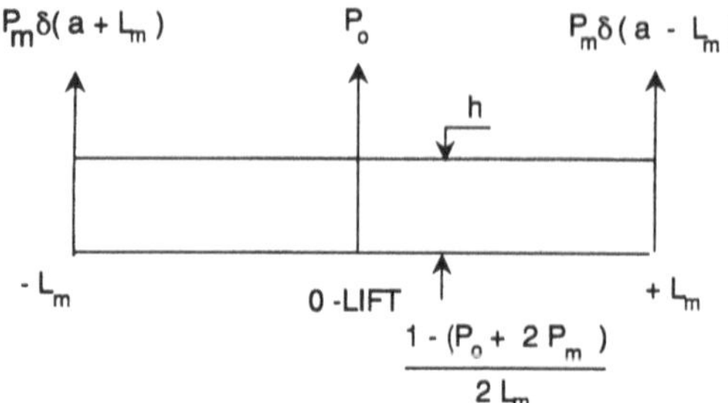

Fig. 13.9 Lift probability and density function.

Our first concern is how to represent the probability density function of Eq. (13.69). First, we note that there is a discrete probability P_m that the lift coefficient will be at the maximum positive value; there is the equal probability that the lift coefficient will be at the maximum negative value. There will be the additional (discrete) probability P_0 that there will be no lift. Finally, there is the continuous probability that the lift will be transitioning between positive, negative, and zero-lift conditions. We will use this information in setting the probability density functions. First, however, we must review the properties of the delta function, which is useful in combining discrete and continuous functions.

$$\int_{-\infty}^{+\infty} \delta(a - a^*)\mathrm{d}a = \int_{-L_m}^{+L_m} \delta(a - a^*)\mathrm{d}a = 1 \qquad -L_m \leq a^* \leq L_m$$

$$= 0 \qquad |a^*| > L_M$$

$$\int_{-\infty}^{+\infty} g(a)\delta(a - a^*)\mathrm{d}a$$

$$= \int_{-L_m}^{+L_m} g(a)\delta(a - a^*)\mathrm{d}a = g(a^*) \qquad -L_m \leq a^* \leq L_m$$

$$= 0 \qquad |a^*| > L_m$$

Now we may represent the probability density function as a composition of three discrete probabilities—P_m (at L_m), P_m (at $-L_m$), and P_0 (at zero lift)—and a continuous distribution of density h. We will see that assigning values to P_m and P_0 will set the value of h in the expression below:

$$f(a) = P_m\delta(a + L_m) + P_0\delta(a) + P_m\delta(a - L_m) + h \tag{13.70}$$

This probability density function is shown in Fig. 13.9.

Thus we have two conditions that must be met: first, the definition of the standard deviation (or variance here) in Eq. (13.71a); and second, in Eq. (13.71b) the probability of 1.0 that the lift will be found at the positive maximum, at the negative maximum, at the no-lift condition, or transitioning among these three conditions. These two conditions are described mathematically as

$$\sigma_m^2 = \int_{-\infty}^{+\infty} a^2[P_m\delta(a+L_m) + P_0\delta(a) + P_m\delta(a-L_m) + h]\mathrm{d}a \qquad (13.71a)$$

$$1 = \int_{-\infty}^{+\infty} [P_m\delta(a+L_m) + P_0\delta(a) + P_m\delta(a-L_m) + h]\mathrm{d}a \qquad (13.71b)$$

It is up to the analyst to set levels for P_m and P_0 based upon an assessment of the performance of the re-entry vehicle. Qualitatively, a re-entry vehicle with a low lift-to-drag ratio would be assigned a low value of P_m, as one might expect that the lift would be applied sparingly to minimize velocity loss due to induced drag. In any event, the values assigned to the discrete probabilities P_m and P_0 must be made independent of the estimating algorithm. As pointed out, the height h of the continuous part of the distribution is set by Eq. (13.71b) and given as

$$h = (1 - 2P_m - P_0)/2L_m \qquad (13.71c)$$

As a result, h can be eliminated from Eq. (13.71a) to provide the standard deviation of the entire maneuvering process, or

$$\sigma_m^2 = (L_m^2/3)(1 + 4P_m - P_0) \qquad (13.72)$$

The lift coefficient is regarded as a stochastic variable. It is described adequately as a first-order Gauss-Markov process. As a result, two additional states are added to the system, $L1$ and $L2$, one along the ***UA*** direction and the other along the ***UB*** direction. The additional state equations are

$$\frac{\mathrm{d}L_i}{\mathrm{d}t} = -\alpha L_i + W_i \qquad (13.73)$$

where

α = reciprocal of the Gauss-Markov time constant
W_i = white noise sequence, variance = σ_w^2
$\sigma_w^2 = 2\alpha\sigma_m^2$
$\sigma_m^2 = (L_m^2/3)(1 + 4P_m - P_0)$

The defense must select (estimate) values to assign to the lift coefficient L_m and to the probability that the re-entry vehicle is at the maximum/minimum and no-lift conditions. Also the ground-based observer (or the analyst, in the case of a simulation) must select some value for α, or the reciprocal of the time constant. As pointed out earlier, α is a measure of how quickly the maneuvering vehicle can transfer through the maximum angle of attack range, since lift is produced by varying the angle of attack.

There are other interesting adaptations of Kalman filtering to maneuvering vehicles. Bekir[12] proposes the sequence

$$\tilde{\mathbf{Z}} = \mathbf{Z} - \hat{\mathbf{Z}}$$

where Z is the measurement vector and $\hat{Z}$ is a measurement vector consistent with the estimated states. The covariance matrix of the sequence is

$$C = E[\tilde{Z}\tilde{Z}^{\mathrm{T}}] = HPH^{\mathrm{T}} + R$$

which is of course the *denominator* of the Kalman gain matrix. Obviously if $\tilde{Z}$ is large (by some standard), then the diagonal term, or trace, of C will be large. Thus, the Kalman gain will be small. Consequently, the measurements will be emphasized and the transitioned state estimates will be de-emphasized.

We will next rewrite the equations of motion that were originally given in Eqs. (13.30). The new set will include the two lift components, one in the ***UA*** direction ($L1$) and the other in the ***UB*** direction ($L2$) [Eqs. (13.67)], and the two additional differential equations that treat lift as a GM process. We have

$$\frac{\mathrm{d}X1}{\mathrm{d}t} = X4 \qquad \frac{\mathrm{d}X2}{\mathrm{d}t} = X5 \qquad \frac{\mathrm{d}X3}{\mathrm{d}t} = X6$$

$$\frac{\mathrm{d}X4}{\mathrm{d}t} = -K_D e^{-X3/H} VX4 + K_L \left[\left(\frac{X5}{V1}\right)V^2L1 + \left(\frac{X4X6}{V1}\right)VL2\right] e^{-X3/H}$$

$$\frac{\mathrm{d}X5}{\mathrm{d}t} = -K_D e^{-X3/H} VX5 + K_L \left[-\left(\frac{X4}{V1}\right)V^2L1 + \left(\frac{X5X6}{V1}\right)VL2\right] e^{-X3/H} \tag{13.74}$$

$$\frac{\mathrm{d}X6}{\mathrm{d}t} = -K_D e^{-X3/H} VX6 + K_L \left[-\left(\frac{V1}{V}\right)L2\right] e^{-X3/H} - g$$

$$\frac{\mathrm{d}X7}{\mathrm{d}t} = -\alpha X7 \qquad \frac{\mathrm{d}X8}{\mathrm{d}t} = -\alpha X8$$

where

$$V = (X4^2 + X5^2 + X6^2)^{1/2}$$

$$V1 = (X4^2 + X5^2)^{1/2}$$

$$K_D = \frac{\rho_s S C_D g}{2W} \qquad K_L = \frac{\rho_s S g}{2W}$$

$$X7 = L_1 \qquad X8 = L_2$$

Note that the following terms assume an exponential atmosphere:

$$K_D e^{-X3/H} \qquad K_L e^{-X3/H} \tag{13.75}$$

that is, the density varies exponentially with altitude (valid only for an isothermal atmosphere). Also note that in previous representations of the exponential atmosphere ρ_0 was the symbol for the corrected sea level density; however, here we are reserving "0" to indicate the nominal trajectory. Therefore ρ_s replaces ρ_0. Equations (13.74) are used in two ways: first, to generate the nominal trajectory and second, to form the Jacobian or the system matrix [Eqs. (13.35)].

When Eqs. (13.74) are used to generate the nominal trajectory, then the atmospheric terms of Eqs. (13.75) are replaced by

$$K_D e^{-X3/H} \rightarrow \frac{\rho S C_D g}{2W} \qquad K_L e^{-X3/H} \rightarrow \frac{\rho S g}{2W}$$

where the density ρ is taken from an atmosphere model. The next step is to form the various terms in the system matrix. A few of the terms might require some explanation.

Note that the K_D term contains the drag coefficient C_D. Because the development of lift carries an additional drag penalty (lift-induced, or just *induced*, drag), the drag coefficient depends upon the lift coefficient. In Chapter 9 we assumed for analytical purposes that the variation of C_D with C_L is adequately represented by a drag polar, i.e.,

$$C_D = C_{D_0}\left[1 + \left(C_L/C_L^*\right)^2\right] \tag{13.76}$$

In the present usage, we have

$$C_L = \left(L1^2 + L2^2\right)^{1/2}$$

Taking the derivative of C_D with respect to $L1$ or $L2$ and evaluating this derivative at either $L1 = 0$ or $L2 = 0$ would indicate that C_D is insensitive to the lift coefficient C_L. We can rewrite K_D as

$$K_D = \left(\frac{\rho S g C_{D_0}}{2W}\right)\left[1 + \left(\frac{C_L}{C_L^*}\right)^2\right] \tag{13.77a}$$

or, linearized, as

$$K_D = K_{D_0}\left(1 + \frac{1}{K_{D_0}}\frac{\partial K_D}{\partial C_L}\bigg|_{(C_L = \bar{m}C_L^*)} C_L\right) \tag{13.77b}$$

We note that

$$\frac{\partial K_D}{\partial C_L}\bigg|_{(C_L = \bar{m}C_L^*)} = \frac{\partial K_D}{\partial L1} = \frac{\partial K_D}{\partial L2} = \frac{\rho_s S g C_{D_0}\bar{m}}{W C_L^*} \tag{13.78}$$

where we have evaluated the derivative, not at $C_L = 0$ but at $C_L = \bar{m}C_L^*$. As we pointed out, evaluating the derivative at $C_L = 0$ would indicate an insensitivity of K_D (i.e., C_D) to $C_L(L1$ or $L2)$.

According to the drag polar representation of the drag dependency upon lift, at $C_L = C_L^*$

$$K_D \,|_{C_L = C_L^*} = 2K_D \,|_{C_L = 0}$$

For Eq. (13.77b) to be compatible with the preceding requirements, the factor $\bar{m}$ must equal 0.5. Here we will leave $\bar{m}$ as a parameter that can be set to some value in the formulation.

We now can write the coefficients of the system matrix as

$$\frac{\mathrm{d}}{\mathrm{d}t}\begin{bmatrix} X1 \\ X2 \\ X3 \\ X4 \\ X5 \\ X6 \\ X7 \\ X8 \end{bmatrix} = \begin{bmatrix} 0 & 0 & 0 & 1 & 0 & 0 & 0 & 0 \\ 0 & 0 & 0 & 0 & 1 & 0 & 0 & 0 \\ 0 & 0 & 0 & 0 & 0 & 1 & 0 & 0 \\ 0 & 0 & a_{43} & a_{44} & a_{45} & a_{46} & a_{47} & a_{48} \\ 0 & 0 & a_{53} & a_{54} & a_{55} & a_{56} & a_{57} & a_{58} \\ 0 & 0 & a_{63} & a_{64} & a_{65} & a_{66} & a_{67} & a_{68} \\ 0 & 0 & 0 & 0 & 0 & 0 & -\alpha & 0 \\ 0 & 0 & 0 & 0 & 0 & 0 & 0 & -\alpha \end{bmatrix}\begin{bmatrix} X1 \\ X2 \\ X3 \\ X4 \\ X5 \\ X6 \\ X7 \\ X8 \end{bmatrix} + \begin{bmatrix} W1 \\ W2 \\ W3 \\ W4 \\ W5 \\ W6 \\ W7 \\ W8 \end{bmatrix} \tag{13.79a}$$

or in compact form

$$\frac{\mathrm{d}\boldsymbol{X}}{\mathrm{d}t} = A\boldsymbol{X} + \boldsymbol{W} \tag{13.79b}$$

where $\boldsymbol{X}$ is the eight-state vector and $\boldsymbol{W}$ is the corresponding noise vector.

Let us write some of the nonzero terms in the system matrix of Eq. (13.79). For instance,

$$a_{43} = -\frac{K_D}{H}e^{-X3_0/H}V_0X4_0 = -\left(\frac{\rho C_{D_0} S g}{2WH}\right)V_0X4_0 \tag{13.80}$$

The term a_{53} has been given previously in Eq. (13.37a). Obviously the expressions for both a_{53} and a_{63} are identical to that given in Eq. (13.80) if $X4$ is replaced, respectively, by $X5$ and $X6$. Note that the actual density corresponding to that of the nominal trajectory is used rather than the density computed from the exponential model.

Continuing, we have

$$a_{44} = -\left(\frac{\rho S g C_{D_0}}{2W}\right)\left(V + \frac{X4^2}{V}\right) \tag{13.81}$$

The terms a_{55} and a_{66} are equivalent if $X4$ in Eq. (13.81) is replaced, respectively, by $X5$ and $X6$. Of course $X4$, $X5$, and $X6$ all appear in the lift vector [Eq. (13.74)], but the derivatives are evaluated at zero lift conditions ($L1 = L2 = 0$) so that these additional velocity contributions do not appear.

The following six terms are easily derived:

$$a_{45} = a_{54} = -\left(\frac{\rho S g C_{D_0}}{2W}\right)\frac{X4X5}{V} \tag{13.82a}$$

$$a_{46} = a_{64} = -\left(\frac{\rho S g C_{D_0}}{2W}\right)\frac{X4X6}{V} \tag{13.82b}$$

$$a_{56} = a_{65} = -\left(\frac{\rho S g C_{D_0}}{2W}\right)\frac{X5X6}{V} \tag{13.82c}$$

It was shown in Eq.(13.79a) that

$$a_{ij} = 0 \qquad i = 1,3,\ j = 1,3$$

$$a_{i+3,j} = \delta_{i+3,j} \qquad i = 1,3,\ j = 4,6$$

$$a_{17} = a_{18} = a_{27} = a_{28} = 0$$

$$a_{ij} = 0 \qquad i = 7,8,\ j = 1,6$$

Further,

$$a_{77} = a_{88} = -\alpha \tag{13.83a}$$

$$a_{78} = a_{87} = 0 \tag{13.83b}$$

Finally, we must consider the terms a_{47}, a_{48}, a_{57}, a_{58}, a_{67}, and a_{68}. For example, using the drag polar representation [Eq. (13.77)] we have

$$a_{47} = -2\left(\frac{\rho S C_{D_0} g}{2W}\right)\left(\frac{\bar{m}}{C_L^*}\right)VX4 + \left(\frac{\rho S g}{2W}\right)\frac{X5}{V1}V^2 \tag{13.84}$$

The remaining terms may be written by inspection from Eq. set (13.74).

The authors are aware of a certain inconsistency in the evaluation of the coefficients of the system matrix [Eq. (13.79a)]. Coefficients not linked to the additional lift states, $X7$ and $X8$, are evaluated at zero lift conditions, i.e., $C_D = C_{D_0}$. However, coefficients linked to states $X7$ and $X8$ include an induced drag term [Eq. set (13.77)].

With all the coefficients of the system matrix A (now for the eight states) available, it is possible to construct the transition matrix ϕ [Eq. (13.39)]. The error covariance matrix P [Eq. (13.21)] as well as the state estimate vector $\hat{\boldsymbol{X}}$ can then be propagated between measurements [Eq. (13.66)]. The final requirement is to relate the measurement vector $\boldsymbol{Z}$ to the state vector $\boldsymbol{X}$.

We will assume that the radar is capable of making three measurements: R, range; Az, azimuth; E, elevation (Fig. 13.8). Thus,

$$\boldsymbol{Z} = [R, Az, E]^{\mathrm{T}} = [Z1, Z2, Z3]^{\mathrm{T}} \tag{13.85}$$

Each of the three measurements may be related to the positional states, although through nonlinear functions.

$$R = \left(X1^2 + X2^2 + X3^2\right)^{1/2} \tag{13.86a}$$

$$Az = \tan^{-1}(X1/X2) \tag{13.86b}$$

$$E = \tan^{-1}\left[X3/\left(X1^2 + X2^2\right)^{1/2}\right] \tag{13.86c}$$

Since the filter algorithm is linear, the foregoing relationships must be linearized.

First we may rewrite Eqs. (13.86) in matrix form as

$$\boldsymbol{Z} = h(\boldsymbol{X}) = \begin{bmatrix} \left(X1^2 + X2^2 + X3^2\right)^{1/2} \\ \tan^{-1}(X1/X2) \\ \tan^{-1}\left[X3/\left(X1^2 + X2^2\right)^{1/2}\right] \end{bmatrix} + \boldsymbol{V} \tag{13.87}$$

where we have included the measurement noise vector $\boldsymbol{V}$ (assumed to be, at this point, a white noise sequence). Linearizing the preceding expressions, we get

$$\boldsymbol{Z} = \boldsymbol{Z}_0 + \tilde{\boldsymbol{Z}} = \boldsymbol{Z}_0 + \left.\frac{\partial h(\boldsymbol{X})}{\partial \boldsymbol{X}}\right|_{\boldsymbol{X}=\boldsymbol{X}_0} \tilde{\boldsymbol{X}} + \boldsymbol{V} \tag{13.88a}$$

or

$$\tilde{\boldsymbol{Z}}_k = H_k \tilde{\boldsymbol{X}}_k + \boldsymbol{V}_k \tag{13.88b}$$

Since $\tilde{\boldsymbol{X}}_k$ is an 8×1 vector and $\boldsymbol{Z}_k$ is a 3×1 vector, the observation matrix H_k is 3×8 matrix. If we let the matrix H be written in terms of its elements, we have

$$H_k = [H_{ij}]_k \tilde{\boldsymbol{X}}_k = \left[\frac{\partial h_i}{\partial X_j}\right]_k \tilde{\boldsymbol{X}}_k \tag{13.89a}$$

where

$$H_{11} = \frac{X1}{R} \qquad H_{12} = \frac{X2}{R} \qquad H_{13} = \frac{X3}{R}$$

$$H_{21} = \frac{X2}{X1^2 + X2^2} \qquad H_{22} = -\frac{X1}{X1^2 + X2^2} \qquad H_{23} = 0 \tag{13.89b}$$

$$H_{31} = -\frac{X1X3}{R^2(X1^2 + X2^2)^{1/2}} \qquad H_{32} = -\frac{X2X3}{R^2(X1^2 + X2^2)^{1/2}}$$

$$H_{33} = \frac{(X1^2 + X2^2)^{1/2}}{R^2}$$

and

$$H_{ij} = 0 \qquad i = 1, 3, \ j = 4, 8$$

R is defined in Eq. (13.86a). Obviously the elements H_{ij} of matrix H are evaluated at each measurement interval by the positional states of the nominal trajectory.

To use the preceding algorithm we must assign some initial value to the error covariance matrix P. We will assume here that initializing P must be done external to the implementation of the algorithm. The values assigned to the elements of P might come from an earlier process. For example, in modeling the trajectory of a re-entry vehicle from the end of boost, the error covariance matrix reflects errors in launch point as well as those errors accumulated during boost. Once P is initialized it is propagated according to Eq. (13.66).

The matrix Q_k must also be assigned some value. A separate modeling effort can show how well the system matrix A models the actual phenomena. For example, if the re-entry vehicle is assumed capable of a certain level of maneuvering, then the validity of the ballistic assumption can be evaluated. The expectation of the states taken two at a time over an interval equal to that between measurements should provide an estimation of Q_k.

Finally, the remaining term in the matrix Eq. (13.64) is the noise covariance matrix R_k. In our case this is a matrix of order (3×3), representing an error in measurement of the range and the azimuth and elevation angles. Appendix G contains a suggested method for arriving at the elements of the R_k matrix.

13.6 Fixed Coefficient Filtering

The computations associated with the Kalman filter are often too intensive for use in a system that must function in real time. We note that n is the order of the state vector (and of course the order of the state estimation vector). Therefore, there is the requirement of a matrix multiplication of order $n \times n$. If m is the order of the measurement vector, there is the additional requirement of the inversion of a square matrix of order $m \times m$. For a tracking radar, we might expect the state vector to be of order $n = 3$ (position) and the measurement vector to be also of order $m = 3$ (position). The Kalman filter remains very attractive from a performance point of view. Nevertheless, considerable effort has been expended to develop simplifications (or approximations) to the Kalman filter to reduce the computational loading without greatly degrading tracking performance.

One of the attractive properties of the Kalman filter is that it is a recursive fading memory algorithm: all measurements are used to some degree in constructing the gain matrix. However, as the time associated with any given measurement recedes from the time of the current measurement, these data are *forgotten* at an exponential rate. (In a batch processing algorithm, all data are processed together regardless of when such data were obtained.)

One approach to developing a non-Kalman recursive algorithm is first to assume some type of target motion, e.g., constant velocity. Next, develop the Kalman gain matrix associated with such a steady state process. Such an approach is often called "Wiener filtering." The underlying assumption is that there must be a steady state condition, an assumption that may not be realizable in practice. An improvement might be to use the Kalman algorithm first to compute a gain table or a set of gain matrices that change according to a schedule of some kind. However, a gain scheduling process may be little more than a trade-off of computational demands for storage requirements.

In this section we will discuss briefly a fixed coefficient filtering algorithm that is used in radar tracking, particularly when only positional estimates are available. This filter, commonly known as an $\alpha - \beta$ tracker, assumes a constant velocity target. A modification of this tracker, in which a constant acceleration is assumed, is identified as an $\alpha - \beta - \gamma$ filter. Only a summary can be given here; more extensive treatments may be found in the literature.[13–16]

First, we emphasize that only the position vector $\boldsymbol{X}$ is measured. However, both the position and velocity states must be estimated, as velocity is needed for transitioning the position estimation vector. In forming the estimation and transitioning equations of the $\alpha - \beta$ tracker we will use the notation of Ref. 12. Thus,

$$\hat{\boldsymbol{X}}_k(+) = \hat{\boldsymbol{X}}_k(-) + \alpha[\mathbf{Z}_k - \hat{\boldsymbol{X}}_k(-)] \tag{13.90a}$$

$$\hat{\boldsymbol{V}}_k = \hat{\boldsymbol{V}}_{k-1} + \frac{\beta}{qT}[\mathbf{Z}_k - \hat{\boldsymbol{X}}_k(-)] \tag{13.90b}$$

$$\hat{\boldsymbol{X}}_{k+1}(-) = \hat{\boldsymbol{X}}_k(+) + T\hat{\boldsymbol{V}}_k \tag{13.90c}$$

The first equation uses a fixed coefficient, i.e., α, to estimate the position state at the kth observation. Obviously we have a recursive filter where the fixed coefficient weights the innovation sequence $(\mathbf{Z}_k - \hat{\boldsymbol{X}}_k)$. The term T is the sampling period, and $\hat{\boldsymbol{X}}_k(+)$ and $\hat{\boldsymbol{V}}_k$ are the estimates of the position and velocity states at the kth observation.

The second equation provides an estimate of the velocity vector $\hat{\boldsymbol{V}}_k$. The fixed coefficient β weights the innovative sequence $[\mathbf{Z}_k - \hat{\boldsymbol{X}}_k(-)]$. The term q in Eq. (13.90b) is usually set to unity. However, if the probability of detection is less than unity, then $\hat{\boldsymbol{V}}_k$ remains unchanged and, until detection is resumed, it is this last value that continues to be used in Eq. (13.90c) to transition the positional estimation vector to the next observation time. To start the $\alpha - \beta$ filter we could initialize the positional and velocity vectors as follows:

$$\hat{\boldsymbol{X}}_1(+) = \mathbf{Z}_1 \qquad \hat{\boldsymbol{X}}_2(+) = \mathbf{Z}_2 \qquad \hat{\boldsymbol{V}}_2 = \frac{1}{T}[\hat{\boldsymbol{X}}_2(+) - \hat{\boldsymbol{X}}_1(+)] \tag{13.91}$$

In the Kalman filter we did not assume that the measurement vector $\mathbf{Z}$ and the state vector $\boldsymbol{X}$ were identical. In the first example we assumed that the state vector was of order six (three positions, three velocities) [Eq.(13.74)] while the measurement vector was of order three (range, azimuth, and elevation) [Eq.(13.86)]. Here we will continue to assume that we still have only three measurements. We can further assume that the measurement vector has been converted to the same form as the estimation vector [i.e., from spherical coordinates (range, azimuth, and elevation) to Cartesian coordinates].

We will simply state here one method for setting the $\alpha - \beta$ coefficients. First, we define the tracking index Λ as a dimensionless parameter proportional to the ratio of some measure of the positional uncertainty due to maneuvering to a measure of positional uncertainty due to sensor accuracy. That is,

$$\Lambda = T^2 \frac{\sigma_w}{\sigma_n} \tag{13.92}$$

where σ_w is the uncertainty in the maneuvering acceleration (meters/second^2) and σ_n is uncertainty in positional measurement (meters). Kalata[15] shows that for the optimal steady state solution, the coefficients α and β are given as

$$\beta^2 = \Lambda^2(1 - \alpha) \tag{13.93a}$$

$$\beta = 2(2 - \alpha) - 4(1 - \alpha)^{1/2} \tag{13.93b}$$

These two simultaneous equations may be iteratively solved for α and β.

The $\alpha - \beta$ tracker assumes a constant target velocity. The $\alpha - \beta$ tracker will therefore follow a ramp input (constant velocity) with no steady-state mean error. However, if the acceleration is other than zero, there will be a steady state error, as

$$\lim_{k \to \infty} [X_k - \hat{X}_k(+)] = \frac{(1 - \alpha)T^2}{\beta} \ddot{X} \tag{13.94}$$

We will now briefly consider the $\alpha - \beta - \gamma$ tracker. As one might expect, this algorithm provides a zero steady state error for a constant acceleration target. We will give the estimation and transitioning for this algorithm, paralleling Eqs. (13.90). Thus,

$$\hat{\boldsymbol{X}}_k(+) = \hat{\boldsymbol{X}}_k(-) + \alpha[\mathbf{Z}_k + \hat{\boldsymbol{X}}_k(-)] \tag{13.95a}$$

$$\hat{\boldsymbol{V}}_k = \hat{\boldsymbol{V}}_{k-1} + T\hat{\boldsymbol{A}}_{k-1} + \frac{\beta}{qT}[\mathbf{Z}_k - \hat{\boldsymbol{X}}_k(-)] \tag{13.95b}$$

$$\hat{\boldsymbol{A}}_k = \hat{\boldsymbol{A}}_{k-1} + \frac{\gamma}{(qT)^2}[\mathbf{Z}_k - \hat{\boldsymbol{X}}_k(-)] \tag{13.95c}$$

$$\hat{\boldsymbol{X}}_{k+1}(-) = \hat{\boldsymbol{X}}_k(+) + T\hat{\boldsymbol{V}}_k + (T^2/2)\hat{\boldsymbol{A}}_k \tag{13.95d}$$

We can now state the three equations relating the $\alpha - \beta - \gamma$ coefficients, using the tracking index parameter defined earlier in Eq. (13.92).

$$\Lambda^2 = \frac{\gamma^2}{4(1-\alpha)} \tag{13.96a}$$

$$\beta = 2(2-\alpha) - 4\sqrt{1-\alpha} \tag{13.96b}$$

$$\gamma = \frac{\beta^2}{\alpha} \tag{13.96c}$$

References

[1]Wylie, C. R., Jr., *Advanced Engineering Mathematics*, 3rd ed., McGraw-Hill, Hightstown, NJ, 1966, Chap. 11.

[2]Strang, G., *Introduction to Applied Mathematics*, Addison-Wesley, Reading, MA, 1986.

[3]Howard, E. R., Houston, W. B., and Modesitt, M. B., "Generalized Error Analysis Program (GEAP)," Singer Company, Kearfott Division, Little Falls, NJ, 1976.

[4]Gelb, A. (ed.), *Applied Optimal Estimation*, MIT Press, Cambridge, MA, 1974, Chap. 4.

[5]ADVANCED CONTINUOUS SIMULATION LANGUAGE User Guide, Mitchell and Gauthier Associates, Concord, MA, 1985.

[6]Brown, R. G., *Introduction to Random Signal Analysis and Kalman Filtering,* Wiley, Somerset, NJ, 1983, Chap. 5.

[7]Ramesh, A. V., Senol, U., and Garba, J. A., "Computational Complexities and Storage Requirements of Some Riccati Equation Solvers," *Journal of Guidance, Control, and Dynamics*, Vol. 12, No. 4, July–Aug. 1989, pp. 469–479.

[8]Kleinman, D. L., "On the Iterative Techniques for Riccati Equation Computations," *IEEE Transactions on Automatic Control*, Vol. AC-13, 1968, pp. 114–115.

[9]Blackburn, T. R., "Solutions of the Algebraic Matrix Riccati Equation via Newton-Raphson Iteration," *AIAA Journal*, Vol. 6, May 1968, pp. 951–953.

[10]Kailath, T., "A View of Three Decades of Linear Filtering Theory," *IEEE Transactions on Information Theory,* Vol. IT-20, No. 2, March 1976, pp. 146–181.

[11]Singer, R., "Estimating Optimal Tracking Filter Performance for Manned Maneuvering Targets," *IEEE Transactions on Aerospace and Electronic Systems*, Vol. AES-6, No. 4, July 1970, pp. 473–483.

[12]Bekir, E., "Adaptive Kalman Filter for Tracking Maneuvering Targets," *Journal of Guidance and Control,* Vol. 6, No. 5, Sept.–Oct. 1983, pp. 414–416.

[13]Blackman, S. S., *Multiple-Target Tracking with Radar Applications*, Artech House, Norwood, MA, 1986.

[14]Simpson, H. R., "Performance Measures and Optimization Conditions for a Third Order Sample Data Tracker," *IEEE Transactions on Automatic Control,* Vol. AC-8, April 1963, pp. 182–183.

[15]Kalata, P. R., "The Tracking Index: A Generalized Parameter for $\alpha - \beta$ and $\alpha - \beta - \gamma$ Target Trackers," *IEEE Transactions on Aerospace and Electronic Systems,* Vol. AES-20, March 1984, pp. 174–182.

[16]Lefferts, R. E., "Adaptive Correlation Regions for Alpha-Beta Tracking Filters," *IEEE Transactions on Aerospace and Electronic Systems,* Vol. AES-17, Nov. 1981, pp. 738–747.

14
Inertial Guidance

14.1 Introduction

A ballistic re-entry vehicle (BRV), regardless of its intentions, may be compared to a long-range artillery shell. The impact point depends upon the conditions of velocity and position at trajectory initialization. These initial conditions have been designated as the V-γ *conditions*, indicating the setting of velocity magnitude and direction. Of course altitude and geographic location are also part of the initialization process. In Chapter 6 we showed that the subsequent trajectory and particularly the impact point are extremely sensitive to the conditions at trajectory initiation. Of course uncertainties in defining the gravitational field and the atmospheric medium (winds, density, etc.) are of first-order significance in limiting the accuracy of impact point prediction.

Even with extensive gravitational and atmospheric models it may not be possible to meet impact accuracy requirements without some kind of in-flight trajectory "correction." In order to steer out errors, from whatever source, the re-entry vehicle must be able to solve the navigation problem using some kind of on-board navigation equipment. For most re-entry vehicles as well as many atmospheric vehicles, the primary navigation system is some form of inertial navigation. Other navigational procedures may be necessary to update the inertial navigator; however, in this chapter we will concentrate on positional determination using inertial navigation.

To put inertial navigation in perspective, let us first consider its predecessor, celestial navigation. Celestial/chronometer navigation determined position at discrete and widely separated intervals to a precision of something like one second of arc (about 30 or 40 meters). A magnetic compass, together with a speed estimator (ship's log in early maritime application), provides a continuous estimate of position between measurements—a procedure known as *dead* (or *deduced*) *reckoning*. However, the accuracy of celestial navigation in a maritime application is about 1 nautical mile because of a variety of error sources. Inertial navigation provides essentially the function that was served by dead reckoning: a continuous estimation of position and velocity and possibly vehicle attitude.

In this chapter we will discuss various implementations of inertial navigation, the instruments, the computational algorithms, if any, and error analysis, or the limitations imposed on the navigator by the instruments. Inertial navigation is for the most part driven by manufacturing technology and as such has not had widespread academic attention.* Consequently, there are few texts on the subject. One of the most complete books on the background of inertial navigation is that by Britting.[1] However, this book is 20 years old and, therefore, the discussion of instruments is entirely dated. A comprehensive guide to the literature associated with strapdown systems is a paper by Garg et al.[2] The most current survey of inertial measurements has been provided by Savage in various papers.[3–5] Therefore, we will avoid here any attempt at an extensive listing of references.

14.2 Implementations

For convenience we will separate the angular velocity of an aerospace vehicle into two distinct categories: the angular rate of the vehicle relative to some Earth-local or navigation frame, i.e., $\boldsymbol{\omega}_{b/N}$, and the angular rate of the navigation frame relative to inertial space, i.e., $\boldsymbol{\omega}_{N/I}$. (See Appendix H for a definition of the various major reference frames.) Let us first define the navigation and inertial frames.

A convenient navigation frame has one axis aligned with the negative of the local normal to the geoid; this axis is designated as the D-axis. The second axis is directed north (N-axis) and the third axis, directed to the east (E-axis), completes the triad. Another convenient frame is the Earth-centered inertial (ECI) frame, which has one axis along the north-south pole line; the other two are arbitrary at least for now. The inertial frame shares the Earth's linear translation, but not its rotation. (We are ignoring the acceleration effects associated with the translational motion of the center of the Earth which is of course orbital motion about the Sun.)

The important point to be made is that the angular rate of the vehicle relative to the navigation frame can be several orders of magnitude greater than the angular rate of the navigation frame relative to inertial space. Nevertheless, we will here sum the two rates; in most implementations the calculations associated with each rate are carried out in separate computer elements. That is,

$$\boldsymbol{\omega}_{b/I} = \boldsymbol{\omega}_{b/N} + \boldsymbol{\omega}_{N/I} \tag{14.1}$$

where $\boldsymbol{\omega}_{b/N}$ is the rotation rate of the body relative to the navigation frame and $\boldsymbol{\omega}_{N/I}$ is the rotation rate of the navigation frame relative to inertial space.

If an instrument cluster (a grouping of accelerometers and gyroscopes) is mounted on a platform that is isolated from vehicle angular motion relative to the navigation frame, then it must exist in an angular environment set by the angular rate of the navigation frame relative to inertial space, $\boldsymbol{\omega}_{N/I}$. On the other hand, if the instrument cluster is mounted directly to the vehicle structure the instruments must operate in the much more demanding angular rate environment set by the rotation rate of the body relative to inertial space, $\boldsymbol{\omega}_{b/I}$.

---

*There is at least one notable exception: much of the pioneering work in the development of inertial navigation was carried out by the MIT Instrumentation Laboratory, now the independent Charles Stark Draper Laboratory.

Early inertial navigation systems had five or more gimbals. The outer three gimbals isolated the instrument cluster from the rotation of the body relative to the navigation frame. The next inner gimbal indicated latitude; and the innermost gimbal, longitude. This innermost gimbal was torqued at the Earth's rate. The two inner gimbals were then essentially analog computers providing geographical position.

A simple error analysis shows the shortcomings of the five-gimbal system beyond the obvious mechanical complexities. Assume that the goal of the navigator is a 1.0 nautical mile accuracy for the system. If the range angle θ is taken to the extreme of 180 degrees (note: there are 60 nautical miles per degree), then

$$\epsilon\theta \approx 1.0 \text{ nautical mile}$$

or

$$\epsilon = \frac{1.0}{\theta} = \frac{1.0}{180.0 \times 60.0} = \frac{1.0}{10800} \approx 1.0 \times 10^{-4}$$

or ϵ must represent one part in 10,000 or 100 ppm, a precision that is beyond the capability of an analog computer.

As inertial navigation technology progressed, i.e., became more accurate, three gimbals were retained for motion isolation, but the inner two gimbals were omitted. The computations for longitude and latitude were then performed by a digital computer identified as the navigation computer. If gimbal removal is to be taken as an indication of technological progress, then the "ultimate" inertial system would have no gimbals: the so-called *strapdown* system. In the strapdown system, the instruments are mounted directly to the vehicle structure, although perhaps through vibration isolators.

The removal of all gimbals greatly reduces the mechanical complexity of the inertial navigator. However, gimbal removal places two important burdens on the navigator. First, we note that the gimbals isolate the cluster from rotation relative to inertial space, $\boldsymbol{\omega}_{b/I}$ or, in some implementations, rotation relative to the navigation frame, $\boldsymbol{\omega}_{b/N}$. A transformation computer must now be in place to provide the attitude of the cluster or, equivalently, the body relative to either the navigation or inertial frame and do so in real time. Second, the dynamic environment in which the instruments must operate is much more demanding for the strapdown system than it is for the platform. Since either $\boldsymbol{\omega}_{b/N}$ or $\boldsymbol{\omega}_{b/I}$ can be several orders of magnitude greater than $\boldsymbol{\omega}_{N/I}$, the required bandwidth of strapdown instruments must be far greater than is necessary in the platform implementation.

The computational requirements of the strapdown system seem not to be a great obstacle to gimbal removal; however, the bandwidth requirement placed on the strapdown instrument is the reason that the platform implementation remains attractive in certain applications. For a re-entry vehicle spinning at about 500 degrees per second, the dynamic environment may be too demanding for any currently available instrument. We should point out that there are two major implementations of platform systems: the platform that (ideally at least) remains aligned with respect to an inertial frame (space-stabilized) and the platform that is continually aligned with respect to the local navigational frame (locally stabilized). The strapdown or gimbal-less system is obviously the direction of technological development. The strapdown system has more flexibility in packaging, is easier to maintain and check out, has increased reliability and reduced power demands. As a consequence strapdown systems should be less costly to maintain. Presumably adding computational capacity is less costly than fabricating gimbals.

As we pointed out, the drawback to the strapdown system arises primarily from the dynamic environment in which the instruments must function. For example, assume that a gyro cluster can have a navigation error no greater than 3.0 nautical miles per hour, or equivalently 3 arc-min/h or 0.05 deg/h. If the gyro were attached to a re-entry vehicle having a rotational rate of 500 deg/s, then the required precision of the instrument would be

$$[0.05/(500.0 \times 3600)] = 1/(3.6 \times 10^7)$$

or one part in 3.6×10^7.

An alternative means to examine dynamic range requirements is to set the total angle through which the vehicle rotates. A re-entry vehicle, in a not too extreme case, might rotate through 1×10^6 degrees. If the gyroscope instrumenting the axis of rotation must meet the performance requirement of 0.05 deg/h (drift rate of a navigational-quality platform system), then the dynamic range must be

$$0.05/(1.0 \times 10^6) \approx 1/(20 \times 10^6)$$

or one part in 2×10^7.

The preceding comments refer only to performance goals. However, the operation of the instruments in a severe dynamic environment means that there will be rotation-induced errors present. Also, in a platform gyro misalignment errors have a negligible effect because platform rates are orders of magnitude less than vehicle rates.

For some applications where a high rotation rate about one axis would put excessive dynamic range demands upon the instrumenting gyro, a hybrid system called a roll-isolated-platform (RIP) might be used. In this implementation a single gimbal prevents the high rotational rate of the vehicle from penetrating the instrument cluster. For the remainder of this chapter we will assume that the platform and strapdown systems remain distinct. Figures 14.1 and 14.2 summarize the basic operational flow of each implementation.

In Fig. 14.1 we have indicated a *true* inertial platform—the space stabilized platform. In this realization the platform is set to some initial orientation and then (ideally) it maintains this orientation with respect to inertial space throughout the trajectory. Of course spurious or noncompensated gyro drifting causes the platform to deviate from the intended orientation. Since accelerometers are mounted on the platform, their output is the resolution of the specific force vector in inertial space. The gravitational acceleration must be added vectorially to the specific force $\boldsymbol{f}^I$ to obtain $\boldsymbol{a}^I$, the acceleration relative to iner-

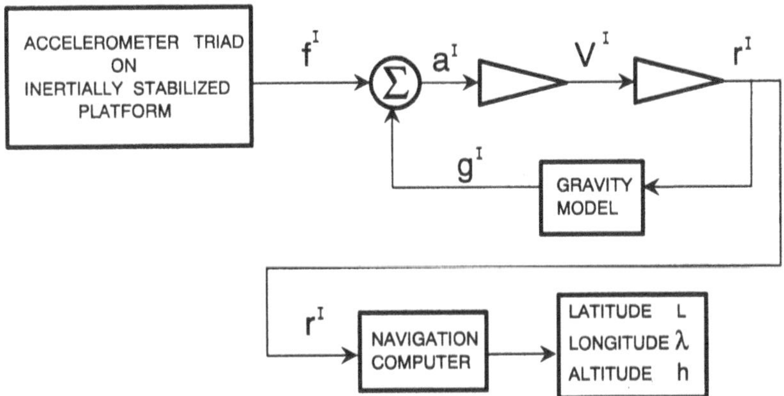

Fig. 14.1 Space stable inertial platform schematic.

tial space. This vector is integrated once for velocity and again for position with components in the inertial frame. The navigation computer then converts the position vector $\boldsymbol{r}^I$, into the more meaningful Earth-fixed components: L, latitude; λ, longitude; h, altitude.

Inertial navigation has been called a *bootstrap* system in that the outputs of L, λ, and h must be returned to the system so as to enter the gravity model and to construct the local to inertial frame DCM. Knowledge of the vehicle's attitude in the navigation frame, say, is not available except through gimbal resolvers.

A variation on the stable platform method is to carry out the computations in a local geographic or navigational frame. In this method, commonly used in the navigation of atmospheric vehicles, the platform is torqued such that one axis is

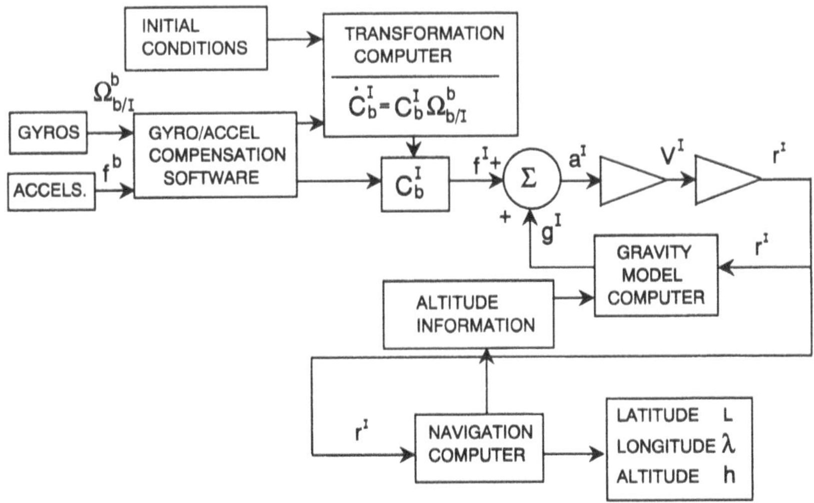

Fig. 14.2 Strapdown schematic.

continuously aligned with the local normal to the geoid. The platform torque rate $\boldsymbol{\omega}^{p}_{N/I}$ is the rotation rate of the navigation frame relative to inertial space, resolved in the platform coordinate system, or

$$\boldsymbol{\omega}^{p}_{N/I} = \boldsymbol{\omega}^{p}_{N/e} + C^{p}_{e}\boldsymbol{\omega}^{e}_{e/I} \tag{14.2}$$

All terms on the right except $\boldsymbol{\omega}^{e}_{e/I}$, which is a constant, must come from the navigation computer. The e-frame is of course the Earth-fixed reference frame. In some implementations, one of the horizontal axes is maintained in a north direction, or equivalently two axes are maintained in the local meridian. A common variation is to relax the north-pointing or preferred direction requirement of either horizontal axis. The local level implementation has not found extensive application to re-entry vehicles and so will not be discussed further. Schmidt has provided a more extensive summary of the various implementations.[6]

Although not universally applied to re-entry vehicle navigation, the strapdown system without doubt will find increasing application as instruments improve. At the present time, however, for re-entry vehicles rolling through a total of 1.0 million degrees, a gyro in the strapdown implementation may not be able to equal the performance of the gimballed (platform) system.

14.3 Instruments (Sensors)

In this section we will discuss a variety of sensors that have or may have application to the inertial navigation of re-entry vehicles. Of course space does not permit an extensive discussion of the engineering details of these devices. Consequently this discussion will be largely qualitative.

The following is a listing of the sensors that must be seriously considered for application to inertial guidance:

Pendulous integrating torque-to-balance accelerometers (PIA)
Single-degree-of-freedom floated rate-integrating gyro (SDFG)
Dry-tuned rotor gyro (DTRG)
Ring laser gyro (RLG)

Pendulous Integrating Accelerometer (PIA)

Contrary to their name, accelerometers do not measure acceleration. In free fall (at 1-g acceleration) an accelerometer will read nothing. On the other hand, on a bench at zero acceleration the device will measure $1g$. An accelerometer measures the difference between inertial kinematic acceleration $\boldsymbol{a}$, and gravitational acceleration $\boldsymbol{g}$ (weight divided by mass or gravitational specific force). The difference is called specific force $\boldsymbol{f}$ and is essentially the sum of all contact forces divided by mass, or

$$\boldsymbol{f} = \boldsymbol{a} - \boldsymbol{g} \tag{14.3}$$

For free fall in a vacuum, clearly $\boldsymbol{a} = \boldsymbol{g}$ and therefore $\boldsymbol{f} = 0$. For the accelerometer at rest in a gravitational field, $\boldsymbol{a} = 0$ so $\boldsymbol{f} = -\boldsymbol{g}$.

Conceptually the simplest accelerometer is a spring/mass/dashpot arrangement that will indicate specific force by the displacement of the seismic mass. A more accurate seismic mass accelerometer would be a balance-to-null device, where the current required to bring the mass to null would be used as a measure of the component of specific force along the axis of sensitivity. Balance-to-null instruments are potentially more accurate than displacement instruments in that they are nearly insensitive to parameter degradation (e.g., changes in spring stiffness).

An inertial grade accelerometer uses balance-to-null circuits. However, the device itself resembles a pendulum rather than a spring/mass instrument. A pendulous accelerometer is shown in Fig. 14.3. The PIA consists of a delicate flexure-hinged pendulum. Some designs have the pendulum suspended in a gas-filled medium (dry), others in a fluid medium (wet). We will not discuss the relative merits of dry versus wet except to note that the wet accelerometer is more complex to manufacture and service but does provide partial buoyancy to relieve hinge loads.

As shown in Fig. 14.3 three axes must be defined for an operational description of the PIA. The XI-axis is taken as the input axis, the YO-axis as the output or hinge axis and the ZP-axis as the pendulous axis. As the case is subjected to an acceleration component along the XI-axis, the pendulum rotates about the YO-axis due to the mass offset from the hinge axis. As the pendulum rotates the YO-axis remains fixed, but both the XI-axis and the ZP-axis rotate. Thus, we need to define a set of reference axes—XR, YR, and ZR; the reference axes remain case fixed. When there is a component of specific force along the XI-axis, the XI, ZI axes develop an angular displacement A_g relative to the reference axes.

An electrical pick-off senses the off-null angle A_g. An electro-magnetic torque generator then returns the pendulum to null (torque is provided by a coil and a case-mounted permanent magnet). A measure of torquer current then provides a signal that is equivalent to the input acceleration. This signal is formed into current-time pulses that are equivalent to the time integral of the sensed acceleration (specific force). Since the output of the inertial-grade accelerometer is acceleration integrated over a fixed time interval, i.e., velocity increments, the device is sometimes called a *velocity meter*. Further details are available from several sources. Savage provides a comprehensive overview.[3–5]

Two intrinsic error sources are associated with accelerometers. The first error source is due to the angular displacement of the input axis relative to the reference axis. Therefore keeping the angle A_g small minimizes the cross-coupling effect that occurs because the input axis XI has rotated through a finite angle from the input reference axis XR. However, if there is a component of specific force along both the input and pendulous axes XR and ZR, respectively, there will be a spurious change in the sensed specific force, or

$$\Delta f_{IA} \approx f_{PA} A_g \tag{14.4}$$

In other words, if A_g is nonzero, a specific force component along the pendulous axis will also cause a rotation about the hinge axis and hence will be

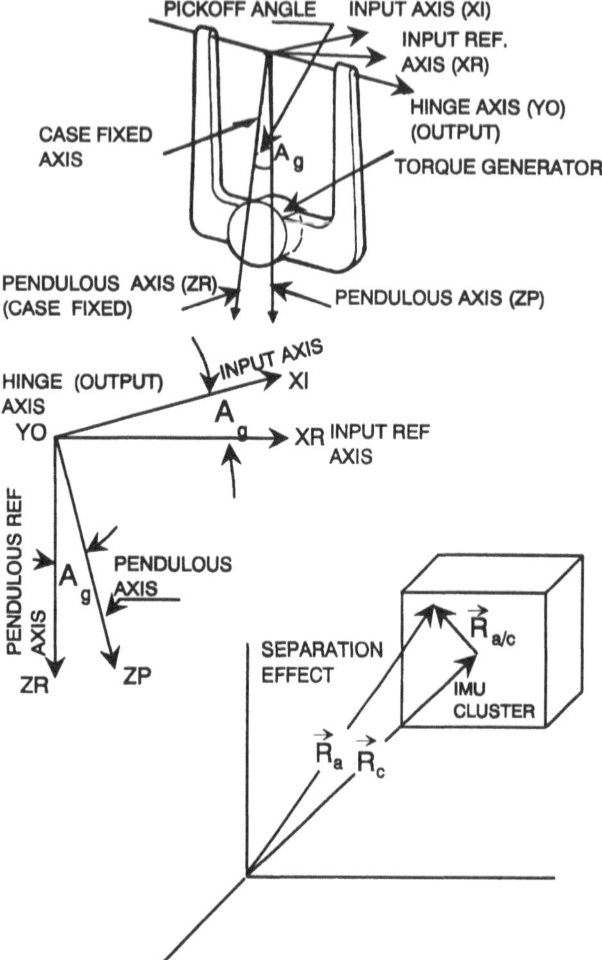

Fig. 14.3 Electrically servoed (torque-to-balance) pendulous integrating accelerometer.

indistinguishable from a component of specific force along the XI or input axis. We will see that a similar cross-coupling effect occurs for mechanical gyros.

The second intrinsic error source experienced by accelerometers, the finite separation effect, does not occur for gyros. When a strapdown IMU (inertial measurement unit) cluster rotates, all instruments are subjected to the same rotation rate but not the same acceleration. From Fig. 14.3 we sum $\boldsymbol{R}_c$, the vector distance of the cluster center from an inertial reference, with $\boldsymbol{R}_{a/c}$, the vector distance to a representative accelerometer from the center, as

$$\boldsymbol{R}_a = \boldsymbol{R}_{a/c} + \boldsymbol{R}_c \tag{14.5}$$

Assuming that the cluster is rotating relative to inertial space at a rate $\boldsymbol{\omega}_{b/I}$, we obtain the derivative of $\boldsymbol{R}_a$ with respect to time as

$$\left.\frac{\mathrm{d}\boldsymbol{R}_a}{\mathrm{d}t}\right|_I = \left.\frac{\mathrm{d}\boldsymbol{R}_c}{\mathrm{d}t}\right|_I + \left.\frac{\mathrm{d}\boldsymbol{R}_{a/c}}{\mathrm{d}t}\right|_I = \left.\frac{\mathrm{d}\boldsymbol{R}_c}{\mathrm{d}t}\right|_I + \left.\frac{\mathrm{d}\boldsymbol{R}_{a/c}}{\mathrm{d}t}\right|_b + \boldsymbol{\omega}_{b/I} \times \boldsymbol{R}_{a/c}$$

$$= \left.\frac{\mathrm{d}\boldsymbol{R}_c}{\mathrm{d}t}\right|_I + \boldsymbol{\omega}_{b/I} \times \boldsymbol{R}_{a/c} \tag{14.6}$$

where $\mathrm{d}\boldsymbol{R}_{a/c}/\mathrm{d}t|_b = 0$ because of rigid body restrictions. The acceleration follows as

$$\left.\frac{\mathrm{d}^2\boldsymbol{R}_a}{\mathrm{d}t^2}\right|_I = \left.\frac{\mathrm{d}^2\boldsymbol{R}_c}{\mathrm{d}t^2}\right|_I + \boldsymbol{\omega}_{b/I} \times (\boldsymbol{\omega}_{b/I} \times \boldsymbol{R}_{a/c}) + \frac{\mathrm{d}\boldsymbol{\omega}_{b/I}}{\mathrm{d}t} \times \boldsymbol{R}_{a/c} \tag{14.7}$$

The second and third terms on the right in Eq. (14.7) are the source of the finite separation effect; that is, the centrifugal effect is

$$\boldsymbol{\omega}_{b/I} \times (\boldsymbol{\omega}_{b/I} \times \boldsymbol{R}_{a/c})$$

and the tangential acceleration effect is

$$\frac{\mathrm{d}\boldsymbol{\omega}_{b/I}}{\mathrm{d}t} \times \boldsymbol{R}_{a/c}$$

We can use the following representative values to estimate both of these effects:

$$|\boldsymbol{\omega}_{b/I}| = 3.0 \text{ radians/s}$$

$$|\boldsymbol{R}_{a/c}| = 0.05 \text{ m}$$

The centrifugal error can be calculated as

$$\omega_{b/I}^2 R_{a/c} = (9.0)(0.05) = 0.45 \text{ m/s}^2 \approx 5.0 \times 10^{-2} g$$

The tangential acceleration error can be best estimated as a velocity error, or

$$\int_0^t a \,\mathrm{d}t = \int_0^t \frac{\mathrm{d}\omega_{b/I}}{\mathrm{d}t} R_{a/c} \,\mathrm{d}t = (3.0)(0.05) = 0.15 \text{ m/s}$$

There are various other error sources associated with an accelerometer. If one assumes that a cubic polynomial relates indicated acceleration to the output signal, uncertainties in the coefficients of the polynomial can lead to an error in the output. Also there are misalignments between the input, output, and pendulous axes as well as misalignments in the instrument as installed. Thus, we have

$$\delta f_I = B + K_{SF} f_I + K_{DF} f_I + K_{2I} f_I^2 + K_{3I} f_I^3 + K_{IP} f_I f_p + K_{IO} f_I f_O + \delta_{IP} f_P - \delta_{IO} f_O \tag{14.8}$$

The terms used in the preceding expression are defined as follows:

B = bias
K_{SF} = scale factor error
K_{DF} = scale factor dissymmetry
K_{2I} = quadratic scale factor
K_{3I} = cubic scale factor
K_{IP} = input-pendulous coupling
K_{IO} = input-output coupling
δ_{IP} = pendulous-to-input axes misalignment
δ_{IO} = output-to-input axes misalignment

Note in the above coefficients that the subscript "I" means "input" not "inertial."

Single-Degree-of-Freedom Floated Rate-Integrating Gyro

The single-degree-of-freedom floated rate-integrating gyro (SDFG) is the seminal inertial-grade gyro. Although this device is still in production and is being continually refined, it has to a large extent been replaced by other gyros—both mechanical and optical—as a high-performance instrument.

A functional illustration of the typical SDFG is given in Fig. 14.4. The instrument case contains a hermetically sealed cylindrical tube called a *float*, which in turn contains the spinning inertial wheel or rotor. The float rotates about a single axis relative to the gyro case. The axis of rotation of the spinning rotor is designated as the SA or spin axis. The axis of rotation of the float is called the OA or output axis. The axis normal to both the SA- and OA-axis is the IA or input axis.

The term float is used in describing the cylinder containing the rotor because this container is submerged in a fluid medium. This fluid serves two purposes. The primary function is to provide a viscous damping moment about the OA-axis. This damping moment is proportional to the rate of change of the relative angular displacement of the float relative to the gyro case. A secondary function is to provide a buoyant force to reduce or eliminate bearing support loads on the OA-axis at the case.

The fundamental mathematical description of gyro operation is given by a vectorial relationship among the applied moment $\boldsymbol{M}$, the angular velocity of the float relative to inertial space $\boldsymbol{\omega}_{F/I}$, and the angular momentum of the wheel $\boldsymbol{H}$, or

$$\boldsymbol{M} = \boldsymbol{\omega}_{F/I} \times \boldsymbol{H} \tag{14.9}$$

As shown in Fig. 14.4, the float has only a single degree of freedom of rotational motion relative to the case. Since the wheel's angular momentum is nominally normal to the case/float axis (OA) of rotation, only the component of angular velocity normal to both the wheel's angular momentum vector and the

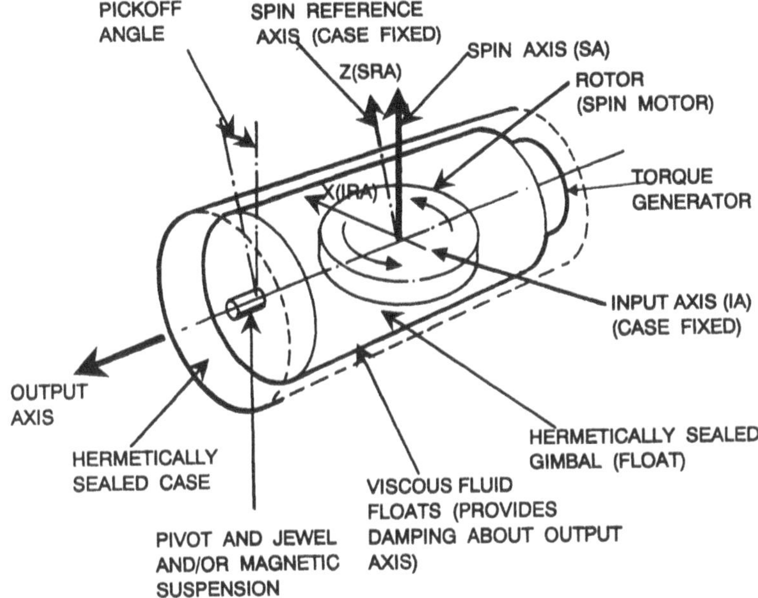

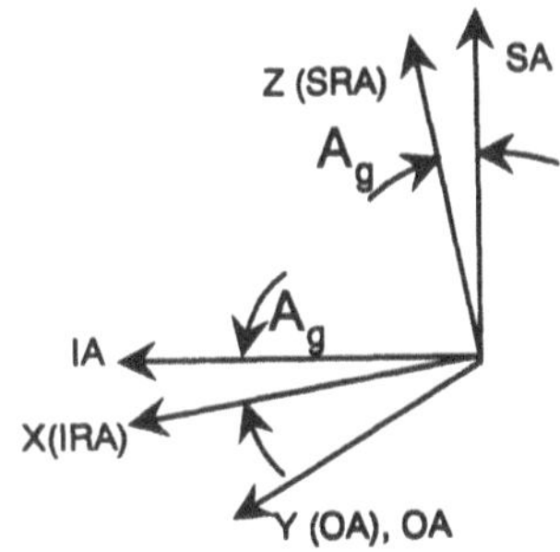

Fig. 14.4 Single-degree-of-freedom floated rate-integrating gyro. (Figure of gyro courtesy of Strapdown Associates, Plymouth, MN.)

float axis of rotation (OA) can result in angular displacement of the float relative to the case. Since the applied moment $\boldsymbol{M}$ (about the OA-axis) is proportional to the rate of change of the case/float relative displacement angle A_g, a measurement of A_g is equivalent to a measurement of the integral of the component of angular velocity along the input or IA-axis.

If ($\boldsymbol{I}$, $\boldsymbol{J}$, $\boldsymbol{K}$) are unit vectors along the IA, OA, and SA axes, we have

$$\boldsymbol{\omega}_{F/I} = [\omega_{\text{in}} \cos(A_g)]\boldsymbol{I} + [\omega_{\text{in}} \sin(A_g)]\boldsymbol{K}$$

$$\boldsymbol{H} = H\boldsymbol{K}$$

$$\boldsymbol{M} = D\,(\mathrm{d}A_g/\mathrm{d}t)\,\boldsymbol{J}$$

where D is a constant of proportionality. We have from Eq. (14.9)

$$A_g \approx \left(\frac{H}{D}\right)\int [\omega_{\text{in}} \cos(A_g)]\mathrm{d}t \approx \left(\frac{H}{D}\right)\int \omega_{\text{in}}\mathrm{d}t \tag{14.10}$$

Essentially, a measure of A_g is equivalent to a measure of the integral of the input angular rate ω_{in}, or the component of $\boldsymbol{\omega}_{F/I}$ normal to both the OA and the spin reference axis, SRA. Clearly one can see from Eq. (14.10) that the angle A_g must be kept as small as possible to avoid *contamination* by the $\cos(A_g)$ term. In strapdown applications we regard the integral of ω_{in} as the incremental angular displacement about the input reference axis, or IRA.

The ratio H/D is called the *gyro scale factor.* In platform applications the scale factor is unimportant since the gyro is used only to detect motion of the platform relative to inertial space and to send the corrective signal to the platform gimbal motors. The stability of the scale factor is of prime concern in setting the limitations of gyro design in a strapdown application.

Let us now consider the fundamental physics of the gyro. In an axis system fixed to the float, unit vectors ($\boldsymbol{I}, \boldsymbol{J}, \boldsymbol{K}$), previously defined, provide $\boldsymbol{H}^F$, or the components of angular momentum in a float-fixed axis system.

$$\boldsymbol{H}^F = \begin{bmatrix} I_{Fx} & 0 & 0 \\ 0 & I_{Fy} & 0 \\ 0 & 0 & I_{Fz} \end{bmatrix}\begin{bmatrix} \omega_{(F/I)x} \\ \omega_{(F/I)y} \\ \omega_{(F/I)z} \end{bmatrix} + \begin{bmatrix} I_{Rx} & 0 & 0 \\ 0 & I_{Ry} & 0 \\ 0 & 0 & I_{Rz} \end{bmatrix}\begin{bmatrix} \omega_{(F/I)x} \\ \omega_{(F/I)y} \\ \omega_{(F/I)z} + \omega_s \end{bmatrix} \tag{14.11}$$

where $\boldsymbol{\omega}_{(F/I)}$, the rotation rate of the float (exclusive of the rotor) relative to inertial space, is resolved in the axis system fixed to the float; the rotor shares the same angular rate as the float except along the Z-axis where the spin rate of the wheel ω_s must be added to the rotation rate of the float relative to inertial space.

Taking

$$I_x = I_{Fx} + I_{Rx} \qquad I_y = I_{Fy} + I_{Ry} \qquad I_z = I_{Fz} + I_{Rz} \qquad I_s = I_{Rz}$$

$$\omega_{(F/I)x} = \omega_{Fx} \qquad \omega_{(F/I)y} = \omega_{Fy} \qquad \omega_{(F/I)z} = \omega_{Fz}$$

we may write Eq. (14.11) as follows:

$$\boldsymbol{H}^F = \begin{bmatrix} I_x & 0 & 0 \\ 0 & I_y & 0 \\ 0 & 0 & I_z \end{bmatrix}\begin{bmatrix} \omega_{Fx} \\ \omega_{Fy} \\ \omega_{Fz} \end{bmatrix} + \begin{bmatrix} 0 \\ 0 \\ I_s\omega_s \end{bmatrix} = \begin{bmatrix} H_x^F \\ H_y^F \\ H_z^F \end{bmatrix} \tag{14.12}$$

Again we note that ω_s is the spin rate of the rotor relative to an axis system fixed in the float. The moment equation is

$$M^F = \left.\frac{\mathrm{d}H^F}{\mathrm{d}t}\right|_F + \Omega^F_{F/I} H^F \tag{14.13}$$

where $\Omega^F_{F/I}$ is the skew-symmetric form of the vector $\omega^F_{F/I}$.

If Eq. (14.12) is inserted into Eq. (14.13), we get

$$M^F = \begin{bmatrix} I_x & 0 & 0 \\ 0 & I_y & 0 \\ 0 & 0 & I_z \end{bmatrix} \begin{bmatrix} \dot{\omega}_{Fx} \\ \dot{\omega}_{Fy} \\ \dot{\omega}_{Fz} \end{bmatrix} + \begin{bmatrix} 0 \\ 0 \\ I_s \dot{\omega}_s \end{bmatrix} + \begin{bmatrix} 0 & -\omega_{Fz} & \omega_{Fy} \\ \omega_{Fz} & 0 & -\omega_{Fx} \\ -\omega_{Fy} & \omega_{Fx} & 0 \end{bmatrix} \begin{bmatrix} H^F_x \\ H^F_y \\ H^F_z \end{bmatrix} \tag{14.14}$$

In component form we have

$$\begin{aligned} M_{Fx} &= I_x \dot{\omega}_{Fx} + (I_z - I_y)\omega_{Fy}\omega_{Fz} + I_s \omega_{Fy}\omega_s \\ M_{Fy} &= I_y \dot{\omega}_{Fy} + (I_x - I_z)\omega_{Fx}\omega_{Fz} - I_s \omega_{Fx}\omega_s \\ M_{Fz} &= I_z \dot{\omega}_{Fz} + (I_y - I_x)\omega_{Fx}\omega_{Fy} + I_s \dot{\omega}_s \end{aligned} \tag{14.15}$$

The angular velocity of the float relative to inertial space with components in the float axis system, or

$$\omega^F_{F/I} = [\omega_{Fx}, \omega_{Fy}, \omega_{Fz}]^{\mathrm{T}}$$

may be separated into two parts: 1) the rotation of the float relative to the case and 2) the rotation of the case relative to inertial space, as

$$\omega^F_{F/I} = \omega^F_{F/c} + \omega^F_{c/I} \tag{14.16a}$$

where the subscripts are self-explanatory. Alternatively,

$$\omega^F_{F/I} = \omega^F_{F/c} + C^F_c \omega^c_{c/I} \tag{14.16b}$$

where C^F_c is the DCM from the case to the float. From Fig. 14.4 we have

$$C^F_c = \begin{bmatrix} \cos A_g & 0 & -\sin A_g \\ 0 & 1 & 0 \\ \sin A_g & 0 & \cos A_g \end{bmatrix} \approx \begin{bmatrix} 1 & 0 & -A_g \\ 0 & 1 & 0 \\ A_g & 0 & 1 \end{bmatrix} \tag{14.16c}$$

In component form we can write the rates given in Eq. (14.16b) as

$$\omega^F_{F/I} = \begin{bmatrix} \omega_{Fx} \\ \omega_{Fy} \\ \omega_{Fz} \end{bmatrix} \qquad \omega^F_{F/c} = \begin{bmatrix} 0 \\ \dot{A}_g \\ 0 \end{bmatrix} \qquad \omega^c_{c/I} = \begin{bmatrix} \omega_{cx} \\ \omega_{cy} \\ \omega_{cz} \end{bmatrix} \tag{14.16d}$$

Consequently, from Eq. (14.16b) we have

$$\boldsymbol{\omega}_{F/I}^{F} = \begin{bmatrix} \omega_{cx} - A_g\omega_{cz} \\ \omega_{cy} + \dot{A}_g \\ A_g\omega_{cx} + \omega_{cz} \end{bmatrix} \tag{14.16e}$$

Equation (14.16e) may be inserted into Eq. (14.15). However, the float has only a single degree of rotation—about the OA- or y-axis. Therefore, we are interested only in moments about this axis: moments about the IA and SA axes must be *taken up* or absorbed by the structure. Thus, we will concentrate only on moments about the OA-axis. Important components of the moment about the OA-axis are given as

$$M_{Fy} = M_T - D\dot{A}_g + U \tag{14.17}$$

where

M_T = moment due to torque motor
$D(\mathrm{d}A_g/\mathrm{d}t)$ = damping moment caused by viscous medium between case and float
U = moments of uncertain or spurious origin

Inserting Eqs. (14.16) and (14.17) into (14.15) and retaining only the components about the OA-axis, we have

$$\begin{aligned} M_T - D\dot{A}_g + U = {} & I_y(\dot{\omega}_{cy} + \ddot{A}_g) + (I_x - I_z)(\omega_{cx} - \omega_{cz}A_g)(\omega_{cz} + A_g\omega_{cx}) \\ & - I_s(\omega_{so} + \Delta\omega_s)(\omega_{cx} - A_g\omega_{cz}) \end{aligned} \tag{14.18}$$

We recognize in the preceding equation that the spin rate of the rotor ω_s can be separated into a nominal value ω_{so} and a deviant term $\Delta\omega_s$.

Next we change the axis notation as follows:

$$Fx = \mathrm{IA} \qquad cx = \mathrm{IRA}$$

$$Fy = \mathrm{OA} \qquad cy = \mathrm{OA} \qquad s = R$$

$$Fz = \mathrm{SA} \qquad cz = \mathrm{SRA}$$

Equation (14.18) now becomes

$$\begin{aligned} I_{\mathrm{OA}}\ddot{A}_g + D\dot{A}_g = {} & M_T + H\omega_{\mathrm{IRA}} - I_{\mathrm{OA}}\dot{\omega}_{\mathrm{OA}} + I_R\omega_{\mathrm{IRA}}\Delta\omega_s \\ & + (I_{\mathrm{SA}} - I_{\mathrm{IA}})\omega_{\mathrm{IRA}}\omega_{\mathrm{SRA}} \\ & + [(I_{\mathrm{SA}} - I_{\mathrm{IA}})(\omega_{\mathrm{IRA}}^2 - \omega_{\mathrm{SRA}}^2) \\ & - H\omega_{\mathrm{SRA}} - I_R\omega_{\mathrm{SRA}}\Delta\omega_s]A_g \\ & + [(I_{\mathrm{IA}} - I_{\mathrm{SA}})(\omega_{\mathrm{IRA}}\omega_{\mathrm{SRA}})]A_g^2 + U \end{aligned} \tag{14.19}$$

where $H = I_s\omega_{so}$. In its simplest representation the major terms of Eq. (14.19) are

$$\dot{A}_g = (H/D)\omega_{\text{IRA}}$$

or

$$A_g = \left(\frac{H}{D}\right)\int_t^{t+\Delta t} \omega_{\text{IRA}} \text{d}t \approx \left(\frac{H}{D}\right)\tilde{\omega}_{\text{IRA}}\Delta t \approx \left(\frac{H}{D}\right)\Delta\theta_{\text{IRA}} \tag{14.20}$$

or the measured value of the angular displacement of the float relative to the case A_g is equivalent to the scale factor H/D times the angular displacement of the case about the IRA (input reference axis) over the sample interval Δt. Obviously there are many other terms in Eq. (14.19), terms that must be regarded as intrinsic instrument errors. Such errors may be listed as follows:

$I_{\text{OA}}\dot{\omega}_{\text{OA}}$	output axis rotation error
$(I_{\text{SA}} - I_{\text{IA}})\omega_{\text{IRA}}\omega_{\text{SRA}}$	aniso-inertia error
$[(I_{\text{SA}} - I_{\text{IA}})(\omega_{\text{IRA}}^2 - \omega_{\text{SRA}}^2)$ $-H\omega_{\text{SRA}} - I_R\omega_{\text{SRA}}\Delta\omega_s]A_g$	cross-coupling error
$[(I_{\text{IA}} - I_{\text{SA}})(\omega_{\text{IRA}}\omega_{\text{SRA}})]A_g^2$	quadrature error
$\frac{1}{H}[(I_{\text{OA}}\ddot{A}_g + D\dot{A}_g)]$	stored error
$I_R\omega_{\text{IRA}}\Delta\omega_s$	spin motor regulation error

In a platform application the output angle A_g is measured using the signal generator. Since a measurement of A_g is equivalent to a measurement of the angular displacement of the case about the IRA axis, A_g is used as a signal to the gimbal motors to rotate the platform (and the case) in the opposite direction, in the process driving the angle A_g back to null. Since the gyro is in a benign angular velocity environment, the various error terms are insignificant.

In the strapdown application, the gyro is now subjected to vehicle angular rates: the various error terms already indicated can be significant. The quadrature, cross-coupling, and stored error terms can be made arbitrarily small by maintaining the angle A_g at an arbitrarily small value.

In order for the gyro to have utility as a rate-integrating instrument, the following must be true:

$$\int_t^{t+\Delta t} \omega_{\text{IRA}} \text{d}t = \Delta\theta_{\text{IRA}} = -\frac{1}{H}\int_t^{t+\Delta t} M_T \text{d}t \tag{14.21}$$

where the torque motor signal must be summed over the sampling time Δt to obtain a measure of the angular displacement.

Equation (14.19) may now be written as a composition of the dominant terms and the error terms. First dividing by H, the angular momentum of the rotor, we have

$$\omega_{\mathrm{IRA}} = \frac{1}{H} M_T + \frac{1}{H} \text{ (Error Terms)} \tag{14.22}$$

Equation (14.21) indicates that the angular displacement of the case is related to the integration of the torque motor output; the presence of the error terms must then result in an error in angular displacement $\Delta\theta_{\mathrm{IRA}}$.

The various error sources can be estimated using representative SDFG data. Table 14.1 provides some data that are useful in obtaining a sense of the relative sizes of the error terms.[7]

We can estimate the size of the rotation error if we assume that the angular acceleration about the OA-axis integrates to 1 radian/s, or

$$\frac{I_{\mathrm{OA}}}{H} \int^{t} \dot{\omega}_{\mathrm{OA}} \mathrm{d}t = \frac{\omega_{\mathrm{OA}} I_{\mathrm{OA}}}{H} = \frac{(1)(225.0)}{1.45 \times 10^5} = 1.6 \times 10^{-3}$$

A misalignment of 1.6 milli-radians (0.09 deg) is considerable, leading to a position error of about 5 nautical miles. Our conclusion is that a gyro axis aligned with the direction of body rotation can cause a significant error in ω_{IRA}.

We could consider the sizes of the other error terms by inserting values from Table 14.1. However, we need only note here that the origin of the aniso-inertia contribution arises from the fact that the moment of inertia around the SA-axis is not equal to that around the IA-axis.

Let us combine the aniso-inertia error and the spin regulation error to get

$$\frac{1}{H}[(I_{\mathrm{SA}} - I_{\mathrm{IA}})\omega_{\mathrm{IRA}}\omega_{\mathrm{SRA}} + I_R \omega_{\mathrm{IRA}} \Delta\omega_s] \tag{14.23}$$

Table 14.1 Representative SDFG parameters

Parameter	Value
ω_{IRA}	1.0 radian/s
A_g	100.0 μ radian
H	1.45×10^5 g-cm^2/s
D	6.0×10^5 dyne-cm-s
I_{OA}	225.0 g-cm^2
I_R	60.0 g-cm^2
$(I_{\mathrm{SA}} - I_{\mathrm{IA}})/H$	9.96 deg/h/(radian/s)^2
m_R	20.0 g
ω_s	2.5×10^3 radian/s
I_{SA}	450.0 g-cm^2
I_{IA}	443.0 g-cm^2

The spin motor has a bandwidth of something like 5 Hz. For cluster rotational rates well below this value, $\Delta\omega_s$ is negligible and the error expression previously given becomes

$$\frac{1}{H}[(I_{\mathrm{SA}} - I_{\mathrm{IA}})\omega_{\mathrm{IRA}}\omega_{\mathrm{SRA}}] \tag{14.24a}$$

For cluster rotation rates comparable to the spin motor bandwidth we can write

$$\Delta\omega_s = -\omega_{\mathrm{SRA}}$$

so that Eq. (14.23) becomes

$$\frac{1}{H}[(I_{\mathrm{SA}} - I_R - I_{\mathrm{IA}})]\omega_{\mathrm{IRA}}\omega_{\mathrm{SRA}} \tag{14.24b}$$

The term $(I_{SA} - I_R)$ is the moment of inertia of the float minus the moment of inertia of momentum wheel about the SA-axis. Clearly if this quantity equals the moment of inertia about the IA-axis (axis normal to the momentum wheel spin axis), then the aniso-inertia contribution is zero. To summarize: to avoid an aniso-inertia error, the following must be true for low angular velocities about the SA-axis:

$$I_{\mathrm{SA}} = I_{\mathrm{IA}}$$

and for high angular velocities,

$$(I_{\mathrm{SA}} - I_R) = I_{\mathrm{IA}}$$

Since the magnitude and direction of the angular velocity will vary during operational conditions, a compromise is set by requiring

$$(I_{\mathrm{SA}} - I_R) < I_{\mathrm{IA}} < I_{\mathrm{SA}} \tag{14.25}$$

The aniso-inertia error is an identifiable cross-coupling of the angular rates along the input and spin axes. We will now consider the remaining cross-coupling errors.

The next major cross-coupling error contribution is

$$\frac{1}{H}[(I_{\mathrm{SA}} - I_{\mathrm{IA}})(\omega_{\mathrm{IRA}}^2 - \omega_{\mathrm{SRA}}^2)]A_g \tag{14.26}$$

which works out to about 1.0×10^{-3} deg for a one-hour flight.

Another cross-coupling error source is

$$\frac{1}{H}(H\,\omega_{\mathrm{SRA}})A_g = \omega_{\mathrm{SRA}}A_g \tag{14.27a}$$

which amounts to 1.0×10^{-4} radian/s or 20 deg/h. This error occurs only in a strapdown application where the gyro is subjected to body angular rates. The only solution for a strapdown application is to keep the angle A_g as small as possible, say 1.0 μ radian rather than 100.0 μ radian.

Finally let us consider the last term from the cross-coupling error contributor, that is,

$$\frac{1}{H}(I_R \omega_{\mathrm{SRA}} \Delta\omega_s) A_g = \frac{\Delta\omega_s}{\omega_s}(A_g \omega_{\mathrm{SRA}}) \qquad (14.27\mathrm{b})$$

With motor regulation at 0.1% the preceding error contribution is about 0.02 deg/h. This is a fairly significant contribution to gyro drift in comparison to that associated with a platform implementation.

The serious errors are these:

1) Output axis rotation error $= (I_{\mathrm{OA}}/H)(\mathrm{d}\omega_{\mathrm{OA}}/\mathrm{d}t)$.

2) Low frequency aniso-inertia error $= (1/H)(I_{\mathrm{SA}} - I_{\mathrm{IA}})\omega_{\mathrm{IRA}}\omega_{\mathrm{SRA}}$.

3) Cross-coupling error $= \omega_{\mathrm{SRA}} A_g$.

The final sources of error are lumped into the U (for uncertain) torque. This term includes noncompensated effects due to manufacturing tolerances, mass offset from gimbal axes, and departure from perfect elastic isotrophy in various supports. This last attribution refers to variations in gimbal stiffness with load direction.

The uncertainty torques U may be divided into some general categories: 1) bias drift (thermal gradients, manufacturing tolerances), 2) g-sensitive drift (mass imbalance), and 3) g^2-sensitive drift (mostly aniso-elasticity).

The bias drift is an output where there is no input. [We will later observe for a ring laser gyro (RLG) a somewhat reverse phenomenon: laser *lock-in*—no output in the presence of an input.] Thermal gradients in the support fluid or across the support of an air-filled gyro, resulting in spurious movement of the float about the gyro input axis, interpreted as the presence of an input rate. Of course, measurements from thermocouples can be input into a software compensation algorithm. Unfortunately, if the thermocouple must be placed other than at the critical position, an additional estimation algorithm must be included. Spurious outputs or gyro drift cannot be completely compensated; the U moment then contains, or more accurately is an estimation of, such noncompensated errors.

The g-sensitive drift is primarily due to an offset of the wheel's mass from the elastic axis. Variations in the thermal environment will result in corresponding variations in the degree of offset. Obviously under an acceleration normal to the output axis there will be a moment about the output axis resulting in a spurious drift rate. This drift rate will be interpreted as a measured input rate. This spurious output, identified as a drift rate, is given as

$$\Delta\omega_{\mathrm{ind}} = \frac{1}{H}(a_{\mathrm{IA}} m_R \Delta Z) \qquad (14.28\mathrm{a})$$

where the symbols are identified as H: wheel momentum; a_{IA}: acceleration component along the input axis; m_R: wheel mass; and ΔZ: mass offset from the elastic axis. In Fig. 14.5a we note that in the presence of a mass offset linear

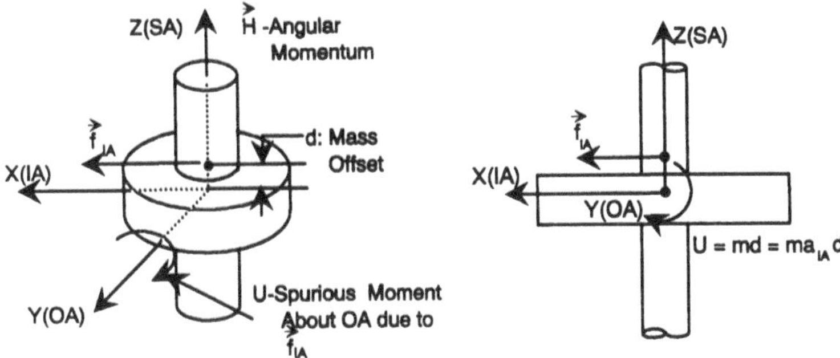

Fig. 14.5a Spurious moment caused by mass offset from output axis.

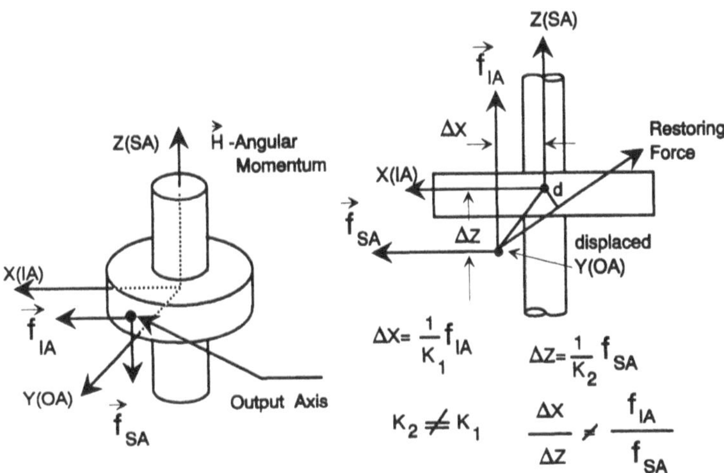

Fig. 14.5b Spurious moment caused by aniso-elastic mass offset from output axis.

acceleration along the input axis causes a moment about the output axis. This moment is interpreted (erroneously) as an input angular rate.

To get an idea of the required tolerance, we obtain from Eq. (14.28a) the following:

$$\Delta Z = \frac{(\Delta\omega_{\text{ind}}/g)H}{(980.0)m_R} \tag{14.28b}$$

Let us take as a required performance goal a drift rate per gram $(\Delta\omega_{\text{ind}}/g)$ of 0.01 deg/h/g or 4.85×10^{-8} radian/s/g. Using Eq. (14.28b), we get, with the data from Table 14.1,

$$\begin{aligned}\Delta Z &= \left[(4.85\times10^{-8})(1.45\times10^{5})\right]/\left[(980.0)(20.0)\right]\\ &\approx 3.588\times10^{-7}\text{cm}\\ &\approx 36.0\ \text{Å}\end{aligned}$$

Thus the lateral mass offset must be restricted to about 36 Å to achieve a performance level of 0.01 deg/h/g.

The g^2 error source is the aniso-elastic error. This effect is caused by a difference in structural compliance in two orthogonal directions. As shown in Fig. 14.5b, the restoring force F_R (equal to the negative of the applied force F_A) does not pass through the elastic axis of the output axis, OA. Consequently there is a moment about the output axis that is interpreted as an input angular rate.

We note the error caused by gyro misalignment both in the instrument itself (output axis not exactly orthogonal to the input axis) and in the installation of the gyro in cluster. One way to describe misalignment is through the angle ϕ_{ij} where the intent of the subscripts is the misalignment of the i-axis about the j-axis. Because of the misalignment, there is a body rate error given as

$$\Delta\boldsymbol{\omega}^b_{b/I} = C^b_c \boldsymbol{\omega}^c_{c/I} \tag{14.29}$$

where

$$C^b_c = \begin{bmatrix} 0 & -\phi_{yz} & \phi_{zy} \\ \phi_{xz} & 0 & -\phi_{zx} \\ -\phi_{xy} & \phi_{yx} & 0 \end{bmatrix} = \begin{bmatrix} 0 & -\phi_{\mathrm{OS}} & \phi_{\mathrm{SO}} \\ \phi_{\mathrm{IS}} & 0 & -\phi_{\mathrm{SI}} \\ -\phi_{\mathrm{IO}} & \phi_{\mathrm{OI}} & 0 \end{bmatrix}$$

and I, O, S refer to the Input Axis, the Output Axis and the Spin Axis. If the misaligned axes remain orthogonal then

$$C^b_c = \begin{bmatrix} 0 & -\phi_{\mathrm{S}} & \phi_{\mathrm{O}} \\ \phi_{\mathrm{S}} & 0 & -\phi_{\mathrm{I}} \\ -\phi_{\mathrm{O}} & \phi_{\mathrm{I}} & 0 \end{bmatrix}$$

A final consideration is the internal bending moments that must be absorbed by the gyro structure. The SDFG will rotate about the Y- or OA-axis in response to an angular rate about the X- or IA-axis. Of course the gyro motor provides a torque that opposes the angular motion about the OA-axis.

Let us assume that the float is subjected to angular rates and acceleration about the IA and SA axes as

$$\omega_{Fx} = \omega_{Fy} = \omega_{Fz} = 1.0 \text{ radian/s}$$

$$\dot{\omega}_{Fx} = \dot{\omega}_{Fz} = \dot{\omega}_s = 1.0 \text{ radian/s}^2$$

We can now rewrite Eqs. (14.15), first replacing x, y, z by IA, OA, SA, respectively, to get

$$\begin{aligned} M_x &= M_{\mathrm{IA}} = I_{\mathrm{IA}}\dot{\omega}_{Fx} + (I_{\mathrm{SA}} - I_{\mathrm{OA}})\omega_{Fy}\omega_{Fz} + (I_s\omega_s)\omega_{Fy} \\ M_z &= M_{\mathrm{SA}} = I_{\mathrm{SA}}\dot{\omega}_{Fz} + (I_{\mathrm{OA}} - I_{\mathrm{IA}})\omega_{Fy}\omega_{Fx} + I_s\dot{\omega}_s \end{aligned} \tag{14.30}$$

Using values from Table 14.1, we get

$$M_x = [443.0 + 225.0 + 1.45 \times 10^5] \text{ dyne-cm}$$

$$M_z = [225.0 - 218.0 + 60.0] \text{ dyne-cm}$$

Clearly there is an enormous torque of 1.45×10^5 dyne-cm about the X- or IA-axis. In a strapdown system where the cluster must be subjected to body rates, the gyro structure must sustain a large bending moment about the IA-axis. For a gyro to be used in a strapdown application the structure must be much more robust than is required in a platform application.

Two Degrees of Freedom Gyro (TDFG), or Dry-Tuned Rotor Gyro (DTRG)

A mechanical gyro that has become the successor to the SDFG discussed in the last section is known by a variety of names. We will use here the acronym DTRG. The term *two degrees of freedom* is appropriate in that the DTRG measures the angular rate (or its integral) about two mutually perpendicular axes at the same time. *Dry* is also applicable in that the gyro operates in an inert atmosphere rather than in a *wet* suspension fluid as does the SDFG. The descriptor *tuned* is also significant in that the momentum element or rotating wheel can be effectively isolated from its support by spinning the wheel at a rate that satisfies a *tuned* condition.

Other names for the DTRG are *gyro-flex*, as well as more restricted manufacturer's descriptors or trademarks. The use of the term gyro-flex refers to the use of flexures to meet the tuning condition.

An ideal gyro would be one in which the reference angular momentum vector is somehow supported or contained in the case without introducing spurious torques associated with the support mechanism. Spurious torques cause the spinning element to acquire precession rates. When precession rates occur, there is no way to distinguish these rates from actual input rates to the gyro.

In Fig. 14.6 we have presented a simplified view of the DTRG emphasizing the essential elements: the rotor, the gimbal, the flexures, and the rotor shaft. The entire structure (gimbal, rotor, flexure) is rotated about the concentric drive shaft. The drive shaft is connected to the gimbal (or inner ring) by means of two flexible couplings. The gimbal in turn is connected to an outer ring or rotor by two more flexures, whose torsion axes are orthogonal to the inner flexures. Ideally the flexures are rigid with respect to lateral motion but permit elastically restricted angular motion.

A two-axis angular pick-off, although not shown in the figure, measures the angular deviation of the rotor. A two-axis permanent magnet torque generator then torques the rotor relative to the case. The torquer magnets are attached to the rotor and the torquer coils to the gyro case.

The *tuning* condition mentioned before refers to setting the rotor spin rate such that the flexure spring constant K is related to the rotor spin rate ω_s and mass properties as follows:

$$K = I_G \omega_s^2 \tag{14.31}$$

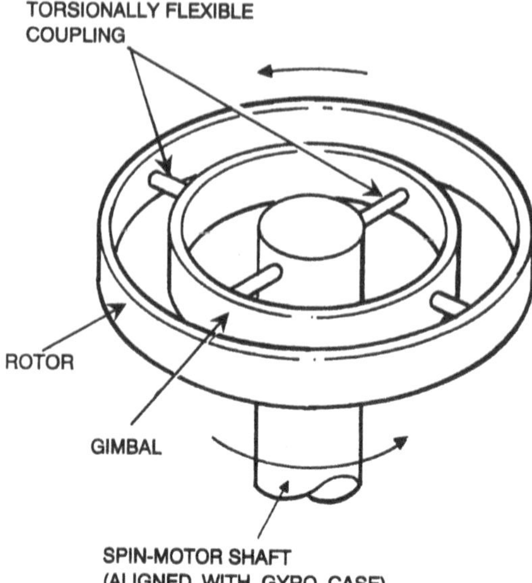

Fig. 14.6 Dry-tuned rotor gyro concept. (Courtesy of Strapdown Associates, Plymouth, MN.)

where I_G is the gimbal in-plane moment of inertia. Meeting the above requirement then allows the gyro to act like a free gyro, i.e., to have infinite compliance relative to its support.

Assume that the rotor's spin vector is at an angle relative to the drive shaft. The gimbal is then driven into oscillatory motion in and out of the rotor plane at twice the spin frequency of the rotor. The reaction torque on the rotor has a component along the angular displacement vector. This torque is proportional to angular displacement of the rotor but acts as a spring with a negative flexure constant. The flexible connection between the rotor and the gimbal provides a similar spring torque on the rotor but of opposite sign. Therefore, to free the rotor from spurious torques, it is necessary to set the gimbal pivot springs so that they cancel the inverse spring effect of the gimbals.[3] This coupled motion is shown in Fig. 14.7.

To analyze the DTRG, we must first define three coordinate systems. The first system is a de-spin or d-frame: the Z-axis is normal to the plane of the rotor, but the X,Y axes, although in the plane of the rotor, do not share the rotor spin. The second axis system is the gimbal or g-frame whose Z-axis is normal to the gimbal plane; and finally a case system whose Z-axis is along the rotor shaft or the spin reference direction.

The DCM between the case and gimbal system follows from Fig. 14.8a as

$$C_c^g = \begin{bmatrix} 1 & 0 & 0 \\ 0 & 1 & A_x \\ 0 & -A_x & 1 \end{bmatrix} \tag{14.32a}$$

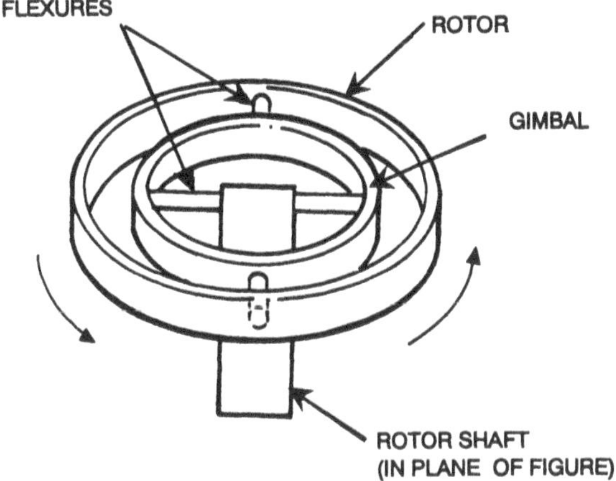

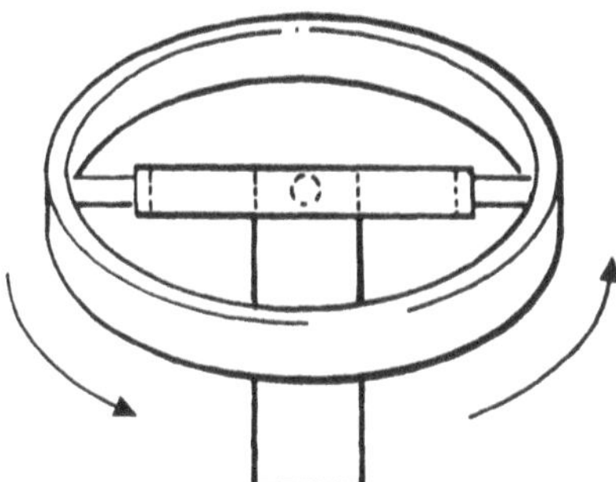

Fig. 14.7 Assembly dynamics of two degrees of freedom dry-tuned rotor gyro. (Courtesy of Strapdown Associates, Plymouth, MN.)

Next we define the DCM between the d-frame and the g-frame as

$$C_g^d = \begin{bmatrix} 1 & 0 & -A_y \\ 0 & 1 & 0 \\ A_y & 0 & 1 \end{bmatrix} \tag{14.32b}$$

Figure 14.8b provides a geometric rendition of frame alignment. All three axis systems are shown in Fig. 14.8c.

Next we must obtain an expression for the angular momentum of the rotor. The angular momentum of the rotor relative to inertial space (by definition) is

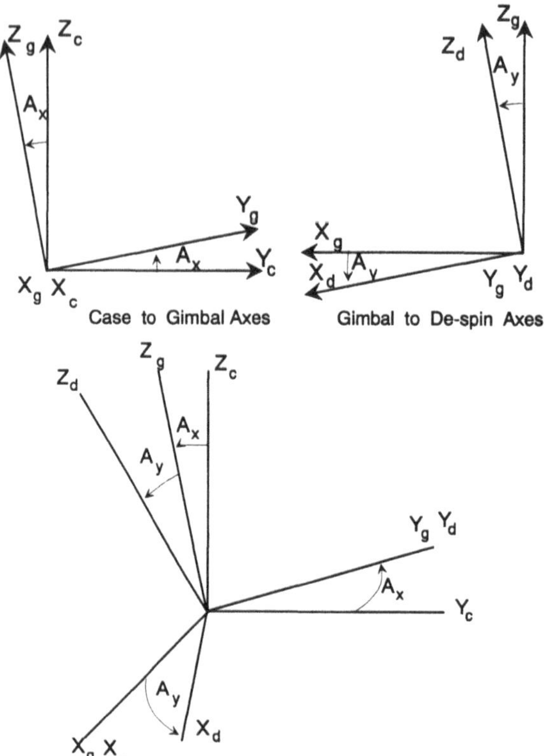

Fig. 14.8 Combined axis transformation for dry-tuned rotor gyro.

the product of the inertia tensor and the angular velocity of the rotor relative to inertial space, or

$$\boldsymbol{H} = I \cdot \boldsymbol{\omega}_{r/I} = \begin{bmatrix} I_T & 0 & 0 \\ 0 & I_T & 0 \\ 0 & 0 & I_S \end{bmatrix} \begin{bmatrix} \omega_{(r/I)x} \\ \omega_{(r/I)y} \\ \omega_{(r/I)z} \end{bmatrix} \tag{14.33}$$

where I_T and I_S are, respectively, the transverse and polar moment of inertia of the rotor. The angular velocity of the rotor relative to inertial space is $\boldsymbol{\omega}_{r/I}$.

We must now separate the angular velocity into two contributions: 1) the rotation of the rotor relative to the de-spun axes (d-axes) and 2) the rotation of the de-spun axes relative to inertial space. This operation is expressed as

$$\boldsymbol{\omega}_{r/I} = \boldsymbol{\omega}_{r/d} + \boldsymbol{\omega}_{d/I} = \begin{bmatrix} 0 \\ 0 \\ \omega_s \end{bmatrix} + \begin{bmatrix} \omega_{dx} \\ \omega_{dy} \\ \omega_{dz} \end{bmatrix} \tag{14.34}$$

The moment on the rotor is then

$$M = \left.\frac{\mathrm{d}H}{\mathrm{d}t}\right|_d + \omega_{d/I} \times H \tag{14.35a}$$

Substituting into Eq. (14.35a) from both Eqs. (14.33) and (14.34) gives

$$M^d = \begin{bmatrix} M_x \\ M_y \\ M_z \end{bmatrix} = \frac{\mathrm{d}}{\mathrm{d}t}\begin{bmatrix} I_T\omega_{dx} \\ I_T\omega_{dy} \\ I_S\omega_{dz} + I_S\omega_s \end{bmatrix} + \begin{bmatrix} 0 & -\omega_{dz} & \omega_{dy} \\ \omega_{dz} & 0 & -\omega_{dx} \\ -\omega_{dy} & \omega_{dx} & 0 \end{bmatrix}\begin{bmatrix} I_T\omega_{dx} \\ I_T\omega_{dy} \\ I_S\omega_{dz} + I_S\omega_s \end{bmatrix} \tag{14.35b}$$

After some manipulation we obtain

$$\begin{aligned} M_x &= I_T\dot{\omega}_{dx} + (I_S - I_T)\omega_{dy}\omega_{dz} + I_S\omega_s\omega_{dy} \\ M_y &= I_T\dot{\omega}_{dy} + (I_T - I_S)\omega_{dx}\omega_{dz} - I_S\omega_s\omega_{dx} \\ M_z &= I_S(\dot{\omega}_{dz} + \dot{\omega}_s) \end{aligned} \tag{14.36}$$

Now let us write $\omega_{d/I}$, the angular velocity of the d-frame (de-spun) relative to inertial space, as

$$\omega^d_{d/I} = \omega^d_{d/g} + C^d_g\omega^g_{g/c} + C^d_g C^g_c \omega^c_{c/I} \tag{14.37}$$

where we have expressed the components of $\omega_{d/I}$ in the d-frame.

We may now use Eqs. (14.36) as appropriate to get

$$\begin{bmatrix} \omega_{dx} \\ \omega_{dy} \\ \omega_{dz} \end{bmatrix} = \begin{bmatrix} 0 \\ \dot{A}_y \\ 0 \end{bmatrix} + \begin{bmatrix} 1 & 0 & -A_y \\ 0 & 1 & 0 \\ A_y & 0 & 1 \end{bmatrix}\begin{bmatrix} \dot{A}_x \\ 0 \\ 0 \end{bmatrix} + \begin{bmatrix} 1 & 0 & -A_y \\ 0 & 1 & A_x \\ A_y & -A_x & 0 \end{bmatrix}\begin{bmatrix} \omega_{cx} \\ \omega_{cy} \\ \omega_{cz} \end{bmatrix} \tag{14.38}$$

where we have simplified the subscript notation by letting

$$\omega^c_{c/I} = [\omega_{cx}, \omega_{cy}, \omega_{cz}]^{\mathrm{T}}$$

In Eq. (14.38) the angles A_x and A_y are measured quantities.

We recognize that the moment about the Z-axis of the d-frame is provided by the regulated spin motor: the motor will provide whatever torque is necessary to spin the rotor at fixed speed [to meet the tuning requirements indicated in Eq. (14.31)]. However, from the utility point of view we are interested in the moments about the X- and Y-axes of the d-frame, or

$$M_x = M_{\mathrm{TM}_x} - K_d \dot{A}_x + V_x \tag{14.39a}$$

$$M_y = M_{\mathrm{TM}_y} - K_d \dot{A}_y + V_y \tag{14.39b}$$

where M_{TM} is the moment applied by the torque motor, required since the DTRG is a drive-to-null instrument and is electrically caged when used in the strapdown mode. The terms $K_d \mathrm{d}A_x/\mathrm{d}t$ and $K_d \mathrm{d}A_y/\mathrm{d}t$ are included to represent the damping moment caused by the inert gas. The terms V_x and V_y are spurious torques that have no readily identifiable source.

As we pointed out earlier, although the spin motor is regulated, there will be some variation in the spin rate of the momentum wheel. To account for such deviations, we separate the spin rate ω_s into nominal and off-nominal components, or

$$\begin{aligned} \omega_s &= \omega_{so} + \Delta\omega_s \\ H &= I_S \omega_{so} \end{aligned} \tag{14.40}$$

Inserting Eqs. (14.38), (14.39), and (14.40) into Eqs. (14.36) and ignoring higher order terms (H.O.T.), we get

$$\begin{aligned} I_T \ddot{A}_x + K_d \dot{A}_x + H \dot{A}_y = {} & M_{\mathrm{TM}_x} + H\omega_{cy} - I_T \dot{\omega}_{cx} - I_S \Delta\omega_s \omega_{cy} \\ & + (I_T - I_S)\omega_{cy}\omega_{cz} - H\omega_{cz} A_y + V_x + \mathrm{H.O.T.} \end{aligned} \tag{14.41a}$$

$$\begin{aligned} I_T \ddot{A}_y + K_d \dot{A}_y + H \dot{A}_x = {} & M_{\mathrm{TM}_y} - H\omega_{cx} - I_T \dot{\omega}_{cy} + I_S \Delta\omega_s \omega_{cx} \\ & + (I_S - I_T)\omega_{cx}\omega_{cz} - H\omega_{cz} A_x + V_y + \mathrm{H.O.T.} \end{aligned} \tag{14.41b}$$

Since the gyro is electrically caged, the ideal output is

$$\tilde{\omega}_{cy} \approx -M_{\mathrm{TM}_x}/H \tag{14.42a}$$

$$\tilde{\omega}_{cx} \approx M_{\mathrm{TM}_y}/H \tag{14.42b}$$

As we did in Eqs. (14.24), we can divide all terms by H to get the errors in the indicated angular velocity. A listing of the intrinsic errors associated with the DTRG includes 1) output axis rotation errors, or

$$-(I_T/H)\dot{\omega}_{cx}, \qquad -(I_T/H)\dot{\omega}_{cy}$$

2) Aniso-inertia errors, or

$$(1/H)(I_S - I_T)\omega_{cy}\omega_{cz}, \qquad (1/H)(I_S - I_T)\omega_{cx}\omega_{cz}$$

3) cross-coupling errors, or

$$-\omega_{cz}A_x, \qquad -\omega_{cz}A_y$$

and 4) drift errors, or

$$V_x/H, \qquad V_y/H$$

Momentum wheel gyros, such as the SDFG and the DTRG instruments that we have already discussed as well as their many variations, no doubt have certain advantages in comparison to the more recently developed instruments. As Savage[4] points out, the momentum wheel gyro has a long production history and its limitations are well understood, particularly in the gimbal or platform application. (By *understood* we mean that compensation procedures have been developed that can minimize such limitations.) However, in a strapdown implementation, which most certainly is the primary application of the coming decades, the mechanical gyro has serious competition from various optical gyros. Again, using the work of Savage,[4] we should note some of the drawbacks of the momentum wheel gyro in the strapdown application.

The mechanical or momentum wheel gyro is bandwidth limited because of the torque-to-balance requirement. Consequently only angular rates with a frequency up to the bandwidth limit can be measured. In addition errors arise in the rectification of vibrations. The resulting error torques produce false precessional rates that are interpreted as case rotations.

The aniso-inertia effect results when the moments of inertia of the case about the SRA-axis and IA-axis are unequal: the difference between these moments of inertia multiplied by the component of angular velocity of the case (and hence the vehicle) about the SRA provides an additional component of angular momentum that must be added to that of the wheel [see Eq. (14.24)]. The additional torque required to cage the gyro electrically is then interpreted as an input rate. In the next section we will consider a typical optical gyro, called the ring laser gyro, which seems to be particularly well suited for the strapdown role.

Ring Laser Gyro (RLG)

Both the SDFG and DTRG, discussed in the previous sections, are balance-to-null instruments (at least in the strapdown mode). In such instances the measurement of the rebalance torque is equivalent to the measurement of the body angular rate. The off-null operation of a mechanical gyro can lead to significant errors due to cross-coupling effects. The ring laser gyro (RLG) considered in this section is ideally suited to the strapdown application since it is not a balance-to-null instrument.

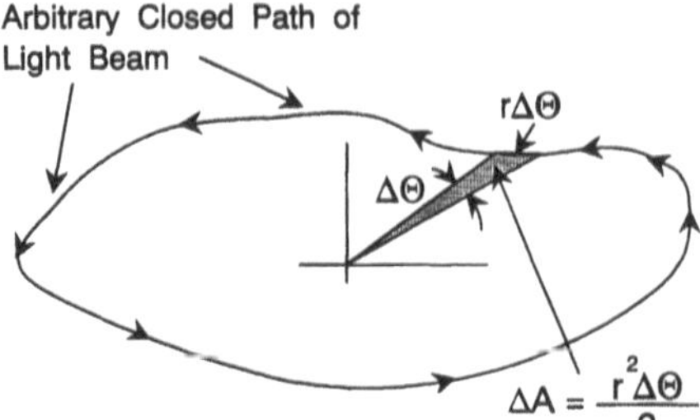

Fig. 14.9 Light beam path.

In Fig. 14.9 we have depicted a closed path of arbitrary shape around which a light beam is traversing. First we will assume that this closed but arbitrary shape is nonrotating with respect to inertial space. The time ΔT for the light to traverse the distance $r\Delta\theta$ is

$$\Delta T = r\Delta\theta/c \tag{14.43}$$

where c is the speed of light.

Now assume that the path is rotating counterclockwise at a rate ω. The incremental angle $\Delta\theta$ increases by an amount $\delta\theta$ during the period ΔT or

$$\delta\theta = \omega\Delta T = r\omega\Delta\theta/c \tag{14.44}$$

The area of the elementary triangle ΔA is $\frac{1}{2}r^2\Delta\theta$. Now from Eq. (14.44) the elementary length $r\Delta\theta$ changes due to path rotation. This change can be written as

$$\delta L = r\delta\theta = \frac{r^2\omega\Delta\theta}{c} = \frac{2\omega}{c}\Delta A$$

The total change in path length, ΔL, is

$$\Delta L \approx \int \delta L = \frac{2\omega}{c}\int \Delta A = \frac{2A\omega}{c} \tag{14.45}$$

Now we can estimate the path length change by inserting the following typical values:

$$
\begin{aligned}
\ell &= 0.1 \text{ m (assumed length of a side of an equilateral triangle)}\\
c &= 3 \times 10^8 \text{ m/s (speed of light propagation)}\\
\omega &= 1 \text{ radian/s (rotation rate of case)}\\
A &= \left(\sqrt{3}/4\right)\ell^2 = 4.33 \times 10^{-3} \text{ m}^2\\
\Delta L &= 2.88 \times 10^{-11} \text{ m} = 0.288 \text{ Å}
\end{aligned}
$$

Therefore, it would seem that an unacceptably small distance would have to be measured to detect a typical angular rate in a packagable device.

If the distances of the order of a few tenths of an angstrom must be measured, then such a requirement would seem to make an RLG impracticable. However, as we shall see, it is not necessary to make such measurements directly.

First we note that time and frequency are two quantities that can be measured with great precision. Assume that we have a gyro in the shape of an equilateral triangle of total length L, and each leg has a length of $\frac{1}{3}L$. Next assume that a laser light source supplies a beam of coherent light with a fixed number N of waves in the circuit of length L. The path length L and the wavelength are related as

$$\lambda = L/N$$

The change in wavelength, $\Delta\lambda$, is therefore

$$\Delta\lambda = \frac{\Delta L}{N} = \frac{2A\omega}{cN} \tag{14.46a}$$

where we have made use of Eq. (14.45). Since $\lambda f = c$, where f is the frequency, we have

$$(\Delta\lambda)f + \Delta f(\lambda) = \Delta c = 0 \tag{14.46b}$$

In Eq. (14.46b) we have indicated that the speed of light is independent of the rotation rate ω. We thus get

$$\begin{aligned} \Delta f &= -(f/\lambda)\Delta\lambda = -(f/\lambda)(2A\omega/cN) \\ &= -2A\omega f/Lc = -(2A\omega/\lambda L) \end{aligned} \tag{14.47}$$

Note that we have made use of the expression $\lambda f = c$.

Equation (14.47) is the basic equation for the RLG. However, it is not in the most useful form. First, let us use some typical numbers: we will assume a path length of 0.3 m (0.1 m on each side of an equilateral triangle) to get

$$\frac{\Delta f}{f} = -\frac{2A\omega}{cL} = -\frac{(2)(4.33 \times 10^{-3})(1)}{(0.3)(3.0 \times 10^{8})} = 9.62 \times 10^{-11}$$

To detect a rotation rate of 1.0 radian/s, apparently we are required to measure frequency to one part in 10^{10}. However, the designers of the RLG neatly sidestep this extremely demanding requirement by using two beams—one clockwise and the other counterclockwise.

The difference in frequencies between the two beams, a difference we will indicate by δf, may be obtained from Eq. (14.47) as

$$\begin{aligned} \delta f &= f_{\text{cw}} - f_{\text{ccw}} \\ &= (f + \Delta f) - (f - \Delta f) \\ &= 2\Delta f = 4A\omega/\lambda L \end{aligned} \tag{14.48}$$

(We have arbitrarily defined the clockwise beam as positive.) The frequency difference δf is formed from an interference pattern between the two beams. Roughly, this pattern is developed by placing a partially silvered mirror at one of the vertices of the ring (triangular in this case).

In the upper portion of Fig. 14.10 is illustrated the partially silvered mirror that combines the clockwise and counterclockwise beams to form an interference pattern. The fringe pattern is detected by detectors A and B set one-quarter of a wavelength apart (see Fig. 14.11). If the two beams have frequencies that are different, that is, the ring is rotating, then the fringe patterns move. The signals from the detectors are digitized so that what the instrument is doing is counting fringes. A threshold is set and the detector is either "on" or "off," depending upon whether the fringe pattern is above or below the pre-set threshold. Since the detectors are set one-quarter of a wavelength apart, motion direction is determined by recording which detector comes on first. For motion to the right the A detector comes on first; the situation is reversed for left-moving fringe patterns.

We will now define $\Delta\phi$ as the phase change in the fringe pattern (cycles) and $\Delta\theta$ as the angle through which the case rotates about the gyro's input axis. The phase change in the fringe pattern may be written as

$$\Delta\phi = \int \delta f \mathrm{d}T = \int (4A/\lambda L)\,\omega\;\mathrm{d}T = \frac{4A}{\lambda L}\Delta\theta \qquad (14.49)$$

The quantity $(\lambda L/4A)\Delta\phi$ is called the *scale factor*. The units are μ radians per pulse. The logic circuit from the A-B detectors will produce a pulse every time a fringe pattern passes. A typical magnitude for the scale factor S.F., using

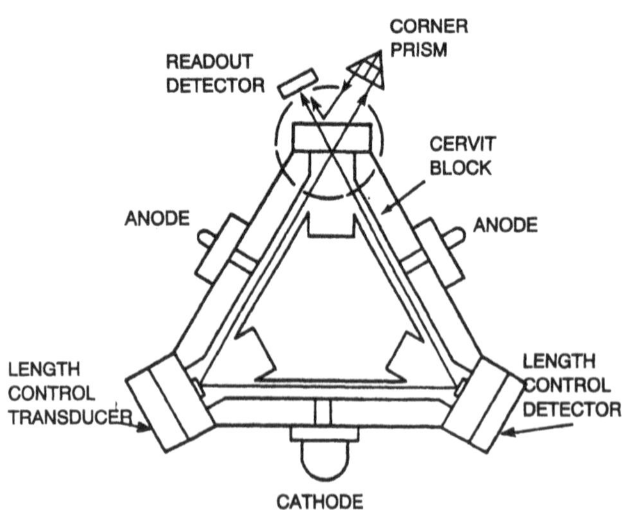

Fig. 14.10 Ring laser gyro elements.

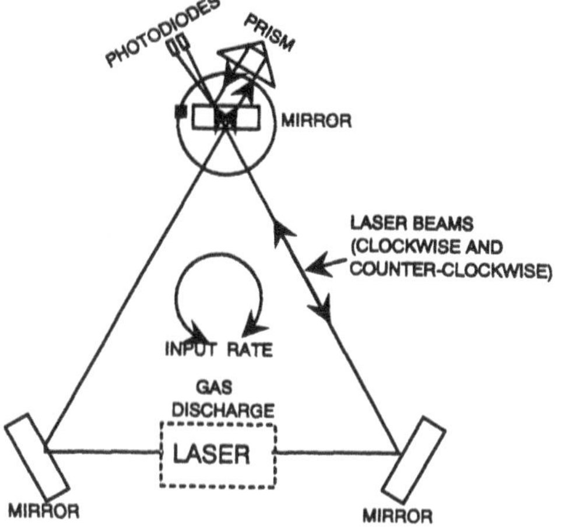

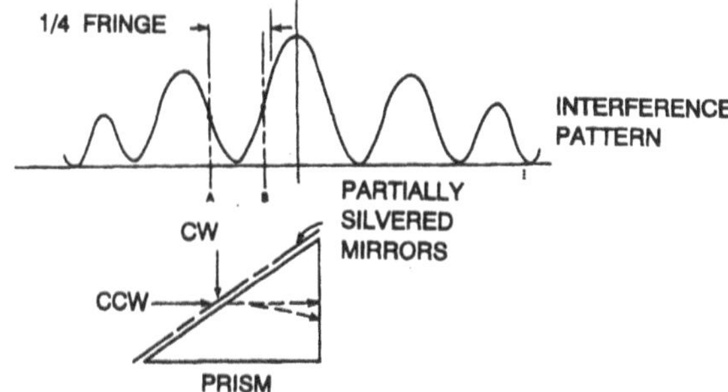

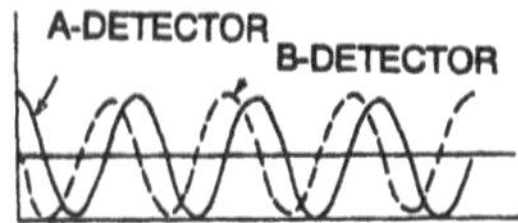

Fig. 14.11 Laser operating elements and fringe patterns.

$L = 0.3$ m, $A = 4.33 \times 10^{-3}$ m^2, $\lambda = 6.0 \times 10^{-7}$ m, and $\Delta\phi = \frac{1}{4}$, can be calculated as

$$\text{S.F.} = \frac{(0.3)(6.0 \times 10^{-7})}{(4.33 \times 10^{-3})(4)} \frac{1}{4} = 2.6 \times 10^{-6} = 2.6 \ \mu \text{ radian/pulse}$$

Thus we get a pulse every time the case rotates through an angle of 2.6×10^{-6} radians, or 2.6 μ radians. In other words, if the case is rotating about the input axis at a rate of 1.0 radian/s, the gyro will produce 1/S.F. or 3.8×10^5 pulses per second.

The RLG has certain advantages as a strapdown rate-measuring instrument. Among these are good quantization and no rotation-induced errors. However, there is one major problem associated with RLGs that must be addressed before the instrument has any utility: this problem is *lock-in*.

The term *lock-in* refers to the coupling between the clockwise and counter-clockwise beams. As a result, for rotational rates that are below a certain value, the beams lock together and produce no measurable output. Depending upon the design, lock-in can occur for rotation rates below 1 to 500 deg/h. Onc solution to the lock-in phenomenon is to dither the case mechanically about the input axis. Dither amplitude may be about 1500 μ radians, at a frequency between 20 and 40 Hz.

For example,† let us take a dead zone of 500 deg/h, a dither frequency of 30 Hz, and an amplitude of 1500 μ radians. We can calculate that

$$500 \text{ deg/h} = 0.0024 \text{ radian/s}$$

$$f_d = 30\text{Hz} \qquad \omega_d = 188.5 \text{ radian/s} \qquad T_d = 0.0333 \text{ s}$$

$$d\theta/dt = \omega_d \theta = (188.5)(1500 \times 10^{-6}) = 0.2827$$

$$\text{Error} = \frac{\text{time in dead zone}}{\text{total period of dither}} = \frac{(4)(4.54 \times 10^{-5})}{0.0333} = 0.00546 \approx 1/2\%$$

where the time in dead zone, $t = \sin^{-1}(.0024/.2827)/188.5 \approx 4.54 \times 10^{-5}$s. There are four dead zone times per dither cycle. Mechanical dither removes all but one-half of one percent of the potential 500 deg/h drift rate.

Another solution to lock-in is to use a Faraday cell to induce an apparent shift in frequency. This essentially biasing mechanism superimposes a frequency on the beams to remove them from the dead zone of lock-in. Since a strapdown implementation is expected to see rates on the order of 1.0 radian/s, the Faraday cell must bias the beam at this rate.

Other considerations in evaluating an RLG are concerns of bias and scale factor stability. Gyro bias, from the performance point of view, is something like the reverse of lock-in. Whereas lock-in might be described as a zero output for a nonzero input, RLG bias is a nonzero output for a zero input. Physically, bias is caused by circulating flow in the laser cavity that causes forward scattering due to mirror imperfections, among other sources.

The scale factor was evaluated earlier. Clearly, variations in scale factor will result in a quantization error in the measurement of case rotations. This error source can be attributed to a variety of phenomena. Here we need only indicate scale factor stability requirements for strapdown applications. Earlier, we mentioned the dynamic range requirements of the strapdown laser gyro. If the gyro case rotates 1.0×10^6 degrees in an hour and provides performance comparable to inertial grade gyros in a platform implementation (0.05 deg/h),

---

†The numerical values used in this example are from a discussion with Dr. G. T. Schmidt of Charles Stark Draper Laboratory.

then the instrument must be able to measure one part in 20 million. To realize this performance level, the gyro must maintain scale factor stability to better than one part in 20 million. Such performance is certainly attainable in a laboratory setting where long warm-up precedes any testing. Whether a practical RLG implementation can meet such a demanding performance level depends upon a critical examination of current technology.

14.4 Instrument Error Analysis

In this section we will present error models for both the accelerometer and the gyro. We will assume that three mutually orthogonal axes are instrumented by the IMU cluster, i.e., that the specific force vector and angular velocity vector, regardless of direction, are completely resolved in the cluster axes. In addition we will limit each instrument to a single output axis. However, for each instrument there are two other orthogonal axes that must be identified to describe the operation completely.

The accelerometer has an input axis normal to both the pendulum and the hinge, an output or hinge axis, and a pendulum axis along the pendulum. Using the right-hand rule, we can show that an acceleration input along the positive input axis will result in a positive rotation about the output axis (or hinge axis).

A similar geometry is in place for the gyro. The gyro has a spin axis S, an input axis I, and an output O (sometimes identified as the quadrature axis, Q). For an angular momentum vector along the positive S-axis and an input along the positive I-axis, we have an output along the positive output axis.

We will not provide any numerical values for the various error parameters since such assignments are often subjected to industrial proprietary restrictions. Here we will simply assume each error-characterizing parameter represents one standard deviation of noncompensated error.

For the accelerometer the error vector, $\delta \boldsymbol{a}$, is

$$\delta \boldsymbol{a} = \begin{bmatrix} \delta a_x \\ \delta a_y \\ \delta a_z \end{bmatrix} = \begin{bmatrix} \mathrm{BIASX} \\ \mathrm{BIASY} \\ \mathrm{BIASZ} \end{bmatrix} + \begin{bmatrix} \mathrm{SFX} & \mathrm{MXY} & \mathrm{MXZ} \\ \mathrm{MYX} & \mathrm{SFY} & \mathrm{MYZ} \\ \mathrm{MZX} & \mathrm{MZY} & \mathrm{SFZ} \end{bmatrix} \begin{bmatrix} \mathrm{AX} \\ \mathrm{AY} \\ \mathrm{AZ} \end{bmatrix}$$

$$+ \begin{bmatrix} \mathrm{DSFX} & 0 & 0 \\ 0 & \mathrm{DSFY} & 0 \\ 0 & 0 & \mathrm{DSFZ} \end{bmatrix} \begin{bmatrix} \mathrm{AX} \\ \mathrm{AY} \\ \mathrm{AZ} \end{bmatrix}$$

$$+ \begin{bmatrix} \mathrm{K2X} & 0 & 0 \\ 0 & \mathrm{K2Y} & 0 \\ 0 & 0 & \mathrm{K2Z} \end{bmatrix} \begin{bmatrix} \mathrm{AX}^2 \\ \mathrm{AY}^2 \\ \mathrm{AZ}^2 \end{bmatrix} + \begin{bmatrix} \mathrm{K3X} & 0 & 0 \\ 0 & \mathrm{K3Y} & 0 \\ 0 & 0 & \mathrm{K3Z} \end{bmatrix} \begin{bmatrix} \mathrm{AX}^3 \\ \mathrm{AY}^3 \\ \mathrm{AZ}^3 \end{bmatrix} \tag{14.50}$$

Note that we have indicated the components of the specific force vector by $[\mathrm{AX}, \mathrm{AY}, \mathrm{AZ}]^T$. The definition and units of the various error parameters are given in Table 14.2.

Table 14.2 Accelerometer error parameters

Symbol	Description	Commercial Units
BIASi	i-accelerometer bias	μg (micro-g)
SFi	i-accelerometer scale factor error	ppm (parts per million)
Mij	Nonorthogonal misalignment of input axis of i-accelerometer to input axis of j-accelerometer	arc-seconds
DSFi	i-accelerometer differential scale factor error	ppm (parts per million)
K2i	Nonlinearity of quadratic scale factor error	$\mu g/g^2$ (micro-g per g^2)
K3i	Nonlinearity of cubic scale factor error	$\mu g/g^3$ (micro-g per g^3)

The gyro may be represented by the following composite error model:

$$\begin{bmatrix} \delta\omega_x \\ \delta\omega_y \\ \delta\omega_z \end{bmatrix} = \begin{bmatrix} \text{GBDX} \\ \text{GBDY} \\ \text{GBDZ} \end{bmatrix} + \begin{bmatrix} \text{GADIX} & \text{GADSX} & \text{GADOX} \\ \text{GADOY} & \text{GADIY} & \text{GADSY} \\ \text{GADOZ} & \text{GADSZ} & \text{GADIZ} \end{bmatrix} \begin{bmatrix} \text{AX} \\ \text{AY} \\ \text{AZ} \end{bmatrix}$$
$$+ \begin{bmatrix} \text{GAADSIX} & \text{GAADSOX} & 0 \\ 0 & \text{GAADSIY} & \text{GAADSOY} \\ \text{GAADSOZ} & 0 & \text{GAADSIZ} \end{bmatrix} \begin{bmatrix} \text{AXAY} \\ \text{AYAZ} \\ \text{AXAZ} \end{bmatrix}$$
$$+ \begin{bmatrix} \text{GSFX} & \text{ALXY} & \text{ALXZ} \\ \text{ALYX} & \text{GSFY} & \text{ALYZ} \\ \text{ALZX} & \text{ALZY} & \text{GSFZ} \end{bmatrix} \begin{bmatrix} \omega_x \\ \omega_y \\ \omega_z \end{bmatrix} + \begin{bmatrix} \text{GDSFX} & 0 & 0 \\ 0 & \text{GDSFY} & 0 \\ 0 & 0 & \text{GDSFZ} \end{bmatrix} \begin{bmatrix} \omega_x \\ \omega_y \\ \omega_z \end{bmatrix} \tag{14.51}$$

The definitions and units of the various error parameters are given in Table 14.3.

14.5 Strapdown Algorithms

In this section we will outline some algorithms that can have some utility in modeling strapdown systems. As pointed out earlier, the gyro measurements are used differently in the two major implementations: strapdown and platform. In the platform application, the gyro senses whatever angular motion has penetrated the gimbal cluster. The measurement of this motion is then sent to the gimbal motors to restore the attitude of the platform. In the strapdown system the gyro output (angular rate or integral of angular rate) is used to update the computer-held Directional Cosine Matrix between a body and inertial frames. In this section we will assume that the gyro provides either $\boldsymbol{\omega}_{b/I}$ or the rate integrated over a small time interval.

Table 14.3 Gyro error parameters

Symbol	Definition	Commercial units
GBD*i*	Bias drift rate of *i*-gyro	deg/h
GADI*i*	Input axis acceleration-sensitive drift rate of *i*-gyro	deg/h/*g*
GADS*i*	Spin axis acceleration-sensitive drift rate of *i*-gyro	deg/h/*g*
GADO*i*	Output axis acceleration-sensitive drift rate of *i*-gyro	deg/h/*g*
GAADSI*i*	Spin-input axis compliance drift rate of *i*-gyro	deg/h/g^2
GAADSO*i*	Spin-output axis compliance drift rate of *i*-gyro	deg/h/g^2
GSF*i*	Scale factor error of *i*-gyro	ppm (parts per million)
GDSF*i*	Differential scale factor of *i*-gyro	ppm (parts per million)
AL*ij*	Nonorthogonality misalignment of input axis of *i*-gyro to input axis of *j*-gyro	arc-s

In Chapter 4 we considered various ways of relating two frames, in particular a body or moving m-frame and an inertial or fixed f-frame. Let us first use the angle-axis vector to describe the rotation of one frame (the m-frame) with respect to a reference frame (the f-frame). Recall that the angle-axis vector consisted of a line of rotation (defined by three directional cosines) and an angle ϕ indicating the angle of rotation about the vector $\boldsymbol{a}$. Thus, the angle-axis vector is a four-parameter descriptor, as

$$\boldsymbol{\phi} = \phi \boldsymbol{a}$$

$$\boldsymbol{a} = [a_1, a_2, a_3]^{\mathrm{T}} \tag{14.52a}$$

where a_1, a_2, a_3 are the direction cosines of the vector $\boldsymbol{a}$. We can write the derivative of $\boldsymbol{\phi}$ as

$$\frac{\mathrm{d}\boldsymbol{\phi}}{\mathrm{d}t} = \frac{\mathrm{d}\phi}{\mathrm{d}t}\boldsymbol{a} + \phi \frac{\mathrm{d}\boldsymbol{a}}{\mathrm{d}t} \tag{14.52b}$$

From Eqs. (4.71) and (4.77) we may write

$$\frac{\mathrm{d}\phi}{\mathrm{d}t} = (\boldsymbol{\omega}^m_{m/f})^{\mathrm{T}}\boldsymbol{a} = \boldsymbol{a}^{\mathrm{T}}\boldsymbol{\omega}^m_{m/f} \tag{4.77}$$

$$\frac{\mathrm{d}\boldsymbol{a}}{\mathrm{d}t} = \left\{\frac{1}{2}a^* - \frac{1}{2}\left[\frac{\sin(\phi)}{1-\cos(\phi)}\right](\boldsymbol{a}\boldsymbol{a}^{\mathrm{T}} - I)\right\}\boldsymbol{\omega}^m_{m/f} \tag{4.71}$$

where the asterisk indicates the skew-symmetric form of a vector. Thus,

$$\frac{\mathrm{d}\boldsymbol{\phi}}{\mathrm{d}t} = \boldsymbol{a}\boldsymbol{a}^{\mathrm{T}}\boldsymbol{\omega}^{m}_{m/f} + \phi\left\{\frac{1}{2}a^{*} - \frac{1}{2}\left[\frac{\sin(\phi)}{1-\cos(\phi)}\right](a^{*})^{2}\right\}\boldsymbol{\omega}^{m}_{m/f} \tag{14.52c}$$

where we have used the identities

$$\boldsymbol{a} \times (\boldsymbol{a} \times \boldsymbol{\omega}) = \boldsymbol{a}(\boldsymbol{a}\cdot\boldsymbol{\omega}) - \boldsymbol{\omega}(\boldsymbol{a}\cdot\boldsymbol{a})$$

$$a^{*}a^{*}\boldsymbol{\omega} = (\boldsymbol{a}\boldsymbol{a}^{T} - I)\boldsymbol{\omega}$$

We can now use this identity as

$$\boldsymbol{a}\boldsymbol{a}^{\mathrm{T}}\boldsymbol{\omega}^{m}_{m/f} = (a^{*})^{2}\boldsymbol{\omega}^{m}_{m/f} + \boldsymbol{\omega}^{m}_{m/f}$$

and rewrite Eq. (14.52c) as

$$\frac{\mathrm{d}\boldsymbol{\phi}}{\mathrm{d}t} = \boldsymbol{\omega}^{m}_{m/f} + \frac{1}{2}\boldsymbol{\phi}\times\boldsymbol{\omega}^{m}_{m/f} + \frac{1}{\phi^{2}}\left\{1 - \frac{\phi}{2}\frac{\sin(\phi)}{[1-\cos(\phi)]}\right\}\boldsymbol{\phi}\times(\boldsymbol{\phi}\times\boldsymbol{\omega}^{m}_{m/f}) \tag{14.53}$$

We must emphasize that $\boldsymbol{\phi} = \phi\boldsymbol{a}$ acts like a vector in that it obeys the dot product and cross product vector operations. However, $\boldsymbol{\phi}$ does not follow the rules of vector addition/subtraction. We also note that no superscript need be applied to $\boldsymbol{\phi}$ since the components of $\boldsymbol{\phi}$ in both the m-frame and the f-frame are identical.

From Eq. (4.65) we can use the $\boldsymbol{\phi}$ in vector transformations in place of a DCM. That is,

$$\begin{aligned} \boldsymbol{V}^{f} &= C^{f}_{m}\boldsymbol{V}^{m} \\ &= \left\{\cos(\phi)I + \frac{[1-\cos(\phi)]}{\phi^{2}}\boldsymbol{\phi}\boldsymbol{\phi}^{\mathrm{T}} + \frac{\sin(\phi)}{\phi}\boldsymbol{\phi}\times(\ )\right\}\boldsymbol{V}^{m} \end{aligned}$$

which results in the following equivalency between the DCM and the axis-angle parameters:

$$C^{f}_{m} = \cos(\phi)I + \frac{[1-\cos(\phi)]}{\phi^{2}}[\boldsymbol{\phi}\boldsymbol{\phi}^{\mathrm{T}}] + \frac{\sin(\phi)}{\phi}[\boldsymbol{\phi}\times(\ )] \tag{14.54a}$$

or, after some manipulation,

$$C^{f}_{m} = I + \frac{\sin(\phi)}{\phi}[\boldsymbol{\phi}\times(\ )] + \left[\frac{1-\cos(\phi)}{\phi^{2}}\right][\boldsymbol{\phi}\times(\boldsymbol{\phi}\times(\ )] \tag{14.54b}$$

We can make the almost trivial observation that if ϕ, the rotation angle, is zero, then the DCM reduces to the identity matrix. Also if $\boldsymbol{\phi}$ and $\boldsymbol{\omega}_{m/f}$ are aligned, then

$$\frac{\mathrm{d}\boldsymbol{\phi}}{\mathrm{d}t} = \boldsymbol{\omega}_{f/m}$$

Obviously, if $\boldsymbol{\phi}$ is known, the DCM is easily computed. The reverse step, obtaining the axis/angle parameters from the DCM, is equally straightforward, that is,

$$a_1 = \frac{\phi_1}{\phi} = (C_{32} - C_{23})/2\sin(\phi)$$

$$a_2 = \frac{\phi_2}{\phi} = (C_{13} - C_{31})/2\sin(\phi)$$

$$a_3 = \frac{\phi_3}{\phi} = (C_{21} - C_{12})/2\sin(\phi)$$

The rotation angle ϕ is easily found from the trace of the DCM, and the relationship is

$$\cos(\phi) = \tfrac{1}{2}(C_{11} + C_{22} + C_{33} - 1)$$

Next, the quaternions (or more properly the four parameters of Euler) are derived from the axis/angle parameters as follows:

$$\boldsymbol{q} = [\sin(\phi/2)]\boldsymbol{a} = [q_1, q_2, q_3]^{\mathrm{T}}$$

$$q_4 = \cos(\phi/2)$$

Clearly,

$$\boldsymbol{q}^{\mathrm{T}}\boldsymbol{q} + q_4^2 = 1 = q_1^2 + q_2^2 + q_3^2 + q_4^2$$

From Eqs. (14.54) we have

$$C_m^f = (1 - 2\boldsymbol{q}^{\mathrm{T}}\boldsymbol{q})I + 2q_4[\boldsymbol{q} \times (\ )] + 2\boldsymbol{q}\boldsymbol{q}^{\mathrm{T}} \tag{14.55a}$$

or

$$C_m^f = I + 2q_4[\boldsymbol{q} \times (\ )] + 2\{\boldsymbol{q} \times [\boldsymbol{q} \times (\ )]\} \tag{14.55b}$$

The update equation for the DCM is given in Eq. (4.70a) as

$$\frac{\mathrm{d}C_m^f}{\mathrm{d}t} = C_m^f \Omega_{m/f}^m = C_m^f[\boldsymbol{\omega}_{m/f}^m \times (\ )] \tag{14.56a}$$

or, in essence,

$$\dot{C} = \frac{\mathrm{d}C}{\mathrm{d}t} = C\Omega = C[\boldsymbol{\omega} \times (\ )]$$

The integral is symbolically,

$$C(t) = C(0)e^{\Omega t} = C(0)\left(I + \Omega t + \tfrac{1}{2}\Omega\Omega t^2 + \tfrac{1}{3\cdot 2}\Omega\Omega\Omega t^3 + \cdots\right) \tag{14.56b}$$

If we let $\Delta t = t$ be the duration of a pulse from the gyro, we have a vector displacement $\delta\boldsymbol{\theta}$, which can be written as

$$\delta\boldsymbol{\theta} = \boldsymbol{\omega}\Delta t$$

so

$$\begin{aligned} C_{k+1} &= C_k\big\{I + [\delta\boldsymbol{\theta}_k \times (\ )] \\ &\quad + \tfrac{1}{2}[\delta\boldsymbol{\theta}_k \times (\delta\boldsymbol{\theta}_k \times (\ ))] \\ &\quad + \tfrac{1}{6}[\delta\boldsymbol{\theta}_k \times (\delta\boldsymbol{\theta}_k \times (\delta\boldsymbol{\theta}_k \times (\ )))]\big\} \\ &= C_k M_k \end{aligned} \tag{14.57}$$

Since the infinite series represented by M_k must be truncated at some finite number of terms, the conflict is obvious: the more terms that are retained the more accuracy is enhanced, but at the expense of an increase in computational load. An important question at this point is, Can the computations be completed during the gyro pulse cycle Δt?

First- and second-order algorithms follow, respectively, from Eqs. (14.57), or

$$M_k^{(1)} = I + [\delta\boldsymbol{\theta}_k \times (\ )] \tag{14.58a}$$

$$M_k^{(2)} = I + [\delta\boldsymbol{\theta}_k \times (\ )] + \tfrac{1}{2}\{\delta\boldsymbol{\theta}_k \times [\delta\boldsymbol{\theta}_k \times (\ )]\} \tag{14.58b}$$

If we restrict consideration at this point to the first-order algorithm, then the error must be something like the second-order term that is omitted, that is,

$$\tfrac{1}{2}\delta\boldsymbol{\theta}_k \times [\delta\boldsymbol{\theta}_k \times (\ )] = \tfrac{1}{2}\delta\theta_k^2 \times (\ )$$

If we assume that $|\boldsymbol{\omega}| \approx 0.5$ radians per second and require that the error rate should be no greater than 0.05 degrees per hour, then

$$\delta\theta = 0.5\Delta t$$

The error per step is approximately

$$\tfrac{1}{2}\delta\theta^2 = \tfrac{1}{2}(0.5\Delta t)^2 = 0.125\Delta t^2$$

The error rate equals the error per step per Δt, or

$$\tfrac{1}{2}(0.5)^2\Delta t = \tfrac{1}{2}(0.25)\Delta t \leq 0.05 \text{ deg/h} = 2.424 \times 10^{-7} \text{ radian/s}$$

Solving for Δt, we see that Δt must be less than 2 μs. For a rotation rate of 0.1 radian/s, Δt must be about 48 μs. Reducing the drift requirement to 0.01 deg/h will lower these values by one-fifth to 0.4 μs and about 10 μs. Whether this period is too short for computation depends upon the computer, but a magnitude of angular velocity of 0.5 radian/s and a maximum drift rate of 0.01 deg/h seem to indicate the need for a second-order algorithm. For the second-order algorithm [see Eq. (14.58b)] we note that an estimation of the truncation error per step Δt is something like

$$\tfrac{1}{6}(\delta\theta)^3/\Delta t = \tfrac{1}{6}\omega^3\Delta t^2 < 2.424 \times 10^{-7} \text{radian/s}$$

where again we have set the maximum drift rate at 0.05 deg/h. Retaining the angular velocity ω at 0.5 radian/s we have for the computational interval

$$\tfrac{1}{6}(0.5)^3(\Delta t)^2 = 2.424 \times 10^{-7}$$

requiring a Δt of 3.4 ms. The additional computation associated with the second-order algorithm should be easily carried out in 3.4 ms. There seems to be no valid reason to go to a higher order algorithm. Of course, the decision must be based upon a trade-off between computational speed and computational complexity.

If higher order algorithms are required, let us consider the following observations. First, we note that

$$\begin{aligned} \boldsymbol{\delta\theta} \times [\boldsymbol{\delta\theta} \times (\ )] &= [\boldsymbol{\delta\theta}\boldsymbol{\delta\theta}^{\mathrm{T}} - \delta\theta^2 I] \\ &\doteq \delta\theta^2 \times (\ ) \end{aligned} \tag{14.59a}$$

Then we have

$$\begin{aligned} [\boldsymbol{\delta\theta} \times (\ )]^3 &= \boldsymbol{\delta\theta} \times (\boldsymbol{\delta\theta}\boldsymbol{\delta\theta}^{\mathrm{T}} - \delta\theta^2 I) \\ &= -\delta\theta^2 \boldsymbol{\delta\theta} \times (\ ) \end{aligned} \tag{14.59b}$$

since

$$\boldsymbol{\delta\theta} \times \boldsymbol{\delta\theta} = 0$$

and

$$[\boldsymbol{\delta\theta} \times (\ )]^4 = -\delta\theta^2[\boldsymbol{\delta\theta} \times (\ )]^2 \tag{14.59c}$$

If we were to use a fourth-order algorithm we would require a fourth-order cross product. However, a fourth-order cross product reduces to a second-order cross product. Consequently, as an extension of Eqs. (14.58) for $M^{(1)}$ and $M^{(2)}$ to M^{∞} we essentially reproduce Eq. (14.54b) as

$$M_k^{\infty} = I + \frac{\sin(\delta\theta)}{\delta\theta}[\delta\boldsymbol{\theta} \times (\ )] + \left[\frac{1-\cos(\delta\theta)}{(\delta\theta)^2}\right]\{\delta\boldsymbol{\theta} \times [\delta\boldsymbol{\theta} \times (\ )]\} \tag{14.60}$$

Over the time interval Δt, we assume that the body rotation can be represented by the angle/axis vector $\delta\boldsymbol{\theta}$. A reasonable approximation of the derivative of $\delta\boldsymbol{\theta}$ is, from Eq. (14.53),

$$\frac{\mathrm{d}\delta\boldsymbol{\theta}}{\mathrm{d}t} = \boldsymbol{\omega}_{m/f} \tag{14.61}$$

Across the time interval Δt, we can also update the DCM C_m^f from Eqs. (14.56) and (14.57), that is,

$$C_{m_{k+1}}^f = C_{m_k}^f\left\{I + \frac{\sin(\delta\theta)}{\delta\theta}[\delta\boldsymbol{\theta} \times (\ )] + \frac{1-\cos(\delta\theta)}{\delta\theta}[\delta\boldsymbol{\theta} \times (\delta\boldsymbol{\theta} \times (\ ))]\right\} \tag{14.62}$$

Of course we could have replaced M_k^{∞} by a first- or second-order algorithm [see Eq. (14.58a)].

If we assume that the angular velocity is an analytical function of time, we may represent the angular velocity as a Taylor series, or

$$\boldsymbol{\omega}(t) = \boldsymbol{\omega}_k + \dot{\boldsymbol{\omega}}_k t + \ddot{\boldsymbol{\omega}}_k\left(t^2/2\right) + \cdots \tag{14.63a}$$

If we take

$$\Omega \doteq \boldsymbol{\omega} \times (\ )$$

then

$$\Omega(t) = \Omega_k + \dot{\Omega}_k t + \tfrac{1}{2}\ddot{\Omega}_k t^2 + \cdots \tag{14.63b}$$

or, from Eq. (14.56b),

$$C(t) = C_k + \dot{C}_k t + \tfrac{1}{2}\ddot{C}_k t^2 + \tfrac{1}{3!}\dddot{C}_k t^3 + \cdots \tag{14.64}$$

but

$$\dot{C}_k = C_k\Omega_k$$

$$\ddot{C}_k = \dot{C}_k\Omega_k + C_k\dot{\Omega}_k = C_k\Omega_k\Omega_k + C_k\dot{\Omega}_k = C_k\Omega_k^2 + C_k\dot{\Omega}_k$$

$$\dddot{C}_k = C_k\Omega_k^3 + C_k\dot{\Omega}_k\Omega_k + 2C_k\Omega_k\dot{\Omega}_k + C_k\ddot{\Omega}_k$$

so that Eq. (14.64) becomes

$$C(t) = C_k[I + \Omega_k t + \tfrac{1}{2}(\Omega_k^2 + \dot{\Omega}_k)t^2 + \tfrac{1}{6}(\Omega_k^3 + \dot{\Omega}_k\Omega_k + 2\Omega_k\dot{\Omega}_k + \ddot{\Omega}_k)t^3 + \cdots \tag{14.65}$$

We should emphasize that the shorter symbolism, or

$$\Omega_k\Omega_k \doteq \Omega_k^2$$

does not imply that the products are commutative, e.g.,

$$\frac{\mathrm{d}(\Omega_k^2)}{\mathrm{d}t} \neq 2\Omega_k\dot{\Omega}_k$$

$$\frac{\mathrm{d}(\Omega_k^2)}{\mathrm{d}t} = \dot{\Omega}_k\Omega_k + \Omega_k\dot{\Omega}_k$$

We will now compare a second-order update with Eq. (14.65). The second-order update can be written, after Eq. (14.58b), as

$$\begin{aligned} M_k^{(2)} &= I + [\delta\boldsymbol{\theta} \times (\ )] + \tfrac{1}{2}[\delta\boldsymbol{\theta} \times (\ )]^2 \\ C_{k+1} &= C_k M_k^{(2)} \end{aligned} \tag{14.66}$$

From Eqs. (14.61) and (14.63a) we have

$$\begin{aligned} \delta\boldsymbol{\theta}_k &= \int_0^{\Delta t} \boldsymbol{\omega}_k \mathrm{d}t \\ &= \boldsymbol{\omega}_k \Delta t + \frac{1}{2}\frac{\mathrm{d}\boldsymbol{\omega}_k}{\mathrm{d}t}\Delta t^2 + \frac{1}{3!}\frac{\mathrm{d}^2\boldsymbol{\omega}_k}{\mathrm{d}t^2}\Delta t^3 + \cdots \end{aligned} \tag{14.67}$$

where $\boldsymbol{\omega}_k = \boldsymbol{\omega}_{m/f}$ over the time interval Δt. Consequently,

$$\begin{aligned} \delta\boldsymbol{\theta}_k \times (\ ) &= \Omega_k \Delta t + \frac{1}{2}\frac{\mathrm{d}\Omega_k}{\mathrm{d}t}\Delta t^2 + \frac{1}{3!}\frac{\mathrm{d}^2\Omega_k}{\mathrm{d}t^2}\Delta t^3 + \cdots \\ [\delta\boldsymbol{\theta}_k \times (\ )]^2 &= \Omega_k^2 \Delta t^2 + \frac{1}{2}\left(\Omega_k\frac{\mathrm{d}\Omega_k}{\mathrm{d}t} + \frac{\mathrm{d}\Omega_k}{\mathrm{d}t}\Omega_k\right)\Delta t^3 + \cdots \end{aligned} \tag{14.68}$$

The second-order algorithm then becomes

$$\begin{aligned} M_k^{(2)} = \Bigg[& I + \Omega_k \Delta t + \frac{1}{2}\left(\frac{\mathrm{d}\Omega_k}{\mathrm{d}t} + \Omega_k^2\right)\Delta t^2 \\ & + \left(\frac{1}{4}\Omega_k\frac{\mathrm{d}\Omega_k}{\mathrm{d}t}\Omega_k + \frac{1}{4}\frac{\mathrm{d}\Omega_k}{\mathrm{d}t}\Omega_k + \frac{1}{6}\frac{\mathrm{d}^2\Omega_k}{\mathrm{d}t^2}\right)\Delta t^3 \Bigg] \end{aligned} \tag{14.69}$$

Now we may compare the second-order algorithm with the "exact" update by forming the difference in the DCM update, as

$$e[C(\Delta t)] = C(\Delta t)^{(2)}_{\text{algorithm}} - C(\Delta t)_{\text{exact}} \tag{14.70}$$

The error e follows from Eqs. (14.69) and (14.65), that is,

$$e[C(\Delta t)] = C_k\left[-\tfrac{1}{6}\Omega_k^3 + \tfrac{1}{12}\dot{\Omega}_k\Omega_k - \tfrac{1}{12}\Omega_k\dot{\Omega}_k\right] \tag{14.71}$$

The first term in Eq. (14.71) is obviously the third-order term that is neglected in the second-order algorithm. The second and third terms are more interesting, as they represent the non-commutative terms of the rotation over the time interval Δt.

Obviously if the terms

$$\dot{\Omega}_k\Omega_k \qquad \text{and} \qquad \Omega_k\dot{\Omega}_k \tag{14.72}$$

were commutative, then the difference between the second and third terms is zero. We can write the foregoing difference as

$$\dot{\Omega}_k\Omega_k - \Omega_k\dot{\Omega}_k = \begin{bmatrix} 0 & \omega_2\dot{\omega}_1 - \omega_1\dot{\omega}_2 & \omega_3\dot{\omega}_1 - \omega_1\dot{\omega}_3 \\ \omega_1\dot{\omega}_2 - \omega_2\dot{\omega}_1 & 0 & \omega_3\dot{\omega}_2 - \omega_2\dot{\omega}_3 \\ \omega_1\dot{\omega}_3 - \omega_3\dot{\omega}_1 & \omega_2\dot{\omega}_3 - \omega_3\dot{\omega}_2 & 0 \end{bmatrix} \tag{14.73}$$

Each element in this matrix is of the form $\omega_1\dot{\omega}_2 - \omega_2\dot{\omega}_1$. Obviously the term is zero if

$$\omega_1\dot{\omega}_2 = \omega_2\dot{\omega}_1$$

or

$$\frac{\dot{\omega}_2}{\dot{\omega}_1} = \frac{\omega_2}{\omega_1}$$

which is true if

$$\begin{aligned} \omega_2 &= K_1\omega_1 \qquad & \dot{\omega}_2 &= K_1\dot{\omega}_1 \\ \omega_1 &= K_2\omega_3 \qquad & \dot{\omega}_1 &= K_2\dot{\omega}_3 \end{aligned} \tag{14.74}$$

Clearly, Eqs. (14.74) show that the noncommutative error vanishes if the angular velocity is fixed in the body.

If the angular velocity is fixed in the body, then the body is in essence rotating about a shaft (or the rotational motion is equivalent to rotation about a shaft). Under such circumstances the time rate of change of the angle variable equals the angular velocity variable [see Eq. (14.52c)]. Otherwise the noncommutative error will be present. VanderVelde offers the following ideas for at least reducing the contribution of the noncommutative error.[8]

Let us assume that the angular velocity displacement $\delta\boldsymbol{\theta}$ over the time interval Δt may be separated into two parts, $\delta\boldsymbol{\theta}_-$ and $\delta\boldsymbol{\theta}_+$. Then

$$\boldsymbol{\omega}_k = \frac{1}{\Delta t}[\delta\boldsymbol{\theta}_- + \delta\boldsymbol{\theta}_+] \propto \delta\boldsymbol{\theta}_- + \delta\boldsymbol{\theta}_+$$
$$\frac{\mathrm{d}\boldsymbol{\omega}_k}{\mathrm{d}t} = \frac{1}{(\Delta t/2)}\left[\frac{\delta\boldsymbol{\theta}_+}{(\Delta t/2)} - \frac{\delta\boldsymbol{\theta}_-}{(\Delta t/2)}\right] \propto \delta\boldsymbol{\theta}_+ - \delta\boldsymbol{\theta}_- \tag{14.75}$$

We have let $\delta\boldsymbol{\theta}_-$ be the change in the angle $\delta\boldsymbol{\theta}$ over the first half of the interval and $\delta\boldsymbol{\theta}_+$ the change over the remainder, or second half of the interval. From Eqs. (14.75) we see that the angular velocity is proportional to the sum of the two angles, whereas the angular acceleration is proportional to the difference between the two angles.

From Eqs. (14.71) [or (14.74)] the noncommutative contribution may now be written as follows:

$$\begin{aligned}\Omega_k\dot{\Omega}_k - \dot{\Omega}_k\Omega_k \propto & \{[\delta\boldsymbol{\theta}_- \times(\;)] + [\delta\boldsymbol{\theta}_+ \times(\;)]\}\{[\delta\boldsymbol{\theta}_+ \times(\;)] \\ & - [\delta\boldsymbol{\theta}_- \times(\;)]\} - \{[\delta\boldsymbol{\theta}_+ \times(\;)] \\ & - [\delta\boldsymbol{\theta}_- \times(\;)]\}\{[\delta\boldsymbol{\theta}_- \times(\;)] + [\delta\boldsymbol{\theta}_+ \times(\;)]\} \\ \propto & \, 2\{[\delta\boldsymbol{\theta}_- \times(\;)][\delta\boldsymbol{\theta}_+ \times(\;)] - [\delta\boldsymbol{\theta}_+ \times(\;)][\delta\boldsymbol{\theta}_- \times(\;)]\}\end{aligned} \tag{14.76}$$

Thus, the third-order algorithm is

$$\begin{aligned}M_k^{(3)} = & \, I + [\delta\boldsymbol{\theta}_k \times(\;)] + \tfrac{1}{2}[\delta\boldsymbol{\theta}_k \times(\;)]^2 + \tfrac{1}{6}[\delta\boldsymbol{\theta}_k \times(\;)]^3 \\ & + C\left\{[\delta\boldsymbol{\theta}_- \times(\;)][\delta\boldsymbol{\theta}_+ \times(\;)] - [\delta\boldsymbol{\theta}_+ \times(\;)][\delta\boldsymbol{\theta}_- \times(\;)]\right\}\end{aligned} \tag{14.77}$$

We have omitted the subscript k from the parts $\delta\boldsymbol{\theta}_-$ and $\delta\boldsymbol{\theta}_+$. Note that we have inserted a constant C in the preceding expression. The goal now is to evaluate C numerically, subject to the condition that up to the third order the noncommutating error is minimized.

The angular velocity $\boldsymbol{\omega}(t)$ was expanded into a Taylor series in Eq. (14.63a). Thus, we may write for $\delta\boldsymbol{\theta}_-$ and $\delta\boldsymbol{\theta}_+$, truncating after the third term,

$$\delta\boldsymbol{\theta}_- = \int_0^{t/2} \boldsymbol{\omega}(t)\mathrm{d}t = \boldsymbol{\omega}_k\frac{t}{2} + \frac{\mathrm{d}\boldsymbol{\omega}_k}{\mathrm{d}t}\frac{t^2}{8} + \frac{\mathrm{d}^2\boldsymbol{\omega}_k}{\mathrm{d}t^2}\frac{t^3}{48} \tag{14.78a}$$

$$\delta\boldsymbol{\theta}_+ = \int_{t/2}^{t} \boldsymbol{\omega}(t)\mathrm{d}t = \boldsymbol{\omega}_k\frac{t}{2} + \frac{\mathrm{d}\boldsymbol{\omega}_k}{\mathrm{d}t}\frac{3t^2}{8} + \frac{\mathrm{d}^2\boldsymbol{\omega}_k}{\mathrm{d}t^2}\frac{7t^3}{48} \tag{14.78b}$$

It then follows from Eqs. (14.78) that

$$[\delta\boldsymbol{\theta}_- \times (\,)] = \Omega_k\,(t/2) + \dot{\Omega}_k\left(t^2/8\right) + \ddot{\Omega}_k\left(t^3/48\right) \tag{14.79a}$$

$$[\delta\boldsymbol{\theta}_+ \times (\,)] = \Omega_k\,(t/2) + \dot{\Omega}_k\left(3t^2/8\right) + \ddot{\Omega}_k\left(7t^3/48\right) \tag{14.79b}$$

Next we must evaluate the noncommutative term in Eq. (14.77). That is,

$$[\delta\boldsymbol{\theta}_1 \times (\,)][\delta\boldsymbol{\theta}_2 \times (\,)] \approx \Omega_k^2\left(t^2/4\right) + \Omega_k\dot{\Omega}_k\left(3t^3/16\right) + \dot{\Omega}_k\Omega_k\left(t^3/16\right) \tag{14.80a}$$

$$[\delta\boldsymbol{\theta}_2 \times (\,)][\delta\boldsymbol{\theta}_1 \times (\,)] \approx \Omega_k^2\left(t^2/4\right) + \Omega_k\dot{\Omega}_k\left(t^3/16\right) + \dot{\Omega}_k\Omega_k\left(3t^3/16\right) \tag{14.80b}$$

To evaluate the third term in Eq. (14.77), we subtract Eq. (14.80b) from Eq. (14.80a) to get

$$\left(\tfrac{1}{8}\Omega_k\dot{\Omega}_k - \tfrac{1}{8}\dot{\Omega}_k\Omega_k\right)t^3 \tag{14.81}$$

Now we must return to Eq. (14.71). As we recall, this equation represents the residual error between the second-order algorithm [Eq. (14.69)] and the "exact" expansion of the DCM. If we use a third-order algorithm, we will remove the first term in Eq. (14.71) since the Ω_k^3 term is included in the third-order algorithm. However, the noncommutative terms remain. Thus, the error between the third-order algorithm and the "exact" expansion is the difference between Eqs. (14.81) and (14.71), with the first term in Eq. (14.73) omitted for the reason given. Thus, we have the third-order error residual as follows:

$$e[C(t)] = C_k\left[\left(\tfrac{1}{8}C - \tfrac{1}{12}\right)\Omega_k\dot{\Omega}_k + \left(\tfrac{1}{12} - \tfrac{1}{8}C\right)\dot{\Omega}_k\Omega_k\right] \tag{14.82}$$

To make $e[C(t)]$ equal to zero, the constant C is chosen as $\frac{2}{3}$. Thus, we split the time interval into two displacements, $\delta\boldsymbol{\theta}_-$ and $\delta\boldsymbol{\theta}_+$. The following third-order algorithm will result in zero error between the algorithm and the "exact" solution:

$$\begin{aligned} M_k^3 &= I + [\delta\boldsymbol{\theta}_k \times (\,)] + \tfrac{1}{2}[\delta\boldsymbol{\theta}_k \times (\,)]^2 + \tfrac{1}{6}[\delta\boldsymbol{\theta}_k \times (\,)]^3 \\ &\quad + \tfrac{2}{3}\left\{[\delta\boldsymbol{\theta}_- \times (\,)][\delta\boldsymbol{\theta}_+ \times (\,)] - [\delta\boldsymbol{\theta}_+ \times (\,)][\delta\boldsymbol{\theta}_- \times (\,)]\right\} \end{aligned} \tag{14.83}$$

The preceding third-order algorithm was based upon two angular increments $\delta\boldsymbol{\theta}_-$ and $\delta\boldsymbol{\theta}_+$ that divided a single interval into two equal parts.

The next, and final, algorithm, is also third order. We will still consider two angular increments, but this time over two time intervals, each of duration t.

Again we will expand the angular velocity in a Taylor series, truncated at third order [Eq. (14.63a)]. Thus,

$$\begin{aligned} \delta\boldsymbol{\theta}_{k-} &= \int_0^t \left(\boldsymbol{\omega}_k + \frac{\mathrm{d}\boldsymbol{\omega}_k}{\mathrm{d}t}t + \frac{1}{2}\frac{\mathrm{d}^2\boldsymbol{\omega}_k}{\mathrm{d}t^2}t^2\right)\mathrm{d}t \\ &= \boldsymbol{\omega}_k t + \frac{1}{2}\frac{\mathrm{d}\boldsymbol{\omega}_k}{\mathrm{d}t}t^2 + \frac{1}{6}\frac{\mathrm{d}^2\boldsymbol{\omega}_k}{\mathrm{d}t^2}t^3 \end{aligned} \tag{14.84a}$$

and

$$\delta\boldsymbol{\theta}_{k+} = \int_t^{2t}\left(\boldsymbol{\omega}_k + \frac{\mathrm{d}\boldsymbol{\omega}_k}{\mathrm{d}t}t^2 + \frac{1}{2}\frac{\mathrm{d}^2\boldsymbol{\omega}_k}{\mathrm{d}t^2}t^2\right)\mathrm{d}t$$

$$= \boldsymbol{\omega}_k t + \frac{3}{2}\frac{\mathrm{d}\boldsymbol{\omega}_k}{\mathrm{d}t}t^2 + \frac{7}{6}\frac{\mathrm{d}^2\boldsymbol{\omega}_k}{\mathrm{d}t^2}t^3 \tag{14.84b}$$

In skew-symmetric form we have

$$[\delta\boldsymbol{\theta}_{k-} \times (\ )] = \Omega_k t + \frac{1}{2}\frac{\mathrm{d}\Omega_k}{\mathrm{d}t}t^2 + \frac{1}{6}\frac{\mathrm{d}^2\Omega_k}{\mathrm{d}t^2}t^3 \tag{14.85a}$$

$$[\delta\boldsymbol{\theta}_{k+} \times (\ )] = \Omega_k t + \frac{3}{2}\frac{\mathrm{d}\Omega_k}{\mathrm{d}t}t^2 + \frac{7}{6}\frac{\mathrm{d}^2\Omega_k}{\mathrm{d}t^2}t^3 \tag{14.85b}$$

Paralleling the work done in developing Eq. (14.80), we have the following products:

$$[\delta\boldsymbol{\theta}_{k-} \times (\ )][\delta\boldsymbol{\theta}_{k+} \times (\ )] = \Omega_k^2 t^2 + \tfrac{3}{2}\Omega_k\dot{\Omega}_k t^3 + \tfrac{1}{2}\dot{\Omega}_k\Omega_k t^3 \tag{14.86a}$$

$$[\delta\boldsymbol{\theta}_{k+} \times (\ )][\delta\boldsymbol{\theta}_{k-} \times (\ )] = \Omega_k^2 t^2 + \tfrac{1}{2}\Omega_k\dot{\Omega}_k t^3 + \tfrac{3}{2}\dot{\Omega}_k\Omega_k t^3 \tag{14.86b}$$

Subtracting Eq. (14.86b) from (14.86a), we get

$$[\delta\boldsymbol{\theta}_{k-} \times (\ )][\delta\boldsymbol{\theta}_{k+} \times (\ )] - [\delta\boldsymbol{\theta}_{k+} \times (\ )][\delta\boldsymbol{\theta}_{k-} \times (\ )] = \Omega_k\dot{\Omega}_k t^3 - \dot{\Omega}_k\Omega_k t^3 \tag{14.87}$$

Following Eq.(14.82), we find the error term is

$$e[C(t)] = C_k\left[\left(C - \tfrac{1}{12}\right)\Omega_k\dot{\Omega}_k + \left(\tfrac{1}{12} - C\right)\dot{\Omega}_k\Omega_k\right] \tag{14.88}$$

Clearly we can make the error vanish up to the third order if we set $C = \frac{1}{12}$.

Thus, a third-order algorithm using two intervals becomes

$$M_k^{(3)} = I + [\delta\boldsymbol{\theta}_k \times (\ )] + \tfrac{1}{2}[\delta\boldsymbol{\theta}_k \times (\ )]^2 + \tfrac{1}{6}[\delta\boldsymbol{\theta}_k \times (\ )]^3$$
$$+ \tfrac{1}{12}([\delta\boldsymbol{\theta}_{k-} \times (\ )][\delta\boldsymbol{\theta}_{k+} \times (\ )] - [\delta\boldsymbol{\theta}_{k+} \times (\ )][\delta\boldsymbol{\theta}_{k-} \times (\ )]) \tag{14.89}$$

The update of the DCM from the beginning of the kth interval to the beginning of the $(k+1)$th interval is

$$C_{k+1} = C_k M_k^{(i)} \tag{14.90}$$

where the superscript (i) is the order of the algorithm.

Although space does not permit an extensive development of updating algorithms, let us examine at least briefly the quaternion approach. The fundamental equation for quaternion update is given by Eq. (4.82c), repeated below, as

$$\dot{Q} = \tfrac{1}{2}AQ \tag{4.82c}$$

where A is given in Eq. (4.82b) and where

$$Q = [q_1, q_2, q_3, q_4]^{\mathrm{T}} = [\boldsymbol{q}, q_4]^{\mathrm{T}}$$

VanderVelde[8] suggests expanding the quaternion Q in a Taylor series as

$$Q(t) = Q_k + \frac{\mathrm{d}Q_k}{\mathrm{d}t} + \frac{1}{2}\frac{\mathrm{d}^2Q_k}{\mathrm{d}t^2}t^2 + \frac{1}{3!}\frac{\mathrm{d}^3Q_k}{\mathrm{d}t^3}t^3 + \cdots$$

as was done for the DCM in Eq. (14.64). Making use of Eq. (4.82c), we have

$$\frac{\mathrm{d}Q_k}{\mathrm{d}t} = \frac{1}{2}A_kQ_k$$

$$\frac{\mathrm{d}^2Q_k}{\mathrm{d}t^2} = \frac{1}{2}A_k\frac{\mathrm{d}Q_k}{\mathrm{d}t} + \frac{1}{2}\frac{\mathrm{d}A_k}{\mathrm{d}t}Q_k = \frac{1}{2}\left[\left(\frac{1}{2}A_k^2\right) + \left(\frac{\mathrm{d}A_k}{\mathrm{d}t}\right)\right]Q_k$$

$$\frac{\mathrm{d}^3Q_k}{\mathrm{d}t^3} = \frac{1}{2}\left(\frac{1}{4}A_k^3 + \frac{1}{2}A_k\frac{\mathrm{d}A_k}{\mathrm{d}t} + \frac{\mathrm{d}A_k}{\mathrm{d}t}A_k + \frac{\mathrm{d}^2A_k}{\mathrm{d}t^2}\right)Q_k$$

Therefore,

$$Q(t) = \left[I + \tfrac{1}{2}A_kt + \tfrac{1}{4}\left(\tfrac{1}{2}A_k^2 + \dot{A}_k\right)t^2 + \tfrac{1}{12}\left(\tfrac{1}{4}A_k^3 + \tfrac{1}{2}A_k\dot{A}_k + \dot{A}_kA_k + \ddot{A}_k\right)t^3\right]Q_k$$

First define

$$A_\theta = \begin{bmatrix} 0 & \delta\theta_3 & -\delta\theta_2 & \delta\theta_1 \\ -\delta\theta_3 & 0 & \delta\theta_1 & \delta\theta_2 \\ \delta\theta_2 & -\delta\theta_1 & 0 & \delta\theta_3 \\ -\delta\theta_1 & -\delta\theta_2 & -\delta\theta_3 & 0 \end{bmatrix} \tag{14.91a}$$

and

$$\delta\theta = (\delta\theta_1^2 + \delta\theta_2^2 + \delta\theta_3^2)^{1/2} \tag{14.91b}$$

from Eq. (4.82b). VanderVelde[8] then develops the following third-order quaternion updating algorithm:

$$Q_{k+1} = [(1 - \tfrac{1}{8}\delta\theta^2)I + \tfrac{1}{2}(1 - \tfrac{1}{24}\delta\theta^2)A_\theta - \tfrac{1}{6}(A_{\theta_-}A_{\theta_+} - A_{\theta_+}A_{\theta-})]Q_k \tag{14.92}$$

where $\delta\theta$ is the full count over the interval, $\delta\theta_-$, counts over the first half of the interval, and $\delta\theta_+$ counts over the second half of the interval.

As we have indicated, there are various parametric representations of frame orientation. The term *frame updating* means computation of the change in orientation of the frame over an interval. The often tacit operative assumption is the orthogonality of the computed moving frame at the beginning of each interval. Experience has shown that for various reasons the moving frame ceases to maintain orthogonality when more than three descriptive parameters are used. The three-parameter method of Euler angles retains frame orthogonality. (Of course, Euler angles have the serious drawback of requiring extensive computations of transcendental functions.)

The four-parameter quaternions and the nine-parameter directional cosine elements can allow the computationally held moving frame to depart from orthogonality. Departure from orthogonality is easily checked: the norm of the quaternion must be unity, and the product of a DCM with its transpose must equal the identity matrix. However, if nonorthogonality is detected, a special algorithm must be invoked to bring the moving frame back into orthogonality.

As pointed out, the norm of the quaternion must be unity, as

$$Q^{\mathrm{T}}Q = 1 = q_1^2 + q_2^2 + q_3^2 + q_4^2 \tag{4.87c}$$

If numerical round-off errors violate this constraint, we must calculate a new quaternion $Q^{(u)}$, that is,

$$Q^{(u)} = Q/(Q^{\mathrm{T}}Q)^{1/2} \tag{14.93}$$

This reorthogonalization need not take place at every computational step but should be applied at regular intervals.

Of course once the quaternions are reorthogonalized, it is a straightforward matter to construct the DCM. Nevertheless, the DCM may be independently orthogonalized according to the following algorithm. For a representative DCM, or C, the requirement for orthogonality is

$$C^{\mathrm{T}}C = CC^{\mathrm{T}} = I \tag{14.94}$$

Clearly, the nine elements of the DCM, or $C = C_m^f$, may be thought of as the inner product of the unit vectors of the f-system with the unit vectors of the m-system, or

$$C_m^f = \begin{bmatrix} C_1^{\mathrm{T}} \\ C_2^{\mathrm{T}} \\ C_3^{\mathrm{T}} \end{bmatrix} \tag{14.95a}$$

where C_1, C_2, C_3 are the unit vectors of the f-system resolved in the m-system. Also,

$$(C_m^f)^{\mathrm{T}} = C_f^m = [C_1 \quad C_2 \quad C_3] \tag{14.95b}$$

To meet the requirements of Eq. (14.94), we have

$$CC^{\mathrm{T}} = \begin{bmatrix} C_1^{\mathrm{T}}C_1 & C_1^{\mathrm{T}}C_2 & C_1^{\mathrm{T}}C_3 \\ C_2^{\mathrm{T}}C_1 & C_2^{\mathrm{T}}C_2 & C_2^{\mathrm{T}}C_3 \\ C_3^{\mathrm{T}}C_1 & C_3^{\mathrm{T}}C_2 & C_3^{\mathrm{T}}C_3 \end{bmatrix} = I = \begin{bmatrix} 1 & 0 & 0 \\ 0 & 1 & 0 \\ 0 & 0 & 1 \end{bmatrix} \tag{14.96a}$$

Thus,

$$\begin{aligned} C_i^{\mathrm{T}}C_j &= 1 \qquad \text{if} \qquad i = j \\ C_i^{\mathrm{T}}C_j &= 0 \qquad \text{if} \qquad i \neq j \end{aligned} \tag{14.96b}$$

We must now separate the corrections into *stretching* (unit vector no longer of unit magnitude) and *rotating* (unit vectors no longer orthogonal) components. For the stretching corrections, we have

$$C_i^{(u)} = \frac{1}{(C_i^{\mathrm{T}}C_j)^{1/2}}C_i \qquad \text{if} \qquad i = j \tag{14.97a}$$

where $C_i^{(u)}$ is the new or corrected value of $\mathbf{C}_i$.

Next, for the second correction we have

$$C_i^{(u)} = C_i - (C_i^{\mathrm{T}}C_j)C_j \qquad \text{if} \qquad i \neq j \tag{14.97b}$$

Obviously if the axes remain orthogonal, then the term in the parentheses, $C_i^{\mathrm{T}}C_j$, vanishes and $C_i^{(u)} = C_i$. When the term in the parentheses is not zero, then there is a component of C_i along C_j; the second term on the right then is a vector that is subtracted from C_i so that orthogonality is once again true. Note that Eq. (14.97a) applies one condition to C_i because only one condition is necessary to change the length. However, Eq. (14.97b) applies two conditions to C_i since a unit vector can change orientation in two orthogonal directions. Since the preceding corrections are applied to both length and direction, they may be carried out separately.

14.6 System Error Analysis

We will adopt a formalism that is frequently used in system analysis. Let the vector X represent the system states (those entities that are obtained from integration). For a re-entry vehicle we will assume that the state vector consists of a linear velocity vector $\boldsymbol{V}$ and angular velocity $\boldsymbol{\omega}$, a positional vector $\boldsymbol{R}$, and an orientation vector $\boldsymbol{\phi}$. The orientation vector $\boldsymbol{\phi}$ does not obey the laws of vector multiplication, as we indicated in Chapter 4. Thus, the state vector is represented as

$$\boldsymbol{X} = [\boldsymbol{V}, \boldsymbol{\omega}, \boldsymbol{R}, \boldsymbol{\phi}]^{\mathrm{T}} \tag{14.98a}$$

The state $\boldsymbol{X}$ is propagated by integrating the following differential equation:

$$\frac{\mathrm{d}\boldsymbol{X}}{\mathrm{d}t} = \boldsymbol{f}(\boldsymbol{X}, \boldsymbol{U}, t)$$
$$\boldsymbol{X}_0 = \boldsymbol{X}\big|_{t=0} \tag{14.98b}$$

The vector $\boldsymbol{U}$ represents the various control channels.

The state equations [Eqs.(14.98b)] are formed from three force equations and three moment equations, both of which are second order. The first integration of the force equations provides three components of linear velocity; the first integral of the moment equations produces the three components of angular velocity. The integration of six or more geometric equations produces three positional coordinates and a set of three or more variables that characterizes frame orientation, i.e., a body frame relative to a navigation or inertial frame. The form of these equations depends upon frame choice such as Earth-centered inertial (ECI), Earth-centered rotating (ECR), local, etc. Since the introduction of specific frames would at this point mire the discussion, we will retain the general form given in Eq. (14.98b).

The error in a navigation system may be represented as a deviation from a *perfect* or nominal path. We will designate the actual or off-nominal path as the *indicated* path. The presence of noncompensated errors means that the cluster measurements are resolved in the indicated frame, which has drifted from the nominal frame. In a simulation of a navigational system, for example, linear and angular acceleration components are obtained from the force and moment equations; then these *IMU measurements* are corrupted by the accelerometer and gyro error models. Clearly, an error in acceleration will integrate into an error in position. The gyro error (drift) will cause the nominal and indicated body frames to become increasingly misaligned. Consequently the acceleration is resolved in an incorrect frame resulting in a positional error in addition to that attributed to acceleration measurement error. A positional error then produces an incorrect gravity correction to the accelerometer output, resulting in an additional positional error.

To express such navigational errors more precisely, let us first represent state deviations as

$$\boldsymbol{X}(t) = \boldsymbol{X}_n(t) + \delta\boldsymbol{X}(t)$$
$$\boldsymbol{U}(t) = \boldsymbol{U}_n(t) + \delta\boldsymbol{U}(t) \tag{14.99a}$$

The state transition equation then becomes

$$\frac{\mathrm{d}\boldsymbol{X}}{\mathrm{d}t} = \frac{\mathrm{d}\boldsymbol{X}_n}{\mathrm{d}t} + \frac{\mathrm{d}\delta\boldsymbol{X}}{\mathrm{d}t} = \boldsymbol{f}(\boldsymbol{X}_n + \delta\boldsymbol{X}, \boldsymbol{U}_n + \delta\boldsymbol{U}, t)$$

$$\boldsymbol{X}(0) = \boldsymbol{X}_{n_0} + \delta\boldsymbol{X}_0 \tag{14.99b}$$

One method for obtaining the off-nominal cases is to solve the differential equations once for the nominal condition and again (maybe several times) for off-nominal conditions. The nominal solution is then subtracted from the perturbed or off-nominal solution to obtain $\delta\boldsymbol{X}$ as a function of time. The disadvantage of such a procedure is that the nominal and perturbed solutions are usually of the same magnitude. Subtraction then means a loss of numerical significance. On the other hand, if the perturbations are large, then the perturbed solutions tend to introduce system nonlinearities.

The approach that we will take here is to linearize the equations of motion directly, as

$$\boldsymbol{f}(\boldsymbol{X}_n + \delta\boldsymbol{X}, \boldsymbol{U}_n + \delta\boldsymbol{U}, t) = \boldsymbol{f}(\boldsymbol{X}_n, \boldsymbol{U}_n, t) + \left.\frac{\partial \boldsymbol{f}}{\partial \boldsymbol{X}}\right|_n \delta\boldsymbol{X} + \left.\frac{\partial \boldsymbol{f}}{\partial \boldsymbol{U}}\right|_n \delta\boldsymbol{U} \tag{14.100a}$$

The vector gradients are defined as

$$F \doteq \left.\frac{\partial \boldsymbol{f}}{\partial \boldsymbol{X}}\right|_n \qquad G \doteq \left.\frac{\partial \boldsymbol{f}}{\partial \boldsymbol{U}}\right|_n \tag{14.100b}$$

where the subscript n indicates evaluation at the nominal conditions. In subscript notation we have

$$F_{ij} = \frac{\partial f_i}{\partial X_j} \qquad G_{ij} = \frac{\partial f_i}{\partial U_j}$$

The state equations in the perturbed variable then become

$$\frac{\mathrm{d}}{\mathrm{d}t}(\delta\boldsymbol{X}) = F\delta\boldsymbol{X} + G\delta\boldsymbol{U}$$

$$\delta\boldsymbol{X}(0) \doteq \delta\boldsymbol{X}_0 \tag{14.101}$$

Let us now define a b-, or body, frame and two navigation frames, an N_n-, or nominal, frame and an N_i-, or indicated, frame. The indicated frame will drift relative to the nominal frame due to uncompensated gyro drift. To simplify the notation, we will represent the DCM between the N_n and the b-frame as

$$C_n \doteq C_b^{N_n} \tag{14.102a}$$

and between the N_i and the b-frame as

$$C_i \doteq C_b^{N_i} \tag{14.102b}$$

Let

$$C_{N_i}^{N_n} = C(\boldsymbol{\phi}) \qquad C_{N_n}^{N_i} = C(-\boldsymbol{\phi}) \tag{14.102c}$$

where we have indicated the misalignment between the nominal frame and the indicated navigation frames by the rotational vector $\boldsymbol{\phi}$. (Of course $\boldsymbol{\phi}$ could be replaced by three Euler angles, four Euler parameters, or nine DCM elements.)

Clearly, we have

$$C_b^{N_i} = C_{N_n}^{N_i} C_b^{N_n} \tag{14.103a}$$

or

$$C_i = C(-\boldsymbol{\phi}) C_n \tag{14.103b}$$

We may now use Eq. (14.54a) to represent the misalignment after replacing the frame designator f, m with N_i, N_n. That is,

$$C(-\boldsymbol{\phi}) = [\cos(\phi)]I - \frac{\sin(\phi)}{\phi}[\boldsymbol{\phi} \times (\ )] + \frac{[1 - \cos(\phi)]}{\phi^2}\boldsymbol{\phi}\boldsymbol{\phi}^{\mathrm{T}} \tag{14.104}$$

where we replaced $\boldsymbol{\phi}$ by $-\boldsymbol{\phi}$. To the first order, Eq. (14.104) becomes

$$C(-\boldsymbol{\phi}) = I - [\boldsymbol{\phi} \times (\ )]$$

or, from Eq. (14.103b),

$$C_i = \{I - [\boldsymbol{\phi} \times (\ )]\} C_n \tag{14.105}$$

Next, we note that the time derivative of C_n may be given as

$$\frac{\mathrm{d}}{\mathrm{d}t}(C_n) = C_n \Omega_n \tag{14.106}$$

where

$$\Omega_n \doteq \Omega_{b/N_n}^b \doteq [\boldsymbol{\omega}_{b/N_n}^b \times (\ )]$$

and the angular rate $\boldsymbol{\omega}_{b/N_i}^b$ may be represented as a perturbation of $\boldsymbol{\omega}_{b/N_n}^b$, or

$$\begin{aligned} \boldsymbol{\omega}_{b/N_i}^b &= \boldsymbol{\omega}_{b/N_n}^b + \delta\boldsymbol{\omega}_{b/N_n}^b \\ \delta\boldsymbol{\omega}_{b/N_n}^b &= \boldsymbol{\omega}_{N_i/N_n}^b \end{aligned} \tag{14.107}$$

In other words, the error in the angular rate of the body frame relative to the navigation frame is equal to the rotation rate of the indicated frame relative to the nominal frame.

The corresponding skew-symmetric form for the angular velocity follows directly from Eq. (14.107).

Next, return to Eq. (14.105), which we may differentiate as follows:

$$\begin{aligned}\frac{\mathrm{d}}{\mathrm{d}t}(C_i) &= -\left[\frac{\mathrm{d}\boldsymbol{\phi}}{\mathrm{d}t}\times(\,)\right]C_n + \{I - [\boldsymbol{\phi}\times(\,)]\}\frac{\mathrm{d}}{\mathrm{d}t}(C_n)\\ &= -\left[\frac{\mathrm{d}\boldsymbol{\phi}}{\mathrm{d}t}\times(\,)\right]C_n + \{I - [\boldsymbol{\phi}\times(\,)]\}C_n\Omega_n \qquad (14.108\text{a})\end{aligned}$$

or, alternatively,

$$\begin{aligned}\frac{\mathrm{d}}{\mathrm{d}t}(C_i) &= C_i\Omega_i\\ &= \{I - [\boldsymbol{\phi}\times(\,)]\}C_n\Omega_i\\ &= \{I - [\boldsymbol{\phi}\times(\,)]\}C_n\left(\Omega_n + \delta\Omega^b_{b/N_n}\right) \qquad (14.108\text{b})\end{aligned}$$

Again we must emphasize that N refers to the navigation frame (which may be an inertial frame or a local level frame) and i and n refer to *indicated* and *nominal*, respectively.

Next, equating $\mathrm{d}C_i/\mathrm{d}t$ in Eqs. (14.108a) and (14.108b), we get

$$-\left[\frac{\mathrm{d}\boldsymbol{\phi}}{\mathrm{d}t}\times(\,)\right]C_n + \{I - [\boldsymbol{\phi}\times(\,)]\}C_n\Omega_n = \{I - [\boldsymbol{\phi}\times(\,)]\}C_n\left(\Omega_n + \delta\Omega^b_{b/N_n}\right)$$

Assuming that $[\boldsymbol{\phi}\times(\,)]\delta\Omega^b_{b/N_n}$ is of higher order, we get

$$-\left[\frac{\mathrm{d}\boldsymbol{\phi}}{\mathrm{d}t}\times(\,)\right]C_n = C_n\delta\Omega^b_{b/N_n} = C_n\left[\delta\boldsymbol{\omega}^b_{b/N_n}\times(\,)\right] \qquad (14.109\text{a})$$

Next we post-multiply both sides of the above expression by $(C_n)^{\mathrm{T}}$ and obtain

$$\left[\frac{\mathrm{d}\boldsymbol{\phi}}{\mathrm{d}t}\times(\,)\right] = -C^{N_n}_b\delta\Omega^b_{b/N_n}C^b_{N_n} \qquad (14.109\text{b})$$

The right-hand side of Eq. (14.109b) will be recognized as a transformation of a skew-symmetric matrix [Eq. (4.34)] from the b-frame to the N_n-frame, that is,

$$\left[\frac{\mathrm{d}\boldsymbol{\phi}}{\mathrm{d}t}\times(\,)\right] = -\delta\Omega^{N_n}_{b/N_n} = -\left[\delta\boldsymbol{\omega}^{N_n}_{b/N_n}\times(\,)\right]$$

or, equivalently,

$$\frac{\mathrm{d}\boldsymbol{\phi}}{\mathrm{d}t} = \left.\frac{\mathrm{d}\boldsymbol{\phi}}{\mathrm{d}t}\right|_{N_n} = -\delta\boldsymbol{\omega}^{N_n}_{b/N_n} \qquad (14.110)$$

Next we note that

$$\boldsymbol{\omega}^b_{b/I} = \boldsymbol{\omega}^b_{b/N_n} + \boldsymbol{\omega}^b_{N_n/I}$$

which can also be written as

$$\boldsymbol{\omega}^b_{b/N_n} = \boldsymbol{\omega}^b_{b/I} - \left(C^{N_n}_b\right)^{\mathrm{T}} \boldsymbol{\omega}^{N_n}_{N_n/I} \tag{14.111a}$$

Also we may write for the indicated or i-frame

$$\boldsymbol{\omega}^b_{b/N_i} = \boldsymbol{\omega}^b_{b/I} - \left(C^{N_i}_b\right)^{\mathrm{T}} \boldsymbol{\omega}^{N_i}_{N_i/I}$$

From Eqs. (14.110) and (14.111a) we have

$$\begin{aligned}\boldsymbol{\omega}^b_{b/N_n} + \delta\boldsymbol{\omega}^b_{b/N_n} &= \boldsymbol{\omega}^b_{b/I} + \delta\boldsymbol{\omega}^b_{b/I} - \left(\{I - [\boldsymbol{\phi} \times (\ )]\}\, C^{N_n}_b\right)^{\mathrm{T}} \left(\boldsymbol{\omega}^{N_n}_{N_n/I} + \delta\boldsymbol{\omega}^{N_n}_{N_n/I}\right) \\ &= \boldsymbol{\omega}^b_{b/I} + \delta\boldsymbol{\omega}^b_{b/I} - \left(C^{N_n}_b\right)^{\mathrm{T}} \{I + [\boldsymbol{\phi} \times (\ )]\} \left(\boldsymbol{\omega}^{N_n}_{N_n/I} + \delta\boldsymbol{\omega}^{N_n}_{N_n/I}\right)\end{aligned} \tag{14.111b}$$

The nominal solution given in Eq. (14.111a) may be removed from the preceding expression to give

$$\delta\boldsymbol{\omega}^b_{b/N_n} = \delta\boldsymbol{\omega}^b_{b/I} - \left(C^{N_n}_b\right)^{\mathrm{T}} \delta\boldsymbol{\omega}^{N_n}_{N_n/I} - \left(C^{N_n}_b\right)^{\mathrm{T}} [\boldsymbol{\phi} \times (\ )]\boldsymbol{\omega}^{N_n}_{N_n/I} \tag{14.112}$$

where second-order terms such as $[\boldsymbol{\phi} \times (\ )]\delta\boldsymbol{\omega}^{N_n}_{N_n/I}$ are ignored.

Equation (14.112) may now be written as

$$\begin{aligned}\delta\boldsymbol{\omega}^{N_n}_{b/N_n} &= C^{N_n}_b \delta\boldsymbol{\omega}^b_{b/N_n} \\ &= C^{N_n}_b \delta\boldsymbol{\omega}^b_{b/I} - \delta\boldsymbol{\omega}^{N_n}_{N_n/I} - [\boldsymbol{\phi} \times (\ )]\boldsymbol{\omega}^{N_n}_{N_n/I} \\ &= \left[\boldsymbol{\omega}^{N_n}_{N_n/I} \times (\ )\right]\boldsymbol{\phi} + C^{N_n}_b \delta\boldsymbol{\omega}^b_{b/I} - \delta\boldsymbol{\omega}^{N_n}_{N_n/I}\end{aligned} \tag{14.113a}$$

or from Eq. (14.110)

$$\frac{\mathrm{d}\boldsymbol{\phi}}{\mathrm{d}t} = -\Omega^{N_n}_{N_n/I}\boldsymbol{\phi} - C^{N_n}_b \delta\boldsymbol{\omega}^b_{b/I} + \delta\boldsymbol{\omega}^{N_n}_{N_n/I} \tag{14.113b}$$

Again we must emphasize that $\boldsymbol{\phi}$ is a measure of the misalignment between the nominal and indicated navigational frames. Also, when $\mathrm{d}\boldsymbol{\phi}/\mathrm{d}t$ is nonzero, the N_n- and N_i-frames become increasingly misaligned as time increases.

Note in Eq. (14.113b) that three distinct sources contribute to a nonzero value for $\boldsymbol{\phi}$ or misalignment between the two navigation frames. The first term indicates that if the nominal and indicated frames are initially misaligned, the angular rate $\Omega^{N_n}_{N_n/I}$ will be resolved in the wrong frame. In other words, if there is an initial misalignment, then the misalignment will continue to grow. The second term indicates that if there is an uncertainty in the measurement of $\boldsymbol{\omega}_{b/I}$, i.e., gyro drift, then the nominal and indicated navigational frames will become misaligned.

The third source of misalignment is due to an error in the computation of $\boldsymbol{\omega}_{N_n/I}$, indicated by the term $\delta\boldsymbol{\omega}_{N_n/I}$. The term $\boldsymbol{\omega}_{N_n/I}$ is obtained from the navigation computer. For example,

$$\begin{aligned}\boldsymbol{\omega}_{N_n/I} &= \boldsymbol{\omega}_{e/I} + \boldsymbol{\omega}_{N_n/e} \\ &= \begin{bmatrix} \omega_e \cos(L) \\ 0 \\ -\omega_e \sin(L) \end{bmatrix} + \begin{bmatrix} \dot{\ell}\cos(L) \\ -\dot{L} \\ -\dot{\ell}\sin(L) \end{bmatrix} \\ &= \begin{bmatrix} (\dot{\ell} + \omega_e)\cos(L) \\ -\dot{L} \\ -(\dot{\ell} + \omega_e)\sin(L) \end{bmatrix}\end{aligned}$$

Clearly, any positional error will produce errors in longitude ℓ and latitude L, resulting in the error in $\boldsymbol{\omega}^{N_n}_{N_n/I}$ or the third term in Eq. (14.113b).

Next we consider velocity errors. The specific force $\boldsymbol{f}$ is measured in the body frame by the accelerometers (at least for the strapdown system). The measurement may be resolved in the navigation frame as

$$\boldsymbol{f}^N = C^N_b \boldsymbol{f}^b \tag{14.114}$$

Next we treat the errors as a linear deviation from the measurement and using Eq. (14.105), we get

$$\boldsymbol{f}^N_n + \delta\boldsymbol{f}^N = \{I - [\boldsymbol{\phi} \times (\ )]\}\, C^{N_n}_b (\boldsymbol{f}^b_n + \delta\boldsymbol{f}^b) \tag{14.115a}$$

The nominal equations are

$$\boldsymbol{f}^N_n = C^{N_n}_b \boldsymbol{f}^b_n \tag{14.115b}$$

After removing the nominal equations [i.e., subtracting Eq. (14.115b) from (14.115a)], we get

$$\begin{aligned}\delta\boldsymbol{f}^{N_n} &= C^{N_n}_b \delta\boldsymbol{f}^b - [\boldsymbol{\phi} \times (\ )] C^{N_n}_b \boldsymbol{f}^b_n \\ &= C^{N_n}_b \delta\boldsymbol{f}^b - [\boldsymbol{\phi} \times (\ )] \boldsymbol{f}^{N_n}_n \\ &= C^{N_n}_b \delta\boldsymbol{f}^b + [\boldsymbol{f}^{N_n}_n \times (\ )]\boldsymbol{\phi}\end{aligned} \tag{14.116}$$

From this expression we note that

$$\delta\boldsymbol{f}^{N_n} \neq C^{N_n}_b \delta\boldsymbol{f}^b$$

which may or may not be surprising.

Returning to Eq. (14.116), we note that two error sources are combined. The first term on the right accounts for measurement errors. In a simulation these errors come from an accelerometer error model. The second error source is due to frame misalignment in that the indicated frame has drifted from the nominal frame [as expressed by Eq. (14.113b)].

As noted several times earlier, the accelerometer measures only specific force (contact forces divided by mass); the gravitational acceleration must be added to the specific force to provide kinematic acceleration. This gravitational contribution is in error because of a positional error. In other words, because position is in error, the gravitational acceleration as obtained from the gravity model will be in error, resulting in an incorrect gravity compensation to the accelerometer error. The kinematic acceleration will then integrate into a further position error.

We will now write the error equations in the Earth-centered rotating (ECR) frame. As before, we will designate the re-entry vehicle by the geocentric position vector $\boldsymbol{R}$. The velocity of interest will be $\boldsymbol{V}_{b/e}$, the velocity of the RV relative to the Earth. In the ECR frame we have, from Coriolis' Theorem,

$$\left.\frac{\mathrm{d}\boldsymbol{R}}{\mathrm{d}t}\right|_I = \boldsymbol{V}_{b/I} = \left.\frac{\mathrm{d}\boldsymbol{R}}{\mathrm{d}t}\right|_e + \boldsymbol{\omega}_{e/I} \times \boldsymbol{R}$$

$$\boldsymbol{V}_{b/I} = \boldsymbol{V}_{b/e} + \boldsymbol{\omega}_{e/I} \times \boldsymbol{R} \tag{14.117a}$$

when $\boldsymbol{V}_{b/I}$ is the linear velocity of the re-entry vehicle, i.e., b-frame, relative to inertial, I-frame.

It follows from Newton's Second Law that

$$\left.\frac{\mathrm{d}\boldsymbol{V}_{b/I}}{\mathrm{d}t}\right|_I = \boldsymbol{f} + \boldsymbol{g}_e$$

$$= \left.\frac{\mathrm{d}}{\mathrm{d}t}[\boldsymbol{V}_{b/e} + \boldsymbol{\omega}_{e/I} \times \boldsymbol{R}]\right|_e + \boldsymbol{\omega}_{e/I} \times (\boldsymbol{V}_{b/e} + \boldsymbol{\omega}_{e/I} \times \boldsymbol{R})$$

which reduces to

$$\left.\frac{\mathrm{d}\boldsymbol{V}_{b/e}}{\mathrm{d}t}\right|_e = \boldsymbol{f} + \boldsymbol{g} - 2\boldsymbol{\omega}_{b/e} \times \boldsymbol{V}_{b/e}$$

where

$$\boldsymbol{g} = \boldsymbol{g}_e - \boldsymbol{\omega}_{e/I} \times (\boldsymbol{\omega}_{e/I} \times \boldsymbol{R})$$

To summarize, we have for the state equations in the ECR frame

$$\left.\frac{\mathrm{d}\boldsymbol{R}}{\mathrm{d}t}\right|_e = \boldsymbol{V}_{b/e} \tag{14.118a}$$

$$\left.\frac{\mathrm{d}\boldsymbol{V}_{b/e}}{\mathrm{d}t}\right|_e = \left.\frac{\mathrm{d}^2\boldsymbol{R}}{\mathrm{d}t^2}\right|_e = \boldsymbol{f} + \boldsymbol{g} - 2\boldsymbol{\omega}_{e/I} \times \boldsymbol{V}_{b/e} \tag{14.118b}$$

For simplicity we will ignore the Coriolis term in Eq. (14.118b) and represent the gravitational acceleration by an inverse square spherical field, or

$$\left.\frac{\mathrm{d}^2\boldsymbol{R}}{\mathrm{d}t^2}\right|_e = \frac{\mathrm{d}^2\boldsymbol{R}}{\mathrm{d}t^2} = \boldsymbol{f} - \frac{\mu\boldsymbol{R}}{R^3} \tag{14.119}$$

where we have used Eq. (3.93) to describe the central field. Next we represent

$$\boldsymbol{R} = \boldsymbol{R}_n + \delta\boldsymbol{R}$$

$$\boldsymbol{f} = \boldsymbol{f}_n \tag{14.120}$$

where we are ignoring any deviation of the specific force from nominal to simplify the expression.

Inserting Eq. (14.120) into Eq. (14.119), we get

$$\frac{\mathrm{d}^2\boldsymbol{R}}{\mathrm{d}t^2} = \frac{\mathrm{d}^2\boldsymbol{R}_n}{\mathrm{d}t^2} + \frac{\mathrm{d}^2}{\mathrm{d}t^2}(\delta\boldsymbol{R}) \approx \boldsymbol{f}_n - \left(\frac{\mu\boldsymbol{R}_n}{R_n^3} - \frac{2\mu\delta\boldsymbol{R}}{R_n^3}\right) \tag{14.121}$$

By removing the nominal condition of Eq. (14.119), we have

$$\frac{\mathrm{d}^2}{\mathrm{d}t^2}\delta\boldsymbol{R} = \frac{2\mu}{R_n^3}\delta\boldsymbol{R} \tag{14.122}$$

In going from Eq. (14.119) to (14.121), we have made the assumption that $\delta\boldsymbol{R}$ is along $\boldsymbol{R}$, or

$$\boldsymbol{R}\delta R = R\delta\boldsymbol{R}$$

Equation (14.122) admits of the solution

$$\delta\boldsymbol{R} = \tfrac{1}{2}\delta\boldsymbol{R}(0)\left(e^{t/T} + e^{-t/T}\right) \tag{14.123}$$

where

$$T = \left(R_n^3/2\mu\right)^{1/2}$$

Thus, we see that an initial error in position of $\delta\boldsymbol{R}(0)$ will lead to a divergence with a time constant of about 570 seconds due to a gravity gradient effect.

In general the gravity gradient might be generalized by the following matrix:

$$G = \frac{\mathrm{d}\boldsymbol{g}}{\mathrm{d}\boldsymbol{R}}$$

For a central spherical gravitational field, this expression becomes

$$G = -\left(\mu/R^3\right)\left[I - \left(3/R^2\right)\boldsymbol{R}\boldsymbol{R}^{\mathrm{T}}\right] \tag{14.124}$$

This matrix has an additional correction ΔG for oblateness, as given by Eq. (3.100a) and repeated here:

$$\Delta G = -\frac{3}{2}\mu J_2 \frac{R_e^2}{R^5}\left(I - \frac{5\boldsymbol{R}\boldsymbol{R}^{\mathrm{T}}}{R^2}\right) \tag{14.125}$$

We may now write for the total error equations in the ECR frame, replacing N_n by E_n to specify the frame,

$$\frac{\mathrm{d}}{\mathrm{d}t}(\delta \boldsymbol{R}) = \delta \boldsymbol{V} \tag{14.126a}$$

$$\frac{\mathrm{d}}{\mathrm{d}t}(\delta \boldsymbol{V}) = C_b^{E_n}\delta \boldsymbol{f}^b + \left[\boldsymbol{f}_n^{E_n} \times (\ )\right]\boldsymbol{\phi} + G\delta \boldsymbol{R}$$

$$- \boldsymbol{\omega}_{e/I} \times (\boldsymbol{\omega}_{e/I} \times \mathbf{R}) - 2\boldsymbol{\omega}_{e/I} \times \delta \boldsymbol{V} \tag{14.126b}$$

If the Z-axis is coincident with the Earth's polar axis, then the last two terms may be written as

$$\boldsymbol{\omega}_{e/I} \times (\boldsymbol{\omega}_{e/I} \times \delta \boldsymbol{R}) - 2\boldsymbol{\omega}_{e/I} \times \delta \boldsymbol{V} = \begin{bmatrix} \omega_{e/I}^2 \delta X & + & 2\omega_{e/I}\delta V_y \\ \omega_{e/I}^2 \delta Y & - & 2\omega_{e/I}\delta V_x \\ & 0 & \end{bmatrix}$$

The first term on the right-hand side in Eq. (14.126b) accounts for error in the accelerometer. In a simulation this error would come from the accelerometer error model. The second term couples the specific force $\boldsymbol{f}^b$ with the angular displacement between the nominal and indicated frames. The angle $\boldsymbol{\phi}$, representing the nominal/indicated frame misalignment, comes from the integration of the gyro drift rate (gyro error model). The third term indicates that an error in position results in the application of an incorrect gravity correction to the accelerometer output. The final two terms indicate that integration in the ECR system causes errors in position and velocity to feed back into the kinematic acceleration error.

14.7 Summary

In this chapter we have discussed the various instruments that might be used in re-entry vehicle inertial navigation system. We then discussed the types of navigation systems, putting some emphasis on the strapdown implementation. Since the strapdown system is computationally intensive, we discussed at some length the various algorithms for updating the orientation of the body with respect to some chosen inertial frame. Finally we considered the error contributors that compromise the accuracy of the navigator. We showed, for example, that gyro drift contributes not only to frame orientation but also to accelerometer errors. We also showed that positional and velocity errors are contributors to acceleration errors.

References

[1]Britting, K. R., *Inertial Navigation Systems Analysis*, Wiley Interscience, Somerset, NJ, 1971.

[2]Garg, S. C., Morrow, L. D., and Mamen, R., "Strapdown Navigation Technology: A Literature Survey," *Journal of Guidance and Control*, Vol. 1, No. 3, May/June 1978, pp. 161–172.

[3]Savage, P. G., "Introduction to Strapdown Inertial Navigation Systems," Strapdown Associates, Inc., Plymouth, MN, Feb. 1990.

[4]Savage, P. G., "Laser Gyros in Inertial Strapdown Systems," IEEE Position, Location and Navigation Symposium, San Diego, CA, Nov. 1976.

[5]Savage, P. G., "Strapdown Sensors," AGARD Lecture Series 95—Strapdown Inertial Systems, AGARD 7 Rue Ancelle, Neuilly sur Seine, France, June 1978.

[6]Schimdt, G. T., "Strapdown Inertial Systems—Theory and Applications—Introduction and Overview," AGARD Lecture Series 95—Strapdown Inertial Systems, AGARD 7, Neuilly-sur-Seine, France, June 1978.

[7]Gilmore, J. P., and Feldman, J., "Gyro in Torque-to-Balance Strapdown Application," *Journal of Spacecraft and Rockets*, Vol. 7, No. 9, Sept. 1970, pp. 1076–1082.

[8]VanderVelde, W., "Notes for a Course in Strapdown Navigation," Massachusetts Institute of Technology, Cambridge, MA, 1980.

Appendices

Appendix A: Atmosphere Model

Appendix B: Gravitational Model of the Earth

Appendix C: Direction Cosine Matrix for Euler Angles

Appendix D: Re-Entry Trajectory with Diveline Guidance

Appendix E: Solution Formalism for First- and Second-Order Linear Ordinary Differential Equations

Appendix F: Gravity Gradient Moments

Appendix G: Radar Error Model

Appendix H: Establishment of Axis Systems and Directional Cosine Matrices

Appendix A: Atmosphere Model

This appendix contains an atmosphere model that is based upon the 1976 Standard Atmosphere up to 86 km; for altitudes above 86 km up to an altitude of 700 km the model is based upon the 1962 Standard Atmosphere. However, the model may easily be changed to meet another standard or to accommodate different views on the value that should be set for the exospheric temperature. The model specifies that thermal layers be identified and that within each layer the temperature varies at most linearly with altitude. The altitude of the lower boundary of each stratum must be entered along with the appropriate value of molecular temperature and molecular weight. These quantities are entered as subscripted variables in the DATA statement. Within the restriction of linearly varying temperature within each stratum, the model may be generalized by changing the temperature and altitude of each stratum.

The TRUEBASIC™ program in Table A.1 is of main use as a subroutine and is presented as such here.

Table A.1 Atmosphere model

```
! THIS IS A PROGRAM THAT CALLS THE SUBROUTINE ATMOS TO
! CALCULATE ATMOSPHERIC PROPERTIES.
!*************************************************************
!*************************************************************
ALT = 1000
TO = 300
PO = 101325
CALL ATMOS(ALT,TO,PO,PR,DE,TM,C,L)
N1$ = "GEOM-ALT      PRESSURE       DENSITY       KIN TEMP"
N2$ = "KILO-M        NEWT/SQ-M      KG/CU-M       DEG-K"
N3$ = "GEOM-ALT        SPEED OF SOUND         MEAN FREE PATH"
N4$ = "KILO-M          METERS/SECOND          METERS"
M1$ = "###.##        #.####^^^^     #.####^^^^    ###.#"
M2$ = "###.##          ####.###               #.####^^^^"
PRINT N1$
PRINT N2$
PRINT USING M1$: ALT/1000,PR,DE,TM
```

(continued on next page)

Table A.1 (continued) Atmosphere model

```
PRINT
PRINT N3$
PRINT N4$
PRINT USING M2$: ALT/1000,C,L
END
!************************************************************
! SUBROUTINE FOR THE CALCULATION OF ATMOSPHERIC PROPERTIES

SUB ATMOS(ALT,T0,P0,PR,DE,TM,C,L)
! THIS SUBROUTINE CALCULATES ATMOSPHERIC TEMPERATURE,
! DENSITY, PRESSURE, SPEED OF SOUND AS A FUNCTION OF
! ALTITUDE
!
! INPUT: ALTITUDE - ALT (METERS)
!        GROUND LEVEL TEMPERATURE - T0 (DEG-C)
!        GROUND LEVEL PRESSURE - P0 (N/SQ-M)
!
! OUTPUT: PRESSURE AT ALTITUDE - PR (N/SQ-M)
!         DENSITY AT ALTITUDE - DE (KG/CU-M)
!         TEMPERATURE AT ALTITUDE - TM (DEG-C)
!         SPEED OF SOUND - C (METERS/SEC)
!         MEAN FREE PATH LENGTH - L (METERS)
!
! PROGRAM PARAMETERS (REAL VARIABLES)
!        RSTAR - UNIVERSAL GAS CONSTANT (JOULES/KG)
!        P0 - SEA LEVEL ATMOSPHERIC PRESSURE (N/SQ-M)
!        N - AVOGADRO'S NUMBER
!        M0 - SEA LEVEL MOLECULAR WEIGHT OF ATMOSPHERIC GAS
!             (KG/MOLE)
!        S - EFFECTIVE DIAMETER OF ATMOSPHERIC GAS MOLECULE
!            (METERS)
!        KT - THERMAL CONDUCTIVITY (WATTS/M-DEG K)
!        F - COLLISION FREQUENCY (1/SEC)
!        V - PARTICLE AVERAGE VELOCITY (M/SEC)
!        L - MEAN FREE PATH (METERS)
!        RP - PLANETARY RADIUS (M)
! PROGRAM PARAMETERS (SUBSCRIPTED VARIABLES)
!        Z(I) - ALTITUDES OF THERMAL STRATA BREAK POINTS (M)
!        T(I) - MOLECULAR TEMPERATURE AT STRATA BREAK POINTS
!               (DEG-K)
!        D(I) - DENSITY AT STRATA BREAK POINTS (KG/CU-M)
!        P(I) - PRESSURE AT STRATA BREAK POINTS (N/CU-M)
!        LZ(I) - THERMAL LAPSE OVER STRATA (DEG K/M)
!        M(I) - MOLECULAR WEIGHT AT STRATA BREAK POINTS
```

(continued on next page)

Table A.1 (continued) Atmosphere model

```
RSTAR = 8313.432
GO = 9.7803
N = 6.0221E+26
S = 3.65E-10
MO = 28.9664
RP = 6.3781E6
B = 2.0/RP
R = RSTAR/MO
DIM Z(25),T(25),D(25),P(25),M(25),LZ(25)
Z(1) = 0.00
T(1) = TO
M(1) = MO
P(1) = PO
D(1) = P(1)/(R*T(1))
FOR I = 2 TO 21
READ Z(I),T(I),M(I)
NEXT I
DATA 11.0191,        216.65,       28.964
DATA 20.0631,        216.65,       28.964
DATA 32.1619,        228.65,       28.964
DATA 47.3501,        270.65,       28.964
DATA 51.4125,        270.65        28.964
DATA 71.8020,        214.65,       28.962
DATA 86.00,          186.946,      28.962
DATA 100.00,         210.65,       28.880
DATA 110.00,         260.65,       28.560
DATA 120.00,         360.65,       28.070
DATA 150.00,         960.65,       26.920
DATA 160.00,        1110.60,       26.660
DATA 170.00,        1210.65,       26.500
DATA 190.00,        1350.65,       25.850
DATA 230.00,        1550.65,       24.690
DATA 300.00,        1830,65,       22.660
DATA 400.00,        2160.65,       19.940
DATA 500.00,        2420.65,       17.940
DATA 600.00,        2590.65,       16.840
DATA 700.00,        2700.00,       16.170
FOR I = 1 TO 21
Z(I) = Z(I)*1000.0
NEXT I
FOR I = 1 TO 20
LZ(I) = (T(I+1)-T(I))/(Z(I+1)-Z(I))
NEXT I
```

(continued on next page)

Table A.1 (continued) Atmosphere model

```
FOR I = 1 TO 20
  IF ABS(LZ(I))>0.001 THEN
    Q1 = 1 + B*((T(I)/LZ(I))-Z(I))
    Q2 = (Q1*G0/(R*LZ(I)))
    Q3 = T(I+1)/T(I)
    Q4 = Q3^((-1.0)*Q2)
    Q5 = EXP((G0*B*(Z(I+1)-Z(I)))/(R*LZ(I)))
    P(I+1) = P(I)*Q4*Q5
    Q7 = Q2 + 1
    D(I+1) = D(I)*Q5*Q3^((-1.0)*Q7)
  ELSE
    Q8 = (-1.0)*G0*(Z(I+1)-Z(I))*(1-(B/2)*(Z(I+1)+Z(I)))/(R*T(I))
    P(I+1) = P(I)*EXP(Q8)
    D(I+1) = D(I)*EXP(Q8)
  END IF
NEXT I
FOR I = 1 TO 20
    IF ALT < Z(I+1) THEN
        IF ABS(LZ(I))>= 0.001 THEN
           Q1 = 1+B*((T(I)/LZ(I))-Z(I))
           TM = T(I)+LZ(I)*(ALT-Z(I))
           Q2 = (Q1*G0/(R*LZ(I)))
           Q3 = TM/T(I)
           Q4 = Q3^((-1.0)*Q2)
           Q5 = EXP((B*G0*(ALT-Z(I)))/(R*LZ(I)))
           PR = P(I)*Q4*Q5
           Q7 = Q2+1.0
           DE = D(I)*(Q3^((-1.0)*Q7))*Q5
        ELSE
           TM = T(I)
           Q8 = (-1.0)*G0*(ALT-Z(I))*(1.0-(B/2)*)ALT+Z(I)))/(R*T(I))
           PR = P(I)*EXP(Q8)
           DE = D(I)*EXP(Q8)
        END IF
  C = SQR(1.4*R*TM)
  MOL = M(I)+(M(I+1)-M(I))/(Z(I+1)-Z(I))*(ALT-Z(I))
  TM = MOL*TM/M0
  MU = 1.458E-6*(TM^(3/2))/(TM+110.4)
  NU = MU/DE
  KT = 2.64638E-3*(TM^(3/2))/(TM+245.4*(10^(-12/TM))
  V = SQR(8*R*TM/PI)
  F = SQR(2)*PI*N*(S^2)*V
  L = V/F
  EXIT FOR
    END IF
NEXT I
END SUB
```

Appendix B: Gravitational Model of the Earth

This appendix contains a gravitational model of the Earth that incorporates both the spherical and nonspherical contributions. The program can easily be adapted to model the gravitational field of other planets by setting the appropriate geophysical parameters, that is,

R_E = planetary equatorial radius
R_p = planetary polar radius
e = ellipticity (difference between the equatorial radius and the pole radius), divided by equatorial radius, or

$$(R_E - R_p)/R_E$$

J_2, J_3, J_4 = Jeffery constants

The program requires as input the geographical latitude L and the altitude or distance along the geoid normal. Latitude is in degrees and altitude is in meters. The output is components of the gravitational vector (m/s^2) in geographic and geocentric coordinates.

The TRUEBASIC™ program in Table B.1 is intended to be used as a subprogram. In such a case all the comment statements (REM or !) and PRINT statements could be omitted. The INPUT requiring L (latitude) and H (altitude) would be omitted or replaced by a call sequence.

Table B.1 Gravity program

```
! THIS PROGRAM IS A CALL TO THE SUBROUTINE GRAVITY WHICH
! CALCULATES THE COMPONENTS OF THE GRAVITATIONAL ACCELERATION
! IN A GEOCENTRIC AND A GEOGRAPHIC REFERENCE FRAME.
L = 35
H = 1000
CALL GRAVITY(L,H,GCX,GCZ,GGN,GGD,D,R0)
N1$ = "ALTITUDE        LATITUDE         VERTICAL DEF        EARTH
RAD"
N2$ = "METERS          DEGREES          DEGREES           KILO-M"
N3$ = "#.####^^^^      ###.###          ##.####           ######.#"
```

(continued on next page)

Table B.1 (continued) Gravity program

```
N4$ = "GEOCENTRIC GRAVITY            GEOCENTRIC GRAVITY"
N5$ = "EARTH RADIUS COMPONENT          LATITUDE COMPONENT"
N6$ = "METERS/SECOND-SQ                METERS/SECOND-SQ"
N7$ = "##.####                         ##.####^^^^"
M1$ = "GEOGRAPHIC GRAVITY             GEOGRAPHIC GRAVITY"
M2$ = "DOWN COMPONENT                  NORTH COMPONENT"
M3$ = "METERS/SECOND-SQ                METERS/SECOND-SQ"
M4$ = "##.####                         ##.####^^^^"
PRINT N1$
PRINT N2$
PRINT
PRINT USING N3$: H,L*180/PI,D*180/PI,R0/1000
PRINT
PRINT
PRINT N4$
PRINT N5$
PRINT N6$
PRINT
PRINT USING N7$: GCZ,GCX
PRINT
PRINT
PRINT M1$
PRINT M2$
PRINT M3$
PRINT
PRINT
PRINT USING M4$: GGD,GGN
END
!*******************************************************
SUB GRAVITY(L,H,GCX,GCZ,GGN,GGD,D,R0)
! THIS PROGRAM COMPUTES THE GRAVITATIONAL ACCELERATION IN
! GEOCENTRIC AND GEOGRAPHIC FRAMES
!
!              VARIABLE  DEFINITION
! GEOCENTRIC-X:  GCX - POSITIVE FOR INCREASING LATITUDE
! GEOCENTRIC-Z:  GCZ - POSITIVE TOWARDS EARTH CENTER
! GEOGRAPHIC-N:  GGN - POSITIVE NORTH
! GEOGRAPHIC-D:  GGD - POSITIVE DOWN
!
! INPUT          L: LATITUDE (DEGREES)
!                H: ALTITUDE (METERS)
! OUTPUT:        GCX: X-GRAVITATIONAL ACCELERATION (M/SEC-SQ)
!                GCZ: Z-GRAVITATIONAL ACCELERATION (M/SEC-SQ)
!                GGN: N-GEOGRAPHIC ACCELERATION (M/SEC-SQ)
!                GGD: D-GEOGRAPHIC ACCELERATION (M/SEC-SQ)
!
!
```

(continued on next page)

Table B.1 (continued) Gravity program

```
GM = 3.986005E+14
J2 = 1.08263E-03
J3 = 2.532153E-03
J4 = 1.61098E-06
RE = 6.378135E+06
E = 1/298.257
L = (PI/180.0)*L
PHI = (PI/2.0)-L
D0 = E*SIN(2*L)-((E^2)/2)*SIN(2*L)+2*(E^2)*SIN(2*L)*((SIN(L))^2)
K2 = 2*E*(1-(E/2))
LC0 = L-D0
R0 = SQR(1-K2*((COS(LC0))^2))
R0 = RE*(1-E)/R0
R = (R0+H)*SQR(1-(2*H*R0*(1-COS(D0))/((R0+H)^2)))
GR = (1.5)*J2*((RE/R)^2)*(3*(COS(PHI)^2)-1)
GR = GR+2*J3*((RE/R)^3)*COS(PHI)*(5*(COS(PHI)^2)-3)
GR = GR+(5/8)*J4*((RE/R)^4)*(35*(COS(PHI)^4)-30*(COS(PHI)^2)+3)
GR = (-1)*GM*(1-GR)/(R^2)
G0 = J2+(5/6)*J4*((RE/R)^2)*(7*(COS(PHI)^2)-3)
G0 = 3*(GM/(R^2))*((RE/R)^2)*SIN(PHI)*COS(PHI)*G0
G0 = G0+(3/2)*(GM/(R^2))*((RE/R)^3)*J3*SIN(PHI)*(5*(COS(PHI)^2)-1)
GCX = G0
GCZ = (-1)*GR
D = E*(1-(H/R0))*SIN(2*L)
SD = (K2/2)*(R0/(R0+H))*SIN(2*L)*COS(D0)
SD = SD+K2*(R0/(RO+H))*(SIN(L)^2)*SIN(D0)
CD = SQR(1-SD^2)
GGN = (-1)*G0*CD-GR*SD
GGD = G0*SD-GR*CD
END SUB
```

Appendix C: Direction Cosine Matrix for Euler Angles

In Section 4.4 we considered one of twelve possible Euler schemes of rotation. In this appendix we will present the Direction Cosine Matrix (DCM) for all twelve schemes of Euler Angles. The symbolism is the same as that used in Chapter 4 and must be understood before the following presentation is of use.

The Euler angle approach to developing a (DCM) involves three successive rotations about axes that are mutually orthogonal. In the symbolism used here, θ_3 is the first angle of rotation, θ_2 is the second angle of rotation, and θ_1 is the last angle. The subscripts on the symbol for the rotation matrices refer to the axes about which rotation takes place. For example,

$$C_3(\theta_1)C_1(\theta_2)C_2(\theta_3) \tag{C.1}$$

means that the first axis of rotation is the 2-axis, the second axis of rotation is the intermediate 1-axis and the third axis of rotation is the new 3-axis. Obviously there are 12 combinations of three objects taken three at a time such that no two adjacent objects are equal. $C_1C_1C_3$ is meaningful, but does not fully span three space. Such a rotation means rotate about the 3-axis, then rotate about the 1-axis and finally rotate again about the 1-axis. Obviously the last two rotations could be combined into a single rotation about the 1-axis. All 12 combinations are listed in Table C.1.

The combination $C_1C_2C_1$ is given in Eq. (4.39), where $\theta_1 = \phi, \theta_2 = \alpha, \theta_3 = \psi$.

In Table C.2 we have the matrices that transform body-fixed components of angular velocity into the time derivatives of the Euler angles. In general,

$$\begin{aligned} \frac{\mathrm{d}\boldsymbol{\theta}}{\mathrm{d}t} &= [\dot{\theta}_1, \dot{\theta}_2, \dot{\theta}_3]^{\mathrm{T}} \\ &= E_{123}^{-1}\boldsymbol{\omega} \\ &= E_{123}^{-1}(\theta_1, \theta_2)\boldsymbol{\omega}_{b/I}^{b} \end{aligned} \tag{C.2}$$

The Euler angle subscripted with 3 is the first angle of rotation, that subscripted with 2 is the second angle of rotation, and that subscripted with 1 is the third angle of rotation.

The transformation matrices to transform Euler angle rates into components of body-fixed angular velocity are given in Table C.3. For example,

Table C.1 Twelve schemes of Euler angles

$$C_1(\theta_1)C_2(\theta_2)C_1(\theta_3) = \begin{bmatrix} c_2 & s_2 s_3 & -s_2 c_3 \\ s_1 s_2 & c_1 c_3 - s_1 c_2 s_3 & c_1 s_3 + s_1 c_2 c_3 \\ c_1 s_2 & -s_1 c_3 - c_1 c_2 s_3 & -s_1 s_3 + c_1 c_2 c_3 \end{bmatrix}$$

$$C_1(\theta_1)C_3(\theta_2)C_1(\theta_3) = \begin{bmatrix} c_2 & s_2 c_3 & s_2 s_3 \\ -c_1 s_2 & c_1 c_2 c_3 - s_1 s_3 & c_1 c_2 s_3 + s_1 c_3 \\ s_1 s_2 & -s_1 c_2 c_3 - c_1 s_3 & -s_1 c_2 s_3 + c_1 c_3 \end{bmatrix}$$

$$C_1(\theta_1)C_3(\theta_2)C_2(\theta_3) = \begin{bmatrix} c_2 c_3 & s_2 & -c_2 s_3 \\ -c_1 s_2 c_3 + s_1 s_3 & c_1 c_2 & c_1 s_2 s_3 + s_1 c_3 \\ s_1 s_2 c_3 + c_1 s_3 & -s_1 c_2 & -s_1 s_2 s_3 + c_1 c_3 \end{bmatrix}$$

$$C_1(\theta_1)C_2(\theta_2)C_3(\theta_3) = \begin{bmatrix} c_2 c_3 & c_2 s_3 & -s_2 \\ -c_1 s_3 + s_1 s_2 c_3 & c_1 c_3 + s_1 s_2 s_3 & s_1 c_2 \\ s_1 s_3 + c_1 s_2 c_3 & -s_1 c_3 + c_1 s_2 s_3 & c_1 c_2 \end{bmatrix}$$

$$C_2(\theta_1)C_1(\theta_2)C_2(\theta_3) = \begin{bmatrix} c_1 c_3 - s_1 c_2 s_3 & s_1 s_2 & -c_1 s_3 - s_1 c_2 c_3 \\ s_2 s_3 & c_2 & s_2 c_3 \\ s_1 c_3 + c_1 c_2 s_3 & -c_1 s_2 & -s_1 s_3 + c_1 c_2 c_3 \end{bmatrix}$$

$$C_3(\theta_1)C_2(\theta_2)C_1(\theta_3) = \begin{bmatrix} c_1 c_2 & c_1 s_2 s_3 + s_1 c_3 & -c_1 s_2 c_3 + s_1 s_3 \\ -s_1 c_2 & -s_1 s_2 s_3 + c_1 c_3 & s_1 s_2 c_3 + c_1 s_2 c_3 \\ s_2 & -c_2 s_3 & c_2 c_3 \end{bmatrix}$$

$$C_2(\theta_1)C_3(\theta_2)C_2(\theta_3) = \begin{bmatrix} c_1 c_2 c_3 - s_1 s_3 & c_1 s_2 & -c_1 c_2 s_3 - s_1 c_3 \\ -s_2 c_3 & c_2 & s_2 s_3 \\ s_1 c_2 c_3 + c_1 s_3 & s_1 s_2 & -s_1 c_2 s_3 + c_1 c_3 \end{bmatrix}$$

$$C_2(\theta_1)C_3(\theta_2)C_1(\theta_3) = \begin{bmatrix} c_1 c_2 & c_1 s_2 c_3 + s_1 s_3 & c_1 s_2 s_3 - s_1 c_3 \\ -s_2 & c_2 c_3 & c_s s_3 \\ s_1 c_2 & s_1 s_2 c_3 - c_1 s_3 & s_1 s_2 s_3 + c_1 c_3 \end{bmatrix}$$

$$C_2(\theta_1)C_1(\theta_2)C_3(\theta_3) = \begin{bmatrix} c_1 c_3 - s_1 s_2 s_3 & c_1 s_3 + s_1 s_2 c_3 & -s_1 c_2 \\ -c_2 s_3 & c_2 c_3 & s_2 \\ s_1 c_3 + c_1 s_2 s_3 & s_1 s_3 - c_1 s_2 c_s & c_1 c_2 \end{bmatrix}$$

$$C_3(\theta_1)C_1(\theta_2)C_3(\theta_3) = \begin{bmatrix} c_1 c_3 - s_1 c_2 s_3 & c_1 s_3 + s_1 c_2 c_3 & s_1 s_2 \\ -s_1 c_3 - c_1 c_2 s_3 & -s_1 s_3 + c_1 c_2 c_3 & c_1 s_2 \\ s_2 s_3 & -s_2 c_3 & c_2 \end{bmatrix}$$

$$C_3(\theta_1)C_2(\theta_2)C_3(\theta_3) = \begin{bmatrix} c_1 c_2 c_3 - s_1 s_3 & c_1 c_2 s_3 + s_1 c_3 & -c_1 s_2 \\ -s_1 c_2 c_3 - c_1 s_3 & -s_1 c_2 s_3 + c_1 c_3 & s_1 s_2 \\ s_2 c_3 & s_2 s_3 & c_2 \end{bmatrix}$$

$$C_3(\theta_1)C_1(\theta_2)C_2(\theta_3) = \begin{bmatrix} c_1 c_3 + s_1 s_2 s_3 & s_1 c_2 & -c_1 s_3 + s_1 s_2 c_3 \\ -s_1 c_3 + c_1 s_2 s_3 & c_1 c_2 & s_1 s_3 + c_1 s_2 c_3 \\ c_2 s_3 & -s_2 & c_2 c_3 \end{bmatrix}$$

E_{121} is given in Eq. (4.52). The particular sequence of axis rotations is designated by the sequence of subscripts associated with E or E^{-1}: E_{ijk} means that the k-axis is the first axis of rotation, the j-axis is the second axis of rotation and the i-axis is the third axis of rotation. For example, the sequence (3,1,2) means rotate first about the 2-axis through the angle θ_3, then about the new 1-axis through the angle θ_2, and finally about the new 3-axis through the angle θ_1. Of course, we could associate 3 with Z, 1 with X, and 2 with Y, but we could as well associate 3 with X, 1 with Y, and 2 with Z.

In all of these tables, the terms c and s refer to the cosine and sine functions, respectively, and the subscript identifies the associated angle [e.g., $s_1 \equiv \sin(\theta_1)$].

Table C.2 Transformation of angular velocity in body frame to Euler angle rates

$$E_{121}^{-1} = \begin{bmatrix} 1 & -s_1c_2/s_2 & -c_1c_2/s_2 \\ 0 & c_1 & -s_1 \\ 0 & s_1/s_2 & c_1/s_2 \end{bmatrix} \qquad E_{232}^{-1} = \begin{bmatrix} -c_1c_2/s_2 & 1 & -s_1c_2/s_2 \\ -s_1 & 0 & c_1 \\ c_1/s_2 & 0 & s_1/s_2 \end{bmatrix}$$

$$E_{131}^{-1} = \begin{bmatrix} 1 & c_1c_2/s_2 & -s_1c_2/s_2 \\ 0 & s_1 & c_1 \\ 0 & -c_1/s_2 & s_1/s_2 \end{bmatrix} \qquad E_{231}^{-1} = \begin{bmatrix} c_1s_2/c_2 & 1 & s_1s_2/c_2 \\ -s_1 & 0 & c_1 \\ c_1/c_2 & 0 & s_1/c_2 \end{bmatrix}$$

$$E_{132}^{-1} = \begin{bmatrix} 1 & -c_1s_2/c_2 & s_1s_2/c_2 \\ 0 & s_1 & c_1 \\ 0 & c_1/c_2 & -s_1/c_2 \end{bmatrix} \qquad E_{213}^{-1} = \begin{bmatrix} s_1s_2/c_2 & 1 & -c_1s_2/c_2 \\ c_1 & 0 & s_1 \\ -s_1/c_2 & 0 & c_1/c_2 \end{bmatrix}$$

$$E_{123}^{-1} = \begin{bmatrix} 1 & s_1s_2/c_2 & c_1s_2/c_2 \\ 0 & c_1 & -s_1 \\ 0 & s_1/s_2 & c_1/c_2 \end{bmatrix} \qquad E_{313}^{-1} = \begin{bmatrix} -s_1c_2/s_2 & -c_1c_2/s_2 & 1 \\ c_1 & -s_1 & 0 \\ s_1/s_2 & c_1/s_2 & 0 \end{bmatrix}$$

$$E_{212}^{-1} = \begin{bmatrix} -s_1c_2/s_2 & 1 & c_1c_2/s_2 \\ c_1 & 0 & s_1 \\ s_1/s_2 & 0 & -c_1/s_2 \end{bmatrix} \qquad E_{323}^{-1} = \begin{bmatrix} c_1c_2/s_2 & -s_1c_2/s_2 & 1 \\ s_1 & c_1 & 0 \\ -c_1/s_2 & s_1/s_2 & 0 \end{bmatrix}$$

$$E_{321}^{-1} = \begin{bmatrix} -s_2c_1/c_2 & s_1s_2/c_2 & 1 \\ s_1 & c_1 & 0 \\ c_1/c_2 & -s_1/c_2 & 0 \end{bmatrix} \qquad E_{312}^{-1} = \begin{bmatrix} s_1s_2/c_2 & c_1s_2/c_2 & 1 \\ c_1 & -s_1 & 0 \\ s_1/c_2 & c_1/c_2 & 0 \end{bmatrix}$$

Table C.3 Transformation of Euler angular rates to body-fixed components of angular velocity

$$E_{121} = \begin{bmatrix} 1 & 0 & c_2 \\ 0 & c_1 & s_2 s_1 \\ 0 & -s_1 & c_1 s_2 \end{bmatrix} \qquad E_{232} = \begin{bmatrix} 0 & -s_1 & c_1 s_2 \\ 1 & 0 & c_2 \\ 0 & c_1 & s_1 s_2 \end{bmatrix}$$

$$E_{131} = \begin{bmatrix} 1 & 0 & c_2 \\ 0 & s_1 & -c_1 s_2 \\ 0 & c_1 & s_1 s_2 \end{bmatrix} \qquad E_{231} = \begin{bmatrix} 0 & -s_1 & c_1 c_2 \\ 1 & 0 & -s_2 \\ 0 & c_1 & s_1 c_2 \end{bmatrix}$$

$$E_{132} = \begin{bmatrix} 1 & 0 & s_2 \\ 0 & s_1 & c_1 c_2 \\ 0 & c_1 & -s_1 c_2 \end{bmatrix} \qquad E_{213} = \begin{bmatrix} 0 & c_1 & -s_1 c_2 \\ 1 & 0 & s_2 \\ 0 & s_1 & c_1 c_2 \end{bmatrix}$$

$$E_{123} = \begin{bmatrix} 1 & 0 & -s_2 \\ 0 & c_1 & s_1 c_2 \\ 0 & -s_1 & c_1 c_2 \end{bmatrix} \qquad E_{313} = \begin{bmatrix} 0 & c_1 & s_1 s_2 \\ 0 & -s_1 & c_1 s_2 \\ 1 & 0 & c_2 \end{bmatrix}$$

$$E_{212} = \begin{bmatrix} 0 & c_1 & s_2 s_1 \\ 1 & 0 & c_2 \\ 0 & s_1 & -c_1 s_2 \end{bmatrix} \qquad E_{323} = \begin{bmatrix} 0 & s_1 & -c_1 s_2 \\ 0 & c_1 & s_1 s_2 \\ 1 & 0 & c_2 \end{bmatrix}$$

$$E_{321} = \begin{bmatrix} 0 & s_1 & c_1 c_2 \\ 0 & c_1 & -s_1 c_2 \\ 1 & 0 & s_2 \end{bmatrix} \qquad E_{312} = \begin{bmatrix} 0 & c_1 & s_1 c_2 \\ 0 & -s_1 & c_1 c_2 \\ 1 & 0 & -s_2 \end{bmatrix}$$

Appendix D: Re-Entry Trajectory with Diveline Guidance

In Chapter 9 one of the guidance algorithms discussed at length was identified as "diveline guidance." Diveline guidance is a method for directing the lift vector to bring the velocity vector in coincidence with some diveline emanating from an intended impact point. This appendix includes a particle trajectory program that incorporates the method of diveline guidance. A listing of the TRUEBASIC™ program is given in Table D.1. The output from the listed program is given graphically in Fig. D.1.

Table D.1 Particle re-entry trajectory

```
!*****************************************************************
! THIS PROGRAM CALCULATES THE TRAJECTORY OF A MANEUVERING RE-ENTRY
! VEHICLE (MaRV) OF CONSTANT BALLISTIC COEFFICIENT. THE USER MUST
! ENTER VEHICLE CHARACTERISTICS, INITIAL CONDITIONS, THE DIVELINE
! COORDINATES, SWITCHING ALTITUDES, AZIMUTHS, ELEVATIONS, AND THE
! ATMOSPHERIC SURFACE TEMPERATURE AND PRESSURE.
! ****************************************************************
!                          INPUTS
!                     VEHICLE CONSTANTS
! CD0 - ZERO LIFT DRAG COEFFICIENT
! LDMAX - MAXIMUM LIFT-TO-DRAG RATIO
! FRACLDM - FRACTION OF MAXIMUM LIFT-TO-DRAG RATIO
! REFL - REFERENCE LENGTH (MAXIMUM DIAMETER) (M)
! MASS - MASS (KG)
! MAXACC - MAXIMUM PERMISSIBLE SIDE ACCELERATION (G'S)
! GDL - DIVELINE GAIN
!                   ATMOSPHERIC CONSTANTS
! TEMPSL - TEMPERATURE SEALEVEL (DEG-C)
! PRESSL - PRESSURE SEALEVEL (N/M^2)
!
!                    INITIAL CONDITIONS
! GAMEV0 - FLIGHT PATH ANGLE (DEG)
! AZEV0  - AZIMUTH ANGLE (DEG)
! VELEV0 - VELOCITY MAGNITUDE (M/S)
```

(continued on next page)

Table D.1 (continued) Particle re-entry trajectory

```
!
!                  OUTPUT
! XEV() - EVADER STATES (SIX):
! XEV(1,..) - POSITION EAST (M)
! XEV(2,..) - POSITION NORTH (M)
! XEV(3,..) - POSITION UP (M)
! XEV(4,..) - VELOCITY MAGNITUDE EAST (M/S)
! XEV(5,..) - VELOCITY MAGNITUDE NORTH (M/S)
! XEV(6,..) - VELOCITY MAGNITUDE UP (M/S)

PUBLIC GRAVAC, CLSTAR, REFL, REFA, WEIGHT, BETA
PUBLIC TEMPSL, PRESSL

N1$ = "###.#    ######.#     ######.#   ######.#  #####.#"
N2$ = "TIME       EAST          NORTH      ALTITUDE      VEL-MAG"
N3$ = "SEC        METERS        METERS     METERS        MET/SEC"
N4$ = "ESTIMATED TIME OF IMPACT  ###.### SECONDS"
N5$ = "TIME       ALTITUDE        DRAG        LIFT"
N6$ = "SEC        METERS          G'S         G'S"
N7$ = "###.#     ######.#     ####.##     ####.##"
OPEN #1: NAME "EVSAVE.TRU", CREATE NEWOLD ! CREATE OUTPUT FILE
ERASE #1

PRINT #1: N2$
PRINT #1: N3$
PRINT #1:

! READ EVADER DATA
CD0 = 0.1   ! ZERO LIFT DRAG COEFFICIENT
LDMAX = 2.5   ! MAXIMUM LIFT-TO-DRAG RATIO
REFL = 0.4   ! REFERENCE LENGTH - DIAMETER (M)
MASS = 140.0   ! EVADER MASS (KG)
GRAVAC = 9.7803   ! GRAVITATIONAL ACCELERATION (M/S^2)
MAXACC = 100   ! MAXIMUM PERMISSIBLE SIDE ACCELERATION (G'S)
FRACLDM = 0.95   ! FRACTION OF MAXIMUM LIFT-TO-DRAG COEFFICIENT
GDL = 0.5   ! DIVELINE GAIN
REFA = PI*(REFL^2)/4
BETA = MASS/(CD0*REFA)
CLSTAR = 2.0*CD0*LDMAX
CLGOOD = ((1+SQR(1-FRACLDM^2))/FRACLDM)*CLSTAR
```

(continued on next page)

Table D.1 (continued) Particle re-entry trajectory

```
! READ IN ATMOSPHERIC CONDITIONS
TEMPSL = 20.0    ! SEA LEVEL TEMPERATURE (DEG-C)
PRESSL = 1.01325E+5   ! SEA LEVEL PRESSURE (N/M^2)

! DIMENSION MARB (EVADER) ARRAYS
DIM XEV(1:6,1:500), XDL(1:3,1:6),AZDL(6),ELDL(6),T(1:500)
DIM F(1:6,1:4), F1(6), XE(1:6),XD(1:6), AC(1:3,1:500)
DIM DRAG(1:500), LIFT(500)
!
! INITIALIZE MARB (EVADER) VELOCITY VECTOR
GAMEVO = 30.0    ! INITIAL FLIGHT PATH ANGLE (DEG)
AZEVO =  45.0    ! INITIAL VELOCITY AZIMUTH ANGLE (DEG)
VELEVO = 5.0E+3  ! INITIAL VELOCITY MAGNITUDE (M/S)
!
! INITIALIZE MARB (EVADER) POSITION STATES
XEV(1,1) = 0.0   ! INITIALIZE POSITION EAST (M)
XEV(2,1) = 0.0   ! INITIALIZE POSITION NORTH (M)
XEV(3,1) = 2.5E4 ! INITIALIZE POSITION UP (M)
!
! INITIALIZE MARB (EVADER) VELOCITY STATES
XEV(4,1) = VELEVO*COS(RAD(GAMEVO))*SIN(RAD(AZEVO))
XEV(5,1) = VELEVO*COS(RAD(GAMEVO))*COS(RAD(AZEVO))
XEV(6,1) = (-1.0)*VELEVO*SIN(RAD(GAMEVO))

! INITIALIZE TIME
MAT T = 0.0

! SET DIVELINE GROUND COORDINATES AND SWITCHING ALTITUDES (M)
MAT READ XDL
DATA 3.0E3, 3.0E3, 3.0E3, 3.0E3, 3.0E3, 3.0E3
DATA 3.0E3, 3.0E3, 3.0E3, 3.0E3, 3.0E3, 3.0E3
DATA 20.0E3, 15.0E3, 12.0E3, 8.0E3, 6.0E3, 0.0

! READ IN DIVELINE AZIMUTH ANGLES (DEGREES)
MAT READ AZDL
DATA 45.0, 45.0, 45.0, 45.0, 45.0, 45.0

! READ DIVELINE ELEVATION ANGLES (DEGREES)
MAT READ ELDL
DATA 60.0, 60.0, 60.0, 60.0, 60.0, 60.0
```

(continued on next page)

Table D.1 (continued) Particle re-entry trajectory

```
DT = 0.1   ! SET INTEGRATION TIME STEP (SECONDS)

! DEFINE EVADER STATE EQUATIONS
DEF EVADX(A)   ! SEND OVER EVADER X-VELOCITY
    EVADX = A
END DEF

DEF EVADY(A)   ! SEND OVER EVADER Y-VELOCITY
    EVADY = A
END DEF

DEF EVADZ(A)
    EVADZ = A   ! SEND OVER EVADER Z-VELOCITY
END DEF

DEF EVADVX(RHO,CL,XE(),Q)
    VM = SQR(XE(4)^2+XE(5)^2+XE(6)^2)
    Z9 = (-1.0)*(RHO/(2.0*BETA))*VM*XE(4)*(1+(CL/CLSTAR)^2)
    Z9 = Z9+(RHO*REFA/(2.0*MASS))*CL*(VM^2)*Q
    EVADVX = Z9
END DEF

DEF EVADVY(RHO,CL,XE(),Q)
    VM = SQR(XE(4)^2+XE(5)^2+XE(6)^2)
    Z9 = (-1.0)*(RHO/(2.0*BETA))*VM*XE(5)*(1+(CL/CLSTAR)^2)
    Z9 = Z9+(RHO*REFA/(2.0*MASS))*CL*(VM^2)*Q
    EVADVY = Z9
END DEF

DEF EVADVZ(RHO,CL,XE(),Q)
    VM = SQR(XE(4)^2+XE(5)^2+XE(6)^2)
    Z9 = (-1.0)*(RHO/(2.0*BETA))*VM*XE(6)*(1+(CL/CLSTAR)^2)
    Z9 = Z9+(RHO*REFA/(2.0*MASS))*CL*(VM^2)*Q-GRAVAC
    EVADVZ = Z9
END DEF

MAT F = 0   ! INITIALIZE THE ADAMS-BASHFORD-MOULTON MATRICES

! BEGIN INTEGRATION LOOP
FOR I = 1 TO 480
```

(continued on next page)

Table D.1 (continued) Particle re-entry trajectory

```
FOR L = 1 TO 6
   XE(L) = XEV(L,I)
NEXT L

   FOR J = 1 TO 6
      IF XEV(3,I) > XDL(3,J) THEN
         K = J
         EXIT FOR
      END IF
    NEXT J

! CALCULATE THE DIRECTION OF THE LIFT VECTOR

RDE = SQR((XEV(1,I)-XDL(1,K))^2+(XEV(2,I)-XDL(2,K))^2+XEV(3,I)^2)
VELEV = SQR(XEV(4,I)^2+XEV(5,I)^2+XEV(6,I)^2)
UR1 = (XEV(1,I)-XDL(1,K))/RDE
UR2 = (XEV(2,I)-XDL(2,K))/RDE
UR3 = XEV(3,I)/RDE
UD1 = COS(RAD(ELDL(K)))*SIN(RAD(AZDL(K)))
UD2 = COS(RAD(ELDL(K)))*COS(RAD(AZDL(K)))
UD3 = SIN(RAD(ELDL(K)))
DOT1 = UR1*UD1+UR2*UD2+UR3*UD3
W1 = UR1*DOT1-GDL*UD1
W2 = UR2*DOT1-GDL*UD2
W3 = UR3*DOT1-GDL*UD3
DOT2 = XEV(4,I)*W1+XEV(5,I)*W2+XEV(6,I)*W3
L1 = XEV(4,I)*DOT2-W1*(VELEV^2)
L2 = XEV(5,I)*DOT2-W2*(VELEV^2)
L3 = XEV(6,I)*DOT2-W3*(VELEV^2)
DOT3 = L1^2+L2^2+L3^2
UL1 = L1/SQR(DOT3)
UL2 = L2/SQR(DOT3)
UL3 = L3/SQR(DOT3)
E = ACOS((XEV(4,I)*UD1+XEV(5,I)*UD2+XEV(6,I)*UD3)/VELEV)
ALT = XEV(3,I)
CALL ATMOS(ALT,TEMPSL,PRESSL,PR,RHO,TEMP)
ATMOSD = RHO/(2.0*BETA)
ATMOSL = RHO*REFA/(2.0*MASS)
LM1 = ATMOSL*CLGOOD*(VELEV^2)
LM2 = MAXACC*GRAVAC
```

(continued on next page)

Table D.1 (continued) Particle re-entry trajectory

```
IF LM1 < LM2 THEN   ! CHECK IF LIFT EXCEEDS MAX ACCELERATION
   CL = CLGOOD
   LIFT(I) = LM1/GRAVAC
ELSE
   CL = LM2/(ATMOSL*(VELEV^2))
   LIFT(I) = LM2/GRAVAC
END IF

! GRADUALLY DECREASE LIFT FOR SMALL LIFT OFFSET FROM DIVELINE
IF ABS(E) < 0.2 THEN
    CL = CL*ABS(E)/0.2
END IF

! BEGIN PREDICTOR PART OF THE ADAMS-BASHFORD-MOULTON INTEGRATOR
F(1,1) = EVADX(XEV(4,I))
F(2,1) = EVADY(XEV(5,I))
F(3,1) = EVADZ(XEV(6,I))
F(4,1) = EVADVX(RHO,CL,XE(),UL1)
F(5,1) = EVADVY(RHO,CL,XE(),UL2)
F(6,1) = EVADVZ(RHO,CL,XE(),UL3)

! STORE VEHICLE ACCELERATION
AC(1,I) = F(4,1)
AC(2,I) = F(5,1)
AC(3,I) = F(6,1)

IF I < 2 THEN   ! EULER INTEGRATOR TO JUMPSTART INTEGRATION
    FOR L = 1 TO 6
      XEV(L,I+1) = XEV(L,I)+F(L,I)*DT
    NEXT L
ELSEIF I < 5 THEN   ! TRAPEZOIDAL INTEGRATOR
     FOR L = 1 TO 6
       XEV(L,I+1) = XEV(L,I)+(F(L,I-1)+F(L,I))*(DT/2.0)
     NEXT L
ELSE   ! BEGIN PREDICTOR PART OF ADAMS-BASHFORD-MOULTON ALGORITHM
  FOR L = 1 TO 6
     Z9 = (DT/24.0)*(55.0*F(L,1)-59.0*F(L,2)+37.0*F(L,3)-9.0*F(L,4))
     XEV(L,I+1) = XEV(L,I)+Z9
  NEXT L
```

(continued on next page)

Table D.1 (continued) Particle re-entry trajectory

```
! BEGIN CORRECTOR PART OF ADAMS-BASHFORD-MOULTON ALGORITHM
ALT = XEV(3,I+1)
CALL ATMOS(ALT,TEMPSL,PRESSL,PR,RHO,TEMP)

F1(1) = EVADX(XEV(4,I+1))
F1(2) = EVADY(XEV(5,I+1))
F1(3) = EVADZ(XEV(6,I+1))
F1(4) = EVADVX(RHO,CL,XE(),UL1)
F1(5) = EVADVY(RHO,CL,XE(),UL2)
F1(6) = EVADVZ(RHO,CL,XE(),UL3)

FOR L = 1 TO 6
   Z9 = (9.0*F1(L)+19.0*F(L,1)-5.0*F(L,2)+F(L,3))*(DT/24.0)
   XEV(L,I+1) = XEV(L,I) + Z9
NEXT L

! RESEQUENCE THE TERMS IN THE ABM CORRECTOR ALGORITHM
FOR L = 1 TO 6
    FOR M = 1 TO 3
      F(L,5-M) = F(L,4-M)
    NEXT M
NEXT L

T(I+1) = T(I) + DT   ! INCREMENT TIME ARRAY

IF XEV(3,I+1) < 10.0 THEN EXIT FOR

END IF
T(I+1) = T(I) + DT   ! INCREMENT TIME ARRAY

NEXT I

FOR J = 1 TO 480
VM = SQR(XEV(4,J)^2+XEV(5,J)^2+XEV(6,J)^2)
Z8 = AC(1,J)*XEV(4,J)+AC(2,J)*XEV(5,J)+AC(3,J)*XEV(6,J)
Z9 = Z8/(VM*GRAVAC)
Z8 = Z8/(VM^2)
A1 = AC(1,J)-XEV(4,J)*Z8
A2 = AC(2,J)-XEV(5,J)*Z8
A3 = AC(3,J)-XEV(6,J)*Z8
! LIFT (J) = SQR(A1^2+A2^2+A3^2)/GRAVAC
DRAG(J) = ABS(Z9)
```

(continued on next page)

Table D.1 (continued) Particle re-entry trajectory

```
IF XEV(3,J) < 10 THEN
   TSTAR = XEV(3,J)/XEV(6,J)
   TIMP = TSTAR+T(J)
   EXIT FOR
END IF

PRINT #1, USING N1$: T(J),XEV(1,J), XEV(2,J), XEV(3,J),VM
NEXT J
PRINT #1:
PRINT #1, USING N4$: TIMP
PRINT #1:
PRINT #1: N5$
PRINT #1: N6$
PRINT #1:
FOR J = 1 TO 480
IF XEV(3,J) < 10.0 THEN EXIT FOR
PRINT #1, USING N7$: T(J),XEV(3,J),DRAG(J),LIFT(J)
NEXT J

END
!***************************************************************
```

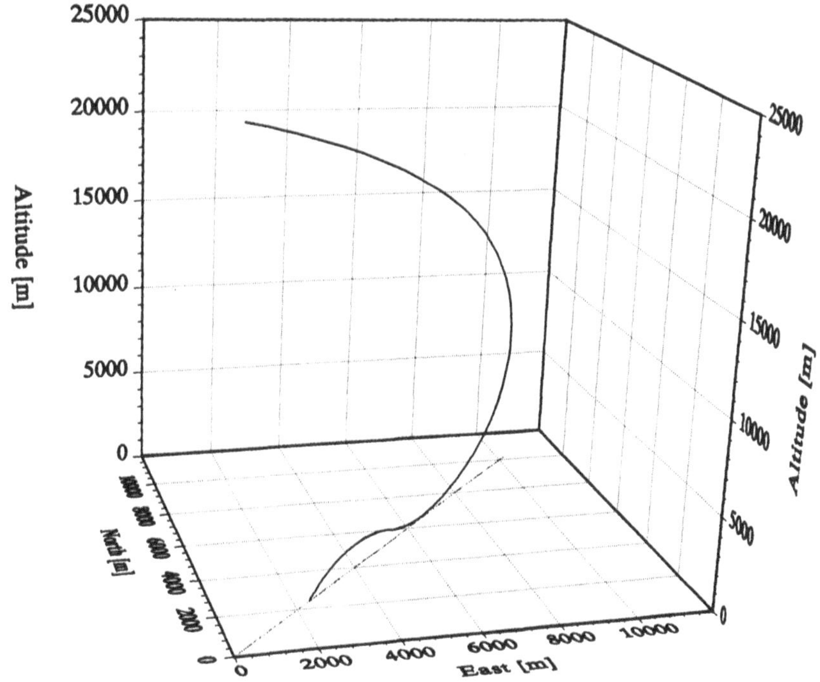

Fig. D.1 Sample diveline trajectory.

The formulation of this program is based upon the analysis given in Chapter 9. We will therefore discuss only the main features of the program and little of the analytical underpinnings. First we note a call to an atmosphere subroutine (ATMOS(..)). This subroutine is given in Appendix A and need not be considered further here except to note that sea-level temperature (TEMPSL) and sea-level pressure (PRESSL) are required in the argument list as "sent over" variables. The position and velocity components are contained in a six-state vector XEV(I,J) where I is the state identifying counter-three positions [east (1), north (2), up (3)] followed by three velocities [east (4), north (5), up (6)]; and J is the integration step counter.

The aerodynamic/mass characteristics are the next input. The drag varies with lift in the form of a drag polar although the operational lift coefficient may be offset (FRACLDM) from the lift coefficient (CLSTAR) corresponding to maximum lift-to-drag ratio (LDMAX).

Next the velocity states are initialized as flight path angle (GAMEV0), azimuth (AZEV0), and magnitude (VELEV0); the position coordinates are then initialized as XEV(1,1), XEV(2,1), and XEV(3,1). Next the diveline ground coordinates and switch altitudes are contained in the matrix XDL, followed by the azimuth and elevation angles of the diveline, AZDL and ELDL respectively.

The state equations are given as defined functions according to the Eq. (9.25). The lift vector is then set by the diveline guidance algorithm according to the development leading up to Eq. (9.35). The Adams-Bashford-Moulton algorithm is used for state equation integration.

The total acceleration is of course computed from the state equations. This acceleration vector is then resolved into drag and lift components as follows: if $\boldsymbol{A}$ is the total acceleration, then the drag acceleration is

$$\text{DRAG} = -(\boldsymbol{A} \cdot \boldsymbol{V})/V = -\left[\, \boldsymbol{A}^{\mathrm{T}}\boldsymbol{V}/(\boldsymbol{V}^{\mathrm{T}}\boldsymbol{V})^{1/2} \,\right] \tag{D.1}$$

where $\boldsymbol{V}$ is the velocity vector; DRAG is of course in the negative direction of the velocity vector; the lift acceleration is then the remaining after the drag vector has been removed from the total acceleration vector, as

$$\text{LIFT} = \boldsymbol{A} - \left[(\boldsymbol{A} \cdot \boldsymbol{V})/V^2\right]\boldsymbol{V} = \boldsymbol{A} - (\boldsymbol{A}^{\mathrm{T}}\boldsymbol{V}/\boldsymbol{V}^{\mathrm{T}}\boldsymbol{V})\boldsymbol{V} \tag{D.2}$$

Appendix E: Solution Formalism for First- and Second-Order Linear Ordinary Differential Equations

E.1 First-Order Differential Equations

Consider the following first-order differential equation with variable coefficients:

$$\frac{dY}{dX} + P(X)Y = Q(X) \tag{E.1}$$

This equation is linear and of first order with variable coefficients: first order because that is the highest order of the derivative, linear because the dependent variable Y appears only to the first order, and of variable coefficient because the coefficients $P(X)$ and $Q(X)$ are not constants but functions of the independent variable.

If $Q(X)$ is zero, the equation is also homogeneous and under such circumstances the solution to Eq. (E.1) is

$$Y = A\exp\left[\int^{X} -P(\tau)\,d\tau\right] \tag{E.2a}$$

where τ is a dummy variable of integration and A is the constant of integration.

The solution of the nonhomogeneous equation is of the form

$$Y = F(X)\exp\left[-\int^{X} P(\tau)\,d\tau\right] \tag{E.2b}$$

The function $F(X)$ is an integrating function. Equation (E.2b) may be rewritten as

$$F(X) = Y\exp\left[\int^{X} P(\tau)\,d\tau\right] \tag{E.3}$$

Differentiating Eq. (E.3), we get

$$\frac{dF(X)}{dX} = \frac{dY}{dX}\exp\left[\int^{X} P(\tau)\,d\tau\right] + YP(X)\exp\left[\int^{X} P(\tau)\,d\tau\right]$$

Next, using Eq. (E.1) we find that the preceding expression becomes

$$\frac{\mathrm{d}F(X)}{\mathrm{d}X} = [Q(X) - P(X)Y]\exp\left[\int^X P(\tau)\,\mathrm{d}\tau\right] + P(X)Y\exp\left[\int^X P(\tau)\,\mathrm{d}\tau\right]$$

$$= Q(X)\exp\left[\int^X P(\tau)\,\mathrm{d}\tau\right] \tag{E.4}$$

Integration of Eq. (E.4) gives

$$F(X) = \int^X Q(\tau)\exp\left[\int^\tau P(\nu)\,\mathrm{d}\nu\right]\mathrm{d}\tau + C \tag{E.5}$$

Replacing $F(X)$ in Eq. (E.2) by Eq. (E.5) provides the integral of the nonhomogeneous equation, or

$$Y = \left\{\int^X Q(\tau)\exp\left[\int^\tau P(\nu)\,\mathrm{d}\nu\right]\mathrm{d}\tau + C\right\}\exp\left[-\int^X P(\tau)\,\mathrm{d}\tau\right] \tag{E.6}$$

E.2 Second-Order Differential Equations

Assume that we have the following second-order differential equation with variable coefficients, i.e., coefficients that are at most functions of the independent variable X:

$$\frac{\mathrm{d}^2Y}{\mathrm{d}X^2} + P(X)\frac{\mathrm{d}Y}{\mathrm{d}X} + Q(X)Y = R(X) \tag{E.7}$$

We wish to find the particular solution, that is, the solution for the case where $R(X)$ is not identically zero. The procedure to be outlined below is called the *method of variation of parameters*. As in the case of the first-order equation, we must find the homogeneous solution first; since Eq. (E.7) is linear, we might expect the homogeneous solution to be formed from two separate functions that are added to form the most general homogeneous solution. Furthermore, since Eq. (E.7) is second order, we also anticipate two constants of integration. Thus, the homogeneous solution can only be written as

$$Y_h(X) = C_1Y_1(X) + C_2Y_2(X) \tag{E.8}$$

Two functions (in this case two solutions) $Y_1(X)$ and $Y_2(X)$ in the domain $(a < X < b)$ are linearly independent if

$$D_1Y_1(X) + D_2Y_2(X) = 0$$

This expression implies that

$$D_1 = D_2 = 0$$

Otherwise, the two solutions are linearly dependent.

The method of variation of parameters states that the particular solution $Y_p(X)$ may be written in terms of the two homogeneous solutions $Y_1(X)$ and $Y_2(X)$ and two as yet undetermined functions $U_1(X)$ and $U_2(X)$ as

$$Y_p(X) = U_1(X)Y_1(X) + U_2(X)Y_2(X) \tag{E.9}$$

It is assumed that $Y_1(X)$ and $Y_2(X)$ are known. Differentiating $Y_p(X)$, we obtain

$$Y_p' = U_1(X)'Y_1(X) + U_1(X)Y_1(X)' + U_2(X)'Y_2(X) + U_2(X)Y_2(X)' \tag{E.10}$$

Since we have two conditions that we may impose upon $Y_p(X)$, we can arbitrarily set the following expression to zero:

$$U_1(X)'Y_1(X) + U_2(X)'Y_2(X) = 0 \tag{E.11}$$

The second condition on $Y_p(X)$ is obvious: $Y_p(X)$ must satisfy the original differential Eq. (E.7). Substituting Eq. (E.9) into Eq. (E.7) gives

$$\begin{aligned} &U_1(X)[Y_1(X)'' + P(X)Y_1(X)' + Q(X)] \\ &\quad + U_2(X)[Y_2(X)'' + P(X)Y_2(X)' + Q(X)] \\ &\quad + U_1(X)'Y_1(X)' + U_2(X)'Y_2(X)' = R(X) \end{aligned}$$

Obviously the terms in the two brackets must vanish and do so independently to leave

$$U_1(X)'Y_1(X)' + U_2(X)'Y_2(X)' = R(X) \tag{E.12}$$

A simultaneous solution of Eqs. (E.11) and (E.12) gives

$$U_1(X) = -\int^X \frac{R(\tau)Y_2(\tau)\,\mathrm{d}\tau}{Y_1(\tau)Y_2(\tau)' - Y_1(\tau)'Y_2(\tau)} \tag{E.13a}$$

$$U_2(X) = \int^X \frac{R(\tau)Y_1(\tau)\,\mathrm{d}\tau}{Y_1(\tau)Y_2(\tau)' - Y_1(\tau)'Y_2(\tau)} \tag{E.13b}$$

Finally, inserting Eqs. (E.13) into Eq. (E.9) yields

$$Y_p = \int^X \left[\frac{Y_2(X)Y_1(\tau) - Y_1(X)Y_2(\tau)}{Y_1(\tau)Y_2(\tau)' - Y_1(\tau)'Y_2(\tau)}\right] R(\tau)\,\mathrm{d}\tau \tag{E.14}$$

Obviously if the functions $Y_1(X)$, $Y_2(X)$, and $R(X)$ are other than the most simple expressions, the utility of Eq. (E.14) for finding the closed-form solutions to Eq. (E.7) is quite limited.

Appendix F: Gravity Gradient Moments

In this appendix we will consider moments applied to a re-entry vehicle due to changes in the gravitational force field across the structure of the vehicle. As might be expected, moments due to gravitational gradient are completely negligible in comparison to the aerodynamic moments of atmospheric entry. However, for a vehicle comparable in size to the Space Shuttle gravitational moments can significantly affect vehicle attitude during orbital motion and therefore at the initiation of atmosphere entry.

Figure F.1 depicts a shuttle-like vehicle in orbit. The vector $\boldsymbol{R}_0$ locates the center of mass of the vehicle from the Earth's center; the vectors $\boldsymbol{R}_i$ and $\boldsymbol{r}$ locate the mass element $\rho \mathrm{d}\forall$ from the center of the Earth and the vehicle center of mass, respectively. Thus, we have

$$\boldsymbol{R}_i = \boldsymbol{R}_0 + \boldsymbol{r} \tag{F.1}$$

The moment due to the gravitational force is given as

$$\boldsymbol{M} = \int_{\forall} \rho(\boldsymbol{r} \times \boldsymbol{g})\, \mathrm{d}\forall \tag{F.2a}$$

The gravitational specific force $\boldsymbol{g}$ acting on the mass element $\mathrm{d}m$, i.e., $\rho \mathrm{d}\forall = \mathrm{d}m$ follows from Newton's law of gravitational attraction as

$$\begin{aligned} \boldsymbol{g} &= -\frac{\mu}{R_i^2}\frac{\boldsymbol{R}_i}{|\boldsymbol{R}_i|} \\ &= -\frac{\mu}{R_i^2}\boldsymbol{U}_R \end{aligned} \tag{F.2b}$$

where we have indicated that the direction of the $\boldsymbol{g}$ vector is along the negative of the unit vector $\boldsymbol{U}_R$ from the center of the Earth to the mass element.

Next we replace the vector $\boldsymbol{R}_i$ according to Eq. (F.1); after some manipulation we obtain

$$\boldsymbol{M} = -\mu \int_{\forall} \left\{ \frac{\boldsymbol{r} \times \boldsymbol{R}_0}{[(\boldsymbol{R}_0 + \boldsymbol{r}) \cdot (\boldsymbol{R}_0 + \boldsymbol{r})]^{3/2}} \right\} \rho\, \mathrm{d}\forall \tag{F.3}$$

where the body density ρ depends in general upon the vector $\boldsymbol{r}$.

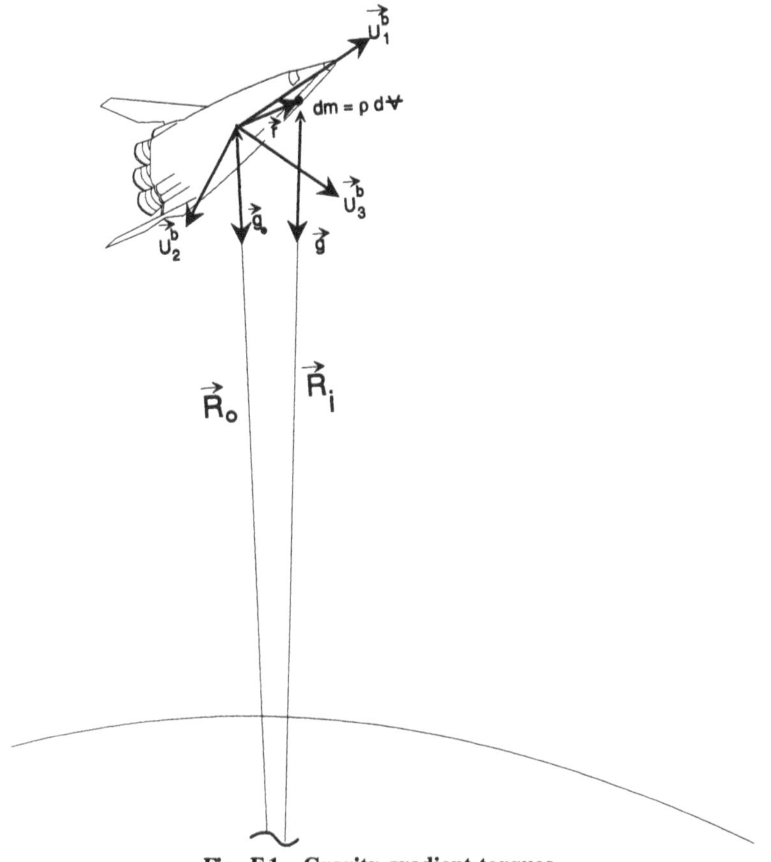

Fig. F.1 Gravity gradient torques.

In carrying out this integration we must write the vectors $\boldsymbol{R}_0$ and $\boldsymbol{r}$ in the body frame as

$$\begin{aligned} \boldsymbol{R}_0 &= X_1\boldsymbol{U}_1^b + X_2\boldsymbol{U}_2^b + X_3\boldsymbol{U}_3^b \\ \boldsymbol{r} &= x_1\boldsymbol{U}_1^b + x_2\boldsymbol{U}_2^b + x_3\boldsymbol{U}_3^b \end{aligned} \tag{F.4}$$

We may now simplify the denominator of Eq. (F.3), subject to the condition that

$$|\boldsymbol{r}| \ll |\boldsymbol{R}_0|$$

$$\frac{|\boldsymbol{r}|}{|\boldsymbol{R}_0|} = \mathcal{O}(-5)$$

Thus, the denominator becomes

$$\begin{aligned}|(\boldsymbol{R}_0 + \boldsymbol{r}) \cdot (\boldsymbol{R}_0 + \boldsymbol{r})|^{-3/2} &= [R_0^2 + 2\boldsymbol{R}_0 \cdot \boldsymbol{r} + r^2]^{-3/2} \\ &\approx [R_0^2 + 2\boldsymbol{R}_0 \cdot \boldsymbol{r}]^{-3/2} \\ &\approx R_0^{-3}\left[1 - (3/R_0^2)(\boldsymbol{R}_0 \cdot \boldsymbol{r})\right] \\ &\approx R_0^{-3}\left[1 - (3/R_0^3)(X_1x_1 + X_2x_2 + X_3x_3)\right] \quad \text{(F.5)}\end{aligned}$$

where the binomial theorem has been used.

Equations (F.4) and (F.5) may now be substituted into Eq. (F.3) to get

$$M_1 = -\frac{\mu}{R_0^3}\int_{V}(x_2X_3 - x_3X_2)\left[1 - \frac{3(X_1x_1 + X_2x_2 + X_3x_3)}{R_0^2}\right]\rho \; \mathrm{d}V \quad \text{(F.6a)}$$

$$M_2 = -\frac{\mu}{R_0^3}\int_{V}(x_3X_1 - x_1X_3)\left[1 - \frac{3(X_1x_1 + X_2x_2 + X_3x_3)}{R_0^2}\right]\rho \; \mathrm{d}V \quad \text{(F.6b)}$$

$$M_3 = -\frac{\mu}{R_0^3}\int_{V}(x_1X_2 - x_2X_1)\left[1 - \frac{3(X_1x_1 + X_2x_2 + X_3x_3)}{R_0^2}\right]\rho \; \mathrm{d}V \quad \text{(F.6c)}$$

We may now examine the second of the preceding equations in detail. The parallel development of the other two equations will be left as an exercise.

First we will expand equation (F.6b) into eight separate integrals, as

$$\begin{aligned}M_2 = &-\frac{\mu}{R_0^3}\left(X_1\int_{V} x_3\rho \; \mathrm{d}V - X_3\int_{V} x_1\rho \; \mathrm{d}V\right) \\ &+ \frac{3\mu}{R_0^5}\left(X_1^2\int_{V} x_1x_3\rho \; \mathrm{d}V + X_1X_2\int_{V} x_2x_3\rho \; \mathrm{d}V + X_1X_3\int_{V} x_3^2\rho \; \mathrm{d}V\right) \\ &- \frac{3\mu}{R_0^5}\left(X_1X_3\int_{V} x_1^2\rho \; \mathrm{d}V + X_2X_3\int_{V} x_1x_2\rho \; \mathrm{d}V + X_3^2\int_{V} x_1x_2\rho \; \mathrm{d}V\right)\end{aligned} \quad \text{(F.7)}$$

To use Eq. (F.7), we first choose an orthogonal body-fixed axis system (U_1^b, U_2^b, U_3^b) that has the following properties: 1) it is located at the center of mass; and 2) it is a principal axis system. We have made use of axis orthogonality in carrying out the cross-product operation in Eq. (F.3). Locating the axis system at the center of mass means that the first two integrals in Eq. (F.7) vanish. Finally the requirement that the axes be principal means that third, fourth, seventh and eighth integrals (the products of inertia) vanish. Now with the fifth and sixth integrals combined, we may write for Eq. (F.7)

$$M_2 = \frac{3\mu}{R_0^5}X_1X_3\left[\int_{V}(x_2^2 + x_3^2)\rho \; \mathrm{d}V - \int_{V}(x_1^2 + x_2^2)\rho \; \mathrm{d}V\right]$$

According to the definition of the moments of inertia [Eqs. (5.54)], the preceding expression may be written as

$$M_2 = (3\mu X_1 X_3 / R_0^5)(I_{11} - I_{33}) \tag{F.8a}$$

The reader can readily show that

$$M_1 = (3\mu X_2 X_3 / R_0^5)(I_{33} - I_{22}) \tag{F.8b}$$

$$M_3 = (3\mu X_1 X_2 / R_0^5)(I_{22} - I_{11}) \tag{F.8c}$$

To discuss vehicle dynamics in which the moment due to the gravitational gradient is significant, we will need three systems of axes. The first is an inertial or I-frame with origin at the center of the Earth. We will arbitrarily set the 2-axis of this system normal to the trajectory or orbital plane. A second axis system has its origin at the center of mass of the vehicle and shares the vehicle's linear velocity associated with orbital motion. The 1-axis of this system defines the local horizontal and the 2-axis is normal to the orbital plane. This system will be identified as the e-frame. The third axis system is the b-frame, so called because it is fixed to the body of the vehicle. The 1-axis is approximately along the thrust direction. The 2-axis is along the right wing and the 3-axis completes the right-handed system. These axes are illustrated in Fig. F.2.

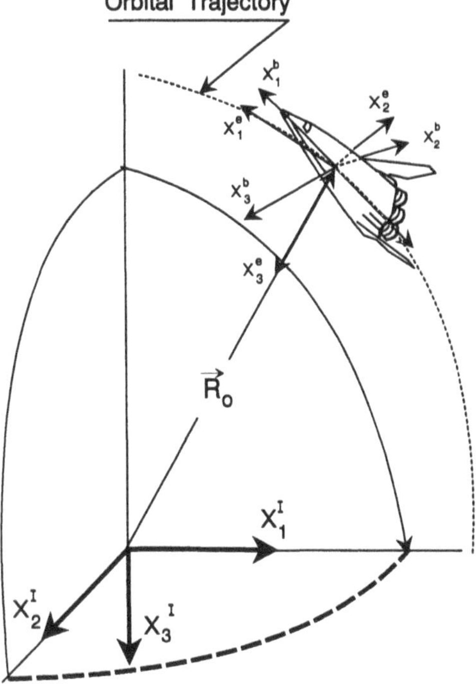

Fig. F.2 Axis system defintion.

The vector representing $\boldsymbol{\omega}^b_{b/I}$, the angular velocity of the body relative to inertial space with components in the body frame, may be separated into two parts: 1) the rotation of the e-frame relative to inertial space due to orbital motion and 2) the rotation of the body or b-frame relative to the e-frame. That is,

$$\begin{aligned}\boldsymbol{\omega}^b_{b/I} &= \boldsymbol{\omega}^b_{b/e} + \boldsymbol{\omega}^b_{e/I} \\ &= \boldsymbol{\omega}^b_{b/e} + C^b_e \boldsymbol{\omega}^e_{e/I}\end{aligned} \tag{F.9}$$

Let us assume that the b-frame makes only small angle displacements relative to the e-frame. We will identify these angles as successive rotations about the 1-, 2-, and 3-axes. We will assume that these angular displacements are small enough that the order of rotation is insignificant. Thus, we will represent these angular displacements by the vector $[\theta_1, \theta_2, \theta_3]^{\mathrm{T}}$. The DCM C^b_e may be written in terms of these angles. For example, refer to Eq. (4.39) where ν is replaced by e and $\theta_1 = \phi, \theta_2 = \alpha$, and $\theta_3 = \psi$. In keeping with the small-angle approximation, we may write Eq. (F.9) as

$$\begin{bmatrix}\omega_1 \\ \omega_2 \\ \omega_3\end{bmatrix} = \begin{bmatrix}\dot{\theta}_1 \\ \dot{\theta}_2 \\ \dot{\theta}_3\end{bmatrix} + \begin{bmatrix}1 & \theta_3 & -\theta_2 \\ -\theta_3 & 1 & \theta_1 \\ \theta_2 & -\theta_1 & 1\end{bmatrix}\begin{bmatrix}0 \\ -\Omega \\ 0\end{bmatrix} \tag{F.10}$$

The rate of change of the central angle is Ω, or the first time derivative of the true anomaly [see Eq. (F.15)].

Carrying out these multiplications gives, for the components of the angular velocity of the body relative to inertial space in b-frame components,

$$\begin{aligned}\omega_1 &= \dot{\theta}_1 - \theta_3\Omega \\ \omega_2 &= \dot{\theta}_2 - \Omega \\ \omega_3 &= \dot{\theta}_3 + \theta_1\Omega\end{aligned} \tag{F.11}$$

The positional vector $\boldsymbol{R}_0$ may be written in the body or b-frame as

$$\begin{bmatrix}X_1 \\ X_2 \\ X_3\end{bmatrix} = C^b_e \boldsymbol{R}^e_0 = \begin{bmatrix}1 & \theta_3 & -\theta_2 \\ -\theta_3 & 1 & \theta_1 \\ \theta_2 & -\theta_1 & 1\end{bmatrix}\begin{bmatrix}0 \\ 0 \\ -R_0\end{bmatrix}$$

The components of the positional vector in the body frame are then

$$\begin{aligned}X_1 &= \theta_2 R_0 \\ X_2 &= -\theta_1 R_0 \\ X_3 &= -R_0\end{aligned} \tag{F.12}$$

We now refer to Eq. (5.56b), where we identify the applied moment with Eqs. (F.8); furthermore, we replace f with I and m with b. Thus we may write

$$I_{11}\dot{\omega}_1 + (I_{33} - I_{22})\omega_3\omega_2 = \frac{3\mu}{R_0^5}X_2X_3(I_{33} - I_{22}) \tag{F.13a}$$

$$I_{22}\dot{\omega}_2 + (I_{11} - I_{33})\omega_1\omega_3 = \frac{3\mu}{R_0^5}X_1X_3(I_{11} - I_{33}) \tag{F.13b}$$

$$I_{33}\dot{\omega}_3 + (I_{22} - I_{11})\omega_1\omega_2 = \frac{3\mu}{R_0^5}X_1X_2(I_{22} - I_{11}) \tag{F.13c}$$

Replacing $\omega_1, \omega_2, \omega_3$ and X_1, X_2, X_3 according to Eqs. (F.11) and (F.12) and neglecting products of perturbational variables and their derivatives gives the following approximations:

$$\omega_2\omega_3 = (\dot{\theta}_2 - \Omega)(\dot{\theta}_3 + \theta_1\Omega) \approx -\dot{\theta}_3\Omega - \theta_1\Omega^2$$

$$\omega_1\omega_3 = (\dot{\theta}_1 - \theta_3\Omega)(\dot{\theta}_3 + \theta_1\Omega) \approx 0$$

$$\omega_1\omega_2 = (\dot{\theta}_1 - \theta_3\Omega) \approx -\dot{\theta}_1\Omega + \theta_3\omega^2$$

$$X_2X_3 = (-\theta_1R_0)(-R_0) = R_0^2\theta_1$$

$$X_1X_3 = (\theta_2R_0)(-R_0) = -R_0^2\theta_2$$

$$X_1X_2 = (\theta_2R_0)(-\theta_1R_0) \approx 0$$

Thus, Eqs. (F.13) may be written as

$$I_{11}\ddot{\theta}_1 + (I_{22} - I_{33} - I_{11})\Omega\dot{\theta}_3 - 4(I_{33} - I_{22})\Omega^2\theta_1 = 0 \tag{F.14a}$$

$$I_{22}\ddot{\theta}_2 + (I_{11} - I_{33})(3\Omega^2\theta_2) = 0 \tag{F.14b}$$

$$I_{33}\ddot{\theta}_3 + (I_{33} - I_{22} + I_{11})\Omega\dot{\theta}_1 + (I_{22} - I_{11})\Omega^2\theta_3 = 0 \tag{F.14c}$$

In the development of the foregoing equations we made use of Kepler's third law, which provides a relationship between the orbital angular rate Ω and the geocentric distance of the satellite as

$$\Omega^2 = \mu/R_0^3 \tag{F.15}$$

This equation of course follows from the fact that the centrifugal acceleration is opposed by the gravitational acceleration. For example, if we replace r by R_0 in Eq. (6.66) we get

$$R_0\left(\frac{\mathrm{d}\theta}{\mathrm{d}t}\right)^2 = \frac{\mu}{R_0^2} = R_0\Omega^2$$

Recall that the 2-axis of the b-frame (pitch axis) is initially normal to the orbital plane. This orientation, which brings the 3-axis along the negative of the geocentric vector and the 1-axis more or less along the velocity vector, is the orientation of a lifting body during re-entry. Thus, Eq. (F.14b) indicates that the pitch angle θ_2 undergoes harmonic motion at a frequency ω_p, given as

$$\omega_p = \Omega\,[3(I_{11} - I_{33})/I_{22}]^{1/2} \tag{F.16a}$$

The frequency given in Eq.(F.16a) is appropriate provided that I_{11} is greater than I_{33}. However, for most aerodynamic configurations I_{33}, the yaw moment of inertia, is larger than that of I_{11}, the roll moment of inertia. Thus, Eq. (F.14b) indicates that when $I_{33} > I_{11}$, the angular motion will be characterized by exponential divergence. The time constant τ for the divergence is given as

$$\tau = \frac{1}{\Omega}\left[\frac{I_{22}}{3(I_{33} - I_{11})}\right]^{1/2} = \frac{T_0}{2\pi}\left[\frac{I_{22}}{3(I_{33} - I_{11})}\right]^{1/2} \tag{F.16b}$$

where T_0 is the orbital period.

A typical single-stage-to-orbit (SSTO) vehicle might have the following inertial properties[1]:

$$I_{11} = 7.53 \times 10^5 \text{ kg-m}^2 \qquad \text{(roll axis)}$$

$$I_{22} = 1.15 \times 10^8 \text{ kg-m}^2 \qquad \text{(pitch axis)}$$

$$I_{33} = 6.60 \times 10^7 \text{ kg-m}^2 \qquad \text{(yaw axis)}$$

$$I_{13} = 7.53 \times 10^5 \text{ kg-m}^2 \qquad \text{(inertial roll-yaw coupling)}$$

Let us assume that all moments of inertia are principal (although clearly I_{11} and I_{33} are not). The divergence time constant is calculated to be

$$\tau = 0.1220 T_0$$

If the SSTO vehicle is oriented such that the 1-axis (thrust axis) is along the velocity vector and the 2-axis (pitch) is normal to the trajectory or orbital plane, the vehicle's attitude will diverge with a time constant something like one-tenth of the orbital period. Clearly, if the vehicle were placed in an attitude usually associated with atmospheric flight at orbital insertion, the vehicle might diverge significantly from that attitude at the time of re-entry.

We may now return to Eqs. (F.14a) and (F.14c) and examine the dynamics associated with the 1-axis and 3-axis. The work in the remainder of this appendix follows that given by Wiesel,[2] although the notation has been altered slightly. Equations (F.14a) and (F.14c) are coupled, but the set is uncoupled from Eq. (F.14b). If we make the following variable substitutions

$$K_{11} = (I_{22} - I_{33})/I_{11} \qquad K_{33} = (I_{22} - I_{11})/I_{33}$$

we may rewrite Eqs. (F.14a,c) as

$$\frac{d^2\theta_1}{dt^2} + (K_{11} - 1)\Omega\frac{d\theta_2}{dt} + K_{11}\Omega^2\theta_1 = 0 \tag{F.17a}$$

$$\frac{d^2\theta_3}{dt^2} + (1 - K_{33})\Omega\frac{d\theta_1}{dt} + K_{33}\Omega^2\theta_3 = 0 \tag{F.17b}$$

Now if we let $\boldsymbol{\theta} = [\theta_1, \theta_2]^{\mathrm{T}}$, we may rewrite the preceding equation set in matrix form as

$$\begin{bmatrix} 1 & 0 \\ 0 & 1 \end{bmatrix}\frac{d^2\boldsymbol{\theta}}{dt^2} + \begin{bmatrix} 0 & \Omega(K_{11} - 1) \\ \Omega(1 - K_{33}) & 0 \end{bmatrix}\frac{d\boldsymbol{\theta}}{dt} + \begin{bmatrix} 4K_{11}\Omega^2 & 0 \\ 0 & K_{33}\Omega^2 \end{bmatrix}\boldsymbol{\theta} = 0 \tag{F.18}$$

Let us assume a solution of the form $\boldsymbol{\theta} = \boldsymbol{\theta}_0 e^{\lambda t}$ where λ and $\boldsymbol{\theta}_0$ are constants. Equation (F.18) then becomes

$$\begin{bmatrix} \lambda^2 & 0 \\ 0 & \lambda^2 \end{bmatrix}\boldsymbol{\theta}_0 + \begin{bmatrix} 0 & \lambda\Omega(K_{11} - 1) \\ \lambda\Omega(1 - K_{33}) & 0 \end{bmatrix}\boldsymbol{\theta}_0 + \begin{bmatrix} 4K_{11}\Omega^2 & 0 \\ 0 & K_{33}\Omega^2 \end{bmatrix}\boldsymbol{\theta}_0 = 0 \tag{F.19}$$

and the characteristic determinant is given as follows:

$$\begin{vmatrix} \lambda^2 + 4K_{11}\Omega^2 & \lambda\Omega(K_{11} - 1) \\ \lambda(1 - K_{33})\Omega & \lambda^2 + K_{33}\Omega^2 \end{vmatrix} = 0$$

which results in the following quadratic in λ^2:

$$\lambda^4 + \lambda^2[3K_{11} + K_{11}K_{33} + 1]\Omega^2 + 4K_{11}K_{33}\Omega^4 = 0 \tag{F.20}$$

If we wish to have an oscillatory motion, then all four values of λ must be imaginary or equivalently both of the roots of Eq. (F.20), quadratic in λ^2, must be negative. We should note in passing that both roots of the quadratic form of Eq. (F.20) cannot be negative if the system has no dissipation, i.e., if the coefficient of λ^2 is zero. For the λ^2 roots to be negative requires that the discriminant of Eq. (F.20) be positive, or

$$(3K_{11} + K_{11}K_{33} + 1) > 4(K_{11}K_{33})^{1/2} \tag{F.21a}$$

or, equivalently, that

$$K_{11}K_{33} > 0 \tag{F.21b}$$

In terms of the moments of inertia, we may rewrite this expression as

$$[(I_{22} - I_{33})/I_{11}][(I_{22} - I_{11})/I_{33}] > 0 \tag{F.21c}$$

Equation (F.21c) requires that the moment of inertia about the 2-axis be either the major or minor principal axis.

For gravity gradient stabilization, the vehicle moments of inertia must be either

$$I_{22} > I_{11} > I_{33}$$

or

$$I_{11} > I_{33} > I_{22}$$

For the preceding derivation to take meaning, the e- and the b-axes are initially aligned, and the b-frame then undergoes either harmonic or divergent motion with respect to the e-frame. The axes of the e-frame, and initially the b-frame, are aligned as follows: the 1-axis must be in the direction of the velocity vector, the 2-axis normal to the orbital plane (or re-entry trajectory vertical plane) and the 3-axis geocentric, along the negative of the position vector. We need not have made the 2-axis the pitch axis or the 3-axis the yaw axis, although this is the orientation that might be expected at the beginning of re-entry.

We note that for the typical SSTO example examined earlier, the pitch moment inertia is the maximum, and so for stable or oscillatory motion the vehicle pitch axis may be aligned normal to the orbital plane; the minimum moment of inertia must be aligned with the 3-axis or in the geocentric direction. Thus, the vehicle roll axis must be in the orbital plane pointing toward the Earth and the yaw axis must be aligned with the velocity vector. Another stable configuration will have the roll axis aligned normal to the orbital plane, since the roll moment of inertia is the minimum. In either event the vehicle would enter the atmosphere with its lift-generating surfaces normal to the flow direction. Obviously, it is necessary to alter orientation prior to entry.

References

[1]Freeman, D. C., and Powell, R. W., "The Results of Studies to Determine the Impact of Far-Aft Center of Gravity Locations on the Design of a Single-Stage-to-Orbit Vehicle System," AIAA Paper 79-0892, Hampton, VA, May 1979.

[2]Wiesel, W. E., *Space Dynamics*, McGraw-Hill, Hightstown, NJ, 1989.

Appendix G: Radar Error Model

In Chapter 13 we discussed the contributions of measurement errors in estimating the state of a re-entry vehicle. In particular we noted in Eq. (13.63) that in setting the Kalman gain matrix we must have available the measurement noise covariance matrix R. In this appendix we will consider how numerical values can be assigned to the diagonal elements of the R matrix (we will ignore off-diagonal or cross-correlation elements).

Figure G.1a represents some of the relevant geometry associated with the radar system. The target is detected by the reflection of the radio frequency energy originally transmitted by the radar antenna. The radar antenna illuminates objects contained within the main lobe or beam of the radar, defined by a cone of solid angle ψ. Not all of the transmitted energy is placed in this main lobe—some energy goes into side lobes that illuminate other objects. The reflected energy from the side lobes, returned to the radar receiver, causes a spurious signal. The target, located in the main lobe, reflects or, in essence, retransmits a portion of the main lobe energy. A fraction of this retransmitted energy is acquired by the radar-receiving antenna. As the target—the re-entry body in this

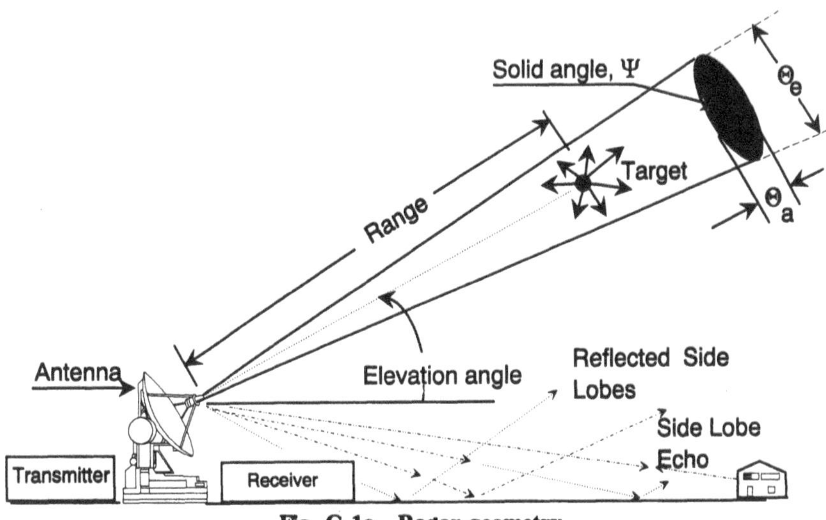

Fig. G.1a Radar geometry.

case—maneuvers and changes orientation relative to the radar axis, the reflected energy varies with time.

Our concern here is not with the operation of radars. Extensive information concerning radar engineering is available in such texts as Toomay[1] and Barton.[2,3] Our interest will be confined to making preliminary estimations of radar measurement errors and to relating such errors to radar characteristics (e.g., transmitted power, antenna gain), and target range, and effective radar cross section.

Some representative error sources are depicted in Fig. G.1b. Obviously a detailed modeling of most of these contributions is beyond the scope of this appendix. However, we will obtain an estimation of the combined errors in the measurement of range R, azimuth Az, and elevation E.

Radar is regarded here as a device for making measurements of range, azimuth and elevation at discrete time intervals. The central utility of a radar is to make range and possibly Doppler velocity measurements; the two angle measurements, although essential for locating the target, can be obtained from angular pick-offs. Therefore, the pivotal equation in assessing the capability of a given radar is the *radar range equation*. One form of this equation expresses the power S of the returned radar signal as a function of the range to target R, the transmitted power P_t, and various geometric characteristics of the radar, or

$$S = P_t G_t G_r \lambda^2 \sigma_{cs} / \left[(4\pi)^3 R^4\right] \tag{G.1}$$

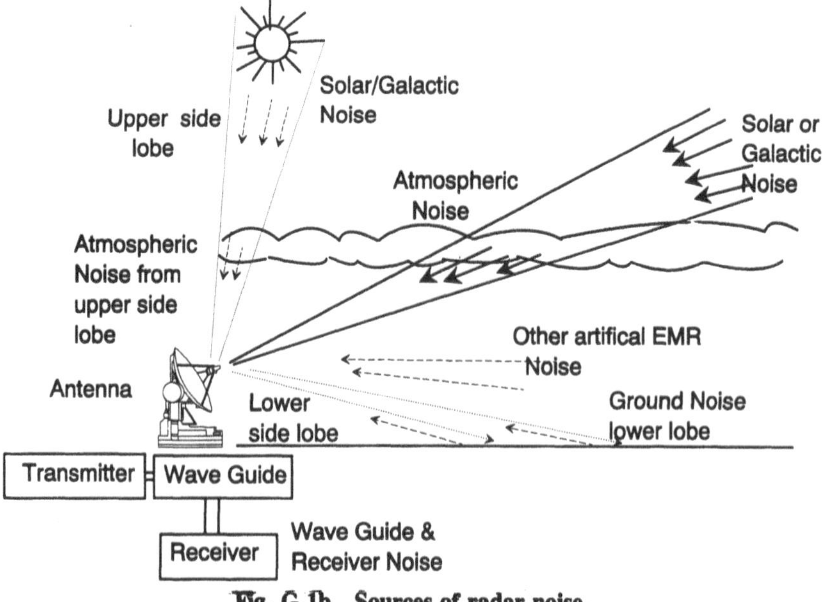

Fig. G.1b Sources of radar noise.

where

S = power returned, watts
P_t = transmitted power, watts
G_t = power gain of the transmitting antenna
G_r = power gain of the receiving antenna
λ = wavelength, m
σ_{cs} = radar cross section, m^2
R = range to target, m

A full derivation of Eq. (G.1) may be found in the Refs. 1, 2, and 3.

Extensive and varied sources for radar errors are illustrated in Fig. G.1. Here we will consider only two categories of noise: 1) the totality of background noise that is received by the antenna and 2) the totality of losses associated with all radar subsystems.

Background noise (solar, atmospheric, galactic, etc.) is often expressed as an equivalent temperature, or

$$N_0 = KT_a \tag{G.2}$$

where K is Boltzmann's constant [1.38×10^{-23} watts/(Hz-K)] and T_a is obtained from the following equation:

$$T_a = 290.0 + [(0.88T_{\text{as}} - 254.0)/L_a] \tag{G.3}$$

where T_{as} is the apparent temperature of the sky as viewed at the operating frequency, and L_a is the dissipative loss within the antenna. Of course, assigning numerical values to the terms L_a and T_{as} requires some knowledge of the medium and the antenna. On page 15 of Ref. 3, a chart (reproduced in Fig. G.2) is given for computing the noise temperature as a function of the radar signal frequency. Barton[3] (and most other radar texts) gives the frequency range for the various bands. For the tracking of re-entry vehicles, we will be primarily interested in the S (2–4 GHz) and the C (4–8 GHz) bands. Typically, T_{as} may be between 10 K for a 30-deg beam elevation angle to 100 K for a 0-deg beam elevation angle.

The term L_a, representing antenna dissipative loss, is around 0.1 dB (1.023) to 0.4 dB (1.096). For example, with L_a = 0.2 dB (1.047) we obtain for T_a the following:

$$\begin{aligned} T_a &= 290.0 + (0.88(65.0) - 254.0/(1.047)) \\ &= 102.0 \text{ K} \end{aligned}$$

The other losses are due to the RF components, as

$$T_r = T_{\text{tr}}(L_r - 1) \tag{G.4}$$

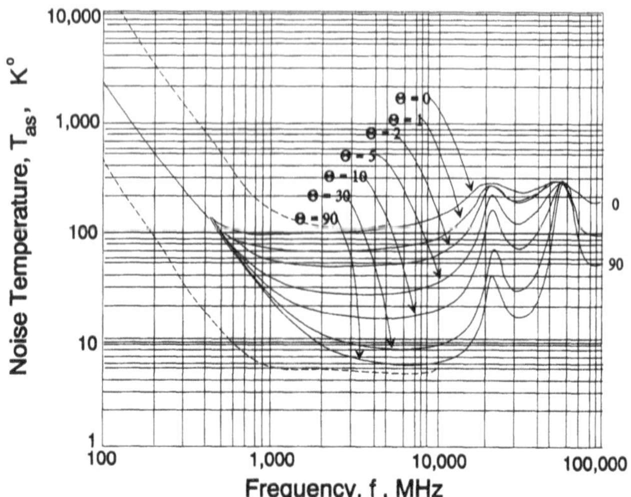

Fig. G.2 Noise temperature of an idealized antenna at Earth's surface as a function of frequency for a number of beam elevation angles. (Reprinted from D. K. Barton, *Modern Radar Systems Analysis*, ARTECH House, Inc., Norwood, MA, 1980.)

For example, we might let T_{tr}, the ambient temperature, be 290 K and L_r, the RF component loss, be 1.0 dB or 1.259. Thus,

$$T_r = 290.0(1.259 - 1.0) = 75 \text{ K}$$

The losses due to the receiver are given as

$$T_e = T_0(F_n - 1.0) \tag{G.5}$$

where T_0 is the reference temperature of 290.0 K and F_n is the noise factor of the receiver. A typical value for F_n, 5.5 dB (3.548), would give

$$T_e = 290.0(3.548 - 1.0) = 738.96 \text{ K}$$

The entire loss of the system then is given as

$$T_s = T_a + T_r + L_r T_e \tag{G.6}$$

where L_r is the loss attributed to the input RF components (typical value about 1.0 dB or 1.259), or

$$\begin{aligned} T_s &= 102.0 + 75.0 + (1.259)(738.9) \\ &= 1107.0 \text{ K} \end{aligned}$$

The effective noise input power is given as

$$N = KT_sB_n \tag{G.7}$$

where K is Boltzmann's constant, T_s is defined in Eq. (G.6), and B_n is the noise bandwidth, defined as

$$B_n = \frac{1}{|H(f_0)|^2} \int_{-\infty}^{+\infty} |H(f)|^2 \, \mathrm{d}f \tag{G.8}$$

where $H(f)$ is the frequency response of the receiving filter and f_0 is the central frequency of $H(f)$. The noise bandwidth B_n is approximately the half-power bandwidth. If the filter is matched to take a pulse of duration τ in seconds, then the noise bandwidth is $1/\tau$. If $\tau = 10^{-6}$ seconds, B_n is 1 MHz. Thus, from Eq. (G.7)

$$N = 1.38 \times 10^{-23} \times 1107.0 \times 10^6$$

$$N = 1.5277 \times 10^{-14} \qquad \text{(watts)}$$

By way of summary we have for the noise power

$$N = KB_n(T_a + T_r + L_rT_e) \tag{G.9}$$

Equation (G.1), the radar range equation, and Eq. (G.9) permit the calculation of the signal-to-noise ratio S/N. The signal-to-noise ratio is important for the calculation of the threshold of target detection, although in this appendix we will confine our attention to calculating the error standard deviation in range, elevation, and azimuth and the Doppler velocity magnitude.

Barton[3] provides a tabulated summary of the root mean square (RMS) or standard deviation of each of the three basic radar measurements, representing the minimum achievable errors according to theoretical limitations. The standard deviation of the basic radar measurements—delay time, angle, and frequency—are as follows.

Delay time (s):

$$\sigma_t = 1/\left(\beta\sqrt{\mathscr{E}}\right) \tag{G.10a}$$

Angle (rad):

$$\sigma_\theta = \lambda/\left(L\sqrt{\mathscr{E}}\right) \tag{G.10b}$$

Frequency for the coherent case (Hz):

$$\sigma_{f_{co}} = 1/\left(\alpha\sqrt{\mathscr{E}}\right) \tag{G.10c}$$

Frequency for the noncoherent case (Hz):

$$\sigma_{f_{nc}} = 1/\left(\alpha_1 \sqrt{\mathscr{E}}\right) \tag{G.10d}$$

The symbol $\mathscr{E}$ represents twice the ratio of the energy of the reflected signal to the noise energy, β is the RMS bandwidth of the signal, λ is the wavelength, L is the RMS value of the aperture width, α is the RMS value of the total duration of the signal and α_1 is the RMS value of the pulse width.

We can alter the expressions given in Eqs. (G.10) to provide more useful forms. First we can obtain the range and Doppler velocity errors as follows:

$$\sigma_r = c\sigma_t \qquad \sigma_v = \lambda\sigma_f \tag{G.11}$$

where c is the speed of propagation of the electromagnetic waves.

In Eq. (G.1) we used the radar equations to provide an expression for the signal strength S in terms of range and radar geometrical and operating characteristics. Therefore, in estimating the magnitude of the various error sources, the signal-to-noise ratio S/N is more useful than the energy ratio $\mathscr{E}$.

If we let τ be the pulse width (seconds) and β the bandwidth of the receiver, we can relate $\mathscr{E}$, the energy ratio, to the signal-to-noise ratio, as

$$\mathscr{E} = 2B\tau(S/N) \tag{G.12}$$

This expression was developed by Barton[2] using the approximation

$$\beta^2 \approx 2B/\tau \tag{G.13}$$

for rectangular pulse widths that satisfy the relationship $\tau \geq 1/\beta$.

Equations (G.10) may now be rewritten in terms of the signal-to-noise ratio S/N, the signal receiver bandwidth B, and the pulse width τ. The standard deviation for range, angle, and Doppler velocity are as follows:

Range (m):

$$\sigma_r = c/\left[2B(S/N)^{1/2}\right] \tag{G.14a}$$

Angle (rad):

$$\sigma_\theta = \lambda/\left[\pi\left(\tfrac{2}{3}B\tau\right)^{1/2} D(S/N)^{1/2}\right] \tag{G.14b}$$

Doppler velocity (coherent):

$$\sigma_{v_{co}} = \lambda/\left[\pi\left(\tfrac{2}{3}B\tau\right)^{1/2} n_p\tau(S/N)^{1/2}\right] \tag{G.14c}$$

Doppler velocity (noncoherent):

$$\sigma_{v_{nc}} = \lambda/\left[\pi\left(\tfrac{2}{3}B\tau\right)^{1/2} \tau(S/N)^{1/2}\right] \tag{G.14d}$$

where D is the maximum diameter of the receiving antenna, n_p is the number of pulses, and L in Eq. (G.10b) is replaced by

$$L = \left(\pi/\sqrt{3}\right)D \tag{G.15a}$$

which assumes a rectangular aperture. Also under the assumption of rectangular pulses,

$$\alpha = \left(\pi/\sqrt{3}\right)\tau n_p \tag{G.15b}$$

$$\alpha_1 = \left(\pi/\sqrt{3}\right)\tau \tag{G.15c}$$

Equations (G.14) are the suggested form for assessment of the uncertainties in range, angle and Doppler velocity. In Chapter 13 a three-state model was used. The noise covariance matrix of Eq. (13.59) might then be written as follows:

$$R = \begin{bmatrix} \sigma_r^2 & 0 & 0 \\ 0 & \sigma_\theta^2 & 0 \\ 0 & 0 & \sigma_\theta^2 \end{bmatrix} \tag{G.16}$$

References

[1]Toomay, J. C., *Radar Principles for the Non-Specialist,* Lifetime Learning Publications, Belmont, CA, 1982.

[2]Barton, D. K., *Radar System Analysis*, ARTECH House, Inc., Norwood, MA, 1976.

[3]Barton, D. K., *Modern Radar System Analysis*, ARTECH House, Inc., Norwood, MA, 1988.

Appendix H: Establishment of Axis Systems and Directional Cosine Matrices

Re-entry vehicle trajectory shaping, impact area control, navigation system evaluation, etc., require the use of reference frames. The selection of any particular reference frame is based upon convenience. For example, in simulating navigation, a useful frame has one axis normal to the geoid and the other two in orthogonal compass directions. On the other hand, aerodynamic loads are usually expressed in a body-fixed frame, with one axis along an axis of symmetry, or a zero lift line or in some cases two axes in a plane of symmetry.

Our concern in this appendix will be limited to defining the various axis systems and the Directional Cosine Matrix (DCM) between any of several frame pairs. If six frames are identified, then 15 DCMs would be needed to complete transformations between any two frames. Such an extensive listing is not necessary since the chain rule permits the evaluation of most nonlisted DCMs.

The most fundamental reference frame is the inertial frame. An inertial, or Newtonian, frame is any frame in which observed positional change can be explained entirely by Newton's laws of motion. If, for example, acceleration is observed, then there must be a physically justifiable force or set of forces to account for such acceleration. Let us consider the following definition: "A reference frame is said to be inertial in a certain region of space and time when, throughout that region and within some specified accuracy, every test particle that is initially at rest remains at rest and every test particle that is initially in motion continues that motion without change in speed or direction."[1]

As Taylor and Wheeler[1] indicate, a frame in accelerated free fall, whose fall is due entirely to a gravitational field, may be considered a locally inertial frame. The *local* restriction means that the frame is restricted to a spatial region over which particle acceleration due to a gravity gradient is not experimentally detectable. If the space surrounding the inertial frame were expanded to include detectable gravity gradient effects there would be accelerated motion without any physically justifiable force. There is a measurable gravity gradient across the Earth's diameter due to the field of the Sun so that in this context any Earth-centered frame cannot be inertial.

Obviously the definition of Taylor and Wheeler is of no direct use in dealing with practical navigation problems. If the gravitational field could be mapped to account for the variations in magnitude and direction of the gravity field, then the gravitational force could be regarded as an applied force in the context of Newton's laws of motion.

Another more important source of spurious motion is caused by frame angular velocity and angular and linear acceleration. Thus, a frame fixed to the Earth's surface is non-Newtonian because such a frame shares the rotation of the Earth. Corrections must be applied in such noninertial frames to remove the effects caused by spurious accelerations. For our purposes we will regard a geocentric frame, nonrotating with respect to the stars, as inertial. Since the gravitational fields of the Earth, Moon, Sun, etc., are not uniform over a sphere of 1.5 R_E, say, we will still regard such a frame as inertial since the gravitational field is adequately defined in such a region. (The gravitational gradient of the Sun has been used by experimenters to demonstrate that the acceleration of an object due to gravity is independent of the material composition of the object.)[2]

Several useful frames are shown in Fig. H.1. The inertial frame, or *I*-frame, does not rotate but its origin remains fixed at the Earth's center of mass. In addition the $X^I Z^I$ plane by definition contains the re-entry vehicle at the moment of trajectory initiation. We can define this point of trajectory initiation as termination of boost or of atmospheric entry.

An Earth-fixed geocentric frame designated as the initial or *e*-frame is coincident with the *I*-frame at trajectory initiation. Since the Earth rotates at a rate

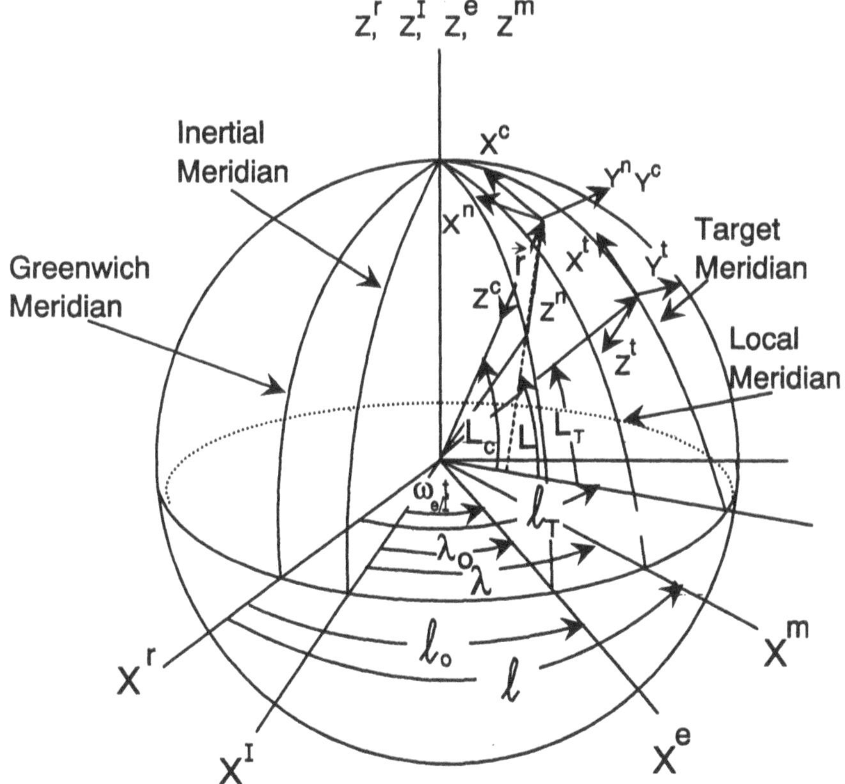

Fig. H.1 Definition of coordinate frames.

$\omega_{e/I}$, the I- and e-frames become increasingly misaligned at the rotational rate of the Earth. The m-frame, or the meridian frame, contains the re-entry vehicle at any time during the trajectory.

There are two related frames, both of which have their origin at some reference point (e.g., the center of mass) of the re-entry vehicle. The navigation or n-frame has the Z^n-axis down along the normal to the geoid. The X^n-axis is north and the Y^n is east. The geocentric or c-frame also has its origin at the RV. The Z^c-axis is down along the negative of the geocentric vector. The X^c is in the meridian and the Y^c completes the triad.

Both the X^nZ^n and the X^cZ^c planes lie in the meridian; the Z^n and the Z^c axes subtend the angle D, known as the deflection of the vertical. The target or t-frame is a navigation frame (Z^t down, X^t north and Y^t east) located at the intended impact or landing point.

A third frame whose origin is at the RV is a body-fixed or b-frame. For a rigid body of constant mass, the origin is usually at the center of mass. One axis is usually directed along an axis of aerodynamic symmetry (zero lift line). For a body having a plane of symmetry, two axes are contained in this plane.

In Fig. H.1 we have indicated the celestial longitude λ and the terrestrial longitude ℓ. These are obviously related, as

$$\ell = \ell_0 + \lambda - \omega_{e/I} t \tag{H.1}$$

where ℓ_0 locates the meridian containing the re-entry vehicle at burnout of the booster, or trajectory initiation. We also note that there are two slightly different definitions of latitude: the geocentric latitude L_c and the geographic latitude L. Further discussion of these two angles is given in Chapter 3. An equatorial view of the angles is given in Fig. H.2; the meridian planes are shown in edge view.

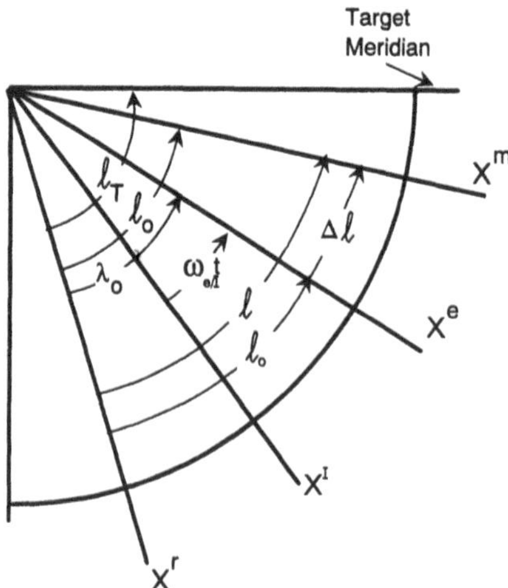

Fig. H.2 Angles in equatorial plane.

H.1 Transformations Among Inertial, Navigation, Earth, Geocentric, and Target Frames

We will now consider some transformations between the various frames, following the work of Britting.[3]

Inertial Frame to Navigation Frame

$$C_I^n = \begin{bmatrix} -\sin(L)\cos(\lambda) & -\sin(L)\sin(\lambda) & \cos(L) \\ -\sin(\lambda) & \cos(\lambda) & 0 \\ -\cos(L)\cos(\lambda) & -\cos(L)\sin(\lambda) & -\sin(L) \end{bmatrix} \tag{H.2}$$

where L is the geographic latitude and the celestial longitude is given after Eq. (H.1) as

$$\lambda = (\ell - \ell_0) + \omega_{e/I} t \tag{H.3}$$

The angular velocity of the navigation frame relative to the inertial frame is

$$\boldsymbol{\omega}_{n/I}^n = \begin{bmatrix} (\frac{d\ell}{dt} + \omega_{e/I})\cos(L) \\ -\frac{dL}{dt} \\ -(\frac{d\ell}{dt} + \omega_{e/I})\sin(L) \end{bmatrix} \tag{H.4}$$

$$\boldsymbol{\omega}_{n/I}^I = \begin{bmatrix} \frac{dL}{dt}\sin[(\ell - \ell_0) + \omega_{e/I} t] \\ -\frac{dL}{dt}\cos[(\ell - \ell_0) + \omega_{e/I} t] \\ \left(\frac{d\ell}{dt} + \omega_{e/I}\right) \end{bmatrix} \tag{H.5}$$

Inertial Frame to Earth Frame

$$C_I^e = \begin{bmatrix} \cos(\omega_{e/I} t) & \sin(\omega_{e/I} t) & 0 \\ -\sin(\omega_{e/I} t) & \cos(\omega_{e/I} t) & 0 \\ 0 & 0 & 1 \end{bmatrix} \tag{H.6}$$

The angular velocity of the e-frame relative to the i-frame is

$$\boldsymbol{\omega}_{e/I}^I = \boldsymbol{\omega}_{e/I}^e = \begin{bmatrix} 0 \\ 0 \\ \omega_{e/I} \end{bmatrix} \tag{H.7}$$

Inertial Frame to Geocentric Frame

$$C_I^c = \begin{bmatrix} -\sin(L_c)\cos(\lambda) & -\sin(L_c)\sin(\lambda) & \cos(L_c) \\ -\sin(\lambda) & \cos(\lambda) & 0 \\ -\cos(L_c)\cos(\lambda) & -\cos(L_c)\sin(\lambda) & -\sin(L_c) \end{bmatrix} \tag{H.8}$$

The angular velocity of the c-frame relative to the I-frame is

$$\boldsymbol{\omega}_{c/I}^c = \begin{bmatrix} \left(\frac{d\ell}{dt} + \omega_{e/I}\right)\cos(L_c) \\ -\frac{dL_c}{dt} \\ -\left(\frac{d\ell}{dt} + \omega_{e/I}\right)\sin(L_c) \end{bmatrix} \tag{H.9}$$

$$\boldsymbol{\omega}_{c/I}^I = \begin{bmatrix} \frac{dL_c}{dt}\sin[(\ell - \ell_0) + \omega_{e/I}t] \\ -\frac{dL_c}{dt}\cos[(\ell - \ell_0) + \omega_{e/I}t] \\ \left(\frac{d\ell}{dt} + \omega_{e/I}\right) \end{bmatrix} \tag{H.10}$$

Geocentric Frame to Navigation Frame

$$C_c^n = \begin{bmatrix} \cos(D) & 0 & \sin(D) \\ 0 & 1 & 0 \\ -\sin(D) & 0 & \cos(D) \end{bmatrix} \tag{H.11}$$

where D (deflection of the vertical) is the difference between the geocentric and geographic latitude, given as

$$D = L - L_c \tag{H.12}$$

The angular velocity of the c-frame relative to the n-frame is

$$\boldsymbol{\omega}_{c/n}^c = \boldsymbol{\omega}_{c/n}^n = \begin{bmatrix} 0 \\ \frac{dD}{dt} \\ 0 \end{bmatrix} \tag{H.13}$$

Navigation Frame to Earth Frame

$$C_n^e = \begin{bmatrix} -\sin(L)\sin(\ell - \ell_0) & -\sin(\ell - \ell_0) & -\cos(L)\cos(\ell - \ell_0) \\ -\sin(L)\sin(\ell - \ell_0) & \cos(\ell - \ell_0) & -\cos(L)\sin(\ell - \ell_0) \\ \cos(L) & 0 & -\sin(L) \end{bmatrix} \tag{H.14}$$

The angular velocity of the n-frame with respect to the e-frame is

$$\boldsymbol{\omega}^{e}_{n/e} = \begin{bmatrix} \frac{dL}{dt}\sin(\ell - \ell_0) \\ -\frac{dL}{dt}\cos(\ell - \ell_0) \\ \frac{d\ell}{dt} \end{bmatrix} \tag{H.15}$$

$$\boldsymbol{\omega}^{n}_{n/e} = \begin{bmatrix} \frac{d\ell}{dt}\cos(L) \\ -\frac{dL}{dt} \\ -\frac{d\ell}{dt}\sin(L) \end{bmatrix} \tag{H.16}$$

Inertial Frame to Target Frame

$$C^t_I = \begin{bmatrix} -\sin(L_t)\cos(\omega_{e/I}t) & -\sin(L_t)\sin(\omega_{e/I}t) & \cos(L_t) \\ -\sin(\omega_{e/I}t) & \cos(\omega_{e/I}t) & 0 \\ -\cos(L_t)\cos(\omega_{e/I}t) & -\cos(L_t)\sin(\omega_{e/I}t) & -\sin(L_t) \end{bmatrix} \tag{H.17}$$

The angular velocity of the t-frame relative to the I-frame is as follows:

$$\boldsymbol{\omega}^{t}_{t/I} = \begin{bmatrix} \omega_{e/I}\cos(L_t) \\ 0 \\ -\omega_{e/I}\sin(L_t) \end{bmatrix} \tag{H.18}$$

Target Frame to Navigation Frame

$$C^n_t = \begin{bmatrix} \sin(L)\sin(L_t)\cos(\ell - \ell_t) + \cos(L)\cos(L_t) & -\sin(L)\sin(\ell - \ell_t) & \sin(L)\cos(L_t)\cos(\ell - \ell_t) - \sin(L_t)\cos(L) \\ \sin(L_t)\sin(\ell - \ell_t) & \cos(\ell - \ell_t) & \cos(L_t)\sin(\ell - \ell_t) \\ \sin(L_t)\cos(L)\cos(\ell - \ell_t) - \sin(L)\cos(L_t) & -\cos(L)\sin(\ell - \ell_t) & \cos(L)\cos(L_t)\cos(\ell - \ell_t) + \sin(L)\sin(L_t) \end{bmatrix} \tag{H.19}$$

The angular velocity of the n-frame relative to the t-frame with coordinates in the n-frame is given as

$$\boldsymbol{\omega}^{n}_{n/t} = \begin{bmatrix} \frac{d\ell}{dt}\cos(L) \\ -\frac{dL}{dt} \\ -\frac{d\ell}{dt}\sin(L) \end{bmatrix} \tag{H.20}$$

H.2 Transformations Among Body, Local, and Velocity Frames

During re-entry one of the most important frames is the body-fixed or b-frame, since aerodynamic loads are usually available in this frame. The moment equation or the strapdown gyros provide the vector, which is

$$\boldsymbol{\omega}^b_{b/I} = [P \;\; Q \;\; R]^{\mathrm{T}} \tag{H.21}$$

It is fairly straightforward to integrate $\boldsymbol{\omega}^b_{b/I}$ to obtain the elements of the DCM C^I_b since

$$C^I_b = \int^t C^I_b \Omega^b_{b/I} \mathrm{d}t$$

or

$$C^I_b|_{j+1} = C^I_b|_j \exp\left[\Omega^b_{b/I}(t_{j+1} - t_j)\right] \tag{H.22}$$

Some of the algorithms for carrying out the integration are discussed in Chapter 14.

We will now introduce two other frames to bring the body frame into the navigation frame. These frames are shown in Fig. H.3 and are designated the local or ℓ-frame and the velocity or v-frame. In the local frame the X^ℓ and Y^ℓ axes are horizontal, with the X^ℓ-axis in the direction of the ground track. The X^v-axis of the velocity frame is in the direction of the velocity vector and the Y^v-axis remains horizontal.

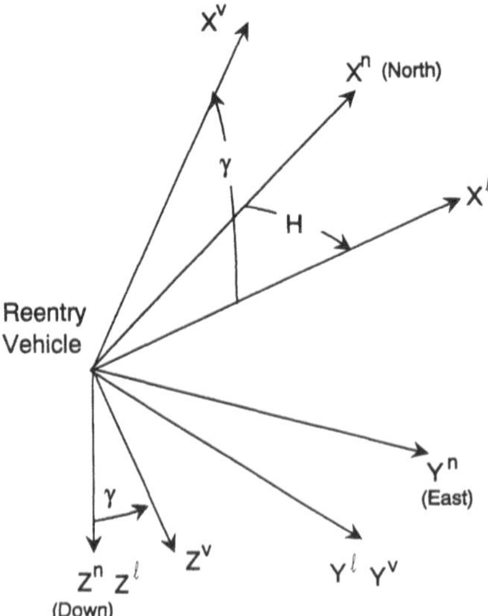

Fig. H.3 Navigation frame to local frame to velocity frame.

To obtain the local frame from the navigation frame requires a rotation about the Z^n-axis through the angle H, which is the heading or azimuth angle. The velocity frame is then obtained by rotating about the Y^ℓ-axis through the flight path angle γ. Thus, the DCM from the n-frame to the ℓ-frame is

$$C_n^\ell = \begin{bmatrix} \cos(H) & \sin(H) & 0 \\ -\sin(H) & \cos(H) & 0 \\ 0 & 0 & 1 \end{bmatrix} \tag{H.23}$$

The angular velocity of the ℓ-frame relative to the n-frame is

$$\boldsymbol{\omega}_{\ell/n}^\ell = \boldsymbol{\omega}_{\ell/n}^n = \begin{bmatrix} 0 & 0 & \frac{dH}{dt} \end{bmatrix}^T \tag{H.24}$$

As pointed out previously, the ν-frame is formed by rotating about the Y^ℓ-axis through the angle γ. The DCM between the ℓ-frame and the ν-frame is as follows:

$$C_\ell^\nu = \begin{bmatrix} \cos(\gamma) & 0 & -\sin(\gamma) \\ 0 & 1 & 0 \\ \sin(\gamma) & 0 & \cos(\gamma) \end{bmatrix} \tag{H.25}$$

The angular velocity of the ν-frame relative to the ℓ-frame is given as

$$\boldsymbol{\omega}_{\nu/\ell}^\ell = \boldsymbol{\omega}_{\nu/\ell}^V = [0 \quad \tfrac{d\gamma}{dt} \quad 0]^T \tag{H.26}$$

We must now relate the body axis to the ν-frame. We will select an Euler axis scheme that is often used when dealing with the dynamics of bodies of revolution. The three successive rotations are as follows: rotation about the X^ν-axis through the angle ψ, then about the Y intermediate axis through the total angle of attack $\bar{\alpha}$, and finally about the X intermediate axis through the roll angle ϕ.

The DCM C_ν^b has already been given in Eq. (4.39) and is repeated here for completeness:

$$C_\nu^b = \begin{bmatrix} \cos(\bar{\alpha}) & \sin(\bar{\alpha})\sin(\psi) & -\sin(\bar{\alpha})\cos(\psi) \\ \sin(\phi)\sin(\bar{\alpha}) & \begin{array}{c}\cos(\phi)\cos(\psi) \\ -\sin(\phi)\cos(\bar{\alpha})\sin(\psi)\end{array} & \begin{array}{c}\cos(\psi)\sin(\psi) \\ +\sin(\phi)\cos(\bar{\alpha})\cos(\psi)\end{array} \\ \cos(\phi)\sin(\bar{\alpha}) & \begin{array}{c}-\sin(\phi)\cos(\psi) \\ -\cos(\bar{\alpha})\sin(\psi)\cos(\phi)\end{array} & \begin{array}{c}-\sin(\psi)\sin(\phi) \\ +\cos(\phi)\cos(\bar{\alpha})\cos(\psi)\end{array} \end{bmatrix} \tag{H.27}$$

The angular velocity of the b-frame relative to the v-frame is

$$\boldsymbol{\omega}^b_{b/v} = \begin{bmatrix} \frac{d\phi}{dt} + \frac{d\psi}{dt}\cos(\bar{\alpha}) \\ \frac{d\bar{\alpha}}{dt}\cos(\phi) + \frac{d\psi}{dt}\sin(\phi)\sin(\bar{\alpha}) \\ -\frac{d\bar{\alpha}}{dt}\sin(\phi) + \frac{d\phi}{dt}\cos(\phi)\sin(\bar{\alpha}) \end{bmatrix} \tag{H.28}$$

The angle-of-attack plane is defined by the X^b-axis (usually the zero-lift line) and the velocity vector. The angle ψ is a measure of the tilt from vertical of the angle-of-attack plane. The angle $\bar{\alpha}$ (total angle of attack) is the angle between the X^b-axis and the velocity vector. The angle ϕ is a measure of the roll of the body about the X^b-axis. Aerodynamic loads are a function of $\bar{\alpha}$ and, for a configuration with asymmetries, the roll angle ϕ as well. Such asymmetries include the presence of fins, body cutouts, spallation in the heat shield, etc. Regardless of the configuration, aerodynamic loads are never a function of ψ.

References

[1]Taylor, E. T., and Wheeler, J. A., *Spacetime Physics*, W. H. Freeman & Co., San Francisco, CA, 1963.

[2]Dicke, R. H., *The Theoretical Significance of Experimental Relativity*, Gordon & Breach, New York, 1964.

[3]Britting, K. R., *Inertial Navigation System Analysis*, Wiley-Interscience, New York, 1971.

Index

www.ingramcontent.com/pod-product-compliance
Lightning Source LLC
LaVergne TN
LVHW090946250526
838985LV00005B/1

* 9 7 8 1 5 6 3 4 7 0 4 8 6 *